Physical organic chemistry

Physical organic chemistry

Neil S. Isaacs

Senior Lecturer in Chemistry,
University of Reading

Longman
Scientific &
Technical

Copublished in the United States with
John Wiley & Sons, Inc., New York

Longman Scientific & Technical

Longman Group UK Limited
Longman House, Burnt Mill, Harlow
Essex CM20 2JE, England
and Associated companies throughout the world

Copublished in the United States with
John Wiley & Sons, Inc., 605 Third Avenue, New York, NY 10158

© Longman Group UK Limited 1987

First published 1987

British Library Cataloguing in Publication Data
Isaacs, Neil S.
  Physical organic chemistry.
  1. Chemistry, Organic   2. Chemistry,
  Physical and theoretical
  I. Title
  547.1'3   QD476

  ISBN 0-582-46366-1   ppr
  ISBN 0-582-00474-8   csd

Library of Congress Cataloging in Publication Data
Isaacs, Neil S., 1934–
  Physical organic chemistry.
  Bibliography: p.
  Includes index.
  1. Chemistry, Physical organic.   I. Title.
  QD475.I846   1987   547.1'3   87-3848

  ISBN 0-470-20787-6 (USA only)

Printed in Northern Ireland by
The Universities Press (Belfast) Ltd.

"Through doubting we come to questioning and through questioning we come to the truth"

Peter Abelard, Paris, 1122

"Seek for simplicity — and then distrust it"

Alfred North Whitehead (1861–1947)

Foreword

Physical organic chemistry, the study of the underlying principles and rationale of organic reactions is over eighty years of age. During this period of development, much has been learned which is now enshrined within the permanent fund of chemical knowledge. At the same time the process of refinement of chemical theory continues, new techniques are developed and viewpoints shift their emphasis. A crucial issue of one decade becomes resolved in another. This then underlies the reason for offering another text on the subject of physical organic chemistry, continuing the series of accounts which began with the notable and still useful book of the same title of 1940 written by Professor Hammett. It is hoped that the present work will help to fill the increasingly large gap between present knowledge and practice and the status of the subject as treated in earlier texts. In particular, the last decade has witnessed the increasing use of sophisticated instrumentation, particularly nuclear magnetic resonance which can probe the structures and even the shapes of molecules in solution. Other trends have been the adoption throughout every branch of the subject of computational techniques including molecular orbital theory both of the simple Hückel type and also at high levels and of molecular mechanics. These aids to understanding are increasing in importance as the reliability of the results is improved and as fast computers become more available to chemists. The trend is likely to continue and computer graphics (cover design) as an aid to making educated guesses as to molecular properties seems likely to make a major contribution to (as Woodward put it) 'the armamentarium of the chemist'. As a result of this, our understanding of chemical processes is shifting more towards the framework of quantum mechanics. The present text has been written with the object of presenting to the senior undergraduate, graduate student and research worker an account of the more important organic reactions including both the traditional evidence – for it is a subject dependent on observation and inference – and modern approaches.

Considerable amounts of data have been included since a firm grasp of a subject is better aided by perusal of collected information than by single representative values. Information up to 1986 is included. Chapters 1 to 9 deal with underlying principles of reaction pathways, of the physical forces

Foreword

which shape bonding between atoms and of the changes of bonding which are chemical reactions. Chapters 10 to 16 describe present knowledge and understanding of the various reaction types which comprise organic chemistry and discuss the ingenious techniques which have been devised for mechanistic investigations. Space rather than choice has prevented the inclusion of certain topics including the organic chemistry of sulfur, phosphorus, silicon and metals, now of great importance but requiring a further book to do them justice.

Gratitude is extended to those colleagues who have advised me on the contents and who have read and criticized this text, notably Professors J. B. Lambert, L. K. Montgomery and N. Turro, and to Dr. A. Gilbert for his help on the photochemical chapter.

University of Reading, November, 1986.

Contents

Chapter 1 Models of chemical bonding   **1**
  1.1   Covalency and molecular structure   1
      1.1.1   The Valence Bond model   2
      1.1.2   The Molecular Orbital model   3
  1.2   Approximate Molecular Orbital theory   3
      1.2.1   The Hückel molecular orbital method   3
      1.2.2   Properties of Hückel molecular orbitals   12
      1.2.3   The relationship between MO and VB models   20
      1.2.4   Advanced MO methods   20
  1.3   Properties of covalent bonds   24
      1.3.1   Bond lengths   24
      1.3.2   Interbond angles   25
      1.3.3   Force constants   28
      1.3.4   Bond and molecular dipole moments   31
      1.3.5   Molecular and bond polarizabilities   33
      1.3.6   Bond dissociation enthalpies   36
      1.3.7   Group additivities to bond enthalpies   38
  1.4   Intermolecular forces   46
      1.4.1   Electrostatic forces   46
      1.4.2   Ion-pairs   49
      1.4.3   Short-range intermolecular forces   56
      1.4.4   The hydrogen bond   62
      1.4.5   Charge-transfer complexes   67
  Problems   70

Chapter 2 Kinetics and thermodynamics   **77**
  2.1   Enthalpy   77
      2.1.1   Endothermic reactions   79
  2.2   Entropy   79
  2.3   The Gibbs function, G 81
  2.4   Factors which contribute to entropy   83
  2.5   Chemical equilibrium   84
  2.6   Some useful thermodynamic relationships   87
      2.6.1   Temperature dependence   87

Contents

2.7 The application of thermodynamics to rate processes 88
    2.7.1 Activation 88
    2.7.2 The potential energy surface 91
    2.7.3 The transition state model 93
2.8 Properties of the transition state 93
    2.8.1 Activation parameters 93
    2.8.2 Heat capacity of activation 94
    2.8.3 Variation of rate with pressure 95
2.9 The uses of activation parameters 96
    2.9.1 The empirical treatment of rates 96
    2.9.2 The rate-determining step 99
    2.9.3 Relative rates 100
    2.9.4 Entropies and volumes of reaction 102
    2.9.5 The isokinetic relationship 102
2.10 The location of the transition state 104
    2.10.1 The Hammond Postulate 104
    2.10.2 Reactivity and selectivity 106
    2.10.3 Kinetic and thermodynamic control of products 107
    2.10.4 The principle of least motion 108
    2.10.5 The principle of microscopic reversibility 109
Problems 110

Chapter 3 Reagents and reaction mechanisms **114**
3.1 Polar and radical pathways 114
    3.1.1 Polar reactions 114
    3.1.2 Nucleophiles 115
    3.1.3 Electrophiles 116
    3.1.4 Radicals 116
3.2 A classification of fundamental reaction types 117
    3.2.1 Bond formation and bond breaking 117
    3.2.2 Transfer reactions 118
    3.2.3 Elimination and addition 119
    3.2.4 Pericyclic reactions 120
3.3 Reaction mechanism 120
    3.3.1 The advantages of synchronous reactions 121
3.4 Electron supply and demand 121
3.5 Transition state properties and structural change 122
Problems 127

Chapter 4 Correlation of structure with reactivity **129**
4.1 Electronic demands 129
4.2 The Hammett equation 131
4.3 Substituent constants 133
4.4 Theories of substituent effects 135

4.4.1	The resonance effect	136
4.4.2	The inductive effect	137
4.5	Interpretation of σ-values	140
4.5.1	Unshared-pair (n) substituents, $-X$	140
4.5.2	Alkyl groups	143
4.5.3	Acceptor groups, $-Z$	144
4.5.4	Cationic centers	144
4.6	Reaction constants, ρ	144
4.7	Deviations from the Hammett equation	146
4.7.1	Random deviations	146
4.7.2	Mechanistic change	146
4.7.3	Enhanced resonance	148
4.7.4	Variable resonance interactions	152
4.8	Dual-parameter correlations	153
4.8.1	Inductive substituent constants	154
4.8.2	The Taft model	154
4.8.3	Other chemical model systems	156
4.9	Molecular orbital considerations	163
	Problems	166
Chapter 5	**Solvent effects**	**171**
5.1	The structure of liquids	172
5.2	Solutions	173
5.3	Solvation	176
5.4	Thermodynamic measures of solvation	177
5.4.1	Free energies of solution and transfer functions	177
5.4.2	Activities of solutes	179
5.4.3	Solvation in the gas phase	181
5.5	The effects of solvation on reaction rates and equilibria	182
5.5.1	Solvent effects on rates	183
5.6	Empirical indexes of solvation	184
5.6.1	Scales based on physical properties	184
5.6.2	Scales based on solvent-sensitive reaction rates	190
5.6.3	Scales based on spectroscopic properties	192
5.6.4	Scales for specific solvation	196
5.7	Relationships between empirical solvation scales	199
5.8	The use of solvation scales in mechanistic studies	200
5.8.1	Multiparameter solvation analysis	202
	Problems	205
Chapter 6	**Acids and bases, electrophiles and nucleophiles**	**210**
6.1	Acid–base dissocation	210
6.2	The strengths of oxygen and nitrogen acids	211
6.2.1	The interpretation of K_A	214
6.3	Linear free-energy relationships	216

Contents

6.4	Rates of proton transfers	216
6.5	Structural effects on amine dissociations	218
	6.5.1 Linear free-energy relationships	219
6.6	Acidities of carbon acids	219
	6.6.1 The measurement of weak acidity	221
6.7	Factors which influence carbon acidity	222
6.8	Rates of ionization of carbon acids	225
6.9	Gas-phase acidity and basicity	225
6.10	Theories of proton transfer	229
6.11	Highly acidic and highly basic solutions	231
	6.11.1 Highly acidic solutions	232
	6.11.2 Highly basic media	236
6.12	Nucleophilicity and electrophilicity	236
	6.12.1 Nucleophilicity and basicity	236
	6.12.2 Hard and soft acids and bases	239
	6.12.3 Nucleophilicity scales	242
	6.12.4 The relationship between nucleophilicity and nucleo-fugacity	246
	6.12.5 The 'α-effect'	247
	6.12.6 Ambident nucleophiles	248
6.13	The measurement of electrophilicity	250
	Problems	251
Chapter 7	**Kinetic isotope effects**	**255**
7.1	Isotopic substitution	255
7.2	Theory of isotope effects	256
7.3	Transition-state geometry	263
7.4	Secondary kinetic isotope effects	263
	7.4.1 'Inductive' and 'steric' isotope effects	268
7.5	Heavy atom isotope effects	268
7.6	The tunnel effect	269
7.7	Solvent isotope effects	272
	7.7.1 Fractionation factors	273
	7.7.2 The proton inventory technique	275
	7.7.3 Examples	277
	Problems	279
Chapter 8	**Steric and conformational properties**	**282**
8.1	The origins of steric strain	282
8.2	Examples of steric effects upon reactions	285
	8.2.1 *Ortho* effects	285
	8.2.2 *F*-strain effects	287
	8.2.3 Bond-angle strain	289
	8.2.4 Steric inhibition of resonance	290
	8.2.5 Steric acceleration	290

8.2.6	Steric enhancement of resonance	291
8.2.7	Calculation of steric effects: the molecular mechanics method	292
8.3	Measurement of steric effects upon rates	295
8.3.1	The Taft–Ingold hypothesis	297
8.3.2	Other steric parameters	297
8.3.3	Examples of steric LFER	299
8.4	Conformational barriers to bond rotation	301
8.4.1	Spectroscopic detection of individual conformers	304
8.4.2	Acyclic compounds	306
8.4.3	Cyclic compounds	309
8.5	Rotations about partial double bonds	313
8.5.1	Inversion at Group V elements	314
8.6	Chemical consequences of conformational isomerism — the Winstein–Holness–Curtin–Hammett principle	315
Problems		324
Chapter 9 Homogeneous catalysis		**330**
9.1	Acid and base catalysis	331
9.1.1	Specific and general catalysis	331
9.1.2	Mechanisms of acid catalysis	334
9.1.3	Methods of distinguishing between A1 and A2 reactions	337
9.1.4	Linear free-energy relationships; the Brønsted Catalysis Law	340
9.1.5	Interpretation of Brønsted coefficients	342
9.1.6	Nucleophilic catalysis	345
9.1.7	Potential-energy surfaces for proton transfers	348
9.1.8	Solvent isotope effects	350
9.1.9	Electrophilic catalysis	351
9.2	The mechanisms of some acid-catalyzed reactions	353
9.2.1	Substitutions α- to a carbonyl group	353
9.2.2	Keto–enol equilibria	355
9.2.3	Hydrolyses of acetals, ketals, orthoesters and related compounds	356
9.2.4	Dehydration of aldehyde hydrates and related compounds	358
9.2.5	The formation of oximes, semicarbazones and hydrazones	358
9.2.6	Decarboxylation	358
9.2.7	Acid-catalyzed alkene–alcohol interchange	359
9.2.8	Some acid-catalyzed rearrangements	361
9.2.9	Rate-limiting proton transfers	365
9.3	Catalysis by non-covalent binding	367
Problems		370

Contents

Chapter 10   Substitution reactions at carbon   **375**
  10.1   Substitutions at saturated carbon   375
      10.1.1   Nucleophilic substitution (S_N2)   375
      10.1.2   The bimolecular reaction, S_N2 376
      10.1.3   Solvolytic reactions   386
      10.1.4   Measurement of solvent participation   387
      10.1.5   Kinetic isotope effects   394
      10.1.6   Structures of intermediates in S_N1 reactions   397
      10.1.7   The phenomenon of 'return'   399
      10.1.8   Rearrangement criteria for return   400
      10.1.9   The 'special' salt effect   402
      10.1.10 Structural effects upon ionization   403
      10.1.11 Leaving group effects   404
      10.1.12 Bridgehead systems   407
      10.1.13 Linear free-energy relationships   407
      10.1.14 Intramolecular assistance in ionization   411
      10.1.15 Activation parameters   414
      10.1.16 The $S_N i$ reactions   415
      10.1.17 Aliphatic S_N2 reactions in the gas phase   416
  10.2   Electrophilic substitutions at saturated carbon   418
      10.2.1   The S_E1 mechanism   418
      10.2.2   The S_E2 mechanism   419
      10.2.3   Electrophilic substitution via enolization   423
  10.3   Nucleophilic displacements at vinyl carbon   424
  10.4   Electrophilic displacements at an aromatic carbon   428
      10.4.1   Timing of bond breaking and making   428
      10.4.2   The general mechanism for electrophilic aromatic
              substitution   430
      10.4.3   The nature of the electrophilic reagents   431
      10.4.4   Kinetic isotope effects   434
      10.4.5   Kinetics of S_E2–Ar reactions   436
      10.4.6   Structural effects on rates   440
      10.4.7   The *ortho–para* selectivity ratio   446
      10.4.8   The nature of the intermediate   448
      10.4.9   *Ipso* attack   450
      10.4.10 The MO interpretation of aromatic reactivity   450
  10.5   Nucleophilic substitution at an aromatic center   453
      10.5.1   The addition–elimination pathway   453
      10.5.2   The unimolecular mechanism   458
      10.5.3   The aryne mechanism   459
      10.5.4   Nucleophilic substitution via ring opening; the S_N
              (ANRORC) route   461
  10.6   Nucleophilic substitutions at carbonyl carbon   462
      10.6.1   Basic hydrolysis of carboylic esters   467
      10.6.2   Acidic hydrolysis of esters   474

10.6.3 Stereoelectronic factors in the decomposition of the tetrahedral intermediate   475
10.6.4 Other mechanisms for ester hydrolysis   477
10.6.5 Hydrolysis of amides, acyl halides and anhydrides   484
10.6.6 Properties of tetrahedral intermediates   488
10.6.7 Nucleophilic catalysis in carbonyl substitutions   490
Problems   492

Chapter 11   Elimination reactions   504
11.1 Base-promoted eliminations in solution   504
     11.1.1 Kinetic criteria of mechanisms   507
     11.1.2 Structural effects on rates of elimination   508
     11.1.3 Kinetic isotope effects   514
     11.1.4 Variation of the base-solvent system   518
     11.1.5 Competition between elimination and substitution   520
     11.1.6 Orientation of product formation   523
     11.1.7 Stereochemistry of E2 reactions   524
     11.1.8 Frontier orbital considerations   529
     11.1.9 E1cb reactions   530
     11.1.10 Ester hydrolysis by the E1cb mechanism   532
11.2 Intramolecular pyrolytic eliminations (the E_i reaction)   533
     11.2.1 Ester pyrolysis   534
     11.2.2 The Chugaev reaction   536
     11.2.3 Amine oxide, sulfoxide and selenoxide pyrolyses   537
     11.2.4 Pyrolysis of alkyl halides   538
11.3 Eliminations forming carbenes   539
11.4 Oxidative eliminations   541
     11.4.1 Oxidations of alcohols by chromium (VI)   541
     11.4.2 The Moffatt oxidation   543
Problems   543

Chapter 12   Polar addition reactions   548
12.1 Electrophilic additions to alkenes   549
     12.1.1 Kinetics   549
     12.1.2 Effect of structure   551
     12.1.3 Isotope effects   555
     12.1.4 Orientation and stereochemistry   556
     12.1.5 The nature of the intermediates in Ad_E reactions   558
12.2 Miscellaneous additions   561
     12.2.1 Hydroboration   561
     12.2.2 Addition with ring closure   565
     12.2.3 Addition of carbocations   565
     12.2.4 Additions to dienes, alkynes and allenes   566
12.3 Nucleophilic additions to multiple bonds   567
     12.3.1 Michael addition   568

Contents

	12.3.2	Carbonyl additions	570
	12.3.3	Additions to heterocumulenes	576
12.4		Frontier orbital considerations	578
12.5		Vinyl substitution via addition and elimination	579
	12.5.1	Examples	581
	12.5.2	Stereochemistry	583
Problems			584

Chapter 13   Intramolecular reactions   **589**

13.1		Neighboring-group participation	589
	13.1.1	The scope of neighboring-group effects	593
	13.1.2	Methods for recognizing neighboring group participation	593
	13.1.3	The kinetic criterion	593
	13.1.4	Linear free-energy relationships	596
	13.1.5	Kinetic isotope effects	599
	13.1.6	Solvent effects	600
	13.1.7	Participation in carbonyl reactions	601
	13.1.8	The stereochemical criterion	605
	13.1.9	The rearrangement criterion	605
	13.1.10	Factors influencing neighboring-group participation	608
	13.1.11	Observation and isolation of cyclic intermediates	613
	13.1.12	σ- and π-participation; the question of non-classical ions	616
13.2		Enzymic reactions	624
	13.2.1	The structures of enzymes	624
	13.2.2	A model for enzyme action	626
	13.2.3	Mechanisms of some enzyme-catalyzed reactions	631
	13.2.4	Enzymes which use cofactors	635
	13.2.5	Enzyme model systems	640
Problems			641

Chapter 14   Pericyclic reactions   **648**

14.1		Classification of pericyclic reactions	648
14.2		The theory of pericyclic reactions	649
	14.2.1	Conservation of orbital symmetry	649
	14.2.2	The frontier orbital concept	652
	14.2.3	The aromaticity concept	654
	14.2.4	Suprafacial and antarafacial geometries	655
14.3		Thermal cycloadditions	658
	14.3.1	The Diels–Alder reaction	658
	14.3.2	Stereo- and regio-specificity	665
	14.3.3	Retro Diels–Alder reactions	670
	14.3.4	The nature of the Diels–Alder transition state	672
	14.3.5	Related six-electron cycloadditions	674

14.4   Thermal $(2 + 2)$ cycloadditions   676
    14.4.1   Cycloadditions of cumulenes   677
    14.4.2   Two-step cycloadditions   680
    14.4.3 $(2 + 2)$ Cycloreversions   682
14.5   1,3-Dipolar cycloadditions   685
14.6   Electrocyclic reactions   689
14.7   Chelotropic reactions   692
14.8   Sigmatropic reactions   696
    14.8.1   Concertedness in sigmatropic rearrangements   700
14.9   Acid catalysis of the Diels–Alder reaction   704
Problems   706

Chapter 15   Reactions via free radicals   **714**
15.1   The generation of radicals   714
    15.1.1   Primary processes   714
    15.1.2   Secondary routes   719
15.2   The detection of radicals   720
    15.2.1   Direct observation   720
    15.2.2   Indirect methods   725
    15.2.3   By chemical characteristics   731
15.3   Reactions of radicals   735
    15.3.1   Radical coupling   735
    15.3.2   Displacement (abstraction, transfer) reactions   737
    15.3.3   Additions to π-systems   740
    15.3.4   Fragmentation of radicals   743
    15.3.5   Rearrangements   744
    15.3.6   Preparatively useful radical chain reactions   747
    15.3.7   Linear-free energy relationships   748
    15.3.8   Electron transfer catalysis   748
15.4   The reactivities of radicals   752
    15.4.1   Radical stability   752
    15.4.2   Polar influences   757
    15.4.3   Solvent effects on radical reactions   759
    15.4.4   Steric effects in radical reactions   759
    15.4.5   Frontier orbital considerations   763
15.5   The stereochemistry of radicals   766
Problems   767

Chapter 16   Organic photochemistry   **777**
16.1   Excited electronic states   777
    16.1.1   Absorption of light by molecules   777
    16.1.2   Vertical and horizontal excitation   777
    16.1.3   Spin multiplicity: singlet and triplet states   779
    16.1.4   Sensitization and quenching   780
    16.1.5   Techniques of photochemistry   781

Contents

16.2 Photochemistry of the carbon–carbon double bond 782
    16.2.1 Geometrical isomerization 782
    16.2.2 Photochemical pericyclic reactions 784
    16.2.3 The Di-π-methane rearrangement 789
    16.2.4 Photoadditions to alkenes 790
16.3 Photoreactions of carbonyl compounds 791
    16.3.1 Carbon–carbon bond cleavage 791
    16.3.2 Cycloadditions 793
16.4 Photochemistry of aromatic compounds 795
    16.4.1 Photosubstitutions at the aromatic ring 795
    16.4.2 The photo-fries rearrangement 796
    16.4.3 Valence isomerization 797
    16.4.4 Photocycloadditions 798
    16.4.5 Photo-oxidations with oxygen 799
Problems 799

Symbols and abbreviations **803**
    Mechanistic designations 808

Index **815**

1 Models of chemical bonding

1.1 COVALENCY AND MOLECULAR STRUCTURE

An understanding of chemical reactivity begins with an understanding of chemical bonding, the forces which render certain aggregates of atoms (i.e. the familiar molecules) more stable than others[1-3]. It is on this basis that chemical reactions — changes in bonding — may be approached and a rational and consistent theory of organic chemistry devised. Two milestones in the understanding of bonding may be quoted. The first, the recognition of the electron-pair covalent bond by Lewis[4] and by Langmuir[5] in 1919, still provides a model for the description of molecular structure adequate for most purposes and will be extensively employed in the following text. According to this concept, valence electrons are shared so as to create filled shell configurations and are regarded as essentially localized in the internuclear space. For the first row elements of which organic compounds are almost entirely composed, this is the octet ($2s^2$, $2p^6$); for hydrogen, $1s^2$. The second leap in understanding was made by the introduction of quantum mechanics to chemistry following the molecular orbital description of bonding in the hydrogen molecule by Heitler and London[6], in 1929. This approach superseded the concept of localized electrons and paved the way to quantitative understanding of bonding, the satisfactory calculations of bonding energies, optimum bond lengths and geometries. It will be necessary to turn to these methods, despite the necessity of somewhat lengthy computation, when the need arises to consider specific molecular orbital properties (for example, in the theory of pericyclic reactions, Chapter 14). Nonetheless, quantum concepts permeate any descriptions of chemical bonding, though two rather distinct models may be used which will now be briefly described.

1.1.1 The Valence Bond (VB) model[1]

We know the structure of a molecule in that it contains defined atoms located precisely in space. One begins with this determinate part of molecular structure (which can be obtained accurately, by X-ray crystal diffraction for instance) by setting all nuclei in their correct spatial positions.

The indeterminate part, the disposition of the bonding electrons, is then accomplished by adding these in pairs such that no atom exceeds its closed shell number. There are inevitably many ways in which this can be achieved, each localized structure (known as a 'contributing' structure or, in the older literature, a 'canonical' structure) being regarded as contributing, in some measure determined by its energy, to the true structure. The molecule is conceived as a 'resonance hybrid' of many contributing structures whose contribution can be expressed as a fraction entering into the whole. The relationship between contributing structures, *which differ only in the distribution of valence electrons,* is expressed by the double arrow, ↔. For the purposes of exact calculations of molecular energies, for instance, many contributing structures are needed even for a molecule such as H_2; it is frequently found that a single VB structure suffices to describe the structure adequately for qualitative purposes. For example, methane may be represented as **1** and contributions from such structures as **2** ignored for the interpretation of reaction mechanisms. If, however, we need to express the dipolar nature of chloromethane by this formalism, structure **3** is not sufficient; a contribution from **4** needs to be considered, a VB expression of the inductive effect, i.e. the tendency of electronegative elements to attract bonding electrons towards themselves. The interactions between neighboring π-systems is frequently expressed by 'resonance', the inclusion of two or more VB structures in the molecular description. Contributing structures **5b**, **5c** express the slight shortening and double-bond character of the C2–C3 bond while the many contributing structures included in the VB description of benzene, **6**, are an expression of its six-fold symmetry, not apparent in any single structure. The disadvantage of

2

this system for the qualitative description of molecular structure is that it lacks compactness. Consequently, it is usual to write a single principle VB structure, unless it is desired to emphasize certain properties, with the implicit understanding that it represents the real molecule, the superposition of many such structures.

1.1.2   The Molecular Orbital (MO) model[7-10]

Starting again with the framework of nuclei correctly located, the problem of electron distribution is attacked by calculating the permitted solutions to the quantum-mechanical Schrödinger equation, each of which is known as a molecular orbital (MO) and corresponds to a definite energy state and distribution for a pair of electrons. The valence electrons are then permitted to occupy MOs from the lowest energy upward until all are accounted for. The total energies and electron distributions can be obtained by summation of the individual values for the occupied MOs. The method will be discussed in more detail in Section 1.2. The individual MOs are a combination of the constituent atomic orbitals (AOs) and may be approximately localized as in the σ-bonds of methane or, in the case of a conjugated π-system, completely delocalized over all its atoms, as shown for example in Fig. 1.1 for buta-1,3-diene.

1.2   APPROXIMATE MOLECULAR ORBITAL THEORY

Many of the underlying principles of organic chemistry may now be rationalized by the application of molecular orbital theory. It is a remarkable and fortunate fact that whereas the quantitative application of quantum principles to molecules requires very large and elaborate computational facilities, qualitative understanding is best served by a very approximate approach. Calculations may now be carried out routinely on quite large molecules with the aid of a microcomputer. This section outlines the approximations to be made for obtaining energies and electron distributions according to the Hückel method first introduced in 1931.

1.2.1   The Hückel Molecular Orbital (HMO) method[11-17]

In principle, we wish to calculate the molecular orbitals of large molecules, i.e. the wavefunctions of all electrons moving within the attractive field of all nuclei in the molecule and the average repulsive field of all other electrons. Such wavefunctions are solutions to the Schrödinger equation for the molecule and are of enormous complexity. For example, to evaluate the MOs of benzene we must take into account H-1s and C-1s, 2s, 2p_x, 2p_y, 2p_z orbitals — altogether 36 AOs leading to 666 interaction terms [$n(n+1)/2$, where n = number of AOs]. If H-2s and C-3s, 3p AOs are included this increases the total to 2211 individual terms. The problem was attacked by making some drastic assumptions; this procedure may be justified if the results obtained in this way are seen to relate to experimental physical and chemical properties.

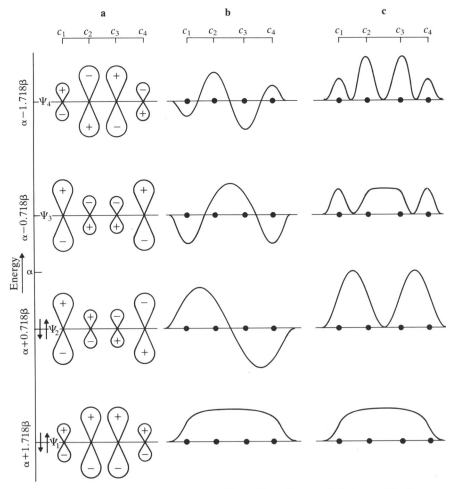

Fig. 1.1 Representations of the molecular orbitals of butadiene. **a**, Relative contributions and phases of each C-2p_z AO to the four MOs; **b**, as for **a**, in graphical form; $c(\psi)$ against a distance parameter; **c**, electron densities: $c^2(\psi)$ against a distance parameter; **d**, sketches of the four MOs.

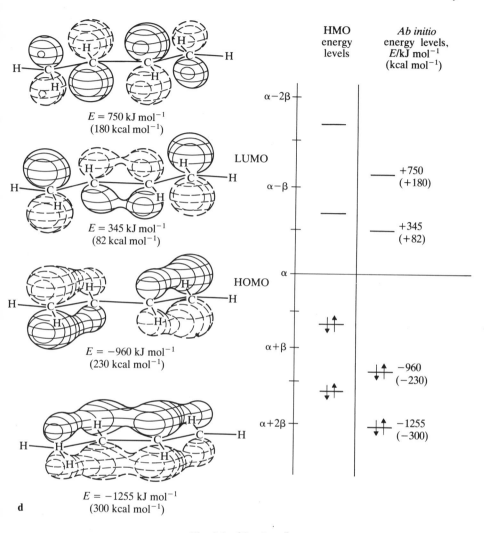

HMO
energy
levels

Ab initio
energy levels,
E/kJ mol^{-1}
(kcal mol^{-1})

$E = 750$ kJ mol^{-1}
(180 kcal mol^{-1})

LUMO

$E = 345$ kJ mol^{-1}
(82 kcal mol^{-1})

HOMO

$E = -960$ kJ mol^{-1}
(230 kcal mol^{-1})

$E = -1255$ kJ mol^{-1}
(300 kcal mol^{-1})

d

$\alpha - 2\beta$

$\alpha - \beta$

α

$\alpha + \beta$

$\alpha + 2\beta$

$+750$
($+180$)

$+345$
($+82$)

-960
(-230)

-1255
(-300)

Fig. 1.1   (*Continued*)

Assumption 1

Electrons which occupy σ-orbitals are regarded as being localized in the region of the two atoms bound and to have properties (energies, electron distributions, etc.) which are constant and unaffected by other parts of the molecule. This relative constancy of σ-bond properties will be justified in Section 1.3. Thus, the molecular wavefunction is separated into two parts, Ψ_σ and Ψ_π and the calculations will focus on the second, the interaction of carbon $2p_z$ AOs to form the appropriate π-MOs. This reduces the basis set to one AO per conjugated atom. In the case of benzene this will now include only the six C-$2p_z$ orbitals with an interaction matrix of only 21 terms. The development of the theory will be carried through for the case of buta-1,3-diene.

Buta-1,3-diene

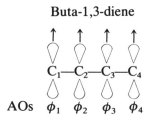

AOs ϕ_1 ϕ_2 ϕ_3 ϕ_4

The interaction matrix, permutating all pairs of AOs, is:

$$
\begin{vmatrix}
A_{11} & A_{12} & A_{13} & A_{14} \\
A_{21} & A_{22} & A_{23} & A_{24} \\
A_{31} & A_{32} & A_{33} & A_{34} \\
A_{41} & A_{42} & A_{43} & A_{44}
\end{vmatrix}
$$

where the terms A_{ij} take account of interaction between atomic orbitals i, j. Since $A_{ij} = A_{ji}$, the number of independent terms by symmetry is 10 rather than 16.

Now, A_{ij} takes the form:

$$A_{ij} = (\mathbf{H}_{ij} - E\mathbf{S}_{ij}), \tag{1.1}$$

where

$$\mathbf{H}_{ij} = \int \phi_i \mathcal{H} \phi_j \, \partial\tau \tag{1.2}$$

and

$$\mathbf{S}_{ij} = \int \phi_i \phi_j \, \partial\tau. \tag{1.3}$$

The Hamiltonian operator, $\mathcal{H}$, contains the sum of kinetic and potential energy terms. The next point to be considered is the values which may be assigned to these integrals.

Assumption 2

Terms $\mathbf{H}_{ii}$ (i.e. those terms which occur on the diagonals of the interaction matrix) physically represent the energy of an electron in the field of its own nucleus and are known as *Coulomb integrals*. If we assume we are dealing with an all-carbon π-system, then this term represents the energy of an electron in a C-$2p_z$ orbital. This may be regarded as a constant and could be assigned a value (e.g. the first ionization potential of carbon = 1085 kJ mol^{-1}

(259 kcal mol^{-1})) but, since we are interested in relative rather than absolute values of energy this integral will be denoted α.

For butadiene, $H_{11} = H_{22} = H_{33} = H_{44} = \alpha$.

Assumption 3

Terms H_{ij} $(i \neq j)$ refer to the energy of electron i in the field of nucleus j or the interaction between AOs i and j: hence they are known as *resonance integrals*. It will be assumed that:

For adjacent (i.e. σ-bonded) i, j, H_{ij} = constant = β.

For non-adjacent i, j, $H_{ij} = 0$.

Hence for butadiene,

$$H_{12} = H_{23} = H_{34} = H_{21} = H_{32} = H_{43} = \beta$$

and

$$H_{13} = H_{14} = H_{24} = H_{31} = H_{41} = H_{42} = 0.$$

Assumption 4

Terms S_{ij} physically signify the spatial distribution which is common to AOs i and j, i.e. the amount of interpenetration or overlap. Consequently, they are known as *overlap integrals*. The Hückel treatment makes the assumptions that

$$S_{ii} = 1, \qquad S_{ij} \, (i \neq j) = 0.$$

The zero-overlap might be considered too severe an approximation. In practice the results are not greatly affected by the inclusion of values of S_{ij} for overlap between contiguous atoms, where $0 < S_{ij} < 1$.

Now, the interaction matrix for C-$2p_z$ orbitals of butadiene may be written as follows:

Butadiene

$$\begin{bmatrix} (\alpha - E) & \beta & 0 & 0 \\ \beta & (\alpha - E) & \beta & 0 \\ 0 & \beta & (\alpha - E) & \beta \\ 0 & 0 & \beta & (\alpha - E) \end{bmatrix} = 0.$$

Each such expression is an $i \times i$ symmetric matrix whose determinant is a polynomial of order i and whose roots, the eigenvalues, are the energies of the molecular orbitals which we seek. The computation is simplified by dividing through by β and setting

$$(\alpha - E)/\beta = x, \qquad (1.4)$$

whence

$$\begin{bmatrix} x & 1 & 0 & 0 \\ 1 & x & 1 & 0 \\ 0 & 1 & x & 1 \\ 0 & 0 & 1 & x \end{bmatrix} = 0.$$

Multiplying out the determinant into polynomial form one obtains

$$x^4 - 3x^2 + 1 = 0,$$

the roots of which are

$$x = +1{\cdot}618, \ +0{\cdot}618, \ -0{\cdot}618, \ -1{\cdot}618,$$

and, since from Eq. (1.4) $E = \alpha - x\beta$, the eigenvalues may be expressed as

$$E = \left.\begin{array}{l} \alpha - 1{\cdot}618\beta \quad \psi_4 \\ \alpha - 0{\cdot}618\beta \quad \psi_3 \end{array}\right\} \text{antibonding}$$

$$E = \left.\begin{array}{l} \alpha + 0{\cdot}618\beta \quad \psi_2 \\ \alpha + 1{\cdot}618\beta \quad \psi_1 \end{array}\right\} \text{bonding}$$

Since both α and β are negative, the solutions which correspond to the most negative energies (i.e. the most stable MOs) are those with the most negative x.

Attention is now turned to electron densities at each atom. Each MO is the linear combination of its constituent AOs:

$$\psi = c_1\phi_1 + c_2\phi_2 + c_3\phi_3 \ldots$$

in which the coefficients c_i express the contribution of the ith AO. Each line of the interaction matrix relates to one of a set of simultaneous equations (*secular equations*) containing the MO coefficients, c_i, as follows:

$$\begin{array}{l} c_1x + c_2 + 0 + 0 = 0, \\ c_1 + c_2x + c_3 + 0 = 0, \\ 0 + c_2 + c_3x + c_4 = 0, \\ 0 + 0 + c_3 + c_4x = 0, \end{array}$$

together with the normalization condition,

$$c_1^2 + c_2^2 + c_3^2 + c_4^2 = 1.$$

These equations are solved to obtain values of the four coefficients for each orbital, i.e. for each of the four values of x.

	ψ_1	ψ_2	ψ_3	ψ_4
c_1	0·3717	0·6015	0·6015;	0·3717
c_2	0·6015	0·3717	−0·3717	−0·6015
c_3	0·6015	−0·3717	−0·3717	0·6015
c_4	0·3717	−0·6015	0·6015	−0·3717
Symmetry about mirror plane	S	A	S	A

The values of c or c^2 in Fig. 1.1 signify the extent to which each of the four AOs contributes to the four derived MOs. The MO parameters for this compound are now fully characterized. Before progressing to use this information to correlate interesting chemical properties, the following general procedure may be adopted to determine Hückel Molecular Orbital (HMO) parameters for any conjugated π-system.

HMO recipe

(1) Write down the formula of the system to be analyzed and number each atom. The order does not matter but it is necessary to identify atom i with coefficient c_i. All atoms must be part of a single conjugated system.
(2) Write down the interaction matrix such that all diagonal elements (Coulomb integrals) are x and all off-diagonal elements are either 1 if they refer to an adjacent pair of atoms or 0 if they refer to a non-adjacent pair. Note that the matrix must be symmetrical about the diagonal.
(3) Find the eigenvalues of the matrix by a suitable diagonalization routine with the aid of a computer. Programs which do this job are usually available in computer libraries; they may also print out the eigenvectors (coefficients). Eigenvalues of some conjugated systems are given in Table 1.1.

Further examples: C_6 isomers

Hexa-1,3,5-triene   Fulvene   Benzene

$$\begin{bmatrix} x & 1 & 0 & 0 & 0 & 0 \\ 1 & x & 1 & 0 & 0 & 0 \\ 0 & 1 & x & 1 & 0 & 0 \\ 0 & 0 & 1 & x & 1 & 0 \\ 0 & 0 & 0 & 1 & x & 1 \\ 0 & 0 & 0 & 0 & 1 & x \end{bmatrix} \quad \begin{bmatrix} x & 1 & 0 & 0 & 0 & 0 \\ 1 & x & 1 & 0 & 0 & 1 \\ 0 & 1 & x & 1 & 0 & 0 \\ 0 & 0 & 1 & x & 1 & 0 \\ 0 & 0 & 0 & 1 & x & 1 \\ 0 & 1 & 0 & 0 & 1 & x \end{bmatrix} \quad \begin{bmatrix} x & 1 & 0 & 0 & 0 & 1 \\ 1 & x & 1 & 0 & 0 & 0 \\ 0 & 1 & x & 1 & 0 & 0 \\ 0 & 0 & 1 & x & 1 & 0 \\ 0 & 0 & 0 & 1 & x & 1 \\ 1 & 0 & 0 & 0 & 1 & x \end{bmatrix}$$

Table 1.1 Hückel solutions for energy levels of some conjugated systems: Values of $(-x)$ where $E = (\alpha - x)\beta$

	System	ψ_1	ψ_2	ψ_3	ψ_4	ψ_5	ψ_6	ψ_7
1		−1·414	0	1·414				
2		−2·000	1·000	1·000				
3		−1·618	−0·618	0·618	1·618			
4		−1·732	0	0	1·732			
5		−2·000	0	0	2·000			
6		−2·170	−0·311	1·000	1·480			
7		−2·214	−1·000	0·539	1·000	1·675		
8		−1·732	−1·000	0	1·000	1·732		
9		−2·000	−0·619	−0·619	1·618	1·618		
10		−1·802	−1·247	−0·445	0·445	1·247	1·802	
11		−1·932	−1·000	−0·518	0·518	1·000	1·932	
12		−1·90	−1·18	0	0	1·18	1·90	
13		−2·000	−1·000	−1·000	1·000	1·000	2·000	
14		−2·115	−1·000	−0·618	0·254	1·618	1·860	
15		−2·247	−0·802	−0·555	0·555	0·802	2·247	
16		−2·175	−1·126	0	0	1·126	2·175	
17		−2·414	−0·618	−0·618	0·414	1·618	1·618	

Table 1.1   (*Continued*)

	System	ψ_1	ψ_2	ψ_3	ψ_4	ψ_5	ψ_6	ψ_7
18	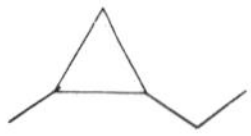	−2·334	−1·099	−0·274	0·594	1·374	1·740	
19	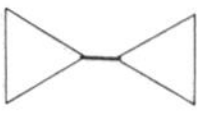	−2·414	−1·732	0·414	1·000	1·000	1·732	
20	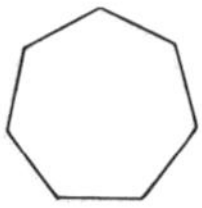	−2·000	−1·247	−1·247	0·445	0·445	1·801	1·801
21	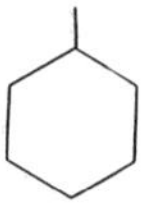	−2·101	−1·259	−1·000	0	1·000	1·259	2·101
22		−1·848	−1·414	−0·765	0	0·765	1·414	1·848
23	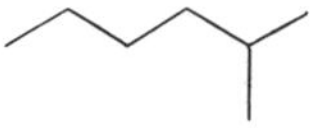	−1·931	−1·414	−0·518	0	0·518	1·414	1·931
24	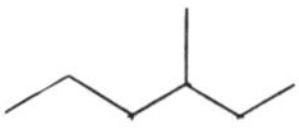	−1·970	−1·285	−0·684	0	0·684	1·285	1·970
25	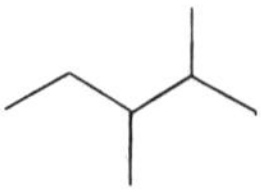	−2·053	−1·209	−0·570	0	0·570	1·209	2·053
26	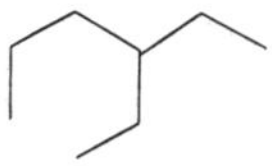	−2·000	−1·000	−1·000	0	1·000	1·000	2·000

(4)   The orbitals are now populated with the available π-electrons in pairs, beginning with the orbital of lowest energy and working upwards (*Aufbau* principle). Those orbitals lower in energy than α are bonding orbitals (e.g. **11**) while those of higher energy than α are antibonding orbitals. Molecular orbitals for which $E = \alpha$ are non-bonding orbitals (NBMOs), which will be discussed below.

Depiction of the MOs

Figure 1.1, showing the MOs of butadiene, illustrates various approaches to depicting the geometries and nodal properties of the MOs. The indication of individual C-2p AOs on each atom, drawn to scale relative to c or c^2 (Fig. 1.1b,c), is useful when symmetry properties are important. A plot of c^2 along

11

the molecular framework helps to visualize the electron distribution for each MO (Fig. 1.1c).

Alternant and non-alternant systems

If we 'mark' with an asterisk alternate atoms of a conjugated system, two types emerge. *Alternant* systems are those for which no pair of adjacent atoms bears the same designation; *non-alternant* systems are those for which it is impossible to avoid one pair of adjacent atoms bearing the same designation. It is conventional to 'star' the larger set.

The following examples will clarify this principle; the categories are further divided into even- and odd-membered types.

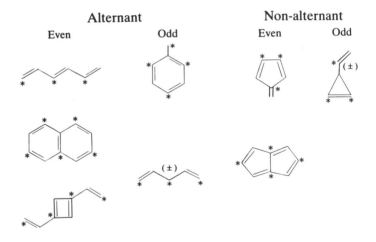

1.2.2 Properties of Hückel molecular orbitals

The following generalizations can be made.

(1) Alternant hydrocarbon systems (AHs) have MOs symmetrically arranged about $E = \alpha$ on the energy scale. A further division may be made into odd- and even-numbered systems. An even AH will have an equal number of bonding and antibonding orbitals only, while an odd AH will have, in addition, one non-bonding MO ($E = \alpha$). Non-alternant systems will not have a symmetric arrangement of orbitals.

(2) Electron densities of even AHs are 1·000 on each carbon. Odd AHs must have a cationic or anionic charge or an odd electron; the charge or odd electron density is equally divided between starred atoms (the larger set), e.g. **7**. Non-AHs will in general show a marked disparity of charge at each carbon (for example, **8**) which means the compound has a dipole moment, in accordance with observations.

(3) Cyclically conjugated hydrocarbons (annulenes) can be divided into two

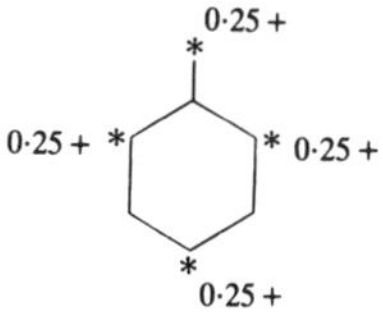

0·25 +
*

0·25 + *   * 0·25 +

0·49 −

0·12 +

*
0·25 +

0·18 +   0·18 +

Benzyl cation

7

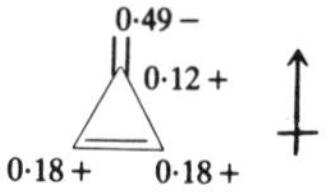

Methylenecyclopropene

8

sets by two different approaches:

(a) Even-numbered rings, as for example in benzene (**10**), have a symmetric set of MOs, the lowest and highest having energies ($E = \alpha \pm 2\beta$) while the others occur in degenerate pairs (pairs of the same energy though of different symmetry).

Odd-numbered rings, necessarily with ionic or radical character, have the lowest-energy MO at ($\alpha - 2\beta$), the others being arranged in non-symmetric pairs upwards. The cyclopentadienide anion, **9**, is an example. The roots of

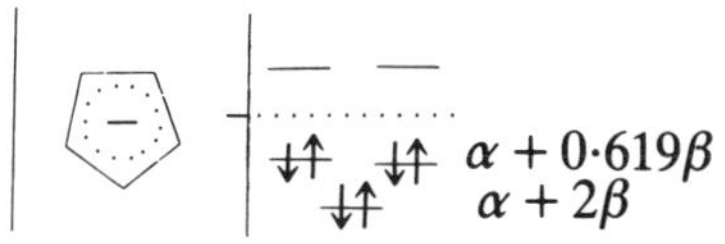

$\alpha + 0.619\beta$
$\alpha + 2\beta$

9

the secular equations for the annulenes (total number of carbons $= 2k$), are given by Eq. (1.5):

$$x = -2\cos(l\pi/k), \quad (l\pi/k) \text{ is in radians} \qquad (1.5)$$

where

$$l = 0, +1, +2 \ldots k.$$

Hence, energies of MOs are given by Eq. (1.6):

$$E_l = \alpha + 2\beta \cos(l\pi/k). \qquad (1.6)$$

(b) A more fundamental division from the viewpoint of chemistry is made between annulenes with ($4n + 2$) π-electrons and those with $4n$ electrons, n being a positive integer including 0. The ($4n + 2$) series, having 2, 6, 10, 14, 18 ... delocalized π-electrons, have all their bonding orbitals filled. They are therefore closed-shell molecules of more than usual stability, and are denoted *aromatic,* benzene being the prime example (**10, 11**).

The $4n$ series, on the other hand, have two electrons sharing a pair of non-bonding MOs. They are therefore of low delocalization energy and stability and their chemistry is that of highly reactive molecules. They are

13

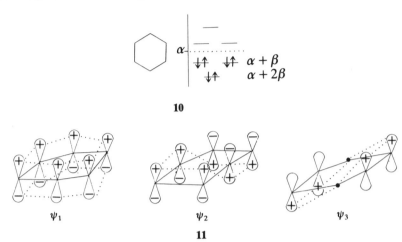

10

ψ_1 ψ_2 ψ_3

11

denoted *anti-aromatic*. The transient molecule cyclobutadiene, **12**, exemplifies this type. (The Hückel MO energies would indicate zero resonance energy for cyclobutadiene and the term 'non-aromatic' is sometimes applied. Refined calculations suggest its stability to be even lower still while the ring currents of circulating π-electrons in $4n$ systems are in the opposite sense to those of $4n + 2$ (aromatic) molecules. Hence the designation 'anti-aromatic' is appropriate.) It is a remarkable fact that the highly approximate solutions to a wave equation rationalize the utterly different chemistry of benzene and cyclobutadiene.

(4) It is possible to include heteroatoms within the scope of simple HMO theory. While the calculations performed above concern implicitly all-carbon systems, the same solutions are appropriate for all homo-atomic systems. The calculated MO energies for the molecule N_2 would take the same form as those for ethene; however, α and β would take different values from those appropriate to carbon. This suggests that if we could differentiate the Coulomb and resonance integrals in the interaction matrix, heteroatom systems could be treated. This is achieved in a purely empirical way. The computation is as before, with the following adjustments.

12

The Coulomb integral appropriate to the heteroatom is given by $(\alpha + \delta\beta)$, where δ is a constant which depends upon the atom, and a further modification may be made by setting the Coulomb integral at carbon adjacent to the heteroatom at $(\alpha + 0 \cdot 1\delta\beta)$.

Table 1.2 Suggested parameter values for use with heteroatom calculations $\alpha_M = \alpha + \delta_M \beta$; $\beta_{C-M} = \rho_{C-M} \nu$

Heteroatom, M	Coulomb integral increment, δ_M	Resonance integral coefficient, ρ_{C-M}
—B<	−1	0·7
—C=C—	0	1·1
—N<	0·5	0·8
—N=	0·5	1·0
=N⁺<	2	1·0
—Ö—	2	0·8
=O	1	1·0
=O⁺—	2·5	0·8
—F	3	0·7
—Cl	2	0·4
—Br	1·5	0·3
—CH₃	2	0·7

Resonance integrals for bonds between carbon and the heteroatom become $\rho\beta$, where ρ depends upon the heteroatom.

Atomic constants δ and ρ recommended for use in heteroatom calculations are given in Table 1.2.

(5) Other useful derived quantities are defined as follows.

$$\text{Total } \pi\text{-energy} = \sum_{}^{occ_i} nE_i,$$

i.e. the sum of the energies (in α and β units) of each occupied orbital multiplied by its occupancy, n (1 or 2 electrons).

$$\text{Electron density at atom } i = \sum_{}^{occ_i} nc_i^2,$$

i.e. the sum of the squares of the coefficients c_i at that atom over each occupied orbital multiplied by its occupancy.

π-bond order of the bond between atoms r and s,

$$p = \sum_{}^{occ_i} 2c_{ir}c_{is}.$$

Total bond order, $P = p + 1$

(i.e. 1 for the σ-bond).

(6) Chemical reactivity indices can be derived from HMO parameters. Reaction rates in a series of structurally related compounds depend upon changes in the energies of activation (Chapter 2), whereas transition-state properties are not calculable by the HMO approximation. Nonetheless, several indices of reactivity have been proposed and depend upon several quite different approaches.

Reagent state properties

Intermolecular attractions depend upon the charges upon bonding atoms and upon the amount of overlap of the two interacting orbitals; either term (see Eq. (1.6)) may be dominant. Thus, a cationic reagent would be attracted toward the site of highest charge density (e.g. C4 in the methylenecyclopropene example) and an anion to that of lowest. Such a reaction might be described as under charge control.

Alternatively, the greatest interaction might be governed by orbital overlap, which is greater the larger the coefficients of the frontier orbitals at the atoms which become bound. This is likely to be the case when both reacting molecules are uncharged and is the principle governing frontier orbital control, a reactivity index of wide application exemplified in later sections. Fukui has defined a reactivity index called *superdelocalizability*,

$$s_r = \sum_{}^{occ_i} c_{jr}^2 / E_j,$$

where s_r refers to atom r for which c_{jr} is the coefficient at r of the jth MO and E_j is its energy.

Both these approaches view the reaction at the initial stages of interaction and are applications of perturbation MO theory.

A related index is *free valence, F.* This is defined as the difference between total bond order, N, at an atom of a conjugated system and the maximum value possible, N_{max}. The latter is the total bond order at trimethylenemethane, 4·732, no larger one being known. Free valence is a measure of the additional bonding capacity of an atom.

Product properties

Though the activated complex, by virtue of the partial bonding it contains, is not amenable to simple MO calculations, it may closely resemble the product (a 'late' transition state), which thereby serves as a model. Bond formation at a conjugated center will remove that center from conjugation so the π-energy of the system will be reduced. For example, benzene when accepting an electrophile to form a benzenium ion (Section 10.2) loses 2·54β

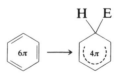

of π-energy. This is known as *localization energy, L_+,* the energy cost of removing an atom from the π-system. The lower the value of L_+, the more reactive the compound. Examples of localization energies in electrophilic displacements (L_+) are as follows.

Free valence, F	0·536	1 10·545 2 20·447	1 10·608 2 20·422 9 90·811	o 0·502 m 0·407 p 0·518 α 0·517 β 1·055
Localization energy, L_+	2·536	1 12·299 2 22·480	1 12·252 2 22·400 9 92·010	o 2·370 m 2·546 p 2·424 α 2·424 β 1·704

The change in π-energy for coordination of a nucleophile would involve two more electrons in the product and is denoted L_-.

Worked example: Hückel MO characteristics of methylenecyclopropene (NB: a non-AH)

$$\begin{array}{c} CH_2 \\ \| \\ C \\ \diagup \backslash \\ HC = CH \end{array}$$

(1) Secular determinant

$$\begin{bmatrix} x & 1 & 1 & 1 \\ 1 & x & 1 & 0 \\ 1 & 1 & x & 0 \\ 1 & 0 & 0 & x \end{bmatrix} = 0$$

(2) Eigenvalues and eigenvectors

Symmetry (xz-plane) x	S 2·1700	S 0·3111	A −1·0000	S −1·4812
	ψ_1	ψ_2	ψ_3	ψ_4
c_1	0·6116	0·2536	0·0000	0·7494
c_2	0·5227	−0·3682	0·7071	−0·3020
c_3	0·5227	−0·3682	−0·7071	−0·3020
c_4	0·2818	0·8152	0·0000	−0·5059

17

(3) Energy level diagram

$$
\begin{array}{ll}
\text{—} & \alpha - 1 \cdot 4812\beta \\[4pt]
\text{—} & \alpha - 1 \cdot 0000\beta \\[10pt]
\uparrow\downarrow & \alpha + 0 \cdot 3111\beta \\[4pt]
\uparrow\downarrow & \alpha + 2 \cdot 1700\beta
\end{array}
$$

π-energy $= (2 \times 2 \cdot 1700) + (2 \times 0 \cdot 3111) = 4 \cdot 962\beta$

Delocalization energy (DE) $= 4 \cdot 962 - 4 \cdot 000 = 0 \cdot 962\beta$

(4) MO geometries

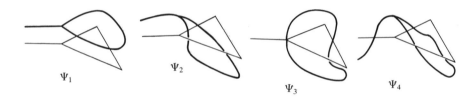

$\Psi_1 \qquad \Psi_2 \qquad \Psi_3 \qquad \Psi_4$

(5) Charge densities

$$q_i = \sum^{\text{occ}_i} nc_i^2$$

C1: $q_1 = (2 \times 0 \cdot 6116^2) + (2 \times 0 \cdot 2536^2) \quad = 0 \cdot 8767$ Predicted

C2: $q_2 = (2 \times 0 \cdot 5227^2) + (2 \times (-0 \cdot 3682^2)) = 0 \cdot 8176$ dipole

C3: $q_3 = (2 \times 0 \cdot 5227^2) + (2 \times 0 \cdot 3682^2) \quad = 0 \cdot 8176$

C4: $q_4 = (2 \times 0 \cdot 2818^2) + (2 \times 0 \cdot 8152^2) \quad = \underline{1 \cdot 4880}$

$\qquad\qquad\qquad\qquad\qquad\qquad\qquad 3 \cdot 9996 = 4e$

(6) Bond orders

$$p_{rs} = \sum_{i=1} 2c_{ir}c_{is} \quad (= \pi\text{-bond order; add 1 for total bond order}).$$

C1—C2 = C1—C3:
$$p_{12} = p_{13} = (2 \times 0 \cdot 6116 \times 0 \cdot 5227) + (2 \times 0 \cdot 2536 \times (-0 \cdot 3682)) = 0 \cdot 4526.$$

C2—C3:
$$p_{23} = (2 \times 0 \cdot 5227 \times 0 \cdot 5227) + (2 \times (-0 \cdot 3682) \times (-0 \cdot 3682)) = 0 \cdot 8176.$$

C1—C4: $p_{14} = (2 \times 0 \cdot 6116 \times 0 \cdot 2818) + (2 \times 0 \cdot 2536 \times 0 \cdot 8152) = 0 \cdot 7582.$

Total bond orders:

$$P_{12} = P_{13} = 1 \cdot 4526;$$
$$P_{23} = 1 \cdot 8176;$$
$$P_{14} = 1 \cdot 7582.$$

(7) Bond lengths

$$R = 1 \cdot 54 - \frac{1 \cdot 54 - 1 \cdot 32}{1 + 0 \cdot 765\left(\dfrac{2 - P}{P - 1}\right)}.$$

Hence,

$$R_{12} = R_{13} = 1 \cdot 54 - \frac{0 \cdot 22}{1 + 0 \cdot 765\left(\dfrac{2 - 1 \cdot 4526}{1 \cdot 4526 - 1}\right)} = 1 \cdot 426 \text{ Å}$$

$$R_{23} = 1 \cdot 54 - \frac{0 \cdot 22}{1 + 0 \cdot 765\left(\dfrac{2 - 1 \cdot 8176}{1 \cdot 8176 - 1}\right)} = 1 \cdot 352 \text{ Å}$$

$$R_{14} = 1 \cdot 54 - \frac{0 \cdot 22}{1 + 0 \cdot 765\left(\dfrac{2 - 1 \cdot 7582}{1 \cdot 7582 - 1}\right)} = 1 \cdot 363 \text{ Å}$$

(8) Free valence

$$F_r = N_{\text{max}} - N_r,$$

where N_r = bond number, $N_{\text{max}} = 4 \cdot 732$.

C1: $F_1 = 4 \cdot 732 - (P_{12} + P_{13} + P_{14})$

$$= 4 \cdot 732 - (1 \cdot 4526 + 1 \cdot 4526 + 1 \cdot 7582) = 0 \cdot 0686$$

C2, C3: $F_2 = F_3 = 4 \cdot 732 - (1 + 1 \cdot 8176 + 1 \cdot 4526) = 0 \cdot 4618$

C4: $F_4 = 4 \cdot 732 - (2 + 1 \cdot 7582) = 0 \cdot 9738$ (most reactive position).

(9) Localization energy

         C2, C3   C4

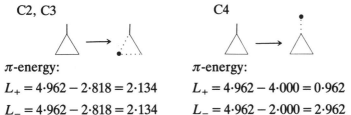

    π-energy: π-energy:

    $L_+ = 4 \cdot 962 - 2 \cdot 818 = 2 \cdot 134$ $L_+ = 4 \cdot 962 - 4 \cdot 000 = 0 \cdot 962$

    $L_- = 4 \cdot 962 - 2 \cdot 818 = 2 \cdot 134$ $L_- = 4 \cdot 962 - 2 \cdot 000 = 2 \cdot 962$

Nucleophilic attack at C4 is thus predicted to be facile.

1.2.3 The relationship between MO and VB models[18]

Both these approaches at their simplest tend to give an extreme picture of molecular structure. Molecular orbital structures emphasize the delocalized nature and the lack of correlation between electron motions — the latter due to calculation of the properties of each MO without regard to the occupancy of the others, the 'one-electron approximation'. Valence bond theory, on the other hand, emphasizes the localized nature of electron pairs and the extreme correlation of electronic motions implicit in a single VB structure. The truth lies somewhere in between. This can be shown diagrammatically by linking corresponding states, as in Fig. 1.2. A more precise MO theory would predict energy levels to be somewhere between the two extremes in accordance with experimental observations of, for example, photoelectron spectroscopy, which can probe the energies of valence electrons in a molecule.

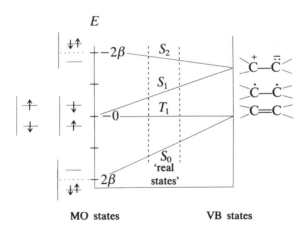

Fig. 1.2 Diagram showing correlation between molecular states as described by simple MO and VB models.

1.2.4 Advanced MO methods[6,19]

Hückel MOs are highly approximate and limited in application to alternant hydrocarbons so that much more exact solutions to the wave equation of a molecule are needed for the investigation of energy levels and reactivity. The starting point for *ab initio* calculations is the same, the Hamiltonian or energy operator, which contains several terms — electronic kinetic energy, electron–nuclear attractions, electron–electron and nuclear–nuclear repulsions — all of which are summed over all possible combinations of nuclei and electrons. Simplification can be achieved by prescribing a particular molecular geometry and by limiting the basis set of AOs used. For first row elements $2s$ and $2p$ orbitals are seldom sufficient and the participation of 3*s*, 3*p* and 3*d* orbitals may be included. Each AO is given explicit mathematical form which is computationally convenient and no separation into σ- and π-types is made initially. Solutions of the wave equation are then obtained iteratively so as to minimize the energy of the system, for which the aid of a very large computer is essential. Programs such as GAUSSIAN 76 are widely available so that computations at this level are within the capabilities of many non-specialists. Programs of this type are under continual revision and are becoming more and more accurate as computers become more powerful. A further constraint on acceptable solutions is the requirement that MOs are either symmetrical or anti-symmetrical with respect to symmetry elements of the molecule. Having minimized the energy with one particular geometry, the calculation may be repeated with small changes in bond lengths and angles until a true minimum is reached which should correspond to the real molecule. The resulting MOs look quite different from those to which one is accustomed from the application of localized bond models. All are spread over the whole molecule and may be designated σ- or π-types according to whether they possess an axis or a plane of symmetry. For a molecule such as methane, Fig. 1.3a, the four MOs appear as four combinations for the localized C–H σ-bonds, three of which are of equal energy and higher than that of the fourth. Experimental measurements show there are two distinct energy levels; this is not apparent from the localized bond model. A few other examples will suffice for the present; others will be introduced in later chapters. Ethene (Fig. 1.3b) has a total of 12 valence electrons: therefore there are 12 MOs of which six are bonding and are illustrated together with their actual energies. The lowest MOs are too deep-seated to be much concerned in chemical reactions. The important ones for this purpose are the HOMO and LUMO — the *frontier orbitals* — which look much like the conventional π and π^* MOs. It would be too space-consuming to show here the whole set of MOs for a large molecule; most of them are not relevant to its chemistry, so for most purposes attention will be drawn to the energies and electronic distributions of the frontier orbitals. A ketone such as acetone (Fig. 1.3c) for instance, has a HOMO (constructed largely from an unshared pair) in which the electron density is mainly on oxygen and a LUMO mainly located on carbonyl carbon. Electron-donating capacity is expected to be at oxygen and electron-accepting capacity at carbon, and this is the reactivity pattern observed. Acrolein (Fig. 1.3d) shows a large lobe of its LUMO both at carbonyl carbon and at β-carbon in accordance with the principle of homology making both of these electrophilic centers.

Semi-empirical MO methods

Full self-consistent field (SCF) treatment involves inclusion of terms which contain three and four AOs (three- and four-center integrals) known as

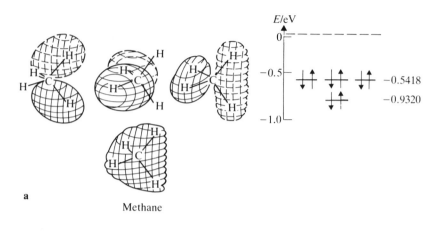

a

Methane

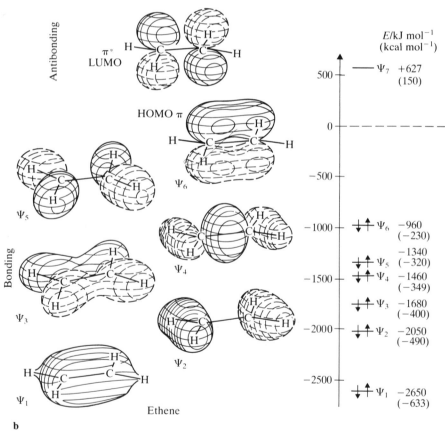

b

Fig. 1.3 Examples of all-valence electron *Ab initio* MOs. **a**, Methane; **b**, ethene (bonding MOs and LUMO); **c**, acetone (frontier orbitals); **d**, acrolein (frontier and adjacent MOs).

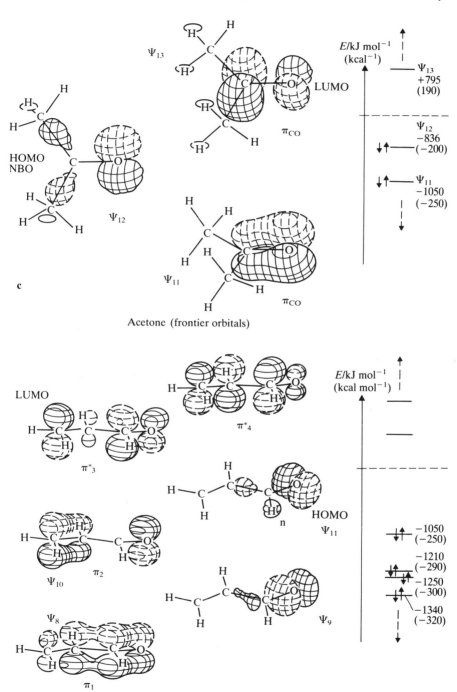

Acetone (frontier orbitals)

c

Propenal (acrolein) (frontier orbitals)

d

differential overlap. A three-centre term, for instance, would signify interaction between the overlap region of AOs *i, j* and *k*. Such terms are troublesome to evaluate and many computational methods adopt simplifying procedures yielding MO information of varying degrees of approximation. The CNDO (Complete Neglect of Differential Overlap) method ignores all such terms. The INDO version (Intermediate Neglect of Differential Overlap) includes some terms, as does MINDO (Modified INDO). Such methods are computationally easier and can yield satisfactory predictions of properties such as dipole moments, relative energies and optimum geometries.

1.3 PROPERTIES OF COVALENT BONDS[3-16]

1.3.1 Bond lengths

The equilibrium internuclear distance of a covalent bond, the bond length, is a resultant of the attractive and repulsive forces depending, respectively,

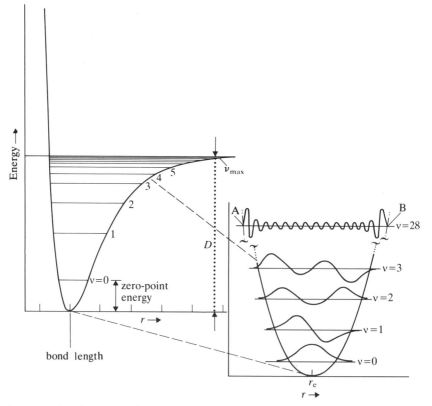

Fig. 1.4 The Morse potential energy curve of energy against interatomic distance for a covalent bond showing vibrational levels: the magnified diagram on the right indicates the vibrational wave functions for a few levels.

upon the degree of overlap of AOs and internuclear electrostatic interactions. These together dictate the shape of the potential energy–bond length function (Fig. 1.4). It turns out that a hard-sphere model works quite well for the computation of bond lengths, each atom being assigned a characteristic 'covalent radius' from which bond lengths may be predicted by summation. Covalent radii (Table 1.3) vary with bond order but are reasonably constant over a wide range of structures. Bond lengths are affected by the state of hybridization of the atoms bonded. The more s-character in a hybrid AO (Table 1.4) the closer it lies to the nucleus and the more 'stubby' its geometry so that effective overlap will require a closer approach. The additional repulsive interaction is compensated by π-overlap, **13**, which increases monotonically as the two atoms approach.

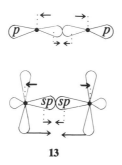

13

Multiple bonds tend therefore to be shorter than single bonds. Another major source of perturbation of bond length is conjugation. Since bond orders are not necessarily integral, the degree of double- or triple-bond character of a given bond may vary widely, giving rise to bond lengths which lie between those of the extreme types. The relationship between bond order and length is shown in Fig. 1.5; a semi-empirical relationship due to Coulson (Eq. (1.7)), may also be used:

$$r = 154 - \frac{20}{1 + 0.765[(2 - P)/(P - 1)]}, \tag{1.7}$$

where $P =$ bond order, obtainable from HMO calculations: see Section 1.2.2, worked example).

Covalent radii increase with atomic number within a group but decrease across a period on account of the greater attraction of an increased nuclear charge (electronegativity) on the valence electrons.

1.3.2 Interbond angles[24]

Both p-orbitals and hybrids are directional in character: the interorbital angles between pairs of AOs are: p, 90°; sp^3, 109·9° (tetrahedral); sp^2, 120° (trigonal); sp, 180° (linear).

Table 1.3 Covalent radii of some elements, r/pm (1 Å = 100 pm)

Bond type					
	H				
—X	28				
		B	C	N	O	F
—X		88	77	70	66	64
=X			66·5	60	55	
≡X			60·2	55		
			Si	P	S	Cl
—X			117	110	104	99
			Ge	As	Se	Br
—X			122	121	117	114
			Sn	Sb	Te	I
—X			140	140	137	133

Average lengths, r, of some covalent bonds X—Y

X	Y	Context	r/pm
C (sp^3)	C (sp^3)	alkane	154
C (sp^3)	C (sp^3)	>C—C—Ar	153
C (sp^3)	C (sp^3)	>C—C—C=	153
C (sp^3)	C (sp^2)	>C—C=	146
C (sp^2)	C (sp^2)	>C=C<	134
C (sp^2)	C (sp^2)	>C=C=C<	131
C (sp^2)	C (sp^2)	=C—C=	142
C (sp^2)	C (sp^2)	arene	140
C (sp)	C (sp)	—C≡C—	120
C (sp)	C (sp)	≡C—C≡	136
C (sp^3)	O (sp^3)	ether	143
C (sp^2)	O (sp^2)	>C=O	123
C (sp^2)	O (sp^2)	>C=C—C=O	121
C (sp^2)	O (sp^2)	ketene	117
C (sp^3)	N (sp^3)	amine, ammonium	148
C (sp^3)	N (sp^2)	>C—N=	147
C (sp^2)	N (sp^3)	Ar—N<	143
C (sp^2)	N (sp^2)	imine, pyridine	135
C (sp^2)	N (sp^2)	isocyanate	132
C (sp)	N (sp)	cyanide	116
C (sp^3)	S (sp^3)	thiol, sulfide	181
C (sp^2)	S (sp^3)	=C—S	173
C (sp^2)	S (sp^2)	thione	171

Table 1.3   (*Continued*)

Some average bond lengths, r/pm			
H—H	74·13	C—As	196
D—D	74·14	N≡N	110
C—H	109	N=N	126
O—H	97	N—N	143
N—H	104	O=O	120·7
F—H	91·7	O—O	148
Cl—H	127·5	S=O	145
Br—H	140·8	C—F	136
I—H	160·0	=C—F	132
P—H	143	C—Cl	177
S—H	134	C—Br	193
Si—H	146	C—I	214
Ge—H	152		

Bond angles of exactly these values are usually found only in extremely symmetric compounds though deviations are usually only a few degrees in a wide range of structures (Table 1.5). Excluding the demands of ring size, this is usually attributable to electrostatic forces between ligands including unshared pairs and also, occasionally, to fractional hybridization (Table 1.4).

Table 1.4   *s*-Character of some carbon atoms[20-23]

Compound	s-Character/%	Compound	s-Character/%
CH$_4$	25	X—CH$_3$	
CH$_2$=CH$_2$	33	X = CH$_3$	23·4
HC≡CH	50	X = OH	25
		X = SH	25·7
		X = F	25·9
CH$_2$ / \ CH$_2$—CH$_2$	30	X = Cl	26·3
		X = Br	26·9
		X = I	27·1
CH$_2$ / \ HC=CH	44	X = N(CH$_3$)$_2$	24·1
(triangle) a b	a 29 b 40		

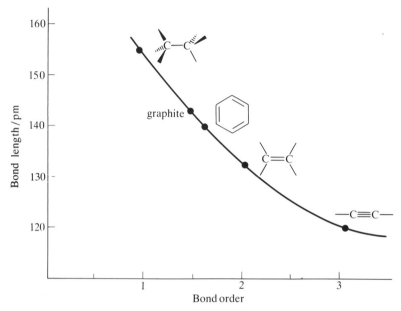

Fig. 1.5 The relationship between bond length and bond order.

1.3.3 Force constants[2-29]

A covalent bond, at least in the lower vibrational states, acts as a harmonic oscillator whose frequency, v, and the masses of the two fragments bonded, m_1, m_2, are connected by the classical (Hooke's Law) equation:

$$v = \frac{1}{2\pi} \left(\frac{f}{\mu}\right)^{\frac{1}{2}} \qquad (1.8)$$

where $\mu = m_1 m_2/(m_1 + m_2)$, the reduced mass of the system.

The force constant, f, is a measure of the restoring force resulting from unit bond extension or bending and so characterizes the stiffness of the covalent bond.

$$f = [\text{force}] \times [\text{length}^{-1}]$$

and, since energy = [force] × [length],

$$f = [\text{energy}] \times [\text{length}^{-2}]; \text{ units, m dyne Å}^{-1} = \text{aJ Å}^{-2}.$$

Force constants may be obtained with great precision from infrared or microwave spectra; typical values are given in Table 1.6. To put these values in perspective, we may calculate the energy required to stretch a

Table 1.5   Examples of interbond angles

H, H, C, CH_3, CH_3 — $111°$	(chair cyclohexane) $110°$	$\equiv C-C$, H, H — $109°$
H, H, C, Cl, Cl — $112°$	$H-C$, Cl, Cl, Cl — $112°$	H, H — $118°$; $60°$ (cyclopropene-type)
H, H, $C=C$, H, H — $110°$	(cyclopentadiene) $110°$, $101°$	H, H, H, H, H — $122°$, $125°$
H, H, $C=O$ — $120°$	$H-C$, O, OH — $121°$	O, CH_3, CH_3 — $111°$
	S, CH_3, CH_3 — $100°$	Cl, Cl, $C=O$ — $114°$
CH_3, CH_3, CH_3, N — $110°$	CH_3-N, O, O — $130°$	

single bond $(f = 5 \times 10^{-4}\,aJ\,nm^{-2})$ by 10 pm (0·1 Å), i.e. by some 7% of its length.

$$\text{Energy} = fl^2 = 5 \times 0{\cdot}1^2 \times 6{\cdot}02 \times 10^{23} \times 10^{-3} \times 10^{-18}$$
$$= 30{\cdot}1\,kJ\,mol^{-1}\ (7{\cdot}19\,kcal\,mol^{-1}).$$

(The bond dissociation energy will be about 400 kJ mol^{-1} [96 kcal mol^{-1}].) The energy required to bend a bond $(f = 0{\cdot}5\,aJ\,(\text{radian})^{-2})$ by 10° (0·175 rad) is calculated as

$$0{\cdot}5 \times 0{\cdot}175 \times 0{\cdot}175 \times 6{\cdot}03 \times 10^{23} \times 10^{-3} \times 10^{-18}$$
$$= 9{\cdot}2\,kJ\,mol^{-1}\ (2{\cdot}2\,kcal\,mol^{-1}).$$

This is comparable with thermal energies at ordinary temperatures so we can see that bond bending by appreciable angles costs very little energy in comparison with bond stretching. This has a considerable bearing on molecular geometry. Thus, it is possible for compounds with highly distorted bond angles, such as those with three- and four-membered rings, to exist and be quite stable although the strain inherent in these structures may be a driving force towards unusual reactivity. The potential energy

Table 1.6   Force constants, f, of some covalent bond vibrations [25-29]
Carbon-stretching modes

Bond	Context	$f/aJ\ Å^{-2}$ [a]
C—H	R—CH$_3$	4·70
C—H	R—CH$_2$—R	4·53
C—H	—O—CH$_3$	4·74
C—H	—S—CH$_3$	4·66
C—H	O=C—CH$_3$	4·80
C—H	R—CH$_2$—F	4·76
C—H	R—CH$_2$—Cl	4·87
C—H	R—CH$_2$—Br	4·91
C—H	R—CH$_2$—I	4·92
C—H	=C—H	5·9
C—H	≡C—H	5·3
C—C	R—CH$_2$—CH$_2$—R	4·45
C—C	R—CH$_2$—CH$_2$—O	4·51
C—C	R—CH$_2$—CH$_2$—S	4·42
C—C	CH$_3$—C=O	5·08

[a] 1 aJ Å^{-2} = 1 mdyne Å^{-2}

Other stretching modes

Bond	$f/aJ\ Å^{-2}$ [a]	Bond	$f/aJ\ Å^{-2}$ [a]	Bond	$f/aJ\ Å^{-2}$ [a]
Si—H	2·8	C—N	4·7	C—F	5·7
Ge—H	2·6	C—O	5·1	C—Cl	3·4
N—H	6·2	C—Si	2·8	C—Br	2·9
P—H	3·1	C=O	9·6	C—I	2·3
As—H	2·9	C≡C	10·8	H—F	8·8
Sb—H	2·3	C≡C	16	H—Cl	4·8
		C≡N	18	H—Br	3·9
				H—I	2·9

[a] 1 aJ Å^{-2} = 1 mdyne Å^{-2}

Bending modes

Bonds	$f/aJ\ rad^{-2}$	Bonds	$f/aJ\ rad^{-2}$
—CH$_3$ (S)	0·58–0·68	—CH$_2$ (S)	0·55–0·59
—CH$_3$(A)	0·53–0·55	—CH$_2$ (A)	0·76–0·80
—CH$_3$ (rock)	0·63–0·85		
C—CH$_2$—C	0·94		
C—O—C	1·21		
C—S—C	1·36		

Torsional modes

Bonds	$f/aJ\ rad^{-2}$ [a]
CH$_3$—CH$_2$—	0·107
R—CH$_2$—OCH$_2$—	0·098
R—CH$_2$—SCH$_2$—	0·045

[a] 1 aJ Å^{-2} = 1 mdyne Å^{-2}.

curve for a bond, Fig. 1.4 shows that the force constant will diminish as the bond is stretched. The values quoted refer to the situation near the equilibrium position.

1.3.4 Bond and molecular dipole moments[30,31]

The asymmetry of electronic charge in a covalent bond may be expressed as a dipole moment. An electric dipole exists when two opposite charges, $+q$ and $-q$, are separated by distance r. The dipole moment, **p**, is given by $\mathbf{p} = qr$. The SI units are coulomb meters (C m) but the older units of Debye (D) are normally used (1 D = 3.336×10^{-30} C m).

To put matters in perspective, if two atoms each bearing one electronic

Table 1.7 Pauling–Allred[32] electronegativity values, χ. The scale was originally derived from the relationship

$$D_{(A-B)} = 1/2[D_{(A-A)} + D_{(B-B)}] + 23.06(\chi_A - \chi_B)^2$$

where D = bond dissociation energy.

H						
2·2						
Li	Be	B	C	N	O	F
0·98	1·57	2·04	2·55	3·04	3·44	3·98
Na	Mg	Al	Si	P	S	Cl
0·93	1·31	1·61	1·90	2·19	2·58	3·16
			Ge	As	Se	Br
			2·01	2·18	2·55	2·96
						I
						2·66

Bond dipole moment contributions, **p**, and ionicities

Bond	p/D	Ionicity/%	Bond	p/D	Ionicity/%
C—H	0·35	6	C—O	0·89	17
O—H	1·24	25	C—N	0·49	9
N—H	0·84	16	C—F	1·43	30
F—H	1·78	39	C—Cl	0·61	11
Cl—H	0·96	18	C—Br	0·41	7
Br—H	0·76	14	C—I	0·11	1·8
I—H	0·46	8	C—S	0·03	0·5

Group contributions

Group	p/D	Group	p/D
>C=O	2·25	—C≡N	3·44
—OCH₃	0·81	—CONH₂	3·4
>C=O	ca 3·2	—SO₂—	4·25
—ONO₂	2·73		

unit of charge ($1 \cdot 6022 \times 10^{-19}$ C) are separated by 150 pm,

$$\mathbf{p} = \frac{(1 \cdot 6022 \times 10^{-19}) \times (150 \times 10^{-12})}{3 \cdot 336 \times 10^{-30}} = 7 \cdot 2 \, \text{D}.$$

Bond dipole moments are reasonably constant for a given type of covalent bond and may be conveniently calculated from Pauling electronegativity values[32], χ (Table 1.7), as follows.

(a) The negative end of the dipole will be on the atom with the greater χ.
(b) The bond dipole contribution, $\mathbf{p}$, is given by

$$\mathbf{p}/D = (\chi_A - \chi_B). \tag{1.9}$$

(c) The ionicity of the bond, i.e. the percentage contribution of the ionic structure, A^+B^-, in VB terms, is given by

$$\text{Ionicity} = 16\mathbf{p} + 3 \cdot 5\mathbf{p}^2. \tag{1.10}$$

Molecular dipole moments, $\mathbf{P}$, may be obtained experimentally (Table 1.8); they are the vector sum of the bond moments, a fact which may be of use in

Table 1.8 Selected values of experimental molecular dipole moments
Values are for the gas phase but often differ significantly if measured in solution or by different workers.

Compound	p/D	Compound	p/D
$CHHal_3$		CH_3X	
Hal = F	1·6	X = F	1·8
Hal = Cl	1·0	X = Cl	1·94
Hal = Br	1·0	X = Br	1·79
Hal = I	0·8	X = I	1·6
CH_3COX		X = OH	1·71
X = F	2·96	X = OMe	1·3
X = Cl	2·71	X = SMe	1·45
X = Br	2·45	X = SH	1·26
X = H	2·68	X = SeMe	1·32
X = OMe	1·77	X = CN	4·0
C_6H_5X		X = NO_2	3·4
X = F	1·35	X = NH_2	1·29
X = Cl	1·75	X = SCN	3·34
X = Br	1·7		
X = I	1·7		
X = NO_2	4·0		

ortho	1·72	ortho	4·63
meta	2·5	meta	3·4
para	0	para	2·6

structure determination. All molecules which are centrosymmetric must have $\mathbf{P} = 0$ and conversely all those with $\mathbf{P} \neq 0$ are not centrosymmetric. For example, dichloromethane has a molecular dipole moment, $\mathbf{P} = 1\cdot6\,\text{D}$; this clearly excludes a planar centrosymmetric structure; on the other hand, vector addition of the bond dipoles would predict a value $\mathbf{P} = 1\cdot2\,\text{D}$, showing a discrepancy due both to experimental error and the approximate nature of the additivity relationship. Dipolar properties give rise to electrostatic forces of attraction or repulsion between molecules and make an important contribution to the forces by which reactions are initiated; see Section 1.4.

1.3.5 Molecular and bond polarizabilities [33–35]

Electrons, being charged particles, respond to an external electric field. Since the bonding electrons in a molecule have a degree of mobility, the charge distribution will be affected by an electric field which creates an induced dipole in addition to any permanent dipole which may be present. The direction of the response in the molecule is such as to tend to diminish the electric field gradient and is known as *polarizability*. The phenomenon is of great importance in chemistry since ionic reagents produce in their vicinity very large field gradients which can induce dipoles in neighboring molecules and thereby bring about additional attractive forces which may lead to reaction. Many other physical properties depend upon polarization; the most familiar is the refraction of light. The passage of a light wave is accompanied by an oscillating electric field at right angles to the direction of propagation which produces a corresponding oscillating dipole in a nearby molecule. This interaction reduces the velocity of propagation of the wave, which is to say that the refractive index, n, of the material medium is greater than 1. The more easily polarized the molecule, the higher is the refractive index. This shows that polarizability is a dynamic property and that electrons, because of their very small mass, can respond to very high frequencies ($>10^{16}\,\text{Hz}$ for visible light). Polarizability, $\mathbf{b}$, of a molecule is defined as the induced dipole produced by unit field strength (units: dipole moment/field gradient $= [qr/qr^{-2}] = [r^3]$, i.e. volume). In general, if the field is located at an arbitrary position with respect to the molecule, the action will occur at some angle to this direction but there will be three orientations for which this angle is zero, the three principal polarizabilities, $\mathbf{b}_1, \mathbf{b}_2, \mathbf{b}_3$, which together give the total electronic polarizability $\mathbf{P}_E$, of the molecule:

$$\mathbf{P}_E = 4\pi L(\mathbf{b}_1 + \mathbf{b}_2 + \mathbf{b}_3)/9, \qquad (1.11)$$

where L is Avogadro's number. The molecular polarizability, $\mathbf{P}_E$, is obtained experimentally from the molar refraction at wavelength λ, R_λ, such that

$$\mathbf{P}_E = R_\infty = R_{\lambda(\lambda^{-2} \to 0)},$$

and

$$R_\lambda = \frac{M(n^2-1)}{d(n^2+2)} \qquad (1.12)$$

(the Lorentz–Lorentz equation; M is molecular weight, d density). Since refractive index varies with the wavelength used for measurement, R_λ is obtained from measurement of n at two wavelengths, when

$$R_\lambda = \frac{R_1 R_2(\lambda_2^2 - \lambda_1^2)}{(\lambda_2^2 R_2 - \lambda_1^2 R_1)}. \qquad (1.13)$$

Polarizabilities may be factored into atomic contributions but, more successfully, into bond contributions (Table 1.9). Furthermore the polarizability of a bond has three directional components, one longitudinal, $\mathbf{b}_L$, and two transverse, $\mathbf{b}_T$ and $\mathbf{b}_V$, which are frequently identical. The three may be separated and values are shown in Table 1.9. The partition of polarizability into bond contributions enables estimates to be made of molecular

Table 1.9   Bond contributions to molecular polarizabilities/10^{-3} nm$^{-3\,a}$ and refractivities, R

Bond	Context	b_L	b_T	b_V	R
C—H	alkane	0·65	0·65	0·65	1·644
C—C	alkane	0·97	0·26	0·26	1·254
C—C	cyclopropane				1·437
C—C	cyclobutane				1·323
C=C	alkene	2·80	0·73	0·77	1·254
C≡C	alkyne	3·79	1·26	1·26	5·670
Ar—Ar	biphenyl				2·546
C—F	MeF	1·2	0·4	0·4	
C—Cl	MeCl	3·18	2·2	2·2	6·361
C—Cl	t-BuCl	3·94	1·81	1·81	
C—Cl	PhCl	4·2	1·9	1·5	
C—Br	MeBr	4·65	3·1	3·1	9·064
C—Br	t-BuBr	5·98	2·6	2·6	
C—Br	PhBr	6·4	2·4	2·2	
C—I	MeI	6·7	4·8	4·8	13·92
C—I	t-BuI	9·2	3·7	3·7	
C—I	PhI	9·1	5·3	3·3	
C—O	ether	0·9	0·46	0·46	1·49
C—O	acetal				1·43
C=O	ketone	2·3	1·4	0·5	3·24
N—H	ammonia	0·5	0.83	0·83	1·76
C—N	amine	0·57	0·7	0·7	1·49
N—N	hydrazine				1·80
N≡N	azo				3·97
C=N	imine				3·509
C≡N	cyanide				4·718
C—S	sulfide	1·9	1·7	1·7	
C—CN	t-BuCN	4·0	1·5	1·5	
S—S	disulfide				7·72

^{a} 10^{-6} Å^{-3}

Table 1.9   (*Continued*)

Molecular polarizabilities, $\alpha/10^{-24}$ cm^3

Molecule	$\alpha/10^{-24}\,cm^3$	Molecule	$\alpha/10^{-24}\,cm^3$
H_2	0·790	$CHCl_3$	—
N_2	1.76	CH_2Cl_2	6·48
CO_2	2·65	CH_3Cl	8·23
CO	1·95	CH_4	2·60
HF	[0·51]	CH_3OH	3·23
HCl	2·63	CH_3CH_2OH	—
HBr	3·61	C_6H_6	10·32
HI	5·45	$C_6H_5CH_3$	—
H_2O	1·48	$o\text{-}C_6H_4(CH_3)_2$	—
NH_3	2·21	He	0·2
CCl_4	10·1	Ar	1·6

polarizabilities for a given molecular geometry. The values may be compared with experimental ones, thus providing a useful probe into conformation (Section 8.4).

It will be seen that bond polarizability varies somewhat with the chemical environment but especially high contributions are due to π-bonds, especially aromatic structures, and to atoms towards the bottom right hand side of the periodic table, paralleling the ease of removal of an electron. Being a sum of bond contributions, refractivity will increase with the molecular complexity of the compound.

Exaltation of refractivity

Molecules with extended conjugation often show refractivities in excess of the values calculated from bond additivities. This difference is known as the *exaltation* of refractivity and is a measure of additional electron mobility in the molecule. Exaltation is especially large for aromatic molecules and serves as an experimental criterion of aromaticity. Examples are shown in Table 1.10

Ab initio calculations have been used to estimate the polarizations of π-systems by adjacent dipoles. For the generalized geometry (**A**) the polarization energy, E_{pol}, of the C$=$Y bond by the C—X dipole is given[36]

Table 1.10   Exaltation of refractivity

	R_{obs}	R_{calc}	*Exaltation*
Hexa-1,3,5-triene	30·58	28·52	2·06
p-Benzoquinone	27·98	26·93	1·05
Naphthalene	91·9	61·4	30·5
Anthracene	130·3	81·7	48·6

Table 1.11 Polarizations of π-systems (Eq. (1.14))[36]

X:	F	OH	CN	Me	OMe	CHO	NO_2
(a) $CH_2{=}CH_2$ H—X ($r = 400$ pm) $10^4\,dq^a$ (electrons)	−122	−75	−120	2	−75	−56	−171
(b) H—X ($r = 400$ pm) $10^4\,dq$ at C1	−42	—	−52	−3	−31	−21	−73

^{a} dq is the change in electronic charge brought about by approach of the two molecules to the specified distance r.

by:

$$E_{pol} = \varepsilon(2 \cos \theta \cos \phi - \sin \theta \sin \phi)/r^3 \qquad (1.14)$$

(A)

Some examples of this relationship are given in Table 1.11.

1.3.6 Bond dissociation enthalpies (BDE)[37,38]

The standard enthalpy (heat) change, ΔH (see Section 2.1) for the process

$$A{-}B(gas) \rightarrow A^{\cdot} + B^{\cdot}$$

is known as the bond dissociation enthalpy of the bond A—B, expressed as $D_{A{-}B}$, and is the vertical distance between the lowest vibrational state and the energy zero on the potential curve, Fig. 1.4. This is one of the most important bond characteristics since the balance of energy gained and lost in the bond reorganizations which constitute a chemical reaction may determine whether it occurs at all. In order to bring order to a large body of BDE data, it is necessary to break them down into contributions from small structural units. Additivity does not apply to atomic contributions to bonding. If it did, a reaction such as

$$H{-}H + F{-}F \rightarrow 2H{-}F$$

should have zero enthalpy of reaction instead of the $535\,kJ\,mol^{-1}$ ($128\,kcal\,mol^{-1}$) exothermicity which is observed: i.e. $2D_{H{-}F} \gg D_{H{-}H} +$

Table 1.12   Average bond dissociation energies, $D/\text{kJ mol}^{-1}$ (kcal mol^{-1})
Single bonds X—Y

X	C	H	O	N	S	F	Cl	Br	I
					Y				
C	330	410	330	275	235	425	325	270	240
	(78·9)	(98·0)	(78·9)	(65·7)	(56·1)	(101·6)	(77·7)	(64·5)	(57·4)
H		431	455	385		564	425	362	295
		(103)	(109)	(92·0)		(135)	(101)	(86·5)	(70·5)
O		142							

Single bonds X—X

F—F	Cl—Cl	Br—Br	I—I
155	239	189	149
(37·0)	(57·1)	(45·1)	(35·6)

Double bonds

C=C	C=O
585	750
(140)	(179)

Triple bonds

C≡C	C≡N	N≡N
806	898(?)	841
(193)	(215)	(201)

$D_{\text{F—F}}$, a fact attributable to increased bond strengths between atoms of differing electronegativity[39]. Additivity is more applicable to specific bond types and some average values are given in Table 1.12 which are found to apply to a wide variety of compounds with an uncertainty of perhaps $\pm 10\,\text{kJ mol}^{-1}$ ($\pm 2\cdot4\,\text{kcal mol}^{-1}$), still too approximate for quantitative applications.

Summation of all the BDEs of a molecule gives the enthalpy of formation from the constituent atoms, normally referred to as the heat of atomization. For example, for ethane:

$$2C + 6H \rightarrow C_2H_6.$$

$$\Delta H = D_{\text{C—C}} + 6D_{\text{C—H}}$$

$$= -367 - (6 \times 409) = -2824\,\text{kJ mol}^{-1}\ (-674\cdot9\,\text{kcal mol}^{-1}).$$

Experimental value $= -2814\ (-672\cdot6)$

However, heats of formation are not normally based on atoms but on the elements in their (arbitrary) standard states — carbon as graphite, and hydrogen, oxygen, nitrogen as diatomic gases at STP. Calculation of heats of formation, ΔH_f, must include the addition of the heats of atomization of the elements, constants that are given in Table 1.13.

Table 1.13 Heats of formation of some atoms, small molecules and radicals

	$\Delta H_f/kJ\,mol^{-1}$	$\Delta S_f/J\,K^{-1}\,mol^{-1}$		$\Delta H_f/kJ\,mol^{-1}$	$\Delta S_f/J\,K^{-1}\,mol^{-1}$
C	+714	+158	$:CH_2$	+360	+193
F	+79	+165	$CH_2{=}\dot{C}H$	+289	+235
Cl	+121	+165	$:CF_2$	−181	+240
H	+218	+114	$:CCl_2$	+196	+266
D	+221	+123	$\cdot CH_3$	+143	+194
I	+107	+181	$\cdot CF_3$	−470	+266
O	+249	+161	$\cdot CCl_3$	+77	+295
S	+279	+168	$\cdot C_2H_5$	+110	+242
NO	+90·2	+210	$\cdot C_6H_5$	+328	+290
CO	−110·3	+197	$C_6H_5CH_2^{\cdot}$	+188	+314
O_3	+142	+238	$CH_3O^{\cdot}$	+14·5	+15
NO_2	+33	+240	$CH_3S^{\cdot}$	+140	+136
N_2O	+82	+220	CH_3NH	+190	+240
CO_2	−393	+213	$C_6H_5CO^{\cdot}$	+111	+326
CS_2	+117	+237	$HCO^{\cdot}$	+37·6	+224
N_2H_4	+95	+238	$H_2O(l)$	−285·8	
H_2O_2	−136	+234	$H_2O(g)$	−241·8	+189
SO_3	−395	+256			
NH_3	−46	+192			
C_6H_6	+82·7	+269			
$C_6H_5CH_3$	+50	+320			

Now,

$$2C \text{ (graphite)} \rightarrow 2C \quad \Delta H = +2\times714$$
$$3H_2 \rightarrow 6H \qquad\qquad +6\times217$$
$$2C + 6H \rightarrow C_2H_6 \qquad \underline{-2814}$$
$$\Delta H_f = -84\,\text{kJ mol}^{-1}\,(-20\cdot0\,\text{k cal mol}^{-1}),$$

or $14\,\text{kJ mol}^{-1}$ ($3\cdot35\,\text{kcal mol}^{-1}$) per C—H bond, which should be contrasted with $D_{C-H} = 435\,\text{kJ mol}^{-1}$ ($104\,\text{kcal mol}^{-1}$).

1.3.7 Group additivities to bond enthalpies

The variation of bond dissociation energies in the context of different structures is too great for values in Table 1.12 to be used for calculation of unknown heats of formation which can match the precision of experimental values, often obtainable to $\pm1\,\text{kJ mol}^{-1}$ ($\pm0\cdot2\,\text{kcal mol}^{-1}$). The values in Table 1.14 illustrate this point. Even within this small selection of values, variation of some 15% on either side of the mean is apparent. Additivity may be improved by collating contributions to ΔH_f from each type of bond, taking into account the variation which exists within different bonding situations, that is with the ligands attached to the bonding atoms. This type of analysis has been carried out by Benson and coworkers, some of whose

Table 1.14   Calculated bond dissociation energies for the reaction $R—H \rightarrow R^{•} + H^{•}$

R	$D_{R—H}/kJ\,mol^{-1}\,(kcal\,mol^{-1})$
CH_3	= 435 (104·0)
CF_3	444 (106·0)
$CHMe_2$	394 (94·2)
CMe_3	387 (92·5)
$CH{=}CH_2$	453 (108·3)
$C{\equiv}CH$	500 (119·5)
$HC{=}O$	363 (86·7)

data are appended in Tables 1.13 and 1.15[40,41]. Tabulations such as these are necessarily larger than Table 1.12 since, for any given atom contributing to ΔH_f of a molecule, a large series of ligand permutations must be considered separately. However, the inconvenience is offset by improved accuracy. The Tables allow the computation of accurate heats of formation

Table 1.15   Group contributions to ΔH_f and ΔS_f[40,41]

Key   Each group is specified as a central polyvalent atom followed by the appropriate ligands.

Central atoms

A	Tetrahedral carbon	**E**	Carbonyl carbon
B	Olefinic carbon	**F**	Oxygen, —O—
C	Acetylenic carbon	**G**	Nitrogen, —N<
D	Aryl carbon		

Ligands

1	H	9	NO_2
2	Saturated carbon	10	CN
3	Olefinic carbon	11	F
4	Aryl carbon	12	Cl
5	Acetylenic carbon	13	Br
6	Oxygen, —O—	14	I
7	Carbonyl	15	Sulfur, —S—
8	Nitrogen, —N—		

Each group is designated by a letter indicating the central atom and up to 15 digits denoting the numbers of ligands of the types 1–15 in the above order. The total number of ligands possible depends on the central atom: **A**, 4; **B**, 2; **C**, 1; **D**, 1; **E**, 2; **F**, 2; **G**, 3. When all have been specified, the numbers of any successive ligands must be zero and the designating digits are omitted.

A–3, 1 indicates the unit

A–2, 1, 0, 0, 1 indicates

E–1, 0, 1 indicates

Table 1.15   (*Continued*)

A–0, 3, 0, 0, 0, 1 indicates

B–1, 0, 0, 0, 0, 0, 0, 0, 0, 1 indicates

Central atom: tetrahedral carbon

1	2	3	4	5	6	7	8	9	10	11	12	13	14	15	ΔH_f/kJ mol^{-1}	ΔS_f/J K^{-1} mol^{-1}
A–3	1														−42·6	+127
2	2														−20·6	+39
1	3														−7·9	−50
0	4														+2·1	−147
2	1	1													−19·8	+41
2	0	2													−17·9	+43
2	0	1	1												−17·9	+43
2	1	0	0	1											−19·8	+43
2	1	0	1												−20·3	+39
1	2	1													−6·2	−49
1	2	0	0	1											−7·2	−47
1	2	0	1												−4·1	−51
0	3	1													+7·0	−145
0	3	0	1												+11·7	−147
3	0	0	0	0	1										−42	+127
2	1	0	0	0	1										−33·8	+41
2	0	1	0	0	1										−27	
2	0	0	1	0	1										−34	
2	0	0	0	1	1										−27	
2	0	0	0	0	2										−67	
1	2	0	0	0	1										−30·1	−46
1	1	0	0	0	2										−68	
0	3	0	0	0	1										−27	−140
3	0	0	0	0	0	1									−42	+127
2	1	0	0	0	0	1									−21·7	+40
2	0	1	0	0	0	1									−15·8	
2	0	0	1	0	0	1									−22·6	
2	0	0	0	0	0	2									−31·8	
1	2	0	0	0	0	1									−7·1	−50
0	3	0	0	0	0	1									+5·8	
3	0	0	0	0	0	0	1								−42	+127
2	1	0	0	0	0	0	1								−27·6	+41
1	2	0	0	0	0	0	1								−21·7	−50
0	3	0	0	0	0	0	1								−13·3	−142
2	1	0	0	0	0	0	0	1							−63·1	+202
1	2	0	0	0	0	0	0	1							−66·0	+112
2	1	0	0	0	0	0	0	0	1						+94·0	+168
1	2	0	0	0	0	0	0	0	1						+107	+83
0	3	0	0	0	0	0	0	0	1						+121	−12
2	1	0	0	0	0	0	0	0	0	1					−215	+148
1	2	0	0	0	0	0	0	0	0	1					−205	+60
2	1	0	0	0	0	0	0	0	0	0	1				−69	+158
1	2	0	0	0	0	0	0	0	0	0	1				−61·8	+73
1	2	3	4	5	6	7	8	9	10	11	12	13	14	15	ΔH_f/kJ mol^{-1}	ΔS_f/J K^{-1} mol^{-1}

Table 1.15   (*Continued*)

1	2	3	4	5	6	7	8	9	10	11	12	13	14	15	$\Delta H_f/kJ\,mol^{-1}$	$\Delta S_f/J\,K^{-1}\,mol^{-1}$
0	3	0	0	0	0	0	0	0	0	0	1				−53·5	+22
2	1	0	0	0	0	0	0	0	0	0	0	1			−22·5	+170
1	2	0	0	0	0	0	0	0	0	0	0	1			−14·2	
0	3	0	0	0	0	0	0	0	0	0	0	1			−1·6	−8
2	1	0	0	0	0	0	0	0	0	0	0	0	1		+33	+179
1	2	0	0	0	0	0	0	0	0	0	0	0	1		+44	+89
0	3	0	0	0	0	0	0	0	0	0	0	0	1		+54	0
3	0	0	0	0	0	0	0	0	0	0	0	0	0	1	−42	+127
2	1	0	0	0	0	0	0	0	0	0	0	0	0	1	−23·6	+41
1	2	0	0	0	0	0	0	0	0	0	0	0	0	1	−2·3	−144

Central atom: olefinic carbon

1	2	3	4	5	6	7	8	9	10	11	12	13	14	15	$\Delta H_f/kJ\,mol^{-1}$	$\Delta S_f/J\,K^{-1}\,mol^{-1}$
B–2															+26·2	+115
1	1														+35·9	+33
0	2														+43	−53
1	0	1													+28·3	+27
0	1	1													+37·1	−61
1	0	0	1												+28·3	+27
0	1	0	1												+36	−62
1	0	0	0	1											+28	+27
0	0	2													+19·2	
0	1	0	0	0	1										+43	
0	1	0	0	0	0	1									+31·5	
1	0	0	0	0	0	0	0	0	0	1					−157	+137
1	0	0	0	0	0	0	0	0	0	0	1				−5·0	+147
1	0	0	0	0	0	0	0	0	0	0	0	1			+45	+160
1	0	0	0	0	0	0	0	0	0	0	0	0	1		+102	+169

Central atom: acetylenic carbon

1	2	3	4	5	6	7	8	9	10	11	12	13	14	15	$\Delta H_f/kJ\,mol^{-1}$	$\Delta S_f/J\,K^{-1}\,mol^{-1}$
C–1															+112	+103
0	1														+115	+26
0	0	1													+122	+27
0	0	0	1												+122	+27

Central atom: allenic carbon (no ligands)   +142   +25

Central atom: aryl carbon

1	2	3	4	5	6	7	8	9	10	11	12	13	14	15	$\Delta H_f/kJ\,mol^{-1}$	$\Delta S_f/J\,K^{-1}\,mol^{-1}$
D–1															+13·8	+48
0	1														+23	−32
0	0	1													+23·7	−32
0	0	0	1												+23·7	−32

Central atom: carbonyl carbon

1	2	3	4	5	6	7	8	9	10	11	12	13	14	15	$\Delta H_f/kJ\,mol^{-1}$	$\Delta S_f/J\,K^{-1}\,mol^{-1}$
E–2															−109	219
1	1														−122	146
1	0	0	1												−122	
1	0	1													−122	
1	0	0	0	1											−122	

1	2	3	4	5	6	7	8	9	10	11	12	13	14	15	$\Delta H_f/kJ\,mol^{-1}$	$\Delta S_f/J\,K^{-1}\,mol^{-1}$

Table 1.15   (*Continued*)

Central atom: carbonyl carbon

1	2	3	4	5	6	7	$\Delta H_f/kJ\,mol^{-1}$	$\Delta S_f/J\,K^{-1}\,mol^{-1}$
1	0	0	0	0	0	1	−106	
1	0	0	0	0	1		−134	
0	2						−131	63
0	1	0	1				−129	
0	0	0	2				−108	
0	1	0	0	0	1		−147	62
0	1	0	0	0	0	1	−122	
0	0	1	0	0	1		−134	
0	0	0	1	0	1		−153	
0	0	0	1	0	0	1	−112	
0	0	0	0	0	2		−125	
0	0	0	0	0	1	1	−123	

Central atom: divalent oxygen

	1	2	3	4	5	6	7	$\Delta H_f/kJ\,mol^{-1}$	$\Delta S_f/J\,K^{-1}\,mol^{-1}$
F–2								−241	+188
	1	0	1					−158	+120
	1	0	0	0	0	0	1	−242	+102
	0	1	1					−127	+40
	0	1	0	0	0	1		−18·8	+42
	0	0	2					−137	
	0	0	0	0	0	0	2	−194	

	1	2	3	4	5	6	7	$\Delta H_f/kJ\,mol^{-1}$	$\Delta S_f/J\,K^{-1}\,mol^{-1}$
F–1	1							−158	+121
	1	0	0	0	0	1		−68	+115
	0	2						−97	+36
	0	1	0	1				−192	
	0	1	0	0	0	0	1	−189	+35
	0	0	1	0	0	1			

Central atom: trivalent nitrogen

	1	2	3	4	5	6	7	$\Delta H_f/kJ\,mol^{-1}$	$\Delta S_f/J\,K^{-1}\,mol^{-1}$
G-2	1							+20·0	+124
	0	3						+102	−96
	0	2	1					+102	
	0	2	0	1				+109	

	1	2	3	4	5	6	7	$\Delta H_f/kJ\,mol^{-1}$	$\Delta S_f/J\,K^{-1}\,mol^{-1}$
G–1	2							+64·4	+38
	2	0	1					+20	
	2	0	0	1				+20	

1	2	3	4	5	6	7	$\Delta H_f/kJ\,mol^{-1}$	$\Delta S_f/J\,K^{-1}\,mol^{-1}$

Structural contributions to be used for radicals

A distinction must now be made for both central atoms and ligands which bear an unpaired electron.

Key

                Central atoms
                A′ Tetrahedral carbon   **H′** Radical carbon, C˙
                B′ Olefinic carbon   **I′** Radical nitrogen, N˙
                D′ Aryl carbon

Table 1.15   (*Continued*)

Ligands

1	H	6 Radical O·
2	Tetrahedral C	7 Radical S·
3	Olefinic =C—	8 Radical N·
4	Aryl C	9 Radical —·C=O
5	Radical C·	

Central atom: tetrahedral C

	1	2	3	4	5	6	7	8	9	$\Delta H_f/kJ\ mol^{-1}$	$\Delta S_f/J\ K^{-1}\ mol^{-1}$
A′	-3	0	0	0	1					-42	+127
	2	1	0	0	1					-20·7	+39
	1	2	0	0	1					-7·9	-50
	0	3	0	0	1					+6·3	-147
	2	1	0	0	0	1				+25·4	+152
	1	2	0	0	0	1				+32·6	+61
	0	3	0	0	0	1				+35·9	-31
	2	1	0	0	0	0	1			+135	+163
	1	2	0	0	0	0	1			+148	+74
	0	3	0	0	0	0	1			+156	-22
	2	1	0	0	0	0	0	1		-27·5	+41
	1	2	0	0	0	0	0	1		-21·7	-49
	0	3	0	0	0	0	0	1		-13·3	+142
	3	0	0	0	0	0	0	0	1	-22·6	+278
	2	1	0	0	0	0	0	0	1	-4·2	+190

Central atom: olefinic carbon

	1	2	3	4	5	$\Delta H_f/kJ\ mol^{-1}$	$\Delta S_f/J\ K^{-1}\ mol^{-1}$
B′	-1	0	0	0	1	+35·9	+33
	0	1	0	0	1	+43	-51

Central atom: aryl carbon

	1	2	3	4	5	6	7	8	9	$\Delta H_f/kJ\ mol^{-1}$	$\Delta S_f/J\ K^{-1}\ mol^{-1}$
D′	-0	0	0	0	1					+23	-32
	0	0	0	0	0	0	0	1		—	-40

Central atom: radical carbon

	1	2	3	4	$\Delta H_f/kJ\ mol^{-1}$	$\Delta S_f/J\ K^{-1}\ mol^{-1}$
H′	-2	1			+150	+128
	1	2			+156	+45
	0	3			+159	-45
	2	0	1		+97	+115
	1	1	1		+106	+29
	0	2	1		+104	-63
	1	1	0	1	+103	+27
	0	2	0	1	+106	-65
	2	0	0	1	+96	+112

Central atom: radical nitrogen

	1	2	3	4	$\Delta H_f/kJ\ mol^{-1}$	$\Delta S_f/J\ K^{-1}\ mol^{-1}$
I′	-1	1			+230	+126
	0	2			+244	+43
	0	1	0	1	+178	+27
	1	0	0	1	+158	+114

Note: 1 cal = 4·184 J

of a wide variety of compounds and also of BDEs by comparison of ΔH_f for the compound and for the two fragments resulting from bond fission. The following examples will clarify the method of use.

Examples of BDE calculation

(a) $\qquad\qquad CH_3-CH_3 \qquad\qquad\qquad \longrightarrow \qquad 2CH_3^{\bullet}$

$\qquad \Delta H_f \qquad 2 \times \mathbf{A}$–3, 1 (C(H$_3$, C)) $\qquad\qquad 2 \times 143$ (from Table 1.13)

$\qquad\qquad = 2 \times (-42 \cdot 6) = -85 \cdot 2 \qquad\qquad = 286$

$\qquad\qquad BDE = 286 - (-85 \cdot 2) = +371 \text{ kJ mol}^{-1} (88 \cdot 7 \text{ kcal mol}^{-1})$

(b) $\qquad\qquad CH_2{=}CH-CH{=}CH_2 \qquad \longrightarrow \qquad 2CH_2{=}CH^{\bullet}$

$\qquad \Delta H_f \quad 2 \times \mathbf{B}$–2 ($=$C(H$_2$)) $= 52 \cdot 4 \qquad\qquad 2 \times 289$

$\qquad 2 \times \mathbf{B}$–1, 0, 1 ($=$(H, C$=$)) $= 56.2$

$\qquad\qquad\qquad\qquad\qquad\qquad\overline{108 \cdot 6} \qquad\qquad \overline{578}$

$\qquad\qquad BDE = 578 - 109 = +469 \text{ kJ mol}^{-1} (112 \text{ kcal mol}^{-1})$

(c) $\quad CH_3-O-CH_3 \qquad\qquad\qquad \longrightarrow \quad CH_3^{\bullet} + {}^{\bullet}OCH_3$

$\Delta H_f \quad 2 \times \mathbf{A}$–3, 0, 0, 0, 0, 1 (C(H$_3$, O))

$\qquad\qquad\qquad\qquad\qquad = -84 \qquad\quad +143 \qquad\quad +14 \cdot 6$

$\qquad \mathbf{F}$–0, 2(O(C$_2$)) $\qquad\qquad = -97$

$\qquad\qquad\qquad\qquad\qquad\overline{-181} \qquad\quad \overline{+158}$

$\qquad\qquad BDE = +158 - (-181) = +339 \text{ kJ mol}^{-1} (81 \text{ kcal mol}^{-1})$

(d) Fragmentation of *n*-butane:

$\qquad CH_3-CH_2-CH_2-CH_3 \qquad\qquad\qquad \longrightarrow 2CH_3-CH_2^{\bullet}$

$\Delta H_f \quad 2 \times \mathbf{A}$–3, 1 $\qquad\qquad = -85 \cdot 2 \qquad 2 \times \mathbf{A}'$–3, 0, 0, 0, 1 $\quad = -84$

$\qquad 2 \times \mathbf{A}$–2, 2 $\qquad\qquad = -41 \cdot 2 \qquad 2 \times \mathbf{H}'$–2, 1 $\qquad\qquad = +300$

$\qquad\qquad\qquad\qquad\quad \overline{-126} \qquad\qquad\qquad\qquad\qquad \overline{+216}$

$\qquad\qquad BDE = +216 - (-126) = +342 \text{ kJ mol}^{-1} (81 \cdot 7 \text{ kcal mol}^{-1})$

$\qquad CH_3-CH_2-CH_2-CH_3 \qquad \longrightarrow \quad CH_3-CH_2-CH_2^{\bullet} + CH_3^{\bullet}$

$\Delta H_f \qquad\qquad\qquad -126 \qquad\qquad\qquad \mathbf{A}$–3, 1 $\qquad\quad = -42 \cdot 6$

$\qquad\qquad\qquad\qquad\qquad\qquad\qquad\qquad \mathbf{A}'$–2, 1, 0, 0, 1 $= -20 \cdot 7$

$\qquad\qquad\qquad\qquad\qquad\qquad\qquad\qquad \mathbf{H}'$–2, 1 $\qquad\quad = 150$

$\qquad\qquad\qquad\qquad\qquad\qquad\qquad\qquad\qquad\qquad \overline{86.7 + 143}$

$\qquad BDE = (86 \cdot 7 + 143) - (-126 \cdot 4) = +356 \text{ kJ mol}^{-1} (85 \text{ kcal mol}^{-1})$

This would predict that butane would tend to fragment thermally at the C2–C3 bond in preference to the C1–C2 bond.

(e) Stability order of the isomeric butyl radicals.

1-Butyl, CH_3—CH_2—CH_2—$\dot{C}H_2$

$$\Delta H_f \quad \begin{array}{ll} \text{A–3, 1} & = -42 \cdot 6 \\ \text{A–2, 2} & = -20 \cdot 6 \\ \text{A'–2, 1, 0, 0, 1} = & -20 \cdot 7 \\ \text{H'–2, 1} & = +150 \\ \hline & +66 \cdot 1 \text{ kJ mol}^{-1} \text{ (15·8 kcal mol}^{-1}\text{)} \end{array}$$

2-Butyl, CH_3—CH_2—$\dot{C}H$—CH_3

$$\Delta H_f \quad \begin{array}{ll} \text{A–3, 1} & = -42 \cdot 6 \\ \text{A'–2, 1, 0, 0, 1} = & -20 \cdot 7 \\ \text{A'–3, 0, 0, 0, 1} = & -42 \\ \text{H'–1, 2} & = +156 \\ \hline & +50 \cdot 7 \text{ kJ mol}^{-1} \text{ (12·1 kcal mol}^{-1}\text{)} \end{array}$$

2-Methylpropyl (isobutyl), $(CH_3)_2CH$—$\dot{C}H_2$

$$\Delta H_f \quad \begin{array}{ll} 2 \times \text{A–3, 1} & = -85 \cdot 2 \\ \text{A'–1, 2, 0, 0, 1} = & -7 \cdot 9 \\ \text{H'–2, 1} & = +150 \\ \hline & +56 \cdot 9 \text{ kJ mol}^{-1} \text{ (13·6 kcal mol}^{-1}\text{)} \end{array}$$

2-Methyl-2-propyl (*tert*-butyl), $(CH_3)_3\dot{C}$.

$$\Delta H_f \quad \begin{array}{ll} 3 \times \text{A'–3, 0, 0, 0, 1} = -126 \\ \text{H'–0, 3} \quad = +159 \\ \hline +33 \text{ kJ mol}^{-1} \text{ (7·0 kcal mol}^{-1}\text{)} \end{array}$$

The order of stability is:

$$\textit{tert}\text{-butyl} > \text{2-butyl} > \text{isobutyl} > \text{1-butyl.}$$

The sign of BDE is here given as positive, i.e. it is the enthalpy change for breaking the bond. Values are sometimes given with a negative sign signifying bond formation.

Entropy contributions

The entropy of formation of both compounds and radicals is also subject to group contributions. These values will be of interest later (Section 2.4) and are included in Table 1.13.

1.4 INTERMOLECULAR FORCES[42-50]

All molecules are able to condense into liquid or solid states — an indication that intermolecular attractive forces exist universally. The approach of two molecules undergoing reaction is brought about initially by a process of random collision, but is fostered by mutual attractions which develop as their separation diminishes. The attractive forces discussed in this section are weaker and less specific than covalent bonds. They can be equally effective by virtue of their number, e.g. the four hydrogen bonds which hold a water molecule in the ice structure. Weak interactions may conveniently be divided into long- and short-range types. The former occur at intermolecular distances at which the molecules retain their own identity and electron exchange between them can be ignored. Short-range forces become important at very close approach so that quantum mechanical phenomena come into play and electrons tend to lose their identity as belonging to one or other molecule. The attractive forces may be directional and tend to orient reacting molecules but will eventually, at very small distances, become balanced by repulsive interactions. The nature of intermolecular forces, of paramount importance in the early stages of reaction, is now becoming clearer as a result of theoretical approaches, of which the perturbation approach to reactivity is important.

1.4.1 Electrostatic forces

The attraction between charges of opposite sign is the underlying force in both covalent and ionic bonding and may also be a major component of intramolecular forces. The strongest attractions of this type will be between oppositely charged ions, a situation which is frequently encountered in reacting systems. Electrostatic theory gives as the force, F, between two point charges q, q' separated by distance r,

$$F = \frac{1}{4\pi\varepsilon_m} \cdot (qq'/r^2). \tag{1.15}$$

The constant of proportionality in this inverse square relationship contains the permittivity of the medium, ε_m, a measure of its ability to transmit an electric field. For a vacuum,

$$\varepsilon_m = \varepsilon_0 = (10^7/4\pi c^2) = 8\cdot85 \times 10^{-12}\ (C^2\,N^{-1}\,m^{-2}),$$

where ε_0 is the permittivity of free space and c is the velocity of light. For other media, the relative permittivity, $\varepsilon_m/\varepsilon_0 = D$, is known as the dielectric constant (Section 5.6.1), a quantity appearing in the measurement of solvation energy.

An order-of-magnitude calculation for the attractive energy, E, between two oppositely and unit-charged ions at 200 pm distance *in vacuo*, gives

46

(since $E = Fr$),

$$E = -\frac{(1 \cdot \times 10^{-19})^2 (6 \cdot 02 \times 10^{23})(10^{-3})}{(4\pi)(8 \cdot 85 \times 10^{-12})(2 \times 10^{-10})} \frac{C^2 \times kJ\,J^{-1} \times mol^{-1}}{C^2\,N^{-1}\,m^{-2} \times m}$$

$$= -694\,kJ\,mol^{-1}\,(-166\,kcal\,mol^{-1})\quad (N \times m = J).$$

The negative sign implies an attractive force, the zero of energy being at infinite separation. In an isotropic medium this would be reduced by a factor $1/D$. The energy of this electrostatic bond is greater than that associated with a covalent single bond and is responsible for the lattice energy of an ionic crystal such as Na^+Cl^-. In solution, however, electrostatic forces will be considerably less than this. In the first place ions are often complex and do not behave as point charges. The charge is frequently spread over many atoms by inductive and conjugative effects which reduce the electric field gradient in the vicinity of the ion. Secondly, the dielectric constant of the medium may be high; water, for example, has $D = 80$, which would reduce the energy in the example above to about $8 \cdot 6\,kJ\,mol^{-1}$ ($2 \cdot 05\,kcal\,mol^{-1}$) — the reason for the solubility of ionic solutes in water. The value is only approximate since the dielectric constant is a macroscopic property whereas the permittivity of solvent on the microscopic scale — a quantity not really accessible — is really required. This is a common limitation to the quantitative application of electrostatic theory to reactivity. The effect of the medium on interionic forces is demonstrated by the values of association constants for ion-pairs in various solvents (Table 1.16), which undoubtedly vary inversely with dielectric constant: see also Section 1.4.2.

Table 1.16   Some properties of ion-pairs [75–77,79]

(a) Association constants, K_a, for tetra(n-butyl)ammonium perchlorate in various solvents

$$M^+ + Y^- \rightleftarrows M^+Y^-;\quad K_a = [M^+Y^-]/[M^+][Y^-]$$

$(M = Bu^n{}_4N^+, Y = ClO_4^-)$

Solvent	Dielectric constant	K_a	$-\Delta G/kJ\,mol^{-1}$	$(k\,cal\,mol^{-1})$
Benzene	2	3×10^{17}	100	(23·9)
Dioxane	4·9	$ca\ 10^{19}$	109	(26·0)
Anisole	4·35	10^9	52	(12·4)
o-Dichlorobenzene	9·8	93,400	28	(6·7)
1,2-Dichloropropane	9·2	91,600	28	(6·7)
1,2-Dichloroethane	10·1	6,500	22	(5·3)
Propan-2-ol	18·0	1950	19	(4·5)
Butan-2-one	18·4	302	14	(3·3)
Butan-1-ol	20·4	2200	19	(4·5)
Propanone	20·7	200	13	(3·1)
Phenyl cyanide	25	91	11	(2·6)
Ethyl cyanide	31	91	11	(2·6)
Methyl cyanide	36	53	10	(2·4)

Table 1.16 *(Continued)*

(b) Thermodynamic parameters for some ion-pair equilibria

$$(1) \qquad [DQ^-]^+ \rightleftarrows DQ^- + M^+$$
$$\qquad\qquad\quad \text{contact} \qquad \text{free}$$
$$\qquad\qquad\quad \text{ion-pair} \qquad \text{ions}$$

Solvent, dimethoxyethane. DQ^- = durosemiquinone radical anion,

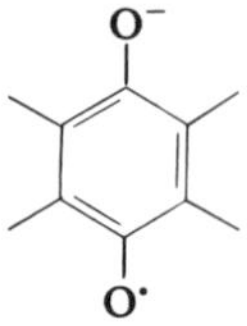

M^+	$\Delta G/kJ\,mol^{-1}$ ($kcal\,mol^{-1}$)	$\Delta H/(kJ\,mol^{-1})$ ($kcal\,mol^{-1}$)	$\Delta S/J\,K^{-1}\,mol^{-1}$ ($cal\,K^{-1}\,mol^{-1}$)
Na^+	+36	−11	−160
	(+8·6)	(−2·6)	(−38)
K^+	+48	−15	−210
	(+11)	(−3·6)	(−50)
Cs^+	+37	−31	−230
	(+8·8)	(−7·4)	(−55)

$$(2) \qquad MDPA^-M^+ \rightleftarrows MDPA^- \,(solv)M^+$$
$$\qquad\qquad\quad \text{contact} \qquad \text{solvent-separated}$$
$$\qquad\qquad\quad \text{ion-pair} \qquad\quad \text{ion-pair}$$

Solvent, tetrahydrofuran. MDPA = 2-methyl-1,3-diphenylallyl anion,

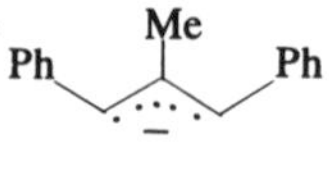

Li^+	+3·3	−28	−105
	(+0·79)	(−6·7)	(−25)
Na^+	+5·7	−45	−170
	(+1·36)	(−10·7)	(−40·6)

$$(3) \qquad DPA^-M^+ \rightleftarrows DPA^-\,M^+ \rightleftarrows DPA(solv)M^+$$
$$\qquad\quad \text{contact} \qquad \text{contact} \qquad \text{solvent-}$$
$$\qquad\quad \text{pair 1} \qquad\;\; \text{pair 2} \qquad \text{separated}$$
$$\qquad\quad \text{(poorly} \qquad \text{(more} \qquad \text{ion-pair SS}$$
$$\qquad\quad \text{solvated)} \qquad \text{solvated)}$$

Solvent, tetrahydrofuran. DPA = 1,3-diphenylallyl anion,

$1 \rightleftarrows 2$

Na^+	−2·8	−12	−31
	(−0·67)	(−2·9)	(−7·4)
Ki^+	+7·6	−30	−13
	(+1·8)	(−7·2)	(−3·1)

Table 1.16 (*Continued*)

$2 \rightleftharpoons SS$			
Na$^+$	+5·9	−26	− 107
	(+1·41)	(−6·2)	(−25)
Li$^+$	+0·3	−6·3	−22
	(+0·07)	(−1·5)	(−5·26)

(c) Partial molar volume changes for ion-pair formation[a]

Solvent[79]	Ion-pair	$\Delta \bar{V}/cm^3\ mol^{-1}$
Tetrahydrofuran	Na$^+$Fl$^-$; contact → loose	−16
	Li$^+$Fl$^-$; contact → loose	−10
Propanol	Me$_4$N$^+$Br$^-$; contact → free ions	−16
Water	Na$^+$SO$_4^{2-}$; contact → free ions	−16

[a] Fl = fluorenyl,

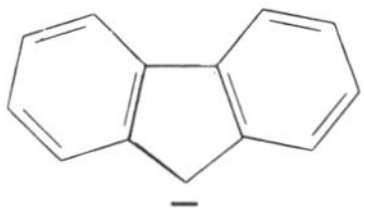

Forces involving dipoles

A dipole in the electric field of an ion experiences a couple tending to rotate it since one end of the dipole will experience an attraction while the other will be repelled.

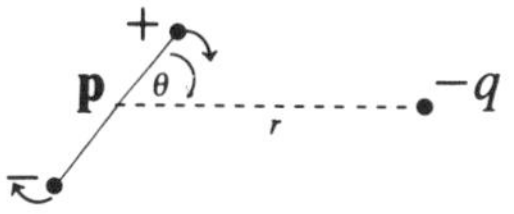

$$\text{Force} = \mathbf{p} \cdot (\partial E/\partial r) \propto \frac{1}{4\pi\varepsilon}(\mathbf{p} \cdot q/r^3). \qquad (1.16)$$

The force varies as the inverse cube of distance and so falls off more rapidly than does an ion–ion interaction. Forces experienced between two dipoles are more difficult to analyze but these vary as the inverse fourth power of distance (Table 1.17). Two dipoles will tend to rotate so that their oppositely charged ends are adjacent and mutually attracted. A molecule in the field of an ion or dipole may acquire an induced dipole moment depending on its polarizability.

1.4.2 Ion-pairs[51–53]

An ion-pair may be said to exist when a cation and an anion are sufficiently close for their electrostatic attraction to exceed thermal energies and for the lifetime of the complex (X^+Y^-) to be in considerable excess of a vibrational

Table 1.17 Summary of electrostatic interactions

		Force $\propto$	Energy $\propto$
(a) Ion–ion	$+q \bullet \cdots_r \cdots \bullet -q$	q^2/r^2	q^2/r
(b) Ion–dipole	$\mathbf{p} \diagup \overset{\theta}{\cdots_r \cdots} \bullet q$	$\mathbf{p} \cos \theta \cdot q/r^3$	$\mathbf{p} \cos \theta \cdot /r^2$
(c) Dipole–dipole	$\mathbf{p} \overset{\theta}{\diagup} \cdots_r \cdots \overset{\theta'}{\diagdown} \mathbf{p'}$	$\mathbf{p} \cos \theta \cdot \mathbf{p'} \cos \theta'/r^4$	$\mathbf{p} \cos \theta \cdot \mathbf{p'} \cos \theta'/r^3$
(d) Ion–induced dipole	$\overset{\alpha}{(+-)} \cdots_r \cdots \bullet -q$	$q\alpha/r^5$	$q\alpha/r^4$

Examples: interaction energies for some specific cases $(\varepsilon = 1, r = 400 \text{ pm})$

(a) $+q \quad_r \quad -q$ $\qquad -E = 320 \text{ kJ mol}^{-1} \, (76 \cdot 5 \text{ kcal mol}^{-1})$

(b) $+q \quad_r \quad - \quad + \atop \mu = 1D$ $\qquad -E = 65 \, kJ \, mol^{-1} \, (15 \cdot 5 \, kcal \, mol^{-1})$

(c) $\underset{\mu = 1D}{+ \quad -} \quad \underset{\mu = 1D}{+ \quad -} \atop \cdots_r \cdots$ $\qquad -E = 15 \text{ kJ mol}^{-1} \, (3 \cdot 6 \text{ kcal mol}^{-1})$

(d) $\pm q \quad_r \quad \oslash \atop \alpha = 10^{-23} \text{ cm}^3$ $\qquad -E = 25 \text{ kJ mol}^{-1} \, (60 \text{ kcal mol}^{-1})$

period $(ca \ 10^{-11} \text{s})$. Clearly, the electrostatic energy to be gained by this association will favor enormously the ion-pair over separated ions in the gas phase, but to a lesser extent in solution in which a measurable equilibrium may be set up. Since most organic reactions are carried out in solvents of low to medium dielectric constant in which ion-pairing is greatly favored, this species is frequently encountered as a reactive intermediate, as, for instance, in the following situation;

$$\text{R—Br} \underset{}{\overset{\Delta H + ve}{\rightleftharpoons}} \text{R}^+\text{Br}^- \underset{}{\overset{\Delta H + ve}{\rightleftharpoons}} \text{R}^+ \ \text{Br}^-$$
$$\text{covalent} \qquad \text{ion-pair} \qquad \text{free ions}$$
$$\text{Products}$$

Ionization takes place in two endothermic steps but the product may be formed very rapidly from either the ion-pair or the free ion. It is likely, then, that the free ion will not form at all in the most rapid pathway to products. The ionic intermediates involved in many mechanisms discussed in later sections are doubtless ion-pairs rather than dissociated ions.

The forces stabilizing ion-pairs are primarily long-range electrostatic attractions as shown by their association constants, K_a, which vary inversely with the dielectric constant of the solvent (Table 1.16). The correlation is not exact, however, because simple electrostatic theory does not account for all of the energy of interaction, nor does the macroscopic dielectric constant adequately measure the microscopic medium effect. Other intermolecular forces — charge transfer and hydrogen bonding, for example — between ions and solvent molecules are also important. Ions in solution are heavily solvated and the organization of solvent molecules in the vicinity of the charge costs a considerable entropy penalty. The formation of an ion-pair, a dipolar rather than a charged species, is probably accompanied by some release of solvent which is thermodynamically favorable. Ion-pair formation can therefore be seen as competitive with solvation in lowering the free energy of the system[54] (Section 2.3). The stability of an ion-pair (association constant) with respect to free ions will, in general, diminish as the temperature is raised (Fig. 1.6a). Also, since an ion-pair in solution has a larger molar volume than the separated ions (see Table 1.16c), due to the reduction in solvation, its stability will diminish with an increase in pressure[55–58].

Evidence for the existence of ion-pairs rests on sound experimental evidence which may be briefly summarized as follows.

(a) The molar conductivity of electrolytes decreases with concentration, an effect very large in organic solvents (Fig. 1.6a). This was first recognized and interpreted by Bjerrum in 1926[59] and by Fuoss[60] as a result of the electrical neutrality of the ion-pair and consequently of its inability to carry a current. Conductivity measurements may be used to determine K_a and often reveal, through deviations of a simple equilibrium constant with concentration, the presence of higher ion aggregates[61]. Variation of $\log K_a$ with $1/T$ should be linear, the slope yielding the free energy change for association (van't Hoff relationship, Section 2.6). Again, these plots are frequently curved: from the extremes, two energy processes may be discerned (Fig. 1.6b), evidently due to the occurrence of more complex equilibria. It is postulated that different types of $1:1$ ion-pair exist known as *contact ion-pairs* (also *tight* or *intimate*), consisting of ions at their closest distance of approach, and *loose ion-pairs* (*solvent-separated*, denoted $M^+ \parallel Y^-$) indicating that the two ions are separated by one or more molecules of solvent with which additional attractive interactions are possible. These two are, as a rule, in equilibrium with the solvent-separated species more favoured at low temperature though relative stabilities are very sensitive to the solvent and the nature of the ions (Fig. 1.7).

(b) The ultraviolet or visible spectra of certain organic ions show changes attributable to ion-pairing and to the type of ion-pair[62]. Loose ion-pairs have spectra which are independent of the counterion, whereas the spectra of contact ion-pairs tend to show shifts of λ_{max} which are related to

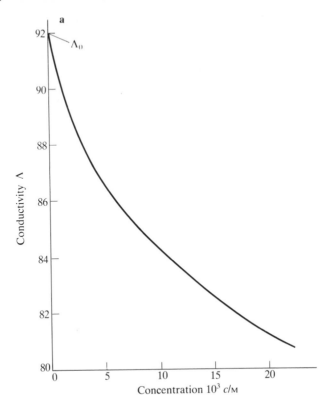

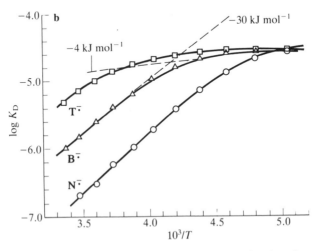

Fig. 1.6 (a) Molar conductivity of sodium picrate in ethanol as a function of concentration[61]. (b) Van't Hoff plots for Na^+ salts of radical anions of naphthalene (N^-), biphenyl (B^-) and triphenylene (T^-), in tetrahydrofuran. Binding energies for contact- and solvent-separated ion-pairs of $[T^- Na^+]$ are indicated[52].

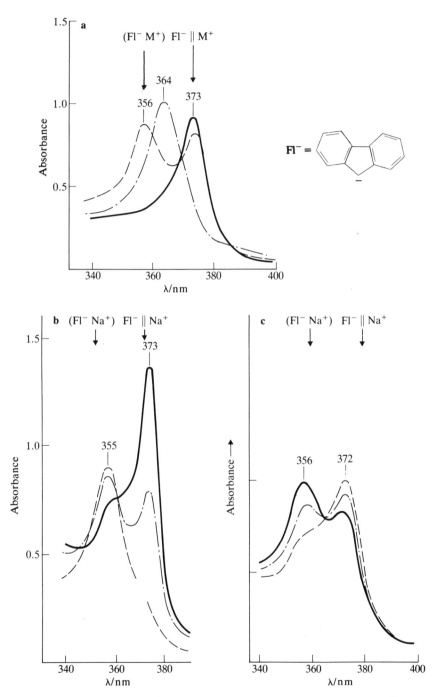

Fig. 1.7 Ion-pairing from ultraviolet spectra of sodium fluorenyl in tetrahydrofuran[51]: **a**, effect of cation (—, Fl^-, Li^+; - - -, Fl^-, Na^+; - · - ·, Fl^-, Cs^+; −30°C); **b**, effect of temperature (—, −50°C; - · - ·, −30°C; - - -, 25°C); **c**, effect of addition of glyme, a polar solvent (glyme concentration, M: —, $1·3 \times 10^{-3}$; - · - ·, $2·6 \times 10^{-3}$; - - -, $4·3 \times 10^{-3}$).

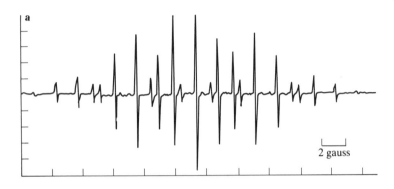

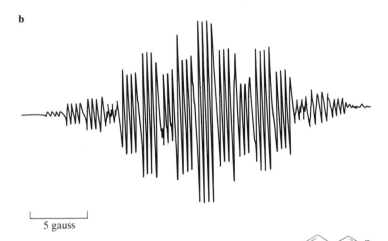

5 gauss

Fig. 1.8 Electron spin resonance spectra of naphthalene radical anion, ⬡⬡⁻ : **a**, free
ion showing 25-line spectrum; **b**, ion pair with $^{23}Na^+$ ($I = 3/2$); all resonances are split into four
($2I + 1$) by the magnetic ion in close contact[68].

counterion size ($\propto 1/r$). Much-studied examples include fluorene salts of
alkali metals (Fig. 1.7).

(c) Magnetic resonance spectra are very sensitive to the environment of the
magnetic nucleus or unpaired electron which is resonating. Proton resonance spectra of organic anions and electron paramagnetic resonance
spectra of radical anions have received much attention and the proximity of
magnetic ions such as $^{23}Na^+$ or $^{87}Rb^+$ can be discerned from direct coupling
(Fig. 1.8) or from line-broadening and the chemical shifts induced[63–66]. The
temperature dependence of these effects can yield information concerning
the energetics and dynamic properties of the ion-pairs present, i.e. their
lifetimes, and may indicate the presence of more than one type of loose
ion-pair[67–69]. Greater complexity may be apparent[70]. Sodium diphenylketyl,
14[71], exists as a diamagnetic dimeric species in equilibrium with various

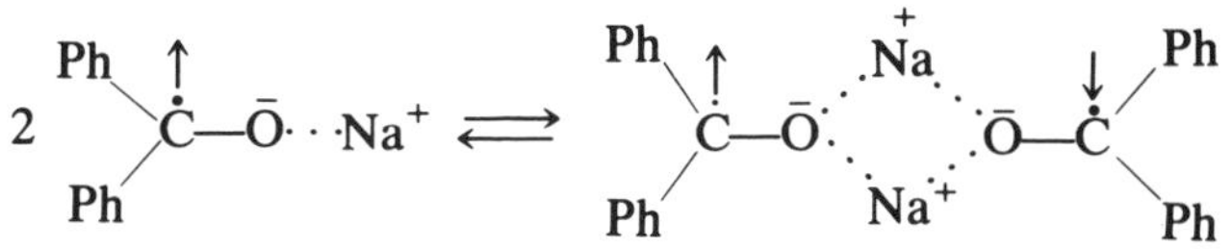

types of ion-pair, and from solutions of the pyridinium iodide, **15**, triple ions, $M^+Y^-M^+$, are detected[72]. These observations are a reminder that ion association is a complex matter which, though of great importance, is at present imperfectly understood.

(d) Abundant chemical evidence both from kinetic and from product studies is available which implicates ion-pairs as intermediate species during the course of chemical reactions. This evidence will be discussed particularly under solvolysis, Section 10.1.3.

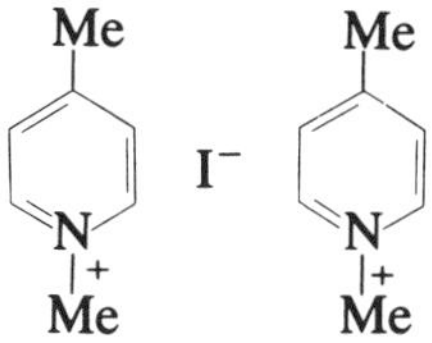

15

Reactivity effects

Association tends to reduce the reactivity of an ion, giving rise to the order:

$$\text{Free ion} \gg \text{loose ion-pair} \gg \text{contact ion-pair.}$$

As an example of the enormous effect which ion association may have on reactivity, rates of anionic polymerization of styrene[16], **16**, are shown in Table 1.18[73-74].

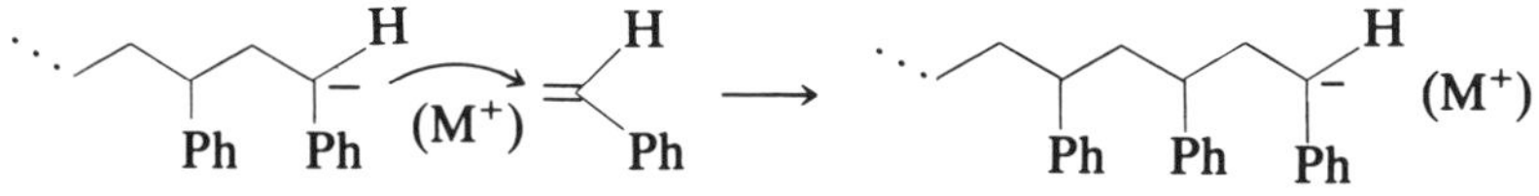

16

Table 1.18   Anionic polymerization of styrene

M^+ ion-pair	Solvent	Reactivity$/M^{-1}s^{-1}$
Free carbanion	Tetrahydrofuran	65 000
Cs^+ion-pair (contact)	Tetrahydrofuran	25
Li^+ion-pair (contact)	Tetrahydrofuran	180
Na^+ion-pair (contact)	Dioxane	3
Na^+ ion-pair (solvent-separated)	Dimethoxyethane	180

It is, evidently, contact ion-pairs whose properties are cation-dependent. The reaction shows an increase in the rate constant upon dilution, an unusual situation but due to the increase in concentration of the highly reactive free anions. The role of the solvent in determining the type of ion-pair formed is also reflected in a rate effect. It must therefore be borne in mind that intrinsic reactivities of ions, whether anions or cations, may be overshadowed by the effects of ion-pairing. Large, polarizable anions such as alkoxides may aggregate with two alkali metal cations forming ion triplets, $M^+A^-M^+$. This is indicated by the inhibition of the cyclization of 29 by Na^+ or by Li^+ according to a rate law which contains a term in $(1/[M^+]^2)$[119].

29

1.4.3   Short-range intermolecular forces[80]

At separations comparable with a covalent bond ($<200\,pm$), MOs of neighboring molecules begin to interact, resulting in a mixture of attractive and repulsive forces. Between two neutral molecules A, B a net weakly attractive force results with binding energy ΔE_b;

$$\Delta E_b = E_{A,B} - (E_A + E_B).$$

We may treat this complex system A, B as a 'supermolecule' and attempt to calculate its total energy, $E_{A,B}$, as well as those of the individual molecules, E_A and E_B, to obtain ΔE_b. This is a formidable task and needs the use of powerful computers and high-level MO programs, but it may now be accomplished in simple cases to give a quantitative description of the various components of the interaction. In connection with VB theory it was mentioned that the complete description of even the hydrogen molecule needs the inclusion of contributions from many electronic configurations, including states in which the electrons occupy orbitals other than the lowest-lying. This mixing-in of excited states stems from the indistinguishability of the electrons at close range and results in stronger binding. The same principles apply to the analysis of the 'supermolecule' and the following analysis of contributing terms (denoted 'energy decomposition analysis') has been used by Morokuma and others[81,82]. Four types of interaction are considered (Fig. 1.9).

(a) Electrostatic interaction energy, ΔE_q (Fig. 1.9a), is the energy of interaction between the undistorted charge distributions of the molecules A and B. This includes all electrostatic forces between permanent charges, dipoles and higher multipoles present in the two molecules as described in Section 1.4.1, and could be attractive or repulsive overall.

(b) Polarization interaction energy, ΔE_{pol}, is the energy change resulting

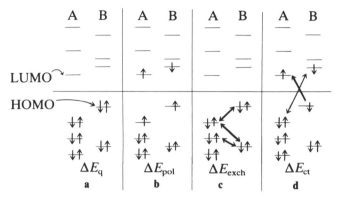

Fig. 1.9   Schematic MO interactions between two closed-shell molecules.

from the distortion of the charge distribution of A and B and results from the mixing-in of electronic configurations such as that shown in Fig. 1.9b in which electrons from either or both molecules occupy antibonding MOs of their own. Since the contribution of such terms depends upon the proximity of the other molecule it can be seen as a distortion of the electron distribution induced by the other molecules, i.e. polarization. This term is always attractive.

(c) Exchange energy, ΔE_{exch}, is the interaction resulting from the loss of identity of the electrons with respect to A and B. This is expressed by mixing-in configurations in which electrons from each molecule have 'changed places' (Fig. 1.9c). It is a result of the total wavefunction satisfying the Pauli exclusion principle. The result of this term is to cause a depletion of electron density in the intermolecular region resulting in a repulsion. The exchange interaction is a result of the mixing of filled orbitals in A and B which, as we saw, gives rise to no net bonding but only to further electron–electron repulsion.

(d) Charge-transfer energy, ΔE_{ct}, is the contribution resulting from mixing-in electronic configurations (Fig. 1.9d) expressing electron transfer between filled and vacant MOs of the two molecules. This results in an attractive force, the magnitude increasing the smaller the energy difference between the two interacting MOs. The most important terms will be those expressing electron transfer between the highest occupied MO (HOMO) of one to the lowest unoccupied MO (LUMO) of the other (the *frontier orbitals*). HOMO–LUMO interactions are important especially in the early stages of reaction and will be frequently encountered in discussions of chemical reactivity.

The energy of the interaction between A and B is given by:

$$\Delta E_b = \Delta E_q + \Delta E_{pol} + \Delta E_{exch} + \Delta E_{ct} \qquad (1.17)$$

and can be calculated with the succesive exclusion of terms so that the magnitude of each contribution may be separately determined. The results of such calculations, referring necessarily to the gas phase, provide a guide to the underlying principles of intermolecular interactions, as the following examples show.

NH₃–BH₃

Although monomeric borane, BH_3, does not exist, its alkyl derivatives do and are known to form quite strong complexes with amines. The interaction is usually written

$$R_3B + :NR_3 \rightarrow R_3\bar{B} \leftarrow \overset{+}{N}R_3,$$

signifying that a two-electron coordinate bond (i.e. a covalent bond) is formed, in which both electrons originate from the same atom rather than one from each. Energy analysis gives the values shown in Table 1.19.

This analysis reveals several interesting features. Firstly, at large separation the predominant force is electrostatic attraction which increases continually as the molecules approach more closely. Repulsive exchange forces begin to compete seriously with the attraction at separations under 200 pm and grow steeply but are compensated by growing polarization and charge-transfer forces. It is the different rates at which these forces grow as the molecules approach which results in a minimum total energy being reached at 180 pm at which the binding energy, -187 kJ mol^{-1} ($-44\cdot7 \text{ kcal mol}^{-1}$), is the resultant of the delicate balance between much larger component forces. The borane–amine type of interaction may be characterized as (q > ct, pol) specifying the principal attractive forces in order. The effect of distance on the components of the interaction are shown graphically in Fig. 1.10a.

Table 1.19 Energy of the system BH_3–NH_3 as a function of distance, r

Energy/kJ mol⁻¹ (*kcal mol*⁻¹)	*B–N separation, r_{opt}/pm*					
	146	156	166	170	186	270
ΔE_b	−141	−175	−187	−209	−174	−63
	(−33·7)	(−41·8)	(−44·7)	(−50)	(−41)	(−15)
ΔE_q	−672	−540	−430	−388	−271	−50
	(−160)	(−129)	(−102)	(−93)	(−65)	(−12)
ΔE_{exch}	+606	+586	+422	+363	+213	+9
	(145)	(142)	(100)	(87)	(51)	(2·1)
ΔE_{pol}	−225	−139	−88	−72	−38	−4
	(−54)	(−33)	(−21)	(−17)	(−9·1)	(−1)
ΔE_{ct}	−282	−191	−132	−133	−73	−17
	(−67)	(−46)	(−31)	(−27)	(−17)	(−4)

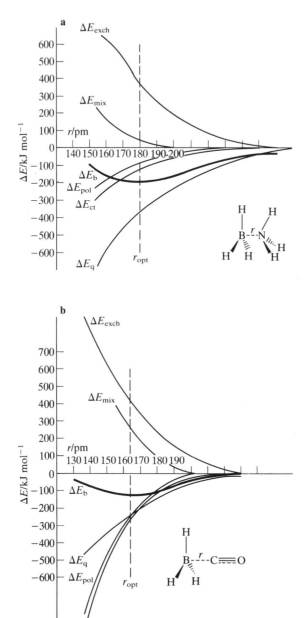

Fig. 1.10 Components of interaction energy as a function of distance: **a**, BH_3–NH_3; **b**, BH_3–CO.

The CO–BH₃ system

This interaction is weaker than that of the borane–amine system and the binding forces are almost equally derived from (ct, pol, q); see Fig. 1.10b, Table 1.20. Of the charge-transfer terms it is calculated that the components $CO \rightarrow BH_3$ and $CO \leftarrow BH_3$ contribute, respectively, 66 and 34% to E_{ct}. Furthermore, the molecular orbitals involved are of σ-type for the former and π-type for the latter (*back-donation* to CO). This information can also be expressed in electron-density contours, Fig. 1.11.

Table 1.20 Dissection of some intermolecular energies

Molecules		r_{opt}/pm	$\Delta E/kJ\,mol^{-1}$ $(kcal\,mol^{-1})$				
			ΔE_q	ΔE_{exch}	ΔE_{pol}	ΔE_{ct}	ΔE_b
NH_3	BH_3	170	−388	+363	−71	−113	−209
			(−92)	(+87)	(−17)	(−27)	(−50)
CO	BH_3	163	−62	+413	−259	−284	−385
			(−15)	(+98·7)	(−62)	(−68)	(−92)
H_2O	$O{=}C(CN)_2$	270	−41	+18	−4·2	−7·5	−30
			(−9·8)	(+4·3)	(−1)	(−1·8)	(−7)
C_6H_6	$O{=}C(CN)_2$	360	−11·7	+7·5	−7·1	−6·7	−18
			(−2·8)	(+1·8)	(−1·7)	(−1·6)	(−4.3)
C_6H_6	Cl_2	360	−2·1	+2·9	−0·4	−3·3	−3
			(−0·5)	(+0·7)	(−0·1)	(−0·8)	(−0·7)
NH_3	Cl_2	293	−16·7	+16·3	−3·3	−9·6	−13
			(−4)	(+3·9)	(−0·8)	(−2·3)	(−3·1)
HCHO	C_2H_4	375	−2·1	+1·7	−0·4	−2·1	−2·9
			(−0·5)	(+0·41)	(−0·1)	(−0·5)	(−0·7)
F^-	CH_3F	180	−26	+11	−7·8	−19	−42
			(−6·2)	(+2·6)	(−1·8)	(−4·5)	(−10)

The benzene–halogen system

Aromatic hydrocarbons and halogen molecules are well known to interact forming what are popularly known as charge-transfer complexes (Section 1.4.5). Model calculations on the benzene–fluorine and benzene–chlorine interactions have been carried out and indicate the following properties. Of five geometries studied, the most favorable energetically is that with halogen located perpendicularly above a carbon relative to the ring, **17**. The

17

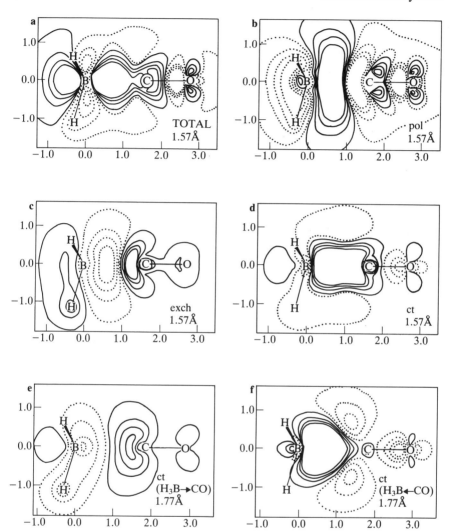

Fig. 1.11 Electron density contours of the BH_3–CO complex (**a**), showing individual contributions from polarization (**b**), exchange (**c**), and charge transfer (**d**). The last of these is partitioned further into forward- (**e**), and back-bonding (**f**) components. Solid contours surround areas of increased electron density corresponding to attractive interactions; dotted contours surround areas of decreased electron density and repulsive interactions. The C–B distance is shown on each map.

interaction energy is even weaker than that of the previous example and is made up as indicated in Table 1.20. The charge-transfer interaction comprises the same contribution to E_{ct} from $C_6H_6 \rightarrow Hal_2$ for both chlorine and fluorine but there is a much larger contribution from Cl_2 than from F_2 to back-donation ($C_6H_6 \leftarrow Hal_2$) as a result of its lower ionization potential.

The CH₃F–F⁻ system

The molecular interactions analyzed in Table 1.20 are representative of strong to very weak complexation which share common origins. The last entry, for FCH$_3$–F⁻, has been frequently analyzed as a model of the bimolecular substitution reactions to be discussed later (Section 10.1.2). Interchange of the fluorines constitutes a chemical reaction.

1.4.4 The hydrogen bond[83-88]

In a famous paper of 1920, Latimer and Rodebush[89] concluded that 'a hydrogen nucleus held between two octets constitutes a weak bond'. The 'hydrogen bond' has proven of widespread importance in phenomena ranging from the abnormally high boiling point of water to the stability of biopolymers. It consists of an attractive force which exists between a hydrogen atom covalently bound to an electronegative atom (e.g. in —O—H) and a second electronegative atom. The two components may occur in two different molecular species (**18**), two molecules of the same species (**19**), or intramolecularly (**20**). The same molecule (e.g. water, ROH) may simultaneously act as H-donor and -acceptor for intermolecular hydrogen bonding, building up two- and three-dimensional structures typical of hydroxylic compounds (**21**). A hydrogen bond is expressed by the dotted line, —O—H⋯O⟨, which conveys the asymmetry of the attachment of hydrogen to each of the electronegative atoms: a covalent bond to one,

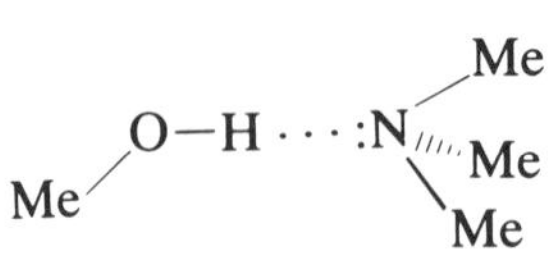

18

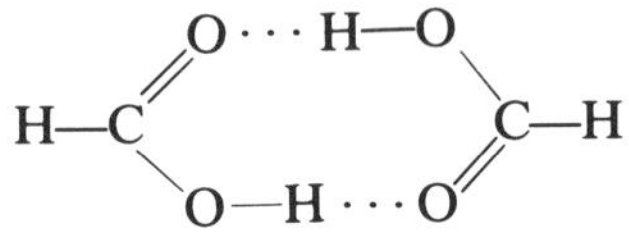

19

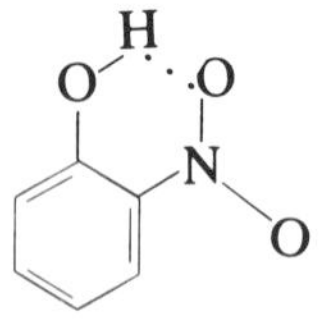

20

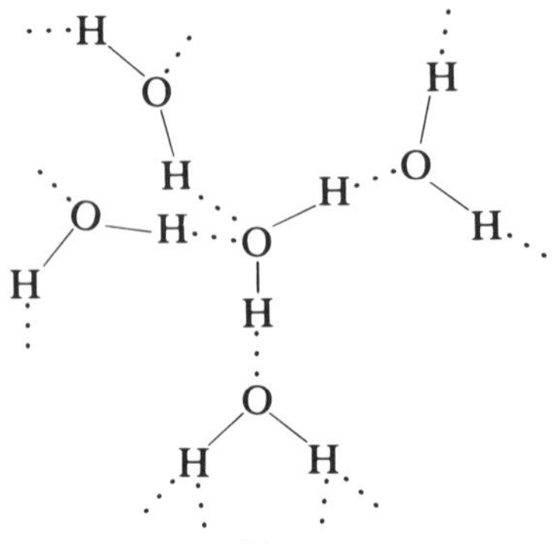

21

and a weak interaction to the other. Its strength depends on several factors; it is strongest when the three-atom system X—H⋯X′ is linear and the elements X, X′ more electronegative. As Table 1.20 shows, $F > O > N > Cl$, Br, $S > C$, P. The following modifications of properties may be used to recognize the existence of hydrogen bonding in the liquid phase.

(a) There are departures from ideal behavior[90,91]. The vapor pressure of a mixture of two liquids whose molecules do not mutually interact is linearly proportional to composition (Raoult's law). Deviations resulting in vapor pressures below the predicted value indicate intermolecular association which may be due to hydrogen bonding. Additional striking properties include the vapor pressures of some individual compounds which show gross abnormalities due to this cause. Water, HF and alcohols, for instance, have boiling points very much higher than homologues with higher molecular weight but no capacity for self-association:

	Boiling pt/°C, 1 bar		
HF 20	H_2O 100	MeOH 60	NH_3 −33
HCl −85	H_2S −61	MeSH −78	PH_3 −87

Hydrogen bonding in general reduces molecular mobility so that coefficients of viscosity of hydrogen-bonded liquids are increased and diffusion decreased; phenomena dependent on these properties will be affected, including density, thermal conductivity and even rates of reaction; heat capacities are increased since it takes additional energy to break the hydrogen bonds while increasing the temperature.

(b) The thermodynamics of hydrogen bonds may be measured from heats of mixing of H-donor and -acceptor species[87,92,93] (Table 1.21c). Enthalpies, ΔH, are usually within the range 10–40 kJ mol^{-1} (2–10 kcal mol^{-1}), values which fall far short of the enthalpies of formation of covalent bonds although quantity can often compensate for quality in making the hydrogen bond an effective force. Furthermore, entropies of formation (Section 2.3) tend to be negative (i.e. unfavorable) due to the loss of translational freedom so that the free energy of hydrogen bond formation, ΔG, is often comparable with thermal energies at room temperature [$RT \sim$ 2·5 kJ mol^{-1} (0·6 kcal mol^{-1})]. Hydrogen bonds in fluid systems, therefore, are likely to be transient features of the molecular scene, rapidly formed and broken.

(c) Infrared absorption frequencies of modes involving hydrogen are lowered by hydrogen bonding since the effective mass is increased; the bands are intensified as the dipole moment of the vibrational transition is increased, and broadened since the degree of bonding present at any instant in a sample varies over a range of values[94,95], (Table 1.21c).

Table 1.21 Properties of some hydrogen bonds

(a) Energy decomposition analysis of some dimeric systems

Acceptor···donor	r_{opt}/pm	$\Delta E/kJ\,mol^{-1}\,(kcal\,mol^{-1})$					
		ΔE_b	ΔE_q	ΔE_{exch}	ΔE_{pol}	ΔE_{ct}	ΔE_{mix}
F—H···NH$_3$	268	−68	−107	+67	−8	−17	−3
		(−16)	(−25)	(+16)	(−2)	(−4)	(−0·7)
F—H···OH$_2$	262	−56	−79	+44	−7	−13	−2
		(−13)	(−19)	(+10·5)	(−1·7)	(−3·1)	(−0·5)
F—H···F—H	271	−32	−34	+19	−2	−13	−1
		(−7·6)	(−8·1)	(+4·5)	(−0·5)	(−3)	(−0·2)
HO—H···OH$_2$	288	−33	−44	+26	−2	−10	−2
		(−7·9)	(−10·5)	(+6·2)	(−0·5)	(−2·4)	(−0·5)
H$_2$N—H···NH$_3$	330	−17	−24	+15	−2	−5	−1
		(−4)	(−5·7)	(−3·5)	(−0·5)	(−1·2)	(−0·2)
HO—H···NH$_3$	293	−38	−58	+38	−5	−10	−1
		(−9)	(−14)	(+9)	(−1·2)	(−2·4)	(−0·2)
H$_2$N—H···OH$_2$	322	−17	−19	+10	−1	−6	−1
		(−4·1)	(−4·5)	(+2·4)	(−0·2)	(−1·4)	(−0·2)
H$_3$C—H···OH$_2$	380	−4·6	−2·1	+2·1	−0·4	−3·7	0
		(−1)	(−0·5)	(0·5)	(−0·1)	(−0·9)	(0)
H$_3$N—H···OH$_2$	268	−114	−142	+25	−17	−21	0
		(−27)	(−34)	(+6)	(−4)	(5)	(0)
F—H···F$^-$	229	−262	−354	+324	−24	−107	−36
		(−63)	(−84)	(+77)	(−5·7)	(−25)	(−8·6)

(b) Thermodynamics of hydrogen bond formation

Type	System[a]	State[b]	$-\Delta G/$ kJ mol^{-1} (kcal mol^{-1})	$-\Delta H/$ kJ mol^{-1} (kcal mol^{-1})	$-\Delta S/$ J K^{-1} mol^{-1} (cal K^{-1} mol^{-1})
Intermolecular					
O—H···O	Phenol–dioxane	l	3·9 (0·9)	21 (5·0)	58 (14)
	Methanol	l		39 (9·3)	117 (28)
O—H···O=C	Water–acetone	l	1·5 (0·36)	10·5 (2·5)	30 (7·2)
	Acetic acid	g		30 (7·2)	75 (18)
O—H···N	Methanol–DMF	l	4·2 (1)	15·5 (3·7)	38 (9·1)
	Methanol–pyridine	l	12 (2·9)	38 (9·1)	95 (23)
	Phenol–NEt$_3$	l	10 (2·4)	35 (8·4)	84 (20)
O—H···F$^-$	Phenol–fluoroalkane	l		9 (2·1)	
N—H···N	Aniline–pyridine	c	1·2 (0·3)	14 (3·3)	44 (10·5)
N—H···O=C	2-Pyridone	l	11 (2·6)	18 (4·3)	25 (6)

[a] HMB = hexamethylbenzene; THF = tetrahydrofuran; DMF = dimethylformamide.
[b] g, Gas phase; l, CCl$_4$; c, cyclohexane.

Table 1.21   (*Continued*)

Type	System[a]	State[b]	$-\Delta G/$ kJ mol^{-1} (kcal mol^{-1})	$-\Delta H/$ kJ mol^{-1} (kcal mol^{-1})	$-\Delta S/$ J K^{-1} mol^{-1} (cal K^{-1} mol^{-1})
N—H$\cdots$S	*N*-Methylacetamide	l	10·2 (2·4)	15 (3·6)	16 (3·8)
N—H$\cdots$O	Aniline–THF	c	0·2 (0·05)	12 (2·9)	42 (10·0)
N—H$\cdots$S	NCNH–H$\cdots$SBu$_2$	l	2·5 (0·6)	15 (3·6)	41 (9·8)
O—H$\cdots$(π-system)	Phenol–HMB	l	1·2 (0·3)	7 (1·7)	23 (5·5)
Intramolecular					
O—H$\cdots$N	2-Cyanoethanol	l		0·6 (0·1)	
O—H$\cdots$I	2-Iodophenol	l		3 (0·7)	
O—H$\cdots$O	2-Methoxybenzoic acid	l		14 (3·3)	

(c) Infrared spectroscopic shifts, $\Delta \bar{\nu}$

CH$_3$O—H stretch

Solvent	$(\bar{\nu}_{\text{solvent}} - \bar{\nu}_{\text{gas}})/cm^{-1}$
CH$_3$CONMe$_2$	160
MeCOMe	112
NEt$_3$	430
Pyridine	286
Benzene	23
C$_4$H$_9$Cl	30

(CF$_3$)$_2$CHO—H stretch

Et$_3$N	1040

Ph$_2$N—H stretch

Pyridine	150
Et$_2$O	94

(d) Proton NMR spectra show a down-field shift of hydrogen-bonded protons indicating decreased shielding (electron density around the proton) due to the H-acceptor atom which may be thought of as repelling the σ-electrons of the A–H bond. The opposite trend may be seen in resonances of the heavy atom[96,97].

Theory of the hydrogen bond[98]

The hydrogen bond does not appear to fit the expected valence properties of hydrogen and for many years electrostatic forces were held responsible.

Application of energy decomposition analysis reveals its origin to be none other than the normal quantum mechanical forces responsible for other intermolecular interactions. The water dimer, $H_2O \cdots H$—O—H, has often been used as a model for computation (Table 1.21a). There seems to be agreement that the attractive force is primarily electrostatic with smaller contributions from charge transfer and polarization and may thus be denoted (q > ct > pol). The relatively large magnitude of the interaction is due to the special characteristics of hydrogen, its small size possibly being the crucial factor.

Relationship of hydrogen bonding to proton transfer

The strength of a hydrogen bond depends on the same factors that determine the ability of an H-donor or -acceptor to act as an acid or base respectively (Chapter 6). The strongest acid–base pairs form the strongest hydrogen bonds which merge with no sharp dividing line into a proton-transfer situation in which two energy minima, corresponding to two stable locations of the proton, exist. Interactions of this kind are important in maintaining the shapes of protein molecules, often referred to as 'salt bridges', **22**. This property is illustrated also by NMR chemical shifts of hydroxylic protons which may correlate with their acid dissociation constants, pK_A; for example, for phenols:

$$pK_A = -0.56\delta + 12.40 \text{ (in } CCl_4);$$

$$pK_A = -1.492\delta + 23.64 \text{ (in dimethyl sulfoxide)}.$$

Similar correlations are found for weak bases.

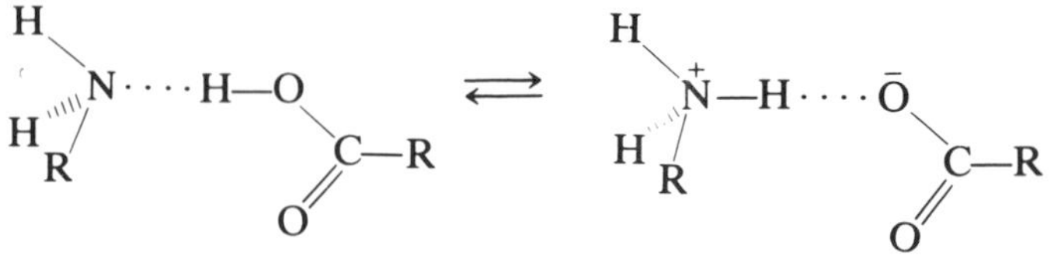

22

Molecular geometry plays a part in determining hydrogen-bond strength. The intramolecular interactions in the 2-hydroxyalkylpyridine series, **23**,

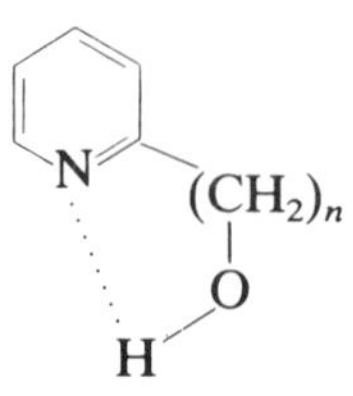

23

Table 1.22 Intramolecular interactions in 2-hydroxyalkylpyridine

	Compound 23		
	$n = 1$	$n = 2$	$n = 3$
Change in —OH stretch/cm^{-1}	192	203	357
O—H···N angle/deg.	125	145	180
O—N distance/pm	260	220	160

illustrate the importance of a linear O—H···N arrangement which is possible only in the larger rings and is accompanied by stronger hydrogen bonding (Table 1.22).

The solvent plays an important part in determining hydrogen-bond equilibria. If it can itself act as a hydrogen-bond acceptor, donor or both (as can hydroxylic solvents) it will compete with solutes for interaction. This can be seen in the dimerization of carboxylic acids, **19**, which is essentially complete in a solvent such as benzene but very slight in ethanol. Hydrogen bonding can bring about stabilization of a species, the best example being the enolic form of a carbonyl compound. This is normally less stable than the keto isomer by ~50 kJ mol^{-1} (12 kcal mol^{-1}) on account of the greater heat of formation of C=O compared with C=C. β-Dicarbonyl compounds exist largely in the enolic forms, **24**, on account of their ability to form an intramolecular hydrogen bond.

24

1.4.5 Charge-transfer complexes[99–107]

It is often observed that on mixing a compound which possesses a high-lying HOMO (i.e. a good electron-pair donor) with one which has a low-lying LUMO (hence an acceptor), intense colors instantly form[108], clearly indicating intramolecular interaction though without chemical reaction. For example, tetracyanoethene (**25**), and hexamethylbenzene (**26**), both initially colorless, form an intensely purple solution on mixing in solution, from which both may be recovered unchanged. Iodine in the gas phase and in solution in 'inert' solvents, such as carbon tetrachloride, is purple, whereas in alcohol and many other solvents containing oxygen or nitrogen (donor solvents) it is brown on account of new absorption at the blue end of the spectrum. The colored species are denoted *charge-transfer complexes* though it is the absorption of light which is accompanied by transference of an electron due to the creation of *intermolecular orbitals*, **25**, **26**, whose separation fortuitously corresponds to a transition in the visible region. Table 1.23 sets out examples of donor and acceptor compounds, the frontier orbitals of which may either be of σ-type as in the case of iodine[116] (σ^*) and ethanol (unshared pair) or of a π-type, e.g. benzene (π) or tetracyanoethene (π^*)[115]. The addition of donor to acceptor in solution may be accompanied by a small volume contraction $(ca\ -4\,cm^3\,mol^{-1})$ and negative enthalpy change (Table 1.23)[109–113].

Association constants, K_a, and molar extinction coefficients, ε, may be estimated graphically from the Benesi–Hildebrand equation (Eq. (1.18))[117,118].

NC   CN   NC   CN   Me   Me Me

NC   CN   NC   CN   Me   Me Me

(A)   ct   (D)

25   (C)   (**26**)

LUMO   ct

$h\nu_{ct}$ HOMO

$$[A]l/\mathrm{Abs} = 1/N_D K_a \varepsilon + 1/\varepsilon, \tag{1.18}$$

where Abs is the absorbance of the solution and N_D is the mole fraction of D. Nonetheless, although solutions of a strong donor and acceptor pair such as dimethylaniline and tetracyanoethene are of an intense blue color, their colligative properties (e.g. vapor pressure) exhibit no evidence of complex formation. It appears that the energy of the interaction is very weak and the lifetime of the complex very short, perhaps only of the order of a vibrational period $(\sim10^{-11}\,s)$, sufficient for electronic excitation to occur $(10^{-16}\,s)$ but not such as to influence solution properties. Such complexes are best regarded as the result of 'sticky collisions' between the two species. Very large dipole moments for the complexes are inferred, consistent with a good deal of electron transfer between donor and acceptor[114]. Nevertheless it is now apparent that the attractive force is the same mixture of quantum mechanical forces which characterizes all intermolecular interactions. Calculations on the benzene–chlorine system (Table 1.20) suggest that charge transfer is the largest contributor, followed by electrostatic forces. It seems likely that this will be found to be broadly true

Table 1.23   Thermodynamics of charge-transfer complexation[109–113]

Donor	Acceptor	$-\Delta G/kJ\,mol^{-1}$ $(kcal\,mol^{-1})$	$-\Delta H/kJ\,mol^{-1}$ $(kcal\,mol^{-1})$	$-\Delta S/J\,K^{-1}\,mol^{-1}$ $(cal\,K^{-1}\,mol^{-1})$
Benzene	iodine	1·1 (0·26)	5·5 (1·3)	15 (3·6)
HMB[a]	iodine	6·8 (1·6)	16 (3·8)	30 (7·1)
1,4-Dioxane	iodine	5·5 (1·3)	14 (3·3)	28 (6·7)
Trimethylamine	iodine	27 (6·5)	51 (12)	79 (19)
HMB[a]	TCNE[b]	18 (4·3)	32 (7·6)	50 (12)
HMB[a]	chloranil[c]	11 (2·6)	22 (5·2)	38 (9·1)
HMB[a]	picric acid[d]	9·5 (2·2)	17 (4·1)	25 (7)

Dipole moments of charge-transfer complexes[114]

Donor	Acceptor	$\mathbf{P}_D/D$	$\mathbf{P}_A/D$	$\mathbf{P}_{DA}/D$
Benzene	iodine	0	0	1·8
Pyridine	iodine	2·2	0	4·5
Trimethylamine	iodine	0·63	0	11·3
HMB	trinitrobenzene	0	0	0·9
HMB	TCNE[b]	0	0	1·3
Durene	TCNE[b]	0	0	1·26
Aniline	picric acid	0	1·5	6·75

[a] HMB, hexamethylbenzene.
[b] Tetracyanoethene, **25**.
[c] Tetrachloro-*p*-benzoquinone.
[d] 2,4,6-Trinitrophenol.

for this class of interactions as a whole which may be designated more precisely the (ct, q) type. Molecules acting as good electron-pair donors (*Lewis bases*, Chapter 6) include species with unshared pairs (particularly on nitrogen or oxygen), or with extended π-orbitals such as aromatic systems; in these donor ability is enhanced by the presence of electron-donating groups such as amino, alkoxy or alkyl. The most powerful electron-pair acceptors (*Lewis acids*) have an unsaturated structure (alkene or aromatic nucleus) to which are attached a number of strongly electron-withdrawing groups such as $-NO_2$, $-CN$, $-C{=}O$. In addition molecules with vacant *p*-orbitals such as boranes (BR_3) or low-lying π^* orbitals such as the halogens also qualify.

Charge-transfer (perhaps better termed 'donor–acceptor') complexes can sometimes be isolated in the crystal, for example bromine–dioxane, while the delocalization of electrons between donor and acceptor molecules can

28

27

give rise to semi-conducting properties. The complex between TCNQ (**27**) and the tetrathiadifulvalene, **28**, is the best example.

PROBLEMS

1 Eigenvalues and eigenvectors for the secular determinant appropriate to trimethylenecyclopropane (numbered as in **I**) are given below:

Eigenvalues:	−2·414	−0·618	−0·618	0·414	1·618	1·618
Eigenvectors	c_1 0·221	−0·162	−0·675	−0·533	−0·006	0·424
	c_2 0·533	−0·100	−0·417	0·221	0·104	−0·687
	c_3 0·533	0·412	0·121	0·221	0·542	0·434
	c_4 0·221	0·666	0·196	−0·533	−0·335	−0·268
	c_5 0·533	−0·311	0·296	0·221	−0·647	0·253
	c_6 0·221	−0·503	0·478	−0·533	0·400	−0·156

I

Sketch the geometries of the molecular orbitals and draw an energy diagram indicating orbital occupancy. Calculate the delocalization energy and charges on each carbon. Predict the stability with respect to benzene and the position of attack by an electrophilic reagent.

2 Write out and solve (using a microcomputer) secular determinants for the following isomeric C_6 systems.

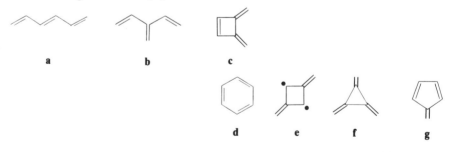

a   b   c

d   e   f   g

(a) Compare the π-energies of each and place in order of stability.

(b) Compare the charge densities at each carbon in **g** and any other isomer; is **g** unique in having non-integral values?

(c) Determine which has the greater π-energy, a linear or branched conjugated system (i.e. **a** or **b**).

3 Give a molecular orbital explanation as to why reaction **A** will not take place while **B** occurs with extreme ease.

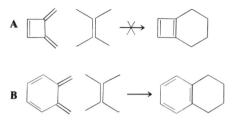

4 Use Hückel MO properties of linear conjugated polyenes to explain why the ultraviolet absorption moves to longer wavelengths with increasing chain length. Why is naphthalene colorless ($\lambda_{max} = 310$–320 nm) while the isomeric azulene is blue ($\lambda_{max} = 540$–700 nm)?

Naphthalene   Azulene

5 Suggest experimental methods which would enable a numerical value of the resonance integral, β, to be determined.
6 Explain the following observations in MO terms.
(a) Cyclopropenone, **1**, and cycloheptatrienone (tropone, **2**) are stable species which do not readily show ketone properties (e.g. they will not form semicarbazones or oximes) whereas cyclopentadienone, **3**, is highly reactive and dimerizes very rapidly.

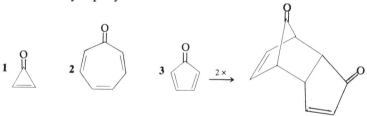

(b) The compound **4** is a very strong dibasic acid completely dissociated in water

O   O—H
          4
O   O—H

(c) The ketone **5** will not enolize whereas its saturated analogue behaves normally.

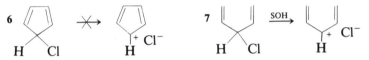

(d) Cyclopentadienyl chloride, **6**, will not undergo S_N1 solvolysis (ionization) whereas the acyclic analogue, **7**, reacts very rapidly and behaves as a normal allylic halide.

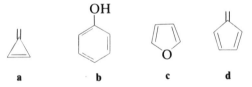

(e) The acidity of cyclopentadiene is very high ($pK_a \sim 12$) whereas its benzo analogues indene, **8**, and fluorene, **9**, are much less acidic ($pK_A \sim 21$ and 24, respectively).

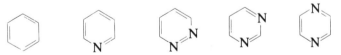

8   **9**

7 Predict the sites of greatest reactivity towards electrophilic attack on the following molecules:

OH

a   b   c   d

8 Determine the effect of substitution on the energy levels, and in particular the frontier orbitals, of compounds C_6H_5—X, where X = —OMe, —NR$_2$, —Cl, —CH=CH$_2$, —C=O, and also of the series

9 Predict, giving reasons, which of the following compounds would be expected to show aromatic character, and which anti-aromatic:

a   b   c

CH$_3$

d 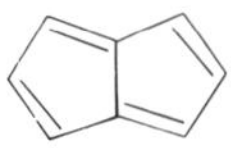   **e** 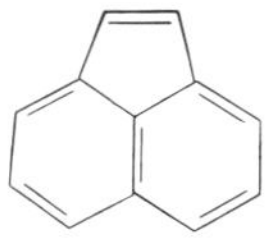   **f** 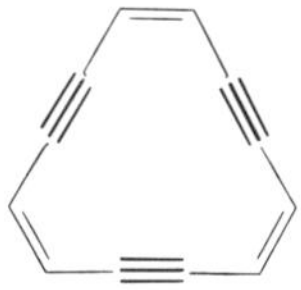   **g**

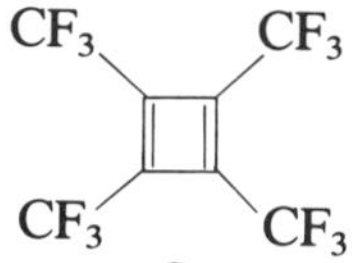

The compound **g** can be obtained stable in solution at $-140°C$; it is a yellow, rectangular and singlet molecule. Comment upon these observations and the Hückel interpretation of cyclobutadiene. [*J. Amer. Chem. Soc.*, **99**, 3524 (1977).]

10 Calculate enthalpies of the following reactions and determine whether they are exo- or endo-thermic.

(a) $CH_2{=}CH_2 + \frac{1}{2}O_2 \longrightarrow CH_2{-}CH_2$
$\qquad\qquad\qquad\qquad\qquad\quad \diagdown\!O\!\diagup$

(b) $CH_2{=}CH_2 + O_2 \longrightarrow CH_2{-}CH_2$
$\qquad\qquad\qquad\qquad\qquad\quad | \qquad |$
$\qquad\qquad\qquad\qquad\qquad\; O{-}\!-\!O$

(c) $2CH_2{=}CH_2 \longrightarrow CH_2{-}CH_2$
$\qquad\qquad\qquad\qquad\qquad\quad | \qquad |$
$\qquad\qquad\qquad\qquad\qquad CH_2{-}CH_2$

(d) $CH_2{=}CH_2 + H_2O \longrightarrow CH_3CH_2OH$

N.B. Obtain strain energies of cyclic products from Table 8.1 and include these in the calculations.

11 The isomerization *n*-butane $\rightleftharpoons$ isobutane can be accomplished on certain catalytic surfaces. Calculate ΔG for the reaction and hence K at atmospheric temperature and at 500°C.

12 Assuming that product stability determines the isomer distribution in the reaction

$$EtCH{=}CH_2 + H{-}Cl \rightarrow EtCH_2{-}CH_2Cl + EtCHCl{-}CH_3,$$

calculate the expected major product. Does this accord with the discussion on orientation in Chapter 12? If not, what may be concluded?

REFERENCES

1 E. Cartmell and G. W. A. Fowles, *Valency and Molecular Structure*, Butterworth, London, 1961.
2 W. Z. Lehmann, *Atomic and Molecular Structure*, Wiley, New York, 1972.
3 J. N. Murrell, S. F. A. Kettle and J. M. Tedder, *The Chemical Bond*, Wiley, Chichester, 1978.
4 G. N. Lewis, *Valence and the Structure of Atoms and Molecules*, Chemical Catalog Co., New York, 1923.
5 I. Langmuir, *J. Amer. Chem. Soc.*, **41**, 868 (1919).

6   M. J. S. Dewar, *The Molecular Orbital Theory of Organic Chemistry*, McGraw-Hill, New York, 1969.
7   D. A. McQuarrie, *Introduction to Quantum Mechanics*, Oxford U.P., 1983.
8   L. Salem, *The Molecular Orbital Theory of Conjugated Systems*, Benjamin, San Francisco, 1966.
9   C. J. Ballhausen and H. B. Gray, *M.O. Theory*, Benjamin, New York, 1964.
10 I. G. Czismadia, *Theory and Practice of M.O. Calculations on Organic Molecules*, Elsevier, Amsterdam, 1976.
11 A. Streitweiser, *Molecular Orbital Theory for Organic Chemists*, Wiley, New York, 1962.
12 A. Liberles, *Introduction to Molecular Orbital Theory*, Holt–Rinehart–Winston, New York, 1966.
13 E. Heilbronner and H. Bock, *The HMO Model*, Vol. 1, Wiley, New York, 1968.
14 P.-O. Lowdin and B. Pullman, *Molecular Orbitals in Chemistry, Physics and Biology*, Academic Press, New York, 1964.
15 K. Higasi, H. Baba and A. Rembaum, *Quantum Organic Chemistry*, Interscience, New York, 1965.
16 M. Karplus and R. N. Porter, *Atoms and Molecules*, Benjamin, Reading, Mass, 1970.
17 C. A. Coulson and A. Streitweiser, *Dictionary of π-Electron Calculations*, Pergamon, London, 1965; C. A. Coulson, *Molecular Orbitals Theory for Organic Chemists*, Academic Press, London, 1978; E. Heilbronner and H. Bock, *The HMO Model*, Vols 2, 3, Wiley, New York, 1968.
18 B. M. Gimarc, *Acc. Chem. Res.*, **7,** 384 (1974).
19 J. L. Whitten, *Acc. Chem. Res.*, **6,** 238 (1973); W. J. Jorgensen and L. Salem, *The Organic Chemist's Book of Orbitals'*, Academic Press, New York, 1973; J. M. Tedder and A. Nechvatal, *Pictorial Orbital Theory*, Pitman, London, 1985.
20 G. L. Closs and L. E. Closs, *J. Amer. Chem. Soc.*, **85,** 2022 (1963).
21 P. Lazzeretti and F. Taddei, *Tetrahedron Letts*, 1453 (1972).
22 N. Muller and D. E. Pritchard, *J. Chem. Phys.*, **31,** 768, 1471 (1959).
23 G. L. Closs, *Proc. Roy. Soc.*, 152 (1962).
24 O. Kennard and D. G. Watson, Eds, *Molecular Structure and Dimensions*, Vols 1–6, International Union of Crystallography, 1935–1974.
25 M. J. S. Dewar, *Hyperconjugation*, Ronald, New York, 1962.
26 M. Fouassier and M.-T. Forel, *Compt. Rend.*, **274B,** 73 (1972).
27 T. Shimanouchi, H. Matsuura, Y. Ogawa and I. Harada, *J. Phys. Chem. Ref. Data,* **7,** 1324 (1978).
28 D. C. McKean, O. Saur, J. Travert and J. C. Lavalley, *Spectrochim. Acta,* **31A,** 1713 (1975).
29 I. M. Mills, *Critical Evaluation of Chemical and Physical Structural Information*, Nat. Acad. Sci., Washington, D.C., 1974.
30 A. L. McClellan, *Tables of Experimental Dipole Moments*, Freeman, San Francisco, 1963.
31 C. J. F. Bottcher and P. Bordewijk, *Theory of Electric Polarisation*, Vol. II, Elsevier, Amsterdam, 1978.
32 L. Pauling, *The Nature of the Chemical Bond*, Cornell U.P., 1945; A. Allred, *J. Inorg. Nucl. Chem.* **17,** 215 (1961).
33 R. J. W. LeFevre, *Adv. Phys. Org. Chem.*, **3,** 1 (1965).
34 J. R. Partington, *An Advanced Treatise on Physical Chemistry*, Longman, London, 1953.
35 J. G. Kirkwood, *J. Chem. Phys.*, **4,** 592 (1936).
36 S. Marriott and R. D. Topson, *J. Chem. Soc., Perkin II*, 113 (1984).
37 J. D. Cox and G. Pilcher, *Thermochemistry of Organic and Organometallic Compounds*, Academic Press, New York, 1970.
38 D. R. Stull, E. F. Westrum and G. C. Sinke, *Chemical Thermodynamics of Organic Compounds*, Wiley, New York, 1969.

39 J. R. Larson, N. D. Epiotis and L. Larson, *Tetrahedron*, **37**, 1557 (1981).
40 S. Benson, *Thermochemical Kinetics*, Wiley, New York, 1976.
41 S. Benson and E. O'Neal, *Free Radicals* (J. K. Kochi, Ed.), Vol. II, Wiley, New York, 1973.
42 G. C. Maitland, M. Rigby, E. B. Smith and W. A. Wakeham, *Intermolecular Forces*, Oxford U.P., 1981.
43 T. Kihara, *Intermolecular Forces*, Wiley, Chichester, 1978.
44 H. Margeneau and N. R. Kestner, *Theory of Intermolecular Forces*, Pergamon, New York, 1971.
45 P. R. Certain and L. W. Bruch, *MTP Int. Rev. Sci., Phys. Chem. I*, 113 (1972); Butterworth, London.
46 W. N. Lipscomb, *MTP Int. Rev. Sci., Phys. Chem. II*, 167 (1974); Butterworth, London.
47 J. O. Hirschfelder, C. F. Curtiss and R. B. Bird, *M.O. Theory of Gases and Liquids*, Wiley, New York, 1954.
48 K. Marokuma, S. Iwata and W. A. Latham, 'Molecular Interactions in Ground and Excited States', in *The World of Quantum Chemistry* (R. Daudel and B. Pullman, Eds), Reidel, Dordrecht, 1974.
49 P. Claverie, *Intermolecular Attractions from Diatomics to Biopolymers*, Wiley, New York, 1978.
50 A. D. Buckingham and R. S. Watts, *Chem. Phys. Letters*, **21**, 186 (1973).
51 M. Szwarc, Ed., *Ions and Ion-Pairs in Organic Reactions*, Wiley, New York, 1972.
52 M. Szwarc, *Acc. Chem. Res.*, **2**, 87 (1969).
53 J. Robbins, *Ions in Solution*, Oxford U.P., 1972.
54 J. W. Barley and R. N. Young, *J. Chem. Soc., Perkin II Trans.*, 835 (1972).
55 N. S. Isaacs, *Liquid Phase High Pressure Chemistry*, Wiley, Chichester, 1981, p. 165.
56 B. Lundgren, S. Claesson and M. Szwarc, *Chem. Scripta*, **3**, 49 (1973).
57 S. Claesson, B. Lundgren and M. Szwarc, *Trans. Farad. Soc.*, **66**, 3053 (1970).
58 T. Asano and W. J. LeNoble, *Chem. Rev.*, **78**, 472 (1978).
59 N. Bjerrum, *Chem. Rev.*, **16**, 287 (1935).
60 R. F. Fuoss and C. A. Kraus, *J. Amer. Chem. Soc.*, **55**, 2387 (1933).
61 J. C. Justice, M. C. Justice and C. Michelotti, *Pure Appl. Chem.*, **53**, 1291 (1981).
62 J. Smid, Ref. *51*, Chapter 3.
63 J. H. Sharp and M. C. R. Symons, Ref. *52*, Chap. 5.
64 E. de Boer and J. L. Sommerdauf, Ref. *52*, Chap. 7.
65 M. C. R. Symons, *Pure Appl. Chem.* **49**, 13 (1977).
66 N. Hirota, R. Carroway and W. Schook, *J. Amer. Chem. Soc.*, **90**, 3606, 3611 (1968).
67 R. F. Adams, *Electron Spin Res.*, **4**, 198 (1977).
68 F. C. Adam and S. I. Weissmann, *J. Amer. Chem. Soc.*, **80**, 1518 (1958).
69 D. H. O'Brien, C. H. Russell and A. J. Hart, *J. Amer. Chem. Soc.*, **101**, 633 (1979).
70 G. Fraenkel and M. P. Haliden-Abberton, *J. Amer. Chem. Soc.*, **103**, 5657 (1981).
71 N. Hirota and S. I. Weissmann, *J. Amer. Chem. Soc.*, **86**, 2538 (1964).
72 E. J. R. Sudhalter and J. B. F. N. Engberts, *Rev. Trav. Chim. Pays-Bas*, **96**, 85 (1977).
73 M. Szwarc, *Pure Appl. Chem.*, **48**, 247 (1976).
74 M. Szwarc, *Carbanions, Living Polymers and Electron-Transfer Processes*, Interscience, New York, 1968.
75 G. J. Janz and R. P. T. Tomkins, *The Non-Aqueous Electrolytes Handbook*, Academic Press, New York, 1972.
76 T. L. Fabry and R. M. Fuoss, *J. Phys. Chem.*, **68**, 1177 (1964).
77 G. C. Greenacre and R. N. Young, *J. Chem. Soc., Perkin II Trans.*, 1661 (1975).
78 T. G. Spiro, A. Revesz and J. Lee, *J. Amer. Chem. Soc.*, **90**, 4000 (1968).
79 R. D. Allendorfer and R. J. Papez, *J. Phys. Chem.*, **76**, 1012 (1972).
80 K. Morokuma, S. Iwata and W. A. Latham, *In the World of Quantum Chemistry* (R. Daudel and B. Pullman, Eds,), Reidel, Boston, 1974.
81 K. Morokuma, *Acc. Chem. Res.*, **10**, 294 (1977).

References

82   K. Morokuma and K. Kitaura, *Mol. Interact.*, **1**, 21 (1980).
83   G. S. Pimentel and A. L. McLellan, *The Hydrogen Bond*, Freeman, San Francisco, 1960.
84   D. Hadzi, Ed., *Hydrogen Bonding*, Pergamon, London, 1959.
85   W. C. Hamilton and J. A. Ibers, *Hydrogen Bonding in Solids*, Benjamin, New York, 1968.
86   A. K. Covington and P. Jones, *Hydrogen-Bonded Solvent Systems*, Taylor and Francis, London, 1968.
87   S. N. Vinogradov and R. H. Linnell, *Hydrogen Bonding*, Van Nostrand, New York, 1971.
88   J. Emsley, *Chem. Soc. Rev.*, **9**, 91 (1980).
89   W. M. Latimer and W. H. Rodebush, *J. Amer. Chem. Soc.*, **42**, 1419 (1920).
90   L. N. Ferguson, *J. Chem. Ed.*, **33**, 267 (1956).
91   L. Pauling, *The Nature of the Chemical Bond*, Cornell U.P., 1945.
92   E. M. Arnett, T. S. S. R. Murty, P.v R. Schleyer and L. Joris, *J. Amer. Chem. Soc.*, **89**, 5955 (1967).
93   S. W. Benson, *J. Chem. Ed.*, **42**, 502 (1965).
94   M. Gorman, *J. Chem. Ed.*, **34**, 304 (1957).
95   Z.-I. Yoshida and E. Osawa, *J. Amer. Chem. Soc.*, **88**, 4019, (1966).
96   D. P. Eyman and R. S. Drago, *J. Amer. Chem. Soc.*, **88**, 1617 (1966).
97   M. F. Rettig and R. S. Drago, *J. Amer. Chem. Soc.*, **88**, 2966 (1966).
98   R. C. Haddon, *J. Amer. Chem. Soc.*, **102**, 1807 (1980).
99   G. Briegleb, *Elektronon-Donator–Acceptor-Komplexe*, Springer, Berlin, 1961.
100 L. J. Andrews and R. M. Keefer, *Molecular Complexes in Organic Chemistry*, Holden–Day, San Francisco, 1964.
101 V. Gutmann, *Pure Appl. Chem.*, **27**, 73 (1971).
102 M. W. Hanna and J. L. Lippert, *Molecular Complexes*, Vol. 1 (R. Foster, Ed.,), Crane Russak, New York, 1973.
103 H. A. Benesi and J. H. Hildebrand, *J. Amer. Chem. Soc.*, **71**, 2703 (1949).
104 R. Foster, *Molecular Complexes*, Vol. 2 (R. Foster, Ed.,), Crane Russak, New York, 1974.
105 M. Tamres and R. L. Strong, *Molecular Association*, Vol. 2 (R. Foster, Ed.,), Academic Press, London, (1979).
106 L. J. Andrews, *Chem. Rev.*, **54**, 713 (1954).
107 V. Gutmann, *The Donor–Acceptor Approach to Molecular Interactions*, Plenum, New York, 1978.
108 E. F. Caldin, D. O'Donnell, D. Smith and J. E. Crooks, *Chem. Comm.*, 1358 (1971).
109 N. M. D. Brown, R. Foster and C. A. Fyfe, *J. Chem. Soc.* (*B*), 406 (1967).
110 R. M. Keefer and L. J. Andrews, *J. Amer. Chem. Soc.*, **77**, 2164 (1955).
111 R. M. Keefer and L. J. Andrews, *J. Amer. Chem. Soc.*, **72**, 4677 (1950); **73**, 462 (1951).
112 A. Zweig, J. E. Lehnsen and M. A. Murrary, *J. Amer. Chem. Soc.*, **85**, 3933 (1963); *Tetrahedron Letts* **89**, (1964).
113 H. E. Winberg and D. D. Coffman, *J. Amer. Chem. Soc.*, **87**, 2776 (1965).
114 S. Kobinata and S. Nagakura, *J. Amer. Chem. Soc.*, **88**, 3905 (1966).
115 E. M. Voiht and C. Reid, *J. Amer. Chem. Soc.*, **86**, 3930 (1964).
116 M. J. Kurylo and N. B. Jurinski, *Tetrahedron Letts*, 1083 (1967).
117 P. Trotter and M. W. Hanna, *J. Amer. Chem. Soc.*, **88**, 3724 (1966).
118 I. D. Knutz, F. P. Gasparro, M. D. Johnston and R. P. Taylor, *J. Amer. Chem. Soc.*, **90**, 4778 (1968).
119 M. Crescenti, C. Galli and L. Mandolini, *J. Chem. Soc., Chem. Comm.* 551 (1986).

2 Kinetics and thermodynamics

The objective of physical organic chemistry is an explanation of the motive forces which prompt spontaneous chemical changes in all their variety, in terms of the fundamental physical properties of the compounds involved. A related problem is an explanation of the factors which determine the rate at which a reaction proceeds. A term to denote the tendency toward chemical reaction is needed: 'driving force' is often used but the outdated word 'affinity' deserves to be reinstated.* What therefore determines reaction affinity? Thermodynamics is the starting point for an examination of this question, though it will not prove an entirely sufficient one. The idea that energy, which may be equated with heat, is released upon formation of a chemical bond or, indeed, upon operation of any of the weak intermolecular forces, has already been made apparent and it seems natural that it should be a spontaneous process to seek the point of lowest potential energy in just the same way as a bicycle spontaneously runs to the point of lowest gravitational potential. It is intuitive that at least a contribution to the driving force comes from the heat of reaction. A familiar example might be

$$CH_4 + 3O_2 \rightarrow CO_2 + 2H_2O + 890\,kJ\ (213\,kcal).$$

The ready tendency for methane to burn in oxygen is seen to be due to the especial stabilities of the products of combustion, carbon dioxide and water. While this is indeed so in this particular case, the argument is incomplete in several senses and must be examined in further detail.

2.1 Enthalpy

Any bulk sample of material, whether element or compound, single substance or mixture, is associated with a total energy content, U, made up of chemical bonds, inter- and intra-molecular forces, vibrational and

---

* Hiftory   The term *affinity*, which is the expreffion of a force by which fubftances of different natures combine with each other, feems to have been pretty early employed by chemical writers.

(*Encyclopaedia Britannica*, 1810)

rotational energy. Changes in U are made up of heat added, q, and work done on the system, w, such that

$$dU = dq + dw, \qquad (2.1)$$

a statement of energy conservation and the First Law of thermodynamics. If heat energy is put into the sample, the work done is a change of volume, pressure or some combination of these (the term pV, which also has units of energy), i.e.

$$dw = -p \, dV$$

and hence

$$dq = dU + p \, dV,$$

or

$$dH = dU + p \, dV \qquad (2.2)$$

since it is convenient to give a name and symbol to the heat term q, which is known as enthalpy, H, and is of great significance in chemistry. Both H and U have units of energy (J or J mol^{-1}, alternatively cal or cal mol^{-1}), they are extensive quantities (i.e. they depend upon the amount of substance considered) and they are state functions (they are independent of the pathway taken to reach a given value). It is not possible, however, to assign absolute values of the enthalpy of any substance, since there is no absolute zero of this quantity from which to make comparisons. Instead, it is conventional to set at zero the enthalpies of all elements in their normal states at 25°C. It is then possible to measure changes in enthalpy — simply as the heat evolved or taken in during some chemical or physical change — and from this basis to build up a consistent set of data expressing the enthalpies of compounds relative to those of their constituent elements in their standard states. This can be done even when direct synthesis from the elements cannot be made to occur. The standard enthalpy of formation of methane, ΔH_f, is, by definition, the heat change for the following (hypothetical) reaction:

C(graphite) + 2H$_2$(gas, 25°C) → CH$_4$(gas, 25°C) + 74·8 kJ (17·9 kcal).

This is expressed as

$$\Delta H_f(\mathrm{CH_4}) = -74\cdot8 \text{ kJ mol}^{-1} \ (-17\cdot9 \text{ kcal mol}^{-1}).$$

The negative sign of ΔH_f indicates that methane is of lower energy (more stable) than the constituent atoms as they occur in the molecules, crystals, etc., of the elements in their standard states. This is information which may be obtained by application of Tables 1.13 and 1.15. ΔH_f must not be confused with the heat change on forming the compound from constituent atoms (heat of atomization, ΔH_{at}) which expresses the energy content of the bonds but is more negative than ΔH_f by an amount corresponding to the heats of atomization of the elements in their standard states:

$$\Delta H_f(\text{compound}) = \Delta H_{at}(\text{compound}) - \Delta H_{at}(\text{elements}).$$

ΔH_f is normally obtained by measurement of the heat of combustion. When all reacting species are in their standard states the energy terms on either side of the equation must balance:

$$CH_4(g) + 3O_2(g) \rightarrow CO_2(g) + 2H_2O(l) + 890\,kJ\,(213\,kcal)$$

ΔH_f: $\quad x \qquad 0 \qquad\quad -393 \qquad 2 \times -286$
$$\qquad\qquad\qquad\qquad (-93\cdot9) \qquad (-68\cdot3)$$

$$x + 0 = [-393 - (2 \times 286) + 890]$$

$$x = \Delta H_f(CH_4) = -75\,kJ\,mol^{-1}\,(-17\cdot9\,kcal\,mol^{-1}).$$

It is an easy matter, therefore, knowing the standard enthalpies of formation of all reagents, to calculate the enthalpy of reaction.

There is a tendency for a spontaneous reaction to occur such that ΔH is negative, i.e. an exothermic process.

2.1.1 Endothermic reactions

Consider now the following reaction used in the production of 'water gas':

$$C(\text{graphite}) + H_2O(g) \rightarrow CO(g) + H_2(g)$$

ΔH_f: $\quad 0 \qquad\quad -242 \qquad -110 \qquad 0$
$$\qquad\qquad (-57\cdot8) \quad (-26\cdot3)$$

$$\Delta H = -110 - (-242) = +132\,kJ\,mol^{-1}\,(31\cdot5\,kcal\,mol^{-1}).$$

This reaction proceeds (providing that the temperature is high enough) despite the fact that heat is absorbed in the process. In this case the reaction is not driven by a reduction in the enthalpy of the system so it is necessary to look for a second contributor to reaction affinity: this turns out to be a quantity known as entropy. That there must be such a quantity is implicit in the First Law since the total energy of a closed system (i.e. the universe) must remain constant; if all processes were solely enthalpy-driven there would be a constant reduction in universal energy.

2.2 ENTROPY

The expansion of a perfect gas into a vacuum and the mixing of two gases are two examples of processes which occur spontaneously but with no heat change; $\Delta H = 0$. Heat flows spontaneously from a hot body to a cold one, and it would seem totally absurd that any such process could occur in reverse without the application of some external agency. The common feature of these examples is an increase in the dispersal of the total energy (which includes the energy of individual molecules) so that it becomes more evenly distributed throughout the available space. It is quite logical that collisions between high- and low-velocity molecules (i.e. hot and cold) should result in an averaging of all velocities: it is of impossibly low probability that the molecules of different velocities should sort themselves in space and create a temperature gradient. There is evidently a universal tendency for dispersal to occur spontaneously, a change from a state of low probability to one of maximum probability — the statistical view of entropy. Since thermal energy, q, is the originator of random motion it is to be expected that entropy, S, increases with temperature (T/K) so that

$$q \propto T$$

or

$$q \equiv ST.$$

Entropy is seen as the constant of proportionality between cause and effect. The thermodynamic definition of entropy expresses the change in S with reversible heat exchange

$$dS = dq_{rev}/T. \tag{2.3}$$

A statistical derivation of entropy may be achieved as follows for a system containing two gases. The total system which we will call the *macrostate*, is defined by the Gas Law, $pV/nRT = $ constant, but this macrostate may be realised by any of a vast number, W, of *microstates* differing in spatial and motional properties of the individual molecules, existing momentarily, and all having equal probability once equilibrium has been reached and the gases are thoroughly mixed. Though very large, the number of microstates is not infinite due to quantization of motion — the 'control of chaos by the quantum laws' as expressed by Hinshelwood[1]. Contributions to W include those due to translational, rotational and vibrational motions and admixture of species. The number W is related to the entropy of the system by Eq. (2.4):

$$S = k \ln W = R/L \ln W \tag{2.4}$$

where $k = $ Boltzmann constant = gas constant per mole. It may be shown that these two viewpoints of entropy are compatible. Entropy is a further extensive and state quantity with the dimension of energy/temperature. The quantity (ST) therefore has the dimension of energy and represents energy present in the system by virtue of its capability of existing in a number of equivalent states. The change of entropy with temperature can be obtained by integrating Eq. (2.3), giving

$$S(T_2) = S(T_1) + \int_{T_1}^{T_2} dH/T$$

in which dH replaces dq_{rev} assuming constant pressure ($p\, dV = 0$). Heat capacity at constant pressure, C_p, the amount of heat to raise the sample by unit temperature, is defined as $C_p = (\partial H/\partial T)_p$, whence

$$S(T_2) = S(T_1) + \int_{T_1}^{T_2} C_p \, \mathrm{d}T/T \qquad (2.5)$$

and changes of entropy are obtained from the areas under the curve of C_p/T between T_1 and T_2. It is usual to assume for all pure crystals that $S(0\mathrm{K}) = 0$ (Third Law); consequently 'absolute' entropies are obtained from Eq. (2.5) with $T_1 = 0$, $S(T_1) = 0$. As with enthalpies, entropies of pure substances may be dissected into structural contributions, which are included in Table 1.13. The Second Law of Thermodynamics is intimately concerned with entropy. It states that 'in a closed system, spontaneous processes occur in the direction of increasing entropy', a principle which can easily be supported by experience. In chemical terms, entropy increases during dissociative processes (solution of crystals, mixing of reagents, breaking of bonds) and, conversely, decreases during associative ones. This does not mean that association cannot occur spontaneously, which would be quite contrary to experience (for example in bimolecular reactions, hydrogen bonding, crystallization). Rather it means that these latter examples must take place despite an unfavorable entropy change, but driven by a compensatingly favorable enthalpy change. Now it can be appreciated that the complete driving force for a spontaneous process, chemical or otherwise, must be sought in the combination of changes in enthalpy and entropy. It is the combination $(\Delta H - T\Delta S)$ which must be negative in order that any unaided change will take place. Although examples have been cited for which one or other of these terms is zero and which can be said to be purely enthalpy- or entropy-driven, it is usual for both enthalpy and entropy to change during a chemical reaction.

2.3 THE GIBBS FUNCTION, G

The sum of energy contributions discussed above is so important that it is given a special name, the Gibbs function (or Gibbs free energy), G:

$$G = H - TS,$$

or

$$\Delta G = \Delta H - T\Delta S;$$

the sign is a result of the requirement that H must decrease but S must increase to create a driving force. The affinity of a chemical reaction may now be equated to a change in G and may only occur spontaneously when ΔG is negative. The energetics of the two gaseous reactions above may now be re-examined in the light of this master principle.

$$CH_4 \quad + \quad 3O_2 \quad \rightarrow \quad CO_2 \quad + \quad 2H_2O(l)$$

ΔH_f:	-75	0	-393	-572
	(-17.9)	(0)	(-93.9)	(-136.7)
ΔS:	0.186	0.615	0.213	0.378
	(0.0444)	(0.147)	(0.051)	(0.903)

$$\Delta H = -890 \text{ kJ mol}^{-1}$$
$$(-212.7 \text{ kcal mol}^{-1})$$

$$\Delta S = -0.210 \text{ kJ K}^{-1} \text{ mol}^{-1}$$
$$(-0.050 \text{ kcal K}^{-1} \text{ mol}^{-1})$$

At $T = 300$ K

$$T\Delta S = 300 \times (-0.21) = -63 \text{ kJ mol}^{-1}$$
$$(-0.05) \quad (-15 \text{ kcal mol}^{-1})$$

$$\Delta G = -890 - (-63) = -828 \text{ kJ mol}^{-1}$$
$$(-212.7) \quad (-15) \quad (-197.9 \text{ kcal mol}^{-1})$$

The entropy change is slightly unfavorable (four gaseous molecules are converted to three) but is amply compensated by the large negative value of ΔH.

$$C \quad + \quad H_2O \quad \rightarrow \quad CO \quad + \quad H_2$$

ΔH_f:	0	-242	-110	0
	(0)	(-57.8)	(-26.3)	
ΔS_f:	0.006	0.189	0.197	0.130
	(0.00143)	(0.045)	(0.047)	(0.031)

$$\Delta H = +132 \text{ kJ mol}^{-1}.$$
$$(+31.5 \text{ kcal mol}^{-1})$$

$$\Delta S = +0.132 \text{ kJ K}^{-1} \text{ mol}^{-1}.$$
$$(+0.0315 \text{ kcal K}^{-1} \text{ mol}^{-1})$$

At $T = 300$ K
$$\Delta G = +132 - (300 \times 0.132) = +92 \text{ kJ mol}^{-1} \quad (+22 \text{ kcal mol}^{-1}).$$
$$(+31.5) \qquad\qquad (0.0315)$$

At $T = 500$
$$\Delta G = +132 - (500 \times 0.132) = +66 \text{ kJ mol}^{-1} \quad (+16 \text{ kcal mol}^{-1}).$$

At $T = 1000$
$$\Delta G = +132 - (1000 \times 0.132) = 0 \text{ kJ mol}^{-1} \quad (0 \text{ kcal mol}^{-1}).$$

(This is somewhat approximate since ΔH is not entirely temperature-independent.)

ΔS alone is favorable (solid going to gaseous products) but cannot overcome the unfavorable enthalpy until a temperature of 1000 K is reached. An endothermic reaction like this may therefore proceed in either direction depending upon the temperature.

Free energies of formation, ΔG_f, can be assigned to pure substances from Eq. (2.6) and calculated from values in Table 1.13.

$$\Delta G_f = \Delta H_f - T\Delta S_f. \tag{2.6}$$

The free energy per mole of substance multiplied by its activity (concentration) is known as its chemical potential, μ. Then by analogy with mechanical and electrical systems a chemical system will tend towards the lowest (chemical) potential.

This brief treatment attempts only to convey a rational impression of the workings of some of the most fundamental natural laws. More rigorous treatments are available in some of the many excellent texts[2-9] on the subject of thermodynamics, which is an exact science even if the subjects to which it is applied are not ideal.

2.4 FACTORS WHICH CONTRIBUTE TO ENTROPY

Absolute entropies of some pure substances are set out in Table 2.1, from which the following trends may be seen.

(a) Entropy is the manifestation of molecular motion and as a consequence diminishes with temperature and with the transition of matter into condensed phases as translational motion is slowed and then completely

Table 2.1   Molar entropies of some pure substances[a] at 298 K

	$S/J\,K^{-1}\,mol^{-1}$ $(cal\,K^{-1}\,mol^{-1})$		$S/J\,K^{-1}mol^{-1}$ $(cal\,K^{-1}\,mol^{-1})$		$S/J\,K^{-1}\,mol^{-1}$ $(cal\,K^{-1}\,mol^{-1})$
Ne	146 (34·9)	HF	173 (41·3)	HCHO	219 (52·3)
H_2	130 (31·0)	HCl	186 (44·4)	CH_4	186 (44·4)
N_2	191 (45·6)	HBr	199 (47·5)	C_2H_6	228 (54·5)
O_2	204 (48·7)	CO	197 (47·1)	$n\text{-}C_3H_8$	270 (64·5)
Cl_2	223 (53·3)	O_3	231 (55·2)	$n\text{-}C_4H_{10}$	310 (74·1)
Br_2	245 (58·5)	H_2O	189 (45·2)	C_2H_4	220 (52·6)
C(diamond)	2·4 (0·57)	SO_2	248 (59·3)	$CH_3CH{=}CH_2$	229 (54·7)
C(graphite)	5·6 (1·3)	NO_2	240 (57·4)	CH_3OCH_3	266 (63·6)
C(gas)	158 (37·7)	HCN	202 (48·3)		
		CO_2	214 (51·1)		

[a] Gaseous unless specified otherwise.

arrested. The sequence $S_{gas} > S_{liquid} > S_{solid}$ is normal, with S tending towards zero at 0 K.

(b) Entropy of similar forms of matter, in the same state and at the same temperature, tends to increase with mass as a result of the greater organization of fundamental particles in a large atom compared with many small ones of similar mass. This can be seen in values for the diatomic gases or for atoms.

(c) Molar entropy tends to increase with greater complexity of the molecule, not only on account of a mass effect but also because more molecular motions become available. However, association of molecules is accompanied by a loss of independent translational motion only partly compensated by additional vibrational modes and hence entropy normally decreases (e.g. because of solvation, intermolecular association and bimolecular reactions). The converse is true for dissociative processes of all kinds. The entropy of a sample of bulk matter depends upon the various energy states which are available to it, such as rotations, vibrations and other forms of molecular motion, and upon the population of such states. A low-energy transition, for which both the upper and lower energy states will be highly populated, will give rise to a larger entropy contribution than a high-energy transition, for which the upper-state will be relatively unpopulated, at least at low temperatures. Thus, a complex molecule of low symmetry will usually have more rotational and vibrational states available to it than a simple one of high symmetry. The higher the temperature, the more populated are the upper states. In a homologous series, each methylene group, $-CH_2-$, contributes about $40\,J\,K^{-1}\,mol^{-1}$ ($9 \cdot 6\,cal\,K^{-1}\,mol^{-1}$).

(d) Entropy increases when substances become mixed or dissolved one in another. This clearly introduces further randomness into the state of matter and will make an important contribution to the properties of solutions (Chapter 5).

It can now be appreciated that the large positive entropy changes associated with both the combustion of methane and the formation of water gas, discussed above, arise from the formation of a greater number of molecules of gaseous products than there were on the reactant side. The ammonia synthesis,
$$N_2 + 3H_2 \rightarrow 2NH_3,$$
is accompanied by a decrease in entropy at 300 K ($\Delta S = -200\,J\,K^{-1}\,mol^{-1}$ ($-47 \cdot 8\,cal\,K^{-1}\,mol^{-1}$)) due to the loss of translational motion. The entropy change becomes positive at high temperatures since ammonia has low vibrational energy levels which can then become populated.

2.5 CHEMICAL EQUILIBRIUM

In thermodynamics much importance is attributed to processes which are reversible, which means in this context that they may be made to proceed in a reverse direction by the (infinitesimal) change of one of the variables. For example, the maximum work output can only be obtained from a change occurring reversibly. Many chemical reactions are demonstrably reversible and in principle all may be considered so from a thermodynamic viewpoint. As will be shown, the thermodynamic analysis of reaction rates stems from the treatment of equilibrium properties. As a simple example, *cis*-but-2-ene, **1**, is a stable and well-behaved compound at ambient conditions; but on heating it above about 500°C or standing it in contact with an acid catalyst (Chapter 9) conversion begins to its isomer, *trans*-but-2-ene, **2**, a spontaneous change which should signal a reduction in free energy. The reaction proceeds at an ever-decreasing rate and finally halts when about half of the *cis* compound is converted. Pure *trans*-but-2-ene would yield the same mixture of isomers under similar conditions. Since it is unreasonable to conclude that some of the molecules are susceptible to isomerization while others are not, it must be inferred that the *cis* and *trans* isomers are being mutually interconverted. Eventually an equilibrium situation is reached at which the two rates are identical so that the system gives the appearance of having a static composition at which the free energy, G, is a minimum. One may measure or calculate free energies of the individual isomers and it might be tempting to conclude that the system should revert completely to the isomer with the lower value of G, i.e. the *trans* form:

	But-2-ene (673 K)	
	cis	*trans*
ΔH_f/kJ mol^{-1}	−13·8	−17·2
(kcal mol^{-1})	(−3·3)	(−4·1)
ΔS_f/kJ K^{-1} mol^{-1}	0·331	0·325
(kcal K^{-1} mol^{-1})	(0·08)	(0·078)
ΔG_f/kJ mol^{-1}	91·7	89·1
(kcal mol^{-1})	(21·9)	(21·3)

The entropy of the mixture of isomers, however, is greater than that of either individually, as discussed above. It is the free energy of the system as a whole that determines the position of equilibrium. Suppose $G(trans) = G(cis)$; then the maximum entropy is achieved with a 1:1 mixture. In fact, $G(trans) < G(cis)$, so the minimum in G occurs at an equilibrium composition containing an excess of the more stable *trans* form; the excess is higher the greater the difference in G. The measure of equilibrium position is the equilibrium constant, K:

$$K = [trans]/[cis].$$

In general,

$$K = \frac{\overset{\text{products}}{\prod} a_i^n}{\underset{\text{reactants}}{\prod} a_j^m} \approx \frac{\overset{\text{products}}{\prod} c_i^n}{\underset{\text{reactants}}{\prod} c_j^m} \qquad (2.7)$$

in which a_i, a_j are activities of species taking part in the equilibrium, although concentration suffices for the present. At equilibrium, the chemical potentials of the two reactants are equal and a small perturbation produces no change in G since the position of equilibrium is located at a minimum on a curve of G against composition, ζ, i.e.

$$\partial G/\partial \zeta = 0.$$

Indeed, Prigogene[19] defines reaction affinity as $\partial G/\partial \zeta$. It may be shown that

$$\Delta G = -RT \ln K \qquad (2.8)$$

where R = gas constant = $8.36\,\text{J K}^{-1}\text{mol}^{-1}$. It is clear that if $\Delta G = 0$, then $\ln K = 0$ and $K = 1$, i.e. a 1:1 mixture exists at equilibrium. This important relationship is known as the reaction isotherm. For the butenes at 400°C (673 K),

$$\Delta G = [G(cis) - G(trans)] = 2.6\,\text{kJ mol}^{-1}\,(0.62\,\text{kcal mol}^{-1})$$

which corresponds to a 61:39 equilibrium mixture in favor of the *trans* isomer.

The relationship between ΔG, K and equilibrium composition for equilibria between equal numbers of reagent and product molecules is given in Table 2.2. When $\Delta G > 20\,\text{kJ mol}^{-1}$ ($4.80\,\text{kcal mol}^{-1}$), equilibrium

Table 2.2 The relationship between ΔG and K for $A \rightleftharpoons B$ at 298 K

$\Delta G/\text{kJ mol}^{-1}$ (kcal mol^{-1})	K	Equilibrium percentage of more stable species
0	1.00	50
1.00 (0.24)	1.5	60
2.1 (0.50)	2.33	70
3.42 (0.82)	4.00	80
5.44 (1.30)	9.00	90
11.35 (2.71)	99	99
17.0 (4.06)	999	99.9
22.8 (5.45)	9999	99.99

amounts of the high-energy component become vanishingly small yet may still be the reactive intermediates on which some chemical process depends, e.g. enols, Section 9.2.1.

2.6 SOME USEFUL THERMODYNAMIC RELATIONSHIPS

Changes in Gibbs function refer to conditions of constant pressure, the usual situation. If constant-volume conditions are considered the pV work term is zero and a different free energy term, A, the Helmholtz free energy, can be defined:

$$A = U - TS. \tag{2.9}$$

Compare

$$G = H - TS = U - TS + pV. \tag{2.10}$$

Hence

$$\underset{\substack{\text{net work}\\\text{available}}}{G} = \underset{\substack{\text{maximum}\\\text{work}}}{A} + \underset{\substack{\text{obligatory}\\\text{work on system}}}{pV} \tag{2.11}$$

The infinitesimal dA can be identified with dw_{rev} for a reversible process and $T\,dS$ with dq_{rev}; hence Eq. (2.1) becomes

$$dU = T\,dS - p\,dV$$

and it follows that

$$dG = V\,dp - S\,dT, \tag{2.12}$$

an important equation which gives

$$(\partial G/\partial T)_p = -S \tag{2.13}$$

and

$$(\partial G/\partial p)_T = V. \tag{2.14}$$

2.6.1 Temperature dependence

Heat capacities at constant pressure, C_p, and volume, C_V, are defined as:

$$C_p = (\partial H/\partial T)_p$$
$$\text{and}\quad C_V = (\partial H/\partial T)_V. \tag{2.15}$$

These quantities are important if we need to determine enthalpies of formation, ΔH_{f}, or of reaction, ΔH_{r}, at temperature other than standard (298 K). This may be done by means of Eqs (2.16):

$$\Delta H_{\text{f}}(T_2) = \Delta H_{\text{f}}(T_1) + (T_2 - T_1)[C_p(T_2) - C_p(T_1)]$$
$$\text{and}\quad \Delta H_{\text{r}}(T_2) = \Delta H_{\text{r}}(T_1) + (T_2 - T_1)[C_p(\text{products}) - C_p(\text{reactants})]. \tag{2.16}$$

Finally, the Gibbs–Helmholtz equation, Eq. (2.17), relates enthalpy to the temperature dependence of G:

$$\partial(G/T)/\partial T = -\Delta H/T^2, \tag{2.17}$$

while the van't Hoff equation, Eq. (2.18), enables one to calculate the enthalpy of reaction, ΔH, from the temperature dependence of the equilibrium constant, K.

$$\mathrm{d}\ln K/\mathrm{d}T = \Delta H/RT^2. \tag{2.18}$$

2.7 THE APPLICATION OF THERMODYNAMICS TO RATE PROCESSES[9,10]

It has been shown that ΔG is the proper function by which to measure reaction affinity, the tendency for reagents to be converted to products. The effects of all manner of variables on ΔG — temperature, pressure, solvent, structure, for example — can be measured and enable one to judge whether a reaction is possible. This is not a sufficient guide as to whether the reaction will actually occur, since it gives information concerning reagents and products only and none concerning the state of affairs during the transformation. Now it is a fundamental postulate that reagents become transformed into products in a gradual process and that, in principle, the geometry and thermodynamic properties of the reacting system are continuously changing functions throughout. The actual changes undergone by the reacting molecules in order to reach the product state may be called the reaction pathway or *reaction coordinate* and upon its nature depends the question whether a thermodynamically-favored reaction will occur at all, and at what rate — matters crucial to chemistry.

2.7.1 Activation

It has been shown that ΔG for the combustion of methane is very favorable and yet a mixture of methane and oxygen can remain in contact without reaction indefinitely unless prompted by a spark, after which the reaction is completed in a very short time.

Hydrogen and ethene can also remain mixed and utterly indifferent to one another unless a platinum surface is introduced, when addition to form ethane occurs smoothly and with the evolution of heat.

$$H_2C=CH_2 + H_2 \rightarrow H_3C-CH_3$$
$$\Delta H = -137 \text{ kJ mol}^{-1} \, (-32.7 \text{ kcal mol}^{-1}).$$

Dibenzoyl peroxide, **3**, is a perfectly stable liquid at room temperature despite the fact that the O—O bond strength is only about 140 kJ mol^{-1} $(33.5 \text{ kcal mol}^{-1})$ and a small fraction of the molecules present actually

Ph $\diagdown$ O—O   Ph   Ph   O· ·O   Ph

$$\text{Ph}\diagdown\!\!\underset{\text{O}}{\overset{\|}{\text{C}}}\!\!\diagup\text{O—O}\diagdown\!\!\underset{\text{O}}{\overset{\|}{\text{C}}}\!\!\diagup\text{Ph}\quad\xrightarrow[\Delta]{(\leftarrow)}\quad \text{Ph}\diagdown\!\!\underset{\text{O}}{\overset{\|}{\text{C}}}\!\!\diagup\text{O· ·O}\diagdown\!\!\underset{\text{O}}{\overset{\|}{\text{C}}}\!\!\diagup\text{Ph}\quad\longrightarrow\quad\text{Products}$$

3

possess kinetic energies in excess of this value according to the Maxwell distribution law (Eq. (2.19)):

$$dn_E = PE^{\frac{1}{2}}e^{-E/kT}\,dE \tag{2.19}$$

where $P = 2\pi N(1/\pi kT)^{\frac{3}{2}}$, dn_E is the fraction of molecules of energy E, N is the total number of molecules present, and k is the Boltzmann constant. As the temperature rises above about 80°C decomposition proceeds at an ever-increasing rate.

These three examples illustrate that a favorable ΔG is not of itself sufficient to promote reaction. The nature of the terrain which lies between reactants and products needs also to be taken into account. The model which leads to an explanation of these characteristics, derived from ideas put forward by Arrhenius (1889), is as follows. Progress along the reaction coordinate requires an initial increase in energy (*energy of activation*) before the final decrease to products (Fig. 2.1). Only those molecules possessing sufficient energy to overcome this *activation barrier* are capable of reaction. Molecules of methane and oxygen at room temperature do not possess sufficient activation energy but once a few fruitful reactions are initiated by a flame or spark, the heat produced provides activation energy for further reaction which spreads at an ever-increasing velocity so that an explosion results. A similar situation exists in the second example but the function of the platinum catalyst is to reduce the amount of activation energy necessary by making available another pathway to products. The final example, the peroxide decomposition, shows that not only must the energy available in a molecule be sufficient — 140 kJ mol^{-1} (33·5 kcal mol^{-1}) in fact — to progress along the reaction coordinate, but it must also be channelled into the appropriate mode, in this case the stretching of the O—O bond to high vibrational levels. It is improbable that, at temperatures below 80°C, sufficient energy will reside in this mode alone. In this case the reaction is endothermic, as in Fig. 2.1c (stage A → I).

Reactions between molecules (bimolecular processes) are initiated by random collisions during which kinetic energy may be transformed into appropriate activation energy. Correct orientation of the colliding molecules will obviously be necessary so that relatively few collisions will be fruitful even if the activation energy is available. Intermolecular attractive forces no doubt also contribute in drawing molecules into collision once they have wandered sufficiently close. Simultaneous collisions between three molecules (termolecular collisions) are very improbable and negligible if some correct orientation of all three is required for reaction to occur.

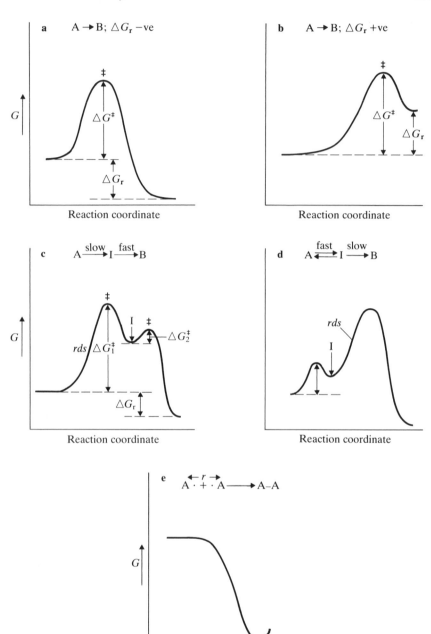

Fig. 2.1 Types of energy profile: free energy of activation $= \Delta G^{\ddagger}$; free energy of reaction $= \Delta G_r$. Transition states are indicated by $\ddagger$ and reactive intermediates by I.

2.7.2 The potential energy surface

The energy (U, G) of a two-atom system such as H—H depends on one spatial variable, the H—H distance, and can be plotted on a graph (Fig. 1.4). A three-atom system (e.g. H—H$\cdots$H $\rightarrow$ H$\cdots$H—H, the simplest possible reaction) needs three parameters (three interatomic distances, or two interatomic distances and an angle) to define the geometry fully. A plot of geometry against energy is a four-dimensional 'surface' already too complex to be represented graphically and just about at the limit of complexity for accurate calculations. The reaction coordinate is the pathway across this surface which requires the least activation energy, analogous to a traveller crossing a mountain range seeking the lowest passes. Any representations of reaction coordinates for more complex systems will require considerable approximations and the implicit understanding that while attention may be focused on the most important geometrical changes such as bond making and breaking, other changes may be occurring simultaneously. For example, in the combination of two methyl radicals to form ethane,

$$H_3C^{\cdot} + {}^{\cdot}CH_3 \rightarrow H_3C\text{—}CH_3,$$

requiring the C–C distance to diminish to 154 pm, H—C—H angles change from 120° to 108° and force constants change simultaneously. Still, it is convenient to use Fig. 2.1e as a means of representation. Traversal of the reaction coordinate is accompanied by a smooth transition of geometries and of molecular orbitals, vibration and rotational levels from those of reagents to those appropriate to products. In more complex examples, potential energy minima (*wells*) may exist which correspond to *reactive intermediates* — species whose reactivities are usually high and corresponding lifetimes short, dependent on the depth of the wells in which they reside. They may or may not lie on the reaction coordinate but a knowledge of the details of these surfaces and of possible pathways to products is a major objective in elucidating reaction mechanism.

Two-coordinate diagrams

It is frequently convenient to express the energy change during the course of a reaction in which two geometrical coordinates are important by plotting them on x and y axes. Contour lines of equal energy then complete a three-dimensional graph (Fig. 2.2). For example, progress of the exchange reaction:

$$I^- + CH_3\text{—}Br \rightarrow I\text{—}CH_3 + Br^-$$

can be represented as changes in the C–Br and C–I bond lengths, other changes being ignored. The reaction coordinate will approximate to the path on the two-dimensional surface which joins reagents and products with

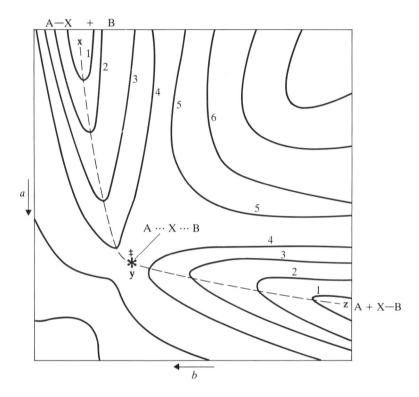

Fig. 2.2 Two-dimensional energy profile for an exchange reaction. Energy contours increase in magnitude 1–6. The coordinates represent *a*, the A–X distance, and *b*, the B–X distance.

the minimum increase in energy. This is seen to pass down the 'valley' **x–y** over the 'saddle point', **y**, and then down the 'valley', **y–z**. Any excursions up the sides of the valley will correspond to vibrations in modes other than those along the reaction coordinate, a situation which presumably occurs but does not lead directly to reaction. The saddle point is of particular interest and is known as the *transition state* or *activated complex* (‡). It corresponds to a structure metastable with respect to any change in its atomic coordinates and has no finite lifetime. The transition state is not a real molecule in that it may have partially formed bonds and higher coordination numbers at reaction centers than are normally permitted by valence bond rules. Now the reaction coordinate is represented on potential energy diagrams such as Fig. 2.2 as the one-dimensional projection of this pathway on the y-axis. The transition state is then the point of highest energy and the difference in energy between this point and that of reagents is the activation energy.

2.7.3 The transition state model [10,11]

Thermodynamics deals with equilibria and reversible processes rather than rates. In order to bring the treatment of rate processes within the scope of thermodynamic arguments, the following model was proposed by Evans and Polanyi [12,13] and by Eyring and others [14-16]. The transition state (or, for that matter, any configuration along the potential surface) is deemed to be a species to which all the usual thermodynamic quantities (H, G, S, etc.) may be ascribed since this must be possible for any assemblage of atoms of definite geometry and bonding. The reagent molecule(s) are then supposed to be in thermodynamic equilibrium with the transition state which passes to products at a frequency appropriate to each system. For example, in a transfer reaction such as that described above,

$$A—X + B \rightarrow A + X—B,$$

this frequency would be that of the pseudo vibration, $v^{\ddagger}$:

$$(A\cdots\overleftarrow{X}\cdots B)^{\ddagger}$$

which naturally leads to products. The reaction rate is the rate at which the transition state progresses to products and will depend on its concentration, [AXB]:

$$\text{Rate} = \frac{-d[AX]}{dt} = \frac{d[XB]}{dt} = k\kappa[AXB] \qquad (2.20)$$

where k is a constant and κ, the *transmission coefficient*, is the fraction of transition state species which pass to product rather than return to starting material; it is often near unity.
    Since $k^{\ddagger} = [AXB]^{\ddagger}/[AX][B]$,

$$\text{Rate} = kK^{\ddagger}\kappa[AX][B], \qquad (2.21)$$

which is clearly of the same form as the empirical expression,

$$\text{Rate} = k_2[AX][B],$$

where k_2 is the second-order specific rate coefficient and $k_2 = kK\kappa$.

2.8 PROPERTIES OF THE TRANSITION STATE

2.8.1 Activation parameters

Since the free energy of the transition state is necessarily higher than that of the reagents, its concentration will increase with temperature according to Eq. (2.18) and as a consequence the rate of reaction will increase.
    The Arrhenius equation, Eq. (2.22), is an empirical expresssion of this principle.

$$\frac{d \ln k}{dT} = \frac{E_A}{RT^2} \quad \text{or} \quad \frac{d \ln k}{d(1/T)} = \frac{-E_A}{R}$$

whence

$$k = Ae^{-E_A/RT}. \tag{2.22}$$

The constants A and E_A are the intercept and slope respectively of the linear plot of $\ln k$ against $1/T$ (Fig. 2.5). A is the extrapolated rate at infinitely high temperature while E_A is interpreted as the *energy of activation*. This may now be derived from thermodynamic principles:

Since $\qquad\qquad E = h\nu \approx kT,$

$$\nu = kT/h.$$

Therefore, $\qquad k = \nu K^{\ddagger} = kT/hK^{\ddagger}$

and, since $\qquad -\Delta G^{\ddagger} = RT \ln K^{\ddagger},$

$$k = kT/he^{-\Delta G/RT} = kT/he^{-\Delta H^{\ddagger}/RT}e^{\Delta S^{\ddagger}/R}, \tag{2.23}$$

which is of the same form as the Arrhenius equation with $\Delta H = E_A$ and $kT/he^{\Delta S^{\ddagger}/R} = A$. The Arrhenius activation energy, $E_A = (\Delta H^{\ddagger} + RT)$ is equated with the enthalpy of activation, strictly speaking the enthalpy change for conversion of a mole of reagents in their standard states into a mole in the transition states. The pre-exponential parameter, A, is linked to the entropy change occurring during the activation process, $\Delta S^{\ddagger}$, which may be obtained from Eq. (2.24):

$$A = \left(\frac{ekT}{h}\right)\exp\left(\frac{\Delta S^{\ddagger}}{R}\right) \tag{2.24}$$

or, at $\sim$300 K,

$$\Delta S^{\ddagger} = 8{\cdot}314 \, (\ln A - 30{\cdot}474). \tag{2.25}$$

The value of $\ln k$ is therefore proportional to the difference in free energy between reagents and transition state, $\Delta G^{\ddagger}$, and is consequently a measure of the affinity of the reaction *as a rate process*. Since rates are so easily measured and are so sensitive to small changes in $\Delta G^{\ddagger}$, kinetic measurements are the single most important method for the investigation of reaction energetics. As will be seen, relative logarithmic rate constants, $\ln k_{rel}$, with respect to some variable provide important data for mechanistic studies.

2.8.2 Heat capacity of activation [17]

The Arrhenius equation (Eq. (2.22)) is linear to a high degree, but over a sufficiently large temperature interval with extremely precise rate constant data slight curvature may sometimes be observed, implying a change in $\Delta H^{\ddagger}$ with temperature that is expressed by Eq. (2.26):

$$\Delta H_1^{\ddagger} - \Delta H_2^{\ddagger} = \Delta C_p^{\ddagger}(T_2 - T_1), \tag{2.26}$$

where $\Delta C_p^{\ddagger}$ is the difference in heat capacity between reagents and products, the *heat capacity of activation,* a parameter which is sometimes ascribed mechanistic significance.

2.8.3   Variation of rate with pressure[18]

In the gas phase, pressure is a measure of concentration. In solution, rates may vary slightly with applied pressure, sometimes being increased and sometimes decreased as pressure is applied. Pressure and volume are conjugate variables and the principle is followed that the application of pressure favors the side of an equilibrium which has the smaller volume. Volume, in this context, for reactions in solution implies partial molar volume — the volume occupied by a mole of solute at infinite dilution. For any equilibrium the pressure dependence of K is given by:

$$\frac{-\boldsymbol{R}T\,\mathrm{d}\ln K}{\mathrm{d}p} = \overset{\text{products}}{\sum} \bar{V} - \overset{\text{reagents}}{\sum} \bar{V} = \Delta V \tag{2.27}$$

where ΔV is the volume change for the reaction. Analogously, via the

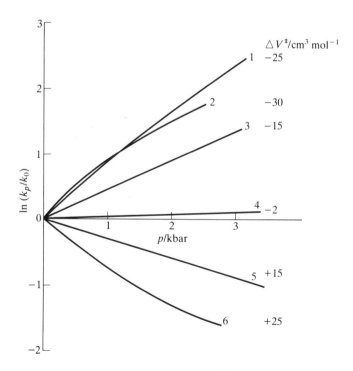

Fig. 2.3 Typical plots of relative rate against pressure[17]: 1, dimerization of cyclopentadiene (Diels–Alder reaction); 2, nitration of benzene; 3, nitration of toluene; 4, acidic hydrolysis of 1,3,5-trioxane; 5, Orton rearrangement of *N*-chloroacetanilide; 6, alkaline hydrolysis of cobalt(II) pentammine sulfate.

transition state assumptions (Section 2.7.3),

$$\frac{-RT\,\mathrm{d}\ln k}{\mathrm{d}p} = \bar{V}^{\ddagger} - \overset{\text{reagents}}{\sum}\,\bar{V} = \Delta V^{\ddagger} \tag{2.28}$$

where $\Delta V^{\ddagger}$ is known as the volume of activation, the difference between the partial molar volume of the transition state, $V^{\ddagger}$, and that of the reagents. The plot of $\ln k$ (or $\ln K$) against p is frequently found to be markedly non-linear such that $\Delta V^{\ddagger}(\Delta V)$ tends to diminish in magnitude with pressure (Fig. 2.3). Again, a second-order term needs to be included and $\mathrm{d}\,\Delta V^{\ddagger}/\mathrm{d}p$ has the significance of a difference in compressibility between reagents and transition state, the *compressibility of activation*, $\Delta K^{\ddagger}$. In practice, $\Delta V^{\ddagger}$ is defined as the value as $p \to 0$ and is obtained from a computed fit of the pressure–rate data to a polynomial:

$$\ln k = a + bp + cp^2 \tag{2.29}$$

where $a = \ln k(p = 0)$, $b = -RT\Delta V^{\ddagger}$ and $c = \Delta \kappa^{\ddagger}/2RT$.

Since partial molar volumes of reagents and products are easily measured from the densities of their solutions, absolute values of transition state volumes can be obtained.

2.9 THE USES OF ACTIVATION PARAMETERS

The following variables are known to affect the rates and products of a reaction and so to provide the means of probing the detailed mechanism:

concentration	temperature	pressure
structure	solvent	isotopy
stereoisomerism		

The most direct method of investigating these multifarious effects is to observe changes in the rate of reaction as one parameter at a time is varied. Many examples of this approach will be encountered.

2.9.1 The empirical treatment of rates

The stoichiometric equation for a reaction is simply a balance sheet of reagents and products whose atoms and charges must be conserved; it gives no mechanistic information. The empirical rate equation (ERE) is of the form;

$$\text{Rate} = \frac{\mathrm{d}c}{\mathrm{d}t}\,(\text{products}) = \frac{-\mathrm{d}c}{\mathrm{d}t}\,(\text{reagents}) = k\prod c_i^n$$

in which terms c are concentrations (activities, to be precise) and term c_i is the concentration of the ith reagent, the exponent n being the order of

Table 2.3 Linear equations for obtaining specific rate coefficients for reactions of various orders

Reactants have initial concentrations (M) A, B, C . . . and concentrations at time t of a, b, c . . .; the amount reacted, x, at time t will therefore be $(A - a)$ etc.

Order, Σn_i	Empirical rate equation (ERE)	Plot t against	Slope	Units of k
0	Rate $= k$	a	$-k$	$concn \times t^{-1}$
1	Rate $= k\,a$	$\ln A/a$ or $\ln 1/a$	$-k$	t^{-1}
2	Rate $= k\,ab$	$\ln (Ab)/(Ba)$	$-1/k(A - B)$	$concn^{-1} \times t^{-1}$
2	Rate $= k\,a^2$	$1/a$	$1/k$	$concn^{-1} \times t^{-1}$
3	Rate $= k\,abc$	$(B - C)\ln \dfrac{a}{A}$ $+ (C - A)\ln \dfrac{b}{B}$ $+ (A - B)\ln \dfrac{c}{C}$	$\begin{array}{l}1/k(A-B)\\ \times (B-C)(C-A)\end{array}$	$concn^{-2} \times t^{-1}$
3	Rate $= k\,a^2 b$	$\dfrac{2x(2B - A)}{A(A - 2x)}$ $+ \dfrac{\ln B(A - 2x)}{A(B - x)}$	$\dfrac{1}{k(A - 2B)^2}$	$concn^{-2} \times t^{-1}$
3	Rate $= k\,a^3$	$1/a^2$	$-1/2k$	$concn^{-2} \times t^{-1}$

reaction with respect to each. The reagent concentrations which appear in the ERE are those which determine the rate, not necessarily those of all species which are included in the stoichiometric equation. It is not possible to predict the form of the ERE by inspection of the stoichiometric equation, this can be done only by varying the concentration of each reagent in turn (and of other species such as catalysts) in order to determine the value of the order, n, for each. It may be sufficient simply to determine the overall order, Σn_i, which can be achieved by fitting the exponential curve of reaction progress to one or other of the linear plots in Table 2.3. The most common situations are reactions of first order $(\Sigma n_i = 1)$ and of second order, $\Sigma n_i = 2$, whose EREs contain one and two concentration terms respectively. The *specific rate coefficient* ('rate constant'), k, is obtained from the slope of the appropriate linear plot and is an absolute measure of rate as a function of concentration under stated reaction conditions (temperature, pressure, solvent, etc.). The importance of the ERE lies in its interpretation: those concentrations which it contains represent the species which together constitute the transition state. The molecularity of the reaction, i.e. the number of molecules which go into the formation of the transition state, is equated with Σn_i with the following proviso. Since the concentration of a reagent which is present in large excess (e.g. the solvent) remains effectively constant as the reaction proceeds $(-dc/dt = 0)$, the effect of its concentration upon the rate and possible involvement in the transition state cannot be determined. Values of n, the order with respect to a reagent, are normally but not necessarily integral (1 or 2 . . .). A fractional order dependence, however, indicates a complex reaction sequence, each case needing its own analysis.

Some examples will clarify the important principles.

(a)
$$R_3N + CH_3I \rightarrow R_3\overset{+}{N}\text{---}CH_3I^-.$$
$$Rate = k_2[R_3N][CH_3I].$$

The reaction is of second order and is implied to be bimolecular, the transition state containing one each of the reagent molecules as in **4**. Note the fractional charges and partial bonds which are normal features of a transition state.

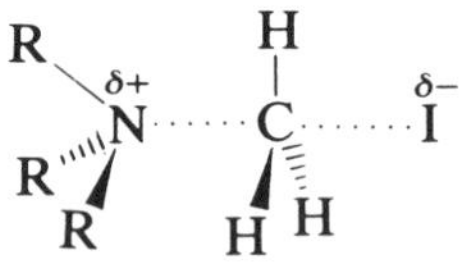

4

(b)

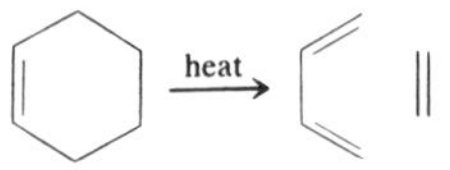

$$Rate = k_1 \text{ [cyclohexene].}$$

The reaction is of first order and is assumed to be unimolecular; the transition state might be represented by **5** or **6**, for instance, but kinetic evidence alone would not distinguish these possibilities.

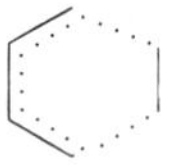

5 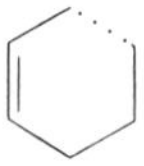   **6**

(c)
$$1\text{-BuBr} + H_2O \rightarrow 1\text{-BuOH} + H^+ + Br^-.$$
$$Rate = k_2[1\text{-BuBr}][H_2O],$$

but since $[H_2O] = $ constant, this becomes

$$Rate = k[1\text{-BuBr}].$$

Water is involved in product formation but it cannot be ascertained whether or not its concentration determines the rate as it is in excess. The reaction is denoted pseudo-first order.

(d)
$$CH_3COCH_3 + Br_2 \xrightarrow[\text{aqueous buffer}]{} CH_3COCH_2Br + HBr.$$

$$Rate = k[\text{acetone}][H^+] \text{ (second-order conditions).}$$

Furthermore, since the medium is buffered, $[H^+]$ and pH are constant.

$$Rate = k'[\text{acetone}] \text{ (first-order conditions).}$$

If the reaction is carried out with a large excess of acetone, [acetone] also is constant and

$$\text{Rate} = k'' \text{ (zero-order conditions)}.$$

In none of these situations does the concentration of bromine appear in the ERE. It is obviously a reagent but does not control the rate of reaction. The inference must be that it enters the reaction subsequently to the transition state. This further requires there to be at least two distinct steps in the reaction. Further discussion of multi-step reactions is necessary before these kinetic results can be interpreted.

2.9.2 The rate-determining step

Many reactions take place by a series of discrete activated steps, each separated by a formation of a real product though one which may be highly reactive and short-lived. *Reactive intermediates,* as they are called, exist within a shallow potential energy well, species indicated as I in Fig. 2.1c and d. In a sequential reaction, the questions arise as to which rate is to be identified with that experimentally observed, and to which process do the activation parameters refer. In general, the overall rate will be that of the slowest step of the sequence (the 'slow' or 'rate-determining' step). Kinetic measurements will give information only concerning events up to the slow step. The following types of complex reaction are often met:

(a)
$$A \xrightarrow[\substack{(1)\\ \text{slow}}]{k_1} B \xrightarrow[\substack{(2)\\ \text{fast}}]{k_2} C \xrightarrow[\substack{(3)\\ \text{fast}}]{k_3} D$$

$k_{\text{obs}} = k_1$ and ΔH_{obs}, ΔS_{obs} refer to this step: rates of steps (2) and (3) do not limit the rate at which A is converted to B.

(b)
$$A \underset{(1)\, k_{-1}}{\overset{k_1}{\rightleftharpoons}} B \xrightarrow[\substack{(2)\,\text{slow}}]{k_2} C.$$

The rate $(= k_2[B])$ depends upon [B], which in turn depends upon the pre-equilibrium by which it is formed so that the characteristics of equilibrium (1) appear in the thermodynamic quantities in addition to those of rate process (2):

$$k_{\text{obs}} = k_1 k_2 / k_{-1} = K \cdot k_1,$$

and

$$\Delta H^{\ddagger}_{\text{obs}} = \Delta H_1 + \Delta H^{\ddagger}_2.$$
$$\Delta S^{\ddagger}_{\text{obs}} = \Delta S_1 + \Delta S^{\ddagger}_2.$$
$$\Delta V^{\ddagger}_{\text{obs}} = \Delta V_1 + \Delta V^{\ddagger}_2.$$

The bromination of acetone fits into this category; the reactive intermedi-

$$CH_3-\overset{\overset{O}{\|}}{C}-CH_3 \xrightarrow[\substack{slow \\ (\leftarrow)}]{(H^+)} CH_3-C\overset{OH}{\underset{CH_2}{\diagdown}} \xrightarrow[fast]{Br_2} CH_3-C\overset{O}{\underset{CH_2Br}{\diagup}} + H^+, Br^-$$

$$\quad\quad 7 \quad\quad\quad\quad\quad\quad\quad 8 \quad\quad\quad\quad\quad\quad 9$$

ate, B, may be identified with the enol tautomer, **8**, and the sequence **7–9** explains the observed kinetics.

2.9.3 Relative rates

An individual rate constant or activation energy carries little mechanistic information and it is generally much more meaningful to examine changes in rate (relative rates, k_{rel}), while systematically varying one of the reaction parameters. A suitable model may then be chosen which affords an explanation of the observed rate sequence. For example, if rates of nitration of a series of substituted benzenes.

$$X-C_6H_5 + (HNO_3) \xrightarrow{k_x} X-C_6H_4-NO_2$$

are measured while X is changed to groups of successively greater power of attracting electrons, the rates diminish strongly. One would then infer that the reaction was facilitated by electrons being released from X and passing in turn to the nitrating agent (Section 2.10.) There is still ambiguity in the interpretation of k ($\equiv \Delta G^{\ddagger}$); firstly a reaction may be facilitated by a more negative $\Delta H^{\ddagger}$ or by a more positive $\Delta S^{\ddagger}$ (or some combination of the two). Different interpretations may hang upon which of these is the controlling factor. For instance, the fact that trimethylamine is a stronger base than ammonia (Section 6.2), i.e. more prone to bond to a proton, could be interpreted as being due to electron release by methyl groups.

$$R_3N: + H^+ \rightleftarrows R_3NH^+$$
$$k: Me_3N > Me_2NH < MeNH_2 > NH_3.$$

An inconsistency would then be revealed by the observation that methylamine is stronger than dimethylamine: the apparent paradox is resolved when it is realised that while values of ΔH decrease steadily the greater the number of methyl groups, while those of ΔS (due to changes in solvation) do not, with the result that ΔG also shows unsystematic behavior. This would need to be probed by measuring the temperature dependence of $\ln K$.

The other ambiguity arises from the fact that k (or K) measures a difference in G and that a perturbation of the rate or equilibrium may arise from a change in G of either reagents or of the transition state (products), or both. The same effect on rate would be observed (i.e. an increase) by a

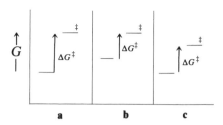

Fig. 2.4   Activation profiles showing that a reduction of activation energy can be achieved by increasing the energy of the reagent (a–b) or by decreasing the energy of the transiton state (a–c). In practice, both are likely to be affected in the same direction but to different extents.

change in the energy profile (Fig. 2.4) from a to b (an initial-state effect) or from a to c (a transition-state effect) each of which may demand a quite different interpretation. It is much more difficult to resolve this problem; one would need to perform lengthy energy summations from thermochemical data to obtain ΔG_f of each reagent and transition state (in solution if need be). As a consequence, while it is common practice to report and interpret values of relative rate constants as such, the reader must be on guard lest assumptions as to the effects brought about by the change of a parameter upon the energetics of reaction prove unwarranted. It may be mentioned here that it is always assumed that perturbations of the energy of the reagents and consequently of the transition state occur in the same sense. That is, if some structural change in the reagent produces a negative change in G(reagent), it will also produce a negative change in $G^{\ddagger}$, though not necessarily of the same magnitude[18]. That this need not be so is shown by the figures of Table 2.4, in which rates of decomposition of some azides are seen to be the resultant of changes in enthalpy of activation compensated by changes in entropy.

Table 2.4   Thermolyses of triphenylmethyl azides, p-X—$C_6H_4 \cdot CPh_2N_3$

X	$10^4\,k/s^{-1}$ (at 169°C)	$\Delta H^{\ddagger}/kJ\,mol^{-1}$ ($kcal\,mol^{-1}$)	$\Delta S^{\ddagger}/J\,K^{-1}\,mol^{-1}$ ($cal\,K^{-1}\,mol^{-1}$)
OMe	1·80	121 (29)	−67 (−16)
Me	1·26	121 (29)	−67 (−16)
H	1·26	134 (32)	−41 (−9·8)
Cl	1·12	141 (34)	−23 (−4·6)
NO$_2$	1·10	143 (34)	−19 (−4·6)

2.9.4 Entropies and volumes of reaction

Individual values of $\Delta S^{\ddagger}$ and of $\Delta V^{\ddagger}$ are capable of interpretation and hold somewhat similar information. In general, association between reagent molecules or between reagent and solvent in or before the rate-determining step will be accompanied by a decrease in both entropy and volume of activation. The change in $\Delta S^{\ddagger}$ is due to the loss of translational freedom and in $\Delta V^{\ddagger}$ it is due to the reduced requirement of *free space,* the vacant space between molecules in the liquid phase. The loss of three degrees of translational freedom (i.e. $A + B \rightarrow AB$) may contribute a loss of entropy amounting to as much as $100 \, J \, K^{-1} \, mol^{-1}$ ($24 \, cal \, K^{-1} \, mol^{-1}$). For example, from Table 2.1, for the reaction

$$H_2 + CH_3CH{=}CH_2 \rightarrow CH_3CH_2CH_3,$$
$$\Delta S_f = -89 \, J \, K^{-1} \, mol^{-1} \; (21 \, cal \, K^{-1} \, mol^{-1}).$$

Volume changes associated with structural change are summarized, highly approximately, in Table 2.5.

Table 2.5   Contributions to $\Delta V^{\ddagger}/cm^3 \, mol^{-1}$

Bond breaking	$+10$
Bond formation	-10
Synchronous displacement	-5
Ionic dissociation	< -20 (solvent-dependent)
Ion recombination	$> +20$ (solvent-dependent)
Charge concentration	-5
Charge dispersal	$+5$

The large volume diminution which accompanies ion formation is due to solvent electrostriction — the tendency of solvent to associate tightly with a charged center by electrostatic forces reducing both entropy and volume. Charge neutralization is accompanied by relaxation of electrostricted solvent. The amount of electrostriction, ΔV_e, varies as the square of the charge, q, on the ion (radius, r) and inversely as the dielectric constant of the medium, ε, according to the Drude–Nernst equation:

$$\Delta V_e = -\frac{Lq^2}{2r} \cdot \frac{1}{\varepsilon} \cdot \left(\frac{d\varepsilon}{dp}\right)_T = -\frac{Lq^2 \Phi}{2r}. \qquad (2.30)$$

2.9.5 The isokinetic relationship [19–21]

It is sometimes found that a linear relationship exists between the enthalpies and entropies of activation of a series of related reactions (or between E_A and $\ln A$):

$$\delta \Delta H^{\ddagger} = \beta \delta \Delta S^{\ddagger}.$$

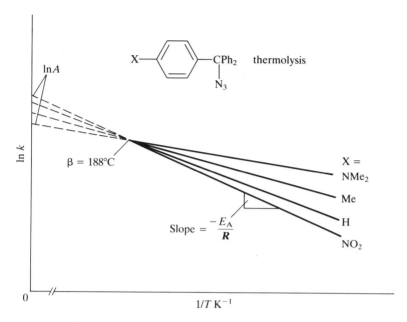

Fig. 2.5 An isokinetic relationship: Arrhenius plots of a series of reactions related by structural change showing compensation of ΔH (E_A) by ΔS (ln A). The isokinetic temperature $\beta = 188°C$.

Hence,

$$\delta\Delta G^{\ddagger} = \delta\Delta H^{\ddagger} - T\delta\Delta S^{\ddagger} = (T - \beta)\delta\Delta S^{\ddagger}. \qquad (2.31)$$

The constant β has the dimensions of temperature and is the actual or virtual temperature at which rates of all members of the series are equal, the *isokinetic temperature*. This implies that the Arrhenius plots for each member of the series will cross at one temperature or close to it (Fig. 2.5). The existence of an isokinetic relationship is usually taken to mean that the reaction parameter which is being varied (structure, solvent, etc.) is operating on only one type of interaction in the system, while the absence of such an effect can point to a complex interplay of effects or to the operation of more than one reaction mechanism. When highly precise rate data are available, the linear relationship may be seen to apply only to a limited part of the series (Fig. 2.6). The isokinetic effect is a tendency for a change in one activation parameter to be compensated by a change in the other in the opposite sense. If a structural change results in a stronger interaction between the reagents and consequent lowering of $\Delta H^{\ddagger}$, it will likely be accompanied by tighter binding in the transition state and a less favorable (more negative) $\Delta S^{\ddagger}$.

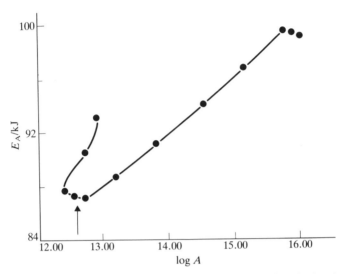

Fig. 2.6 Isokinetic plot of E_A against $\log A$ for the solvolysis of *t*-butyl chloride in ethanol–water of varying composition[28]. The linear relationship only holds over a limited range of solvent composition.

2.10 THE LOCATION OF THE TRANSITION STATE

2.10.1 The Hammond Postulate

While application of the quasi-thermodynamic arguments above enables estimates of the energy and volume of the transition state to be made, no information as to the position of the transition state on the reaction coordinate or of its geometry can be gained. One may argue that in the case of a symmetric reaction the transition state will also have geometrical symmetry. Examples are the transition states for proton transfer between two identical bases, **10**, or for methyl transfer between two iodide ions, **11** (a nucleophilic displacement at carbon).

$$R_3N \cdots H \cdots NR_3$$

10

$$\left(I \cdots \underset{\underset{H \quad H}{|}}{\overset{\overset{H}{|}}{C}} \cdots I \right)^{-}$$

11

There is no restriction *a priori* on the location of the transition state for unsymmetrical reactions; it may closely resemble the reagents (an *early* transition state) or the products (*late*) or possess properties intermediate between the two extremes (*central*). A useful qualitative guide goes by the name of the Hammond Postulate, which states: 'If two states occur

104

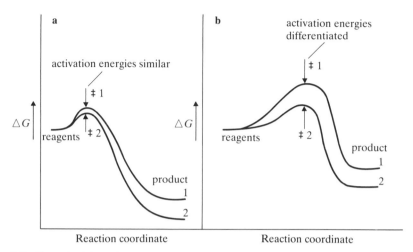

Fig. 2.7 The selectivity relationship; **a**, reaction with a highly reactive reagent leading to an early 'reagent-like' transition state and little selectivity; **b**, analogous reaction with a reagent of low reactivity leading to a 'product-like' transition state of high selectivity.

consecutively during a reaction process and have nearly the same energy content, their interconversion will involve only a small reorganization of molecular structure'. The principle implies, as a corollary, that the transition state of a highly exothermic reaction will resemble reagents (i.e. will be 'early') becoming later as the exothermicity is reduced until, for a highly endothermic reaction, the transition state will be late and product-like (Fig. 2.7). The location of the transition state on the reaction coordinate will therefore vary with the relative energies of the reagent, transition state and product. It will change in a continuous fashion as these energies are perturbed by, for example, changes in structure or solvent (the Leffler–Hammond Principle).

The nitration of benzene illustrates the application of this hypothesis. The activation step of this reaction involves the coordination of a nitronium ion, NO_2^+, to the aromatic ring to form a benzenium ion, **12** (see Chapter 10), a very endothermic step. The transition state should therefore resemble the benzenium ion rather than benzene; so the properties of this type of compound which, though reactive, can be directly observed and studied, may be used as a good approximation to those of the transition state which is inaccessible to direct study.

12

2.10.2 Reactivity and selectivity[19-23]

A reaction frequently leads to the formation of two or more products (for example, *ortho* and *para* isomers from aromatic substitution, alkene isomers from eliminations) by parallel pathways. If two such competing routes to products X, Y have rates k_x, k_y the selectivity S, of the reaction is given by Eq. (2.32):

$$S = \log k_x/k_y. \qquad (2.32)$$

S depends upon the two activation energies, $\Delta G_x^{\ddagger}$ and $\Delta G_y^{\ddagger}$. Now suppose there are two reagents of which one, A, is highly reactive (rates are high) and the other, B, is much less so, both capable of undergoing the same two reaction pathways x and y. The energetics of the reactions might be as shown in Fig. 2.8. The transition states for the reactions of A will be relatively early and reagent-like while those of B will be later and more product-like. Since the reagent is the same species for both pathways x and y there will be little discrimination between their rates for a reagent-like transition state and selectivity will be low. There will however, be a greater difference between the two transition states x and y when they are

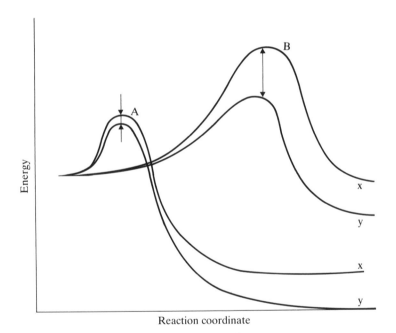

Reaction coordinate

Fig. 2.8 Scheme illustrating the relationship between reactivity and selectivity. A highly reactive reagent attacks two different substrates with a characteristically early (reagent-like) transition states not very dissimilar in energy, A. A reagent of lower reactivity attacking the same two substrates, B, has a later transition state, less exothermic and more differentiated (selective).

product-like since the products to which they lead are different chemical species. The selectivity will then be greater. This is an application of the Hammond Postulate to the situation, frequently experienced, that highly reactive species tend to be non-discriminating and react randomly while unreactive ones are highly selective. The abstraction of hydrogen at primary, secondary and tertiary centers by chlorine atoms (highly reactive) and by bromine atoms (much less reactive) is an example.

$$>C\!-\!H \; + \; \cdot Hal \; \longrightarrow \; >C\cdot \; + \; H\!-\!Hal$$

	$1ry(-CH_3)$	$2ry(>CH_2)$	$3ry(>C\!-\!H)$
k_{rel} Cl·	1	3·5	4·2
Br·	1	80	1700

To take the extremes, the selectivity of tertiary over primary attack, $S = 0·6$ for Cl· and 3·23 for Br·.

For high selectivity to be observed it is necessary for the energies of the two products and of the transition states from which they are derived to be significantly different. In this case the tertiary product radical $R_3C\cdot$ is more stable than a primary, $RCH_2^\cdot$ (Section 15.4). Examples are encountered (Section 6.7) in which a series of reagents may show a wide range of reactivity but similar selectivity; this can be due to fortuitous similarity in product energies.

2.10.3   Kinetic and thermodynamic control of products

The discussion on selectivity above assumes the two rates of product formation determine the product yields, i.e.

$$[X]/[Y] = k_x/k_y. \tag{2.33}$$

When Eq. (2.33) holds, the reaction is said to be under kinetic control. If the energetics of the reaction are such that one or both of the reactions are reversible, then the product ratio will reflect their relative stabilities (Fig. 2.9) so that

$$[X]/[Y] \propto \ln(G_x - G_y), \tag{2.34}$$

when the reaction is said to be under thermodynamic control. If the difference in free energy between the products is $>20\,\text{kJ mol}^{-1}$ ($4·8\,\text{kcal mol}^{-1}$) then the ultimate product will be solely the more stable one. This situation may be reached slowly depending upon the barrier height, E_1, so that the product ratio will change with time. If a product ratio is to be used as a means of assessing relative rates it is essential to ensure that the reaction is under kinetic control.

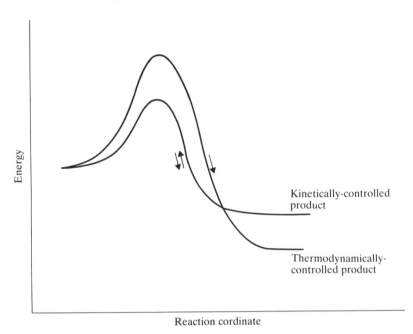

Fig. 2.9 Scheme illustrating the origin of products under kinetic and under thermodynamic control. If two pathways to products are available, one reversible and of low activation energy and the other irreversible and of high activation energy, products arise from the former when reaction times are short and temperatures low but from the latter on prolonged standing or at high temperatures.

2.10.4 The principle of least motion

It was proposed by Rice and Teller[22] that 'those elementary reactions are favored which involve least change in atomic positions and electronic configurations'. This implies that a spontaneous reaction will yield a product which has a similar spatial arrangement of the atoms rather than a very different one even if energetically favorable. It may rationalize the observation that the abstraction of a primary hydrogen atom, **13**, occurs with difficulty in preference to the simultaneous abstraction and rearrangement, **14**, which could in principle yield a more stable product. There is an energy requirement for nuclear motion usually taken as proportional to the square of the nuclear displacement[23-27]. However, least-motion considerations are frequently overshadowed by more powerful directing forces, particularly product stabilities and molecular orbital requirements, so that it is difficult to find examples in which they control product formation. A possible instance is the preference for elimination in cyclohexanes by the diaxial route, **15** (rather than by axial–equatorial, **16**, or diequatorial processes, **17**), which is also calculated to require the least change of nuclear positions.[24]

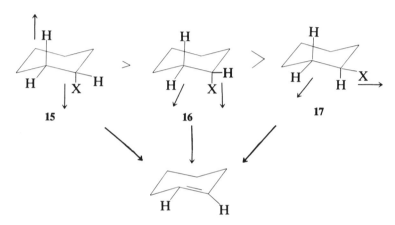

2.10.5 The principle of microscopic reversibility

When building up the 'profile' of a reaction coordinate either in energy or volume terms, it may at times be convenient to examine a reversible reaction from either direction. For example, the energy of activation for the addition of SO_2 to a diene, **18**, may be determined from the temperature dependence of the forward rate, k_1, in the range 20–60°C. The energy of reaction could be determined by thermochemical measurements on all reactants and products but it is more convenient to raise the temperature and study the rates of reaction in reverse, which occurs above 150°C (Fig. 2.7). The question arises whether the mechanism is the same in both directions. The principle of microscopic reversibility asserts that it is. It is logically the case: for suppose a lower-energy pathway were available for the reverse reaction, then it must also be a lower-energy pathway for the forward reaction and it follows that both must traverse exactly identical routes.

$$E_A = 134 \text{ kJ mol}^{-1}$$
$$(32 \text{ kcal mol}^{-1})$$

$$E_A = 67 \text{ kJ mol}^{-1}$$
$$(16 \text{ kcal mol}^{-1})$$

SO_2

18

Kinetics and thermodynamics

PROBLEMS

1 Phenyloxirane (styrene oxide) and piperidine react together in a 1:1 ratio stoichiometrically to give a single product, P. The starting concentrations and product concentration as a function of time are as follows:

$T = 60°C$

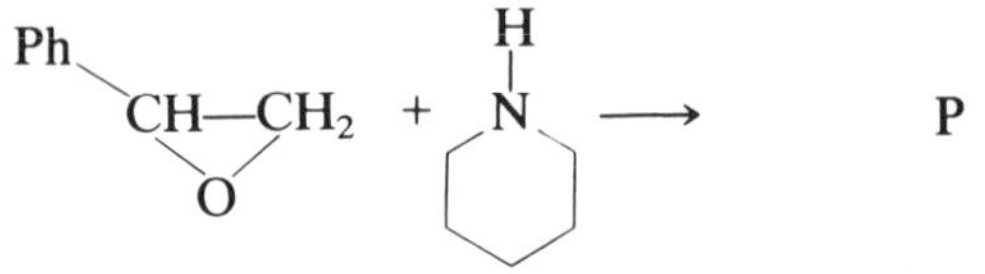

| c_0/M: | 0·1415 | 0·1530 | 0 |

t/s	$10^3[P]/M$	t/s	$10^3[P]/M$
306	11·1	1800	51·2
612	21·8	2112	56·4
900	30·4	2400	63·2
1212	38·5	3000	69·1
1512	45·2	3600	75·7

Show that the reaction obeys second-order kinetics and calculate the specific rate constant. Suggest a structure for P. Rate constants at other temperatures are as follows:

$T/°C$	$10^4k/M^{-1}s^{-1}$
0·0	0·170
24·9	1·50
44·0	7·00
54·7	14·8

Calculate k at 35°C and the activation parameters E_A, $\log A$.

2 Benzyl chloride hydrolyzes to the alcohol and HCl in water, the reaction being followed by the increase in conductance of the solution due to the ionic products. $10^{-3}M$ HCl was found to have a conductance reading of 1000 (arbitrary units). The times to reach various fractions of this value during the hydrolysis of $10^{-3}M$ benzyl chloride in water were as follows:

Conductance (arbitrary units)	Time/s		
	$T/K = 293$	$T/K = 313$	$T/K = 333$
0	0	0	0
100	15030	1379	179
200	31832	2920	380
300	50880	4668	608
400	72870	6686	871
500	98880	9072	1182

110

Determine the order of the reaction and calculate the specific rate constant at each of the three temperatures. From this, calculate the enthalpy, entropy and free energy of activation.

3 Acetone, 0.1M, in aqueous buffer solution was treated with a solution of iodine, 10^{-3}M, and the color of the latter was observed to diminish with time. Absorbance readings at 420 nm at times, t after mixing were as follows:

t/min:	0	1	2	3	4	5	6	7	8	9	10	11	12	15
A:		1·53	1·39	1·26	1·13	0·99	0·86	0·72	0·59	0·45	0·32	0·19	0·12	0·12	0·12

Determine the order of the reaction and comment upon any mechanistic features which it reveals. What further experiments would you perform in order to test any hypotheses which you might put forward?

4 Ethyl acetate and sodium hydroxide both initially at 0.1M concentration in water react to form acetate and ethanol with a second-order rate constant $k_2 = 0.11$M^{-1}s^{-1}.

$$CH_3COOEt + OH^- \rightarrow CH_3COO^- + EtOH$$

What concentrations of reagents and products would remain after (a) 10 s; (b) 10 min; (c) 30 min?

5 The general rate equation, rate $= k \prod c_i^{m_i}$ contains concentration terms $n = \sum m_i$, where $n =$ order of reaction. Show that the ratio $t_{\frac{1}{2}}/t_{\frac{3}{4}}$ is a function only of n and therefore can be used to determine the order of reaction ($t_{\frac{1}{2}} =$ time for 50% of reaction to occur; $t_{\frac{3}{4}} =$ time for 75% of reaction to occur).

6 Derive rate equations for the following pair of parallel reactions:

$$PhCH_2CH_2Cl + OEt^- \xrightarrow{k_s} PhCH_2CH_2OEt + Cl^-$$

$$PhCH_2CH_2Cl + OEt^- \xrightarrow{k_e} PhCH=CH_2 + EtOH + Cl^-$$

What will be the effect on the rate law if $[OEt^-] \gg [PhCH_2CH_2Cl]$? What will be the effect on the rate law if the substitution product undergoes rapid elimination, i.e. $k_e' > k_s$

$$PhCH_2CH_2OEt + OEt^- \xrightarrow{k_{e'}} PhCH=CH_2 + EtOH + OEt^- \quad ?$$

7 The hydrolyses of nitrophenyl esters of p-methoxy- and p-hydroxy-benzoic acids under similar alkaline conditions are dependent upon the

pressure of the reaction. Rates at several pressures are:

Pressure, p/bar:	1	300	600	900
p-Methoxybenzoate, $10^4 k_1/\text{s}^{-1}$:	8·38	10·1	12·0	13·2
p-Hydroxybenzoate, $10^4 k_1/\text{s}^{-1}$:	7·05	6·40	5·17	4·37

Calculate for each reaction the volume of activation and suggest a mechanistic explanation for the difference in sign. [*Chem. Comm.*, 1361 (1984).] N.B. $R = 80\cdot98 \text{ cm}^3 \text{ bar K}^{-1}$.

$$RO\!-\!\!\bigcirc\!\!-\!\!C\!\!\stackrel{O}{\diagdown}\!\!O\!-\!\!\bigcirc\!\!-\!NO_2 \xrightarrow{H_2O} RO\!-\!\!\bigcirc\!\!-\!\!C\!\!\stackrel{O}{\underset{OH}{\diagdown}} + HO\!-\!\!\bigcirc\!\!-\!NO_2$$

8 Calculate the changes in activation parameters necessary to bring about an increase in rate of 2-, 100- and 10^6-fold:
(a) by a change in $\Delta H^{\ddagger}$ only;
(b) by a change in $\Delta S^{\ddagger}$ only at 25° and at 100°C.

9 A Diels–Alder reaction is characteristically associated with a volume of activation $\Delta V^{\ddagger} = -34 \text{ cm}^3 \text{ mol}^{-1}$. What pressure would be required to cause a rate acceleration of 10^6-fold?

REFERENCES

1 C. N. Hinshelwood, *The Structure of Physical Chemistry*, Oxford U.P., 1951.
2 E. F. Caldin, *An Introduction to Chemical Thermodynamics*, Oxford U.P., 1961.
3 E. A. Guggenheim, *Thermodynamics*, North Holland, Amsterdam, 1967.
4 B. H. Mahan, *Elementary Chemical Thermodynamics*, Benjamin, New York, 1963.
5 J. B. Fenn, *Engines, Heat and Entropy*, Freeman, San Francisco, 1982.
6 G. N. Lewis and M. Randall, *Thermodynamics* (revised by K. S. Pitzer and L. Brewer), McGraw-Hill, New York, 1961.
7 E. B. Smith, *Basic Chemical Thermodynamics*, Oxford U.P., 1977.
8 L. K. Nash, *J. Chem. Ed.*, **64**, 42 (1965).
9 I. Prigogene and R. Defuy, *Chemical Thermodynamics*, Longmans, London, 1954.
10 S. W. Benson, *Foundations of Chemical Thermodynamics*, McGraw-Hill, New York, 1960.
11 C. H. Bamford and C. F. H. Tipper (Eds.) *Comprehensive Chemical Kinetics*, Elsevier, Amsterdam, 1969–1980 (22 vols).
12 M. G. Evans and M. Polanyi, *Trans. Farad. Soc.*, 31, 875 (1935); **33**, 448 (1937).
13 M. Polanyi, *J. Chem. Soc.*, 629 (1937).
14 H. Eyring, *J. Chem. Phys*, **3**, 107, (1935).
15 W. F. K. Wynne-Jones, *J. Chem. Phys.*, **3**, 492 (1935).

16 E. A. Guggenheim and J. Weiss, *Trans. Farad. Soc.,* **34,** 57 (1938).
17 S. Singh and R. Robertson, *Can. J. Chem.,* **55,** 2582 (1977); R. Robertson, *Tetrahedron Letters,* **17,** 1489 (1979).
18 N. S. Isaacs, *Liquid Phase High Pressure Chemistry,* Wiley, Chichester, 1981.
19 J. E. Leffler and E. Grunwald, *Rates and Equilibria of Organic Reactions,* Wiley, New York, 1963.
20 W. Good, D. B. Ingham and J. Stone, *Tetrahedron,* **31,** 257 (1975).
21 O. Exner, *Coll. Czech. Comm.,* **29,** 1094 (1964).
22 F. O. Rice and E. Teller, *J. Chem. Phys.,* **6,** 489 (1938); **7,** 199 (1939).
23 J. Hine, *Adv. Phys. Org. Chem.,* **15,** 1 (1977).
24 S. I. Miller, *Adv. Phys. Org. Chem.,* **6,** 185 (1968).
25 O. S. Tee and K. Yates, *J. Amer. Chem. Soc.,* **94,** 3074 (1972).
26 O. S. Tee, J. Altmann and K. Yates, *J. Amer. Chem. Soc.,* **96,** 3141 (1974).
27 O. S. Tee, *J. Amer. Chem. Soc.,* **91,** 7144 (1969).

3 Reagents and reaction mechanism

3.1 POLAR AND RADICAL PATHWAYS [1-4]

Almost all stable molecules of interest to organic chemists possess an even number of valence electrons, spin-paired and filling all bonding orbitals (O_2, NO and NO_2 are notable exceptions). Reactant and product molecules will for the most part conform to this type. It has been mentioned, however, that during the course of a reaction high-energy species such as intermediates (with some stability) or the transition state (with no intrinsic stability) are transiently formed and may have very unusual bonding. Fundamental differences in reaction characteristics are found depending on whether a reaction occurs with retention of spin-pairing throughout (which will be called a *polar* reaction) or whether there is involvement of intermediates with unpaired electrons (*radical* reactions). At the present there seems to be a clear distinction between these reaction types but there remains the possibility that future experimental techniques will be able to discriminate two successive one-electron shifts, **1**, in reactions which today appear to occur by a single two-electron motion, **2**, which for the present discussion will be assumed to take place. The purpose of this section is to classify types of reagent and the fundamental reactions which they will undergo.

$$Nu:^- + E^+ \quad \overset{\textbf{2}}{\underset{(\curvearrowright)}{\curvearrowright}} \quad Nu \overset{\curvearrowleft}{\underset{}{-}}E$$
$$\searrow (Nu \cdot \cdot E) \nearrow$$
$$\textbf{1}$$

3.1.1 Polar reactions

The simplest examples occur with the involvement of a complementary pair of reagents, an electron-pair donor known as the *nucleophile* (Nu:) and an electron-pair acceptor, the *electrophile* (E), designations due to Ingold[1]. The interaction **2** and its reversal represent all reactions of this type, a very large body of organic chemistry closely allied to the combination of base with acid. In principle any nucleophile can react with any electrophile, a curved arrow ($\curvearrowright$) being used to denote to notional motion of the electron pair in the interaction. The position of the implied equilibrium will depend

114

upon the nature of the reagents, (their 'strengths'). Whether a specific molecule is designated a nucleophile or an electrophile may depend upon the context in which it is placed. Some species can show activity of either; for example, ethanol, which acts as a nucleophile (3) towards acetyl chloride, acts as an electrophile (4) towards the hydride ion, upon relinquishing a proton.

3.1.2 Nucleophiles

This term is synonymous with 'Lewis base', the name usually preferred in the field of inorganic chemistry. Nucleophiles will have a readily accessible pair of electrons which therefore lie on a high-energy molecular orbital.* The source of nucleophilic reactivity is usually the HOMO, which may contain an unshared electron pair (:) or π-electrons (5, 6) including

extended and aromatic systems. Typical examples are

$$R_3C:^-   \quad   R_3N: \quad   R_2O: \quad   R—Hal: .$$

Also included are 'virtual' nucleophiles or species which, while not possessing an available electron pair, can make available or transfer a species which does. The ion BH_4^- is an example; it acts as a nucleophile by transferring $H:^-$ (7). Nucleophiles may be anions or neutral species; very rarely they are cations. Ambident nucleophiles have two or more centers of nucleophilic activity: for example, enols, 8, can accept an electrophile at either carbon or oxygen.

---

* The terms 'HOMOgen' and 'LUMOgen' in place of the entrenched 'nucleophile' and 'electrophile' are beginning to make an appearance in the literature, e.g. Tedder and Nachvetal, *Pictorial MO Theory,* Pitman, London, 1985.

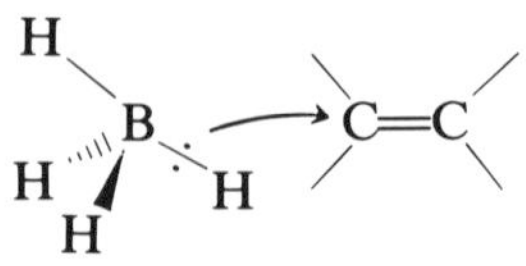

7

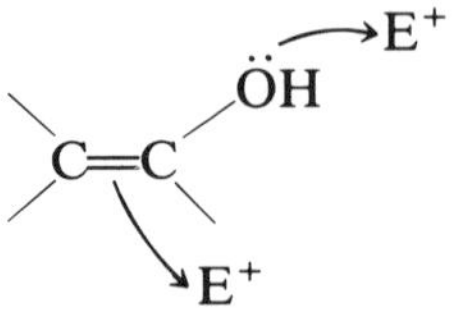

8

3.1.3 Electrophiles

Electrophiles complement nucleophiles, to which they are capable of bonding by accepting an electron pair, making them synonymous with Lewis acids. It would seem at first sight that a low-lying vacant orbital (LUMO) would be a prerequisite to electrophilic reactivity. This would be too limiting, since only species such as R_3C^+ or R_3B (which are, indeed, powerful electrophiles) are endowed in this way. Far more common is the electrophilic center which will accept a pair of electrons while simultaneously releasing another pair. The following examples (**9–12**) illustrate the point. Electrophiles may be neutral or positive but they are rarely negatively charged. They frequently bear a partial positive charge due to polarization or created by the approach of the nucleophile (polarizability, **11**).

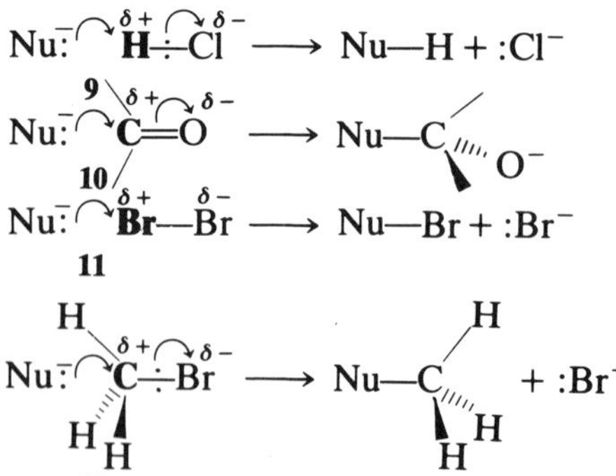

3.1.4 Radicals

These are species which possess an unpaired electron. They are usually highly reactive intermediates which dimerize by pairing of the odd electrons, **13**, or coordinate to a stable species, **14**, resulting in either addition

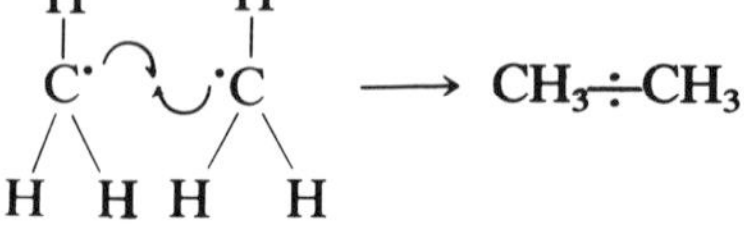

13

116

$$RCH{=\!=}CH_2 + R'' \cdot \longrightarrow R\overset{\cdot}{C}H\overset{\diagup R'}{-\!\!-}CH_2 \longrightarrow \text{further reaction}$$
                                                14

or displacement; in the latter case the reaction may continue since of necessity a further radical species is generated. The singly barbed arrow ($\frown$) is used to represent the movement of single electrons.

3.2 A CLASSIFICATION OF FUNDAMENTAL REACTION TYPES

Though organic reactions may involve overall the making and breaking of many covalent bonds, such complexity is usually the result of a sequence of simple steps. If the circumstances favoring each of these processes is understood, then a complex reaction may become readily comprehensible. Take, for example, the hydrolysis of an acetal:

$$PhCH(OEt)_2 + H_2O \xrightarrow{(H_3O^+)} PhCHO + 2EtOH.$$

At least six atoms are undergoing covalency change involving a correspond-ing number of electron pairs. So complex a reaction cannot be accomplished in a single step. For one thing, the cost in entropy would be extremely high. Instead, a sequence of some seven or eight simple steps, each energetically feasible, will accomplish the process with ease (Section 9.2.3).

Individual steps in virtually all organic reactions are of one of the following four types; there may be polar and radical variants in some cases.

3.2.1 Bond formation and bond breaking

Polar bond formation, the combination of a nucleophile with an elec-trophile, is a simple process of wide occurrence among reactions in solution (the charges are relative):

$$Nu\!:^- + \overset{\frown}{E^+} \xrightarrow{(-)} Nu\overset{\frown}{-}E.$$

The reverse bond breaking process (heterolysis, ionization) typically is favored when the covalent bond occurs between two elements of very different electronegativity and in solvents of high dielectric constant which can stabilize ions. Heterolysis rarely occurs in the gas phase since it requires far more energy than does homolysis.

Radical bond formation, the recombination of two radicals, is normally a very facile process requiring no activation energy and generates the full bond dissociation energy as heat. The reverse, homolysis, is brought about by thermal activation (Δ) and is favored by weak homopolar bonds (bonds between two atoms of the same element). The process is insensitive to the nature of the solvent and is frequently the preferred route in the gas phase.

$$R\cdot + \cdot R' \xrightarrow{\Delta} R\text{---}R'$$

3.2.2 Transfer reactions

In those processes, also known as substitution or displacement reactions, a single atomic center is transferred in one step from covalent binding at one atom to covalent binding at another. They are commonly designated $S_{(N,E, \text{ or } R)}2$ (where S denotes substitution; the subscripts N, E or R indicate nucleophilic, electrophilic or radical respectively; and 2 means bimolecular), according to the mechanistic notation devised by Ingold.

$$X\text{---}Y + Z \rightarrow [X \cdots Y \cdots Z]^{\ddagger} \rightarrow X + Y\text{---}Z.$$

Three possibilities exist provided X and Z are of the same reagent type (Nu, E or R):

(a) Nucleophilic, S_N2 (number of electrons, 4)

$$Nu\text{---}Y + :Nu' \rightarrow Nu: + Y\text{---}Nu'.$$

(This can also be regarded as a transfer of Y^+ between two nucleophilic centers.)

(b) Electrophilic, S_E2 (number of electrons, 2)

$$E\text{---}Y + E' \rightarrow E + Y\text{---}E'.$$

(Transfer of Y^- between two electrophiles.)

(c) Radical, S_R2 (number of electrons, 3)

$$R\text{---}Y + \cdot R' \rightarrow R\cdot + Y\text{---}R'.$$

(Transfer of $Y\cdot$ between two radicals.)

The two complementary ways of regarding these reactions, as substitutions or transfers, focus attention respectively on the reagent or on the central atom. The former is perhaps the more traditional but the latter is a useful concept, for example in emphasizing similarities between proton and methyl transfer:

$$B: H\text{---}B'^+ \rightarrow B\text{---}H^+ + :B'$$
$$Nu: CH_3\text{---}Nu' \rightarrow Nu\text{---}CH_3 + :Nu'.$$

In all these reactions, the bond-forming and bond-breaking steps are considered to occur synchronously. The same products may be formed when two consecutive reactions of type (a) occur sequentially in either order. Sequences of this type are frequently observed to happen, sometimes competing with transfer processes, and will be referred to as addition–elimination (Ad–E) or elimination–addition (E–Ad).

$$\text{Y–Z} \overset{\text{E–Ad}}{\underset{\text{Ad–E}}{\begin{array}{c} \xrightarrow{-Z} \text{Y} \xrightarrow{+X} \\ \xrightarrow{+X} \text{X–Y–Z} \xrightarrow{-Z} \end{array}}} \text{Y–X}$$

(These terms are used since there do not appear to be convenient words to imply 'bond making' and 'bond breaking' which, when put together would not result in clumsy phraseology; they should not be confused with reactions described in **3.2.3**.)

3.2.3 Elimination (E) and addition (Ad)

Elimination here implies the splitting out of an electrophile and a nucleophile from the same atom (α-elimination), from adjacent atoms (β-elimination), or from a more distant pair (γ, δ-elimination). Products will be, respectively, a carbene (**15**) or equivalent, a π-bond (**16**), or a ring (**17**). Addition implies the reverse reaction.

Elimination may be synchronous (E2 — elimination, bimolecular) or sequential (E1 or E1$_{cb}$, Chapter 11), but additions are expected to be sequential since three species are required to combine together. The sequential reactions may involve combination (or loss) of nucleophile and electrophile in either order. Radical additions to olefinic double bonds are more complex (Section 15.3).

3.2.4 Pericyclic reactions

The last category of elementary reaction types contains what are generally referred to as *pericyclic* reactions and consist in the cyclic reorganization of (usually) two or six electrons which may originate either in σ- or in π-bonds. Since the electronic 'movement' is cyclic, the two partners in such a process cannot be designated as behaving either as nucleophiles or electrophiles but it is presumed that the mobile electrons remain spin-paired throughout. The best known example of a reaction of this kind, to be further discussed in Chapter 14, is the Diels–Alder reaction (**18**), the cycloaddition of a 1,3-diene to an alkene. As with other multibond processes, the same products can in principle arise from a synchronous reaction or from a sequence of bond-making and bond-breaking steps of either a polar or radical type (**19**). Examples of both are known.

18

19   intermediate

3.3 REACTION MECHANISM

This term is taken to mean, in its fullest sense, a detailed description of the entire energy surface associated with a reaction, particularly that part which is close to the reaction coordinate and includes the involvement of not only reagent molecules but solvent also. In addition it includes an understanding in terms of physical principles of the features of the energy surface, the existence and stabilities of intermediates, and the differences in reactivity of structurally related reactants. This ideal state of knowledge has never been realized. A more modest aim is to obtain evidence for the principal features of a reaction, namely the constitution of the transition state and its energy characteristics, the presence of intermediates and at least a qualitative understanding of the affinity or driving force of the reaction, the effects of

120

structural variation on rate, and the role of the solvent. Although the reaction coordinate has been discussed as though it were a definite trajectory across the energy surface which is rigorously followed by every molecular event leading to reaction, it is more than likely that there is some latitude in many of the variables and that individual reacting molecules will follow similar but not necessarily identical pathways to product. The energies and energy distributions of individual molecules are not identical but follow a distribution function (Eq. (2.19)) and it is reacting systems of energy above the average that will be able to surmount activation barriers above the minimum available. This comment is a qualitative one and definite information as to variable properties of the transition state is not available.

3.3.1 The advantages of synchronous reactions

In a synchronous transfer reaction, such as a proton transfer or alkyl transfer (S_N2 reaction), a bond is formed simultaneously with that broken. The energy supplied by bond formation tends to cancel that required for bond fission. The result is a very much lower activation energy than that required for a sequential reaction to give the same product by complete bond fission followed by formation of the new bond. The energetics of a synchronous transfer can be viewed as arising from the intersection of two potential energy curves of the type shown in Fig. 1.4 and relating to the two bonds which are undergoing covalency change, Fig. 3.4. A point is reached at which excited vibrational modes of one pass over to excited vibrations of the other at a point well below the bond dissociation energy. In general the activation energy diminishes with the closer approach of the two minima and with increasing exothermicity of the reaction, i.e. the product energy lower than that of the reactant state. On the energy contour diagrams, Fig. 3.2, the concerted pathway is that proceeding directly from reactant to product state diagonally.

3.4 ELECTRON SUPPLY AND DEMAND

The curved arrows which conventionally connect interacting nucleophiles with electrophiles are a guide to a useful principle by which the effects of structural change on reaction affinity may be predicted.

$$Nu: \curvearrowright E^+ \dashrightarrow Nu{-}E$$
$$\begin{array}{cccc} \uparrow & \downarrow & \uparrow & \downarrow \\ X & Z & X & Z \end{array}$$

X and Z are substituent groups attached to the nucleophilic and electrophilic centers which become bound. It is easy to predict the effects which these two groups must have if the reaction is to be made to occur with greater

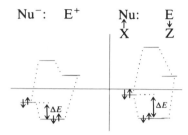

Fig. 3.1 Effect of donor (X) and acceptor (Z) groups on the frontier orbitals of a nucleophile and an electrophile; an electron-donating group tends to raise the HOMO and an electron-withdrawing group tends to lower the LUMO.

facility (i.e. faster or more exothermically). Since the nucleophile is effectively losing one electron, any electron density supplied by X will aid the bonding. Similarly, E has to gain one electron which will be facilitated by electron density being absorbed or removed by Z. The symbols X and Z, respectively, are used throughout to denote substituent groups which tend to donate or withdraw electrons, the point of reference being —H as a substituent group which is assumed not to perturb the surrounding electronic atmosphere. Many aspects of chemistry may be illuminated by the naive application of 'electron-pushing and -pulling' (Chapter 4), which in MO terms means changing the energies of HOMO and LUMO (Fig. 3.1). This may also be viewed as an aspect of delocalization, **20**. A nucleophile containing a donor group is doubly 'electron-rich' while an electrophile bearing an electron-withdrawing substituent is doubly 'electron-poor'. The interaction between the two results in an evening-out of the electronic distribution, a spreading of electron density which is associated with extra stability.

$$Nu \cdots E$$

$$X + Nu \cdots E + Z$$
20

3.5 TRANSITION STATE PROPERTIES AND STRUCTURAL CHANGE

It is desirable to have a reliable guide to the effect on the energy of a transition state wrought by a change in structure of the reagents. In Chapter 2 the Hammond Postulate was mentioned as a useful qualitative principle, but more quantitative theories have been attempted. In the example above, the introduction of groups X and Z will change the energy levels of Nu: and E, but their effect on the transition state will be more than the total initial-state effect. It is more difficult to predict the effect of electronic perturbation by X- or Z-type substituents on the transfer reaction

$$\text{Nu:}^{-} \quad \overset{\overset{\displaystyle \text{X(Z)}}{|}}{\underset{/ \backslash}{\text{C}}}\text{—Nu}' \longrightarrow \text{Nu—}\overset{\overset{\displaystyle \text{X(Z)}}{|}}{\underset{/ \backslash}{\text{C}}} + \text{:Nu}'$$

since electron supply would make bond formation more difficult but bond breaking easier and conversely electron withdrawal would have the opposite effects so that a cancellation would result. Thornton drew attention to the principle that changes in stability of the transition state along the reaction coordinate go in accordance with the Hammond Postulate (greater stability, more reagent-like) but changes which move it perpendicular to the reaction coordinate will have the opposite effect. This can be illustrated with reference to a β-elimination whose two-dimensional energy contour surface is shown schematically in Fig. 3.2a, a type of representation associated with the names of More O'Ferrall, Albery and Jencks. The two coordinates of importance in defining the reaction coordinate and plotted on x- and y-axes are the lengths of the C–H and C–X bonds. The concerted E2 mechanism requires both to increase concurrently, the route being diagonally from origin to top right corner. E1 and E1$_{cb}$ routes will go via the other two corners. An increase in the stability of the product will be reflected in a corresponding though smaller increase in stability of the E2 transition state and a movement away from the product (more reagent-like), a reaction coordinate effect in accordance with the Hammond Postulate. Suppose now the structural change is such as to weaken the C–H bond so that the transition state becomes asymmetrical and 'carbanion-like' (though still the concerted E2 process); this is a movement perpendicular to the reaction coordinate so the transition state moves towards the E–Ad corner (bottom right) where the carbanion intermediate is located, (Fig. 3.2b). Thornton's Rule concerning this aspect states: 'A substituent change which makes progress along the reaction coordinate more difficult will lead to a transition state which is later if the force constant for motion along the coordinate is negative but earlier if it is positive'. A transition state has a negative force constant (i.e a decrease in energy with amplitude) for the vibration which takes it across an energy saddle point, but positive (i.e. of the normal type) if the vibration is of a mode across an energy valley. These principles have been embodied in a form applicable to displacement reactions, indicating shifts in the location of the transition state, in terms of the orders of the partial bonds (Table 3.1). A later transition state means a decrease in order of the C–L bond and an increase in the order of Nu–C. The converse statements also apply.

A similar model has also been proposed to cater for additions to the carbonyl group. The origins of changes in the potential energy surface in the vicinity of the transition state are, of course, quantum-mechanical and it is within the capabilities of current computers to determine by calculation

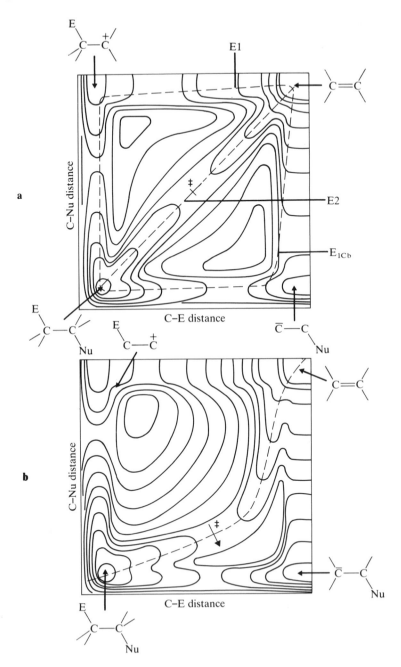

Fig. 3.2 More O'Ferrall–Albery–Jencks diagrams of energy contours for β-elimination reactions as a function of lengths of the two bonds broken. Reaction coordinates follow the dotted lines. **a**, E2 mechanism with 'central' character, simultaneous fission of C_α–Nu and C_β–H bonds. **b**, The result of stabilization of the carbanion, Nu—C—C$^-$; a continuation of this trend would result in a switch to the E1$_{cb}$ mechanism.

Table 3.1   Effects of electron withdrawal on transition state bond orders for

$$\text{Nu:} + \text{C—L} \rightarrow [\text{Nu}\cdots\text{C}\cdots\text{L}]^{\ddagger} \rightarrow \text{Nu—C} + \text{:L}$$

Site of electron withdrawal	Predicted order change	
	Nu$\cdots$C	C$\cdots$X
Nu:	increase	decrease
C	increase	decrease
L	decrease	increase

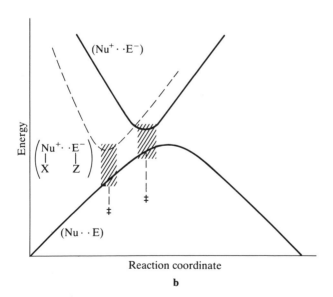

Fig. 3.3 **a**, Electronic states of a nucleophile–electrophile system. **b**, Energies of ground and charge-transfer states as a function of the reaction coordinate. The transition state occurs at the point where energies become close enough for mixing to occur. **c**, Effect of stabilization of the charge-transfer state by suitable substituents. An earlier transition state may occur.

two-dimensional energy plots similar to Figs 2.2 and 3.2. However, some insight into the nature of the interaction during progress along the reaction coordinate can be obtained by considering the energies of separate ground and excited electronic states (Fig. 3.3), and the way they are affected by structural change. The simplest system applicable to a transfer (S_N2) reaction requires the interactions between three MOs to be considered, the HOMO of the nucleophile and the HOMO and LUMO of the electrophile.

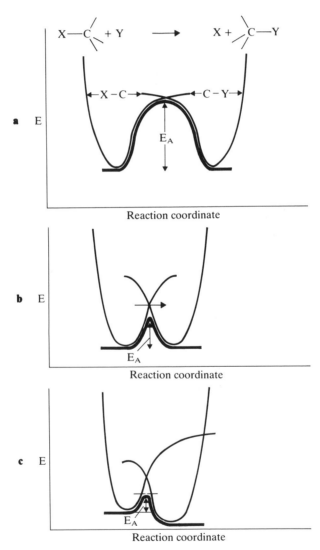

Fig. 3.4 Activation barriers for a displacement reaction shown as resulting from the intersection of potential energy curves of the C—X and C—Y bonds. **a**, Sequential reaction: E_A = bond dissociation energy of C—X, **b** concerted reaction: E_A is very much reduced, **c** the same for an exothermic reaction.

The transition state can then be described in VB terms as a resonance hybrid of the ground-state configuration and a set of excited states (including doubly excited ones) of which the charge-transfer form (Nu^+E^-) is the most important. By using high-level MO calculations the energies of ground and charge-transfer states may be calculated as a function of geometry corresponding to a traversal of the reaction coordinate: it then appears that the former increases in energy while the latter decreases until the two states approach close enough for mixing to occur. Substitutents which stabilize the charge-transfer state (X—Nu: and Z—E) will lower its energy and bring about this mixing earlier; this accords with the Hammond Postulate.

PROBLEMS

1 Classify the following reagents as either nucleophiles or electrophiles indicating the notional electron-pair movement by the curly-arrow convention

(a) $CH_3COOH + OH^- \longrightarrow CH_3COO^- + H_2O$.
(b) $CH_3COO^- + H_2O \longrightarrow CH_3COOH + OH^-$.
(c) $CH_3\!-\!Br + I^- \longrightarrow CH_3\!-\!I + Br^-$.
(d) $CH_3COCH_3 + CN^- \longrightarrow CH_3C(CN)OH\cdot CH_3$.
(e) $CH_3CO\cdot Cl + PhO^- \longrightarrow CH_3CO\cdot OPh + Cl^-$.
(f) $Me_3B + NMe_3 \longrightarrow Me_3B^-\!\!-\!\overset{+}{N}Me_3$.

(g) 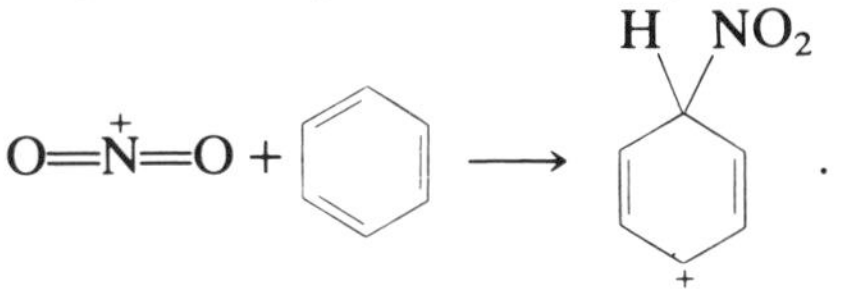 $O=\overset{+}{N}=O +$ [benzene ring] $\longrightarrow$ [cyclohexadiene ring with H, NO$_2$ and +].

(h) [cyclohexadiene ring with H, NO$_2$ and +] $+ HSO_4^- \longrightarrow$ [benzene ring with NO$_2$] $+ H_2SO_4$.

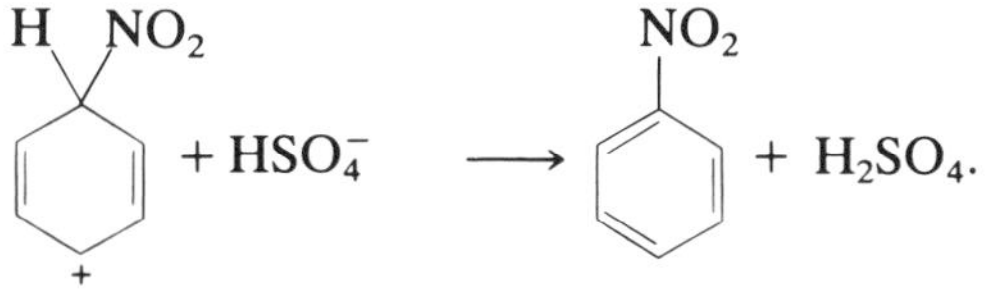

(i) $Br\!-\!Br + CH_2\!=\!CH_2 \longrightarrow BrCH_2\!-\!CH_2^+Br^- \longrightarrow BrCH_2\!-\!CH_2Br$.

2 Classify the reactions in the following scheme and add the necessary reagents and experimental conditions.

$$PhH \xrightarrow{a} PhCOCH_3 \xrightarrow{b} PhCH(OH)CH_3 \xrightarrow{c} PhCH=CH_2$$

$$\downarrow d$$

$$PhC\equiv CH \xleftarrow{f} PhCH=CHBr \xleftarrow{e} PhCHBr\!-\!CH_2Br$$

$$\downarrow g$$

$$PhC\equiv C^- \xrightarrow{h} PhC\equiv C\!-\!CH_2\!-\!CH_2CN \xrightarrow{i} PhC\equiv C\!-\!(CH_2)_3NH_2$$

3 Three pathways for the hydrolysis of acetyl chloride,

$$CH_3CO \cdot Cl + H_2O \rightarrow CH_3CO \cdot OH + H^+, Cl^-$$

might be considered: Ad–E, E–Ad and synchronous. Draw up a More O'Ferrall–Albery–Jencks energy diagram of C–Cl against C–OH bond distances indicating the three pathways. Add energy contours expected for each pathway in turn to be favored.

REFERENCES

1 C. K. Ingold, *Structure and Mechanism in Organic Chemistry*, 2nd edn, Bell, London, 1969.
2 P. Sykes, *A Guidebook to Mechanism in Organic Chemistry*, 5th edn, Longman, London, 1981.
3 T. H. Lowry and K. S. Richardson, *Mechanism and Theory in Organic Chemistry*, 2nd edn, Harper and Row, New York, 1981.
4 H. Maskill, *The Physical Basis of Organic Chemistry*, Oxford U.P., 1985.

4 Correlation of structure with reactivity

Organic compounds uniquely are capable of structural variation in the vicinity of a reaction center, permitting an almost continuous variation in its electrophilic or nucleophilic character. This capacity may then be used as a delicate probe into the effects which electronic perturbation produces upon reaction affinity and from which the electronic demands of the reaction may be inferred. The use of this information in deducing mechanism is a highly developed art, applicable to polar reactions in particular but to a lesser degree to radical or pericyclic processes[1-8].

4.1 ELECTRONIC DEMANDS

Since a polar reaction consists of the interaction between a nucleophile and an electrophile, the electronic demands of the reaction, i.e. the factors which will facilitate this process, may be summarized as supply of electrons to the nucleophilic center and withdrawal of electrons from the electrophilic center. Structural features which enhance nucleophilicity and electrophilicity are (rather naively) described as 'electron-donating' and 'electron-withdrawing', respectively (Fig. 4.1). The degree to which a given reaction responds to electronic perturbation by a substituent depends upon the reaction type and its electronic demands[9].

The structural changes which are customarily used to bring about electronic perturbations are substituent groups which may be introduced suitably near to the reaction center and which do not themselves take part directly in the reaction being considered. Common substituents might include, for example, —OH, —Me, —Cl, —F, —NO$_2$, —CN. It is the changes in reaction affinity brought about by substitution which are of interest, so structural change must be assessed relative to some standard substituent which is electronically neutral. Hydrogen is normally adopted as the 'zero' substituent; others are then reckoned electron-donating or electron-withdrawing relative to it. The reason for this choice lies in the fact that most reactions studied occur at carbon as one reaction center, and the electronegativities of carbon and hydrogen are almost equal so that a C–H bond has no polarity. Furthermore the H-substituent has no unshared pairs or π-electrons and, moreover, it likely to be the most accessible.

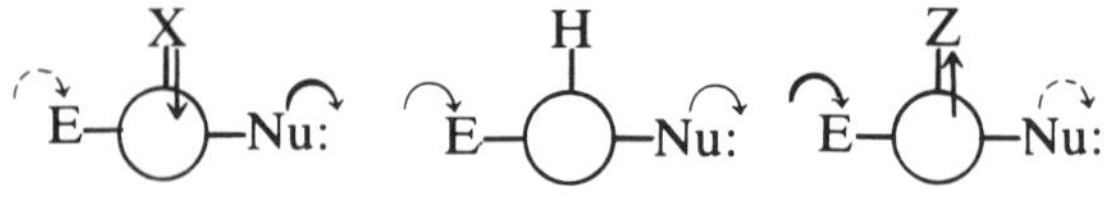

$$\overset{X}{\underset{}{E{-}\bigcirc{-}Nu{:}}} \qquad \overset{H}{\underset{}{E{-}\bigcirc{-}Nu{:}}} \qquad \overset{Z}{\underset{}{E{-}\bigcirc{-}Nu{:}}}$$

Fig. 4.1 Model expressing the effect on reactivities of nucleophilic and electrophilic centers in a molecule by electron-donating and electron-withdrawing substituents relative to hydrogen.

The following example will illustrate the procedure and reveal a problem. Suppose it is desired to investigate the electronic demands of ester hydrolysis by aqueous alkali, to identify the nucleophile and electrophilic centers and the degree to which the ester responds to substituent change.

$$p\text{-}X{\cdot}C_6H_4{\cdot}COOEt \xrightarrow{\ OH^-,k\ } p\text{-}X{\cdot}C_6H_4{\cdot}CO\bar{O} + EtOH$$

The following series of compounds are then devised and synthesized[10,11]:

X:	—NO$_2$	—CN	—Cl	—H	—CH$_3$	—OCH$_3$
$k/\text{M}^{-1}\text{s}^{-1}$:	32·9	15·7	2·10	0·289	0·172	0·143

Rate constants are measured under identical conditions at a single temperature and it is reasonable in a closely related series such as this to assume that the same mechanism is operating throughout and that $\Delta S^{\ddagger}$ is approximately constant, and consequently the series has been ordered by increasing $\Delta G^{\ddagger}$. The problem which remains is that, formally at least, the substituent groups X have not been identified with their electronic character; which are electron-donating and which electron-withdrawing? If the ester is the electrophilic entity, then the order of substituents is one of decreasing electron withdrawal (or increasing electron donation). If the ester is the nucleophilic partner, then the opposite interpretation would be appropriate. It is necessary to begin with a reaction whose electronic demands can be inferred *a priori,* a technique which is credited to Louis Hammett[12–14] (but for which credit should be shared by Burckhardt[15,16]). These workers employed, as model reaction, the ionization in water of substituted benzoic acids, **1**, and measured equilibrium constants, K_A. For the assessment of substituent effects, equilibrium measurements are as valid (more so, perhaps) as rate measurements. There are abundant K_A values of the acids in the literature and, most important, the 'mechanism' of the process is known in that dissociation involves a proton transfer to water. This identifies the carboxylic acid as the electrophile and water as the nucleophile. Consequently K_A will be increased by the substitution of H for an electron-withdrawing group which weakens the O—H bond, and conversely decreased by an electron-donating substituent. The measure of the change in ΔG will be $\Delta(\log K_A)$, or $\Delta(\ln K_A)$.

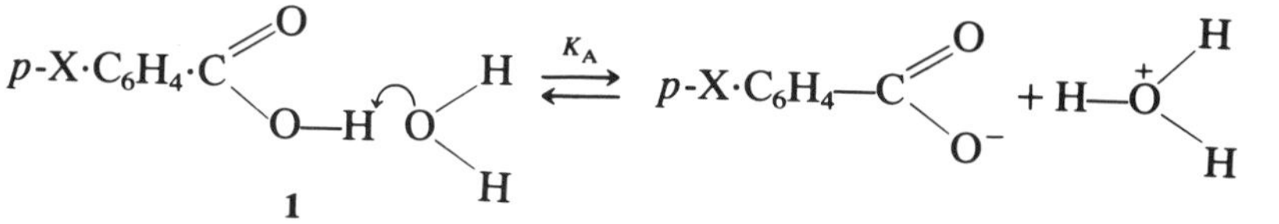

X:	—NO$_2$	—CN	—Cl	—H	—Me	—OMe	
log K_A:	−3·45	−3·56	−4·00	−4·20	−4·37	−4·47	$(pK_A = -\log K_A)$
σ:	+0·75	+0·64	+0·2	0·00	−0·17	−0·27	

Relative to hydrogen in this series, the substituents *p*-Cl, *p*-CN, *p*-NO$_2$ are increasingly acid-strengthening and hence electron-withdrawing, while *p*-Me, *p*-OMe are increasingly electron-donating or acid-weakening. It is highly significant that the ordering of this set of substituents is the same as that in ester hydrolysis.

4.2 THE HAMMETT EQUATION

Hammett drew attention to the fact that a plot of $\log K_A$ for benzoic acid ionization against $\log k$ for ester hydrolysis over many substituents is reasonably linear (Fig. 4.2a), which means that all the substituents are exerting a similar effect in each of these quite dissimilar reactions. There is a proviso, however, that substituents be located at *meta* or *para* positions in the benzene ring. Rates and equilibrium constants for *ortho* substituted compounds do not fall on the line. The reason for this is that changes in k or K_A (i.e. $\Delta G^{\ddagger}$, ΔG) brought about by *m*- or *p*-substituents are virtually changes in $\Delta H^{\ddagger}$ or ΔH since substitution does not greatly affect ΔS. *Ortho* substituents affect both ΔH and ΔS, the latter differently for the two reactions being compared since entropy change depends on the size of both reagents and substituents and on solvation changes which they bring about. The hydrolysis data can now be interpreted. Rates are evidently increased by electron-withdrawing substituents on the ester, which must accordingly be the electrophile, and it is likely that the initial interaction may be expressed as **2**.

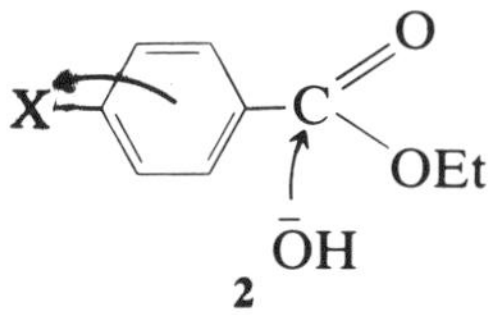

Quantitatively, the effect of each substituent, relative to that of hydrogen, may be obtained by a comparison of ΔG for dissociation constants of substituted benzoic acids (K_X) with that of the parent compounds,

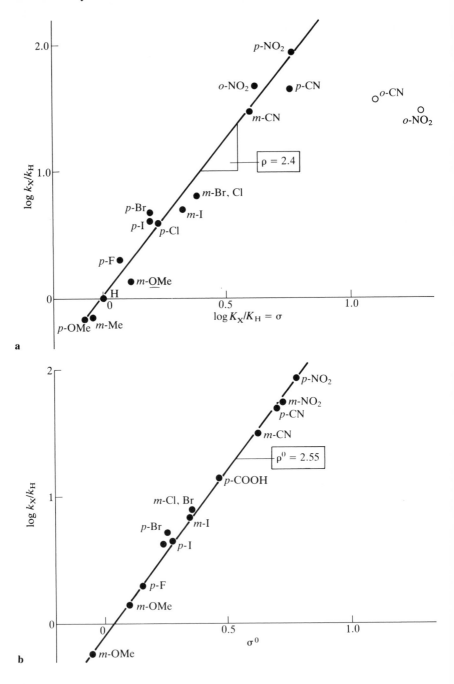

Fig. 4.2 **a**, Log–log plot of relative rates of hydrolysis of substituted ethyl benzoates (ordinate) against acid dissociation constants of benzoic acids (abscissa). **b**, The same, plotting values of σ^0 on the abscissa.

benzoic acid (K_H), thus:

$$\text{Substituent effect} = \Delta G_X - \Delta G_H = \log(K_X/K_H) = \sigma_X$$

in which σ_X is known as the *substituent constant*. Electron-withdrawing substituents are therefore characterized by negative values of σ and electron-donating ones by positive values, their magnitudes being a measure of these qualities. Hydrogen, as reference point, has $\sigma = 0 \cdot 0$. Now it is evident that the linear relationship of Fig. 4.2a implies

$$\log k_X/k_H \propto \log K_X/K_H \propto \sigma.$$

Writing $k_X/k_H = k_{rel}$, $K_X/K_H = K_{rel}$, and introducing a constant of proportionality ρ, known as the *reaction constant*,

$$\log k_{rel} \propto \log K_{rel} = \rho\sigma. \tag{4.1}$$

This expression has become known as the Hammett equation and is obeyed with varying precision by thousands of reactions which take place at or near a benzene ring on which substituents are located at *meta* or *para* positions. A correlation of this type is clear evidence that the changes in structure produce proportional changes in the activation energy ($\Delta G^{\ddagger}$) for all such reactions. It is therefore known as a Linear Free Energy Relationship (LFER). The Hammett equation is the best-known of this type of expression of similarity. It is the rule rather than the exception that the Hammett equation is obeyed with moderate precision for polar reactions. This enables the electronic demands of many reactions to be measured as the constant, ρ, with which mechanistic schemes must accord and models of electronic perturbation may be tested. LFERs truly occupy a central position in the theory of organic chemistry.

4.3   SUBSTITUENT CONSTANTS, σ^0

The plot of $\log k_{rel}$ against σ (or $\log K_{rel}$) of Fig. 4.2a (ignoring *ortho* points) is typical for a comparison between two reactions and still has a considerable scatter of points. There are evidently effects related to solvation changes which detract from an exact proportionality in the relative perturbations produced by each substituent in the two systems. This situation can be greatly improved by using, as substituent constants, values which have been averaged over many reactions. A modified set of substituent constants, known as σ^0, which better express a 'universal' substituent character, is defined by Eq. (4.2) and these will subsequently be employed rather than values defined by Eq. (4.1). Values of σ^0 from the recent compilation by Exner[17] are given in Table 4.1b.

$$\ln k_X/k_H = \rho^0\sigma^0. \tag{4.2}$$

Figure 4.2b shows the great improvement obtainable. The sign and

Table 4.1 (a) Hammett substituent constants from dissociation constants of benzoic acids

$$(m, p)\text{-X—C}_6\text{H}_4\text{COOH} + \text{H}_2\text{O} \xrightleftharpoons{K_A} (m, p)\text{-X—C}_6\text{H}_4\text{COO}^- + \text{H}_3\text{O}^+(\text{H}_2\text{O}, 25°\text{C});$$

$$\log K_A(\text{X}) - \log K_A(\text{H}) = \text{p}K_A(\text{H}) - \text{p}K_A(\text{X}) = \sigma$$

X	pK_A		σ	
	meta	*para*	*meta*	*para*
—NH$_2$	4·60	4·86	−0·40	−0·66
—OH	4·08	4·60	+0·12	−0·40
—OMe	4·09	4·47	+0·11	−0·27
—Me	4·27	4·37	−0·07	−0·17
—H		4·20		0·00
—F	3·86	4·13	0·34	0·07
—Cl	3·83	4·00	0·38	0·20
—Br	3·82	3·97	0·38	0·23
—I	3·85		0·35	
—CN	3·61	3·56	0·59	0·64
—NO$_2$	3·45	3·45	0·75	0·75

(b) Substituent constants[17] for LFER use

Substituent	σ_m^0	σ_p^0	σ_p^+	σ_p^-	σ_I	σ_R
—NMe$_2$	−0·10	−0·32	−1·7		0·06	−0·52
—OMe	0·1	−0·12	−0·78		0·34	−0·45
—NHAc	0·14	0·0	−0·6		0·29	−0·25
—OAc	0·39	0·31			0·38	−0·21
—SMe	0·14	0·06	−0·6	0·04	0·23	−0·20
—SAc	0·39	0·44			0·21	0·01
—Me	−0·06	−0·14	−0·3		0·01	−0·11
—*tert*-Bu	−0·09	−0·15	−0·26		−0·03	−0·12
—Ph	0·05	0·05	−0·18		0·10	−0·11
—F	0·34	0·15	−0·07		0·52	−0·32
—Cl	0·37	0·34	0·11		0·47	−0·23
—Br	0·37	0·26	0·15		0·47	−0·19
—I	0·34	0·28	0·13		0·42	−0·16
—CHO	0·41	0·47		1·04	0·25	0·26
—COMe	0·36	0·47		0·82	0·28	0·16
—COOR	0·35	0·44		0·74	0·31	0·16
—CN	0·62	0·71		0·99	0·57	0·13
—SOMe	0·21	0·17		0·73	0·50	0·00
—SO$_2$Me	0·64	0·73		1·05	0·55	0·12
—NO$_2$	0·71	0·81		1·23	0·65	0·15
—CF$_3$	0·46	0·53		0·65	0·42	0·11
—NMe$_3$	1·04	0·88			0·73	0
—$^+$SMe$_2$	1·00	0·90		1.16	0·89	0·17

Additional values

—NH$_2$	−0·09	−0·30	−1·3		0·12	−0·48
—NPh$_2$	−0·07	−0·19	−1·4			
—OH	0·02	−0·22	−0·92		0·25	−0·43
—OTos	0·36	0·33			0·58	
—OPh	0·26	0·05	−0·5		0·38	−0·34
—CH=CH$_2$	0·08	−0·08	0·18		0·08	−0·05
—C≡CH	0·2	0·23			0·3	0·07
—CH$_2$Ph	−0·05	−0·06	−0·27		0·03	−0·12

Table 4.1 (*Continued*)

Substituent	σ_m^0	σ_p^0	σ_p^+	σ_p^-	σ_I	σ_R
—CH$_2$CN	0·16	0·18	0·12		0·18	−0·09
—CH$_2$NH$_2$	−0·03	−0·11			0·08	−0·10
—CH$_2$OH	0·01	0·01	0·01		0·05	0·0
—CH$_2$F	0·11	0·1			0·18	−0·02
—CH$_2$Cl	0·11	0·12			0·16	0·0
—CF(CF$_3$)$_2$	0·37	0·53			0·48	0·0
—SiMe$_3$	0·11	0·0		0·11	−1·13	0·06
—HgMe	0·43	0·1				
—PR$_2$	0·05	0·03			0·08	−0·05
—P(OR)$_3$	0·12	0·15			0·09	0·06
—COCF$_3$	0·63	0·80			0·58	0·33
—CONH$_2$	0·28	0·31		0·62	0·33	0·0
—COOH	0·35	0·44		0·73	0·32	0·29
—COCl	0·53	0·69			0·38	0·21
—NHSO$_2$Me	0·20	0·03			0·32	
—N≡C	0·48	0·49			0·57	0·02
—NCO	0·3	0·24			0·36	−0·4
—N=NPh	0·29	0·33		0·7	0·19	0·0
—NO	0·49	0·65		1·46	0·33	0·32
—SO$_2$NMe$_2$	0·71	0·9		1·06	0·44	0·0
—CH$_2$NMe$_2$	0·40	0·44				
—(CH$_2$)NMe$_2$	0·16	0·13				
—COO$^-$	0·09	−0·05			−0·15	
—SO$_3^-$	0·05	0·09			0·2	0·07
—D					−0·0021	−0·0003

magnitude of σ^0 express the capability of a substituent to perturb its electronic environment. However, it will be noticed that the Hammett equation requires different values of σ to be assigned to the same groups depending on their orientation, *meta* or *para*. Moreover, even the sign of σ may change according to its location: for *m*-OMe, $\sigma_m = +0\cdot10$ (electron-withdrawing); for *p*-OMe, $\sigma_p = -0\cdot12$ (electron-donating). The σ values are then only characteristic of a substituent in a particular context. These observations can be explained if the electronic perturbation produced by a substituent is the resultant of two (or more) independent effects operating simultaneously. The model will now be examined.

4.4 THEORIES OF SUBSTITUENT EFFECTS

In any discussion of substituent effects on reaction rates and equilibria three components must be borne in mind; the substituent or 'source' of the perturbation, X, the reaction site or 'detector' of the perturbation, Y, and the molecular framework ('core') through which the effect is transmitted.

This is the benzene ring in the original Hammett systems and discussion will for the present center on benzenoid systems. Two basically distinct electronic effects may be generated by a substituent on a reaction site, and are discussed below.

4.4.1 The resonance effect [18]

This is also known as the mesomeric effect, and is denoted *R*. Many substituents give rise to a perturbation which is greater when they are located *para* than when they are *meta*; this suggests the transmission mechanism is of a conjugative nature in which charge is relayed to alternate atoms. This effect is described here as a '+*R* effect' if it results in donation of electrons from substituent to reaction center and a '−*R* effect' if a withdrawal of electrons results. These effects would be characterized by negative and positive values of σ, respectively, so sign convention is slightly confusing.

In order to exercise a resonance effect, a substituent must possess a *p*- or π-orbital which is available to conjugate with the π-MOs of the aromatic system. Two situations are important.

(a) $\ddot{X}$- is a donor group and typically possesses an unshared electron pair or π-electrons on an atom directly attached to the ring. Examples are

$$—NR_2, —OR, —SR, —PR_2, —Hal, —CH{=}CH_2.$$

These groups are all capable of exerting a +*R* effect which stabilizes an acceptor center (e.g. a carbon bearing some degree of carbocation character) when in the *para* or *ortho* positions. The extreme situation is depicted by the VB structures, **3a**, **3b**. Little interaction between donor and

$$\begin{array}{ccccc}
\text{(+R)} & & & \text{(−R)} & \\
\textbf{3a} & \textbf{3b} & \textbf{3c} & \textbf{3d} & \textbf{3e}
\end{array}$$

3f

acceptor centers will occur if they are located *meta* since quinoid resonance structures analogous to **3b** and **3e** cannot be drawn and analogous structures (**3f**) are therefore of high energy and less important.

(b) Substituents Z have a π-acceptor center adjacent to the ring. Since no commonly encountered substituents possess vacant bonding *p*-orbitals, this means in practice groups which can act as electron acceptors by simultaneously releasing π-electrons to adjacent heteroatoms and whose VB contributing structures have a positively charged atom attached to the ring (**3c–e**). Common examples are

All such groups tend to accept electronic charge and stabilize donor centers (e.g. carbon bearing some degree of carbanion character adjacent to the ring), illustrated by contributing structure **3e**. Again, the strongest $-R$ interactions occur when substituent and reaction centers are located *ortho* or *para* (**3e**).

It may be observed that there is no fundamental difference between these two situations; there is in each case a transfer of charge between two centers by conjugation and the differentiation into $+R$ and $-R$ effects merely depends upon which center we designate the substituent and which the reaction center.

4.4.2 The inductive effect

The second type of perturbation evident from substituent constants results in the effect at the *meta* position being greater than that at the *para*. This is clearly observable in substituents such as $-\overset{+}{N}Me_3$ (**4**) and $-CF_3$ and depends for its strength upon proximity. This is called the inductive effect[19,20], designated I; $+I$ means electron-donating, $-I$ electron-withdrawing. The basis of this electronic displacement is probably complex but originates in part from differences in electronegativity which cause

4

polarization of both σ- and π-bonds and also from electrostatic effects experienced at the reaction center due to charges and dipoles resident on the substituent. Two mechanisms for inductive polarization may be considered. The classical inductive effect is a polarization through bonds, both of σ- and π-types, becoming progressively attenuated. The other, known as a field effect, is propagated through space and depends more for its intensity on proximity than on the number of bonds separating source and receptor. In practice it is difficult to separate the two, which may both be components of the I effect[19]. The extent of the attenuation of an inductive effect can be measured by the steadily decreasing value of σ brought about by interposition of a methylene chain[21]:

$$-(CH_2)_n-NMe_3^+$$

n:	0	1	2	3
σ_m^0:	1·04	0·4	0·16	0·06
σ_p^0:	0·88	0·44	0·13	0·02

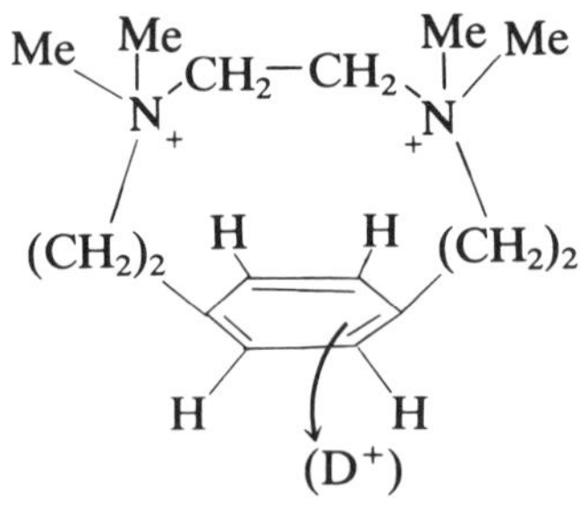

5

The juxtapositioning of a charged or dipolar substituent near a reaction center can have a profound effect on the rate of a reaction involving charge generation since the energy of the transition state will include additional electrostatic terms summarized in Table 1.17. For example, **5** undergoes D-exchange at a rate 10^{-5} times that of benzene[22] and this difference can only be partly accounted for by steric hindrance. Interaction through space must be occurring here, resisting the creation of positive charge on the benzene ring. On the other hand, strong coupling between unshared pairs of the diazabicyclo-octane, **6**, which are too far apart for direct interaction is claimed to occur 'through bonds'[23].

6

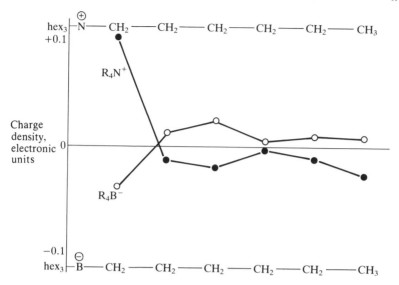

Fig. 4.3 Charge distribution, measured by ^{13}C-NMR chemical shifts, along an alkyl chain perturbed by a charged substituent.

The charges on individual carbons may be measured by ^{13}C-NMR chemical shifts. Some results are shown in Fig. 4.3. The α-carbon apparently takes positive charge more readily than negative charge but the effect is rapidly attenuated so that a slight reversal of charge is apparent on successive carbons[24]. This is in accord neither with simple σ-inductive polarization nor with a field effect. The latter should be experienced more strongly at C4 than at C3 because of the favorable conformation, **7**. High-level MO theory has been applied to the problem and appears to predict an alternation of charge along the carbon chain (one conformation and absence of solvent)[25]: for example,

$$\overset{\delta\delta+}{CH_3}-\overset{\delta\delta-}{CH_2}-\overset{\delta+}{CH_2}-\overset{\delta-}{CH_2}-F.$$

Some experimental data appear to agree with this[26] while other data do not[27]. The nature of the inductive effect therefore seems more complex than was initially thought and evidently contains field effects, bond polarization and extended π-polarization[25-29]. The term 'π-inductive effect' is used to indicate a general polarization of an extended π-system[30-32]. Unlike a resonance effect it is attenuated with distance, though the falloff is less rapid

7

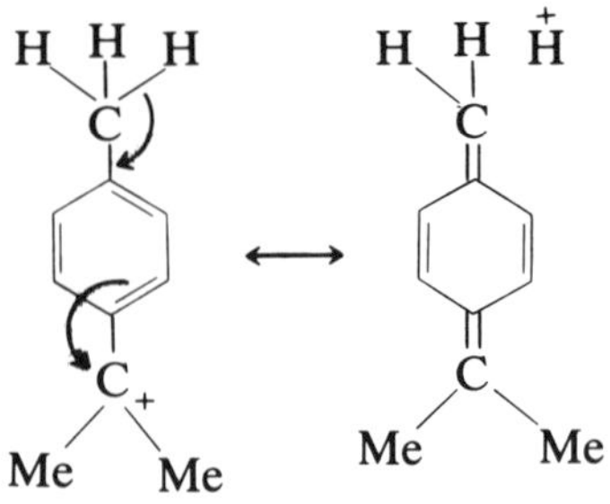

8

than that through a σ-bond framework. Cleavage rates of a series of silanes correlate well with σ^0 and the values of ρ diminish steadily with n; the falloff of sensitivity, measured by ρ_n/ρ_{n-1}, has a constant value[33]. This is a characteristic of neither pure R or σ-inductive effects and may be an example of transmission of electrons by π-polarization.

4.5 INTERPRETATION OF σ-VALUES

Inductive effects tend to be in the electron-withdrawing sense $(-I)$ since most substituent groups available are more electronegative than carbon. Resonance effects can be of either sign. It is convenient to consider four structural types of substituent according to the blend of I and R influences which they are capable of exerting:

Unshared-pair substituents, $-\ddot{X}$	$+R, -I$
Alkyl groups, $-R$	$+R$
Acceptor groups, $-Z$	$-R, -I$
Cationic centers with no vacant orbitals	$-I$

4.5.1 Unshared-pair (n) substituents, —X

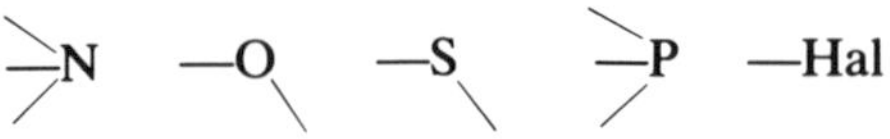

Para (and ortho) position

All are capable of exerting an electron-donating resonance effect (**3a, b**), stabilizing acceptor centers, but the capacity for sharing their n-electrons falls off with increasing electronegativity. At the same time the $+I$ effect becomes increasing strong. The resultant effect leaves σ_p negative for —NR$_2$, —OR, —R and —SR but positive for halogens as the effect changes from $+R > -I$ to $+R < -I$. This is shown graphically in Fig. 4.4a by plotting individual R and I components and the resultant on the σ-scale.

140

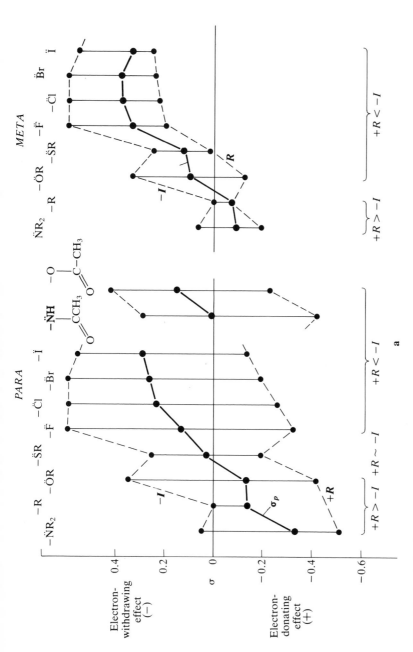

Fig. 4.4   Analysis of total electronic substituent effect, σ^0, as a resultant of I and R components: (**a**) $+R$ groups; (**b**) $-R$ groups.

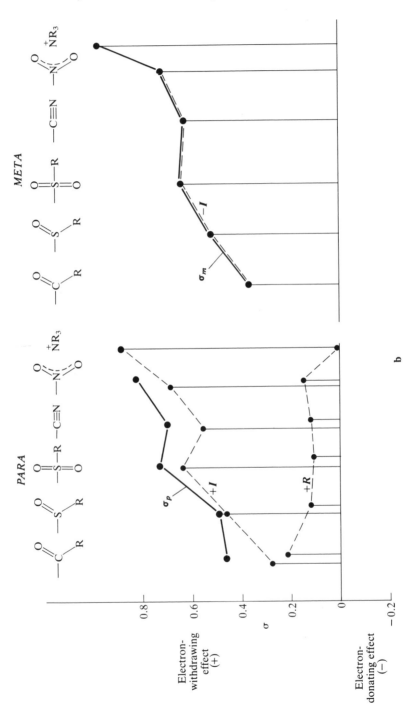

Fig. 4.4 (*Continued*)

Modification of these substituents by a $-R$ group such as carbonyl or sulfonyl ($-NHCOCH_3$, $-OCOCH_3$, $-OSO_2R$) will reduce the electron-donating capacity while anionic charge (e.g. $-O^-$) is accompanied by an immense increase in electron donation ($+R, +I$).

Meta position

The $+R$ effect is weak so σ is dominated by the $+I$ effect, stronger from the *meta* than from the *para* position and increasing with electronegativity. Except for $-NR_2$ and $-R$ all substituents of this type have a positive value of σ_m so are, overall, electron-withdrawing.

4.5.2 Alkyl groups

The electronic effect of alkyl groups such as methyl are experimentally that of mildly $+R$ substituents, i.e. capable of releasing electrons by a resonance mechanism and best able to stabilize a positive center when *ortho* or *para* to it **8**. The effect is known as *hyperconjugation* since the delocalization involves σ-bonded rather than unshared (n-) or π-electrons. The *tert*-butyl cation, $C(Me)_3^+$ (**9**), is far more stable than the methyl cation, $CH_3^{+\,34}$, and both theory and NMR spectra agree that the positive charge is delocalized onto the peripheral hydrogens[35,36]. The VB contributing structures **10** express

this sharing of bonding elecrons which is formally little different from the familiar situation for unshared pairs. Hyperconjugation operates through the framework of the benzene ring, but there is a negligible I effect from such groups[37]. Experimentally, hyperconjugation may be probed by photo-electron spectroscopy, which measures the ionization energy of electrons in individual MOs. Hyperconjugation between $-CZ_3$ and $-CH{=}CH_2$ in allyl compounds raises the energy of the π-orbital (since both components contain all-filled bonding MOs) by an amount proportional to the strength of the interaction. This places the order of σ-bond hyperconjugation as follows[38]:

$$C{-}I > C{-}Br > C{-}H \sim C{-}Cl \sim C{-}C \sim C{-}F.$$

C—C hyperconjugation is slightly more pronounced than C—H hyperconjugation, so the value of σ for *tert*-butyl ($-0{\cdot}15$) is more negative than that for methyl ($-0{\cdot}14$).

4.5.3 Acceptor groups, —Z

Into this category are placed substituents which are capable of accepting electronic charge by a resonance effect. All have a π-system embracing the atom which is directly attached to the core and an important VB contributing structure which gives this atom a positive charge and ligand vacancy (**3d**). Typical examples are the family of carbonyl substituents —CO—A; A = R, OR, NR$_2$, H, etc.), cyano (—C≡N), sulfoxyl and sulfonyl derivatives ($>$SO and $>$SO$_2$) and the nitro group (—NO$_2$). All exhibit a $-R$ effect, stabilizing negative charge in the ring and destabilizing positive charge (**3c–e**). In addition there is an inductive effect in the same sense $(-I)$, often larger than the resonance effect due to the partial charges and the high electronegativity of oxygen or nitrogen in the substituent group. These substituents are doubly electron-withdrawing (Fig. 4.4b).

4.5.4 Cationic centers

Substituents such as —$\overset{+}{N}R_3$, —$\overset{+}{P}R_3$, possess a very powerful electron-withdrawing effect, shown by substituent constants exceeding +1. This may be considered to be entirely due to inductive effects of the charge. Contributions from hyperconjugation must be present but are difficult to separate from the strong $+I$ effect.

Substituent effects are summarized in Table 4.2.

Table 4.2   Summary of substituent effects

Examples	*Effects*	*Sign of σ*
—O$^-$	$+R$, $+I$, electron-donating	− (large)
—NR$_2$, —OR, —SR	$+R > -I$, electron-donating	− (large)
—R	$+R$, electron-donating	− (small)
—Hal	$-R < -I$, electron-withdrawing	+ (small)
—C=O —SO$_2$, —C≡N, —NO$_2$	$-R$, $-I$, electron-withdrawing	+ (large)
—$\overset{+}{N}R_3$	$-I$, electron-withdrawing	+ (large)

4.6 REACTION CONSTANTS, ρ

The slope of a Hammett plot of log k_{rel} against σ is ρ, the *reaction constant*. This is a measure of the sensitivity of a reaction to the effects of electronic perturbation. From the definition of σ and Eq. (4.1), $\rho = 1.000$ for the dissociation of benzoic acids in water at 25°C. This reaction is therefore set arbitrarily as the standard by which to compare susceptibilities or electronic demands of other reactions:

Sign of ρ	Rate increased by:
+	electron withdrawal
−	electron donation

The magnitude of ρ gives a measure of the degree to which the reaction responds to substituents. Since this is on a logarithmic scale, a change in ρ of 1 indicates a 10-fold change in rate. A useful guide to interpreting Hammett constants is as follows: a change in the substituent from p-OMe to p-NO$_2$ ($\Delta\sigma \sim 1$) will be accompanied by a 10-fold change in rate constant if $\rho = 1\cdot0$, 100-fold if $\rho = 2$ (10^n where $\rho = n$), increasing if ρ is positive, and decreasing if it is negative. The value of ρ for the ester hydrolysis, Fig. 4.2b, is 2·54. This reaction is more sensitive than benzoic acid dissociation by a factor of $10^{2\cdot54-1\cdot00} = 35$, consistent with an activation step in which OH$^-$ attacks the carbonyl group, a reaction center two bonds closer to the source

Table 4.3   Reactions which obey the Hammett equation, Eq. (4.2)[5,6,8]

	ρ°	Comment
Ionization of acids		
1 ArCOOH, water, 25°C	1·000	
2 ArCOOH, 50% aq. ethanol	1·52	ρ is solvent-sensitive
3 ArCH$_2$COOH, water, 25°C	0·562 ⎫	Attenuation by alkyl
4 ArCH$_2$CH$_2$COOH, water, 25°C	0·237 ⎭	chain
5 ArCH=CH—COOH (*trans*), 50% aq. ethanol	0·68	π-bond transmits electrons better than σ-bond
6 ArSCH$_2$COOH, water, 25°C	0·30	
7 ArSO$_2$CH$_2$COOH, water, 25°C	0·253	
8 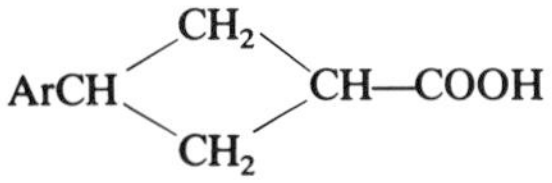 ArCH(CH$_2$CH$_2$)CH—COOH	0·256	
9 ArPO(OH)$_2$, water, 25°C	0·830	
10 ArCH$_2$NH$_3$, water, 25°C	1·05	
Other reactions		
11 ArCOOH + Ph$_2$CN$_2$, PhMe	2·22 ⎫	ArCOOH is the electrophile
12 PhCOOH + Ar$_2$CN$_2$, PhMe	−1·58 ⎭	and Ar$_2$CN$_2$ the nucleophile
13 ArCOOEt + OH$^-$, aq. ethanol	2·55 ⎫	Attack of OH at $>$C=O,
14 ArCH$_2$COOH + OH$^-$, aq. ethanol	0·98 ⎭	attenuated by —CH$_2$—
15 ArCOOEt + H$_2$O(H$^+$), aq. ethanol	−0·577 ⎫	
16 ArCONH$_2$ + H$_2$O(H$^+$), aq. ethanol	−0·502 ⎬	Rather insensitive to substituents
17 ArCH$_2$COPh + Br$_2$(H$^+$), AcOH	−0·22 ⎭	
18 ArCH$_2$CH$_2$I + EtO$^-$ → ArCH=CH$_2$	2·07	α-H$^+$ removal by EtO$^-$
19 ArCH$_3$ + ˙CCl$_3$ → ArCH$_2$ + CHCl$_3$	−1·46	H-atom transfer responds to electronic effect
20 ArCN + H$_2$S → ArCH(SH)NH$_2$	+2·14	Addition at cyano group by a nucleophile

of the perturbation than is the case for acid dissociation. The parameter ρ measures the ability of the core to transmit electronic effects; for example for the acid ionization of the compounds indicated:

K_A of:	ArCOOH	ArCH$_2$COOH	ArCH$_2$CH$_2$COOH	ArCH=CHCOOH	ArC≡CCOOH
ρ:	1·000	0·489	0·212	0·466	1·1

Table 4.3 sets out some typical values of ρ with comments; many more will be quoted in the later text.

4.7 DEVIATIONS FROM THE HAMMETT EQUATION

The Hammett equation, like all linear free-energy relationships, is an expression of similarity of behavior among two or more sets of reactions. As much can be learned of mechanism from reactions which fail to conform as from those which do. Several types of non-linear behavior between rates and σ-values may be experienced.

4.7.1 Random deviations

Apart from experimental error in k-values, one may expect experimental points to lie off the best straight line by some 10–15%, this being the limit to which the relationship holds on account of minor variations in solvation and other extraneous factors. Conversely, this is the limit of reliablility for prediction of an unknown rate constant from a known σ and ρ. Highly random plots, noted in some radical reactions, are an indication that polar factors are in no way controlling reaction rates.

4.7.2 Mechanistic change

Occasionally the energetically-preferred mechanism of a reaction changes upon altering a substituent. This is often the case if two pathways of similar activation energy are available and have very different electronic demands. The result will be a sudden change in the slope of the Hammett plot (Fig. 4.5)[39,40]. A change in the rate-determining step of a multi-step reaction will also result in a discontinuity, or even a sign reversal, of ρ (Fig. 4.6)[41,42]. Non-linear plots may be found in reactions which take place by two concurrent pathways, when the proportion of the two will differ according to the substituent present. Sometimes the presence of a certain substituent makes possible a unique reaction pathway. The reaction of cumyl chlorides, **11** (R = Me), with nucleophiles such as thiophenoxide is greatly retarded by

$$\text{Ar—CR}_2\text{—Cl} \xrightarrow{\text{slow}} (\text{Ar—}\overset{+}{\text{C}}\text{R}_2\bar{\text{C}}\text{l}) \xrightarrow[\text{Nu}^-:]{} \text{products}$$

11

146

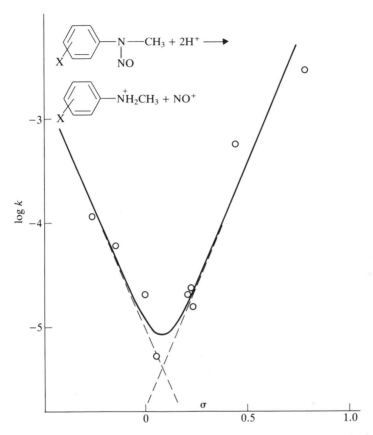

Fig. 4.5 Deviation from linearity of a Hammett plot due to a change in mechanism.

electron-withdrawing groups. Nitro, however, is an exception and a facile displacement is observed against expectations[43]. It appears that *p*-nitrocumyl chloride prefers to react by way of a radical chain process (Section 15.3.7) and therefore its rate is not comparable with those of its

$$PhS^- \curvearrowright O_2N \overset{Cl}{\underset{}{\longrightarrow}} C \longrightarrow \left(PhS^\cdot \ O_2\bar{N} \overset{Cl}{\underset{}{\longrightarrow}} C \right) \longrightarrow products$$

12

$$ArCH{=}NR \underset{(\leftarrow)}{\overset{H^+, k_1}{\longleftarrow}} Ar\overset{+}{C}H{-}NHR \xrightarrow[H_2O]{k_2} ArCHO + RNH_2$$

$$HO^- \curvearrowright HO \overset{O}{\underset{OAr}{\longrightarrow}} C \longrightarrow HOH \ O{=}\langle \rangle{=}C{=}O + \bar{O}Ar$$

13

147

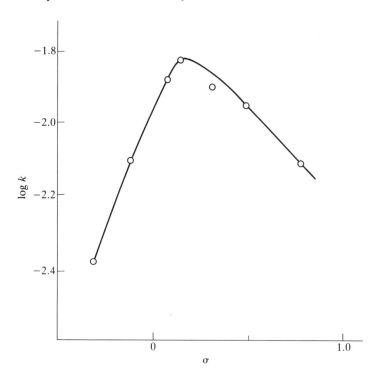

Fig. 4.6 Deviation from linearity of a Hammett plot due to a change in the rate-controlling step; the reaction is the hydrolysis of an imine; electron-donating groups assist step 1 making step 2 slow but electron-withdrawing groups ultimately increase k_2 until it is no longer rate-determining[41,42].

analogues, **12**. The hydrolysis of *p*-hydroxybenzoates, **13**, is unusually facile owing to the possibility of proton removal as part of the driving force[44].

4.7.3 Enhanced resonance

The solvolyses of tertiary halides, **11**, which occur by rate-determining ionization, give linear Hammett plots only for deactivating $-R$ and $-I$ *para* substituents and for other types only when in the *meta* position. Lone-pair and alkyl groups which exert $+R$ effects impart greater reactivity than expected from their values of σ so the plot becomes increasingly curved (Fig. 4.7)[45–47]. The reason lies in the nature of the transition state, in which a vacant *p*-orbital is developing adjacent to the aryl ring. An especially strong conjugative interaction with lone-pair substituents in the *para* position therefore develops, much greater than that measured in dissociations of

148

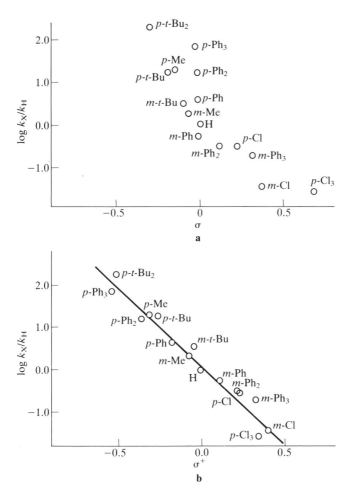

Fig. 4.7 **a**, Hammett plot of the ionization of Ar_3C—Cl in liquid SO_2. **b**, Brown–Okamoto plot of the same reaction.

benzoic acids, and a departure from linearity is the result. This means that the $+R$ effect depends not only on the substituent but also on the reaction. The remedy adopted by Brown and Okamoto[47] was to define a modified LFER applicable to this situation:

$$\log k_X/k_H = \rho^+\sigma^+, \tag{4.3}$$

where σ^+ are a new set of substituent constants expressing enhanced resonance properties (Table 4.1). The scale was defined for the reaction above (cumyl chlorides; **11**. R = Me) solvolyzing at 25°C in aqueous acetone.

149

Table 4.4 Some reactions which obey the Brown–Okamoto equation, Eq. (4.3)
(See also Table 10.7.)

	ρ^+	Ref.
Ionizations		
$ArCMe_2$—Cl (90% aq. acetone, 25°C)	−4·45	47
$ArCMe_2$—Cl (EtOH, 25°C)	−4·67	48
$ArCPh_2$—Cl (EtOH/Et_2O)	−2·57	49
$ArCPh_2$—Cl (liquid SO_2, 0°C)	−3·73	50
$ArCPh_2$—OH (H_2SO_4, 25°C)	−3·64	51
$ArCMe_2$—ODnb (solvent-assisted)	−2·47	52
$ArCMe_2$—ODnb (neighboring-group-assisted)	−7·07	
$ArCOOH + H^+ \rightarrow Ar\overset{+}{C}(OH)_2$	−1·1	49
Electrophilic attack at a π-system		
$ArH + ketene/AlCl_3$	−6·6	53
$ArH + Br_2$ (liquid SO_2)	−9·05	54
$ArCH(OH)—CH{=}CHMe + H^+$ (rearrangement)	−2·97	55
$ArC{\equiv}CH + Cl_2$	−4·19	56
$ArCHROH$ oxidation	−2·14	57
$Ar_2C(OH)N_3$ rearrangement	−2·69	58
$ArCH{=}CH_2$ hydroboration	−0·49	59
Rotation of protonated bithiophenes	+4·8	48
$Ar—\overset{+}{N}{\equiv}N + N_3^-$	+3·7	61
$ArCHO + CH_2(CN)_2$	+1·45	60

	ρ^-	Ref.
Reactions correlating with σ^-		
Acid dissociation of ArOH	+2·23	61
Acid dissociation of $ArNH_3$	+2·89	62
$ArCl + \bar{O}Me \rightarrow ArOMe + Cl^-$ (MeOH, 50°C)	+8·5	63
$ArS^- + PhC{\equiv}C{\cdot}COOEt$	−0·83	63

The linear Hammett plot for *meta* substituents ($\rho = -4·54$) was constructed and the σ^+-values fitted to this line. Equation (4.3), rather than (4.2), should apply to all reactions in which strong acceptor character (in most cases, as here, a carbocation) develops at the ring. Table 4.4 lists reactions which obey the Brown–Okamoto equation and must be of this mechanistic type. They mostly have large negative values of ρ^+ but situations occur in which the reagent is stabilized by strong resonance interactions to a greater extent than is the transition state. In such cases, correlation of rate constants with σ^+ is still found, but with ρ^+ positive. An example is the coordination of a nucleophile to a diazonium ion:

$$Ar—\overset{+}{N}{\equiv}N + N_3^- \rightarrow Ar—N{=}N—N_3; \qquad \rho^+ = +3·73$$

and is, of course, the reverse of the ionization type of reaction by which σ^+ is defined.

An analogous departure from Hammett behavior occurs when a strong

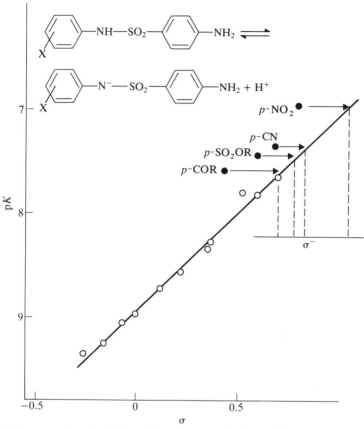

Fig. 4.8 Hammett plot of acid dissociation constants of sulfonamides

$$p\text{-}X\text{—}C_6H_4\text{—}NHSO_2Ph + B \rightleftarrows p\text{-}X\text{—}C_6H_4\bar{N}\text{—}SO_2Ph + BH^+$$

showing deviant points of $-R$ substituents and their fitting to the σ^--scale.

donor center contiguous to the aromatic ring is present as a reactant or generated as a product (**12**). In this case, extra strong conjugation is possible with $-R$ substituents so that non-linear plots are obtained for substituents such as p-nitro and p-cyano. Examples of this behavior occur with many reactions of phenols and anilines (Fig. 4.8). p-Nitrophenol, **14**, for example, is a far stronger acid and *p*-nitroaniline a far weaker base than would be expected from the value of $\sigma^0_{p\text{-NO}_2}$, due to stabilization of the respective conjugate bases by resonance. A scale of substituent constants may be constructed analogously to the σ^+ scale such that reactions of this class obey Eq. (4.4):

$$\log k_X/k_H = \rho^- \sigma^-. \tag{4.4}$$

Values of the nucleophilic substituent constants, σ^-, are given in Table 4.1, and examples of reactions which obey Eq. (4.4) in Table 4.4.

4.7.4 Variable resonance interactions

The correlation of rates with σ^+ and σ^- instead of with σ^0 is an indication that the resonance contribution to a substituent effect is variable and responds more effectively than inductive influences to the electronic demands of a reaction[64]. For practical reasons, the number of substituent scales must be limited and it is for most purposes sufficient to define three types. Then rate behavior may be judged to be 'most like' that of benzoic acid dissociations (σ^0), aniline dissociations (σ^-) or cumyl chloride solvolysis (σ^+). Correlations between or even outside these cases is not

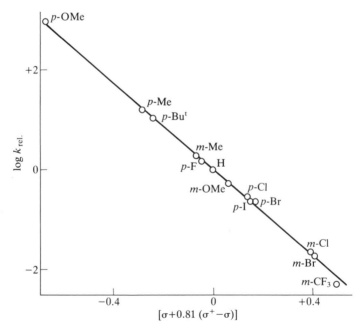

Fig. 4.9 Yukawa–Tsuno plot for the reaction[97]

$$\text{ArC}\!\equiv\!\text{CH} + \text{H}_2\text{O} \xrightarrow{\text{(H}^+)} \text{ArCO}\!-\!\text{CH}_3.$$

Table 4.5   Reaction parameters fitting the Yukawa–Tsuno equation (Eq. (4.5))

Substrate	Reaction	ρ	r
ArCOOH	Ionization (water, 25°C)	1·000	0
ArCMe$_2$Cl	Hydrolysis (90% aq. acetone)	−4·52	1·00
Ar$_2$CHCl	Methanolysis (25°C)	−4·02	1·23
ArN=NPh	Protonation (20% aq. ethanol)	−2·29	0·85
ArH	Nitration (HNO$_3$, in MeCN)	−6·38	0·90
ArH	Bromination (Br$_2$, AcOH)	−10·6	1·2
ArSiMe$_3$	Protodesilylation (H$^+$)	−5·7	0·70
ArSiMe$_3$	Bromodesilylation (Br$_2$)	−6·8	0·79
ArGeMe$_3$	Protodegermylation (H$^+$)	−4·4	0·62
ArPbMe$_3$	Protodeplumbylation (H$^+$)	−2·42	0·63
ArMeCN=NOH	Beckmann reagent (H$^+$)	−1·98	0·43
Ar$_2$N$_2$	Esterification by PhCOOH	−1·57	0·19
ArCOCHN$_2$	Acid-catalyzed decomposition	−0·82	0·56
ArC≡CH	Acid-catalyzed hydration	−4·3	0·81

ruled out. For example, the brominolysis of benzeneboronic acids,

$$Ar—B(OH)_2 + Br_2 \rightarrow Ar—Br + B(OH)_3 + HBr,$$

is so powerfully accelerated by p-OMe that even σ^+ is insufficient to express its effect.

A flexible method of introducing such variability into the resonance effect can be accomplished by the addition of a further parameter. The Yukawa–Tsuno equation, Eq. (4.5), expresses the linear free-energy relationship as containing contributions from 'normal' and 'enhanced' resonance effects[65].

$$\log k_X/k_H = \rho(\sigma^0 + r(\sigma^+ - \sigma^0)). \tag{4.5}$$

Hardly surprisingly, better correlations may be obtained by the adjustment of r than from either simple Hammett or Brown–Okamoto plots (Fig. 4.9, Table 4.5). The constant r expresses the degree of enhanced resonance interaction in relation to benzoic acid dissociations ($r = 0$) and cumyl chloride hydrolysis ($r = 1$). This in turn may be related mechanistically to the extent of development of a vacant orbital at the reaction center in the transition state.

4.8   DUAL PARAMETER CORRELATIONS; THE FLOWERING OF LFER

The evident success of the Hammett equation and the recognition of the need for a model of substituent effects which encompassed at least two physical phenomena—the I and R effects—prompted many workers to undertake a separation of σ-values into their supposed contributing components. If universal inductive (σ_I) and resonance (σ_R) components of σ could be assigned to each substituent, not only would their interaction mechanisms be clarified but the importance of the two effects in stabilizing the transition state could be assessed by the relative values of σ_I and σ_R. If

$$\sigma^0 = \sigma_I + \sigma_R \qquad (4.6)$$

a rate series would be fitted to a two-parameter equation such as Eq. (4.7)[8]:

$$\log k_X/k_H = \rho_I\sigma_I + \rho_R\sigma_R. \qquad (4.7)$$

The extent to which this approach has been successful will be discussed.

4.8.1 Inductive substituent constants

Model systems in which reaction center and substituent are separated by an aromatic ring as core, e.g. the Hammett system, permit a mixture of resonance and inductive interaction to be transmitted, the former only weakly from a *meta* position. The simplest model makes the assumptions

$$\sigma_p^0 = \sigma_I + \sigma_R \quad \text{and} \quad \sigma_m^0 = \sigma_I + \alpha\sigma_R,$$

from which, after assigning a value to α (0·33 has been used) elimination of σ_R is possible. Thus,

$$\sigma_I = (\sigma_m^0 - \alpha\sigma_p^0)/(1 - \alpha). \qquad (4.8)$$

Objections to this scale include the arbitrary value of α and the assumption of the same value of the inductive effect at *meta* and *para* positions.

A different model is required if a real separation of the effects is to be achieved and it is relatively easy to conceive systems in which inductive effects are important but resonance effects are not. This must be the case, for example, for reactions in which substituent and reaction center are separated by a saturated hydrocarbon core.

4.8.2 The Taft model[2,67,68]

Aliphatic esters can be hydrolyzed under either acidic or basic conditions; rates, k^A and k^B respectively, can then be measured for a series of substituted ethyl acetates, but neither gives a satisfactory linear Hammett plot.

$$X-CH_2COOEt \underset{OH^-,\, k^B}{\overset{H^+,\, k^A}{\rightleftharpoons}} X-COOH + EtOH$$

Now each of these reactions is subject to a combination not only of inductive and resonance effects, but also of changes in 'steric' influences of the substituent, E_s, which affect $\Delta S^{\ddagger}$. The latter may include direct hindrance of access to the reaction center, related to the size of X and changes in solvation in its vicinity; all three variables must now be included in an LFER:

$$\log k_X/k_H = \rho_I \sigma_I + \rho_R \sigma_R + E_s. \tag{4.9}$$

It was proposed by Ingold that electronic effects $(I + R)$ are slight in acid-catalyzed hydrolysis (the reasons are discussed in Section 10.3.2) but pronounced in the base-catalyzed reactions, i.e.

$$\log k_X^A/k_H^A = E_s. \tag{4.10}$$

Also, assuming that the —CH$_2$— group will not transmit resonance effects between the substituent and the reaction center (C=O),

$$\log k_X^B/k_H^B = \rho_I \sigma_I + E_s, \tag{4.11}$$

and it follows that

$$\log(k_X^A/k_H^A) - \log(k_X^B/k_H^B) = \rho_I \sigma_I = \rho^* \sigma^*, \tag{4.12}$$

which defines a new substituent constant σ^* for X, mainly of an inductive nature and applicable to reaction series with an aliphatic core. Now assigning $\rho^* = 2\cdot54$ (the reaction constant for basic hydrolysis of benzoate esters, ArCOOEt, from the Hammett equation) permits σ^* to be evaluated (Table 4.6). Equation (4.13) is known as the Taft–Ingold equation, for which Me (rather than H) is the reference substituent:

$$\log k_X/k_{Me} = \rho^* \sigma^* (+\rho_s E_s). \tag{4.13}$$

It gives tolerably linear fits to many aliphatic reactions (Table 4.7). The steric constants, E_s, may be evaluated from Eq. (4.10) and may be incorporated into the correlation equation to improve the fit (Table 8.2). Still, there is usually a far greater scatter of points than with a Hammett plot and it seems clear that σ^* values contain a contribution from 'steric' effects while E_s values themselves contain inductive contributions. The model, in fact, has not produced a complete separation of the two.

Swain and Lupton[2,69,70] measured acid dissociation constants of bicyclo-octanecarboxylic acids, **15**, and other aliphatic systems in which resonance

Table 4.6   Taft inductive substituent constants, σ^{*}[61]

X	$\sigma^*(X)$	$\sigma^*(CH_2X)$	X	$\sigma^*(X)$	$\sigma^*(CH_2X)$
Me	0·00	−0·10	OH		0·555
Et	−0·10	−0·10	OMe	−0·22	+0·52
n-Pr	−0·10	−0·13	OPh		+0·85
n-Bu	−0·13		OEt		
isoPr	−0·19		F		+1·1
IsoBu	−0·125		Cl		+1·05
tert-Bu	−0·300	−0·165	Br		+1·00
neoPe	−0·165		I		+0·85
Ph	+0·60	+0·215	NO$_2$		+1·40
PhCH$_2$	+0·215	+0·08	CN		+1·30
PhCH$_2$CH$_2$	+0·08		CF$_3$		+0·92
			CH$_3$CO		+0·60

Table 4.7 Some reactions correlating with the Taft–Ingold equation, Eq. (4.13)

	Reaction[a]	ρ^*	*log ks*$^{-1}$
1	Alkaline hydrolysis of esters, X—COOEt	+2·48	
2	Alkaline hydrolysis of esters, X—COOEt	+0·97	−4·536
3	Ionization of acids, X—CH$_2$COOH	+1·72	−4·65
4	Hydrolysis of acetals, X—CH(OEt)$_2$	−3·65	−0·73
5	Esterification, X—CH$_2$COOH + Ph$_2$N$_2$	+1·17	−1·94
6	H$^+$ exchange, X—CH$_2$OH + isoPrO$^-$	+1·36	−0·07
7	Decomposition of NH$_2$NO$_2$ by SCH$_2$COO$^-$	−1·42	−0·37
8	Bromination of PhCOCHXX′ (Br$_2$, H$_2$O)	+1·59	−4·63
9	Acidic hydrolysis of X—CH—CH$_2$	−1·83	−2·52
	O		
10	Acetolysis of XX′CHOBros	−3·49	
11	Ethanolysis of XX′X″C—Cl	−3·29	
12	Ethanolysis of XCH$_2$OTos	−0·74	

[a] For identities of X, see Table 4.6.

effects could be discounted, to define an inductive (either field or 'through-bond') substituent constant, $\mathscr{F}$:

$$\log K_X/K_X = \mathscr{F}. \tag{4.14}$$

They then showed that the various values of σ could be expressed consistently as a combination of $\mathscr{F}$ and a resonance substituent parameter, $\mathscr{R}$:

$$\sigma = a\mathscr{F} + b\mathscr{R}. \tag{4.15}$$

$\mathscr{F}$ and $\mathscr{R}$ (Table 4.8) have the same connotation as ρ_I and ρ_R so can be used in the regression equation (4.16) to correlate rates of many chemical and biochemical reactions and spectroscopic data:

$$\log k_X/k_H = f\mathscr{F} + r\mathscr{R}, \tag{4.16}$$

in which f, r are the susceptibility constants of the reaction. The Swain–Lupton analysis shows the contributions of $\mathscr{F}$ and $\mathscr{R}$ which make up the common σ-values:

$$\sigma_m^0 = 0·60\mathscr{F} + 0·27\mathscr{R};$$
$$\sigma_p^0 = 0·56\mathscr{F} + 1·00\mathscr{R};$$
$$\sigma^+ = 0·51\mathscr{F} + 1·59\mathscr{R};$$
$$\sigma^- = 0·75\mathscr{F} + 1·52\mathscr{R}.$$

4.8.3 Other chemical model systems: modern σ_I and σ_R scales

Almost any rigid aliphatic system may be made the basis of an estimate of σ_I. Dissociations of aliphatic carboxylic acids of the series **15–19** have been

Table 4.8 Substituent parameters for use with the Swain–Lupton equation, Eq. (4.16)[70]

X	$\mathscr{F}$	$\mathscr{R}$
$NHCOCH_3$	$0\cdot77 \pm 0\cdot07$	$-1\cdot43 \pm 0\cdot17$
$OCOCH_3$	$0\cdot70 \pm 0\cdot08$	$-0\cdot04 \pm 0\cdot25$
$COCH_3$	$0\cdot50 \pm 0\cdot05$	$0\cdot90 \pm 0\cdot12$
$SCOCH_3$	$0\cdot53 \pm 0\cdot08$	$0\cdot68 \pm 0\cdot24$
NH_2	$0\cdot38 \pm 0\cdot08$	$-2\cdot52 \pm 0\cdot23$
Br	$0\cdot72 \pm 0\cdot03$	$-0\cdot18 \pm 0\cdot07$
$O(CH_2)_3CH_3$	$0\cdot72 \pm 0\cdot10$	$-2\cdot16 \pm 0\cdot32$
$C(CH_3)_3$	$-0\cdot11 \pm 0\cdot05$	$-0\cdot29 \pm 0\cdot11$
CO_2H	$0\cdot44 \pm 0\cdot03$	$0\cdot66 \pm 0\cdot08$
CO_2^-	$-0\cdot27 \pm 0\cdot03$	$0\cdot40 \pm 0\cdot08$
Cl	$0\cdot72 \pm 0\cdot03$	$-0\cdot24 \pm 0\cdot08$
CN	$0\cdot90 \pm 0\cdot03$	$0\cdot71 \pm 0\cdot07$
N_2^+	$2\cdot36 \pm 0\cdot10$	$2\cdot81 \pm 0\cdot27$
$S(CH_3)_2^+$	$1\cdot62 \pm 0\cdot08$	$0\cdot52 \pm 0\cdot13$
OC_2H_5	$0\cdot61 \pm 0\cdot10$	$-1\cdot72 \pm 0\cdot30$
$CO_2C_2H_5$	$0\cdot47 \pm 0\cdot02$	$0\cdot67 \pm 0\cdot07$
C_2H_5	$-0\cdot02 \pm 0\cdot03$	$-0\cdot44 \pm 0\cdot10$
F	$0\cdot74 \pm 0\cdot06$	$-0\cdot60 \pm 0\cdot12$
H	$0\cdot00 \pm 0\cdot00$	$0\cdot00 \pm 0\cdot00$
OH	$0\cdot46 \pm 0\cdot04$	$-1\cdot89 \pm 0\cdot17$
I	$0\cdot65 \pm 0\cdot06$	$-0\cdot12 \pm 0\cdot11$
IO_2	$0\cdot99 \pm 0\cdot08$	$0\cdot99 \pm 0\cdot25$
SH	$0\cdot52 \pm 0\cdot09$	$-0\cdot26 \pm 0\cdot25$
OCH_3	$0\cdot54 \pm 0\cdot03$	$-1\cdot68 \pm 0\cdot16$
CH_3	$-0\cdot01 \pm 0\cdot03$	$-0\cdot41 \pm 0\cdot08$
$SeCH_3$	$0\cdot28 \pm 0\cdot08$	$-0\cdot39 \pm 0\cdot26$
$SOCH_3$	$0\cdot80 \pm 0\cdot05$	$0\cdot45 \pm 0\cdot11$
SO_2CH_3	$0\cdot88 \pm 0\cdot05$	$0\cdot85 \pm 0\cdot12$
SCH_3	$0\cdot68 \pm 0\cdot07$	$-1\cdot30 \pm 0\cdot16$
NO_2	$1\cdot00 \pm 0\cdot00$	$1\cdot00 \pm 0\cdot00$
$O(CH_2)_4CH_3$	$0\cdot75 \pm 0\cdot10$	$-2\cdot27 \pm 0\cdot33$
OC_6H_5	$0\cdot76 \pm 0\cdot07$	$-1\cdot29 \pm 0\cdot15$
C_6H_5	$0\cdot25 \pm 0\cdot05$	$-0\cdot37 \pm 0\cdot11$
HPO_3^-	$0\cdot22 \pm 0\cdot09$	$0\cdot58 \pm 0\cdot27$
$OCH(CH_3)_2$	$0\cdot90 \pm 0\cdot12$	$-2\cdot88 \pm 0\cdot37$
$O(CH_2)_2CH_3$	$0\cdot63 \pm 0\cdot10$	$-1\cdot77 \pm 0\cdot30$
$CH_2Si(CH_3)_3$	$-0\cdot19 \pm 0\cdot08$	$-0\cdot32 \pm 0\cdot26$
SO_2NH_2	$0\cdot55 \pm 0\cdot09$	$1\cdot07 \pm 0\cdot27$
SO_3^-	$-0\cdot05 \pm 0\cdot06$	$0\cdot53 \pm 0\cdot12$
CF_3	$0\cdot64 \pm 0\cdot03$	$0\cdot76 \pm 0\cdot08$
$N(CH_3)_3^+$	$1\cdot54 \pm 0\cdot05$	$0\cdot00 \pm 0\cdot00$
$Si(CH_3)_3$	$-0\cdot10 \pm 0\cdot06$	$0\cdot16 \pm 0\cdot11$
$N(CH_3)_2$	$0\cdot69 \pm 0\cdot13$	$-3\cdot81 \pm 0\cdot42$

variously employed; of these perhaps the bicyclo-octanecarboxylic acids, **15**, and quinuclidines, **16**, provide the best data[71,76]. Clearly, resonance effects are minimal and the geometry of each member similar and rigid to maintain similar 'steric' effects. Moreover the compounds are easier to synthesize than the corresponding cubane derivatives, **17**. In all cases, σ_I and σ_R may be defined according to Eqs. (4.17)–(4.20):

$$\log K_X/K_H = \alpha\sigma_I. \tag{4.17}$$

COOH · COOH COOH

15 16 17 18 19

X X X CH₂X X—CH₂COOH

Each σ_I then needs to be scaled to fit in with σ^0 values. Essentially the same inductive constants result from the application of Eq. (4.17) to acidic dissociation of **15** ($\alpha = 1.464$), **16** ($\alpha = 0.207$) or **17** ($\alpha = 3.95$) or from Eq. (4.18) which uses the ^{19}F-NMR chemical shifts of 3-substituted fluorobenzenes to define inductive effects[77]:

$$\sigma_I = \delta_X/3.71 + 0.084. \qquad (4.18)$$

Minor disagreements between scales can be averaged out to produce a 'universal' scale of inductive substituent constants (Table 4.9)[17]. Resonance substituent constants, σ_R, may then be obtained from Eqs (4.19) and (4.20):

$$\sigma_p = \sigma_I + \sigma_R, \qquad (4.19)$$

$$\sigma_m = \sigma_I + \alpha\sigma_R, \qquad (4.20)$$

in which σ may be σ^0, σ^+, σ^- and α is a scaling constant. Values of σ_R^0 are applicable in the absence of direct conjugation between substituent and

Table 4.9 Substituent constants for dual-parameter correlations, Eq. (4.21)[17]

X	σ_I	σ_R^0	$\sigma_R(BA)$	σ_R^+	σ_R^-	σ_X
NMe₂	0·05	−0·52	−0·83	−1·75	−0·34	0·34
NH₂	0·10	−0·48	−0·82	−1·61	−0·48	0·33
NHAc	0·26	−0·22	−0·36	−0·86		
OPh			−0·55	−0·87		
OH	0·27	−0·44				0·43
OMe	0·26	−0·41	−0·61	−1·02	−0·45	0·44
SMe	0·19	−0·17	−0·32		−0·14	0·10
Me	−0·05	−0·10	−0·11	−1·25	−0·11	0·17
Ph	0·10	−0·11	−0·11	−0·3	−0·04	
H	0·0	0·0	0·0	0·0	0·0	0·0
F	0·51	−0·34	−0·45	−0·57	−0·45	0·52
Cl	0·47	−0·21	−0·23	−0·23	−0·35	0·28
Br	0·45	−0·16	−0·19	−0·19	−0·30	
I	0·39	−0·12	−0·16	−0·11	−0·25	
SOMe	0·25	0·0	0·0		0·0	
COMe	0·28	0·19	0·16	0·16	0·47	0·14
COOR	0·31	0·15	0·14	0·14	0·34	0·19
SO₂Me	0·64	0·12	0·12	0·08	0·17	
CN	0·52	0·14	0·13	0·13	0·33	0·31
NO₂	0·64	0·19	0·15	0·15	0·46	0·40
CF₃	0·41	0·13	0·08	0·08	0·17	0·17
SiH₃						−0·13

reaction center, the type of system which would fit the Hammett equation. Since resonance effects exhibited by a substituent do not respond linearly to electron demand, a feature missing in the Swain–Lupton analysis, there is a need for additional σ_R scales to cope with enhanced resonance in the dual-parameter approach. Resonance scales denoted σ_R^+ and σ_R^- for use with acceptor and donor reaction centers are generated from Eq. (4.19), replacing σ_p by σ^+ and σ^- respectively. A fourth scale, $\sigma_R(BA)$, is also advocated for 'moderate' resonance situations similar to that encountered in benzoic acid (BA) dissociations[68] (Table 4.9). To obtain dual-parameter correlations, it is necessary to fit rate or equilibrium data for a reaction series to Eq. (4.21) using values of σ_I and whichever resonance scale gives the best fit:

$$\log k_X/k_H = \rho_I\sigma_I + \rho_R\sigma_R. \tag{4.21}$$

To summarize:

Scale	σ_R^0	$\sigma_R(BA)$	σ_R^+	σ_R^-
Applicability:				
Resonance interaction at reaction center:	No direct resonance	Weak acceptor	Strong acceptor	Strong donor

Examples of dual-parameter correlations are given in Table 4.10. Two reaction constants for each are obtained, characterizing the sensitivity of their rates to inductive influences and to mesomeric effects. In most cases, the two are of comparable magnitude, i.e. $\rho_R/\rho_I = \lambda \sim 1$, but in some instances I or R effects may dominate (e.g. entries 7, 14 in Table 4.10) or the two reaction constants may even be of opposite sign. An example of the latter behavior is found in the alkaline decomposition of aryldiazonium ions:

$$Ar\!-\!\overset{+}{N_2} \xrightarrow{\ slow\ } Ar^+ + N_2 \rightarrow ArOH.$$

This reaction fails to give a satisfactory correlation with any version of the Hammett equation, but from Eq. (4.21) an adequate fit is obtained, with $\rho_I = -4{\cdot}09$, $\rho_R = +2{\cdot}72$. A single-parameter LFER cannot cope with a reaction having such unusual characteristics (however, it must be admitted that the physical interpretation of this correlation is not clear).

Table 4.10 Examples of reactions correlated by the dual-parameter equation, Eq. (4.21)[2,68,78–84,94]

(a) Reactions correlating with σ_R^0

Reaction	ρ_I	ρ_R^0	λ	$\log k_H$
1 Ionization of ArCH$_2$COOH (water)	0·484	0·433	0·90	0·481
2 Ionization of ArCH$_2$COOH (aq. ethanol)	0·755	0·646	0·86	0·562
3 Ionization of ArCH$_2$NH$_3^+$ (water)	1·082	1·057	0·98	0·908
4 Alkaline hydrolysis of ArCH$_2$COOEt	1·173	0·79	0·67	0·813
5 Alkaline hydrolysis of CH$_3$COOAr	0·714	0·718	1·01	0·732
6 Displacement of Cl by MeO$^-$ in ArCH$_2$Cl	5·961	−8·24	1·38	5·545
7 ^{19}F-NMR chemical shift in ArF	−9·02	−31·2	3·46	−15·0

(b) reactions correlating with σ_R (BA)

Reaction	ρ_I	ρ_R	λ	$\log k_H$
8 Ionization of ArCOOH (dioxane–water)	1·426	1·153	0·81	1·453
9 Ionization of ArCOOH (methanol)	1·358	1·139	0·84	1·382
10 Ionization of ArSO$_2$NHPh (water)	1·140	0·943	0·83	1·062
11 Ionization of ArSe(OH)$_2$ (water)	1·068	0·812	0·76	0·827
12 Ionization of ArC≡C—COOH (aq. ethanol)	0·694	0·664	0·96	0·712
13 Alkaline hydrolysis of ArCOOEt	2·602	2·307	0·89	2·593
14 Transesterification, ArCOOC$_{10}$H$_{19}$, MeO$^-$, methanol	2·578	2·212	0·86	2·608
15 Hydrolysis ArS—CH$_2$Cl	−2·576	−2·302	0·89	−2·618
16 Pyrolysis of ArCOOEt (515°C)	0·215	0·112	0·96	1·361
17 Hydrolysis of (ArCO)$_2$O	3·107	2·925	0·94	1·361
18 ^{19}F-NMR chemical shift, ArCH$_2$F	9·297	16·811	1·81	11·28

(c) Reactions correlating with σ_R^+

Reaction	ρ_I	ρ_R	λ	$\log k_H$
19 Ionization of Ar$_3$C—OH (H$_2$SO$_4$)	−11·10	−9·79	0·88	−9·861
20 Ar—N$_2$ → Ar$^+$ + N$_2$	−4·09	+2·72	−0·66	

(d) Reactions of vinyl compounds

Compound	Reaction	ρ_I	ρ_R	s
trans series				
X, COOH	pK$_A$	−2·26	−2·09	0
X, COOEt	OH$^-$, hydrolysis	+2·83	+4·88	0
X, CH$_2$Cl	Ethanolysis	−5·19	−8·80	0
X, CH$_2$Cl	^{13}C chemical shift	11·9	63·5	0

Table 4.10   (*Continued*)
(d) Reactions of vinyl compounds

Compound	Reaction	ρ_I	ρ_R	s
cis series				
X—CH=CH—COOH	pK_A	−2.28	−1·71	−0·313
X—CH=CH—COOMe	OH⁻, hydrolysis	2·81	1·91	−1·68
X—CH=CH—COOH	+ PhCN₂	1·98	1·97	−0·235
geminal series				
X, COOH (C=CH₂)	pK_A	−4·14	−0·873	0
Ph, X, COOH	pK_A	−3·81	−1·22	0
X, CH₂Cl (C=CH₂)	+I⁻	−0·845	−1·23	0
Reactions at π-bond				
Me—CH=CH—X	+Ag⁺	−3·6	−5·21	0
X (CH₂=CH—X)	+Br—Br/AcOH	−13·0	−3·56	0
Me—CH=CH—X	+H—OCOCF₃	−8·48	−13·7	0

(e) Reactions of *ortho* substituted aryl compounds

Compound	Reaction	ρ_I	ρ_R	s
o-X—ArCOOH	pK_A	−2·41	−2·31	0
o-X—pyH⁺	pK_A	−11·4	−2·30	0
o-X—ArOCH₂COOH	pK_A	−0·386	−0·210	0
o-X—ArCOOMe + OH⁻	Hydrolysis	2·71	0·432	0
o-X—ArCOOH	+ CH₂N₂	1·58	1·05	0
o-X—Ar—CH—CH₃ \\ OAc	Thermal elimination	−1·35	−1·02	0
o-X—Ar—COOMe	+ Me₃N	1·44	1·17	0

Evaluation of dual-parameter reaction constants, ρ_I and ρ_R

Correlation with two independent variables as represented by Eqs (4.16) and (4.21) requires computer analysis of the data; this is usually a routine matter of using a standard multi-correlation analysis program, usually to be found among standard software for any available machine[85,86]. The two-parameter equation is not suitable for graphical display, which is such a useful feature of the Hammett equation, but this may be achieved by means of the following transformations:

$$\log k_X/k_H = \rho_I \sigma_I + \rho_R \sigma_R = \bar{\sigma}\bar{\rho} \tag{4.22}$$

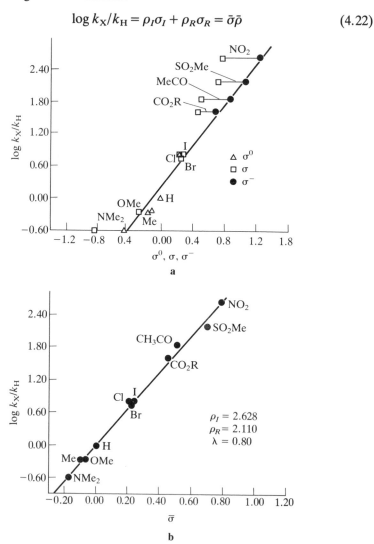

Fig. 4.10 **a**, LFER plots of ionization equilibria of ArSH (EtOH, 25°C) using various σ-values. **b**, The same, plotted against $\bar{\sigma}$ according to Eq. (4.22).

so that

$$\bar{\sigma} = (\sigma_I + \lambda\sigma_R)/(1 + \langle\lambda\rangle).$$

Plotting $\log k_X/k_H$ against $\bar{\sigma}$ gives a linear plot, slope $= \bar{\rho}$, from which

$$\rho_I = \bar{\rho}/(1 + \langle\lambda\rangle) \quad \text{and} \quad \rho_R = \bar{\rho}\lambda/(1 - \langle\lambda\rangle),$$

where $\lambda = \rho_R/\rho_T$. Figure 4.10 gives examples of such a graph compared with a single-parameter plot for the same reaction. It must be stressed that the value of any correlation must be judged by proper statistical criteria of significance[1] (correlation coefficients, T- or F-tests, for example) or by the mathematical formalism of factor analysis[87]. The two-parameter approach, however, may be used to correlate effects of *ortho* substituents[88–91], and reactions through heterocyclic[92,93], olefinic[94] and aliphatic cores.

Computed substituent electronegativity parameters

As has been mentioned, the inductive parameter, σ_I, is an operational measure of the non-resonance effect and is almost certainly a blend of 'through-bond' (true inductive or electronegative) and 'through-space' (field) effects, the latter predominating. It is difficult to conceive of an experimental method capable of measuring the electronegativity effect alone, but recently a scale σ_χ (Table 4.9) has been proposed[95] derived from *ab initio* MO calculations, defined as

$$\sigma_\chi = (1 - q_H),$$

in which q_H is the excess electronic charge on hydrogen in compounds H—X and is a measure of the pure inductive effect of group X. The σ_χ scale differs significantly from the σ_I scale, the differences being related to the field effect. The field effect may be estimated theoretically by calculating the energy change, ΔE, for the system:

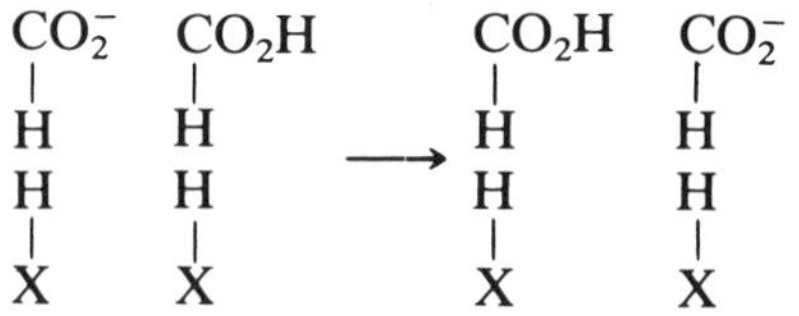

$$
\begin{array}{cc}
CO_2^- & CO_2H \\
| & | \\
H & H \\
H & H \\
| & | \\
X & X
\end{array}
\longrightarrow
\begin{array}{cc}
CO_2H & CO_2^- \\
| & | \\
H & H \\
H & H \\
| & | \\
X & X
\end{array}
\qquad (4.23)
$$

The molecules remain a fixed distance apart and the field effect of substituents, X, is transmitted only through space. Values of ΔE are given in Table 4.11[97].

4.9 MOLECULAR ORBITAL CONSIDERATIONS

Replacement of hydrogen by substituent groups has the effect of perturbing reactivity of a molecule by changing its strength as a nucleophile or electrophile. This in turn is attributable at least in part to a perturbation of the frontier orbitals in a systematic manner. One may examine qualitatively the effect of substitution upon the reactivity of benzene at the *para* position to electrophilic attack. Using heteroatom correction factors, δ_m and ρ_m (Table 1.2), the localization energies, L_+, are obtained for the conversion:

$$L_+ = \Delta E_\pi$$

	δ_M:	0	0·5	1·0	1·5	2·0	
[roughly corresponding to X:		CH_2^-	N	O		Cl]	H
	$E(HOMO)/\beta$:	0	$-0\cdot28$	$-0\cdot50$	$-0\cdot66$	$-0\cdot75$	
	L_+/β:	1·82	2·06	2·21	2·30	2·35	2·536
Coefficient^c in HOMO (ψ_4)	p-C:	0·377	0·397	0·443	0·489	0·520	
	o-C:	0·377	0·382	0·387	0·383	0·372	

Compared with benzene, the localization energy diminishes and hence reactivity should increase in the order

$$p\text{-Cl} < p\text{-OH} < p\text{-NH}_2 < p\text{-CH}_2^-$$

in accordance with observations. The HOMO energy also rises in this order, implying increasing nucleophilicity. The coefficients of the HOMO at the reactive positions (o- and p-) tend to diminish with increasing reactivity and so may be predicted to be of less importance, but their relative magnitudes increasingly converge with reactivity and may be a factor influencing the product ratio.

The origins of substituent constants may be probed by high-level MO theory. Energies of the species involved in the equilibrium[96] may be

calculated, and hence one obtains the overall energy change for the proton exchange, which may be equated with the difference in acidities of the substituted and unsubstituted acids, $\delta\Delta E$. This may be denoted a field/inductive effect parameter, F (Table 4.11).

A similar set of equilibria for proton transfer between anilines can also be computed, substituent effects now being a combination of the field, F, and resonance, R, effects which may then be separated. The values of F and R agree surprisingly well with σ_I and σ_R^0 values, indicating the essential correctness of this approach.

Ion cyclotron resonance (ICR; Section 6.9) permits equilibrium constants in the gas phase to be evaluated for reactions between ions and neutral

Table 4.11 Substituent effects from gas phase equilibria[96]

(a) Computed field/inductive effects, F^a
(from calculated $XCH_2CH_2NH_3^+ + CH_3CH_2NH_2 \rightleftarrows$
$XCH_2CH_2NH_2 + CH_3CH_2NH_3^+$; $\delta\Delta E$)

X	$\delta\Delta E/kcal\,mol^{-1}$
Me	+1·4
H	0
CH=CH$_2$	−0·5
NH$_2$	−2·5
C≡CH	−2·7
OH	−2·8
CF$_3$	−6·4
F	−6·4
CN	−11·2
NO$_2$	−14·9

(b) From ICR equilibria, Eq. (4.24)^a

X	$F/kcal\,mol^{-1}$	$P/kcal\,mol^{-1}$
tert-Bu	2·8	8·7
isoPr	1·7	6·8
n-Pr	1·3	5·7
Et	0·9	3·9
Me	0·0	0·0
CH$_2$Ph	−2·5	7·1
CH$_2$CHF$_2$	−8·7	4·1
CH$_2$CF$_3$	−13·1	2·6

(c) Calculated field effects using Eq. (4.23)^a

X	$\Delta E/kcal\,mol^{-1}$	X	$\Delta E/kcal\,mol^{-1}$
H	0·0	CN	6·31
NH$_2$	1·49	CF$_3$	5·00
NMe$_2$	1·59	CO·Me	2·63
OMe	3·34	COONe	2·59
F	5·51	NO$_2$	8·30
Me	−0·02	CHO	3·01

^{a} The original author's scale, in which energy values are given in kcal mol^{-1}, has been retained.

molecules, and is ideal for proton transfers. For example,

$$R\overset{+}{O}H_2 + MeOH \rightleftarrows ROH + Me\overset{+}{O}H_2 \qquad (A)$$

$$RO^- + MeOH \rightleftarrows ROH + MeO^- \qquad (B)$$

In aliphatic systems, resonance effects are unimportant and structural variation affects ΔG by field, F, and polarizability, P, changes in the structure R. The electronic demands of the two reactions are opposite but

Correlation of structure with reactivity

polarizability assists both. Hence

$$\Delta G(A) \approx F + P$$
$$\Delta G(B) \approx -F + P$$

and $\Delta G(A) - \Delta G(B) = 2F$, corresponding to

$$R\overset{+}{O}H_2 + MeO^- \rightleftarrows RO^- + Me\overset{+}{O}H_2; \qquad \Delta G = 2F \qquad (4.24)$$

from which $P = \Delta G(A) - F$. Values of the F and P constants are given in Table 4.11. Polarizability appears to be dominant for alkyl groups but field effects increase when electronegative substituents are present.

PROBLEMS

1 The following Table sets out values of the pK_A of halogen-substituted benzoic acids and phenols.

	Benzoic acid, 4·76			Phenol, 10·00		
	o-	*m-*	*p-*	*o-*	*m-*	*p-*
F	3·83	4·42	4·70	8·70	9·21	9·91
Cl	3·48	4·39	4·53	8·53	9·13	9·42
Br	3·41	4·37	4·49	8·54	9·03	9·36
I	3·42	4·41	4·46	8·51	9·06	9·30

Construct a Hammett plot from these values and examine the fit for *ortho*, *meta*, and *para* series. Calculate the reaction constant, ρ, and comment upon its magnitude.

2 Using the values above for monochlorophenols and the following pK_a values for polychlorophenols, examine the postulate that substituent effects are additive, i.e. $pK_a \propto \sum \sigma$.

2,3-Dichloro	7·70	2,3,4-Trichloro	7·59
2,4-Dichloro	7·85	2,3,5-Trichloro	7·23
2,5-Dichloro	7·51	2,3,6-Trichloro	6·12
		2,4,6-Trichloro	6·46
		3,4,5-Trichloro	7·74
2,3,4,5-Tetrachloro	5·22		
2,3,4,6-Tetrachloro	5·22		
2,3,5,6-Tetrachloro	5·44		

3 For the elimination reaction,

$$X{-}PhCH{=}N{-}Cl + OH^- \rightarrow X{-}Ph{-}C{\equiv}N + H_2O + Cl^-,$$

rate constants for different X were found to be:

X:	*p*-MeO	*p*-Me	H	*p*-Cl	*m*-Br	*p*-COOEt
$10^3 k/\text{M}^{-1}\,\text{s}^{-1}$:	4·31	8·43	17·3	100	117	168

Do these rates conform with the Hammett equation and, if so, what mechanistic interpretation may be placed upon them? Estimate the rate of reaction when $X = p\text{-}CF_3$.

4 The rate constants for bromination of p-bromophenol and its conjugate base are found to be as follows:

$$k = 3 \cdot 2 \times 10^{-3} \, M^{-1} s^{-1} \qquad 7 \cdot 8 \times 10^9 \, M^{-1} s^{-1}$$

Estimate σ^+ for the $-O^-$ substituent if the reaction is assumed to have $\rho^+ = -10$. [*J. Chem. Soc.*, 63 (1961); cf. *J. Chem. Soc., Perkin II*, 1801 (1984).]

5 Iodine does not react appreciably with substituted benzenes unless very highly activating substituents are present. It will react much more readily under electrolytic conditions:

$$ArH + I_2/MeCN \xrightarrow{\text{electrolysis}} ArI,$$

for which $\rho^+ = -6 \cdot 27$. Suggest a mechanism for this reaction.

6 The rates of chlorination of arylacetylenes,

$$X-C_6H_4C\equiv CH + Cl_2 \rightarrow X-C_6H_4CCl\!=\!CHCl$$

are sensitive to substituents:

X:	p-OMe	p-Me	p-F	H	p-Cl	p-Br	m-NO₂	p-NO₂
k:	19,500	190	14·9	10·6	4·15	2·81	0·0165	0·00325

Determine whether $\log k$ is better correlated with σ or with σ^+ and hence infer the nature of the transition state. [*J. Org. Chem.* **45,** 2377 (1980).]

REFERENCES

1 J. Shorter, *Correlation Analysis in Organic Chemistry*, Oxford U.P., 1973.
2 J. Shorter, 'Multiparameter Extensions to the Hammett Equation' in *Correlation Analysis in Chemistry*, (N.B. Chapman and J. Shorter, Eds), Plenum, London, 1978.
3 N. B. Chapman and J. Shorter (Eds), *Advances in Linear Free Energy Relationships*, Plenum, London, 1972.
4 P. Wells, *The Hammett Equation*, Academic Press, London, 1968.
5 H. H. Jaffe, *Chem. Rev.*, **53,** 191 (1953).
6 N. B. Chapman and J. Shorter (Eds), *Correlation Analysis in Chemistry*, Plenum, London, 1978.
7 S. Ehrenson, R. T. C. Brownlee and R. W. Taft, *Prog. Phys. Org. Chem.*, **10,** 1 (1973).
8 P. Wells, *Prog. Phys. Org. Chem.*, **6,** 111 (1968).
9 C. K. Ingold, *Structure and Mechanism in Organic Chemistry*, 2nd edn, Bell, London, 1969.
10 C. Hinshelwood and E. Tommila, *J. Chem. Soc.*, 1801 (1938).

Correlation of structure with reactivity

11a E. Kivinen and E. Tommila, *Suomen Kemist.*, **14B**, 7 (1941).
11b E. Tommila, L. Brehmer and H. Elo, *Ann. Acad. Sci. Fenn.*, **A59**, 9 (1942).
11c S. Tommila and E. Tommila, *Ann. Acad. Sci. Fenn.*, **A59**, 5 (1942).
11d E. Tommila, *Ann. Acad. Sci. Fenn.*, **A57**, 3 (1941).
12 L. P. Hammett, *Physical Organic Chemistry*, McGraw-Hill, New York, 1940; 2nd Edn, 1970.
13 L. P. Hammett, *J. Chem. Ed.*, **43**, 464 (1966).
14 L. P. Hammett, *Chem. Rev.*, **17**, 225 (1935).
15 G. N. Burckhardt, *Nature*, **136**, 684 (1935).
16 G. N. Burckhardt, W. G. K. Ford and E. Singleton, *J. Chem. Soc.*, 17 (1936).
17 O. Exner, Chapter 10 of Ref. 7.
18 C. K. Ingold and E. H. Ingold, *J. Chem. Soc.*, 1310 (1926).
19 Z. Friedl, H. Hapala and O. Exner, *Coll. Czech. Chem. Comm.*, **44**, 2928 (1979).
20 G. N. Lewis, *Valence and the Structure of Atoms and Molecules*, Chemical Catalog Co., New York, 1923.
21 F. R. Goss, C. K. Ingold and I. S. Wilson, *J. Chem. Soc.*, 2440 (1926); F. R. Goss, W. Hanhard and C. K. Ingold, *J. Chem. Soc.*, 250 (1927); C. K. Ingold and I. S. Wilson, *J. Chem. Soc.*, 810 (1927).
21a P. E. Peterson and C. Casey, *Tetrahedron Letts*, 1569 (1963).
22 A. Danieli, A. Ricci and J. H. Ridd, *J. Chem. Soc., Perkin II*, 1547, 2107 (1972).
23 R. Hoffmann, *Acc. Chem. Res.*, **4**, 1 (1971).
24 D. J. Hart and W. T. Ford, *J. Org. Chem.*, **39**, 363 (1974).
25 J. A. Pople and M. Gordon, *J. Amer. Chem. Soc.*, **89**, 4253 (1967).
26 R. D. Stolow, P. W. Samal and T. W. Giants, *J. Amer. Chem. Soc.*, **103**, 197 (1981).
27 D. J. Sardella, *J. Amer. Chem. Soc.*, **94**, 5206 (1972).
28 J. W. Verhoeven and P. Pasman, *Tetrahedron*, **37**, 943 (1981).
29 E. R. Vorpagel and A. Streitwieser, *J. Amer. Chem. Soc.*, **103**, 3777 (1981); R. Taft, *Prog. Phys. Org. Chem.*, **14**, 247 (1983).
30 R. T. Brownlee and D. J. Craig, *J. Chem. Soc., Perkin II*, 760 (1981).
31 M. Godfrey, *J. Chem. Soc. (B)*, 799 (1967).
32 M. Godfrey, *J. Chem. Soc. (B)*, 751 (1968).
33 C. Eaborn, R. Eastmond and D. R. M. Walton, *J. Chem. Soc. (B)*, 127 (1971).
34 G. A. Olah and P. v. R. Schleyer, *Carbonium Ions*, Interscience, New York, 1968.
35 N. S. Isaacs, *Tetrahedron*, **25**, 3555 (1969).
36 G. A. Olah, *Carbocations and Electrophilic Reactions*, Verlag Chemie, Weinheim, 1973.
37 L. S. Levitt and H. E. Widing, *Prog. Phys. Org. Chem.*, **12**, 119 (1976).
38 H. Schmidt and A. Schweig, *Angew. Chem. Int. Ed.*, **12**, 307 (1973).
39 B. A. Parai-Koshils, E. Yu. Belyaev and E. Shadovskii, *Reacts. Sposobnost. Org. Soedin.*, **1**, 10 (1964).
40 M. Bergon and J. P. Calmon, *Tetrahedron Letts*, **22**, 937 (1981).
41 H. Hart and E. A. Sedor, *J. Amer. Chem. Soc.*, **89**, 2342 (1967).
42 J. Hoffmann, J. Klicnar, V. Sterba and M. Vecera, *Coll. Czech. Chem. Comm.*, **35**, 1387 (1970).
43 H. Feuer, *Tetrahedron Suppl.* **1**, 107 (1967).
44 N. S. Isaacs and T. Najem, *Chem. Comm.*, 1361 (1984).
45 Y. Okamoto, T. Inukai and H. C. Brown, *J. Amer. Chem. Soc.*, **50**, 4972 (1958).
46 A. C. Nixon and G. E. K. Branch, *J. Amer. Chem. Soc.*, **58**, 492 (1936).
47 H. C. Brown and Y. Okamoto, *J. Amer. Chem. Soc.*, **80**, 4979 (1958).
48 D. A. Forsyth and D. E. Vogel, *J. Org. Chem.*, **44**, 3917 (1979).
49 C. D. Ritchie and P. O. I. Virtanen, *J. Amer. Chem. Soc.*, **94**, 1589 (1972).
50 N. Lichtin, *Prog. Phys. Org. Chem.*, **1**, 75 (1963).
51 N. C. Deno and A. Schreisheim, *J. Amer. Chem. Soc.*, **77**, 3051 (1955); N. C. Deno and W. C. Evans, *J. Amer. Chem. Soc.*, **79**, 5804 (1957).

52 H. C. Brown, C. G. Rao and M. Ravindranathan, *J. Amer. Chem. Soc.*, **100,** 7946 (1978).
53 K. R. Fountain, P. Heinze, D. Maddox, G. Gerhardt and P. John, *Can. J. Chem.*, **58,** 1939 (1980).
54 P. Catalanese and P. Villa, *Comptes Rend.*, **289,** 453 (1979).
55 E. A. Braude and E. S. Stern, *J. Chem. Soc.*, 1096 (1947).
56 K. Yates and A. T. Go, *J. Org. Chem.*, **45,** 2377 (1980).
57 J. Mukherjee and K. K. Banerji, *J. Chem. Soc., Perkin II*, 676 (1980).
58 S. N. Ege and K. W. Sherk, *J. Amer. Chem. Soc.*, **75,** 254 (1953).
59 L. C. Vishnakarma and A. Fry, *J. Org. Chem.*, **45,** 5306 (1980).
60 S. Patai and Y. Israeli, *J. Chem. Soc.*, 2020 (1960).
61 R. W. Taft, *Steric Effects in Organic Chemistry* (M. Newman, Ed.), Chapter 13, Wiley, New York, 1956.
62 C. K. Ingold, *J. Chem. Soc.*, 1032 (1930).
63 O. Exner, *Advances in LFER* (N. B. Chapman and J. Shorter, Eds), Plenum, London, 1972.
64 W. F. Reynolds, P. Dais, R. W. Taft and R. D. Topsom, *Tetrahedron Letts*, **22,** 1795 (1981).
65 Y. Tsuno, T. Ibata and Y. Yukawa, *Bull. Chem. Soc. Japan*, 32, 960, 965, 971 (1959); P. R. Young and W. P. Jencks, *J. Amer. Chem. Soc.*, **99,** 8238 (1977); **101,** 3288, 4678 (1979).
66 R. W. Bott, C. Eaborn, R. Baker and D. R. M. Walton, *J. Chem. Soc.*, 2139 (1963); 627 (1964); 384 (1965).
67 R. W. Taft, *J. Amer. Chem. Soc.*, **74,** 3120 (1952); **75,** 4231 (1953).
68 P. R. Wells, S. Ehrenson and R. W. Taft, *Prog. Phys. Org. Chem.*, **6,** 147 (1968).
69 C. G. Swain and E. C. Lupton, *J. Amer. Chem. Soc.*, **90,** 4328 (1968).
70 C. G. Swain, S. H. Unger, N. R. Rosenquist and M. S. Swain, *J. Amer. Chem. Soc.*, **105,** 492 (1983).
71 C. D. Ritchie and E. S. Lewis, *J. Amer. Chem. Soc.*, **84,** 591 (1962).
72 H. D. Holtz and L. M. Stock, *J. Amer. Chem. Soc.*, **86,** 5188 (1964).
73 S. Siegel and J. M. Komarmy, *J. Amer. Chem. Soc.*, **82,** 2547 (1960).
74 N. B. Chapman, J. Shorter and K. J. Toyne, *J. Chem. Soc.*, 1977 (1964).
75 C. F. Wilson and J. S. McIntyre, *J. Org. Chem.*, **30,** 777 (1965).
76 J. Palicek and J. Hlavati, *Z. Chem.*, **9,** 428 (1967).
77 R. W. Taft, E. Price, I. R. Fox, I. C. Lewis, K. K. Andereson and G. T. Davis, *J. Amer. Chem. Soc.*, **85,** 709, 3146 (1963).
78 E. A. Halonen, *Acta Chem. Scand.*, **9,** 1492 (1955).
79 R. H. Wolfe and W. G. Young, *Chem. Rev.*, **56,** 753 (1956).
80 M. Charton, *J. Org. Chem.*, **30,** 974 (1965).
81 K. Bowden, *Can. J. Chem.*, **44,** 661 (1966).
82 M. Charton, *J. Org. Chem.*, **31,** 2991 (1966).
83 P. B. D. De la Mare and R. Bolton, *Electrophilic Additions to Unsaturated Systems*, 2nd edn, Elsevier, Amsterdam, 1982.
84 P. E. Peterson and R. J. Bopp, *J. Amer. Chem. Soc.*, **89,** 1283 (1967).
85 B. R. Kowalsky (Ed.), *Chemometrics, Theory and Application*, ACS Symposium No. 52, 1977.
86 K. B. Wiberg, W. E. Pratt and W. F. Bailey, *J. Org. Chem.*, **45,** 4936 (1980).
87 E. R. Malinovski and D. G. Howery, *Factor Analysis in Chemistry*, Wiley, New York, 1980.
88 M. Charton, *Prog. Phys. Org. Chem.*, **8,** 235 (1971).
89 M. Charton, *Prog. Phys. Org. Chem.*, **10,** 81 (1973).
90 T. Fujita, *Anal. Chim. Acta*, **133,** 667 (1981).
91 M. T. Tribble and J. G. Traynham, *J. Amer. Chem. Soc.*, **91,** 379 (1969).
92 G. P. Ford, A. R. Katritzky and R. D. Topsom, *Correlation Analysis in Chemistry* (N. B. Chapman and J. Shorter, Eds), Chapter 6, Plenum, London 1978.

Correlation of structure with reactivity

93 G. Consiglio, C. Arnone, D. Spinelli, R. Noto and V. Frenna, *J. Chem. Soc., Perkin II*, 388 (1981).
94 M. Charton, *J. Org. Chem.*, **31**, 3745 (1966).
95 S. Marriott, W. F. Reynolds, R. W. Taft and R. D. Topsom, *J. Org. Chem.*, **49**, 959 (1984).
96 R. W. Taft, *Prog. Phys. Org. Chem.*, **14**, 247 (1983).
97 S. Marriott and R. Topsom, *J. Amer. Chem. Soc.*, **107**, 2253 (1985).

5 Solvent Effects

Organic reactions are, for the most part, carried out in the homogeneous liquid phase, more often than not in the presence of a solvent which does not itself enter into the reaction stoichiometry but nonetheless is a vital ingredient of the reaction. The unique properties of a medium which at the same time is a condensed phase and yet possesses high mobility serve to bring together efficiently reagents which may need collisional activation and to break down crystal lattice constraints through solution. Equally important is the complex interaction between solvent and solute molecules which may result in a gross modification of their activities, free energies, and, consequently, reactivities. Many reactions, for instance

$$R—I + NMe_3 \rightarrow R—\overset{+}{N}Me_3\bar{I};$$
$$k(MeCN)/k(CS_2) = 440,$$

are reluctant to occur at all in the gas phase, and in solution their rates may be highly solvent-dependent. Even more dramatic is the proton-transfer reaction [1]:

$$\underset{\underset{Me}{|}}{PhCH·CN} + MeO^- \longrightarrow \underset{\underset{Me}{|}}{Ph\bar{C}—CN} + MeOH,$$
$$k(Me_2SO)/k(MeOH) = 10^9$$

With effects of this magnitude, therefore, the solvent cannot be ignored, and must be considered a true catalyst of the reaction, a fact which has been apparent since the earliest explorations in chemistry and which today is exceedingly well documented [2–11] if still imperfectly understood.

Solvents comprise liquids of all chemical types of which three main categories may be distinguished:

(a) Protic; this term is used to describe solvents which possess a proton-donating function, —OH or —NH, and includes alcohols, amines, carboxylic acids and water. These have both a large dipole moment and a capacity for hydrogen bonding.

(b) Dipolar aprotic; these solvents which possess a large dipole moment

171

and donor properties have no acidic protons. Examples are dimethyl-sulfoxide, alkyl cyanides, secondary amides and ketones.

(c) Non-polar aprotic; these are liquids with only slight dipole moments, no acidic protons or donor/acceptor properties. Consequently there are only weak intermolecular forces. Hydrocarbons, halocarbons and ethers are typical.

Within these categories is found a wide range of solvent behavior, which is difficult to quantify since invariably a mixture of different intermolecular forces, operating with varying susceptibilities upon reagents and transition states, must be considered.

5.1 THE STRUCTURE OF LIQUIDS [12–15]

Compared with gases and crystalline solids, the liquid phase is the most difficult to comprehend, consisting as it does of rapidly changing molecular order while retaining a considerable degree of cohesiveness between molecules. The forces which stabilize the liquid state are the weak intermolecular forces described in Chapter 2, electrostatic (Coulombic) forces, hydrogen bonding and charge-transfer (donor–acceptor) interactions being the principal attractive elements which lower the free energy of the system by their negative enthalpy contributions. At the same time, the entropy of the liquid remains large and positive on account of the freedom of motion. These factors render the liquid state of most compounds stable over a surprisingly large temperature range[14,15]. Standard entropies of liquids are very much influenced by their tendency towards association (Table 5.1).

Water itself is highly associated and its tendency to associate with solute molecules (solvation) is high[16]. Ethanol and other protic solvents have a similar though lesser tendency while CS_2 has a very small capacity for intermolecular attraction.

A satisfactory description of the random and rapidly altering structure of

Table 5.1 Standard entropy of some liquids

Liquid	$S/J\,K^{-1}\,mol^{-1}$ $(cal\,K^{-1}\,mol^{-1})$	Liquid	$S/J\,K^{-1}\,mol^{-1}$ $(cal\,K^{-1}\,mol^{-1})$
Water	69·9 (16·71)	Nitromethane	171 (40·9)
Ethanol	160 (38·2)	1,2-Dibromoethane	223 (53·3)
Acetonitrile	144 (34·4)	1,2-Dichloroethane	208 (49·7)
Acetic acid	160 (38·2)	Carbon disulfide	237 (56·6)

172

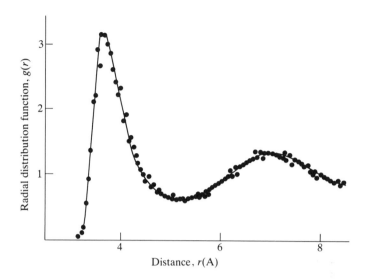

Fig. 5.1 Calculated radial distribution function of liquid argon. The peaks represent successive shells of neighboring atoms.

a liquid needs a statistical approach which for very simple liquids (e.g. inert gases) is beginning to succeed in reproducing experimental observation. Radial distribution functions, showing successive 'shells' of solvent, can be obtained by X-ray diffraction and matched by mathematical models of intermolecular forces and dynamics (Fig. 5.1). Protic liquids are far more complex on account of the additional ordering occasioned by hydrogen bonding. Water, however, has received much attention, consistent with its importance[17,18]. An instantaneous picture of bulk water would seem to include regions of three-dimensional hydrogen-bonded ice structure together with others in which unbonded hydrogens or unshared pairs, 'holes' and free water molecules occur (Fig. 5.2). A similar, although less structured, arrangement may also be expected in alcohols although they have only one hydrogen per functional group available for hydrogen-bond formation. Little is known of the detailed structures of most organic liquids, however.

5.2 SOLUTIONS [18,19]

For the purposes of this discussion, a solution is the homogeneous liquid phase which results from the mixing of a solute (gas, liquid or solid) with a liquid solvent, usually in large excess. If dissolution of the solute is to occur spontaneously it must be accompanied by a reduction in free energy, ΔG_{sol}, of the system. This may be considered to contain the following elements.

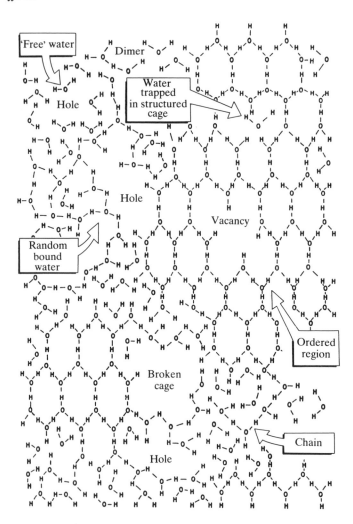

Fig. 5.2 Schematic diagram of water structure showing features present, though rapidly undergoing interchange.

	Contribution to ΔG_{sol}
(a) Creation of a 'cavity' or space within the solvent (reduction of solvent–solvent forces)	+ve
(b) Separation of solute molecule from bulk solute (reduction of solute–solute forces)	+ve
(c) Insertion of solute molecule in cavity (creation of solute–solvent forces (solvation))	−ve
(d) Entropy of mixing of solute and solvent	−ve

Components (a) and (b) are related to latent heats of vaporization of solvent and solute respectively; these quantities need to be balanced by the solvation energy available as expressed by (c). If solvent and solute are similar chemical types, the type and magnitude of forces (a) and (b) will be similar to solvation forces (c), and solubility will be high. This is the basis of the empirical rule 'like dissolves like', known from the earliest times.

The solute and surrounding solvent molecules exert a mutual net attraction (otherwise there would be no solution). The aggregation of solvent molecules around a solute is known as solvation. The presence of a solute molecule affects the properties of its immediate environment, which is known as the *cybotactic region*[20] and may extend well beyond the inner solvation shell though, at present, detailed and exact knowledge of its structure is not available[21]. It is likely that solutes which carry a charge or strong dipole orient contiguous solvent dipoles completely (Fig. 5.3), and solvent molecules beyond this intimate solvation shell less completely. At the same time, thermal motion permits more or less free rotation and rapid interchange of solvent molecules. The extent of the cybotactic region is greater the lower the dielectric constant of the solvent, since the falloff of

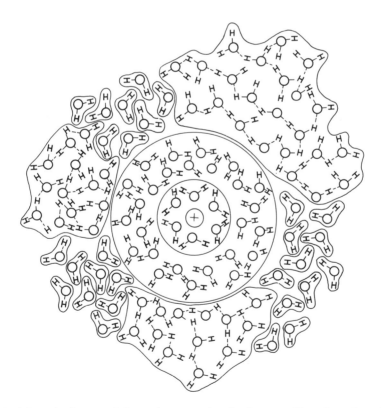

Fig. 5.3 Schematic diagram of the environment of an ion in water. The cybotactic region is enclosed by the large circle and includes the inner solvation shell within the small circle.

electric potential with distance varies inversely with dielectric constant; however, as will be seen, solvation forces will be lower. Also, as a result of the solvation forces, the density of the cybotactic region is higher than that of bulk solvent so that solvation is accompanied by the reduction in volume called *electrostriction*. The effect of intermolecular attractions of all kinds is to generate an 'internal pressure' acting upon all the molecules in the bulk fluid and amounting to several thousand atmospheres.

5.3 SOLVATION

This term describes the total solvent–solute interaction energy (Section 1.4) and is the key to all solvent effects upon chemical and physical phenomena. The free energy of a solute molecule may be modified by any or all of the following components but solvation energy corresponds to the difference between solute–solvent forces and the solvent–solvent forces which they replace.

Polarity

This term is often used in a rather inexact way but it is here taken to cover the non-specific attractive and repulsive forces, essentially of an electrostatic nature, between ionic or dipolar solutes and solvent dipoles. A solvent capable of exerting strong interactions of this type is known as a *polar* solvent (a vague term but one entrenched in the chemical literature). The degree to which a solvent is capable of solvating and stabilizing charges is referred to as its *polarity*. The dipole moment of a solvent, however, is no indicator of its polarity since the ability to solvate will also depend strongly on steric factors, the size and shape of solute and solvent molecules. It is particularly inadequate for the description of solvation by associated liquids such as hydroxylic solvents. The most generally acceptable measures of solvent polarity are those based on the numerous empirical scales described below.

Polarizability

This comprises the attractive interactions between temporary dipoles induced in solvent and solute molecules by their mutual electrostatic fields. The field gradient in the vicinity of an ion is of the order of 10^6 V cm^{-1}, sufficient to distort the electronic distribution of any molecule and provide it with an induced dipole (Section 1.4.3).

Hydrogen bonding

In practice, either the solute or the solvent must possess an hydroxyl function (a hydrogen-bond acceptor (HBA) center) and the other an

176

unshared pair (hydrogen bond donor (HBD) center), e.g. —OH, —OR, $\diagup\!\!\!\diagdown$C=O, —NR$_2$ (Section 1.4.4).

Donor–acceptor interactions

These may occur between contiguous solute and solvent molecules when one possesses a high-lying filled MO, and the other, a low-lying vacant MO. When the two are correctly oriented, a HOMO–LUMO interaction (often referred to as *charge transfer*) can stabilize the system. One may further differentiate the types of orbitals involved — n-, σ- or π- — in donor and acceptor activity (Section 1.4.5).

Solvation frequently includes contributions from all these components in different measure, which creates great difficulties in unraveling gross solvent effects. For example, heats of solvation in mixed solvent systems often vary in a wildly irregular manner with composition[22], which no doubt reflects subtle changes in the pattern of solvation.

The sum total of electrostatic solvation may be expressed as a *reaction field*, $\mathbf{E}_R$, proportional to the dipole moment of the solute molecule and oriented along its dipolar axis. The exact determination of the reaction field is potentially of great interest but would require knowledge of the entire cybotactic region. Instead, model systems have been constructed which are sufficiently simple for the concept to be applied within limits to be discussed below[23,24].

5.4 THERMODYNAMIC MEASURES OF SOLVATION

The energetics of solution and the deviations of solution properties from ideality yield information concerning solvent–solute interactions.

5.4.1 Free energies of solution and transfer functions

Most measurements that are available concern ionic solutes and hence focus attention on polar solvation. The free energy of solution, ΔG_{sol} (= (a) + (b) + (c) + (d) above), and of transfer, ΔG_{tr}, are defined by Eqs (5.1), (5.2) and some typical values are given in Tables 5.2 and 5.3.

$$(X)_{T,c}(\text{pure}) \rightarrow (X)_{T,c}(\text{solution}); \qquad \Delta G = \Delta G_{sol} \qquad (5.1)$$

$$(X)_{T,c}(\text{solvent 1}) \rightarrow (X)_{T,c}(\text{solvent 2}); \qquad \Delta G = \Delta G_{tr}. \qquad (5.2)$$

Although the solubilities vary by a factor of 10^8, ΔG_{sol} is remarkably small and constant since the latent heat terms, (a) + (b), tend to cancel solvation, (c), though these individual terms will be very large for solvents such as water and alcohols, and very small for ether or benzene. The free energy of solution is smaller for the solute considered as an ion-pair than as free ions,

Solvent effects

Table 5.2 Molar solubilities, $c/10^x$M, and free energies of solution, $\Delta G_{sol}/\text{kJ mol}^{-1}$ (kcal mol^{-1}), of dissociated and ion-pair species, $Et_4N^+I^-$ [29]

Solvent	c	x	ΔG_{sol} $Et_4N^+ + I^-$	ΔG_{sol} $[Et_4N^+I^-]$
Water	1·927	0	4·81 (1·15)	3·8 (0·91)
Methanol	0·362	0	12·5 (3·0)	5·4 (1·3)
Ethanol	3·41	−2	22·8 (5·4)	10·7 (2·5)
Propan-1-ol	1·187	−2	28·0 (6·7)	12·8 (3·1)
Propan-2-ol	3·56	−3	33·0 (7·9)	15·6 (3·7)
Butan-1-ol	5·95	−3	32·0 (7·6)	14·0 (3·3)
3-Methylbutan-1-ol	2·75	−3	36·5 (8·7)	15·6 (3·7)
tert-Butanol	2·76	−4	47·4 (11·3)	21·1 (5·0)
Dimethylsulfoxide	0·121	0	13·2 (3·1)	
Nitromethane	0·209	0	12·2 (2·9)	10·5 (2·5)
Acetonitrile	0·115	0	15·1 (3·6)	11·1 (2·6)
Dimethylformamide	0·115	0	15·4 (3·7)	9·2 (2·2)
Propiononitrile	3·00	−2	21·6 (5·2)	12·6 (3·0)
Nitrobenzene	1·96	−2	22·3 (5·3)	14·2 (3·4)
Benzonitrile	1·76	−2	24·0 (5·7)	14·2 (3·4)
Acetone	9·68	−3	27·3 (6·5)	14·9 (3·6)
Butan-2-one	2·03	−3	33·9 (8·1)	19·4 (4·6)
Acetophenone	5·05	−3	30·3 (7·2)	16·6 (4·0)
1,2-Dichloroethane	9·19	−3	35·1 (8·4)	12·2 (2·9)
Dichloromethane	3·77	−2	34·4 (8·2)	8·4 (2·0)
1,1,2-Trichloroethane	1·83	−3	48·7 (11·6)	15·7 (3·7)
Methyl formate	1·20	−3	45·1 (10·8)	17·0 (4·1)
Bromobenzene	3·61	−4	62·3 (14·9)	19·7 (4·7)
Ethyl benzoate	4·96	−5	63·5 (15·2)	24·7 (5·9)
Ethyl acetate	1·56	−5	66·0 (15·8)	7·7 (1·8)
Chlorobenzene	2·21	−5	67·7 (16·2)	26·7 (6·4)
Diethyl ether	4·0	−8	96·6 (23·1)	42·2 (10·0)
Benzene	3·8	−7	136 (32·5)	36·6 (8·7)

since some of the solvation energy is replaced by interionic attraction (Table 5.2). A more useful way of presenting this type of information is as a free energy of transfer, ΔG_{tr} [25-30], the free-energy change for transferring solute (in dilute solution) from solvent 1 to solvent 2. Any standard medium (solvent 1) may be chosen for comparison, including the gas phase. Then ΔG_{tr} is a measure of the change in solvation from one medium to another. While water is consistently associated with the most negative value of ΔG_{tr}, indicative of the strongest solvation, values are surprisingly similar throughout a very wide range of solvents, perhaps because of compensation; the greater the energy of solvation, the more it is offset by energy required for breaking solvent–solvent attractive forces. It is highly instructive to partition free energies of transfer of electrolytes into contributions for the dissociated cations and anions which are shown for Et_4N^+ and I^- in Table 5.3, and for the ion-pairs.

In many solvents, ΔG_{tr} values for the anion and for the cation are of opposite sign. It is found that anions are best solvated by hydroxylic solvents, presumably by hydrogen bonding, while the solvation of cations is

178

Table 5.3 Free energies of transfer, $\Delta G_{tr}/\text{kJ mol}^{-1}$ (kcal mol^{-1}), of $Et_4N^+I^-$ as dissociated ions and ion-pairs, between methanol and other solvents

Solvent	ΔG_{tr} $Et_4N^+ + I^-$	ΔG_{tr} $[Et_4N^+I^-]$	ΔG_{tr} *tert*-BuCl	$\log y_{rel}$ (K^+)	$\log y_{rel}$ (Cl^-)
Water	−7·5 (−1·79)	−1·67 (−0·4)	22·0 (5·26)	−1·5	−2·5
Methanol	0·0 (0·0)	0·0 (0·0)	0·0 (0·0)	0·0	0·0
Ethanol	10·5 (2·51)	5·4 (1·29)	−1·2 (−0·29)		
Propan-1-ol	15·5 (3·70)	7·5 (1·79)	−1·6 (−0·38)		
Propan-2-ol	20·9 (5·0)	10·0 (2·39)	−1·44 (−0·34)		
Butan-1-ol	19·7 (4·71)	8·8 (2·1)	−2·2 (−0·53)		
tert-Butanol	34·7 (8·29)	15·6 (3·73)	−2·2 (−0·53)		
Dimethylsulfoxide	0·8 (0·19)	2·0 (0·48)	−0·5 (−0·12)	−4·5	5·5
Nitromethane	−0·4 (−0·09)	5·0 (1·19)	−0·77 (−0·18)		4·9
Acetonitrile	2·5 (0·59)	5·8 (1·39)	−1·9 (−0·45)	−0·8	6·3
Dimethylformamide	2·9 (0·69)	3·8 (0·91)	−2·6 (−0·62)	−3·7	6·5
Nitrobenzene	10·0 (2·39)	8·8 (2·10)	−3·6 (−0·86)		
Benzonitrile	11·3 (2·70)	8·8 (2·10)			
Acetone	14·6 (3·49)	9·6 (2·29)	−4·0 (−0·95)		
Butan-2-one	21·3 (5·09)	13·3 (3·18)			
Acetophenone	18·0 (4·30)	11·3 (2·70)			
1,2-Dichloroethane	22·6 (5·40)	10·5 (2·51)			
Ethyl acetate	53·5 (12·8)	22·1 (5·28)			
Chlorobenzene	55·2 (13·2)	21·3 (5·09)	−5·4 (−1·29)		
Diethyl ether	84·1 (20·1)	37·2 (8·89)	−4·0 (−0·95)		
Benzene	109 (26·0)	28·8 (6·88)	−5·1 (−1·22)		
Carbon tetrachloride	46 (11·0)				
Cyclohexane	54 (12·9)	58 (13.9)			
Hexane					

strongest by dipolar aprotic solvents. Much less information is available concerning solvation of neutral organic solutes. Heats and entropies of solution of a completely non-polar substance such as ethane are very similar for protic and non-polar solvents (water being somewhat exceptional).

5.4.2 Activities of solutes

An ideal solution is one in which energies of interaction between solvent and solute are essentially the same as solvent–solvent forces and Raoult's Law (Eq. (5.3)) is obeyed.

$$p = p^\circ(1 - x_2) \tag{5.3}$$

(p = vapor pressure of solution; x_2 = mole fraction of solute). For real solutions, concentration terms need to be replaced by activities, a, such that

$$\left.\begin{array}{r} a = yc \\ \text{and} \quad a = fx \\ \text{and} \quad a = \gamma m, \end{array}\right\} \tag{5.4}$$

where c, x and m are concentrations expressed as molarity, mole fraction

Table 5.4 Some polar solvation constants for common solvents

Solvent	ε	$10^4 f_e$	$10^4 f_n$	δ	E_T	Z	π^*	S_{PR}	$\mathscr{S}$
Formic acid	58						0·96		
Acetic acid	6·15	3872	1851	8·9	51·2	79·2	0·62		
Water	78·4	4905	1706	23·4	63·1	94·6	1·09	0·154	
CF₃CH₂OH	26·7				59·5		0·73		
CH₃OH	32·7	4774	1688	14·3	55·5	83·6	0·60	0·0499	−1·89
C₂H₅OH	24·5	4700	1813	12·7	51·9	79·0	0·54	0·000	−2·02
PhCH₂OH	13·1	4448	2389	12·2	50·8				
CH₃CH₂CH₂OH	20·3	4640	1900	11·9	50·7	78·3	0·51	−0·0158	
CH₃CH₂CH₂CH₂OH		4583	1949	11·4	50·2	77·7	0·46		
CH₃CHOHCH₃	19·9	4633	1871	11·5	48·6	76·3	0·46	−0·0413	
CH₃CH₂CHOHCH₃	16·6	4560	1941		47·1				
(Me)₃COH	12·5	4422	1908	10·5	43·9	71·3	0·43	−0·104	
HOCH₂CH₂OH	37·7	4803	2059		56·6	85·1			
MeOCH₂CH₂OH	16·9	4570	1958		52·3				
HCONH₂	111	4932	2110		56·6	83·3	0·98		
HCONHMe	182	4959	2059		54·1				
HCONMe₂	37	4800	2043	11·8	43·8	68·4	0·87	−0·1416	
MeCONHMe	191	4961	2048		52·0	77·9			
MeCONMe₂	37·8	4808	2080	10·8	43·7	66·9	0·88		
CO(NMe₂)₂	29·6	4687	2116		41·0				
N-Methylpyrrolidone	32·0	4374	1748	11·3					
MeCN	37·5	4374	1748	11·7	46·0	71·3	0·76	−0·1039	
EtCN	27·2	4729	1829	10·6	43·7				−0·328
PhCN	25·2	4708	2354		42·0	65·0	0·93		
MeNO₂	35·9	4394	1885	12·6	46·3	71·2	0·85	−0·134	0·041
PhNO₂	34·8	4787	2416	1·1	42·0		1·01	−0·218	
O=P(NMe₂)₃	29·6	4751	2145	8·6	40·9	62·8	0·87		
O=S(Me₂)	46·7	4841	2207	13·0	45·0	71·1	1·00		
(MeCO)₂O	20·7	4646	1918		3·9		0·84		
MeCOMe	20·7	4646	1803	9·6	42·2	65·5	0·68	−0·1748	−0·824
EtCOMe	18·5	4605	1876	9·1	41·3		0·67		
PhCOMe	17·4	4581	2372	10·4	41·3		0·90		
Cyclohexanone	18·3	4601	2121	9·90	40·8		0·75		
MeCOOMe	6·7	3955	1813	9·5	40·0		0·56		
MeCOOEt	6·0	3850	1853	8·90	38·1	59·4	0·545		−1·66
Pyridine	12·4	4419	2303		40·2	64·0	0·87		
NEt₃	2·4	2431	1955	7·45	33·3		0·14		
PhNH₂		3985	2513		44·3				
Glyme	7·2	4026	1879		38·2				
Triglyme	7·5	4062	2030		38·9	61·3			
Tetrahydrofuran	7·6	4072	1976	9·30	37·4	58·8	0·576		
PhOMe	4·3	3447	2323		7·2		0·734		
1,4-Dioxane	2·2	2232	2028	9·8	36·0		0·553		
EtOEt	4·3	3450	1780	7·8	34·6		0·273		−2·92
(Me₂CH)₂O	3·9	3287	1838		34·0		0·27		
(n-C₄H₉)₂O	3·1	2905	1948	7·62	33·4	60·1	0·239		
CH₂Cl₂	8·9	4156	2034	10·0	41·1	64·7	0·802		−0·553
CH₂ClCH₂Cl	10·4	4309	2101	9·9	41·9	63·4	0·807		
CHCl₂CH₃	10·0	4286	2007	9·1	39·4	62·1			
CHCl₃	4·8	3587	2095	9·3	39·1	63·2	0·760	−0·20	−0·886
PhCl	5·6	3774	2338	9·5	37·5	58·0	0·709		
PhI	4·6	3538	2599	10·1	37·9		0·81		
CHCl=CCl₂	3·4	3087	2202		35·9		0·534		−2·85
CCl₄	2·2	2150	2142	8·6	32·5		0·294		−1·74
Benzene	2·3	2303	2276	9·2	34·5	54	0·588	−0·215	
PhMe	2·4	2396	2263	8·9	33·9		0·535		−4·15
Cyclohexane	2·0	2023	2040	8·2	32·1		0·00		−5
n-C₆H₁₄	1·9	1849	1862	7·27	30·9		−0·081	−0·337	
CS₂	2·6	2611	2619	9·9	32·6		0·51		

180

and molality, respectively, and y, f and γ the activity coefficients appropriate to each type of concentration unit. Activity coefficients express the deviation of the substrate from ideality and are based upon an (arbitrary) standard state, e.g. very dilute aqueous solution, for which, for the purposes of comparison, $y(f, \gamma) = 1{\cdot}000$. Values under other conditions then reflect the changing degrees of solvation or association which may cause either an increase or a decrease in y. Activity coefficients for some single ions are given in Table 5.3[27,28]. Their striking features are the very large values in organic solvents compared with those in water. Since an increase in y indicates a reduction in solvation, it may be concluded that ionic solvation decreases with solvent in the order: water > alcohols > dipolar aprotics > non-polar solvents[31]. Single ion activities show this to be a result of differential solvation of cations and anions; cations are solvated by dipolar aprotic solvents to a much greater extent than by hydroxylic solvents whereas the reverse is true for anions. Less information is available for non-ionic solutes but activity coefficients are in general likely to be close to unity in all solvents. These features have profound effects on chemical reactivity.

5.4.3 'Solvation' in the gas phase

Detailed information concerning the most intimately bound solvent shell may be obtained by mass spectroscopic methods[32], by which an ion is

$$M^{\pm}(H_2O)_n + (H_2O) \rightarrow M^{\pm}(H_2O)_{n+1} + \Delta H$$

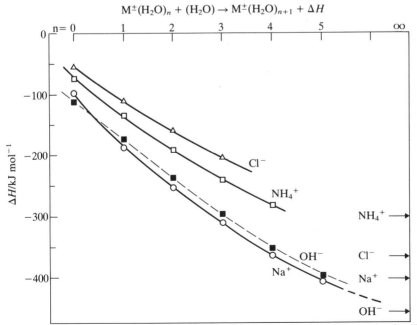

Fig. 5.4   Energetics of hydration of several gas-phase ions by successive water molecules. The extrapolation leads to the energy of hydration of single ions.

created in an atmosphere of solvent vapor at low pressure. The series of solvated ions that form are assumed to be in equilibrium and their abundances are measured by the mass spectrometer. Figure 5.4 shows the energetics of binding of successive water molecules to several ions. Six discrete hydration steps can be discerned with progressively lower energies of addition, binding presumably being electrostatic as in the inner solvation sphere shown in Fig. 5.3. Extrapolation leads to the total solvation energy. This type of study may be adapted to both cations and anions and to solvation by organic molecules such as acetonitrile.

5.5 THE EFFECTS OF SOLVATION ON REACTION RATES AND EQUILIBRIA[8,9,10,33]

The extent to which solvation will reduce the free energies of solute molecules in equilibrium will affect the free energy of reaction, ΔG, and consequently also the equilibrium constant, K. In general, an equilibrium will be shifted by a change in solvent so as to favor the side most stabilized by solvation. For example, in the tautomeric equilibrium between 2-hydroxypyridine, **1a**, and 2-pyridone, **1b**, the latter is favored by an increase in solvent polarity[18](Table 5.5); see also Fig. 5.6.

1a   **1b**

Evidently **1a** is intrinsically the more stable but solvation, particularly by solvents of high polarity, rapidly reverses this situation. One would conclude that **1b** is more susceptible to solvation than **1a** on account of its higher dipole moment since both tautomers are capable of acting as a hydrogen-bond acceptor. It is normal to find severe discrepancies between properties of isolated molecules (e.g. from MO calculations) and those in solution (e.g. Fig. 5.5).

Table 5.5 Solvent effect on the tautomeric equilibrium **1a** $\rightleftarrows$ **1b**

Solvent	K	Percentage of **1b**	$-\Delta G/$ $kJ\,mol^{-1}\,(kcal\,mol^{-1})$
Gas phase	0·4	28	−2·2 (−0·52)
Cyclohexane	1·7	63	1·3 (0·31)
Chloroform	6·0	86	4·5 (1·1)
Acentonitrile	148	99	12·5 (3·0)
Water	910	99·9	17·0 (4·1)

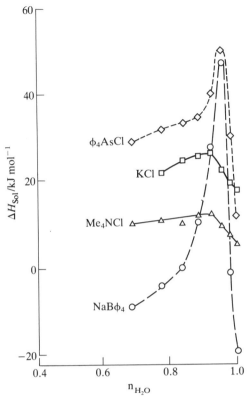

Fig. 5.5 Irregularity in energy of solvation as a function of solvent composition; salts in water–*tert*-butanol mixtures.

5.5.1 Solvent effects on rates

Combining this type of analysis with transition-state assumptions one can see that, if the transition state is capable of greater solvation than are the reagents, then the reaction rate will be increased by a change to a more solvating solvent since the activation energy will be lowered. This is the basis of the Hughes–Ingold Rules of solvent effects which express the principle in terms of the most important contribution to solvation, polar interactions[34].

Three charge types need to be considered.

(1) If the activation step is accompanied by an increase in electrical charges on the reactants, a change to a more polar solvent will cause a large increase in rate.

(2) If the activation step is accompanied by a decrease in electrical charges on the reactants, a change to a more polar solvent will cause a large decrease in rate.

(3) If the activation step is accompanied by a dispersion of electrical charges, a change to a more polar solvent will cause a small decrease in rate.

The kinetic solvent effect on a reaction is determined by its charge type. Examples of the three categories are as follows

Type 1: charge separation $\qquad\qquad$ $k(H_2O)/k(EtOH)$

$$R_3C-Br \longrightarrow [R_3\overset{\delta+}{C}\cdots\overset{\delta-}{Br}]^\ddagger \qquad\qquad 1500$$

Type 2: charge neutralization

$$H\bar{O}\diagdown C-\overset{+}{N}R_3 \longrightarrow [\overset{\delta-}{HO}\cdots \overset{|}{C}\cdots \overset{\delta+}{N}R_3]^\ddagger \qquad\qquad 0{\cdot}001$$

Type 3: charge dispersion

$$H\bar{O}\diagdown C-Br \longrightarrow [\overset{\delta-}{HO}\cdots \overset{|}{C}\cdots \overset{\delta-}{Br}]^\ddagger \qquad\qquad 0{\cdot}2$$

This theme will be developed further after a discussion of the measurement of solvent polarity.

5.6 EMPIRICAL INDEXES OF SOLVATION

Any phenomena, chemical or physical, which are sensitive to the nature of the solvent may be made the basis of a solvation idex. The significance of each such scale will, of course, depend upon the nature of the solvation forces to which the experiment responds. There is no shortage of solvation scales which have been proposed. The problem, rather, is to discover the physical basis for each, so that one may know what is being measured. It is rare that a solvation scale is a measure purely of one type of intermolecular interaction. In many cases the response of the probe appears to be to a blend of polar (non-specific), hydrogen-bonding and donor–acceptor (specific) solvation capabilities, and the further question may then be posed as to the relative importance of each. Hybrid scales of this type are, nonetheless, useful as the basis for comparison of the responses to solvent change by different reactions and they thus lend themselves to a linear free-energy treatment.

5.6.1 Scales based on physical properties

Dielectric constant (relative permittivity), ε

This is the relative capacitance of a simple capacitor with the medium as dielectric, compared with that in vacuum (or air, in practice). It is a measure of the ability of the bulk fluid to store electrical energy by alignment of dipoles and provides to some extent a direct measure of its capability for polar solvation. Dielectric constants range from around 2 for alkanes to 80 for water, up to some exceptionally high values for monosubstituted amides (N-methylacetamide, $\varepsilon = 200$); see Table 5.4. Consequently one finds that ionic solutes dissolve only in solvents of high dielectric constant in which solvation energy can match the crystal lattice energy. Electrostatic theory was applied as early as 1928 to solvent effects, with a model in which the solute was assumed to be a point dipole in a spherical cavity of $\varepsilon = 1$, while the solvent acted as a continuous isotropic dielectric medium, with $\varepsilon =$ bulk dielectric constant as measured[35-37].

It may be shown that the energy associated with a charge or dipole varies according to the dielectric constant function, $(\varepsilon - 1)/(2\varepsilon + 1)$ (Kirkwood function, f_ε). This leads to the Kirkwood–Onsager equation for the electrostatic contribution to ΔG_{tr} for transfer of an isolated spherical dipole (dipole moment, $\mathbf{P}$) to a medium of dielectric constant, ε, from one for which $\varepsilon = 1$;

$$\Delta G_{tr} = \frac{\mathbf{P}^2}{r^3} \cdot \frac{(\varepsilon - 1)}{(2\varepsilon + 1)}. \qquad (5.5)$$

As a very crude measure of solvent polarity, the bulk dielectric constant is often used. The ability to stabilize charges created in the transition state of the Menshutkin reaction (or any other ionogenic process) is accounted for in a general way, as indicated in Table 5.6.

The bulk property, ε, does not, however, give a good quantitative measure of interactions at the molecular level[33,38,39]. The Kirkwood function, depending as it does on a model of solvent as a continuous medium, is little better. Correlation of rates with the Kirkwood function may be satisfactory when variation in the medium is achieved by admixture of two solvents in varying proportions (Fig. 5.6), but is usually poor in comparisons of different solvent types (Fig. 5.7). This type of result is to be expected when a solvation parameter responds principally to one type of force (e.g. polar factors) while the solvents involved can manifest their solvation capability by additional types of force.

Table 5.6   Effect of solvent on rates of the Menshutkin reaction

$$Et_3N + MeI \rightarrow [Et_3\overset{\delta+}{N} \cdots Me \cdots \overset{\delta-}{I}]^{\ddagger} \rightarrow Et_3\overset{+}{N}Me\ \overset{-}{I}$$

Solvent	ε	f_ε	k_{rel}
Hexane	2	0·20	1
Carbon tetrachloride	2·2	0·22	31
Chlorobenzene	5·6	0·38	1200
Acetonitrile	37·5	0·480	12 000
Dimethylsulfoxide	46·7	0·484	50 000

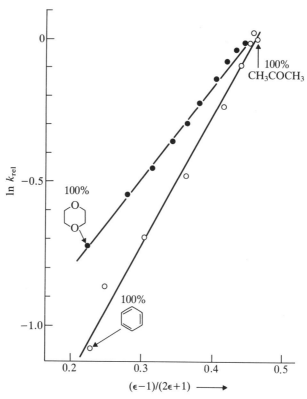

Fig. 5.6 Correlation between log k_{rel} and the Kirkwood function, $f_\varepsilon = (\varepsilon - 1)/(2\varepsilon + 1)$, for the Menshutkin reaction $Et_3N + EtI \rightarrow Et_4\overset{+}{N}\ \overset{-}{I}$ in binary mixtures of acetone–dioxane and acetone–benzene.

Refractive index, n

The square of the refractive index, n^2, or the function $f_n = (n^2 - 1)/(2n^2 + 1)$ (based on the Kirkwood model), is a measure of the response of the medium to the high-frequency alternating electric field of visible light (10^{16} Hz). Since this frequency is much greater than those associated with molecular relaxation (realignment), the molecules can only respond by distortion of valence electron distributions; hence the refractive-index function is a measure of the polarizability of the solvent, within the limits of the model. As will be seen in Section 5.8, polarizability plays a part in solvation of ions but is rarely dominant.

There are examples of phenomena, usually involving light absorption[34], which do correlate well with polarizability, such as the energies of excitation of transient charge-transfer complexes between donor (D) and acceptor (A) species:

$$D + A \xrightarrow[\text{(solvent)}]{h\nu} (D \cdots A); \qquad \bar{\nu}_{max} \propto n^2 \text{ (solvent)}.$$

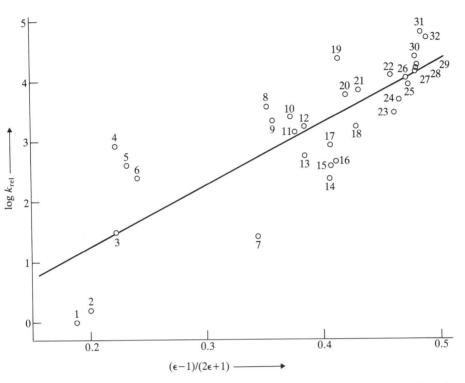

Fig. 5.7 Lack of correlation of the Menshutkin reaction (Fig. 5.6) when a large variety of solvents is used: 1, *n*-hexane; 2, cyclohexane; 3, carbon tetrachloride; 4, 1,4-dioxane; 5, benzene; 6, toluene; 7, diethyl ether; 8, iodobenzene; 9, chloroform; 10, bromobenzene; 11, chlorobenzene; 12, ethyl benzoate; 13, ethyl acetate; 14, 1,1,1-trichloroethane; 15, chlorocyclohexane; 16, bromocyclohexane; 17, tetrahydrofuran; 18, 1,1-dichloroethane; 19, 1,1,2,2-tetrachloroethane; 20, dichloromethane; 21, 1,2-dichloroethane; 22, acetophenone; 23, 2-butan-2-one; 24, acetone; 25, propionitrile; 26, benzonitrile; 27, nitrobenzene; 28, *N,N*-dimethylformamide; 29, acetonitrile; 30, nitromethane; 31, dimethylsulfoxide; 32, propene carbonate.

Reaction field calculations

From the Kirkwood–Onsager model (Eq. (5.5)) an expression for the reaction field E_R can be derived:

$$E_R = 2P/r^3 f_\varepsilon, \qquad (5.6)$$

while modification of the model to allow the dielectric constant of the solute cavity the more reasonable value of 2, and a gradation from this to bulk solvent rather than a sudden boundary change, leads to the analogous expression:

$$E_r = P/r^3 f'_\varepsilon \qquad (5.7)$$

187

where

$$f'_\varepsilon = \frac{3\varepsilon \ln \varepsilon}{\varepsilon \ln \varepsilon - \varepsilon + 1} - \frac{6}{\ln \varepsilon} - 2.$$

Some information is becoming available on the applicability of E_r as a solvation parameter; for example, rate parameters of the Menshutkin reaction have been shown to give an excellent correlation with E_r values for selected solvents (dipolar aprotics in the main: specifically excluding hydroxylic and aromatic solvents).

Solvent cohesive energy — the solubility parameter[40-44]

Hildebrand defined the *cohesive energy density* (i.e. the energy of cohesion of unit volume), D, of a species by Eq. (5.8):

$$D = \Delta H_L / V_m, \tag{5.8}$$

where ΔH_L is its molar latent heat of vaporization and V_m its molar volume. The solubility parameter, δ, (Table 5.4), is defined as

$$\delta = D^{\frac{1}{2}} \tag{5.9}$$

and this was shown to enter into an expression (Eq. (5.10)) for the activity coefficient of a solute, f_2:

$$RT \log f_2 = V_2(\delta_2 - \delta_1)^2, \tag{5.10}$$

in which subscripts 1 and 2 refer respectively to solvent and solute. Since reaction rates depend upon activities of solutes and transition state, the following expression (Eq. (5.11)) is inferred for a bimolecular reaction,

$$A + B \rightarrow (AB)^{\ddagger} \rightarrow \text{products.}$$

$$\text{Rate} \equiv \ln k = \ln k_0 + \ln f_A + \ln f_B - \ln f^{\ddagger}, \tag{5.11}$$

in which k_0 refers to the rate under the hypothetical unit-activity conditions and $f^{\ddagger}$ refers to the activity of the transition state. It follows from Eq. (5.10) that

$$\ln k = \ln k_0 + 1/RT[V_A(\delta_A - \delta_S)^2 + V_B(\delta_B - \delta_S) - V^{\ddagger}(\delta_S - \delta^{\ddagger})^2]. \tag{5.12}$$

where δ_S is the solubility parameter of the solvent. Expansion of Eq. (5.12) gives:

$$RT \ln k/k_0 = 2\delta_S(V^{\ddagger}\delta^{\dagger} - V_A\delta_A - V_B\delta_B) + (V_A\delta_A^2 + V_B\delta_B^2 - V^{\ddagger}\delta^{\dagger 2}), \tag{5.13}$$

from which it may be seen that a plot of $\ln(k/k_0)$ against δ_S should be linear, with slope $= 2(V^{\ddagger}\delta^{\ddagger} - V_A\delta_A - V_B\delta_B)/RT$. Now the initial assump-

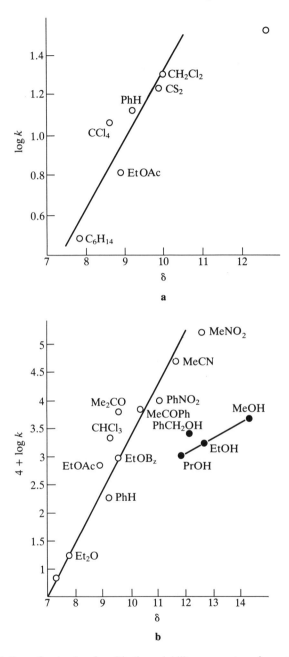

Fig. 5.8   Correlation of rate, log k, with the solubility parameter, δ; **a**, the Diels–Alder reaction, maleic anhydride + butadiene; **b**, the Menshutkin reaction, $Et_3N + p$-nitrobenzyl chloride.

tions on which δ_S is based apply to non-polar solutes, non-polar solvents and the absence of hydrogen bonding (so-called *regular solutions*)[41] and therefore only within the limitations of this type of reaction can we expect a reasonable agreement with Eq. (5.13). Figure 5.8 shows the reasonably linear correlation of ln k with δ for a Diels–Alder reaction, a typically non-polar process. Perhaps surprisingly, a good linear correlation is also found for the polar Menshutkin reaction in dipolar aprotic solvents (though hydroxylic solvents deviate markedly). Values of transition-state properties, $\delta^{\ddagger}$ and $V^{\ddagger}$[45, 46], may be obtained from Eq. (5.13); $V^{\ddagger}$ values can also be determined from pressure effects on rates via the Drude–Nernst equation (Eq. 2.30). Substitution for $\Delta G = -RT \ln k$ in Eq. (5.5) and combining with Eq. (5.11) gives

$$kT \ln k = \left(\frac{\varepsilon - 1}{2\varepsilon + 1}\right) \cdot \left(\frac{P_A^2}{r_A^3} + \frac{P_B^2}{r_B^3} - \frac{P^{\ddagger 2}}{r^{\ddagger 3}}\right), \qquad (5.14)$$

$$\mathrm{Et_3N + EtBr} \rightarrow \left[\overset{\delta+}{\mathrm{Et_3N}} \cdots \underset{\underset{\mathrm{CH_3}}{|}}{\mathrm{CH_2}} \cdots \overset{\delta-}{\mathrm{Br}}\right]^{\ddagger}$$

$$\xleftarrow{\qquad 375 \text{ pm} \qquad}$$

$$P \qquad 0{\cdot}7D \qquad 2{\cdot}0D \qquad\qquad 5{\cdot}1D$$

If the $\mathrm{N}\cdots\mathrm{Br}$ distance in the transition state is 375 pm (10% longer than normal bond lengths), the calculated dipole moment corresponds to charges of $\pm 0{\cdot}28$.

The solvent *internal pressure*, p_i, is derived from these concepts and is defined as the internal energy per unit volume, $(\partial E/\partial V)_T$. Values range from 2000 to 5000 bar for organic liquids[45,47]. Surface tension has also been proposed as a measure of molecular cohesion, relevant to reactivity[46].

5.6.2 Scales based on solvent-sensitive reaction rates

The Y-parameter

The solvolysis of a tertiary halide in hydroxylic media, SOH, occurs by rate-determining ionization:

$$\mathrm{MeC-Cl} \xrightarrow{\mathrm{SOH}} [\overset{\delta+}{\mathrm{Me_3C}} \cdots \overset{\delta-}{\mathrm{Cl}}]^{\ddagger} \longrightarrow \mathrm{Me_3C-OS + Cl^-, H^+}.$$

The separation of charges means that these reactions are very sensitive to solvent polarity. Winstein and Grunwald defined an empirical solvent parameter, Y, from solvolytic rate constants, k, of *tert*-butyl chloride[48–50].

$$\log k/k_0 = m_s Y(+lN), \qquad (5.15)$$

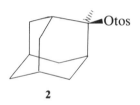

2

where k_0 refers to the rate in standard solvent (taken arbitarily as 80% v/v
ethanol–water) at 25°C; m_s is a constant reflecting the susceptibility of the
reaction to changes in solvent and is determined by the compound
undergoing solvolysis; N is the nucleophilicity of the solvent and l its
susceptibility parameter (see Section 6.12, Table 6.8). (The second term is
often omitted in solvent–reactivity correlations.) The constant m_s is set at
$m_s = 1·000$ for *tert*-butyl chloride although it is now more usual to use
adamantyl tosylate, **2**, as substrate. The form of Eq. (5.15) is that of a linear
free-energy equation which relates the solvent effect to changes in the free
energy of activation of this reaction. Values of Y (Table 5.7) are limited to
hydroxylic solvents in which alone the reaction may take place. A slightly
increased range of solvents can be accommodated by the use of *p*-
methoxyneophyl tosylate ('log k (neophyl)'), **3**, as substrate ($m_s = 0·5$); it
undergoes ionization with rearrangement[51]. The Y-scale measures non-
specific polar solvation with a contribution from solvent basicity.

Table 5.7   Solvation scales for ionization: Y-values and log k (neophyl) (Eq. (5.15))

Solvent	Y	log k (*neophyl*)
Water	3·493	−1·180
Formic acid	2·054	−0·929
CF₃CH₂OH	1·045	
HCONH₂	0·604	
MeOH/H₂O, 80:20	0·381	
EtOH/H₂O, 80:20	0·000	−2·505
n-Hex₄N·PhCOO	−0·39	
NeCOMe/H₂O, 80:220	−0·673	
1,4-Dioxane/H₂O, 80:20	−0·833	
MeOH	−1·090	−2·796
MeCOOH	−1·675	−2·772
EtOH	−2·033	−3·204
MeCHOHMe	−2·73	
Me₃COH	−3·26	
MeSOMe		−3·738
MeNO₂		−3·921
MeCN		−4·221
HCONMe₂	−3·5	−4·298
Pyridine		−4·670
MeCOMe		−5·067
MeCOOEt		−5·947
Tetrahydrofuran		−6·073
EtOEt		−7·000

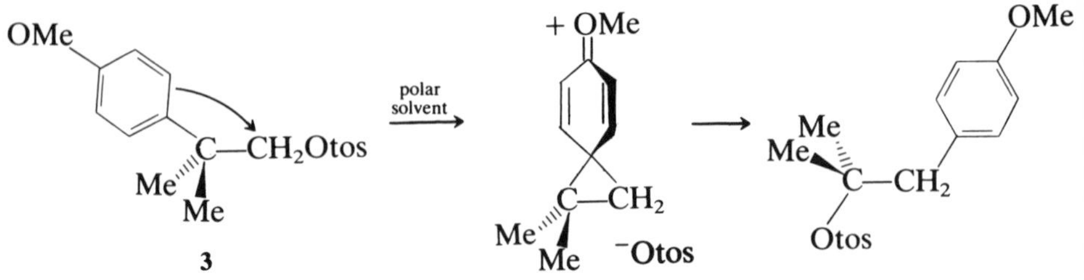

The 𝒮-parameter

As shown above, rates of a Menshutkin reaction are highly solvent-sensitive. An extensive solvation scale, $\mathscr{S}$ (Table 5.4), has been developed[38,39,52], based on rates of reaction between methyl iodide and tri(n-propyl)amine:

$$MeI + nPr_3N \rightarrow nPr_3\overset{+}{N}—MeI^-$$

$$\log k = \mathscr{S}. \tag{5.16}$$

The X-parameter

$$\log k \propto X = pX \tag{5.17}$$

This is analogous to $\mathscr{S}$ but it is based on the brominolysis of tetramethylstannane (4), $p = 1\cdot000$ the reference solvent being acetic acid[53].

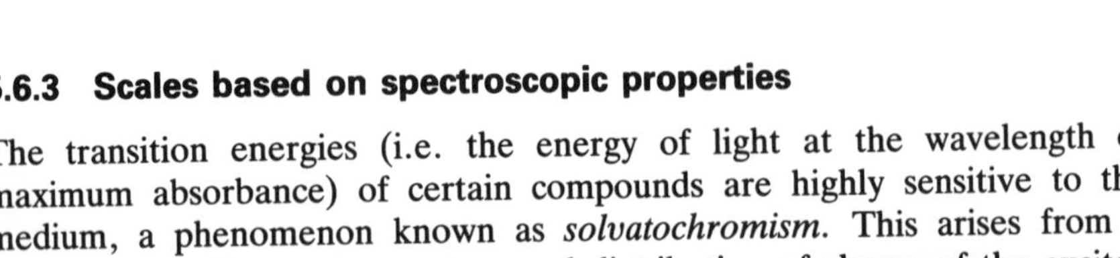

5.6.3 Scales based on spectroscopic properties

The transition energies (i.e. the energy of light at the wavelength of maximum absorbance) of certain compounds are highly sensitive to the medium, a phenomenon known as *solvatochromism*. This arises from a change in the electronic structure and distribution of charge of the excited state as compared with the ground state (Fig. 5.9). If the excited state is more polar than the ground state it will be better solvated and its energy lowered so that the transition will occur at a longer wavelength; i.e. there will be a red-shift with increasing solvent polarity. On the other hand, if the ground state is more polar than the excited state it will be stabilized by polar solvation so that excitation is made more difficult by an increase in solvent

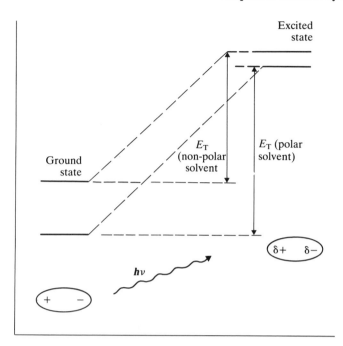

Fig. 5.9 Schematic energy diagram for excitation of a solvatochromic dye of the type **9** in polar and non-polar solvents.

polarity accompanied by a shift of the absorption maximum to a shorter wavelength (blue-shift).

Numerous solvatochromic dyes have been proposed as the basis of solvent parameters. The transition energy, E, is measured as wavelength or frequency at the maximum absorption,

$$E = hv = hc/\lambda = hc\bar{v}. \tag{5.18}$$

The major advantages of such scales include ease of measurement and ability to obtain values for almost all solvents — all that is required is a limited solubility and lack of chemical reaction.

The Z-parameter

The pyridinium iodide, **5**, undergoes a charge-transfer transition with the absorption of visible light to form an excited state, **6**, having a much smaller dipole moment (8·6 D) than that of the ground state (13·9 D)[54]. Transfer from a non-polar to a polar solvent destabilizes the ground state to some extent, but the excited state more so, and the excitation energy is raised in proportion to solvent polarity. Kosower[55–57] defined a solvation parameter,

193

Z (Table 5.4):

$$Z/\text{kcal mol}^{-1} = \frac{2\cdot859 \times 10^4}{\lambda_{max}/\text{nm}}.\qquad (5.19)$$

The χ_B, χ_R-scales

The original suggestion that solvatochromic dyes could be used to measure solvent polarity appears to have originated with Brooker[58] who, at the Kodak laboratories, had access to a variety of highly conjugated novel dyes, many of which showed very solvent-sensitive transitions. Two compounds with more polar excited states than ground states are **7** and **8**, whose transition energies were used to define solvation scales χ_B and χ_R, respectively. It is less desirable to use a dye whose dipole moment increases in excitation than one which shows a decrease, since stabilization of the excited state by polar solvents is required; this needs sufficient time for solvent reorganization, time which may not be available during an electronic transition (10^{-16} s). Hence polarizability of the solvent may be unduly important in the solvatochromism of the former type.

The E_T-parameter

The most solvatochromic dye yet found is the pyridinium betaine, **9**, whose visible transition leads to the less polar excited state, **10**. The solvent parameter, E_T (Table 5.4), based on this compound by Dimroth and

Reichardt[59,60], was defined by Eq. (5.20);

$$E_T/\text{kcal mol}^{-1} = \frac{hc}{\lambda_{max} \text{ for } 9}.$$ (5.20)

Compound **9** is so solvatochromic that λ_{max} shifts from 450 nm (green) in water to around 220 nm (far-UV) in alkanes. One limitation is the exclusion of acidic solvents which protonate the phenolate ion of **9** and prevent the electronic transition.

The S-parameter

Brownstein[61] fitted rates of a number of reactions and absorption maxima of solvatochromic dyes to Eq. (5.21):

$$\log k/k_0 \text{ (or } h\nu) = RS,$$ (5.21)

where S is the solvent polarity parameter and R, the reaction/spectrum sensitivity constant. The expression is equivalent to Eqs (5.15) and (5.17) but S-values were averaged over more than 40 different solvent-sensitive probes (Table 5.4).

The P-scale

[19]F-chemical shifts are very sensitive to electronic perturbation; a compound such as p-nitrofluorobenzene has a solvent-dependent chemical shift since solvation of the nitro group by donor solvents will affect its electron-withdrawing capacity. Solvent P values were defined as

$$P = (\delta - \delta_0)_s - (\delta - \delta_0)_b,$$

where δ, δ_0 are chemical shifts of p-nitrofluorobenzene and fluoroben-zenezene respectively, and s, b refer to test solvent and benzene (standard solvent)[62-63].

The π*-parameter

By an approach similar to that of Brownstein, Taft[64] derived average solvation parameters for 70 solvents by the use of solvatochromic shifts of some 50 indicator compounds. Hydrogen-bonding solvents were excluded and the scale is claimed as a measure of non-specific electrostatic solvation (Table 5.4).

5.6.4 Scales for specific solvation

Hydrogen-bonding capabilities of a solvent frequently affect IR and NMR characteristics of a solute.

Infrared stretching frequencies

An IR-active mode is necessarily associated with a periodic dipole charge, which should be solvent-dependent. The Kirkwood–Bauer–Magat equation, Eq. (5.22), expresses this in terms of electrostatic theory[65]:

$$\frac{\Delta v}{v_0} = \frac{v_0 - v}{v_0} = c\left(\frac{\varepsilon - 1}{2\varepsilon + 1}\right) = c'\left(\frac{n^2 - 1}{2n^2 + 1}\right), \tag{5.22}$$

where c, c' are constants. The expression holds, however, only for a very limited range of non-polar solvents. A more useful scale is derived from the O—H or O—D stretching frequencies of methanol and methanol-d which are lowered by solvents capable of hydrogen-bonding. The basicity (electron-pair donor) solvation parameter, B, is defined by Shorter[66,67] as

$$B/cm^{-1} = (\bar{v}^{\circ}_{O-D} - \bar{v}_{O-D}), \tag{5.23}$$

where $\bar{v}_{O-D}$ refers to the wavenumber of the O—D vibration of MeOD in the basic solvent and $\bar{v}^{\circ}$, that in CCl$_4$. The scale may be extended to protic solvents in which MeOD would exchange, by the use of $\bar{v}_{O-H}$ of phenol.

Values of B are given in Table 5.8[68-70].

Donor Numbers, DN

The capability of electron-pair donation (Lewis basicity, nucleophilicity) may be related to the enthalpy of interaction of donor species with standard acceptor (Lewis acid). Gutmann has set up a scale of Donor Numbers, DN (Table 5.8), defined from the enthalpy of interaction of a donor solvent, S:, with SbCl$_3$ in dilute dichloroethane solution[66,71,72],

$$S:\frown SbCl_3 \rightarrow S\text{-}SbCl_3; \qquad \Delta H/kcal\,mol^{-1} = DN$$

The scale is supplemented by data derived from ^{13}C-NMR chemical shifts of CF$_3$I and the ionization of Ph$_3$C—Cl in the test solvents.

Table 5.8   Solvation scales for acceptor and donor properties

Solvent	Taft-α	Gutmann AN
CF_3COOH		105
$Et_3PO \cdot SbCl_5$		100 (reference)
Acetic acid		52·9
Water	1·13	54·8
MeOH	0·98	41·3
EtOH	0·86	37·1
MeCHOHMe	0·78	33·5
CF_3CH_2OH	1·35	
1-PrOH	0·80	
1-BuOH	0·79	
tert-BuOH	0·62	
$HOCH_2CH_2OH$	0·92	
$PhCH_2CH_2OH$	0·43	
$HCONH_2$	0·66	39·8
HCONHMe		32·1
$HCONMe_2$		16·0
MeSOMe		19·3
$MeNO_2$	0·23	
MeCN	0·15	18·9
PhCN		15·5
MeCOMe	0·07	
CH_2Cl_2	0·22	
$CHCl_3$	0·34	
1,4-Dioxane		10·8
Benzene		8·3
Ether		8·2
Tetrahydrofuran		8·2

Solvent	Taft-β	Gutmann DN	Shorter B
Water	0·18	18·0	
MeOH	0·62	19·0	
EtOH	0·77	31·5	
2-PrOH	0·95		
1-BuOH	0·88		
tert-BuOH	1·01		
Pyridine	0·64	33·1	
$EtNH_2$		55·5	
Et_3N		61·0	314
$PhNO_2$	0·39	4·4	53
PhCN	0·41	11·9	97
MeCN	0·31	14·1	103
$(EtO)_3PO$	0·77	23	
$Me_2S{=}O$	0·76		192
MeCONMe	0·76	27·8	178
$HCONMe_2$	0·69	26·6	166
$O{=}C(NMe_2)_2$	0·78		
MeCOMe	0·48	17·0	123
Cyclohexanone	0·53		132
PhCOMe	0·49		108
EtOEt	0·47	19·2	130
1,4-Dioxane	0·37	14·8	128
Tetrahydrofuran	0·55	20·0	145
Tetrahydropyran	0·54		
Sulfolane		14·8	

Table 5.8 (*Continued*)

Solvent	Taft-β	Gutmann DN	Shorter B
PhOMe	0·22		75
MeCOOEt	0·45	17·1	89
PhCOOEt	0·41		
Cl₃CCOOEt	0·61		
HCOOEt	0·61		
PhMe, PhH	0·0		52

The D_π-scale

Rates of cycloaddition of diphenyldiazomethane, **11**, to tetracyanoethene, **12**, are retarded by solvents capable of acting as donor molecules since this reduces the reactivity of the tetracyanoethene towards the substrate. A solvation scale of donor character towards a π-acid (i.e. electrophile), D_π, has been defined with reference to this reaction [10]:

$$\log k_0/k = D_\pi \tag{5.24}$$

in which k_0 refers to the rate in the standard solvent (benzene).

The β-scale

Basicity towards hydrogen bonding, i.e. capability of acting as a hydrogen bond acceptor, was measured by Taft[67] as the solvatochromic shift of 4-nitroaniline (whose transition is facilitated by hydrogen bonding of the amino group) relative to that of *N,N*-diethyl-4-nitroaniline (which is incapable of hydrogen bonding). Other scales used were based on H-bond formation constants and NMR chemical shifts. These were subjected to an averaging procedure and contributions from solvent polarity and polarizability were subtracted out to give a unified scale of basicity values, β, shown in Table 5.8.

The α-scale of hydrogen-bond acceptors

By an approach similar to that used for construction of the β-scale, Taft has also constructed a scale of hydrogen-bond acidity (capability to provide a proton) by measuring solvatochromic shifts of a number of dyes in protic solvents and subtracting out polarity, polarizability and H-bond donor contributions. After a statistical averaging procedure, a unified scale of α-values remains (Table 5.8)[73].

Acceptor Numbers, AN

Gutmann based a scale of Lewis acidity on the ability of an acceptor solvent to impart a chemical shift to the ^{31}P-NMR of triethylphosphine oxide, proportional to the strength of the interaction (Table 5.8).

$$Et_3P{=}O\mathpunct{:}\overset{\frown}{}S \to Et_3P{=}O\cdots S$$

$$AN = 100 \times \delta/\delta_0 = 2{\cdot}348 \times \delta, \tag{5.25}$$

where δ, δ_0 refer to chemical shifts of $Et_3P{=}O$ in dichloroethane solution in the presence of test solvent and $SbCl_3$, respectively[74].

5.7 RELATIONSHIPS BETWEEN EMPIRICAL SOLVATION SCALES

Many of the multitude of solvation scales which have been proposed respond in a similar fashion to solvent change. Two scales (S_1 and S_2) are linearly related if Eq. (5.26) holds:

$$S_1 = aS_2 + b. \tag{5.26}$$

Establishing such relationships permits many redundant scales to be abandoned. Table 5.9 sets out the regression constants, a and b, and correlation coefficients, r, for relationships between a number of polarity scales.

Table 5.9   Coefficients of Eq. (5.26) for solvation scale correlation[40,67]

S_2	S_1	a	b
E_T	Y	0·357	−19·88
E_T	Y(neophyl)	0·197	−13·26
E_T	$\mathscr{S}$	1·41	−10·69
E_T	Z	1·41	6·92
E_T	S	0·014	−0·733
E_T	Φ^a	0·0126	−0·358
δ	E_T	2·13	20·0
δ	Z	3·60	32·3
δ	S	0·046	−0·624
π^*	δ	0·203	−1·41
π^*	P	4·3	−0·1

^{a} Φ is a solvatochromic solvation parameter using the $n \to \pi^*$ transition of ketones[75].

The high values of $r(\rightarrow 1 \cdot 0)$ indicate strong similarities between these scales. The E_T and π^* scales are very closely related over a select group of solvents, although solvents which participate in hydrogen bonding usually deviate. These two, and perhaps the δ-scale, are therefore recommended for general use as indicators of polarity.

5.8 THE USE OF SOLVATION SCALES IN MECHANISTIC STUDIES

The variation of reaction rate with the properties of the solvent holds mechanistic information. Measurements may be made in a wide range of solvents if it is desired to ascertain the solvation forces responsible for rate variation. Frequently all that is sought is qualitative information as to the sensitivity of a given reaction to solvent polarity. The relative rates just in cyclohexane (very non-polar) and in acetonitrile (polar, aprotic) may give sufficient information to support or reject one or other mechanism, as is the case for the following three cycloadditions.

The solvent effect on the Diels–Alder reaction, **13**, is very small and clearly points to a transition state in which solvation is not much different from that of the reagents. In principle this transition state might be *en route* to the diradical, **13a**, or it might be that of a concerted reaction, **13b**, but it cannot be a dipolar species such as **13c**[45].

The very large solvent effect on the $(2 + 2)$ cycloaddition, **14**, of tetracyanoethene to the vinyl ether equally unambiguously points to a very dipolar transition state which would lead to the intermediate **14a**[46]. The

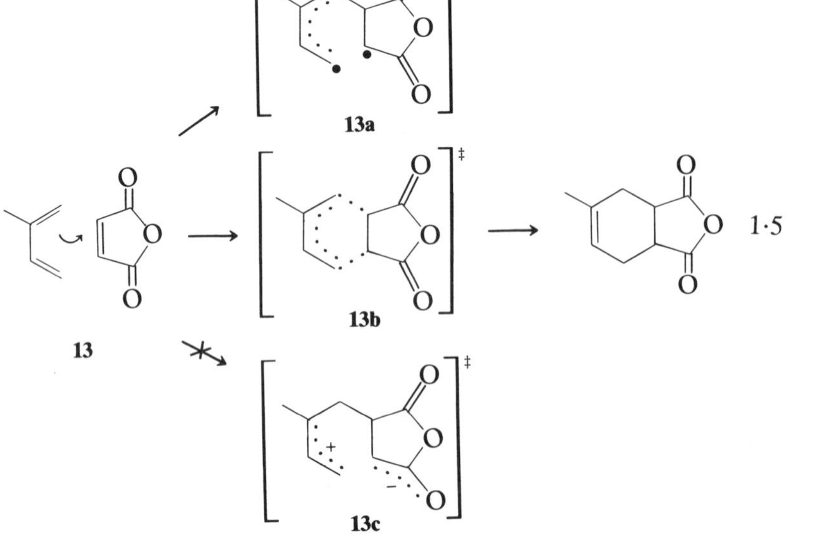

$$\frac{k_{\mathrm{MeCN}}}{k_{\mathrm{C_6H_{12}}}}$$

13a

13 → 13b → 1·5

13c

$$\dfrac{k_{MeCN}}{k_{C_6H_{12}}}$$

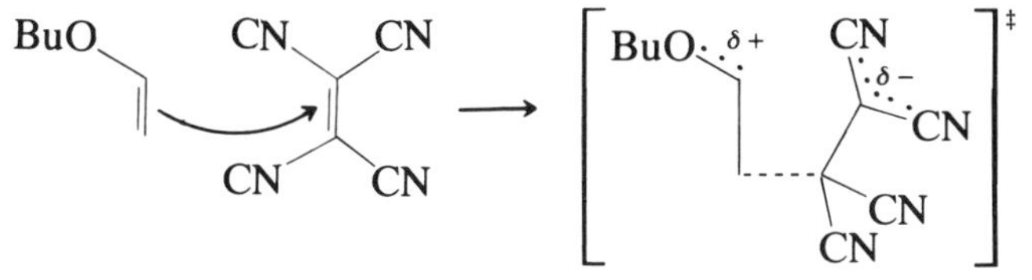

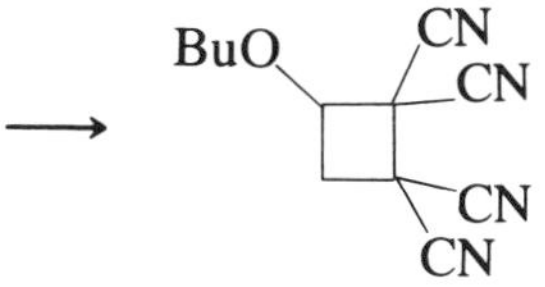

10,800

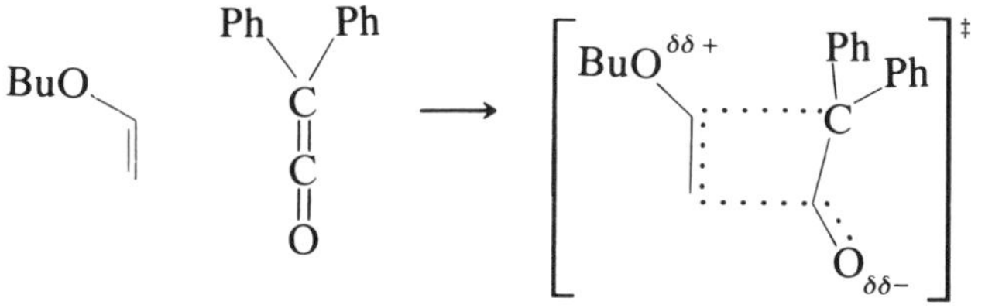

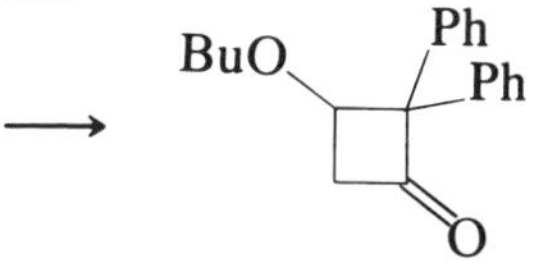

163

moderate solvent effect on the cycloaddition **15**, of diphenylketene to vinyl ether may be interpreted in terms of a concerted cycloaddition with dipolar character via, for instance, a transition state such as **15a**[48] but is rather ambiguous as it could also signify a reaction similar to **14** but with a very early transition state in which little charge is developed.

Some reactions may be favored by a non-polar rather than by a polar solvent. The racemization of the pyramidal allyl sulfoxide, **16**, is a case pointing to a transition state of lower dipole moment than the reagent. A simple inversion would not account for this and it is therefore suggested (though by no means proven on this evidence alone) that the reaction proceeds by allylic rearrangement via the sulfenyl ester, **17**.

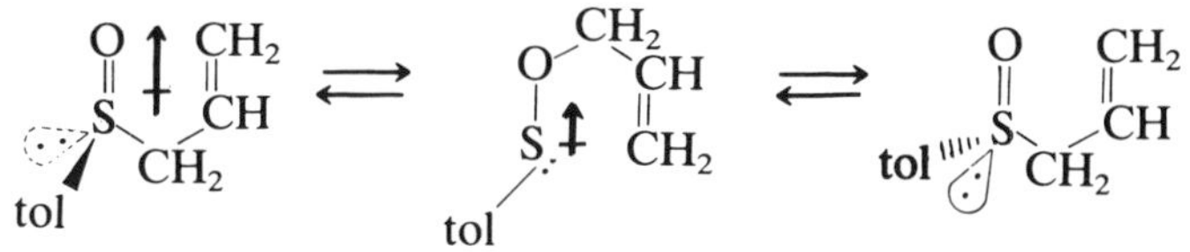

Solvent.	cyclohexane	benzene	dioxane	MeCN	EtOH
k_{rel}:	30	11	7	2·1	1

The correlation of a series of rates with solvent properties requires consideration of the best scale for the purpose. When protic solvents may be excluded, a plot of ln k against E_T, π^* or δ may be used — each scale has its adherents — and, if required, the transition-state properties may be calculated by Eq. (5.13). It seems that the π^*-parameter is a better measure of dipolar interactions, while δ correlates better with non-dipolar reactions, concerted cycloadditions, radical reactions and the like. Figure 5.8b, for example, shows such a plot which qualitatively demonstrates the reaction to be very sensitive to solvent polarity. One could use the slope as an index of sensitivity to solvent polarity (analogous to ρ, the reaction constant of the Hammett equation) but, because of the often-considerable scatter of points, it has not been the practice to do this except in the case of solvolytic reactions whose rates correlate with the Y-scale. The slope of a plot of ln k for a solvolysis against Y (protic solvents only) is the sensitivity constant, m_s, of Eq. (5.15) and is used as a criterion to distinguish ionization (S_N1, $m_s \sim 1$) and synchronous displacement (S_N2, $m_s \sim 0 \cdot 2$) routes; see Section 10.1.2. The difference in behavior is due to a greater development of charge in the S_N1 than in the S_N2 process.

One must select the solvation parameter which best fits the data under examination, to reveal the solvation forces which are of greatest importance. In the examples above, the Y-scale correlates well for solvolysis alone, while δ, E_T or π^* are most suitable where electrostatic forces are important, or perhaps the β-scale is appropriate where hydrogen bonding is a significant influence[76].

A more refined insight into solvation forces may be achieved by multicorrelation analysis using more than one solvation parameter.

5.8.1 Multiparameter solvation analysis

If solvation is due to a summation of contributions from several definite and independent physical processes, each with its unique scale of measurement, it should be possible to relate variations in rate or spectroscopic properties to a multiparameter equation such as Eq. (5.27), analogous to that described for electronic effects.

$$\log k_{rel} = pP + nN + aA + bB + \cdots \tag{5.27}$$

in which $P, N, A, B \ldots$ are values on solvation scales expressing, for example, polarity, polarizability, acidity and basicity. Any other terms which are relevant may be added. Then $p, n, a,$ and $b,$ the regression coefficients, express sensitivity of the system to these components of solvation. The increase in the number of independent variables inevitably improves the correlation so that it is necessary to have data referring to many solvents and to carry out statistical tests for the significance of the fit, e.g. the multicorrelation coefficient, $R,$ all of which require computation. Furthermore, in order for such an analysis to be meaningful, the solvation

Table 5.10   Some recommended sets of independent solvation scales

Solvent property	Preferred scale		
	Koppel and Pal'm[6]	Taft[64]	Krygowski and Fawcett[77]
P (polarity)	f_ε	π^*	E_T
N (polarizability)	f_n	—	—
E (electrophilicity or Lewis acidity)	E^a	α	—
B (nucleophilicity or Lewis basicity)	$\bar{v}_{O-D}(B)$	β	DN

[a] E was defined as $E = E_T - 25 \cdot 57 - 14 \cdot 39 f_\varepsilon - 9 \cdot 08 f_n$ which incidentally analyzes the E_T scale into three components. It may be seen from this that Krygowski and Fawcett's analysis really includes four parameters, three being combined in E_T and that π^* contains both P and N.

scales which are used ideally must each respond to one type of interaction only. Some subjective judgment may be needed to decide how many parameters are really necessary to fit the data. Finally, the magnitudes of the coefficients may be interpreted in terms of mechanism. This approach has been used by Koppel and Pal'm[6] and by Taft and coworkers[64] who have correlated many rate and spectroscopic phenomena with the same four solvent properties using, however, different scales to express these relationships. Krygowski and Fawcett[77] have found two scales sufficient, E_T and basicity. Table 5.10 summarizes these findings.

Some examples of multicorrelation analysis of solvent effects follow.

(a) Solvolysis of tert-butyl chloride

Koppel and Pal'm:

$$\log k/s^{-1} = -19 \cdot 50 + 5 \cdot 67 f_\varepsilon + 17 \cdot 27 f_n + 0 \cdot 379 E; \qquad R = 0 \cdot 982.$$

Taft:

$$\log k/k_0 = 5 \cdot 6 + 8 \cdot 54 \pi^* + 7 \cdot 06 \alpha; \qquad R = 0 \cdot 995.$$

These regressions agree in that the most important effect stabilizing the transition state is non-specific polarity together with a contribution from Lewis acidity, which presumably indicates that solvation of the developing halide ion assists ionization.

(b) Association constants of trialkylammonium ions

$$R_3\overset{+}{N}-H + B \underset{}{\overset{K_f}{\rightleftharpoons}} (R_3N-H\cdots B)^+,$$

in which B is the solvent, correlates with polarity and basicity:

$$\log K_f = -2 \cdot 30 + 1 \cdot 40 \pi^* + 7 \cdot 10 \beta.$$

Free energies of transfer of tetraalkylammonium salts, either as ion-pairs or free ions, correlate with π^*, α and δ_H but are independent of β-values[79]:

$$\Delta G_{tr} = s\pi^* + a\alpha + h\delta_H^2 \,(+ \text{const.}).$$

(c) The reaction of diphenyldiazomethane with benzoic acid

This is a slow proton transfer which correlates also with polarity and basicity:

$$Ph_2CN_2 + PhCOOH \xrightarrow{\text{slow}} (Ph_2\overset{+}{C}HN_2 \cdot Ph\overset{-}{COO}) \rightarrow Ph_2CHOCOPh + N_2$$

$$\log k = 2 \cdot 333 + 6 \cdot 505 f_\varepsilon - 5 \cdot 599 \, DN.$$

In this case the coefficient of the basicity parameter is negative, indicating that donor ability in the solvent stabilized the reagent (presumably benzoic acid, by hydrogen bonding), slowing down the rate.

Swain has developed two solvent parameters, A and B, expressing ability to solvate anions and cations, respectively (by any mechanism)[78]. The scales were generated by statistical analysis of a large number of rate–solvent data setting $A = B = 0$ for *n*-heptane, $A = B = 1$ for water, $A = 0$ for hexamethylphosphoric triamide and $B = 0$ for trifluoroacetic acid (Table 5.11). They are applied by means of Eq. (5.28), as shown in the examples following.

$$\log k_{rel} = aA + bB \qquad (5.28)$$

Example	a	b
1 Solvolysis of MeBr	−4·23	3·44
2 Solvolysis of *tert*-BuBr (Y-scale)	5·63	6·13
3 MeI + NEt$_3$ (Menshutkin reaction)	0·94	4·38
4 Me$_4$Sn + Br$_2$	7·12	6·27

Rates of the S_N1 reaction, 2, are greatly assisted by solvation of both carbocation and departing anion. Solvation, mainly of the cation, is important in 3, since the large I$^-$ ion has little tendency to solvate — unlike Br$^-$ in 4. The opposite signs of a and b in 1 originate in complexing interactions between anion- and cation-solvating centers in the solvent (in this case the hydrogen-bonding centres of the protic solvents).

Analyses of this kind can add greatly to the interpretation of solvent effects and help in mechanistic deductions but they are as yet essays in the science of disentangling solvent effects and it is likely that the future will see further refinement along the same lines[31].

Table 5.11   Swain A and B solvation parameters [78]

Solvent	A	B
CF₃COOH	1·72	0·00
CH₃COOH	0·93	0·13
HCOOH	1·18	0·51
Water	1·00	1·00
MeOH/H₂O, 96:4	0·76	0·61
EtOH, 80:20	0·75	0·65
EtOH/H₂O, 60:40	0·80	0·77
EtOH/H₂O, 50:50	0·82	0·80
MeOH	0·37	0·86
EtOH	0·66	0·45
MeCOMe/H₂O, 80:20	0·62	0·70
Propan-1-ol	0·63	0·44
Propan-2-ol	0·59	0·44
Butan-1-ol	0·61	0·43
Me₃COH	0·45	0·50
HOCH₂CH₂OH	0·78	0·84
BuNH₂	0·15	1·17
Pyridine	0·24	0·96
PhNH₂	0·36	1·19
Et₃N	0·08	0·19
MeCN	0·37	0·86
MeNO₂	0·39	0·92
MeSOMe	0·34	1·08
HCONH₂	0·66	0·99
HCONMe₂	0·27	0·97
MeCONMe₂	0·27	0·97
PhNO₂	0·29	0·86
(Me₂N)₃P=O	0·00	1·07
MeCOMe	0·25	0·81
MeCOEt	0·23	0·74
MeCOPh	0·23	0·90
CH₂Cl₂	0·33	0·80
CHCl₃	0·42	0·73
CCl₂=CCl₂	0·10	0·25
CHCl=CCl₂	0·16	0·54
CH₂ClCH₂Cl	0·30	0·82
CCl₄	0·09	0·34
CS₂	0·10	0·38
MeCOOEt	0·21	0·59
EtOEt	0·12	0·34
PhOMe	0·21	0·74
PhCl	0·20	0·65
PhMe	0·13	0·54
PhH	0·15	0·59
n-C₆H₁₄	0.00	0.00

PROBLEMS

1 Comment upon the observation that $\Delta pK_A = [pK_A(H_2O) - pK_A(MeOH)]$ is $-5\cdot7$ for formic acid but only $-0\cdot4$ for anilinium, $PhNH_3^+$.

2 The tautomeric equilibrium of some β-dicarbonyl compounds shown below is strongly affected by solvent. Suggested factors which contribute to the observed trends. [C. Reichardt, *Solvent Effects in Organic Chemistry*, Verlag Chemie, Weinheim, 1979.]

R = OEt   R = Me

Solvent	K_T	%Enol	K_T	%Enol
Gas phase	0·74	42·6	11·7	92.1
n-Hexane	0·64	39	19	95
Carbon tetrachloride	0·39	28	24	96
Diethyl ether	0·29	22	19	95
Carbon disulfide	0·25	20	16	94
Benzene	0·19	16	8·1	89
1,4-Dioxane	0·12	11	4·6	82
Ethanol	0·11	10	4·6	82
Chloroform	0·081	7·5	6·7	87
Pure solute	0·081	7·5	4·3	81
Methanol	0·062	5·8	2·8	74
Acetonitrile	0·052	4·9	1·6	62
Dimethylsulfoxide	0·023	2·2	1·6	62
Acetic acid	0·019	1·9	2·0	67

3 Effects of solvent on rates of some cycloadditions (Chapter 14) are given below. Comment upon the probable nature of the transition states in each case.

(a) Diels–Alder:

k_{rel}: gas phase, 1; neat liquid, 0·8; benzene, 1·0; CCl$_4$, 1·1; nitrobenzene, 1·9; ethanol, 2·8.

(b) (2 + 2) cycloaddition:

Solvent:	C_6H_{12}	C_6H_5Cl	CH_3COCH_3	C_6H_5CN	CH_3CN
$K_{2,rel}$:	1	13	43	63	163

(c) (2 + 2) cycloaddition:

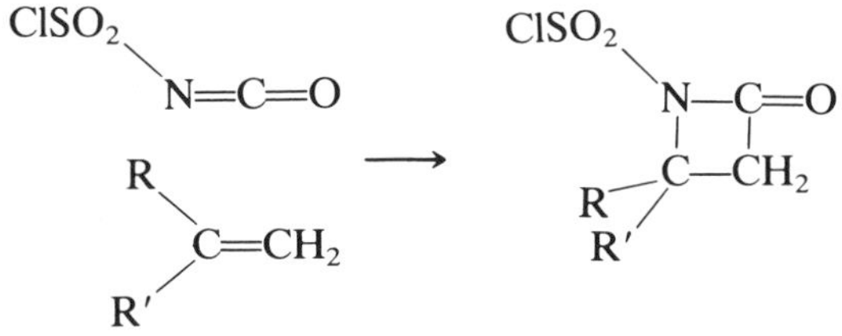

Solvent:	C_6H_{14}	$(C_2H_5)_2O$	$CHCl_3$	CH_2Cl_2	$C_6H_5NO_2$	CH_3NO_2
$k_{2,rel}$:	1	31	250	1700	5000	18 800

[See Chapter 14.]

4 Predict the effect on the rate accompanying a change to a more polar solvent for the following reactions (N.B. the appropriate mechanism must be taken into account in all cases):

(a) $Et_3S^+Br^- \rightarrow Et_2S + EtBr$.
(b) $n\text{-}Pr_3N + MeI \rightarrow n\text{-}Pr_3NMe^+I^-$.
(c) $Me_3N + Me_3S^+ \rightarrow Me_4N^+ + Me_2S$.
(d) $OH^- + Et_4N^+ \rightarrow H_2O + CH_2{=}CH_2 + Et_3N$.
(e) $CH_2{=}CH_2 + ArSCl \rightarrow ArSCH_2CH_2Cl$.
(f) $Me_3N + Et_4N^+ \rightarrow Me_3NH^+ + CH_2{=}CH_2 + Et_3N$.

REFERENCES

1 D. J. Cram, B. Rickborn, C. A. Kingbury and P. Haberfield, *J. Amer. Chem. Soc.*, **83**, 3678 (1961).
2 V. Gutmann, *Inorg Nucl. Chem. Letts*, **2**, 257 (1966).
3 V. Gutmann and R. Schmidt, *Monatsch*, **102**, 798, 806, 1217 (1971).
4 T. Oshima, S. Arokata and T. Nagai, *J. Chem. Res.*, (*S*), 204; (*M*), 2518 (1981).
5 I. A. Koppel and A. I. Pal'm, *Org. React.* (*Tartu*), **11**, 121 (1974).
6 I. A. Koppel and A. I. Pal'm, *Advances in LFER* (N.B. Chapman and J. Shorter, Eds), Plenum, New York, 1972.
7 M. H. Abraham, *Prog. Phys. Org. Chem.*, **11**, 1 (1974).
8 J. March, *Advanced Organic Chemistry*, McGraw-Hill, New York, 1977.
9 E. S. Amis, *Solvent Effects on Reaction Rates and Mechanisms*, Academic Press, New York, 1966.
10 E. S. Amis and J. F. Hinton, *Solvent Effects on Chemical Phenomena*, Vol. 1, Academic Press, New York, 1973.
11 M. J. R. Dack, 'Influence of Solvent on Chemical Reactivity', in *Techniques of Organic Chemistry*, Vol. VIII Part II (A. Weissberger, Ed.), Wiley, New York, 1976, p. 95.
12 D. Henderson 'The Liquid State' in *Physical Chemistry, An Advanced Treatise* (H. Eyring, D. Henderson and W. H. Jost, Eds), Vol. 8, Academic Press, New York, 1971, pp. 377, 414.
13 J. S. Rawlinson, *Liquids and Liquid Mixtures*, Butterworth, London, 1969.

14 F. Kohler, *The Liquid State,* Verlag Chemie, Weinheim, 1972; W. Kohler, E. Wilhelm and H. Posch, *Adv. Molec. Relax. Proces.* 195 (1976).
15 I. R. MacDonald and K. Singer, *Chem. in Brit.,* **9,** 54, (1973).
16 D. Eisenberg and W. Kauzmann, *The Structure and Properties of Water,* Oxford U.P., 1969.
17 G. J. Hills, *Soc. Exptl. Biol. Symp. XXVI,* Chap. 1, Cambridge U.P., 1972.
18 R. A. Horne, 'The Structure of Water and Aqueous Solutions', in *Survey of Progress in Chemistry* (A. F. Scott, Ed.) **4,** 1 (1968).
19 M. J. R. Dack, 'Solutions and Solubilities', in *Techniques of Organic Chemistry* (A. Weissberger, Ed.), Vol. VIII, Wiley, New York, 1976.
20 E. M. Kosower, *Introduction to Physical Organic Chemistry,* Chapter 2, Wiley, New York, 1968.
21 E. S. Amis, 'The Solvation of Ions', in *Techniques of Organic Chemistry* (A. Weissberger, Ed.), Vol. VIII, Part I, Chap. III, Interscience, New York, 1975.
22 E. M. Arnett and D. R. McKelvey, *J. Amer. Chem. Soc.,* **87,** 1393 (1965).
23 M. H. Abraham and R. J. Abraham, *J. Chem. Soc., Perkin II,* 47 (1974).
24 M. H. Abraham and R. J. Abraham, *J. Chem. Soc., Perkin II,* 1677 (1975).
25 M. H. Abraham, *J. Chem. Soc., Perkin II,* 1343 (1972).
26 G. R. Hedwig, D. A. Owensby and A. J. Parker, *J. Amer. Chem. Soc.,* **97,** 3888 (1975); B. G. Cox and A. J. Parker, *J. Amer. Chem. Soc.,* **95,** 402, 408 (1973).
27 A. J. Parker, *Chem. Rev.,* **69,** 1 (1969).
28 A. J. Parker, *Pure Appl. Chem.,* **53,** 1437 (1981).
29 R. Alexander, E. C. F. Ko, A. J. Parker and T. J. Broxton, *J. Amer. Chem. Soc.,* **90,** 5049 (1968).
30 E. Buncel and H. Wilson, *Acc. Chem. Res.,* **12,** 42 (1979).
31 A. Ben-Naim, 'Molecular Origins of Ideal Solutions', in *Techniques of Organic Chemistry* (M. J. R. Dack and A. Weissberger, Eds), Vol. VIII, Wiley, New York, 1976.
32 P. Kebarle, *Ann. Rev. Phys. Chem.,* **28,** 447 (1977).
33 H. Hartmann, *Z. Physik. Chem. (Frankfurt),* **66,** 183 (1969).
34 N. S. Isaacs, *J. Chem. Soc.,* 1361 (1967).
35 M. Born, *Z. Physik,* **1,** 45 (1920).
36 J. G. Kirkwood, *J. Chem. Phys.,* **2,** 351 (1934).
37 L. Onsager, *J. Amer. Chem. Soc.,* **58,** 1486 (1936).
38 Y. Drougard and D. Decroocq, *Bull. Soc. Chim. France,* 2972 (1969).
39 M. Auriel and E. de Hoffmann, *J. Amer. Chem. Soc.,* **97,** 7433 (1975).
40 H. F. Herbrandson and F. R. Neufeld, *J. Org. Chem.,* **31,** 1141 (1966).
41 J. Hildebrand and R. L. Scott, *Regular Solutions,* Prentice Hall, Englewood Cliffs, NJ, 1962.
42 J. H. Hildebrand, J. M. Rousnitz and R. L. Scott, *Regular and Related Solutions,* Van Nostrand–Reinhold, Princeton, 1970.
43 J. L. Hildebrand and R. L. Scott, *The Solubility of Non-Electrolytes,* Reinhold, New York, 1950, p. 419.
44 J. L. Hildebrand and S. E. Wood, *J. Chem. Phys.,* **1,** 817 (1933).
45 N. S. Isaacs, *Liquid Phase High Pressure Chemistry,* Wiley, Chichester, 1981.
46 K. A. Connors and S.-R. Sum, *J. Amer. Chem. Soc.,* **93,** 7239 (1971).
47 M. J. R. Dack, *Chem. Soc. Rev.,* **4,** 211 (1975).
48 E. Grunwald and S. Winstein, *J. Amer. Chem. Soc.,* **70,** 846 (1948).
49 S. Winstein, E. Clippinger, A. H. Fainberg and G. C. Robinson, *Chem. and Ind.,* 664 (1954); *J. Amer. Chem. Soc.,* **76,** 2597 (1954); **74,** 2165 (1954); **78,** 328, 2763, 2767, 2777, 2780, 2784 (1952).
50 P. R. Wells, *Chem. Rev.,* **63,** 171 (1963).
51 S. G. Smith, A. H. Fainberg and S. Winstein, *J. Amer. Chem. Soc.,* **83,** 618 (1961).
52 C. Lassau and J. C. Jungers, *Bull. Soc. Chim. France,* 2678 (1968).
53 M. Gielen and J. Nasielski, *J. Organometall. Chem.,* **1,** 173 (1963); **7,** 273 (1967).

54 J. W. Larsen, A. G. Edwards and P. Dobi, *J. Amer. Chem. Soc.,* **102,** 6780 (1980).

55 E. M. Kosower and M. Mohammed, *J. Amer. Chem. Soc.,* **90,** 3271 (1968); **93,** 2713 (1971); *J. Phys. Chem.,* **74,** 1153 (1970).

56 E. M. Kosower, *J. Amer. Chem. Soc.,* **80,** 3253 (1958); *J. Chem. Phys.,* **61,** 230 (1964).

57 E. M. Kosower, *Physical Organic Chemistry,* Wiley, New York, 1968.

58 L. G. S. Brooker, A. C. Craig, D. W. Heseltine, P. W. Jenkins and L. L. Lincoln, *J. Amer. Chem. Soc.,* **87,** 2443 (1965).

59 C. Reichardt, *Angew. Chem., Int. Ed.,* **3,** 30 (1964).

60 C. Reichardt, *Solvent Effects in Organic Chemistry,* Verlag Chemie, Weinheim, 1979.

61 S. Brownstein, *Can. J. Chem.,* **38,** 1591 (1960).

62 R. W. Taft, G. B. Klingensmith, E. Price and R. Fox, *Symp. LFER Correlations,* Durham, 1964, p. 275.

63 R. E. Uschold and R. W. Taft, *Org. Mag. Res.,* **1,** 375 (1969).

64 J. L. M. Abboud, M. J. Kamlet and R. W. Taft, *Prog. Phys. Org. Chem.,* **13,** 485 (1981).

65 L. J. Bellamy, *Infrared Spectroscopy of Complex Molecules,* Methuen, London, 1960.

66 V. Gutmann, *Chem. Brit.,* **7,** 102 (1971); V. Gutmann and R. Schmidt, *Coord. Chem. Rev.,* **12,** 263 (1974).

67 M. J. Kamlet, J.-L. Abboud, M. E. Jones and R. W. Taft, *J. Chem. Soc., Perkin II,* 342 (1979).

68 M. J. Kamlet and R. W. Taft, *J. Amer. Chem. Soc.,* **98,** 377 (1976).

69 T. Kagiya, Y. Sumida and T. Inoue, *Bull. Chem. Soc., Japan,* **41,** 767 (1968).

70 A. G. Burden, G. Collier and J. Shorter, *J. Chem. Soc., Perkin II,* 1627 (1976).

71 V. Gutmann, *Electrochem. Acta,* **21,** 661 (1976).

72 V. Gutmann and U. Mayer, *Monatsch.,* **100,** 2048 (1969).

73 R. W. Taft and M. J. Kamlet, *J. Amer. Chem. Soc.,* **98,** 2886 (1976).

74 V. Gutmann, *Electrochem. Acta,* **21,** 661 (1976).

75 J. Dubois and A. Bienvenue, *J. Chim. Phys.,* **65,** 1259 (1968).

76 T. Yakoyama, R. W. Taft and M. J. Kamlet, *J. Amer. Chem. Soc.,* **98,** 3233 (1976).

77 T. M. Krygowski and W. R. Fawcett, *J. Amer. Chem. Soc.,* **97,** 2143 (1975); *Austr. J. Chem.,* **28,** 2115 (1975); *Can. J. Chem.,* **54,** 3283 (1976).

78 C. G. Swain, M. S. Swain, A. L. Powell and S. Alunni, *J. Amer. Chem. Soc.,* **105,** 502 (1983).

79 R. W. Taft, M. H. Abraham, R. M. Doherty and M. J. Kamlet, *J. Amer. Chem. Soc.,* 3105, (1975).

6 Acids and bases, electrophiles and nucleophiles

6.1 ACID–BASE DISSOCIATION

In the Brønsted sense [1], an acid–base reaction is the transfer of a proton from a donor (acid) to an acceptor (base), usually under conditions of equilibrium. Although any base and any medium may be considered, there is a special significance attached to dissociations in water which can act as acid, base and solvent and is capable of supporting the partial dissociations of many classes of weak organic acids and bases.

$$H_2\overset{\frown}{O} + \overset{\frown}{H}{-}\overset{\frown}{A} \rightleftharpoons H_2\overset{+}{O}{-}H + A^-$$

The equilibrium constant for this reaction is the acid dissociation constant, K_A, of the acid HA:

$$K'_A = \frac{a_{A^-}a_{H_3O^+}}{a_{HA}a_{H_2O}} = \frac{[A^-][H_3\overset{+}{O}]}{[HA][H_2O]} \cdot \frac{y_{A^-}y_{H_3O^+}}{y_{HA}y_{H_2O}}. \tag{6.1}$$

The activity of water is usually incorporated in K_A, so

$$K_A = K'_A a_{H_2O} = \frac{[A^-]}{[HA]} \cdot a_{H_3O^+} \cdot (y_{A^-}/y_{HA})$$

or

$$pK_A = pH - \log([A]/[HA])$$
$$= pH + \log I, \tag{6.2}$$

where I is the indicator ratio, $[HA]/[A^-]$, assuming the activity coefficients, y, to be unity in dilute aqueous solution. Similarly, if water acts as an acid causing protonation of a dissolved base, B:,

$$B{:}\overset{\frown}{+}H{-}\overset{\frown}{O}H \rightleftharpoons \overset{+}{B}{-}H + \overset{-}{O}H,$$

a base dissociation constant, K'_B, may be defined analogously:

$$K'_B = \frac{a_{B^+H}a_{OH^-}}{a_{B:}a_{H_2O}} = \frac{[\overset{+}{B}H][O\overset{-}{H}]}{[B][H_2O]} \frac{y_{B^+H}y_{OH^-}}{y_{B:}y_{H_2O}} \tag{6.3}$$

and

$$K_B = K'_B a_{H_2O} = \frac{[\overset{+}{B}H]}{[B:]} \cdot a_{O^-H} \cdot \frac{y_{B^+H}}{y_{B:}};$$

$$pK_B = pOH - \log[BH]/[B:]) = pOH - \log I,$$

or, since $pH + pOH = pK_w = 14$ from the autionization of water,

$$pK_B = 14 - pK_A \qquad (6.4)$$

and it is usual to characterize both acidic and basic behavior by pK_A. The larger is the value of pK_A, the weaker the acid and the stronger the conjugate base. These equations apply to any acid–base conjugate pair and any solvent, but beyond the dilute aqueous situation activity coefficients, y, and also the likelihood of ionic products existing as ion-pairs rather than as free ions, would need to be taken into account. In such cases, the apparent K_A would contain the association constant and it would not be a true value. The magnitude of pK_A $(= \Delta G/RT)$ is a measure of the affinity for the proton transfer. Measurement of pK_A requires a knowledge of pH, obtainable from the electrical potential of the glass electrode, and of the ratio $[HA]/[A^-]$, the indicator ratio I, which is obtainable from spectroscopic measurements or assumed to be unity at 'half-neutralization' of the acid by a strong base. Measurement of pK_A in the range 0–14 can be made in water and suitable substrates for this medium include carboxylic acids, phenols and ammonium ions. Consequently acidity has been much used as the basis of reactivity theories[2].

6.2   THE STRENGTHS OF OXYGEN AND NITROGEN ACIDS

The hydroxyl group is an important acidic function strongly affected by its substituents:

	$O_2S(OH)_2$	$CH_3CO \cdot OH$	Ph—OH	H—OH	Me—OH
pK_A	>1	4·76	10·0	14	16
Dissociation in water/%	100	4	3×10^{-4}	10^{-7}	10^{-10}

The acidity of an alcohol is too low to permit appreciable dissociation in pure water but electron-withdrawing groups such as $>C{=}O$, $>SO_2$ greatly weaken the O–H bond towards heterolysis. Ammonium ions are stronger acids than the corresponding oxygen analogs (i.e. $PhNH_3^+$ is stronger than PhOH as an acid) and it follows that the oxygen bases are stronger than the nitrogen analogs (PhO^- is stronger than $PhNH_2$ as a base); see Tables 6.1, 6.2. The pK_A is determined by the difference in free energies of conjugate acid and base forms, and it is to be expected that dissociation will be much

Table 6.1   Effects of media upon acid dissociation equilibria[5]: pK_A

Acid	Solvent				
	H_2O	MeOH	Me_2SO	$H \cdot CONMe_2$	MeCN
HSO_4^-	2·0		14·45	17·1	25·8
CH_3COOH	4·8	9·5	12·6	13·5	
PhCOOH	4·2	9·3	11·1	12·2	20·7
p-Nitrobenzoic	7·2	11·4	11·0	12·6	21
$PhNH_3^+$	4·6		3·2	4·2	

Comparison of acidities in water and in benzene[3,6]

Acid	pK_A	pK_{DPG}[a]
H—C_6H_4—$COOH$	4·20	−5·26
p-NH_2—C_6H_4—$COOH$	4·86	−4·45
p-Me—C_6H_4—$COOH$	4·37	−5·03
p-Cl—C_6H_4—$COOH$	3·98	−5·82
p-NC—C_6H_4—$COOH$	3·55	−6·63
p-NO_2—C_6H_4—$COOH$	3·42	−6·8
	$\rho = 1 \cdot 000$	$\rho = 3 \cdot 0$
CH_3COOH	4·76	−4·45

[a] Association constant with diphenylguanidine, $NH{=}C(NHPh)_2$.

Table 6.2   Thermodynamics of ionization of oxygen and nitrogen acids[3,4,7–9]

Acid	pK_A	$\Delta G/$ kJ mol^{-1} (kcal mol^{-1})	$\Delta H/$ kJ mol^{-1} (kcal mol^{-1})	$\Delta S/$ J K^{-1} mol^{-1} (cal K^{-1} mol^{-1})	$T\Delta S(298K)/$ kJ mol^{-1} (kcal mol^{-1})
PhCOOH	4·213	24·00 (5·7)	0·37 (0·89)	−80 (−19)	−23·8 (−5·68)
X—C_6H_4COOH					
X = m-OMe	4·09	23·3 (5·6)	0·25 (0·06)	−77 (−18)	−22·9 (−5·47)
X = p-OH	4·58	26·1 (6·2)	2·25 (0·54)	−80 (−14)	−23·8 (−6·69)
X = p-Me	4·344	24·7 (5·9)	1·25 (0·29)	−79 (−19)	−23·5 (−5·68)
X = m-Me	4·243	24·1 (5·8)	0·29 (0·07)	−80 (−19)	−23·8 (−5·65)
X = p-Br	4·002	22·8 (5·4)	0·46 (0·11)	−75 (−18)	−22·3 (−5·33)
X = m-Cl	3·827	21·8 (5·2)	0·084 (0·02)	−73 (−17)	−21·7 (−5·18)
X = p-Cl	3·986	22·7 (5·4)	0·83 (0·20)	−73 (−17)	−21·7 (−5·18)
X = p-CN	3·551	20·3 (4·8)	0·12 (0·029)	−68 (−16)	−20·3 (−4·85)
X = o-Me	3·91	22·3 (5·3)	−6·2 (−0·15)	−95 (−22)	−28·3 (−6·76)
X = o-OMe	4·09	23·3 (5·6)	−6·7 (−1·6)	−100 (−24)	−29·8 (−7·12)

Table 6.2   (*Continued*)

Acid	pK_A	$\Delta G/$ kJ mol^{-1} (kcal mol^{-1})	$\Delta H/$ kJ mol^{-1} (kcal mol^{-1})	$\Delta S/$ J K^{-1} mol^{-1} (cal K^{-1} mol^{-1})	$T\Delta S(298K)/$ kJ mol^{-1} (kcal mol^{-1})
HCOOH	3·752	21·5 (5·1)	−0·17 (−0·04)	−73 (−17)	−21·75 (−5·20)
MeCOOH	4·756	27·14 (6·5)	−0·46 (−0·11)	−92 (−22)	−27·4 (−6·55)
EtCOOH	4·875	27·8 (6·6)	−0·96 (−0·23)	−96 (−23)	−28·6 (−6·83)
PrCOOH	4·818	27·5 (6·6)	−2·92 (−0·70)	−102 (−24)	−30·4 (−7·26)
CMe$_3$COOH	5·032	28·8 (6·9)	−3·02 (−0·72)	−107 (−25)	−31·9 (−7·62)
CHCl$_2$COOH	1·30	7·40 (1·8)	−0·42 (−6·1)	−25 (−6)	−7·45 (−1·78)
CCl$_3$COOH	0·64	3·64 (0·87)	4·2 (1·0)	+8·4 (+2)	+2·50 (0·60)
X—C$_6$H$_4$OH					
X = H	10·02	57·2 (13·7)	23·6 (5·6)	−112 (−27)	−33·4 (−8·0)
X = p-Cl	9·38	53·5 (12·8)	24·2 (5·8)	−98 (−23)	−29·2 (−6·98)
X = p-NO$_2$	7·15	40·7 (9·7)	19·7 (4·7)	−70 (−17)	−20·86 (−4·98)
NH$_4^+$	9·245	52·7 (12·6)	51·8 (12·4)	2·9 (0·7)	−0·86 (−0·20)
MeNH$_3^+$	10·62	60·5 (14·4)	54·7 (13·0)	−19·6 (−5)	−5·8 (−1·4)
Me$_2$NH$_2^+$	10·77	61·4 (14·7)	49·6 (11·8)	−40 (−9·5)	−11·9 (−2·84)
Me$_3$NH$^+$	9·80	55·8 (13·3)	36·9 (8·8)	−63 (−15)	−18·7 (−4·47)
Pyridinium	5·21	29·7 (7·10)	18·3 (4·4)	−38·8 (−9·1)	−11·5 (−2·75)
4-Bromopyridinium	3·68	21·0 (5·0)	13·8 (3·3)	24·0 (5·7)	7·1 (1·7)
4-Nitropyridinium	1·23	7·02 (1·7)	−4·52 (−1·08)	39·0 (9·3)	11·6 (2·77)
PhNH$_3^+$	4·60	26·23 (6·27)	30·9 (7·38)	15·5 (3·7)	4·6 (1·10)

affected by solvation, particularly of the charged species, hence K_A values are markedly solvent-dependent (Table 6.1)[3,4]. In general, polar media such as water will be better able to support an increase or concentration of charge than can a less polar solvent. Dissociations of neutral acids (RCOOH) are therefore very much reduced by transferring to media such as methanol or dimethylsulfoxide, and anionic acids (HSO$_4^-$) are even more susceptible. On the other hand the dissociations of cationic acids (ammonium ions) are accompanied by no net change in charge and their pK_A values are much less affected by the solvent. The medium effect therefore depends strongly upon charge type. It is of interest to measure acid/base properties in all types of solvent since organic reactions are mainly carried out in non-aqueous solvents; 'acidity' in benzene, for example, may be measured as association constants with a reference base; but complications due to dimerization of the acid and ion-association need to be taken into account[6].

Dissociation of an acid in a basic solvent gives rise to the 'lyonium' ion (protonated solvent) which is the strongest acid possible in that medium (e.g. H_3O^+ in water). If an acid is sufficiently strong for dissociation to be complete, then all acids of this or greater K_A will appear equal — in all cases the solution contains a molar equivalent of lyonium ions. Hence the solvent has 'leveled' the acidity of such acids. The more basic the solvent, the further the leveling effect spreads to weaker and weaker acids. For example, hydrochloric acid is intrinsically weaker than sulfuric in a solvent such as acetic acid, but in water they are equal. Leveling of bases which are highly dissociated in acidic solvents, such as amines in concentrated sulfuric acid, also occurs.

6.2.1 The interpretation of K_A

One may attempt to rationalize the strengths of acids and bases in terms of the substituent effects previously described. Electron-withdrawing substituents will tend to strengthen acidity and, conversely, electron-donating substituents will weaken it. Frequently anomalies are met and it is necessary to look more deeply into the thermodynamics of proton transfer to obtain an explanation[7,8]. One may partition K_A ($\equiv \Delta G$) into ΔH and $T\Delta S$ components, both of which vary with structure (Table 6.2). Acidic dissociation is favored by a negative ΔH and a positive ΔS. The importance of solvation is shown by the large negative values of ΔS for ionization — of neutral acids, at least. The considerable increase in acidity brought about by substitution of chlorine, e.g. trichloroacetic acid compared with acetic acid ($\Delta pK_A = 4\cdot 12$, $\Delta\Delta G = 23\cdot 5$ kJ mol^{-1} ($5\cdot 6$ kcal mol^{-1})), and usually attributed to the inductive effect of chlorine, is due more to the $T\Delta S$ term ($T\Delta\Delta S = 29\cdot 9$ kJ mol^{-1} ($7\cdot 1$ kcal mol^{-1})) than to $\Delta\Delta H$ ($-4\cdot 66$ ($-1\cdot 1$), which is actually unfavorable.

On the other hand, these two components of $\Delta\Delta G$ are not unconnected. Solvation of the ionic products which results in a favorable enthalpy of ionization also results in association of the solvent with ions and an unfavorable entropy change. There is then a compensating effect and the observed value of pK_A is a resultant of the delicate balance of these two opposing influences. Acetate ion is apparently more highly aquated than is neutral acetic acid, whereas the trichloroacetate ion and trichloroacetic acid must be similarly aquated. The reasons for this may be as follows. The electronic effects exerted by the chlorines, mainly inductive and polarizability, create a dipole in the chloroacetic acids (strongest in trichloroacetic), which results in their becoming more strongly solvated than is acetic acid. Upon ionization this dipole opposes that of the carboxylate group, thereby weakening its solvation by inductively spreading the charge, so the chloroacetate ions tend to be less solvated than is acetate and their dissociations are accompanied by a less unfavorable entropy change, **1**.

$$CH_3-C\overset{O}{\underset{OH}{<}} \longrightarrow CH_3-C\overset{O}{\underset{O}{<}} -$$

increase in solvation

$$Cl_3C-C\overset{O}{\underset{OH}{<}} \longrightarrow Cl_3C-C\overset{O}{\underset{O}{<}} -$$

little change in solvation

1

The dissociation constants of different classes of acids show significant differences in their thermodynamic origins. Ionizations of carboxylic acids are typically accompanied by a small enthalpy term contributing only about 5% to ΔG. This term can account for 50% of the free energy in phenol dissociations and for 80–90% of that of ammonium ions, each being less subject than carboxylic acids to solvation changes.

Steric factors may also be important; formic acid, the least hindered of the simple aliphatic acids to solvation, is considerably stronger than acetic while pivalic (trimethylacetic) acid, highly hindered, is weaker than acetic acid: the entropies of ionization become progressively more negative. It is more plausible to account for this order by steric hindrance to solvation of the carboxylic acid than by the electronic effects of methyl substituents. The acidities of phenols (some seven orders of magnitude greater than those of alcohols) may be partly due to stabilization of phenoxide ion by the delocalization of negative charge into the ring, **2**. As a result, phenoxide is less solvated than a localized oxide ion such as ethoxide would be, and so it is associated with a less negative entropy. Charge delocalization in the conjugate base is particularly important when $-R$ substituents are present (e.g. *p*-nitrophenol, $pK_A = 7.1$). Stabilization of charge is internal in this case, so exerts no cost in entropy.

$$Me-O^{\ominus} \qquad \qquad$$

2

Even in the *meta-* and *para*-substituted benzoic acids, differences in pK_A and consequently in σ values result from changes in both ΔH and ΔS (Table 6.1). The success of the Hammett equation indicates that this blend must be reasonably constant throughout many reactions though much greater fluctuations occur in the *ortho* series.

6.3 LINEAR FREE-ENERGY RELATIONSHIPS [10-12]

The pK_A values of many series of acids can be fitted to linear free-energy regressions — either single- or dual-parameter equations — by means of the appropriate substituent constants, σ. The following equations refer to acid dissociation constants of solutions in water at 25°C.

$$R—CH_2COOH \qquad pK_R = 4\cdot65 - 1\cdot72\sigma^* \qquad (6.5)$$

$$m\text{- or } p\text{-X}—C_6H_4COOH \qquad pK_X = 4\cdot204 - 1\cdot00\sigma \qquad (6.6)$$
$$= 4\cdot204 - 0\cdot992\sigma_I - 0\cdot403\sigma_R \qquad (6.7)$$

$$m\text{- or } p\text{-X}—C_6H_4CH_2COOH \qquad pK_X = 4\cdot297 - 0\cdot489\sigma \qquad (6.8)$$
$$= 4\cdot297 - 0\cdot484\sigma_I - 0\cdot433\sigma_R \qquad (6.9)$$

$$m\text{- or} \qquad pK_X = 4\cdot551 - 0\cdot212\sigma \qquad (6.10)$$
$$p\text{-X}—C_6H_4CH_2CH_2COOH$$

$$m\text{- or} \qquad pK_X = 4\cdot447 - 0\cdot466\sigma \qquad (6.11)$$
$$p\text{-X}—C_6H_4CH{=}CHCOOH$$

$$trans\text{-}CH_2{=}C(X)COOH \qquad pK_X = 4\cdot247 - 2\cdot26\sigma_I - 2\cdot09\sigma_R \qquad (6.12)$$

$$trans\text{-}PhC(X){=}CHCOOH \qquad pK_X = 4\cdot438 - 2\cdot94\sigma_I - 4\cdot09\sigma_R \qquad (6.13)$$

4-Substituted-2-naphthoic acids

$$pK_X = 4\cdot167 - 1\cdot23\sigma_I - 0\cdot75\sigma_R \qquad (6.14)$$

6.4 RATES OF PROTON TRANSFERS [13,14]

The equilibrium constant, K_A is related to the rates of protonation in either direction:

$$HA + B \underset{k_R}{\overset{k_D}{\rightleftharpoons}} A^- + BH^+.$$

$$K_A = k_D/k_R.$$

Table 6.3   Proton transfer rates in water at 25°C

Oxygen acids

$$HA + H_2O \underset{k_R}{\overset{k_D}{\rightleftharpoons}} A^- + H_3O^+$$

HA	k_D/s^{-1}	$k_R/M^{-1}s^{-1}$
H_2O	$2\cdot5 \times 10^{-5}$	$1\cdot4 \times 10^{11}$
D_2O	$2\cdot5 \times 10^{-6}$	$8\cdot4 \times 10^{10}$
HF	$7\cdot0 \times 10^7$	$1\cdot0 \times 10^{11}$
CH_3COOH	$7\cdot8 \times 10^5$	$4\cdot5 \times 10^{10}$
CH_3CH_2COOH	$5\cdot7 \times 10^5$	$4\cdot3 \times 10^{10}$
PhCOOH	$2\cdot2 \times 10^6$	$3\cdot5 \times 10^{10}$
$m\text{-}NH_2\text{---}C_6H_4COOH$	$7\cdot4 \times 10^5$	$4\cdot8 \times 10^{10}$
$o\text{-}NO_2\text{---}C_6H_4OH$	$1\cdot0 \times 10^3$	$1\cdot7 \times 10^{10}$
$m\text{-}NO_2\text{---}C_6H_4OH$	$1\cdot9 \times 10^2$	$4\cdot2 \times 10^{10}$
$p\text{-}NO_2\text{---}C_6H_4OH$	$2\cdot6 \times 10^3$	$3\cdot6 \times 10^{10}$
p-Nitrophenol + triethylamine	$0\cdot31 \times 10^8$	9×10^9
Picric acid + methyl red	$8\cdot1 \times 10^8$	$5\cdot1 \times 10^9$

Nitrogen acids

$$R_3N + H_2O \underset{k_D}{\overset{k_R}{\rightleftharpoons}} R_3NH^+ + OH^-$$

R_3N	k_D/s^{-1}	$k_R/M^{-1}s^{-1}$
NH_3	$6\cdot0 \times 10^5$	$3\cdot4 \times 10^{10}$
CH_3NH_2	$1\cdot6 \times 10^7$	$3\cdot7 \times 10^{10}$
$(CH_3)_2NH$	$1\cdot9 \times 10^7$	$3\cdot1 \times 10^{10}$
$(CH_3)_3N$	$1\cdot4 \times 10^6$	$2\cdot1 \times 10^{10}$

The individual rates of many proton transfers have been measured and are normally extremely fast. Rates of transfer of the aquated protons to anionic bases (i.e. k_R) approach the limit imposed by the time required for reagents to diffuse together (*diffusion control*) of around $10^{11}M^{-1}s^{-1}$ and are aided by mutual electrostatic attractions. It follows that rates of proton transfer from neutral acid to water (k_D) are much slower but they are still fast reactions: for $RCOOH + H_2O \rightarrow RCOO^- + H_3O^+$, k_D is of the order $10^6M^{-1}s^{-1}$ (Table 6.3). These rates are measures of kinetic rather than of thermodynamic acidity. Proton transfers tend to be much slower in non-aqueous solvents than in water, particularly so in non-protic media. Water itself is able to move protons and hydroxide ions with great facility by a 'relay' mechanism, **3**, in which intervening water molecules simultaneously accept and donate these species.

6.5 STRUCTURAL EFFECTS ON AMINE DISSOCIATIONS[9]

Amines and ammonium ions function as important acid–base partners in many reactions.

$$R_3N + HA \rightleftharpoons R_3NH^+ + A^-$$

Aliphatic amines have $pK_A \sim 9$–10 which is not very sensitive to structure. The order of basicity which might be expected, based on the electron-releasing effect of alkyl groups, would be:

$$NH_3 < RNH_2 < R_2NH < R_3N.$$

This is observed in the absence of solvent, but the order found in aqueous solution is:

$$NH_3 < R_3N < RNH_2 < R_2NH.$$

This reversal is due to solvation effects[15]. Dissociation of the tertiary ammonium ion, R_3NH^+, has a more favorable enthalpy than that for ammonium, NH_4^+, but a less favorable entropy change. This is probably due to the numbers of hydrogen bonds each can form. Ammonium can form four tetrahedrally arranged hydrogen bonds similar to those of H_3O^+ in the water structure, so its dissociation is accompanied by a very small ΔS; release of a proton from a tertiary ammonium ion (limited to one hydrogen bond) to water is accompanied by a large increase in the number of hydrogen bonds formed and a decrease in entropy.

Nonetheless, entropies of dissociation of ammonium ions are usually small, much smaller than are those of carboxylic acids, since there is no net change of charge. Cyclic secondary amines are slightly more basic than acyclic analogues, the nitrogen perhaps being sterically more accessible. Adjacent heteroatoms with unshared pairs — oxygen, nitrogen or halogen — tend to weaken basicity. Hydrazine (NH_2NH_2), for instance, has a pK_A two units less than ammonia. This is in contrast to the increase in nucleophilic power of such species (the α-*effect*, Section 6.12.5). Arylamines are much weaker bases than alkylamines. The difference lies mainly in the ΔH of ionization; the protonation of aniline in water is only about half as exothermic as the protonation of methylamine, presumably because of the smaller hydrogen-bonding donor capability and extra stabilization of the amine from conjugation with the aromatic ring. Steric hindrance about the nitrogen results in reduced basicity; for example, 2,6-di(*tert*-butyl)aniline has a pK_A four units less than that of aniline.

6.5.1   Linear free-energy relationships [15]

Attempts to represent pK_A values of aliphatic amines by the Taft equation (Eq. (4.13)) are moderately successful, providing one compares primary, secondary and tertiary amines separately.

Primary: $pK_A = 13 \cdot 71 - 3 \cdot 14\sigma^* - \log n/s$ (6.15)

Secondary: $pK_A = 12 \cdot 43 - 3 \cdot 23 \sum \sigma^* - \log n/s$ (6.16)

Tertiary: $pK_A = 9 \cdot 61 - 3 \cdot 30 \sum \sigma^* - \log n/s$ (6.17)

where σ^* are Taft aliphatic substituent constants, n is the number of protons on nitrogen in the ammonium ion, and s is the number of basic sites in the amine. A single correlation equation, Eq. (6.18), which fits all three series has also been derived [16–18]:

$$pK_A = 9 \cdot 61 + (10 \cdot 92 - 0 \cdot 778n)\log n - (3 \cdot 38 - 0 \cdot 08n) \sum \sigma^* - \log n/s.$$

$$(6.18)$$

N-substituted anilines, $PhNRR'$, need a separate correlation (Eq. (6.19)):

$$pK_A = 5 \cdot 36 + (0 \cdot 75 + 1 \cdot 55n)\log n - (3 \cdot 35 - 0 \cdot 16n)\sigma^* - \log n. \quad (6.19)$$

The need to incorporate n suggests the importance of the number of hydrogen-bonding sites in stabilizing the ammonium ion by solvation. Acidities of anilinium ions correlate with σ^- [19]:

$$pK_A = 4 \cdot 603 - 2 \cdot 89\sigma^-. \quad (6.20)$$

(Some confusion may arise here since ρ^- for acid dissociation of the anilinium ions is $+2 \cdot 89$ — they are strengthened by electron-withdrawing groups. The negative sign originates in the use of pK $(= -\log K_A)$ instead of $\log K$ in Eqs (6.15)–(6.20).)

6.6   ACIDITIES OF CARBON ACIDS [20,21]

Carbon acidity describes the tendency of a C—H bond to transfer a proton to a base with formation of a carbanion:

$$R_3C\!-\!H + :B \rightarrow R_3C^- + BH^+.$$

Since the C–H bond is so nearly homopolar and has little tendency to form hydrogen bonds this is, in general, much less facile than is the acid dissociation of O–H or N–H bonds. Carbon shows considerable reluctance to accept negative charge compared with elements in later groups of the periodic table, to the extent that simple free alkyl anions can be said not to exist in solution. All carbanions which have even a transitory existence require the charge to be delocalized, usually onto heteroatoms. Substituent groups attached to the acidic carbon therefore play a more crucial role in determining acidity than is usual in oxygen acids. An alkane such as ethane ($pK_A > 40$) is less acidic than water by some 26 orders of magnitude. Nevertheless, with suitable structural modification, carbon acidities approaching those of the mineral acids may be reached. Even for very weak acids, chemical properties may depend on ability to ionize: the bromination of acetone ($pK_A \sim 20$) catalyzed by weak bases commences with a rate-determining proton transfer, though the equilibrium amount of enolate is minute.

$$CH_3\overset{\overset{\displaystyle O}{\|}}{C}CH_3 + :B \xrightarrow{\text{slow}} CH_3\overset{\overset{\displaystyle O}{\|}}{C}\bar{C}H_2 \xrightarrow[\text{fast}]{Br_2} CH_3\overset{\overset{\displaystyle O}{\|}}{C}CH_2Br, \text{ etc.}$$
$$+BH^+$$

Table 6.4 gives some examples of pK_A values for carbon acids in which the most striking feature is the enormous range encompassed. Carbon acidities may span 60 orders of magnitude and thus can span the extremes of known acid behavior.

Table 6.4   Acidities of carbon acids

Acid	pK_A	Acid	pK_A
(a) Cyano compounds		**(c) Sulfones and sulfoxides—** *contd.*	
MeCN	25		
$CH_2(CN)_2$	11·2	Sulfolane	>31
$CH(CN)_3$	−5·13	$PhCH_2SO_2CH_2Ph$	22
tert-$BuCH(CN)_2$	13·1	$EtSO_2$—CHMe—SO_2Et	14·6
HCN	9·21	$EtSO_2$—CHBr—SO_2Et	10·7
$BrCH(CN)_2$	5	MeSOMe	28·5
$MeOCO$—$CH(CN)_2$	−2·8		
$CH(CN)_2 \cdot C{=}C(CN)_2$	−8·5	**(d) Ketones**	
$\quad\quad\quad \mid$		MeCOMe	20
$\quad\quad\quad CN$		$MeCOCH_2Cl$	16
		$MeCOCHCl_2$	14·9
(b) Nitro compounds		MeCOPh	19·5
$MeNO_2$	10·2	$RCOCH_2COR$	
$CH_2(NO_2)_2$	3·63	R = Me	8·84
$CH(NO_2)_3$	0·14	R = Ph	8·95
R—$CH(NO_2)_2$		R = tert-Bu	11·57
R = Me	5·30	R = CF_3	5·35
R = Et	5·61		
R = isoPr	6·77	**(e) Cyclopentadienes**	
R = $CH_2{=}CH$—CH_2	4·91	Cyclopentadiene	15·0
R = $PhCH_2$	4·54	1-Cyanocyclopentadiene	9·78
R = H	3·63	2,5-Dicyanocyclopentadiene	2·52
R = Br	3·58	Indene	20·2
R = Ph	3·71	Fluorene	22·74
R = NO_2	0·14	9-Phenylfluorene	18·59
R = $CONH_2$	1·30	9-p-Anisylfluorene	19·01
R = I	3·19	9-Ethylfluorene	22·60
		9-Cyanofluorene	11·41
(c) Sulfones and sulfoxides		**(f) Other weak acids**	
$MeSO_2Me$	28·5	Ph_3CH	31·5
$MeSO_2Ph$	27	Ph_2CH_2	33

Table 6.4   (*Continued*)
Acidities of weak acids measured by the application of Eq. (6.23) in various solvent systems[22]

Compound	Solvent			
	Tetrahydro-furan	*Cyclohexyl-amine*	*Dimethyl-sulfoxide*	*Dimethoxy-ethane*
9-Phenylfluorene	(18·49)[a]	(18·49)[a]	17·9	17·55
2-Phenylindene	19·10		19·4	
3,4-Benzo[c]fluorene	19·90	19·75		
Benzo[a]fluorene	20·54	20·35		
9-Benzylfluorene	20·92	21·27	21·4	20·95
9-Methylfluorene	21·85	22·33	22·3	
Fluorene	22·41	23·04	22·9	22·3
4,5-Methylenephenanthrene	22·42	22·93	22·7	
Benzo[b]fluorene	23·15	23·47	23·5	
9-*tert*-butylfluorene	24·22	24·25	24·3	23·75
1,1,3-Triphenylpropene	26·52	26·59	26·2	
9-Phenyl-10,10-dimethyl-dihydroanthracene	27·86	28·01		
9-Phenylxanthene	28·50	28·49	27·7	27·7
p-Biphenylyldiphenylmethane	29·83	30·17	29·4	29·3
Triphenylmethane	31·02	31·45	30·6	30·75
p-Benzylbiphenyl	31·43	31·82		
Tri(p-tolyl)methane	32·86	33·04		
Diphenylmethane	33·01	33·41	32·6	
Di(o-tolyl)methane	33·95	34·80		

[a] 9-Phenylfluorene is the reference point.

6.6.1 The measurement of weak acidity

While equilibrium acidities of carboxylic acids and phenols can be measured in water, whose acid–base behavior acts as a convenient reference point, the extent of dissociation of acids of $pK_A > 14$ cannot usually be determined in aqueous media. Solvents other than water, and bases other than OH^-, may be chosen in order to obtain a sufficient degree of ionization for measurement of the equilibrium constant, but comparisons of acidity are best made with reference to a common base, preferably water for historical reasons. If two weak acids, HA and HA″, are dissolved in a suitable solvent and a small amount of a very strong base is added, the following equilibrium is set up[15];

$$HA + A''^- \rightleftarrows HA'' + A^-$$

for which

$$K_A = \frac{[HA''][A^-]}{[A''^-][HA]} \cdot \frac{y_{HA''}y_{A^-}}{y_{A''^-}y_{HA}} = \frac{I''}{I}.$$

But

$$pK_A(HA) = pH - [A^-]y_{A^-}/[HA]y_{HA} = pH + \log I, \qquad (6.21)$$

221

and

$$pK_A(HA'') = pH - [A''^-]y_{A''-}/[HA'']y_{HA''} = pH + \log I''. \qquad (6.22)$$

Hence

$$\log K_\Delta = pK_A(HA'') - pK_A(HA). \qquad (6.23)$$

The equilibrium constant yields the difference in pK_A for the two acids: if one is known, the other may be determined. Estimates of activity coefficient ratios must be made and included.

Assuming that $pK_A(HA)$ relative to water is known, then $pK_A(HA'')$ also relative to water may be determined despite the fact that this equilibrium was observed in another medium. Now a similar equilibrium measurement between HA'' and a still weaker acid may be set up and the difference in acidities determined for this pair. In this way, a scale of acidities, all based on the water scale, gradually evolves. After water, several media of increasing basicity have been widely used for relative acidity measurement[23-25]:

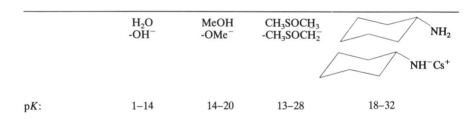

	H$_2$O -OH$^-$	MeOH -OMe$^-$	CH$_3$SOCH$_3$ -CH$_3$SOCH$_2^-$	NH$_2$ / NH$^-$Cs$^+$
pK:	1–14	14–20	13–28	18–32

Cesium cyclohexylamide in cyclohexylamine has been extensively employed by Streitweiser[26-29] as a medium for the measurement of acidity as low as $pK_A = 34$[15]; despite the large cation and relatively high dielectric constant, considerable ion-pairing occurs. A careful choice of acids must be made to ensure that ΔpK_A is no more than 1–2 units, otherwise the equilibrium will lie too far to one side for accurate measurement. The highest accuracy in measuring values of I is required since each acid whose pK_A is determined serves as an indicator for a weaker acid in turn and hence errors accumulate (Table 6.4).

6.7 FACTORS WHICH INFLUENCE CARBON ACIDITY

It is useful to summarize the structural features which facilitate the ionization of carbon acids by stabilizing the carbanions which are their conjugate bases.

(a) *Electronic effects of adjacent +R and +I groups*

All $-R$ substituents ($\diagdown$C=O, $\diagdown$SO$_2$, —C≡N, —NO$_2$) are able to stabilize carbanionic centers by delocalizing negative charge onto the heteroatoms, oxygen and nitrogen, **4**, according to their values of σ^-. A single nitro group has a larger effect than cyano but successive cyano substituents have a greater effect than the equivalent number of nitro groups. This is probably because the cyano groups are linear and even three on one carbon can exert their full resonance effect. Three nitro groups attached to the same carbon twist out of the plane to avoid electrostatic repulsions between adjacent oxygens and so are able to exert only a fraction of their ability to share negative charge. The conjugate bases of carbonyl compounds are enolates whose chemistry is discussed in Section 9.2.2.

4

Inductive effects $(-I)$ are probably also of importance but are difficult to dissect from other electronic perturbations. One might expect strong inductive effects by the —CF$_3$ group to render fluoroform, CHF$_3$, rather acidic, yet it has p$K_A \sim 30$, little different from that of triphenylmethane. The charge on a quaternary nitrogen R$_3\overset{+}{\text{N}}$—CH$_3$ does little to increase the acidity of adjacent protons[30,31].

(b) Stabilization by *d*-orbitals

Elements other than those of the first row of the Periodic Table, and having available vacant *d*-orbitals whose energies are close to those of the valence orbitals, are able to stabilize an adjacent carbanionic center; the *d*-element (S, P, Cl) expands its valence shell. The acidities of chloroform and of phosphonium (**5**) and sulfonium (**6**) ions can be ascribed to this effect and are at least 10 orders of magnitude more acidic than ammonium ions so that this cannot be due to inductive influences alone. The conjugate bases of the 'onium ions are formally neutral and are known as *ylides*[32]. The polarizabilities of the larger atoms may also contribute to stabilization of charge.

$$R_3\overset{+}{P}\text{—CH}_3 \xrightarrow{-\text{H}^+} R_3\overset{+}{P}\text{—}\overset{-}{C}H_2 \longleftrightarrow R_3P\text{=}CH_2$$

5

$$R_2\overset{+}{S}\text{—CH}_3 \xrightarrow{-\text{H}^+} R_2\overset{+}{S}\text{—}\overset{-}{C}H_2 \longleftrightarrow R_2S\text{=}CH_2$$

6

(c) s-Character of carbon hybridization

The proportion of C-$2s$ orbital which is involved in the C–H σ-bond affects its acidity. Orbitals with a high proportion of s-character (Table 1.4) are better able to accept an anionic charge since they experience a greater attractive influence of the carbon nucleus. This is due to the orbital geometry; s-orbitals have finite probability at the nucleus, p-orbitals zero. A familiar example is acetylene with 50% s-character at its carbon and $pK_A \approx 23$, which is low but much greater than acidities of hydrogen on sp^2 or sp^3-hybridized carbon (pK_A *ca* 32, 40, respectively); so acetylene is converted to the acetylide ion, HC≡C⁻, in a medium such as liquid ammonia/amide ion.

(d) Aromaticity

Enormous enhancement of acidity can result if the conjugate base has an aromatic structure when the acid does not. The simplest example is cyclopentadiene, **7**, whose pK_A of 12 is perhaps 20 orders of magnitude greater than that of the corresponding acyclic analogue, penta-1,3-diene.

The cyclopentadienide ion is a $(4n + 2)$ aromatic system with 5-fold symmetry and π molecular orbitals similar to those of benzene. The benzo derivatives, indene and fluorene, show enhanced acidity compared with non-cyclic analogues but are much less acidic than cyclopentadiene since reorganization of the negative charge affects the electron distribution of the aromatic rings. An even more extreme case of the tendency to revert to an aromatic conjugate base is found in benzenium ions (Section 10.4), **8**, whose eagerness to shed a proton and revert to a benzenoid structure is such that they can only exist in a stable state in 'superacid' media (Section 6.11.1) and must be among the strongest acids known, $pK_A < -10$. The converse of this principle is a powerful weakening of acidity if the conjugate base is antiaromatic (a $4n$ system). An example is the cyclobutenones, **9**, which will not enolize since this would lead to a high-energy cyclobutadiene structure.

6.8 RATES OF IONIZATION OF CARBON ACIDS

Whereas rates of ionization of oxygen and nitrogen acids often approach diffusion control, those of carbon acids are much slower and can be very slow indeed. This is connected with the low ability of carbon to form a hydrogen bond, which is evidently a necessary precursor of the transition state for a proton exchange. The measurement of ionization rate by proton exchange with isotopic hydrogen can easily be adapted to the slowest systems by increasing the severity of reaction conditions and extending observations to long times. Shatenshtein[33], for example, has used liquid ND_3/ND_2^- in which benzene exchanges its protons (half-life 2·5 h at 25°C), alkenes exchange allylic protons (100 h, 120°C) and even cyclopentane, probably the least acidic compound of all, exchanges some 30% of its protons (1300 h, 120°C). Detritiation is particularly sensitive and may be applied to reactions whose half-lives are measured in weeks:

$$\overset{\diagdown}{\underset{\diagup}{C}}\!\!-T + B: \xrightarrow{\text{slow}} \overset{\diagdown}{\underset{\diagup}{C}}{}^- \xrightarrow[H^+]{\text{fast}} \overset{\diagdown}{\underset{\diagup}{C}}\!\!-H$$

$$+\ BT^+$$

Rates of ionization (Table 6.5) of carbon acids are a measure of their acidity, in this case kinetic acidity, and it is of interest to enquire what relationship, if any, exists between these values and equilibrium measurements ('thermodynamic acidities'). A plot of log k (ionization) against pK_A, shows a very rough correlation, no more than a trend, if all types of carbon acid are included. Evidently the Brønsted Law, which relates rates with protonation equilibria (Section 9.1.1), holds only poorly. A more linear relationship is observed if only related series of carbon acids are considered, e.g. nitro compounds or ketones. It is found that nitroalkanes ionize much more rapidly than expected for acids of their pK_A while cyano compounds show the opposite trend.

6.9 GAS-PHASE ACIDITY AND BASICITY[34–39]

In order to eliminate the effects of solvent and study intrinsic acidities and basicities, the ideal measurements of proton-transfer equilibria would be made in the gas phase. This possibility has been realized in recent years with the advent of high-pressure mass spectrometry. Ions formed in the source are permitted to interact with molecules introduced along the flight path or by ion cyclotron resonance (ICR), in which a cyclotron is used to maintain organic and inorganic ions in a permanently circulating path in a strong magnetic field.

Reference acids or bases may be admitted so that reaction occurs:

$$A^- + HA' \rightleftharpoons HA + A'^-.$$

Table 6.5   Gas-phase acidities and basicities
(a) ΔH values from the thermochemical cycle

Gas-phase acidities:

$$HA \xrightarrow{D} H^{\cdot} + A^{\cdot} \xrightarrow{EA} A^- + H^+$$

$$\Delta H = \text{Dissociation energy } (D) + \text{electron affinity (EA)}$$

Oxygen acid	$\Delta H / kJ\,mol^{-1}\,(kcal\,mol^{-1})$
MeOH	265 (63·3)
EtOH	257 (61·4)
isoPrOH	250 (59·7)
tert-BuOH	247 (59·0)
PhOH	151 (36·1)
MeCOOH	146 (34·9)
MeCH$_2$COOH	141 (33·7)
HCOOH	132 (31·5)
PhCOOH	112 (26·8)
CH$_2$ClCOOH	91 (21·7)
CHCl$_2$COOH	62 (14·8)
CF$_3$COOH	38 (9·1)
HCl	84 (20·0)

Carbon acid	$\Delta H / kJ\,mol^{-1}\,(kcal\,mol^{-1})$
MeCN	250 (54·7)
MeSOCH$_3$	255 (60·9)
MeCOCH$_3$	236 (56·4)
MeCOCH$_2$CH$_3$	229 (54·7)
MeCHO	224 (53·5)
MeSO$_2$CH$_3$	221 (52·8)
Ph$_2$CH$_2$	206 (49·2)
PhCOCH$_3$	206 (49·2)
MeNO$_2$	184
MeCH$_2$NO$_2$	183 (43·7)
Cyclopentadiene	175 (41·8)
Fluorene	166 (39·7)

Gas-phase basicities (proton affinities): $B_g : + H_g^+ \rightarrow BH_g^+$

Base, B:	$\Delta H / kJ\,mol^{-1}\,(kcal\,mol)$	Base, B:	$\Delta H\,kJ\,mol^{-1}\,(kcal\,mol^{-1})$
NH$_3$	828 (198)	CH$_4$	563 (135)
MeNH$_2$	878 (210)	HCl	575 (138)
EtNH$_2$	891 (213)	HBr	599 (143)
nPrNH$_2$	897 (214)	CO	609 (145)
Me$_2$NH	906 (216)	C$_2$H$_6$	615 (147)
Et$_2$NH	925 (221)	HI	631 (151)
Me$_3$N	923 (220)	SO$_2$	650 (155)
Et$_3$N	952 (227)	C$_2$H$_4$	680 (163)
PhNH$_2$	866 (207)	H$_2$O	696 (166)
Pyridine	907 (217)	CS$_2$	695 (166)
Pyrrole	857 (205)	CF$_3$CH$_2$OH	704 (168)
MeCN	753 (180)	H$_2$S	710 (170)

Table 6.5   (*Continued*)
(b) Gas-phase acidities: $HA \rightarrow H^+ + A^-$. These values differ from those in Table 6.5a by the inclusion of the ionization potential for $H^\cdot$ (1310 kJ mol^{-1} (313 kcal mol^{-1})) and by an entropy correction.

Carbon acid	$\Delta G/kJ\,mol^{-1}\,(kcal\,mol^{-1})$	Oxygen acid	$\Delta G/kJ\,mol^{-1}\,(kcal\,mol^{-1})$
CH_4	1711 (409)	H_2O	1607 (384)
$CH_2{=}CH_2$	1690 (404)	MeOH	1559 (373)
C_6H_6	1656 (396)	PhOH	1441 (344)
PhMe	1558 (372)	MeCOOH	1439 (344)
$HC{\equiv}CH$	1536 (367)	HCOOH	1425 (340)
MeSOMe	1530 (366)	PhCOOH	1404 (335)
$PhC{\equiv}CH$	1517 (362)	HNO_3	1322 (316)
MeCN	1524 (364)	PhSH	1372 (328)
MeCOOMe	1513 (362)	$PhNH_2$	1505 (360)
MeCHO	1504 (359)		
Ph_2CH_2	1502 (359)		
$MeSO_2Me$	1502 (359)		
PhCOMe	1491 (356)		
$MeNO_2$	1473 (352)		
$PhCH_2CN$	1449 (346)		
HCN	1447 (346)		
Fluorene	1445 (345)		
$CH_2(CN)_2$	1384 (331)		

		Cationic acid	$\Delta G/kJ\,mol^{-1}\,(kcal\,mol^{-1})$
		Me_3NH^+	912 (218)
		$PhMe_2NH^+$	903 (216)
		Pyridinium	894 (214)
		$PhNH_3^+$	849 (203)
		NH_4^+	820 (196)
		$Me{-}C{\Big\langle}{\overset{OH}{\underset{OMe}{}}}(+$	793 (189)
		$Me_2C{=}OH^+$	788 (188)
		Me_2OH^+	771 (184)
		PH_4^+	753 (180)
		$C_6H_7^+$	735 (176)
		H_3S^+	696 (166)
		H_3O^+	680 (162)
		H_2Cl^+	531 (127)
		H_2F^+	473 (113)

Simple hydrides

Hydride	$\Delta G/kJ\,mol^{-1}\,(kcal\,mol^{-1})$	Hydride	$\Delta G/kJ\,mol^{-1}\,(kcal\,mol^{-1})$
CH_4	1711 (409)	H_2S	1452 (347)
SiH_4	1520 (363)	HF	1530 (365)
NH_3	1661 (397)	HCl	1369 (327)
PH_3	1519 (363)	HBr	1320 (315)
H_2O	1607 (384)	HI	1281 (306)

Equilibrium is established and product distributions are analyzed by mass-spectrometric detection of the ions present to obtain K_A and hence ΔG. The proton affinity for a base may be calculated, as ΔH for the gas-phase reaction $B: + H^+ \rightarrow BH$ (Table 6.5). Acidity of simple hydrides increases across the Periodic Table and with increasing period number; between CH_4 and SiH_4, $\Delta pK_A = 33$ and between CH_4 and NH_3 it is 8·8. H_2O is less acidic than H_3O^+ by a factor of 10^{163}! A notable result which emerges is the order of basicity of amines which is regular and as expected from the electron-donating capabilities of alkyl groups:

$$NH_3 < RNH_2 < R_2NH < R_3N,$$

which confirms that the anomalies in water are solvation effects. Gas-phase acidities of aliphatic carboxylic acids seem to increase with the size of the alkyl group:

$$MeCOOH < EtCOOH < PrCOOH,$$

possibly because of their greater polarizabilities, but this order is opposite to that found in solution[37]. Acetic acid appears to be comparable in acidity with phenol and much weaker than benzoic acid while the halogenoacetic acids, even chloroacetic, are stronger than HCl, all at variance with pK_A values in water. Among the carbon acids, which in general have a higher enthalpy of heterolysis (i.e. lower acidity) than oxygen acids, the enthalpy of diphenylmethane seems anomalously high but that of fluorene is higher than that of cyclopentadiene. Acidities in the gas phase provide data with which to compare calculations based on high-level MO theory and so to test the origins of structural effects. *Ab initio* calculations can be performed on structures with any prescribed geometry; the energy change $\Delta\Delta E$ for the process

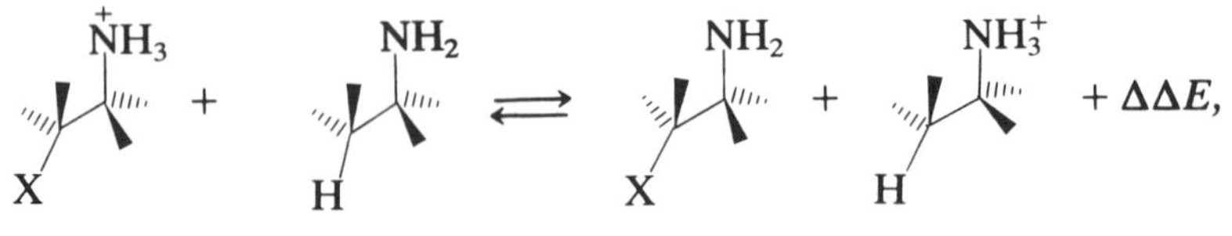

varying X, can be taken as a measure of the inductive (or field) effect of X: a good correlation between these values and σ_I indicates the correctness of this approach. Using *para*-substituted anilinium ions in the equilibrium:

resonance effects can be calculated after separating out the field effect. Values of $\Delta\Delta E$ follow the same order as σ_R.

The calculations can also simulate solvation, since the energy of a molecule in association with any number of solvent molecules of given geometry or in a geometry which minimizes the potential energy of the system, can be computed. Furthermore, the association energy of an ion with solvent molecules may be measured by the ion-cyclotron technique for comparison. A model has proved successful in which it is assumed that a molecule of water accepts a hydrogen bond from each proton on an ammonium ion and one water donates a hydrogen bond to an amine nitrogen. Values of $\Delta\Delta E$ for the equilibrium:

$$\underset{\substack{\\ \text{H}\cdots\text{OH}_2}}{\overset{\substack{..\text{OH}_2 \\ \text{H}}}{\text{XCH}_2-\overset{+}{\text{N}}-\text{H}\cdots\text{OH}_2}} + \text{HCH}_2-\text{N} \rightleftharpoons \text{XCH}_2-\text{N} + \text{HCH}_2-\overset{+}{\text{N}}-\text{H}\cdots\text{OH}_2$$

correlate reasonably with pK_A values. Equilibria of the type:

$$\text{(hexamethyl arenium ion)} + \text{B:} \rightleftharpoons \text{(hexamethylbenzene)} -\text{Me} + \text{B}\overset{+}{\text{H}}$$

may be measured by ICR with and without solvation so that free energies of transfer (gas phase to water) may be measured. Very large values are found for H_3O^+ and simple oxonium ions due to the stabilization of these species by hydrogen bonding. The steric effects of *tert*-butyl groups is also apparent in hindering hydrogen bonding (Table 6.5). Undoubtedly such approaches will become increasingly successful with elaboration of the computations. It is remarkable that the main features of solvation are evident even from the inclusion of only one molecule of water on the acid or base.

6.10 THEORIES OF PROTON TRANSFER

While proton transfer is just about the simplest chemical reaction possible, there appear nonetheless to be complex mechanistic features which may be discerned by careful study. The model put forward by Eigen[40] considers three steps to be essential (10); diffusion together of acid and base to the point at which hydrogen-bond formation occurs; proton transfer via the

$$A—H + B \rightleftharpoons (A—H\cdots B) \rightleftharpoons (A^-\cdots H—\overset{+}{B}) \rightleftharpoons A^- + H—\overset{+}{B}$$

10

hydrogen bridge; diffusion apart of the products. Diffusion rates are determined by the characteristics of the solute molecules according to the Smoluchovsky equation:

$$k_{diff} = 4L(r_{AH} + r_B)(d_{AH} + d_B)\mathscr{E} \qquad (6.24)$$

where r and d are radii and diffusion coefficients of the species, and $\mathscr{E}$ is an electrostatic term, zero for neutral molecules but large for cation–anion interactions. It is this which is responsible for the greatly increased rates of proton transfer between oppositely-charged acid–base pairs compared with rates of transfer between neutral pairs of molecules.

The actual proton-transfer rate in the hydrogen-bonded complex seems to follow the strength of the hydrogen bond (Section 1.4.4):

$$O—H\cdots O > O—H\cdots N > N—H\cdots O > N—H\cdots N > C—H\cdots O > O—H\cdots C$$

and may take place through the intermediacy of other water molecules. Marcus' approach[44], initially developed as a theory of electron-transfer processes, is applicable to proton transfer and possibly to other related reactions such as methyl transfers[42]. The same three stages of the reaction are considered as in the Eigen model. Free-energy changes are as defined in Fig. 6.1 and it follows that the standard free energy change for the proton-transfer step differs from that for the overall process by the work

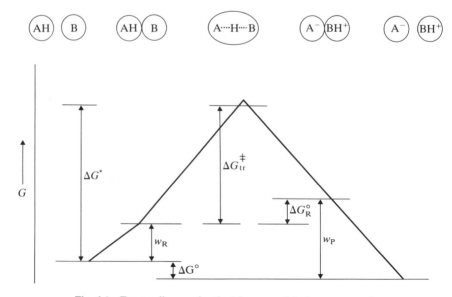

Fig. 6.1   Energy diagram for the Marcus model of proton transfer.

terms:

$$\Delta G_R^{\circ} = \Delta G^{\circ} + w_P - w_R \qquad (6.25)$$

while the free energy of activation for the proton transfer is less than the experimental value by the work of approach:

$$\Delta G_{tr}^{\ddagger} = \Delta G_{tr} + w_R. \qquad (6.26)$$

The free energy of proton transfer, $\Delta G_{tr}^{\ddagger}$, is considered to consist of an intrinsic (internal energy) part and an extrinsic (solvent reorganization) part, proportional to $\lambda/4$, these terms being connected by Eq. (6.27):

$$\Delta G_{tr}^{\ddagger} = \left(1 + \frac{\Delta G_R^{\circ}}{\lambda}\right)^2 \frac{\lambda}{4} + w_R, \qquad (6.27)$$

where ΔG_R° is the free-energy change which drives the proton transfer and differs from the standard free energy of the reaction by the difference in the energies required to bring together reagents and to separate products; λ is related to the electrostatic energy between the ions A^-, HB^+.

This is the equation of a parabola and so a plot of $\Delta G_{tr}^{\ddagger}$ against ΔG° for a series of related acid–base reactions (e.g. varying the base) gives a section of a parabolic curve, a departure from Brønsted behavior which will be discussed in Section 9.1.3.

A thermodynamic argument for the existence of an intermediate in a proton-transfer reaction is derived from the effect of temperature on K_A. The van't Hoff equation (Eq. (2.18)) expresses the linear relationship between $\ln K$ and T for a simple equilibrium. An examination of this relationship for a number of carboxylic acids whose pK_A values are known with great precision shows that these plots are not linear but this discrepancy may be interpreted in terms of a two-step process[43,44]:

$$A—H + :B \underset{\substack{K_1 \\ \Delta H_1}}{\rightleftharpoons} [A^- \cdot H—B^+] \underset{\substack{K_2 \\ \Delta H_2}}{\rightleftharpoons} A^- + H—B^+.$$

A similar analysis has been used for the dissociation of ammonia in water[45].

6.11 HIGHLY ACIDIC AND HIGHLY BASIC SOLUTIONS

Water is an amphiprotic solvent capable of sustaining concentrations of either $H^+(aq)$ or OH^- and providing acidic or basic solutions. Its central role in acid–base equilibrium studies, however, is due to its abundance and its suitability for the study of carboxylic acids, phenols and amines which fortuitously happen to be appreciably but not completely dissociated in this solvent and therefore are not 'leveled'. The range of acidic or basic behavior which is observable is limited by the properties of the solvent since the extent of dissociation of an acid or base is measured by the indicator ratio, I (Eq. (6.2)). In water the hydrogen-ion activity (pH) is limited to a range of approximately 1–14 since

$$pK_W = \log[H^+(aq)] \cdot [OH^-] = 14.$$

Dissociation of a weak acid, $pK_A = 20$, therefore, to its conjugate base in equilibrium in 1M sodium hydroxide solution, $pH = 14$, occurs only to the extent of one part in a million:

$$\log[A^-]/[HA] = (14 - 20) = -6$$

and similarly for a weak base, $pK_A = -6$ in water at $pH = 0$. Such very small degrees of acid–base dissociation do not necessarily preclude reactions occurring in water via the minority species since these species may be very rapidly formed and highly reactive, but equilibrium studies cannot be made with any accuracy when indicator ratios are outside the range 0·01–100 (i.e. <1% or >99% dissociation). The acidity or basicity of aqueous solutions can be considerably raised by using concentrated solutions of mineral acids or alkali-metal hydroxides but the pH scale becomes inappropriate since the acid or base is no longer the dilute, aquated proton or hydroxide ion and the response of the glass electrode by which it is measured becomes non-linear. Non-aqueous media can provide solutions intrinsically more acidic or basic than is possible in water. In general, the higher the pK_A of a solvent the weaker the acids whose dissociations can be supported. New scales of acidity and basicity are, however, needed to replace pH and these are known as *acidity functions*. As will be seen, measurement of acidity depends crucially upon the medium and upon the differential solvation of the acid–base conjugate pair. Hydrogen bonding is probably the single most important solvation force which determines acidity.

6.11.1 Highly acidic solutions

A medium such as 50% sulfuric acid contains acidic species other than the solvated proton — H_2SO_4, $H_3\overset{+}{S}O_4$ for instance — so that its protonating ability is not measurable by pH even could this quantity be determined. The hydrogen-ion activity, a_H, must therefore be assessed by a scale based on the behavior of some suitable base which is incompletely dissociated in this medium.

$$B: + (H^+) \rightleftharpoons BH^+.$$

From Eqs (6.1) and (6.2),

$$I = 1/K_A \cdot a_{H^+} a_B/a_{BH^+}, \tag{6.28}$$

or

$$a_{H^+} = K_A I \cdot a_{BH^+}/a_B. \tag{6.29}$$

If two related weak bases, B(1) and B(2), are simultaneously present in the same protonating solution they will both be partially protonated and a_{H^+} may be eliminated. Hence,

$$\log I(1) - \log I(2) = pK_A(1) - pK_A(2) + \log\left(\frac{a_B(1)a_{BH^+}(2)}{a_B(2)a_{BH^+}(1)}\right). \quad (6.30)$$

It is found that within a range of the medium (e.g. sulfuric acid–water), $[\log I(1) - \log I(2)]$ is constant so the last term in Eq. (6.30) is medium-independent. If this is also assumed to be the case in dilute aqueous solution the last term is zero since all the activities are unity. Therefore

$$\log I(1) - \log I(2) = pK''(1) - pK''(2). \quad (6.31)$$

This equation may be used to determine a pK_A value inaccessible directly in water, and by analogy with Eq. (6.2) was used by Hammett to define an operational measure of protonating ability, H_0 [45, 46]:

$$H_0 = pK_A - \log I$$

$$= -\log\left(a_{H^+} \cdot \frac{a_B}{a_{BH^+}}\right). \quad (6.32)$$

Indicators of different structural types do not necessarily yield the same value of the acidity function in a given medium. There are, in fact, as many acidity functions as there are types of weak base (or indeed, in the limit of precision, individual weak bases, whose activity coefficients in a given medium need not be equal). To obtain a definition which is sufficient for many practical purposes of the protonation behavior of important categories of base, it has been necessary to define a series of complementary acidity functions, H_X. Some of the more important ones are defined in Table 6.6 and their values in the sulfuric acid–water system are shown in Fig. 6.2 and for basic media in Fig. 6.3. Acidity functions M_c are also used and are defined analogously to Eq. (6.32) using activity coefficients of the species [47].

$$M_c = -\log\frac{y_B y_{H^+}}{y_{BH^+}}. \quad (6.33)$$

The necessity for many acidity functions is illustrated by the observation that in a single medium, 80% sulfuric acid say, the extent of protonation of

Table 6.6 Some acidity functions [48–54]

Acidity function	Indicator type	ϕ^a
H_0	Primary anilines	0
H_0'''	N,N-dialkylnitroanilines	-0.33 to -0.48
H_I	Indoles	-0.67 to -0.85
H_A	Amines	$+0.42$ to $+0.55$
H_R	Triarylmethanols ($\rightarrow Ar_3C^+$)	-1.02 to -1.59
H_{ROR}	Ethers	$+0.75$ to $+0.82$

[a] Ref. 55.

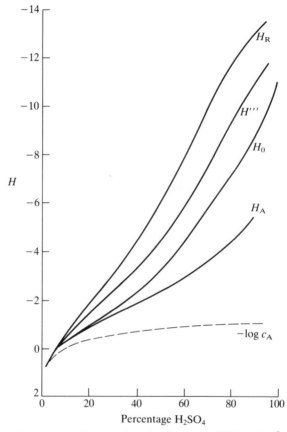

Fig. 6.2 Acidity functions for the system H_2SO_4–H_2O[8].

a benzamide, a nitroaniline and an N,N-dimethylnitroaniline of the same pK_A would be progressively greater. Each acidity function, H_X, may be expressed as the corresponding function J_X by addition of the activity of water:

$$J_0 = H_0 + a_{H_2O}.$$

J_0 is a convenient quantity for relating hydrolytic rates in strongly acidic aqueous solution.

Relationships between acidity functions

To a moderate degree of precision acidity functions are linearly related[56]:

$$H_X = mH_0$$
$$\text{and} \quad M_c(a) = nM_c(b), \tag{6.34}$$

where $M_c(a)$ and $M_c(b)$ are the acidity functions of two different classes of indicator. A more precise function, due to Bunnett, is Eq. (6.35)[55]:

$$H_X + \log c_H = (1 - \phi)(H_0 + c_H). \tag{6.35}$$

Values of the constant, ϕ, are given for various indicator types in Table 6.6 (in principle they will differ for each individual indicator base) and are related to differences in activity-coefficient ratio between two acid–base pairs:

$$\log(y_{H^+}y_X/y_{XH^+}) = (1 - \phi)\log(y_{H^+}y_B/y_{BH^+}),$$

where, if B refers to H_0 indicators (amides),

$$\log \frac{y_X y_{BH^+}}{y_{XH^+} y_B} = \phi(\log y_{BH^+}/y_B y_{H^+}) = \phi(H_0 + \log c_{H^+}).$$

A value of $\phi \neq 0$ means the activity coefficient ratio y_X/y_{XH^+} changes with acidity and differently from that of amides: it will be large and positive for bases whose conjugate acids have highly localized charges or hydrogen-bonding capacity which results in an increase in solvent structuring, e.g. $R_2O \rightarrow R_2OH^+$. Negative values of ϕ are typical of bases which lead to highly delocalized charges on protonation, e.g. carbocations (H_R). Values of ϕ for individual compounds are a useful probe into changes of solvation on protonation, thus:

$(MeO)_3 \cdot C_6H_2 \cdot COMe$	$\phi = -0.11$	
p-Nitroaniline	$\phi = 0$	increasing solvation
Ethyl acetate	$\phi = +0.4$	on protonation
Acetone	$\phi = +0.75$	

Superacid solutions [57–59]

The most protonating media known are solutions of BF_3, PF_5, AsF_5 and especially SbF_5 in liquid hydrogen fluoride, diluted with fluorosulfonic acid, FSO_3F, and sulfonyl chlorofluoride, SO_2ClF. Mixtures with these ingredients are known as *superacids* and are capable of protonating extremely weak bases (e.g. benzene) since acidity functions at least as high as -20 may be reached. This is due to the non-nucleophilic properties of the stable, symmetrical anions BF_4^-, SbF_6^- present in these acids, represented as 'HBF$_4$', 'HSbF$_6$'.

The uses of acidity functions

The acidity functions are measures of protonating ability suitable for use in highly acidic media. They will replace $[H^+]$ in any kinetic expressions in which the proton is involved when reactions take place under highly acidic conditions. All are expressions of similarity. For example, the acid-catalyzed decomposition of trioxane[60] occurs at a rate proportional to H_0; protonation of the ether function here is evidently sufficiently similar to that of the nitroanilines. An even better fit with H_{ROR} would be expected. Rates of hydrolysis of benzamides in sulfuric acid–water relate to H_A (J_A):

$$\log k_{rel} = r \log a_{H_2O} - H_A(\propto J_A).$$

This is indicative of pre-protonation and reaction involving r molecules of water. The rate of decarbonylation of triphenylacetic acid follows H_R — the appropriate scale measuring protonation of a hydroxyl function,

$$Ph_3CCOOH + (H^+) \rightarrow [Ph_3CCO \cdot OH_2^+] \rightarrow Ph_3C^+ + CO, H_2O.$$

6.11.2 Highly basic media

The same formal treatment may be used to assess the deprotonating power of a very basic medium beyond the limitation of the pH scale, which effectively becomes inoperable above pH 14: sodium methoxide in methanol or in dimethylsulfoxide, for example. A series of weak acid indicators is now chosen and their pK_A values measured in the stepwise manner (Eq. (6.30)) which relates them to dissociations in water. In highly basic media equilibrium deprotonation will occur and the indicator ratios are measured spectroscopically. An acidity function H_- can then be defined analogously to H_0 (Eq. (6.32)) applying to these indicators[20,61] (Fig. 6.3). Again, as the medium merges with water, H_- tends towards identity with pH. Separate scales for different classes of weak acid indicators are needed — weak carbon acids have been extensively used. The indicator ratios, $I,$ are also found to depend on the counterion, Fig. 6.3c, since this determines the degree of ion-pairing and affects the spectral changes.

The function H_- may be used as [$\log c_{(base)}$] in kinetic rate laws referring to highly basic solution. The racemization rate of **12**, for example, is given by the linear relationship[64]:

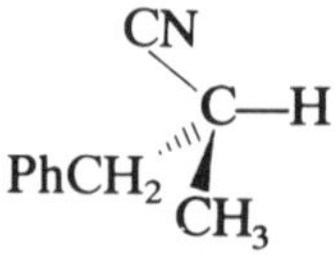

$$\log k = 0.875 H_- - 0.6.$$

12

6.12 NUCLEOPHILICITY AND ELECTROPHILICITY

The term 'base' specifically refers to nucleophilic activity at the proton while 'acid' refers to electrophilic properties of the proton. This section examines the more general concepts of nucleophilic and electrophilic reactivity at other centers.

6.12.1 Measurement of nucleophilicity: nucleophilicity and basicity

Even within a closely related series of nucleophiles, rates of reaction with some common electrophiles are often found to be only very approximately

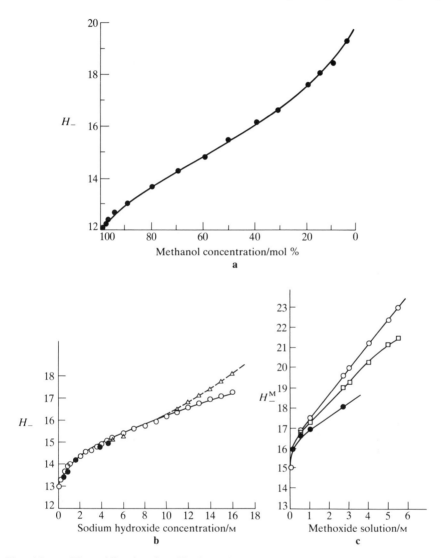

Fig. 6.3 **a**, The acidity function H_- for solutions of sodium methoxide in methanol–dimethylsulfoxide solvent of varying composition[62]. **b**, The acidity function H_- for the system NaOH–H_2O (data of different authors)[63]. **c**, The effect of cation on the acidity function of M^+ OMe^- in methanol at a range of concentrations[64]: O, K^+; □, Na^+; ●, Li^+.

parallel with pK_A values. For example, the displacements by a series of carboxylate ions are shown in Fig. 6.4.

$$RCOO^- \frown CH_2 \overset{\frown}{-} Cl \overset{k}{\longrightarrow} RCOOCH_2 + Cl^-$$
$$\underset{COO^-}{|} \qquad\qquad \underset{COO^-}{|}$$

Even with such a series of similar nucleophiles a plot of $\log k$ versus pK_A

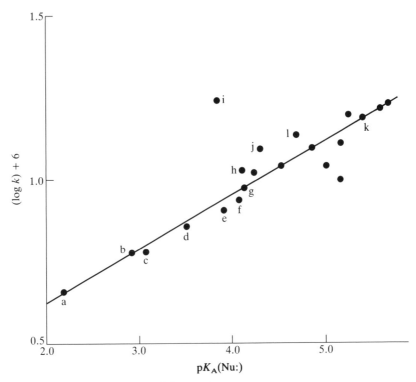

Fig. 6.4 Log–log plot of the rate of the reaction:

$$Nu: + CH_2(COO^-)Cl \rightarrow Nu—CH_2COO^- + Cl^-$$

against $pK_A(Nu:)$. Nucleophiles are as follows: a, *o*-nitrobenzoate; b, salicylate; c, chloroacetate; d, *m*-nitrobenzoate; e, *m*-chlorobenzoate; f, *p*-chlorobenzoate; g, benzoate; h, *p*-nitrophenylacetate; i, formate; j, oxalate; k, malonate; l, acetate.

shows considerable scatter and formate is very anomalous. This is partly because the attempt to relate kinetic and equilibrium phenomena, as has been seen in the case of carbon acidities, is not usually successful, and partly the result of comparing the differing demands of two electrophiles, H^+ and R^+. Despite the formal similarity between proton and methyl transfers,

$$Nu: \longrightarrow H\!-\!\overset{\frown}{O}H \underset{}{\overset{K_A}{\rightleftharpoons}} Nu—H \ + \ OH^-$$

$$Nu: \longrightarrow CH_3\!-\!\overset{\frown}{B}r \overset{k}{\longrightarrow} Nu—CH_3 + Br^-$$

all semblance of a correlation between k and K_A disappears when the range of nucleophiles is widely extended[65,66]. Some nucleophiles are found to be weak bases (i.e. toward H) but act powerfully toward carbon; iodide, azide and cyanide are examples that show this type of behavior. Separate scales of nucleophilicity appropriate to each type of electrophilic center are required in order to measure the capabilities for electron-pair donation at saturated carbon, cationic carbon, phosphorus, sulfur, metals and so on. The reason for this variable behavior is closely connected with the concept of 'hard and soft acids and bases' — the HSAB principle.

6.12.2 Hard and soft acids and bases: frontier orbital interactions

In 1963, Pearson drew attention to a division of nucleophiles and electrophiles each into two classes which he termed *hard* and *soft*[67]. Characteristic of this division, it was observed that strong interactions tended to occur between hard acids and hard bases or between soft acids and soft bases. A combination of a hard with a soft species led to a weak interaction. These concepts are essentially qualitative but point to two distinct types of attractive force involved in general acid–base interactions. Table 6.7 sets out examples of members of these different classes: thus it is seen that the proton and all proton donors are classed as 'hard' electrophiles; saturated carbon, on the other hand, is classed as 'soft'. Hydroxide ion is a hard nucleophile and iodide is soft. Therefore hydroxide ion tends to have a strong interaction with the proton, while iodide interacts weakly with H^+ but strongly at carbon. A semiquantitative measure of softness is given by the availability of the frontier orbitals — the higher the energy of the HOMO or the lower the energy of the LUMO, the softer the nucleophile or electrophile, respectively.

The principle can be illustrated by an explanation of the somewhat paradoxical observation that aqueous HI acts both as a strong acid and a strong nucleophile and is especially capable of cleaving ethers. Protonation on oxygen, a hard base, by HI provides a more facile leaving group and reaction is completed by attack of iodide on carbon, the soft electrophilic center, **11**.

11

Other soft–soft interactions which are well known include the affinity of mercury for sulfur (from which originated the obsolete name 'mercaptan' for thiols) and of silver ion for halogen, making this an effective electrophilic catalyst in substitutions. The position on the hard–soft scale which is occupied by any nucleophile–electrophile pair will determine their tendency to react and hence the disparate relative reactivities of a series of nucleophiles with different electrophilic centers.

The underlying principle on which this division is based may be explained in terms of perturbation MO theory (Section 1.2.2). The forces of

Table 6.7   Hard and soft nucleophiles[68]

Bases (nucleophiles)	Acids (electrophiles)
Hard H_2O, OH^-, F^- $CH_2CO_2^-$, PO_4^{3-}, SO_4^{2-} Cl^-, CO_3^{2-}, ClO_4^-, NO_3^- ROH, RO^-, R_2O NH_3, RNH_2, N_2H_4	*Hard* H^+, Li^+, Na^+, K^+ Be^{2+}, Mg^{2+}, Ca^{2+} Al^{3+}, Ga^{3+} Cr^{3+}, Co^{3+}, Fe^{3+} CH_3Sn^{3+} Si^{4+}, Ti^{4+} Ce^{3+}, Sn^{4+} $(CH_3)_2Sn^{2+}$ $BeMe_2$, BF_3, $B(OR)_3$ $Al(CH_3)_3$, $AlCl_3$, AlH_3 RPO_2^+, $ROPO_2^+$ RSO_2^+, $ROSO_2^+$, SO_3 I^{7+}, I^{5+}, Cl^{7+}, Cr^{6+} RCO^+, CO_2, NC^+ HX (hydrogen-bonding molecules)

Borderline
$C_6H_5NH_2$, C_5H_5N, N_3^-, Br^-, NO_2^-,
   SO_3^{2-}
	Borderline
	Fe^{2+}, Co^{2+}, Ni^{2+}, Cu^{2+}, Zn^{2+}, Pb^{2+}, Sn^{2+},
	   $B(CH_3)_3$, SO_2, NO^+, R_3C^+, $C_6H_5^+$

Soft R_2S, RSH, RS^- I^-, SCN^-, $S_2O_3^{2-}$ R_3P, R_3As, $(RO)_3P$ CN^-, RNC, CO C_2H_4, C_6H_6 H^-, R^-	*Soft* Cu^+, Ag^+, Au^+, Tl^+, Hg^+ Pd^{2+}, Cd^{2+}, Pt^{2+}, Hg^{2+}, CH_3Hg^+, $Co(CN)_5^{2-}$ Tl^{3+}, $Tl(CH_3)_3$, BH_3 RS^+, RSe^+, RTe^+ I^+, Br^+, HO^+, RO^+ I_2, Br_2, ICN, etc. Trinitrobenzene, etc. Chloranil, quinones, etc. Tetracyanoethene, etc. O, Cl, Br, I, N, $RO\cdot$, $RO_2^\cdot$ M^0 (metal atoms) Bulk metals CH_2, carbenes

Corrected experimental energies of frontier orbitals for some nucleophiles and electrophiles

Nucleophile	HOMO energy, $E^{\ddagger}/eV^a$	Electrophile	LUMO energy, $E^{\ddagger}/eV$
H^-	-7.37 *soft*	Ca^{2+}	2.33 *hard*
I^-	-8.31 ↑	Fe^{2+}	0.69 ↑
SH^-	-8.59	Li^+	0.49
CN^-	-8.78	H^+	0.42
Br^-	-9.22	Na^+	0
Cl^-	-9.94	Ag^+	-2.82 ↓
OH^-	-10.45	Hg^{2+}	-4.64 *soft*
H_2O	-10.7 ↓		
F^-	-12.18 *hard*		

^{a} $1\ eV = 96.45\ kJ\ mol^{-1} = 23.05\ kcal\ mol^{-1}$.

interaction between a nucleophile and an electrophile in the early stages of their progress towards bond formation are of three types:

$$E_i = E(\text{core}) + E(\text{electrostatic}) + E(\text{overlap}).$$

The core term is the electron–electron repulsion energy and is positive. The electrostatic term depends upon the charge or dipolar character of the reagents, while the overlap term contains the interaction energy of frontier orbitals (charge transfer and polarizability) and will be large if the reacting partners possess respectively a symmetry-compatible HOMO and LUMO which are close in energy. Both of the last two terms are negative (attractive).

The full expression for the interaction is as follows:

$$\Delta E = -\sum_{ab} (q_a + q_b)\beta_{ab}S_{ab} + \sum_{k<l} \frac{q_k q_l}{\varepsilon r_{kl}} + \sum_{r}^{occ} \sum_{s}^{unocc} - \sum_{s}^{occ} \sum_{r}^{unocc} \frac{2(\sum_{ab} c_{ra} c_{sb})^2}{E_r - E_s}$$

$$\underbrace{\hphantom{AAAAAAAAA}}_{\text{core}} \qquad \underbrace{\hphantom{AAAAAAAAA}}_{\text{electrostatic}} \qquad \underbrace{\hphantom{AAAAAAAAAAAAAAAAA}}_{\text{overlap}}$$

in which q_n, q_e are charges on nucleophile and electrophile, respectively, and c_n, c_e the coefficients of the HOMO and LUMO at the interacting the local dielectric constant; r_{kl} is the separation between atoms k and l; c_{ra} is the coefficient of AO a in MO r in one reactant; and c_{sb} is the coefficient of AO b in MO s in the other reagent.

Since in a related series the core term is constant and the orbital interactions are most important for the frontier orbitals, the simplified expression is used:

$$\Delta E = \frac{-q_n q_e}{\varepsilon r} + \frac{2(c_n c_e \beta)^2}{E_{HOMO} - E_{LUMO}}, \tag{6.36}$$

in which q_n, q_e are charges on nucleophile and electrophile, respectively, and c_n, c_e the coefficients of the HOMO and LUMO at the interacting centers. Values of E_{HOMO} and E_{LUMO} can be experimentally determined from measurements of ionization potentials and electron affinities, respectively. They necessarily refer to the gas phase but, when corrected for solvation, yield values ($E^{\ddagger}$) which are a quantitative measure of hard–soft character (Table 6.7).

Now the attractive force between hard acids and bases is mainly due to the electrostatic term, while that between soft acids and bases is mainly due to the HOMO–LUMO overlap term. It is hardly surprising, therefore, that relative nucleophilicity depends upon the electrophilic center to which it refers. Simple calculations of E_i can duplicate observed reactivity orders, which change according to the LUMO of a reference electrophile; this variation affects the importance of the last term[68], thus:

$E_{LUMO} = -5\,\text{eV}$ (low-lying, easily accessible, e.g. Hg^{2+}).

$E(\text{overlap}) \equiv$ nucleophilicity: $HS^- > CN^- > Br^- > Cl^- > OH^- > F^-$.

This is the order of reactivity at a soft center.

$E_{LUMO} = +1$ eV (high-lying, inaccessible e.g. Ca^{2+})

Nucleophilicity: $OH^- > CN^- > HS^- > F^- > Cl^- > Br^- > I^-$.

This is the order of pK_A (reactivity at a hard center).

6.12.3 Nucleophilicity scales

Scales measuring nucleophilicity embodying the linear free-energy principle have been devised with reference to carbon and also to platinum as electrophilic centers. Others could be created if sufficient rate or equilibrium data were available. Swain and Scott used a linear free-energy approach based on the displacement reaction[69],

$$Nu:^- + CH_3Br \xrightarrow{k_{Nu}} Nu{-\!}CH_3 + Br^-.$$

This is formally a transfer of (CH_3^+) between the incoming nucleophile, $Nu:^-$, and bromide which is quite analogous to proton transfer between bases. The change in free energy of activation is attributed to a parameter, n (nucleophilicity), characteristic of each species Nu:, which may be neutral or anionic (Table 6.8):

$$\log k_{Nu}/k_0 = sn \qquad (6.37)$$

here k_0 refers to that by a standard nucleophile (water), and s is the susceptibility constant expressing sensitivity of the rate to nucleophilicity. Analogies with the Hammett equation (Eq. (4.1)) are obvious. The value of s is defined for methyl bromide diplacement in water as $s = 1$, with other systems relative to this. A more extensive scale (n_0) based on the same reaction series in methanol is now more usually employed. The Swain–Scott equation, Eq. (6.37), best correlates nucleophilic displacements at saturated carbon[70,71] in protic solvents but may be useful in a qualitative sense for reactions at sulfur[73] and phosphorus[74]. It sometimes fails to correlate reactivity at other types of center, for example the carbonyl group[75]. Carbonyl carbon bears a partial positive charge and is therefore 'harder' than saturated carbon with the result that a different order of reactivity holds.

Solvent nucleophilicities, N, of protic solvents with respect to solvolytic reactions are measured by the Grunwald–Winstein relationship, Eq. (5.15) and values are given in Table 6.9. The Edwards ('Oxybase') equation (Eq. (6.38)), which correlates the same type of data, is a two-parameter version of Eq. (6.37)[72]:

$$\log k_N/k_0 = aE_N + bH_N \qquad (6.38)$$

where E_N, H_N may be called 'soft' and 'hard' nucleophilicities (Table 6.9)

Table 6.8   Nucleophilicity parameters

Nucleophile	n^a	$n_0{}^b$	$n_{Pt}{}^c$	$E_N{}^d$	$H_N{}^d$
MeOH			0·0		
H_2O	0			0	0
NO_3^-		1·5	−1·3	0·29	0·4
OAc^-	2·72	4·3	4·19	0·95	6·46
F^-		2·7	ca 2	−0·27	4·9
Cl^-	3·04	4·33	3·04	1·24	−3·0
SO_4^{2-}		3·5		0·59	3·74
Br^-	3·89	5·79	4·18	1·51	−6·0
PhO^-		5·79		1·46	11·7
I^-	5·04	7·42	5·46	2·06	−9·0
Pyridine		5·23	3·19	1·20	7·04
N_3^-	4·0	5·78	3·58	1·58	6·46
OH^-	4·2			1·65	17·5
MeO^-		6·29	ca 2·4		
$PhNH_2$	4·49	5·70	3·16	1·78	6·28
Et_3P		8·72	8·93		
SCN^-	4·77	6·70	5·75	1·83	1·0
NO_2^-		7·42	3·22	1·73	5·09
$S_2O_3^{2-}$	6·36	8·95	7·34	2·52	3·60
SO_3^{2-}		8·53	5·79	2·57	9·0
PhS^-		9·92	7·17		
$S^=$				3·08	14·66
NH_3		5·50	3·07	2·71	11·22
CN^-		6·70	7·14	2·79	10·88

[a] Swain–Scott parameters, solvent water [69].
[b] Swain–Scott parameters, solvent MeOH.
[c] Nucleophilicity to platinum based on

$$Nu: + Ptpy_2Cl_2 \xrightarrow{k_{Nu}} Ptpy_2ClNu + Cl^-; \quad n_{Pt} = \log k_{Nu}/k_{MeOH}.$$

[d] Oxybase scale parameters [72].

defined from oxidation potentials and pK_A values, respectively; *a, b* are the susceptibility constants and, normally $a \gg b$. The following examples are typical [79].

	a	*b*	*a/b*
$Nu:^- + MeBr \rightarrow Nu{-}Me + Br^-$	2·50	0·006	416
$Nu:^- + Cu^{++} \rightarrow Nu{-}Cu^+$	4·95	0·162	30
$Nu:^- + HO{-}OH \rightarrow Nu{-}OH + OH^-$	6·22	−0·43	−14

The different balance of the two components in reactions at hard and at soft electrophilic centers is clearly seen: the negative coefficient in the third example seems to point to a repulsive electrostatic term, due to negative charge on the oxygen at the reaction center. On the whole, however, the

243

Table 6.9   Solvent and anion–solvent nucleophilicities
Grunwald–Winstein solvent nucleophilicities, N[77]

Solvent	N
CF_3COOH	−4·74
CF_3COOH/water, 97:3	−3·93
H_2O	−0·26
CF_3CH_2OH	−2·78
$HCOOH^2$	−2·05
CH_3COOH	−2·05
EtOH	0·09
EtOH/water, 80:20	0·00 (reference)
EtOH/water, 50:50	−0·20
MeCOMe/water, 90:10	−0·43
MeCOMe/water, 50:50	−0·44
p-Dioxane/water, 90:10	−0·65
p-Dioxane/water, 50:50	−0·39

Ritchie anion–solvent nucleophilicities, N_+.[78]

System	N_+
Water	0·00 (reference)
Methanol	0·5
CN^-/water	3·8
$PhSO_2^-$/methanol	3·8
OH^-/water	4·5
N_3^-/water	5·4
CN^-/methanol	5·9
MeO^-/methanol	7·5
N_3^-/methanol	8·5
CN^-/DMSO	8·6
CN^-DMF	9·4
PhS^-/MeOH	10·7
PhS^-	13·1

Molar solubilities (C), ion-pair association constants (K_A), activity coefficients (y), and free energies of solution of dissociated species and ion-pairs as a function of solvent, for $Et_4N^+I^-$ [76]

Solvent	C/M	K_A	y	$\Delta G_e^c/kcal\,mol^{-1}$ $(R_4N^+ + I^-)$	$R_4N^+I^-$
Water	1·927	1·5	0·197	1·15	0·91
Methanol	0·362	18	3·6	3·00	1·29
Ethanol	$3·40 \times 10^{-2}$	133	4·3	5·45	2·55
n-Propanol	$1·187 \times 10^{-2}$	466	4·7	6·70	3·06
Propan-2-ol	$3·565 \times 10^{-3}$	1300	5·0	7·97	3·72
n-Butanol	$5·950 \times 10^{-3}$	1410	6·0	7·65	3·36
3-Methylbutanol	$2·75 \times 10^{-2}$	4500	7·0	8·72	3·74
tert-Butyl alcohol	$2·760 \times 10^{-4}$	$4·00 \times 10^4$	7·0	11·34	5·06
Dimethyl sulfoxide	0·1214	0	4·0	3·16	—
Nitromethane	0·2090	2	3·9	2·92	2·51
Acetonitrile	0·1153	5	3·6	3·61	2·65
Dimethylformamide	0·1150	12·2	3·7	3·69	2·21
Propionitrile	$3·00 \times 10^{-2}$	38	4·5	5·17	3·02
Nitrobenzene	$1·96 \times 10^{-2}$	27	3·9	5·35	3·40

Table 6.9   (*Continued*)

Solvent	C	K_A	γ	$\Delta G_{\varepsilon}^c/kcal\,mol^{-1}$	
				$(R_4N^+ + I^-)$	$R_4N^+ I^-$
Benzonitrile	1.76×10^{-2}	51	4.5	5.73	3.40
Acetone	9.68×10^{-3}	150	5.0	6.53	3.57
Ethyl methyl ketone	2.035×10^{-3}	411	5.0	8.10	4.53
Acetophenone	5.05×10^{-3}	250	5.5	7.25	3.98
1,2-Dichloroethane	9.19×10^{-3}	1.00×10^4	6.0	8.39	2.93
Dichloromethane	3.77×10^{-2}	3.65×10^4	6.0	8.23	2.01
1,1,2,2-Tetrachloroethane	4.82×10^{-2}	1.06×10^5	6.5	8.68	1.83
1,1,2-Trichloroethane	1.83×10^{-3}	5.82×10^5	6.5	11.65	3.76
1,1-Dichloroethane	$\leqslant 1 \times 10^{-4}$	4.50×10^4	6.0	$\geqslant 12.13$	$\geqslant 5.78$
Methyl formate	1.20×10^{-3}	8.44×10^4	6.5	10.79	4.07
Bromobenzene	3.61×10^{-4}	2.92×10^7	7.0	14.89	4.70
Ethyl benzoate	4.96×10^{-5}	6.50×10^6	7.0	15.20	5.91
Ethyl acetate	1.56×10^{-5}	5.00×10^6	7.0	15.77	6.63
Chlorobenzene	2.21×10^{-5}	1.51×10^7	7.0	16.18	6.39
Ether	4.0×10^{-8}	3.14×10^9	7.5	23.10	10.15
Benzene	3.87×10^{-7}	3.41×10^{17}	8.0	32.67	8.75

Swain–Scott and Edwards parameters express nucleophilicity only in a limited context since it is a highly solvent-dependent quantity. Solvation, particularly hydrogen bonding, tends to reduce nucleophilicity since the unshared pairs are less available for bonding. Ritchie[80,81] has proposed a scale, N_+, of reactivities of nucleophile/solvent systems as a whole at carbocation and carbonyl centers:

$$Ar_3C^+ + Nu:^-(solv) \xrightarrow{k_{Nu}} Ar_3C—Nu$$

$$\log k_{Nu}/k_0 = N_+, \tag{6.39}$$

where k_0 refers to the rate when Nu: = water. An interesting feature of this correlation is the absence of a susceptibility parameter. It appears that many series of electrophiles of the carbocation type, such as tropylium ions, diazonium ions, esters, as well as triphenylmethyl cations, obey Eq. (6.39) despite large differences in absolute rates. This remarkable result appears to violate the reactivity–selectivity principle but probably reveals unforeseen complexities; the slow step in this apparently simple reaction may be the formation of an ion-pair with solvent reorganization[82] under electrostatic forces:

$$Ar_3C^+ + :Nu^-(solv) \rightarrow [Ar_3\overset{+}{C}(solv) :Nu^-] \rightarrow Ar_3C—Nu.$$

Values of N_+ in Table 6.9 show how experimental nucleophilic power is influenced by the medium; cyanide ion, for instance, is nearly 10^6-fold more reactive in DMF than in water. Dipolar aprotic solvents such as dimethyl-formamide, due to their poor ability to solvate anions, will detract less from the reactivity of the nucleophile than do hydrogen-bonding solvents such as alcohol or water. This effect is parallel to the changes in activity coefficients of nucleophilic anions due to solvent influences (Table 6.9).

6.12.4 The relationship between nucleophilicity and nucleofugacity

It has been shown that nucleophilicity depends upon the nature of the electrophilic center at which it is tested. Frequently, the reaction of interest is a synchronous displacement (S_N2) in which the leaving group departs as the nucleophile commences to bond. In such cases it may be considered that the nature of the electrophilic center is in a state of change throughout the reaction. For example, if bond breaking is more advanced than bond making it becomes harder as the reaction proceeds. This simply means that the nature of the leaving group (*nucleofuge*) will also affect nucleophilicity assessments in such cases. This is borne out by the observation that the order of reactivity of a series of nucleophiles towards S_N2 reactions may vary according to the leaving group. In general, a highly polarizable leaving group is most easily displaced by a highly polarizable nucleophile, an aspect of the operation of Eq. (6.36) and recognized in the Edwards equation. Consider the following rates:

	$10^5 k_2/\text{M}^{-1}\,\text{s}^{-1}$
$I^- + EtI^* \rightarrow EtI + I^{*-}$	6000
$py + EtI \rightarrow Et\overset{+}{p}y + I^-$	4·17
$I^- + EtBr \rightarrow EtI + Br^-$	195
$py + EtBr \rightarrow Et\overset{+}{p}y + Br^-$	0·725

Using ethyl iodide as electrophile, relative nucleophilicities of iodide and pyridine, $\log k_{I^-}/k_{py} = 3\cdot15$ whereas with ethyl bromide, a similar type of center and identical reaction mechanism but a less polarizable leaving group the corresponding value is 2·42. Similar observations may be made when the slow step consists of coordination of the nucleophile but no fission of the leaving group, for example the $S_N Ar$ reaction (Section 10.5). Rates of attack of PhS^- (a polarizable nucleophile) relative to PhO^- (less polarizable) at 1-halogeno-2,4-dinitrobenzenes increase in the order: $F < Cl < Br < I$, despite the fact that absolute rates for each nucleophile are in the opposite order.

Again, one could not assess the relative nucleophilicity of the two nucleophiles as a single universal value but only with reference to the individual electrophilic center with which it is interacting. For such reactions linear free-energy relationships have been proposed which take into account the polarizability of the leaving group, $\mathbf{P}_{C-X}$. Todesco, for example proposed[83] Eq. (6.40):

$$\log k_p/k_0 = A + B \log \mathbf{P}_{C-X} \tag{6.40}$$

		Polarizability $\longrightarrow$	
$k_{\mathrm{PhS^-}}/k_{\mathrm{PhO^-}}$:	43	700	2200	5000
Hal:	F	Cl	Br	I

in which k_p, k_0 are rates for reactions of a polarizable and a less-polarizable nucleophile, respectively, and B is the susceptibility of the reaction to polarizability differences in the reagents being compared. Both A and B will depend upon the nucleophile, and $\mathbf{P}$ will be a property of the C–X bond; cf. Table 1.9. The reactions above now correlate well with the following values of A and B:

$$\log k_{\mathrm{PhS^-}} = 1{\cdot}3 + 2{\cdot}1 \log \mathbf{P}_{\mathrm{C-X}}$$

and

$$\log k_{\mathrm{PhO^-}} = -1{\cdot}57 + 0{\cdot}1 \log \mathbf{P}_{\mathrm{C-X}}.$$

It is clear that the susceptibility to attack by PhS^- is far more dependent on $\mathbf{P}_{\mathrm{C-X}}$ then is attack by PhO^-. Whereas an equation with so many variables is not of great practical utility, it is an important demonstration of the factors which control nucleophilic displacements and of their complex interrelationships.

6.12.5 The 'α-effect'

Nucleophiles which have an unshared electron pair on the atom immediately adjacent (α) to the nucleophilic center appear to be unusually strong donors, certainly in relation to their weak basicity (which, as has been explained, is no guide). These species include hydrazines, hydroxylamines, and peroxides,

$$-\mathrm{NH}-\mathrm{NH_2} \qquad -\mathrm{O}-\mathrm{NH_2} \qquad -\mathrm{O}-\mathrm{O}^-$$

and this property is in striking contrast to their weakness as bases owing to the inductive effect of the adjacent heteroatom. For example, the hydroperoxide ion, $HO-O^-$, though less basic than OH^- by a factor of 10^4, is more reactive in the displacement of bromide from bromoacetate by a factor of about 20[84,85]. The origin of this hyperactivity appears to lie in interactions between the unshared pairs on adjacent atoms which result in a raising of the HOMO energy and enhancement of 'soft' base behavior[86,87]. Late transition states which enable orbital overlap, the principle driving force, to be maximized are characteristic of these and other soft–soft interactions.

6.12.6 Ambident nucleophiles

Species containing two or more nucleophilic centers of different types are known as *ambident* and, if each center is a part of a conjugated system, the reactivity pattern can be analyzed by considering the distribution of the HOMO and the charge, if any. One of the most important ambident nucleophiles is the enolate ion, an oxygen analogue of the allyl anion. The Hückel MO characteristics may be calculated by solution of the 3×3 determinant containing values of δ_M and ρ_{C-M} recommended for —O⁻ (Table 1.2):

Allyl anion 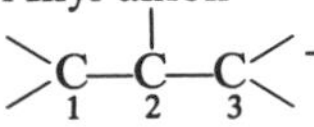
$$\begin{bmatrix} x & 1 & 0 \\ 1 & x & 1 \\ 0 & 1 & x \end{bmatrix} = 0$$

Enolate 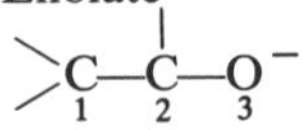
$$\begin{bmatrix} x & 1 & 0 \\ 1 & (x+0.2) & 0.8 \\ 0 & 0.8 & 2 \end{bmatrix} = 0$$

for which the solutions are:

Eigenvalues

	ψ_1	ψ_2	ψ_3	ψ_1	ψ_2	ψ_3
	1·444	0	−1·414	2·366	0.839	−1·00

Eigenvectors, c

	ψ_1	ψ_2	ψ_3	ψ_1	ψ_2	ψ_3
c_1.	0·500	−0·707	0·500	0·173	0·700 ‡	0·692
c_2.	0·707	0	−0·707	0·410	0·588	−0·697
c_3.	0·5	0·707	0·5	0·895 *	−0·405	0·185
		(HOMO)			(HOMO)	

The MOs are depicted opposite and show that the effect of the heteroatom is to increase the charge at oxygen relative to carbon because of the large coefficient c_3 in $\psi_1(*)$ but to decrease the coefficient of the HOMO $\psi_2(\ddagger)$ at oxygen relative to carbon. This makes oxygen a hard nucleophilic center and carbon a soft one so that the position of attack, at O or C, depends upon the hard/soft character of the electrophile; normally, enolates of simple aldehydes and ketones react at carbon with alkyl halides, which are soft electrophiles, but at oxygen with the harder acyl halides.

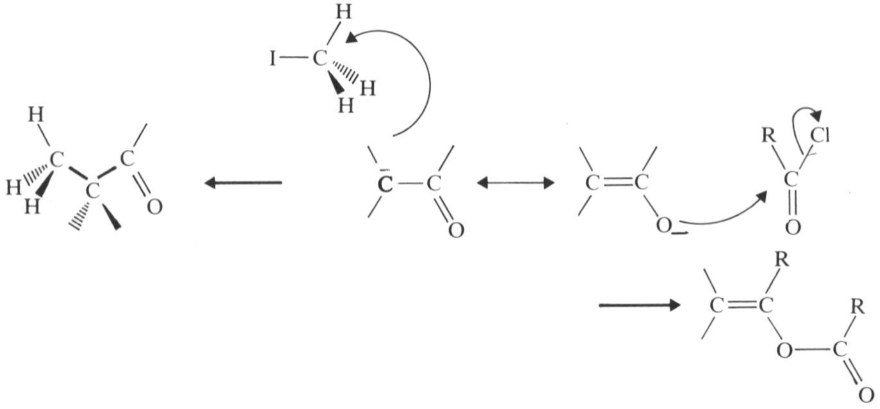

$c^2(\Psi_2)$	0.5	0	0.5	0.49 0.34 0.16
Charge q/e	-0.5	0	-0.5	$-0.04 -0.03 -0.93$

Acetoacetate is a stable enolate on which this theory can be tested; alkylation can lead to *C*-alkyl or *O*-alkyl products, the ratio of which decreases with increasing hardness of the leaving group (and consequently also that of the reaction center):

X	Increasing hardness				
	—I	—Br	—OTos	—OSO₂OEt	—OSO₂CF₃
k_C/k_O	>100	60	6·6	4·8	3·70

Nitrite ion, $O{=}N{=}O^-$, is also ambident and is attacked by soft electrophiles (such as methyl iodide) principally at nitrogen, which bears the largest coefficient of the HOMO (Fig. 6.5). Hard electrophiles (methyl iodide in the presence of silver ion) react at oxygen:

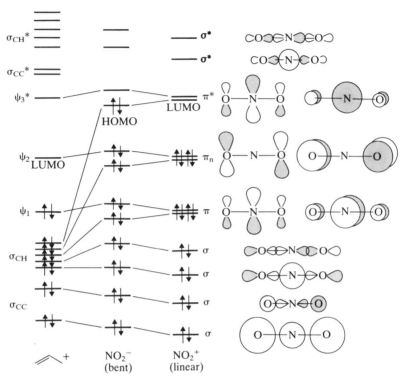

Fig. 6.5 MO's of nitrite and nitronium ions and their relationship with the alkyl cation: σ, π are bonding orbitals; π_n, non-bonding; σ^*, π^*, antibonding.

The related nitronium ion is an electrophile but it also has the principal frontier orbital (LUMO) coefficient at nitrogen at which, for example, it reacts with aromatic molecules.

Aromatic compounds, Ph—X for example, are ambident nucleophiles having *ortho, meta* and *para* positions for attack. This aspect of reactivity is partly frontier-orbital controlled and will be discussed in Section 10.4.10.

Cyanide ion is soft at carbon and relatively hard at nitrogen. Alkylation normally yields alkyl cyanides but in the presence of silver ion the isocyanide forms.

6.13 THE MEASUREMENT OF ELECTROPHILICITY

In a similar manner, electrophilicity may be defined as the capacity to accept an electron pair from a standard donor, for which little in the way of

Table 6.10   Relative electrophilicities[86], water, 25°C[a]

Cl^+	10^7	Br_2	10
Br^+	10^6	$HIOH^+$	10
HN_2O_5	10^5	$HBrOH^+$	10
NO_2^+	10^4	NO^+	1
Cl_2	10^3	Br_3^-	1
$NOHSO_4$	10^3	PhN_2^+	10^{-1}
I^+	10^3	N_2O_4	10^{-1}
$NOBr$	10^3	$HClO$	10^{-2}
$H_2NO_3^+$	10^2	$HBrO$	10^{-2}
H_2ClO	10^2	N_2O_3	10^{-2}
$NOCl$	10	Hg^+	10^{-4}
$H_2NO_2^+$	10^7	I_2	10^{-5}

[a] Highly approximate, not to be taken too seriously.

quantitative information is available except for the proton. One reason for this is that many of the electrophiles of importance in organic chemistry are highly reactive transient species: carbocations of various types, nitronium ions, and halogens or acyl halides in the presence of aluminium halides. It is difficult to devise suitable experimental techniques leading to electrophilicity scales for such species. The list given in Table 6.10 is an attempt to provide 'order of magnitude' electrophilicities which are derived from a variety of sources (but to be taken *cum grano salis!*[78]).

PROBLEMS

1   Using values from Table 6.2, compare the acid dissociation constants of benzoic acid and anilinium ion at 25° and at 100°C. Suggest an explanation for the temperature effect.

2   Rationalize the effects of methyl substitution on the order of acidity of substituted benzoic acids:

Acid:	2,6-Dimethylbenzoic	**Benzoic**	3,5-Dimethylbenzoic	3,4-Dimethylbenzoic
pK_A:	3.24	4·2	4·3	4.4

3   Draw the structures, showing charge delocalization, of the conjugate base of the cyanocarbon acid, the last entry in Table 6.4a.

4   Are the data in Tables 6.2 and 6.3 compatible?

5   Using Tables 6.5, 1.13 and 1.15, compare the energies required to remove a proton and a hydrogen atom from a molecule. What is the relationship between the two quantities?

6   Explain why the inclusion of a covalent bond between the positions

Acids and bases, electrophiles and nucleophiles

marked by an asterisk increases the pK_A of the proton —H by more than 10 units (i.e. increases its acidity by 10^{10}).

7 Ester displacements respond to nucleophilic strength as measured by N_+. How much faster would nitrophenolate be released from a nitrophenyl ester by attack of cyanide in dimethylformamide than in water?

8 The amines

and

have similar pK_A values ($-6\cdot15$). Estimate the percentage protonation for each in 70% $H_2SO_4 : H_2O$ by the use of Fig. 6.2.

REFERENCES

1 J. N. Brønsted, *Chem. Rev.*, **3**, 231 (1928); R. P. Bell, *Ann. Rep. Chem. Soc.*, **31**, 71 (1934).
2 L. P. Hammett, *Physical Organic Chemistry*, 2nd edn, McGraw-Hill, New York, 1970.
3 G. Kortum, W. Vogel and K. Andrussow, *Dissociation Constants of Organic Acids in Aqueous Solution*, Butterworth, London, 1961.
4 D. D. Perrin, *Dissociation Constants of Organic Bases in Aqueous Solution*, Butterworth, London, 1965.
5 A. Covington, *Physical Chemistry of Organic Solvent Systems*, Plenum, London, 1973.
6 M. M. Davis, *Acid Behavior in Aprotic Organic Solvents*, Nat. Bur. Stand. Monograph No. 105, Washington, DC, 1968.
7 J. E. Leffler and E. Grunwald, *Rates and Equilibria of Organic Reactions*, Wiley, New York, 1963.
8 E. M. Arnett and G. Scorrono, *Adv. Phys. Org. Chem.*, **13**, 83 (1976).
9 D. Lemordant and R. Gaboriaud, *J. Chim. Phys, Phys. Chim. Biol.*, **79**, 149 (1982).
10 H. H. Jaffe, *Chem. Rev.*, **53**, 191 (1953).
11 M. Charton, *Prog. Phys. Org. Chem.*, **10**, 81 (1973).
12 P. Wells, *Prog. Phys. Org. Chem.*, **6**, 111 (1968).
13 M. Eigen, W. Kruse, G. Maass and L. de Maeyer, *Prog. React. Kinetics*, **1**, 285 (1964).
14 (a) M. Eigen, *Angew. Chem.*, **3**, 1 (1964); (b) Faraday Soc. Symposium, Vol. 10, The Chemical Society, London, 1975.
15 F. M. Jones and E. M. Arnett, *Prog. Phys. Org. Chem.*, **11**, 263 (1974).
16 H. K. Hall, *J. Amer. Chem. Soc.*, **79**, 5441 (1957).
17 E. Falkers and O. Runquist, *J. Org. Chem.*, **29**, 830 (1964).

18 F. E. Condon, *J. Amer. Chem. Soc.*, **87**, 4481, 4485, 4491, 4494 (1965).
19 C. L. Liotta, E. M. Perdue and H. P. Hopkins, *J. Amer. Chem. Soc.*, **95**, 2439 (1973).
20 J. Jones, *The Ionisation of Carbon Acids*, Academic Press, London, 1973.
21 D. J. Cram, *Fundamentals of Carbanion Chemistry*, Academic Press, New York, 1965.
22 A. Streitweiser, D. A. Bors and M. J. Kaufman, *J. Chem. Soc., Chem. Comm.*, 1394 (1983).
23 E. C. Steiner and J. H. Gilbert, *J. Amer. Chem. Soc.*, **85**, 3054 (1963).
24 A. J. Parker, *Chem. Rev.*, **69**, 1 (1969).
25 M. J. Cook, A. R. Katritzky and A. D. Page, *Tetrahedron*, 2707 (1975).
26 D. W. Boerth and A. Streitweiser, *J. Amer. Chem. Soc.*, **103**, 6443 (1981).
27 A. Streitweiser, J. I. Brauman, J. H. Hammons and A. H. Rudjaatmaka, *J. Amer. Chem. Soc.*, **87**, 384 (1965).
28 A. Streitweiser, C. Perrin, R. G. Lawler, R. A. Caldwell and G. R. Ziegler, *J. Amer. Chem. Soc.*, **87**, 5383, 5388, 5394, 5399 (1965).
29 A. Streitweiser, J. H. Hammons, E. Ciuffarin and J. I. Brauman, *J. Amer. Chem. Soc.*, **89**, 59, 63 (1967); A. Streitweiser and H. F. Koch, *J. Amer. Chem. Soc.*, **86**, 404 (1964).
30 A. I. Dyatkin, *Tetrahedron*, 2991 (1965).
31 H. G. Adolph and M. J. Kamlet, *J. Amer. Chem. Soc.*, **88**, 4761 (1966).
32 A. W. Johnson, *Ylide Chemistry*, Academic Press, New York, 1976.
33 A. I. Shatenshtein, *Adv. Phys. Org. Chem.*, **1**, 153 (1963).
34 P. Kebarle, *Ann. Rev. Phys. Chem.*, **28**, 445 (1977): *J. Amer. Chem. Soc.*, **107**, 2615 (1985).
35 C. G. Pitt, M. M. Bursey and D. P. Chatford, *J. Chem. Soc., Perkin II*, 434 (1976).
36 J. I. Brauman and L. K. Blair, *J. Amer. Chem. Soc.*, **91**, 2126 (1969).
37 Y. Yamdagni and P. Kebarle, *J. Amer. Chem. Soc.*, **95**, 4050 (1973).
38 D. M. Aue, H. M. Webb and M. T. Bowers, *J. Amer. Chem. Soc.*, **98**, 318 (1976); D. H. Aue, *J. Amer. Chem. Soc.*, **97**, 4137 (1975).
39 R. W. Taft, *Prog. Phys. Org. Chem.*, **14**, 247 (1983).
40 M. Eigen, 'Fast React. Primary Processes Chem. Kinet'., in *Proc. Nobel Symp.*, (S. Claessen, Ed.), Almqvist and Wisksell, Stockholm, 1967, p. 245.
41 R. A. Marcus, *J. Phys. Chem.*, **72**, 891 (1968).
42 J. Albery and M. Kreevoy, *Adv. Phys. Org. Chem.*, **16**, 87 (1978).
43 M. J. Blandamer, R. E. Robertson, J. M. W. Scott and P. D. Golding, *J. Amer. Chem. Soc.*, **103**, 5923 (1981).
44 M. J. Blandamer, J. M. W. Scott and R. E. Robertson, *J. Chem. Soc., Perkin II*, 447 (1981); M. J. Blandamer, J. Burgess, P. P. Duce, R. E. Robertson and J. M. W. Scott, *J. Chem. Soc., Faraday I*, **77**, 2281 (1981).
45 L. P. Hammett and A. I. Deyrup, *J. Amer. Chem. Soc.*, **54**, 2721 (1932).
46 M. A. Paul and F. A. Long, *Chem. Rev.*, **57**, 1 (1957).
47 N. C. Marziano, C. M. Cimino, R. C. Passerini, P. G. Traverso and A. Tomasin, *J. Chem. Soc., Perkin II*, 1915 (1973): 306, 309 (1977).
48 K. Yates and J. C. Riorden, *Can. J. Chem.*, **43**, 2328 (1965).
49 K. Yates and J. B. Stevens, *Can. J. Chem.*, **43**, 529 (1965).
50 R. A. Cox, L. M. Druet, A. I. Klausner, T. A. Modro, P. Wan and K. Yates, *Can. J. Chem.*, **59**, 1568 (1981).
51 E. M. Arnett and G. W. Mach, *J. Amer. Chem. Soc.*, **86**, 2671 (1964).
52 K. Yates, J. B. Stevens and A. R. Katritzky, *Can. J. Chem.*, **42**, 1957 (1964).
53 U. Quintily, G. Scorrano and F. Magno, *Gazz. Chim. Ital*, **111**, 401 (1981).
54 N. C. Deno, J. J. Jaruzelski and A. Schreischeim, *J. Amer. Chem. Soc.*, **77**, 3044 (1955).
55 J. F. Bunnett and F. P. Olsen, *Can. J. Chem.*, **44**, 1899 (1966).
56 K. Yates and R. A. McClelland, *J. Amer. Chem. Soc.*, **89**, 2686 (1967); **95**, 3055 (1973); *Prog. Phys. Org. Chem.*, **11**, 323 (1974).
57 G. A. Olah, *Angew. Chem.*, **75**, 800 (1963); *Org. React. Mech. Conf., Cork, 1964* (Special Pub. 19), Chemical Society, London, 1965.

Acids and bases, electrophiles and nucleophiles

58 R. J. Gillespie, *Acc. Chem. Res.*, **1**, 202 (1968).
59 A. F. Clifford, H. C. Beachell and W. H. Jack, *J. Inorg. Nucl. Chem.*, **5**, 57 (1957).
60 M. A. Paul, *J. Amer. Chem. Soc.*, **72**, 3813 (1950); **74**, 141 (1952).
61 A. Streitwieser, *J. Amer. Chem. Soc.*, **89**, 59, 63 (1967).
62 F. Terrier, F. Millott and R. Schaal, *Bull. Soc. Chim. France*, 3002 (1969); F. Terrier, *Ann. Chim.*, **4**, 153 (1969).
63 J. Jones, *The Ionization of Carbon Acids*, Academic Press, London, 1973.
64 F. Terrier, F. Millot and R. Schaal, *Bull. Soc. Chim. France*, 3002 (1969).
65 G. F. Smith, *J. Chem. Soc.*, 521 (1943).
66 R. G. Pearson, H. Sabel and J. Songstad, *J. Amer. Chem. Soc.*, **90**, 319 (1968).
67 R. G. Pearson, *J. Amer. Chem. Soc.*, **85**, 3533 (1963).
68 I. Fleming, *Frontier Orbitals and Organic Chemical Reactions*, Wiley–Interscience, London, 1976.
69 C. G. Swain and C. B. Scott, *J. Amer. Chem. Soc.*, **75**, 141, (1953).
70 K. Bowden and R. S. Cook, *J. Chem. Soc. (B)*, 1529 (1968).
71 E. Yrjanheikki and J. Koskikallio, *Suomen. Kem.*, **42B**, 195 (1969).
72 J. O. Edwards, *J. Amer. Chem. Soc.*, **76**, 1540 (1954).; **78**, 1819 (1956).
73 J. L. Kice and E. Legan, *J. Amer. Chem. Soc.*, **95**, 3912 (1973); J. L. Kice, G. J. Kasperek and D. Patterson, *J. Amer. Chem. Soc.*, **91**, 5516 (1969); O. Rogne, *J. Chem. Soc(B)*, 1056 (1970).
74 H. J. Bestmann, *Angew. Chem. Int. Ed.*, **4**, 645 (1965); S. Trippett, *Quart. Rev.*, **17**, 406 (1954).
75 R. F. Hudson and M. Green, *J. Chem. Soc.*, 1055 (1962).
76 M. H. Abraham, *J. Chem. Soc., Perkin II*, 1343 (1972).
77 F. L. Schadt, T. W. Bentley and P. v. R. Schleyer, *J. Amer. Chem. Soc.*, **98**, 7667 (1976).
78 T. Turvey, private communication.
79 R. E. Davis, S. P. Molnar and R. Nehring, *J. Amer. Chem. Soc.*, **91**, 97, 104 (1969).
80 C. D. Ritchie, *J. Amer. Chem. Soc.*, **94**, 4966 (1972) and other papers.
81 C. D. Ritchie, *Acc. Chem. Res.*, **5**, 348 (1972).
82 A. Pross, *J. Amer. Chem. Soc.*, **98**, 776 (1976); N. S. Isaacs and K. Javaid, *J. Chem. Soc., Perkin II*, 839 (1983).
83 G. Bartoli and P. E. Todesco, *Acc. Chem. Res.*, **10**, 125 (1977).
84 J. E. McIsaac, L. R. Subbaraman, J. Subbaraman, H. A. Mulhausen and E. J. Behrman, *J. Org. Chem.*, **37**, 1037 (1972).
85 M. J. Gregory and T. C. Bruice, *J. Amer. Chem. Soc.*, **89**, 4400 (1967).
86 J. E. Dixon and T. C. Bruice, *J. Amer. Chem. Soc.*, **94**, 2052 (1972).
87 F. Filippini and R. F. Hudson, *Chem. Comm.*, 522 (1972).

7 Kinetic isotope effects

7.1 ISOTOPIC SUBSTITUTION

The replacement of one or more atoms in a reacting system by others of their respective isotopes is one of the most subtle structural perturbations which may be made. Changes in reaction rate which are brought about by isotopic substitution are known as *kinetic isotope effects* and carry mechanistic information [1-4].

Isotopic substitution does not affect the potential energy surface of the molecule nor does it perturb the electronic energy levels. It is only those properties that are dependent upon atomic masses which are affected; for chemical purposes, the perturbation can be considered to be limited to vibrational frequencies.

Each atom in a molecule may move in three dimensions and for a molecule containing N atoms one might consider $3N$ independent modes of motion. However, six of these modes involve motion of the molecule as a whole (three for translation and three for rotation) so are not vibrations. There are, then, $(3N - 6)$ normal vibrational modes present, each one associated with vibrational energy (part of the internal energy, U). Moreover, each vibrational frequency, and hence energy, depends on the masses of the atoms vibrating and will vary with the isotopic species. Vibrational energy will usually change during the course of a reaction or between reagent and transition state since some bonds are in the course of being broken or made and their associated frequencies will be affected. Isotopic substitution should therefore affect reaction rates. The extent to which this occurs depends greatly upon the relative masses of the isotopes, being greatest for those of hydrogen H, D and T (^{1}H, ^{2}H, ^{3}H). Since hydrogen isotope effects are the most thoroughly examined, the following discussion will refer to this case, rates being denoted k_H, k_D, k_T, but the same principles apply to any pair of isotopes.

If $k_H \neq k_D$ a kinetic isotope effect (KIE) exists, expressed as the ratio k_H/k_D; it is described as 'normal' if $k_H/k_D > 1$ and 'inverse' if $k_H/k_D < 1$. The following types of isotope effect are distinguished:

(a) *primary* (PKIE), in which the bond to the isotopic atom is broken in the rate-determining step;

255

(b) *secondary* (SKIE), in which the bond to the isotopic atom(s) remains intact throughout the reaction;
(c) *solvent isotope effects,* which result from isotopic differences in the medium.

These categories are a convenience rather than the result of any fundamental difference in the origins of each type of KIE.

7.2 THEORY OF ISOTOPE EFFECTS [1-4]: THE PRIMARY EFFECT

A large proportion of observed isotope effects concern C–H bond breaking, which is a convenient starting point for discussion. The vibrational energy associated with the stretching of a covalent bond (C—H, C—D) is quantized so that

$$E_v = hc\bar{v} = (V + 1/2)hv; \qquad V = 0, 1, 2, \ldots \qquad (7.1)$$

where V is the vibrational quantum number, v the frequency of the transition from one level to the next and $\bar{v}$ the corresponding wavenumber ($= 1/\lambda$). Values of $\bar{v}$ for C—H and C—D stretching modes (from IR spectra) are around 3000 and 2100 cm^{-1}, respectively, and the corresponding frequencies, v, are 9 and 6·3 (10^{13} s^{-1}). In either case, transitions are associated with energy considerably greater than thermal energies at room temperature (kT) so that almost all molecules will be populating the ground vibrational level at around 25°C. The potential functions (i.e. length–energy relationships) for C–H and C–D bond stretching are essentially identical but the distribution of rotational energy levels in each bond differs. Those of the C–D bond lie at a lower energy than those of the C–H bond for a given value of V because the vibrational frequency of D is lower due to its greater mass (Fig. 7.1). The ground vibrational state is defined by $V = 0$ so that $E_v = \frac{1}{2}hv$; this is known as the *zero-point energy* (vibrational energy remaining even at 0 K) and, as explained, lies lower for the C–D bond. However, the isotopic difference in the energies of corresponding vibrational levels diminishes as the value of V increases until at the dissociation limit it is zero. It is clear, then, that more energy is expended in breaking a C–D bond than a C–H bond since it was originally at the lower potential. In general, the dissociation energy of a bond to a heavy isotope is greater than that to a light isotope of the same element and is associated with a slower rate—a normal PKIE.

Relative vibrational frequencies of C—H and C—D stretching modes (from Eq. (1.8)) are approximately in the ratio:

$$\frac{v_H}{v_D} = \frac{(f_H/\mu_H)^{\frac{1}{2}}}{(f_D/\mu_D)^{\frac{1}{2}}}. \qquad (7.2)$$

The force constants, $f_H = f_D$, are equal and the reduced masses, μ_H, μ_D, can be replaced by the masses of H, D if the remainder of the molecule is

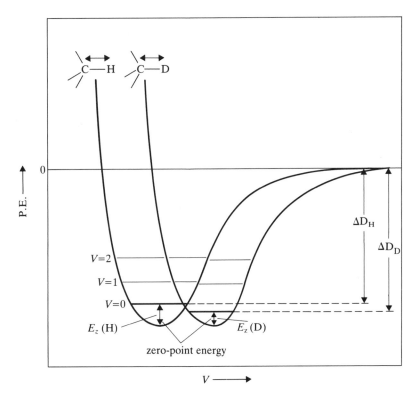

Fig. 7.1 Potential energy curves for C–H and C–D bonds showing the different dispositions of vibrational levels and the origins of their different bond dissociation energies. $E_z = E_v$ at $V = 0$ = zero point energy.

relatively massive. Then

$$\nu_H/\nu_D \approx (m_D/m_H)^{\frac{1}{2}} = 2^{\frac{1}{2}} = 1\cdot41.$$

Similarly, $\nu_H/\nu_T \approx 3^{\frac{1}{2}} = 1\cdot73$ although more exact values are:

$$\nu_H/\nu_D \approx 1\cdot35 \quad \text{and} \quad \nu_H/\nu_T \approx 1\cdot59. \tag{7.3}$$

The thermal dissociation of a C–H bond should proceed faster than that of a C–D bond under similar conditions since the activation energies are simply the bond dissociation energies.

Calculation of the kinetic isotope effect can be achieved by application of statistical thermodynamics, of which only an outline will be attempted here.

A sample of molecules contains quantized energy in many forms— translational, rotational, vibrational and electronic—each associated with energy levels or states. The partition function, Q, is defined by Eq. (7.4);

$$Q = \sum p_i e^{-\varepsilon_i/kT}, \tag{7.4}$$

257

in which the summation is over all energy states (energy ε_i, multiplicity p_i); Q is the product of contributions from each form of internal energy:

$$Q = Q_{tr} \cdot Q_{rot} \cdot Q_{vib} \cdot Q_{elec}, \qquad (7.5)$$

and can define an equilibrium constant—in this case an activation equilibrium,

$$K^{\ddagger} = [Q^{\ddagger}/Q(\text{reagents})] \cdot e^{-E_A/RT}, \qquad (7.6)$$

where $Q^{\ddagger}$ refers to the transition state.

The kinetic isotope effect is given by

$$\frac{k_H}{k_D} = \frac{K_H^{\ddagger}}{K_D^{\ddagger}} = \frac{Q_H^{\ddagger}}{Q_D^{\ddagger}} \cdot \frac{Q_D(\text{reagents})}{Q_H(\text{reagents})}. \qquad (7.7)$$

Expanding the partition functions leads to Eq. (7.8), from which the Bigeleisen equation, Eq. (7.9), may be obtained (in which $u = hv/kT$); the mass terms cancel since the same atoms are present in the reagents as in the transition state.

$$Q = M^{\frac{3}{2}}(ABC)^{\frac{1}{2}} \prod^{3N-6} \left(\frac{e^{-\frac{1}{2}u}}{1-e^{-u}} \right), \qquad (7.8)$$

where A, B, C are the three principal moments of inertia, and M the molecular mass.

$$\frac{k_H}{k_D} = \frac{v_H^{\ddagger}}{v_D^{\ddagger}} \prod^{3N-7} \left(\frac{u_H^{\ddagger}}{u_D^{\ddagger}} \times \frac{e^{-\frac{1}{2}u_H^{\ddagger}}}{e^{-\frac{1}{2}u_D^{\ddagger}}} \times \frac{1-e^{-u_D^{\ddagger}}}{1-e^{-u_H^{\ddagger}}} \right) \prod^{3N-6} \left(\frac{u_D}{u_H} \times \frac{e^{-\frac{1}{2}u_D}}{e^{-\frac{1}{2}u_H}} \times \frac{1-e^{-u_H}}{1-e^{-u_D}} \right). \quad (7.9)$$

transition-state terms   reagent terms

The $v^{\ddagger}$ terms refer to 'imaginary' frequencies corresponding to the hydrogen or deuterium exchange motion along the reaction coordinate.

The important aspect of these equations is that they contain terms from all vibrations of reagents and transition states. A complete analysis of a kinetic isotope effect may be carried out if the frequencies of all $(3N - 6)$ modes of the reagents and $(3N - 7)$ modes of the transition state* are known or can be estimated, e.g. by molecular mechanics (Section 8.2.7).

This is seldom attempted but these cumbersome equations may be approximated. Firstly, for this purpose, translational and rotational partition functions may be ignored since their quanta are so small that they contribute little. The electronic partition function may also be ignored since the quanta are so large that no excited states are normally brought into play. It is frequently assumed that vibrations in parts of the molecule not undergoing chemical change will be the same in reagents and transition state, and so these terms will cancel. In fact, in a simple C–H bond fission, it may be assumed that the C—H stretching frequency is the only important

---

* In the transition state one mode fewer—the stretching of the bond which is breaking—is no longer a true vibration: it is assigned an 'imaginary' frequency, $v^{\ddagger}$.

one to change. Equations (7.7) and (7.8) then combine to become:

$$\frac{k_H}{k_D} = \left(\frac{M_H^{\ddagger} M_D}{M_D^{\ddagger} M_H}\right)^{\frac{3}{2}} \left(\frac{A_H^{\ddagger} B_H^{\ddagger} C_H^{\ddagger}}{A_D^{\ddagger} B_D^{\ddagger} C_D^{\ddagger}} \times \frac{A_D B_D C_D}{A_H B_H C_H}\right)^{\frac{1}{2}} \frac{\sinh \frac{1}{2} u_H}{\sinh \frac{1}{2} u_D}. \qquad (7.10)$$

Furthermore, the mass and moment-of-inertia terms approximately cancel and will contribute little to the right-hand expression. This leaves

$$\frac{k_H}{k_D} = \frac{\sinh \frac{1}{2} u_H}{\sinh \frac{1}{2} u_D}, \qquad (7.11)$$

which comes also from Eq. (7.9) with similar assumptions.

The following calculation applies to breaking of C–H and C–D bonds, $\bar{v}_H = 3000$, $\bar{v}_D = 2100\ \text{cm}^{-1}$, 298 K:

$$\sinh \tfrac{1}{2} u_H = \sinh \left\{ \frac{6 \cdot 626 \times 10^{-34} \times 3 \times 10^{10} \times 3000}{2 \times 1 \cdot 381 \times 10^{-23} \times 298} \right\}$$

$$= \sinh 7 \cdot 2452 = \tfrac{1}{2}(e^{7 \cdot 2452} - e^{-7 \cdot 2452}) = 700 \cdot 72.$$

Similarly,

$$\sinh \tfrac{1}{2} u_D = 79 \cdot 62$$

and

$$k_H / k_D = 8 \cdot 80.$$

From Table 7.1, which summarizes calculations for different temperatures, it emerges that the kinetic isotope effect diminishes with temperature. In addition to loss of a stretching frequency, the loss of two bending frequencies upon fission of a bond ought to be taken into account. Assuming the bending frequencies lie at $\bar{v}_H = 1400\ \text{cm}^{-1}$, $\bar{v}_D = 960\ \text{cm}^{-1}$, the resulting contributions to the kinetic isotope effect can also be added in. The figures in Table 7.1 indicate that very large maximum isotope effects

Table 7.1   Maximum kinetic isotope effects calculated from Eq. (7.11)[a]

$T/°C$	k_H/k_D contributions		Combined isotope effect
	$\overset{\leftarrow}{C}\text{–}\vec{H}$ *stretch*	*bend*	
−100	42	6·2	1620
−50	18·3	4·14	313
0	10·7	3·2	109
25	8·8	2·9	74
100	5·7	2·38	32
200	3·9	2·03	16

[a] N.B. These values, derived from a highly simplified model, are greatly exaggerated but illustrate the temperature effect.

are possible. The experimental observations, however, are in accordance with loss of stretching frequencies only, i.e. there is a change of about 8 at 25°C. The reason is probably that there is compensation. At the transition state, the C–H bond is not completely broken (and may be only slightly stretched); similarly the bending vibrations are not totally lost. Also, reactions involving C–H bond breaking are almost always transfer reactions and the properties of the newly-forming bond should also be considered. Extra high values of k_H/k_D which are sometimes observed may be due to special circumstances which involve loss of bending but also to tunneling (see Section 7.6). In general, the observed isotope effect is treated as an experimental quantity to be interpreted by comparison, but simply establishing its presence or absence is often sufficient for mechanistic deductions to be made.

Calculations involving bond orders to estimate the force field and predict isotope effects have recently been introduced[6], and results similar to those of Table 7.1 may be obtained by considering the energetics of bond vibrations. The total difference in zero-point energy ΔE_z, between C—H and C—D for the three modes, is given by

$$\Delta E_z = \tfrac{1}{2}hcL[(\bar{v}_H - \bar{v}_D)_{str} + 2(\bar{v}_H - \bar{v}_D)_{bend}]$$
$$= 10\,650 \text{ J mol}^{-1} \quad (2545 \text{ cal mol}^{-1}). \tag{7.12}$$

If we equate ΔE_z with the difference in enthalpy of bond fission, $\Delta\Delta H$, then

$$k_H/k_D = e^{\Delta\Delta H/RT} = e^{10\,650/8\cdot36\times298} = 72,$$

which is near the value obtained for 25°C in the earlier calculation.

Actual experimental values of kinetic isotope effects do not approach this figure since in almost all cases the hydrogen (whether as H˙, H$^+$ or H$^-$ is immaterial) is transferred to an acceptor and the transition state is reached without the complete loss of this component of the vibrational energy. Vibrational energies of modes appropriate to the transition state result in at least a partial cancelation of zero-point energies. This, of course, results in activation energies being far less than pure bond dissociation energies, for otherwise most reactions could not occur at atmospheric temperatures. A model due to Westheimer[4] and Melander[1] considers the isotopic effect on energies of the three-atom transition state, A$\cdots$L$\cdots$B, which reduces the maximum value to about 18. More refined calculations[2] put the maximum PKIE for hydrogen transfer from C–L, N–L and O–L bonds at around 7 (L is any hydrogen involved in a kinetic isotope effect; Table 7.2), which is close to the maximum experimentally observed value in the absence of tunneling, diminishing with increased temperature as predicted by Eq. (7.8). Proportionately smaller maximum PKIEs are predicted for breaking of bonds of lower vibrational frequency.

Table 7.2   Maximum normal values of PKIEs

Element	Isotope		$T/°C$	k_A/k_B
	A	B		
H	1	2(D)	0	8·3
H	1	2(D)	25	6·4
H	1	2(D)	100	4·7
H	1	2(D)	200	3·4
H	1	2(D)	500	2·1
H	1	3(T)	25	13
C	12	13	25	1·04
C	12	14	25	1·07
N	14	15	25	1·03
O	16	18	25	1·02
S	32	34	25	1·01
Cl	35	37	25	1·01

Some typical kinetic isotope effects

	Reaction	k_H/k_D
Primary	$Me_2C\!-\!OH + CrO_3 \rightarrow Me_2C\!=\!O$ $\qquad$ | $\qquad$ L	7·0
	$Me_2C\!-\!OH + Br_2 \rightarrow Me_2C\!=\!O$ $\qquad$ | $\qquad$ L	3
	$RCO.CL_2.R' + SeO_2 \rightarrow R.CO.CO.R'$	7
	$RCOCL_3 + Br_2 \rightarrow RCOCL_2Br + LBr$	7
	$CL_3NO_2 + Br_2 \rightarrow BrCL_2NO_2 + LBr$	4·5–6·5

Secondary $ROtos + AcOH \rightarrow ROAc + HOtos$

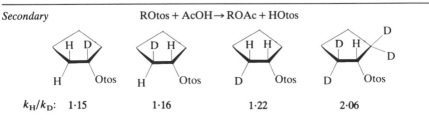

| k_H/k_D: | 1·15 | 1·16 | 1·22 | 2·06 |

The following examples illustrate the uses of isotope effects in elucidating mechanisms.

(a) Elimination mechanisms

$$PhCL_2CH_2\overset{+}{N}Me_3 \xrightarrow{\ OEt^-\ } PhCL\!=\!CH_2 + NMe_3; \qquad k_H/k_D = 4·6 \quad (Ref. 7)$$

Any mechanism proposed for this reaction must therefore feature fission of the β-C–H bond in the activation step consistent with the observation of a PKIE. Transition states **1** and **2** accord with this observation; **3** may be

261

rejected. Furthermore, there is also a nitrogen isotope effect, $k_{14}/k_{15} = 1.009$, showing the C–N as well as the C–H bond to be breaking in the activation step; this is strong evidence for the concerted process, **2**. On the other hand, consider the reaction

$$CL_3\text{—}CMe_2Cl \xrightarrow{\text{EtOH}} CL_2\text{=}CMe_2; \qquad k_H/k_D = 1.1.$$

This k_H/k_D value is too small for a primary kinetic isotope effect: it must be a secondary effect. Consequently transition states **1** and **2** may be rejected and **3** favored since the C–H bond, in this case, must be intact in the transition state.

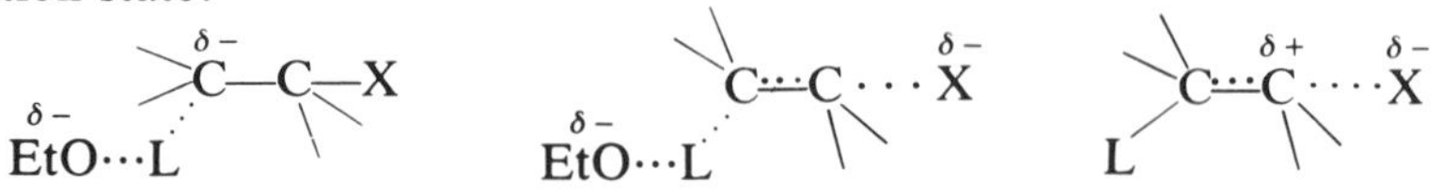

1   **2**   **3**

(b) Aromatic nitration

For reaction **4**, $k_H/k_D = 1.0$. The absence of a PKIE[8] is conclusive evidence that the C–H bond is still unbroken in the transition state of the slow step which rules out the concerted pathway, **5**. The C–H bond is broken at some stage but it must be in a subsequent fast step. This points unambiguously to a two-step route to products, **6**.

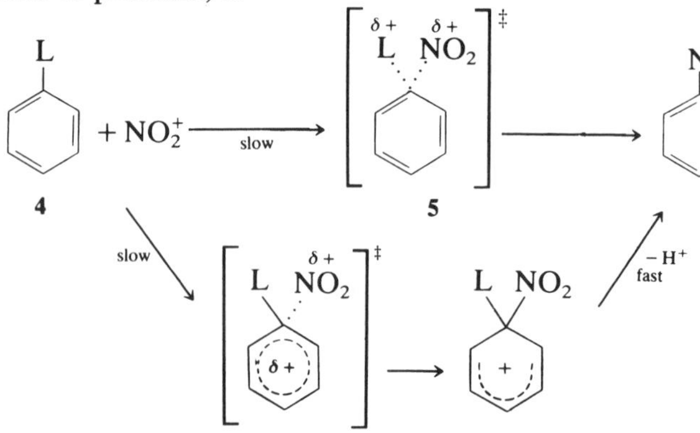

4   **5**

6

(c) Decarboxylation

A carbon isotope effect occurs in the decarboxylation[9] of **7**, $k_{12}/k_{13} = 1.04$, confirming rate-determining loss of CO_2[10,11].

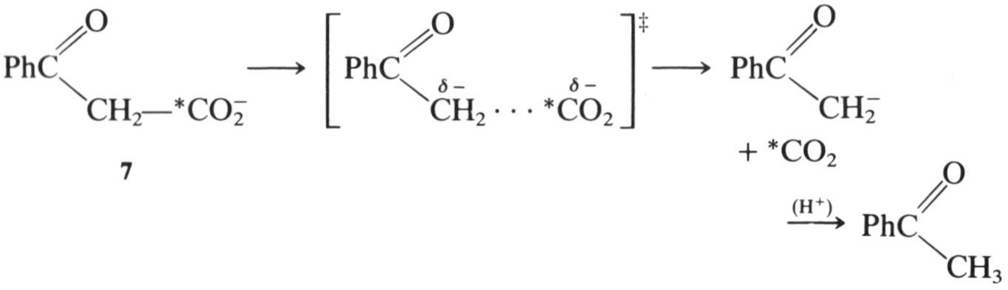

7

7.3 TRANSITION-STATE GEOMETRY

Two aspects of transition-state geometry are important in determining the magnitude of a kinetic isotope effect. These are the position of the transition state along the reaction coordinate

$$C \cdots H \cdots \cdots \cdots X \qquad C \cdots \cdots H \cdots \cdots X \qquad C \cdots \cdots \cdots H \cdots X,$$
        'early'   'central'   'late'

and the linearity of the C—H—X system. The maximum isotope effect is expected from a linear symmetrical transition state with the hydrogen positioned centrally between the two acceptor centres. In this case the symmetrical stretching vibration will result in no motion of H and will be mass-independent. This means that there is no compensating contribution to the energy of the transition state from this mode. Calculated isotope effects for different degrees of hydrogen transfer are shown in Fig. 7.2a and may be compared with observed values for ionization of nitro compounds by bases of varying strength, Fig. 7.2b, for which presumably the position of the transition state varies systematically. The isotope effect reaches a maximum when the two bases are of similar strength, associated with a 'central' transition state.

As an instance of a reaction which must involve a non-linear hydrogen transfer (because of the five-membered ring), the pyrolysis of an amine oxide (Cope Elimination), **8**, may be cited. It is significant that the primary isotope effect is unusually low; $k_H/k_D \sim 3$ [12].

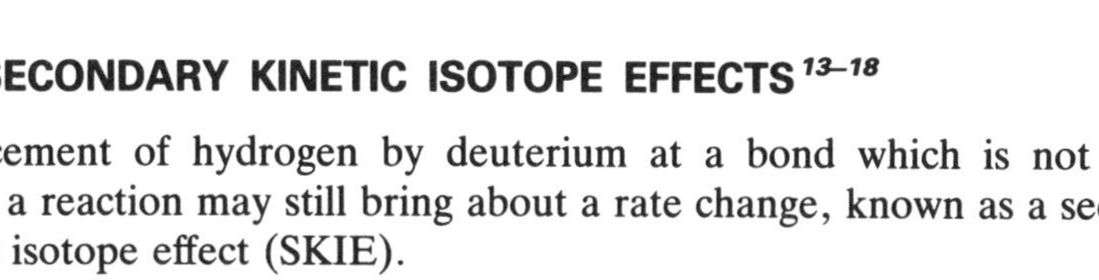

    8

7.4 SECONDARY KINETIC ISOTOPE EFFECTS [13–18]

Replacement of hydrogen by deuterium at a bond which is not broken during a reaction may still bring about a rate change, known as a secondary kinetic isotope effect (SKIE).

The full Bigeleisen equation (Eq. (7.9)) contains frequencies of all vibrational modes and it is only when these are exactly equal in both reagent and transition state that they cancel. When a C–H bond is being broken, the change in frequency of its stretching mode is responsible for the largest term which, in approximation, may be taken as the only one of importance. When no bond fission at the isotopic atom is occurring, smaller isotopic perturbations of the various modes in which it is involved remain; cancellation of the terms of Eq. (7.9) is seen to be incomplete and an effect

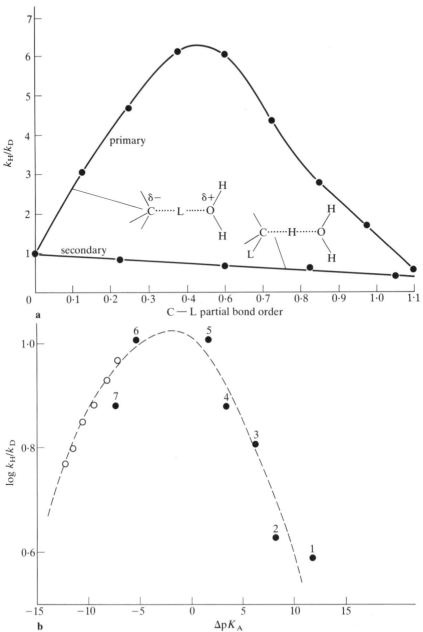

Fig. 7.2 **a,** Calculated curve showing the variation of isotope effect (primary and secondary) for proton transfer from a carbon acid as a function of the degree of bond breaking. **b,** Experimental curve of the observed isotope effect versus difference in pK_A between acid (MeNO$_2$ (I) or Me$_2$CHNO$_2$ (II)) and base for the reaction $\diagdown$CHNO$_2$ + B: $\rightarrow$ $\diagdown$CNO$_2$ + BH$^+$.

The acid–base pairs are: 1, I + H$_2$O; 2, I + CH$_2$ClCOO$^-$; 3, I + OAc$^-$; 4, II + OAc$^-$; 5, II + pyridine; 6, I + OH$^-$; 7, II + OH$^-$.

264

of remote isotopic substitution on the rate, an extremely sensitive probe, is apparent. An understanding of the origins of an SKIE requires a model which connects covalency changes at the reaction center with vibrational frequencies at remote bonds. It is easy to see that isotopic substitution at the reaction center which is undergoing a change in coordination number will involve changes in vibration modes. In a solvolytic reaction, for example, an α-proton both in the sp^3-hybridized reagent and in the product will be associated with one stretching and two bending modes.

$v_{str}(H)$	2900	2800
$v_{bend}(H)$	1340, 1340	1350, 800
$v_{bend}(D)$	957	571

The principal difference in vibration energy between reagents and product is the reduction in frequency of a bending mode, **9**, which will favor the

reaction and will also favor hydrogen over deuterium, for which the change in frequency is greater by a factor of about 1·4. For this change, from Eq. (7.12),

$$\Delta E_z = \tfrac{1}{2}hcL[(1340 - 957) - (800 - 571)] = 873 \text{ kJ mol}^{-1} \ (220 \text{ kcal mol}^{-1}).$$

Hence $k_H/k_D = e^{837/8\cdot34\times298} = 1\cdot4$. The result is a secondary isotope effect, an order of magnitude smaller than the normal range of primary effects, appropriate to any change from tetrahedral to trigonal carbon. Other modes affected by the isotopic substitution need also be taken into account so that the resulting maximum effect is larger than this value; experimental SKIEs are usually smaller, however. For example, $k_H/k_D = 1\cdot15-1\cdot2$ is the range often encountered in solvolytic reactions (Table 10.2). Conversely, a change in the other direction will be associated with an inverse effect, $k_H/k_D = 1/1\cdot15 \sim 0\cdot9$.

    Kinetic effects are observed not only when the isotopic atom is located at the α-carbon, but when it is bonded to more remote centers also. β-SKIEs are at least as large as α-SKIEs (Fig. 7.3). The interaction between a vacant orbital (e.g. in a carbocation) and a β-C–H bond is a familiar phenomenon denoted *hyperconjugation*. The VB interpretation is the inclusion of 'no

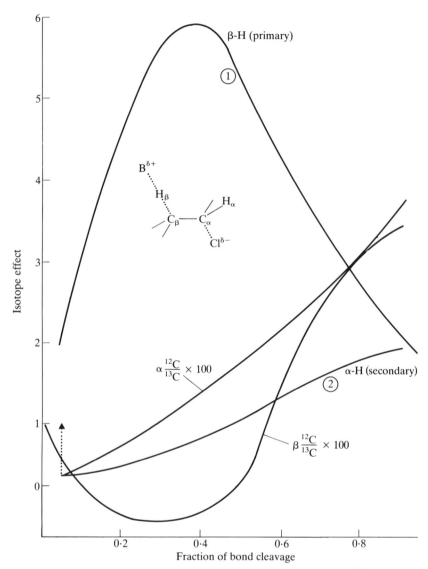

Fig. 7.3 Calculated curves showing the variation of the isotope effect with differing degrees of C–H bond breaking in the E2-elimination transition state, ^HC—C_{Cl}, for the β-H PKIE and for SKIEs at α-H, α-C and β-C.

bond' contributing structures:

The MO interpretation is the π-overlap of the vacant p-orbital with group orbitals of an adjacent methyl (for example), of which two have π-

symmetry and one is capable of effective overlap:

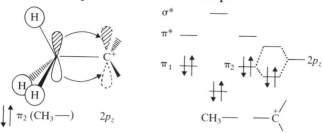

Since only two electrons are involved they may now occupy a delocalized MO of lower energy than the components and extend over the three atoms involved. By either model, the result is the transfer of electron density from the β-C–H bond to the electron-deficient center. This causes weakening of the C–H bond and a lowering of its vibrational frequency, C—H being affected more than C—D. Hence, an increase in hyperconjugative interaction with a β-H will be accompanied by a small SKIE. It is sensitive to substituents on the β-carbon since they affect the energy of the important π-type MO. An electronegative substituent lowers its energy and reduces the hyperconjugative interaction and the isotope effect.

A further important aspect of β-SKIEs is their conformational dependence, which is in accordance with their hyperconjugative origin. π-overlap is dependent on dihedral angle, α, of the interacting orbitals. The SKIE is a maximum when the isotopic β-C–H bond is parallel to the developing $2p$ orbital and a minimum (0) when it is orthogonal (following $\cos^2 \alpha$), as illustrated by the examples in Fig. 7.4. This is because effective hyperconjugation requires maximum overlap with a methyl group orbital of π-symmetry.

The use of secondary isotope effects has been most effective in studies of solvolytic[15] and of pericyclic[19] reactions, from which some examples in Tables 7.2, 10.10 and 14.14 are drawn. Typical ranges of values observed

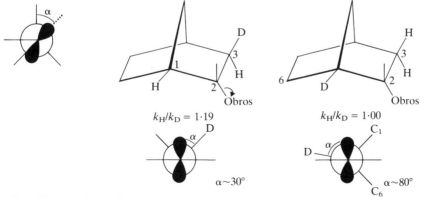

Fig. 7.4 Conformational dependence of secondary isotope effects. Maximum interaction between β-C–L bond and the developing cationic center is at $\alpha = 0$.

are shown below, often expressed as a percentage, i.e. $100 (k_H/k_D - 1)$, per deuterium atom substituted. Values are naturally affected by the degree of bond fission of the leaving group and development of carbocation character.

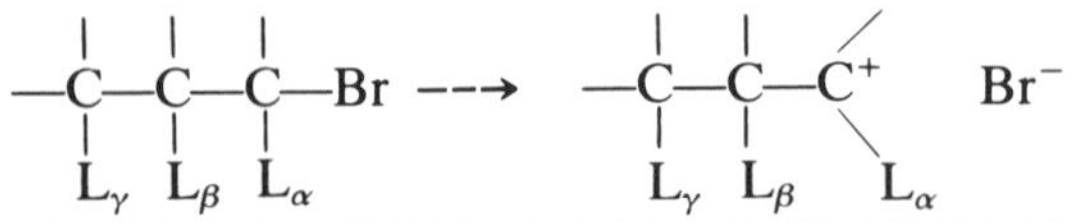

Isotope position	k_H/k_D	Percentage effect per D
α	1·1–1.2	10–20
β	1·15–1·25	15–20
γ	0·92–1·02	−8 to +2

7.4.1 'Inductive' and 'steric' isotope effects

The different vibrational amplitudes of bonds to isotopic hydrogen atoms result in changes in dipole moments, **P**, as shown for example in Table 7.3.

Table 7.3 Dipole moments of H- and D-substituted compounds

Compound	P/Debye	
	$L = H$	$L = D$
$CH_3C\equiv CL$	0·7806	0·7678
$CL_3C\equiv CH$	0·7806	0·7841
CLF_3	1·6460	1·6470
CL_3F	1·8471	1·8583

This is not a normal electronic effect but is due to the slightly shorter length of the C–D bond compared with the C–H bond because of its smaller vibration amplitude. This results in a slightly increased ability to donate electron density by an inductive effect. At the same time, a CD_3 group is less able to supply electron density by hyperconjugation than is CH_3. The reduced amplitude of C—D bending vibrations compared with those of C—H results in deuterium having smaller steric requirements than protium (H). As an example of this, the racemization of the biphenyl, **10**, via the planar transition state (Section 8.2.4) shows a small inverse steric isotope effect, $k_H/k_D = 0·84$[20].

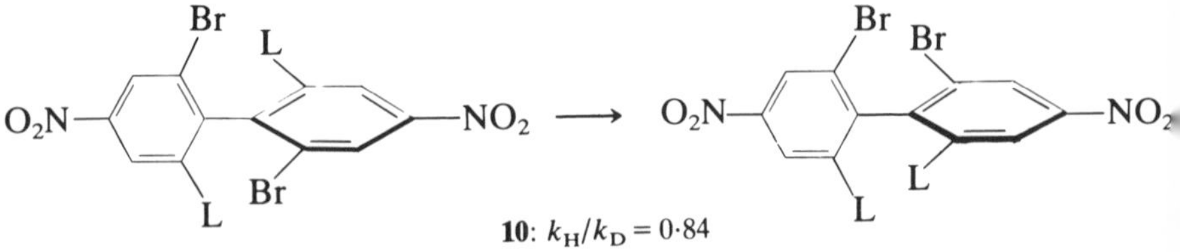

10: $k_H/k_D = 0·84$

7.5 HEAVY-ATOM ISOTOPE EFFECTS[21]

Since the maximum isotopic rate ratio is approximately the square root of the inverse ratio of isotopic masses, the PKIEs involving atoms heavier than hydrogen will be very small and secondary effects will be negligible (Table 7.2). For example, the isotopic discrimination for ionization of ^{79}Br and ^{81}Br from an alkyl bromide, **11**:

$$R\text{—}Br \rightarrow \left[\overset{\delta+}{R} \cdots \overset{\delta+}{Br} \right]^{\ddagger} \rightarrow R^{+} \quad Br^{-} \rightarrow \text{products}$$
11

may be calculated using Eqs. (7.3)–(7.5) and should have a maximum value $k_{79}/k_{81} = (81/79)^{\frac{1}{2}} = 1\cdot044$: in reality it will be far less. Nevertheless such effects can be measured by means of specially adapted mass spectrometers and the mechanistic information that can be deduced resembles that from a hydrogen PKIE.

For example, for **11**, with R = *tert*-Bu and with R = 1-Pr[22], there are primary bromine isotope effects; $k_{79}/k_{81} = 1\cdot00310$ and $1\cdot00169$, respectively. Quite definitely one may conclude that the C–Br bond is being broken in the rate-determining step of each reaction, the process being presumably farther advanced at the transition state in the former example, which has the larger PKIE, although neither ratio appears to be a maximum. Thus the PKIE does not discriminate between the two mechanisms involved (Section 10.13), S_N1 in the first case and S_N2 in the second.

A sulfur isotope effect occurs in the elimination[23]

$$PhCH_2\text{—}CH_2\overset{+}{S}Me_2 \xrightarrow{\ B\ } PhCH\text{=}CH_2 + SMe_2 + BH^{+}; \quad k_{32}/k_{34} = 1\cdot072$$

and a carbon isotope effect in decarboxylations such as[24]

$$CH_3COCH_2COOH \rightarrow CH_3COCH_3 + CO_2: \quad k_{12}/k_{13} = 1\cdot04.$$

In each case the results confirm rate-determining fission of the bond to the isotopic atom (indicated by bold type).

7.6 THE TUNNEL EFFECT[25–28]

Occasionally one encounters a reaction which exhibits a PKIE far in excess of values normally considered to be the maximum at a given temperature. For example, the proton transfer reaction **12** (Table 7.4), when R = Me, has $k_H/k_D = 24$, well above the normal limit of 7. Interestingly, when R = H, the PKIE falls to 9, almost the normal maximum value[32–34]. It is generally considered that in these exceptional cases another factor is operating which is known as the quantum-mechanical *tunnel effect*. A particle is associated with a de Broglie wavelength given by $\lambda = h/mv$ and is consequently mass-dependent. Values of λ for several species are as follows:

Particle:	e^{-}	H	D	T	C	Br
λ/pm:	2690	63	45	36	18	7

When λ is small compared with the motions of the particle, its behavior is essentially classical, but when it becomes comparable in magnitude, then

Table 7.4 Some hydrogen-transfer reactions which show tunneling effects[9,29-31]

Reaction	k_H/k_D	$\Delta V_H^{\ddagger}/\Delta V_D^{\ddagger}$
12 $Me_2CLNO_2 + N$ $\longrightarrow Me_2\bar{C}NO_2 + L\overset{+}{N}$	24	0·7
13 $PhCL_2NO_2 + Et_3N \rightarrow Ph\bar{C}LNO_2 + Et_3\overset{+}{N}L$	(PhCl) 23 (PhCH$_3$) 11 (MeCN) 3	
14 $PhCL_2NO_2 + HN{=}C(NMe_2)_2$	(PhCl) 50	
15 $CL_3OH + H^{\cdot} \rightarrow \dot{C}L_2OH + H{-}L$	(H$_2$O) 20	
16	30	
17 $PhCL(CF_3)OH + MnO_4^-(OH^-) \rightarrow PhCOCF_3$	16	
18	>100	<1
19 $Ar_3C{-}L +$ $\longrightarrow Ar_3\overset{+}{C} +$	18	0·71
20 $\longrightarrow$ $+ LF$	2·5[a]	

[a] But Arrhenius A factors differ: $A_D/A_H = 24$.

quantum effects may be important. A proton moving with kinetic energy E_k and momentum $(2mE_k)^{\frac{1}{2}}$ will be associated with an uncertainty of position according to the Heisenberg Uncertainty Principle, $x \sim \lambda/4\pi$. This amounts to *ca* 5 pm for the proton and is comparable with the width of the energy barrier for a transfer reaction. Therefore there is a substantial probability that a proton, on encountering an energy barrier of this dimension, will be found on the product side without apparently ever having had sufficient energy to pass over the top (Fig. 7.5). This phenomenon is known as tunneling and introduces an additional contribution to the rate. Tunneling is very mass-dependent and is always important in electron motion. The proton and to a lesser extent the deuteron are the only particles for which tunneling can impart significant changes in reactivity in chemical systems but

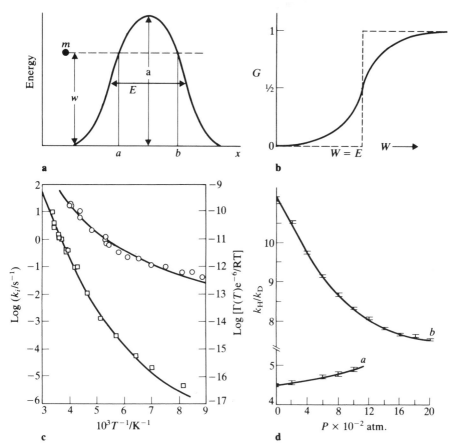

Fig. 7.5 Quantum-mechanical tunneling. **a**, Classical (solid line) and quantum-mechanical (broken line) passage of an energy barrier. **b**, Probability (G) of a particle passing an energy barrier E as a function of its energy, W; classical case (broken line), quantum mechanical case (solid line) showing tunneling ($G > 0$ at $W < E$ and $G < 1$ at $W > E$). **c**, Curved Arrhenius plots (reaction **18**, Table 7.4) due to tunneling; note the PKIE is approximately 10^4 at the lowest temperature ($-150°$C)! **d**, Effect of pressure on kinetic isotope effects; curve a, a normal PKIE ($Ph_2CN_2 + PhCOOH(D) \rightarrow PhCOOCH(D)Ph_2 + N_2$); curve b, a reaction showing tunneling (reaction **19**, Table 7.4) which is reduced with pressure.

because it further discriminates between H and D the tunnel effect can greatly affect a PKIE. Some or all of the following characteristics may be observed when tunneling is an important contributor to reaction rate.

(a) The kinetic isotope effect is abnormally high, particularly at low temperatures (Table 7.4).

(b) The Arrhenius plot of $\log k$ against $1/T$ is non-linear, with the curvature tending towards zero activation energy at low temperatures (Fig. 7.5c).

(c) The isotope effect may diminish with pressure (Fig. 7.4d)[35,36], indicating a difference in volumes of activation $\Delta V^{\ddagger}(D) < \Delta V^{\ddagger}(H)$.

(d) The difference in activation energies, $E_A(H) - E_A(D)$, may be greater than the difference in zero-point energies (*ca* 5 kJ mol^{-1}) and the ratio of A-factors, $A(H)/A(D)$ greater than the theoretical maximum, 1·4, e.g. **20**.

Not all of these criteria are satisfied and, to some extent, they appear independent: For example, abnormal A-factors are found in the enolization reaction, **20**, although the PKIE is only around 2·5.

The question naturally arises as to why tunneling behavior is a rather rare phenomenon while proton transfers are common. It seems likely that the answer lies in solvation of the proton. Normally, the motion of the proton being transferred is coordinated with that of solvent molecules which increases its effective mass. This will render quantum-mechanical behavior negligible. Those instances in which tunneling is observed must be fortuitous cases in which an unsolvated proton is exchanged. It is significant that abnormal isotope effects in **12** are only found when the pyridine base has bulky *ortho* groups attached which could exclude solvent. The pressure effect also has been explained as causing an increase in solvation and hence diminishing quantum-mechanical effects[35]. It is certainly true that the nature of the solvent and of the base enormously affects the magnitude of the tunneling contribution and in a completely capricious way. Analysis of these abnormal isotope effects can lead to an estimation of the width of an activation barrier; for reaction **12** it is estimated at 114 pm.

7.7 SOLVENT ISOTOPE EFFECTS

Kinetic effects carrying mechanistic information may be observed on changing the solvent to one which differs only isotopically, for example from H_2O to D_2O. There are several possible causes of these rate changes[2,3–32,37–43].

(a) The solvent may be a reagent and may be covalently bound to the substrate in the rate-determining step. If this involves a proton transfer, a large (primary) kinetic isotope effect may be observed. It is relatively rare for a proton transfer from solvent to be rate-determining since this is normally a very fast process. Far more common is the equilibrium proton transfer, the initial step in any acid- or base-catalyzed reaction, **21**:

$$A—L + B \underset{}{\overset{slow}{\rightleftharpoons}}\ A^- + BL^+ \xrightarrow{fast} \text{products.}$$

21

Smaller secondary effects may result from changes in vibrational frequencies, even if no proton cleavage occurs, for example if solvent water is acting through oxygen as a nucleophile (**22**).

(b) Protic solvents may exchange protons with acidic centers in the substrate so that the substrates differ isotopically as do the solvents. Either primary or secondary isotope effects may result.

$$L_2O \frown H$$
$$\underset{t\mathrm{Bu}}{\overset{}{\diagdown}} \mathrm{C} \overset{\text{\tiny{wn}}}{\diagup} \overset{\mathrm{CN}}{\underset{\mathrm{CN}}{}} \longrightarrow L_2\overset{+}{\mathrm{O}}H \cdot t\mathrm{Bu}\bar{\mathrm{C}}(\mathrm{CN})_2 \qquad k_H/k_D = 3.7$$

22

(c) The solvation properties of a protic solvent, affecting chemical potentials of solute species, may differ according to isotopic composition[33]. In particular, hydrogen-bonding capability changes, and can in turn alter nucleophilicities of reagents, for instance.

Large solvent isotope effects that are sometimes observed appear to be a combination of primary and secondary effects for the proton transfer, together with solvation effects[33,34], which can in some cases operate simultaneously upon a reaction[44], considerably complicating the analysis.

7.7.1 Fractionation factors[37]

Isotope effects on equilibrium proton exchange between solvent and various types of solute are important in studies of acid and base catalysis. The effect of a change of solvent from H_2O to D_2O may be readily predicted using isotopic fractionation factors. Isotope exchange equilibria of the type

$$RO—D + SH \rightleftarrows RO—H + SD$$

do not necessarily lead to a statistical distribution of the exchangeable protons since the equilibrium constant is the difference between primary isotope effects for forward and reverse reactions and these are not necessarily equal. This is particularly the case if the two basic centers are chemically of different types. When SH is a solute with a single exchangeable hydrogen in a solvent containing RO—H and RO—D, the equilibrium constant, ϕ, is known as the *fractionation factor* and depends upon the nature of S (Table 7.5):

$$\phi = \frac{[\mathrm{SD}][\mathrm{RO—H}]}{[\mathrm{SH}][\mathrm{RO—D}]} = \frac{[\mathrm{SD}]/[\mathrm{SH}]}{[\mathrm{RO—D}]/[\mathrm{RO—H}]}. \tag{7.13}$$

It is a measure of the preference of the solute, relative to water, for combination with deuterium. The value of ϕ for hydroxyl group, —OH, is indeed unity (statistically) due to the similarities of vibrational frequencies with those of water; centers such as —O⁻, —O⁺ and —S⁻ show a marked preference for the lighter isotope ($\phi < 1$). Secondary isotope effects may be estimated from fractionation factors according to Eq. (7.14):

$$K_H/K_D = \overset{\text{reactant sites}}{\underset{i}{\prod}} \phi_i \Big/ \overset{\text{product sites}}{\underset{j}{\prod}} \phi_j, \tag{7.14}$$

Table 7.5 H/D fractionation factors, ϕ, relative to water, for various types of bond

Bond type	ϕ
—O—L	1·0
—CO—OL	1·0
—C—OL	1·25
$\diagdown$O$^+$—L$\diagup$	0·69
O$^-$—L	0·5
$\diagdown$C—L$\diagup$	0·84–1·18
=C—L	0·9
≡C—L	0·69
$\diagdown$N—L$\diagup$	0·92
$\diagdown$N$^+$—L$\diagup$	0·97
—S—L	0·42
H—L	0·29

in which product terms are taken over all exchangeable sites in reagents and products. An important consequence of non-unit values of fractionation factors is the differences which are observed in acid–base dissociation constants in D_2O and H_2O:

$$\diagdown_L \hspace{-0.5em} O \; + \; LOA \; \rightleftharpoons \; \diagdown_L \hspace{-0.5em} \overset{+}{O}\text{—}L \; + \; \bar{O}\text{—}A$$

Only secondary effects need be considered in an equilibrium process and

$$K_H/K_D = \phi_{OL}^2 \phi_{OL}/\phi_{O^+L}^3 = (1/0{\cdot}69^3) = 3{\cdot}04. \tag{7.15}$$

This predicts that all oxygen acids should be stronger by a factor of three in H_2O than in D_2O, and oxygen bases should be correspondingly weaker. Similarly, OD$^-$ is a stronger base than OH$^-$ by a factor of two:

$$L\text{—}O^- + L\text{—}OA \rightleftharpoons L\text{—}OL + OA^-;$$

$$K_H/K_D = \frac{\phi_{OL^-} \cdot \phi_{OL}}{\phi_{OL}^2} = 0{\cdot}5/1 = 2. \tag{7.16}$$

Similar conclusions apply to nitrogen acids and bases. These predictions are often expressed by saying that D_3O^+ in D_2O is a stronger acid than H_3O^+ in H_2O, and DO$^-$ is likewise a stronger base than OH$^-$. There will therefore be rate effects on acid- and base-catalyzed processes since the concentrations of the active species will, in general, be higher in D_2O.

Auto-ionization of water is greater than that of D_2O:

$$2L_2O \xrightleftharpoons{K_W} L_3O^+ + OL^-;$$

$$K_W(H)/K_W(D) = (\phi_{OL}^2 \phi_{OL}^2)/(\phi_{OL^+}^3 \phi_{OL^-}) = 1/0\cdot69^3 \times 0\cdot5$$

$$= 6\cdot1 \quad \text{(experimental value 7)}.$$

$$\text{Hence} \quad pK_W(D) = 14\cdot84,$$

which will have a bearing on pH values. Equation (7.17), analogous to Eq. (7.14), may be used to calculate secondary solvent effects upon rates:

$$k_H/k_D = \prod_i^{\substack{\text{reactant} \\ \text{sites}}} \phi_i \Big/ \prod_j^{\ddagger} \phi_j^{\ddagger}. \tag{7.17}$$

Fractionation factors for the transition state, $\phi^{\ddagger}$, are not experimentally accessible but may be estimated as a weighted mean between those of reagents and products, ϕ_r and ϕ_p respectively:

$$\phi^{\ddagger} = \phi_r^{1-x} \phi_p^x, \tag{7.18}$$

where x is a parameter expressing the location of the transition state on the reaction coordinate, and will vary between 0 (reagents) and 1 (products); it is therefore similar to the Brønsted parameter, α (Section 9.15), which may be used as a measure of x.

7.7.2 Solvent isotope effects in mixed isotopic solvents: the proton inventory technique[40,45]

Interesting mechanistic information may be available from the rates of a reaction as a function of solvent composition using mixtures of H_2O and D_2O. Equilibrium and rate constants are related to the atom-fraction of deuterium, n, by the Gross–Butler equation (Eq. 7.20)[44,46–48].

The rate, k_n, of a reaction in which there is a single exchangeable hydrogen in a water mixture with n atom-fraction of D, is a weighted mean of the rates in isotopically pure solvents, k_H and k_D:

$$k_n = k_H(1 - n) + k_D(n).$$

$$\text{Hence} \quad k_n = k_H(1 - n) + n(k_H/k_D) = k_H(1 - n + n\phi), \tag{7.19}$$

and it may be shown that, in general,

$$k_n = k_H \prod^{\text{products}} (1 - n + n\phi_p) \Big/ \prod^{\text{reagents}} (1 - n + n\phi_r), \tag{7.20}$$

where the product terms are taken over all hydrogens which exchange with solvent. A plot of k_n against n will be of a form characteristic of the number of protons involved in the reaction (Fig. 7.6). The analysis of such curves

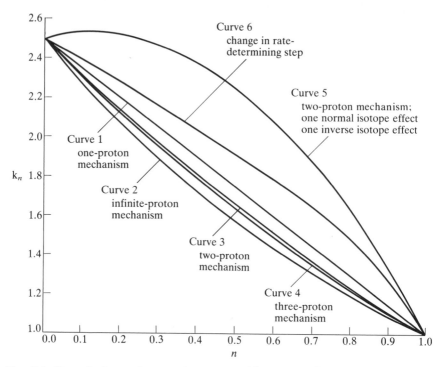

Fig. 7.6 General shape of rate–solvent composition curves (H_2O–D_2O) for reactions proceeding via protonated substrate by different mechanisms.

into the appropriate terms of the Gross–Butler equation is known as a *proton inventory* and may enable transition-state fractionation factors to be calculated and provide evidence of the number of protons involved in a reaction. The following eamples will show something of the power of this approach.

A linear relationship between k_n and n is indicative of a simple solvent isotope effect involving a single exchangeable hydrogen in the transition state (curve 1, Fig. 7.5a) which is isotopically distinct from water (i.e. shows a change in ϕ), thereby contributing to the isotope effect. An example is the hydrolysis of the salicylic ester, **23**, general-base-catalyzed by the adjacent carboxylate group. This is consistent with a transition state in which the carboxylate mediates the transfer of a hydroxyl ion from water to the reaction center. Only the proton undergoing transfer shows a change in fractionation factor between reagent and transition state.

$$k_n = k_H \frac{(1 - n + n\phi^{\ddagger}_{HA})(1 - n + n\phi^{\ddagger}_{HB})}{(1 - n + n\phi_{OH})^2} \tag{7.21}$$

Since $\phi^{\ddagger}_{HB} = \phi_{OH} = 1 \cdot 0$ this becomes

$$k_n = k_H(1 - n + n\phi_{HA})$$

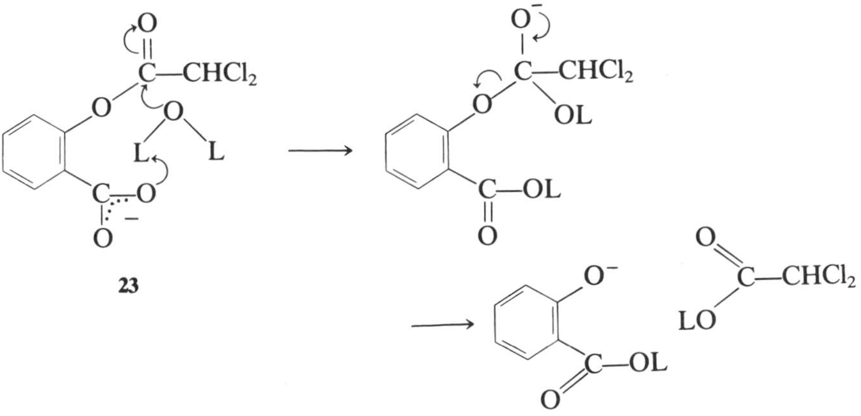

23

and a value of 0·46 can be assigned to the transition state fractionation factor, ϕ_{HA}, applicable to a hydrogen transferred between two oxygens. Downward-curved plots of k_n versus n indicate the participation of greater numbers of isotopically distinct protons, although the precision of the data must be progressively higher to discriminate the larger numbers involved. Upward-curved plots indicate the presence of an inverse isotope effect (i.e. $k_H/k_D < 1$); enzymic examples are known.

7.7.3 Examples of solvent isotope effects

(a)
$$S + H^+ \underset{}{\overset{fast}{\rightleftharpoons}} SH^+$$
$$SH^+ \xrightarrow{slow} product$$

An example is the hydrolysis of acetals and ketals (**24**) in which the pre-equilibrium protonation is followed by slow loss of an alcohol molecule (A1 mechanism, Section 9.1.2). The resulting solvent isotope effect should be $(1/3) \times (1/0\cdot69) = 0\cdot48$, which may be compared with experimental values in the range 0·33–0·55. If the slow step were the displacement of the

24

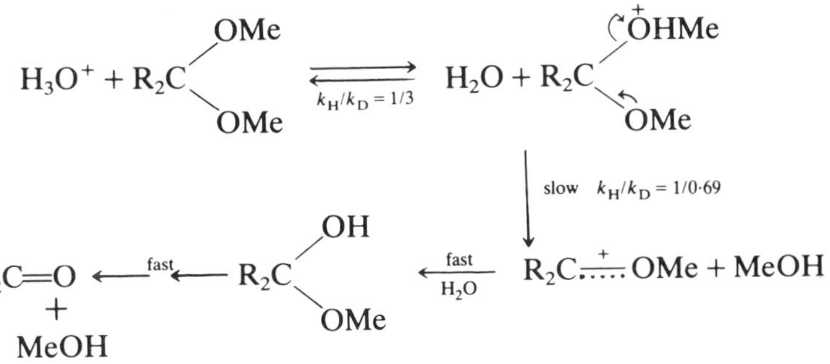

alcohol by attack of a solvent molecule (A2 mechanism), there would be no isotope effect on the latter stage so that the observed value should be lower, around 0·33.

(b)
$$S-H \overset{L_2O}{\rightleftharpoons} S-L$$
$$X + S-L \overset{slow}{\longrightarrow} X-L+S$$

The reaction between diphenyldiazomethane and a carboxylic acid is an example in which rapid and complete exchange of acidic hydrogen with the solvent (in large excess) occurs, followed by a rate-determining proton transfer. The isotope effect, apart from changes in solvation, will be a primary effect, maximum $k_H/k_D \sim 7$ (**25**). Experimental values of 4–5 are reported[49].

$$PhCOOH + L_2O \rightleftharpoons PhCOOL + HOL$$

$$Ph_2CN_2 + PhCOOL \overset{slow}{\longrightarrow} PhCN_2^+ + PhCOO^- \overset{fast}{\dashrightarrow} PhC - O_2CPh$$

(with L below the first product, L below the second, and $+ N_2$)

25

(c) The mutarotation of glucose (α-D-**26** $\rightleftharpoons \beta$-D-**26**) occurs by preprotonation followed by base-assisted ring opening (A2 reaction). Prior to these steps, all hydroxylic protons will exchange in D_2O. The resulting combined effect may be expected to have a maximum value of $k_H/k_D = 0·33 \times 7 = 2·3$. The experimental value, 1·37, supports this analysis but suggests the primary effect is not a maximum.

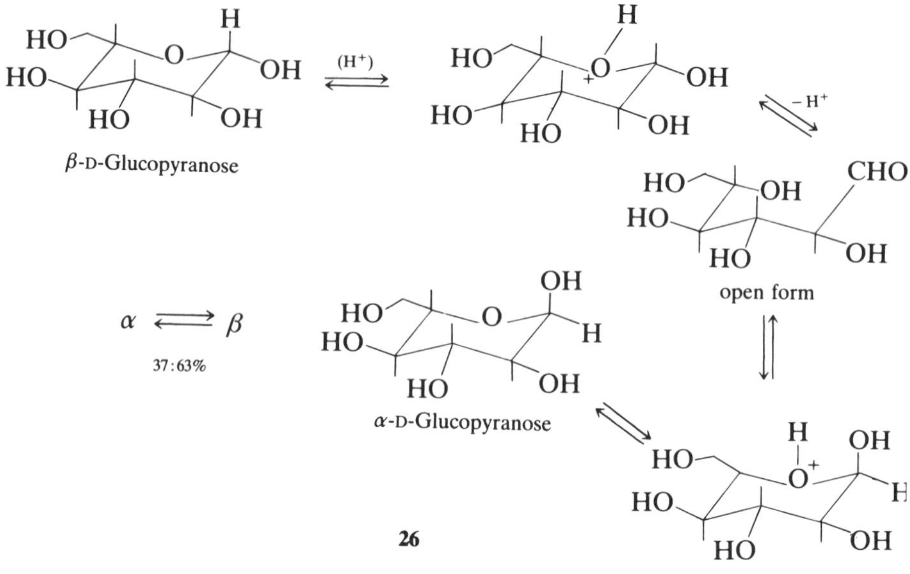

β-D-Glucopyranose

$\alpha \rightleftharpoons \beta$

37:63%

α-D-Glucopyranose

open form

26

PROBLEMS

1 Predict the type and magnitude of kinetic isotope effects associated with the atoms ringed (if two or more positions are ringed consider each separately):

(a)

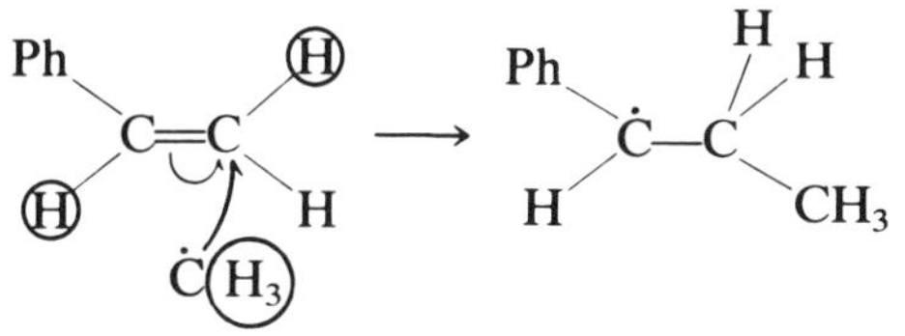

$$Ph\text{—}C\text{=}C\text{—}\textcircled{H} \longrightarrow Ph\text{—}\dot{C}\text{—}C$$

with $\textcircled{H}$, H, $\dot{C}\textcircled{H_3}$ on left; product has H, H, CH₃

(b)

$$Me_2C\text{—}OH, \textcircled{H} \xrightarrow[\text{fast}]{CrO_3} Me_2C\text{—}O\text{—}CrO_2OH, H \xrightarrow{\text{slow}} Me_2C\text{=}O (+C\overset{IV}{r}O_2(OH))$$

(c)

$$\xrightarrow{\Delta,450°C} + CH_3COOH_a$$

with $\textcircled{H}$, $\textcircled{H_a}$, $\textcircled{H_b}$, $O\text{—}C\text{—}CH_3$, O on left; product has H, $H_a H_a$

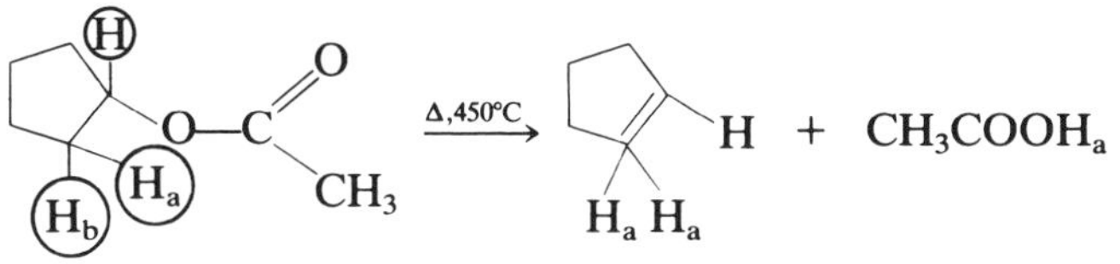

(d) $I^- + \textcircled{H}\text{—}C\text{—}\textcircled{O}\text{—}C\text{—}CF_3 \longrightarrow I\text{—}C + \ddot{O}.COCF_3$

with H, H, H and O

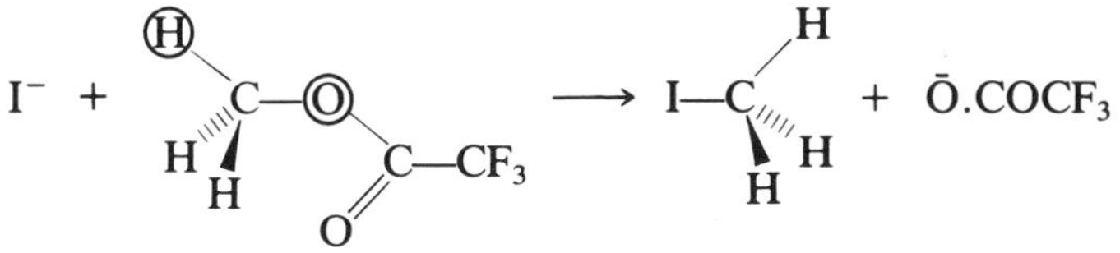

(e) $PhCH_2\text{—}\textcircled{N}\text{=}N\text{—}CH_2Ph \xrightarrow{\Delta} Ph\dot{C}H_2 + N\text{≡}N + \dot{C}H_2Ph$

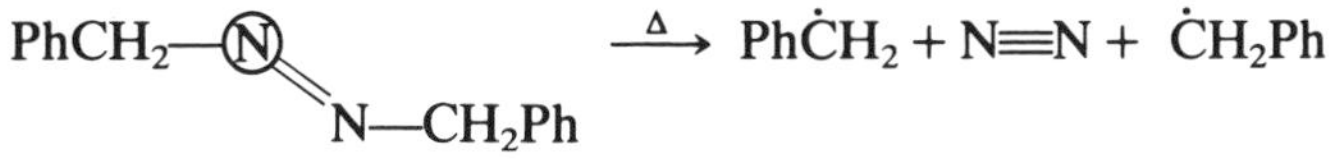

(f)

fluorene structures with $\textcircled{H}$, $\textcircled{C}H_2OH$ → $^-CH_2OH$ → CH_2

(i) $\xrightarrow{\text{slow}}$ $\xrightarrow{\text{fast}}$

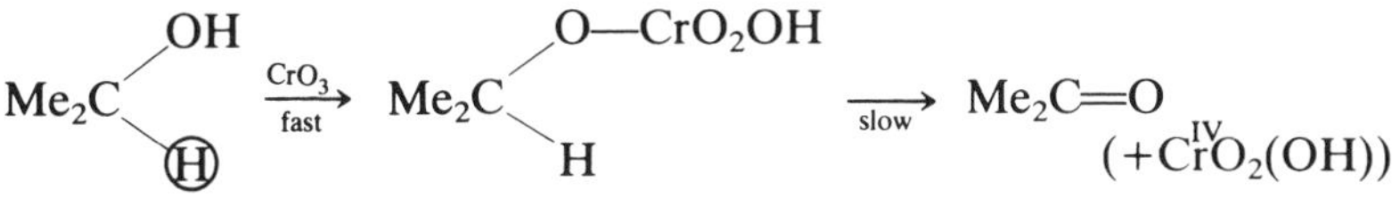

(ii) $\rightleftharpoons$ $\xrightarrow{\text{slow}}$

279

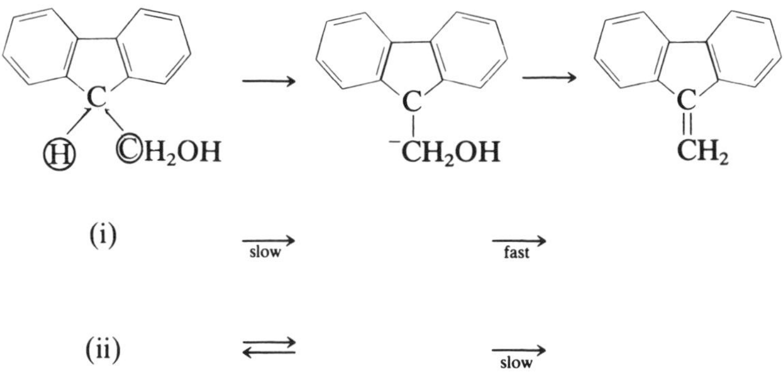

(g)

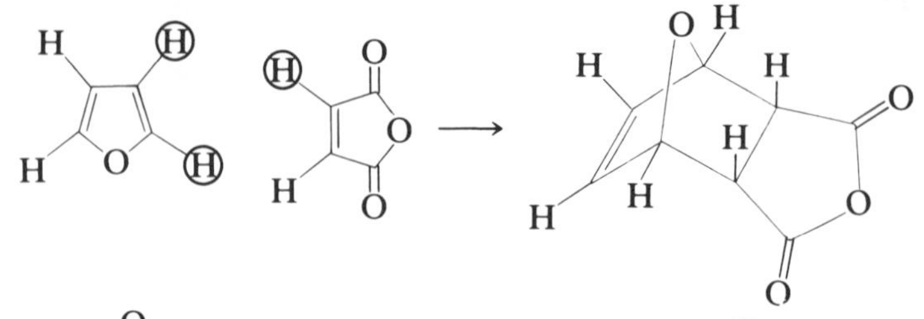

(h)

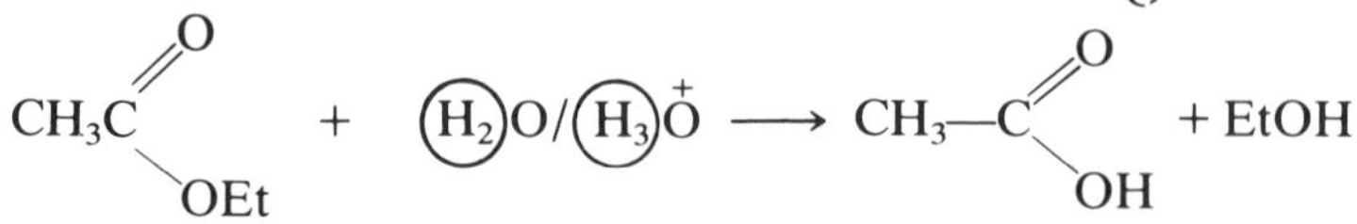

$$CH_3C \overset{O}{\underset{OEt}{}} \quad + \quad \textcircled{H}_2O/\textcircled{H}_3\overset{+}{O} \longrightarrow CH_3{-}C \overset{O}{\underset{OH}{}} \quad + EtOH$$

(i)

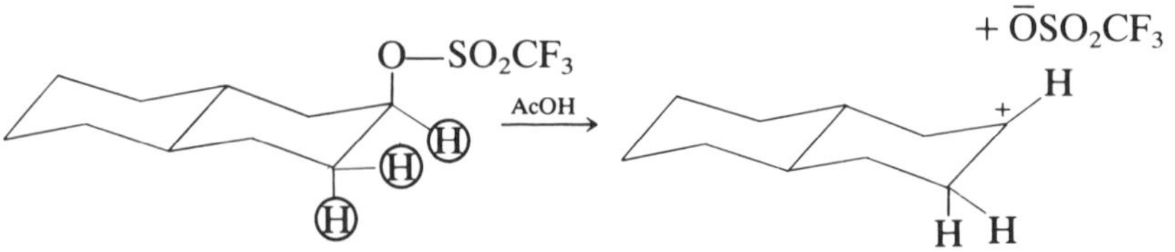

(j)

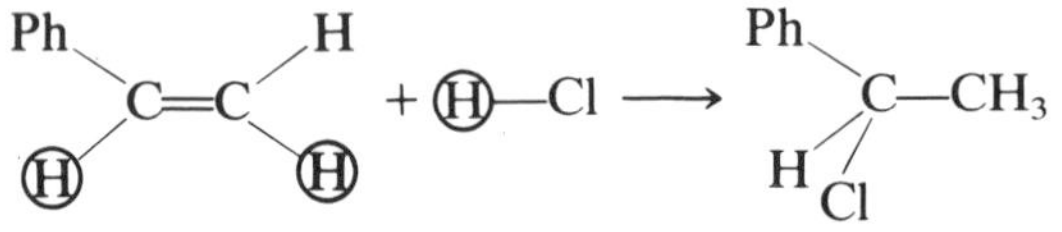

(k)

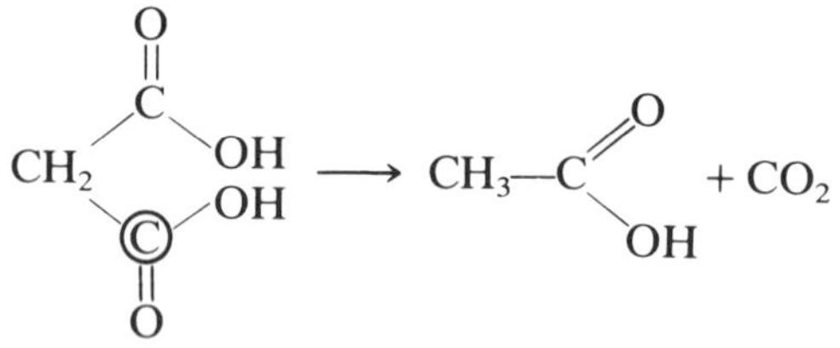

2 Calculate the percentage incorporation of deuterium at equilibrium into the following compounds (exchangeable hydrogens are underlined), when equimolar quantities of D_2O and the compound are mixed: EtOH, CH₃COCH₃, PhSH.

REFERENCES

1 L. Melander, *Isotope Effects on Reaction Rates*, Ronald Press, New York, 1960; K. B. Wiberg, 'The Deuterium Isotope Effect', *Chem. Rev.*, **55**, 713 (1955).

2 L. Melander and W. H. Saunders, *Reaction Rates of Isotopic Molecules*, Wiley, New York, 1980.

3 C. J. Collins and N. S. Bowman, *Isotope Effects in Chemical Reactions*, ACS Monograph 167, Van Nostrand, New York, 1970.

4 F. H. Westheimer, *Chem. Rev.*, **61**, 265 1961.

5 K. Wiberg, *Physical Organic Chemistry*, John Wiley, NY, 1964, p. 358.

6 L. B. Sims and D. E. Lewis, in *Isotopes in Organic Chemistry*, Vol. 16, (E. Buncel and C. C. Lee, Eds), Elsevier, Amsterdam, 1984.

7 W. H. Saunders and A. F. Cockerill, *Elimination Reactions*, Wiley, New York, 1973.

8 L. Melander, *Ark. Kemi.*, **2**, 211 (1950).

9 E. S. Lewis and L. Funderburk, *J. Amer. Chem. Soc.*, **86**, 2531 (1964); **89**, 2322 (1967).

10   G. E. Dunn, in *Isotopes in Organic Chemistry*, Vol. 3, (E. Buncel and C. C. Lee, Eds), Elsevier, Amsterdam, 1978.
11   A. V. Willi, in *Isotopes in Organic Chemistry*, Vol. 3, (E. Bruncel and C. C. Lee Eds), Elsevier, Amsterdam, 1978.
12   W. B. Chiao and W. H. Saunders, *J. Amer. Chem. Soc.*, **100**, 2812 (1978).
13   E. Halevi, *Prog. Phys. Org. Chem.*, **1**, 109 (1963).
14   A. Streitweiser, R. H. Jagow, R. C. Fahey and S. Suzuki, *J. Amer. Chem. Soc.*, **80**, 2326 (1958).
15   D. E. Sunko and W. J. Hehre, *Prog. Phys. Org. Chem.*, **14**, 205 (1983).
16   V. J. Shiner, W. G. Buddenbaum and B. L. Murr and G. Lamarty, *J. Amer. Chem. Soc.*, **90**, 418 (1968).
17   J. A. Llewellyn, R. E. Robertson and J. M. W. Scott, *Can. J. Chem.*, **38**, 222 (1960).
18   K. T. Leffek, J. A. Llewellyn and R. E. Robertson, *Can. J. Chem.*, **38**, 1505 (1960).
19   W. R. Dolbier, in *Isotopes in Organic Chemistry*, Vol. 1, (E. Buncel and C. C. Lee, Eds), Elsevier, Amsterdam, 1975, p. 27.
20   L. Melander and R. E. Carter, *Acta Chem. Scand.*, **18**, 1138 (1964); *J. Amer. Chem. Soc.*, **86**, 295 (1964).
21   A. Fry, 'Heavy Atom Isotope Effects' in Ref. *3*, Chapter 6.
22   J. F. Willey and J. W. Taylor, *J. Amer. Chem. Soc.*, **102**, 2387 (1980).
23   A. F. Cockerill, *J. Chem. Soc., B*, 964 (1967).
24   J. Bigeleisen and L. Friedman, *J. Chem. Phys.*, **17**, 998 (1949).
25   R. P. Bell, *The Tunnel Effect in Chemistry*, Chapman and Hall, London, 1980.
26   E. F. Caldin, *Chem. Rev.*, **69**, 135 (1969).
27   E. F. Caldin, *NATO Adv. Study Inst. Ser. C, 1978*, **C50**, 479 (1979).
28   R. P. Bell, *The Proton in Chemistry*, Methuen, London, 1960.
29   E. S. Lewis and J. K. Robinson, *J. Amer. Chem. Soc.*, **90**, 4337 (1968).
30   G. Brunton, D. Griller and L. R. C. Barclay and K. U. Ingold, *J. Amer. Chem. Soc.*, **98**, 6803 (1976).
31   E. F. Caldin and G. Tomelin, *J. Chem. Soc., Faraday I*, 2823 (1968).
32   E. M. Arnett and D. R. McKelvey, Ref. *43*, p. 344.
33   R. E. Robertson, *Prog. Phys. Org. Chem.*, **4**, 213 (1967); C. G. Swain and E. R. Thornton, *J. Amer. Chem. Soc.*, **84**, 822 (1962).
34   F. Hibbert and F. A. Long, *J. Amer. Chem. Soc.*, **93**, 2836 (1971).
35   N. S. Isaacs, in *Isotopes in Organic Chemistry*, Vol. 6. (E. Buncel and C. C. Lee, Eds), Elsevier, Amsterdam, 1984, p. 67.
36   N. S. Isaacs, K. Javaid and E. Rannala, *Nature*, **268**, 372 (1977).
37   V. Gold, *Adv. Phys. Org. Chem.*, **7**, 259 (1969).
38   V. Gold, in *Hydrogen-Bonded Solvent Systems* (A. K. Covington and P. Jones, Eds), Taylor and Francis, London, 1968, p. 295.
39   J. Albery, in *Proton Transfer Reactions*, (E. Caldin and V. Gold, Eds), Chapman and Hall, London, 1975.
40   R. L. Schowen, *Prog. Phys. Org. Chem.*, **9**, 275 (1972).
41   J. R. Jones, *The Ionisation of Carbon Acids*, Academic Press, London, 1973.
42   J. R. Jones, *Prog. Phys. Org. Chem.*, **9**, 241 (1972).
43   P. M. Laughton and R. E. Robertson, in *Solute–Solvent Interaction* (J. F. Coetzee and C. D. Ritchie, Eds), Marcel Dekker, New York, 1969, p. 400.
44   P. Gross, H. Steiner and F. Krauss, *Trans. Farad. Soc.*, **32**, 877, 879, 883 (1936); N. A. Bergman and L. Baltzer, *Acta Chem. Scand.*, **A31**, 343 (1977).
45   K. B. J. Schowen, in *Transition States of Biochemical Processes* (R. D. Gandour and R. L. Schowen, Eds), Chapter 6, Plenum, New York, 1978.
46   J. C. Hornel and J. A. V. Butler, *J. Chem. Soc.*, 1361 (1936).
47   W. J. C. Orr and J. A. V. Butler, *J. Chem. Soc.*, 330 (1937).
48   W. E. Nelson and J. A. V. Butler, *J. Chem. Soc.*, 957 (1938).
49   R. A. More O'Ferrall, *Prog. Phys. Org. Chem.*, **5**, 331, 1967.

281

8 Steric and conformational properties

Atoms behave as more or less hard spheres having definite spatial requirements which, if unable to be met as a result of particular bonding arrangements in a molecule or transition state, may result in a sharp increase in the energy of the system. This is due to non-bonding repulsions between adjacent atoms or like charges and also to bond-angle strain. Changes in strain energy between reagents and transition state will affect the activation energy and hence rates of reaction. Though all molecules possess some degree of strain in the form of the non-bonding interaction terms, it is convenient to define certain structural features as strain-free. These include the staggered conformations of an acyclic alkane, **1**, and the chair form of cyclohexane, **2**. Strain energy can then be expressed as a difference in the enthalpy of formation of a compound compared with an unstrained analogue (Table 8.1)[1-3].

8.1 THE ORIGINS OF STERIC STRAIN

There are three important sources of steric strain.

(a) Prelog strain: intramolecular van der Waals repulsive forces resulting from crowding together of large atoms or groups, as in **3**. This can result in permanent deformations; for example, the central C–C bond length of **3** is

Table 8.1   Strain energies/kJ mol^{-1} (kcal mol^{-1})[4-6]

(a) Acyclic compounds

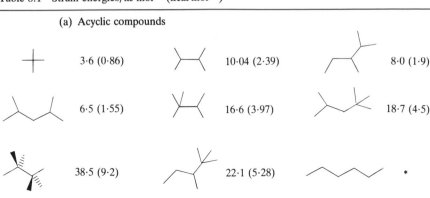

(b) Monocyclic compounds

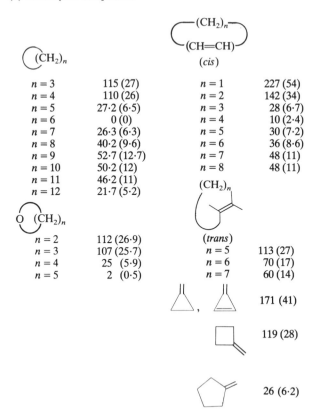

$(CH_2)_n$	
$n = 3$	115 (27)
$n = 4$	110 (26)
$n = 5$	27·2 (6·5)
$n = 6$	0 (0)
$n = 7$	26·3 (6·3)
$n = 8$	40·2 (9·6)
$n = 9$	52·7 (12·7)
$n = 10$	50·2 (12)
$n = 11$	46·2 (11)
$n = 12$	21·7 (5·2)

$O(CH_2)_n$	
$n = 2$	112 (26·9)
$n = 3$	107 (25·7)
$n = 4$	25 (5·9)
$n = 5$	2 (0·5)

(cis)	
$n = 1$	227 (54)
$n = 2$	142 (34)
$n = 3$	28 (6·7)
$n = 4$	10 (2·4)
$n = 5$	30 (7·2)
$n = 6$	36 (8·6)
$n = 7$	48 (11)
$n = 8$	48 (11)

(trans)	
$n = 5$	113 (27)
$n = 6$	70 (17)
$n = 7$	60 (14)

171 (41)

119 (28)

26 (6·2)

Table 8.1 *(Continued)*

(c) Polycyclic compounds[7]

$$(CH_2)_x \overset{\displaystyle CH}{\underset{\displaystyle CH}{\diamondsuit}} (CH_2)_z \quad (CH_2)_y$$

'Bridgehead' compounds

$$(CH_2)_x \overset{\displaystyle CH}{\underset{\displaystyle C=CH}{\diamondsuit}} (CH_2)_{z-1} \quad (CH_2)_y$$

x	y	nz		
1	2	2	63 (15)	209 (50)
1	2	3	43 (16)	163 (39)
1	3	3	38 (9)	101 (24)
2	1	1	172 (41)	277 (66)
2	1	2	63 (15)	209 (50)
2	1	3	43 (10)	163 (39)
2	1	4	66 (16)	125 (30)
1	1	5	162 (39)	236 (56)
2	2	2	44 (10)	211 (50)
2	2	3	62 (15)	114 (27)
2	2	4	85 (20)	119 (28)
2	3	3	85 (20)	106 (25)
3	2	2	62 (15)	144 (34)
3	2	3	86 (20)	106 (25)

Cubane [cube] 656 (157)

stretched to 164 pm from the normal 154 pm[8], and the C=C length in **4** is stretched to 136 pm from 132 pm and twisted from planarity.

t-Bu 136 pm Ph
C=C ≡
t-Bu Ph

t-Bu 24° Ph
Ph *t*-Bu

4

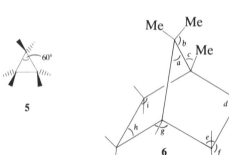

5 60°

Me Me
b
c Me
a
O
d
i
h g
e f
6

Deviations from natural angles / degrees

a	+2
b	−15
c	+10
d	−5
e	−8
f	+3
g	+1
h	−8
i	−6

(b) Baeyer strain: bond-angle distortion, e.g. in **5** and **6**. This may be an inevitable consequence of certain types of structure such as small rings, or of repulsions between neighboring atoms (Table 1.5).

(c) Pitzer strain: torsional deformation by σ-bond rotation from the most stable conformation such as the staggered arrangement of **1**. It includes eclipsing interactions, **7**, as in camphor (**6**) and planar rings **5**, **8**, and steric inhibition of resonance by twisting of a conjugated structure (e.g. **27**, **28**) out of coplanarity[9].

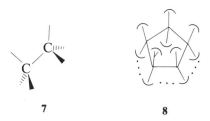

          7   **8**

If any of these components of intramolecular energy increases during the activation step of a reaction, a decrease in rate will result and is known as steric hindrance. In principle, though it is less common, steric acceleration (release of steric strain) can also occur.

8.2 EXAMPLES OF STERIC EFFECTS UPON REACTIONS

8.2.1 *Ortho* effects

Ortho substituents in an aromatic ring usually inhibit substitution reactions in a manner which suggests dependence on size. For example, the proportion of *ortho*:*para* product in the nitration of alkylbenzenes (the alkyl groups having similar electronic effects) is as follows (the statistical ratio would be 2):

$$RC_6H_5 + HNO_3 \rightarrow (o,p)\text{-}RC_6H_4NO_2$$

R:	**Me**	Et	isoPr	*tert*-Bu
o/p:	1·66	0·90	0·41	0·12

Similarly, rates of methylation of pyridines, **9**, by methyl iodide (an S_N2

$$\text{(structure 9)}$$

                           9

reaction) decrease with increasing size of the 2-alkyl substituent[10]:

9; R:	H	Me	Et	isoPr	*tert*-Bu
k_{rel}:	100	41·8	22·8	7·4	0·026

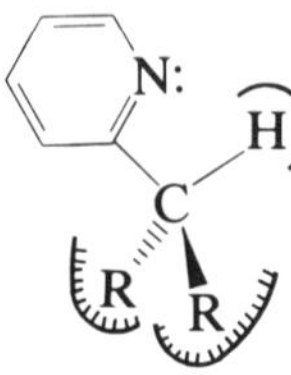

10

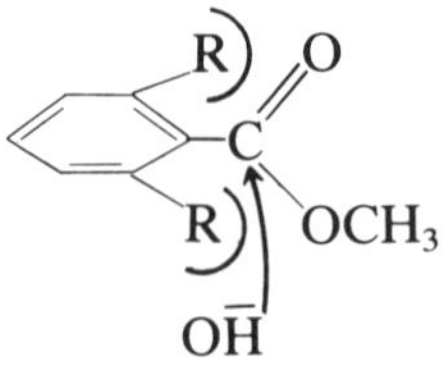

11

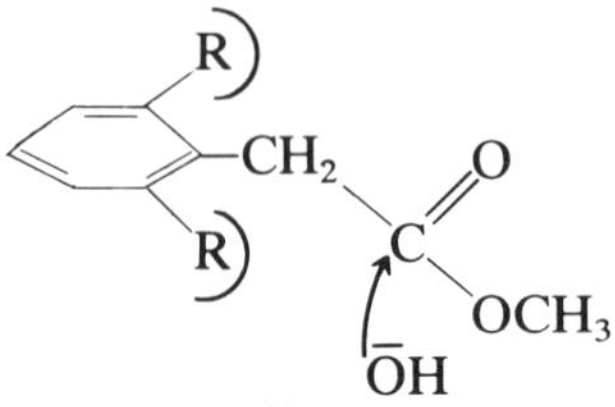

12

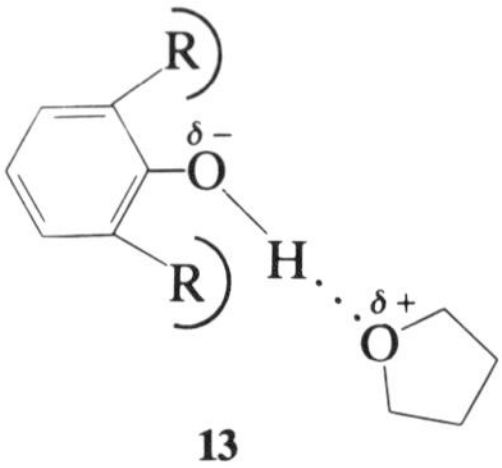

13

The effect is moderate while there is still one α-hydrogen on the *ortho* alkyl group, **10**, which can orient towards the incoming reagent, but increases sharply when that is no longer possible, as with *tert*-Bu. The esterification of 2,6-dialkylbenzoic acids, **11**, and the hydrolyses of the corresponding esters are far more difficult than could be accounted for by electronic effects, while the phenylacetic acid series, **12**, is normal. Clearly, access to the carbonyl group by the incoming nucleophile is denied by bulky *ortho* groups. In general, Prelog strain tends to increase with coordination number which, in these examples, increases during the activation step. The failure of the Hammett equation (Eq. (4.1)) for *ortho* substituents is, of course, a more general result of this effect. It would be tempting to interpret the reduction in equilibrium constant for hydrogen-bond formation by phenols, **13**, in a similar way[11].

13	K	$\Delta G/kJ\,mol^{-1}\,(kcal\,mol^{-1})$	$\Delta S/J\,K^{-1}\,mol^{-1}\,(cal\,K^{-1}\,mol^{-1})$
R = H:	13·6	−16·7 (−4·0)	−35 (−8·4)
R = *tert*-Bu:	0·5	−33·4 (−8·0)	−108 (−26)

However, it appears that the effect brought about by the tertiary butyl group is of a large decrease in entropy of association, the enthalpy actually being more favorable. This suggests that the *ortho* groups tend to reduce the solvation of the phenol much more than that of the complex. It is known that steric inhibition of solvation of phenolate ions can amount to 70 kJ mol^{-1} ($16 \cdot 7 \text{ kcal mol}^{-1}$)[12] ($\Delta S > 200 \text{ J K}^{-1} \text{ mol}^{-1}$ ($48 \text{ cal K}^{-1} \text{ mol}^{-1}$) at 25°C).

8.2.2 F-strain effects

S_N2 *reactions at carbon*

These are greatly hindered by the increasing size and number of alkyl substituents on the α-carbon (**14**) until, with tertiary compounds, rates are so low that a mechanistic change towards ionization becomes favorable (Section 10.1.2). Equally large steric effects are found when the number of alkyl groups on the β-carbon is increased (**15**): see Table 8.2. Both result

14   **15**

from the mechanistic requirement for rearface attack. This type of steric hindrance to approach of the reagent has been denoted F ('front') *strain*[13–15]. It means that bond deformations are necessary in the initial stages of the reaction for the nucleophile to approach the reaction center within bonding distance. This represents additional activation energy.

Table 8.2   Steric effects upon rates of bimolecular displacements (S_N2) at saturated carbon

α-substitution[16] **14**				β-saturation **15**			
R	R'	R''	k_{rel}	R_1	R_2	R_3	k_{rel}
H	H	H	100	H	H	H	1000
Me	H	H	0·6	Me	H	H	28
Me	Me	H	0·0003	Me	Me	H	3
				Me	Me	Me	0·0004

tert-BuC≡CCH$_2$Cl	38 000	

The retarding effect of a *tert*-butyl group is not relayed through a π-system; consequently it cannot be of electronic origin.

Borane–amine complexes

The equilibria for complex formation between boranes and amines are governed by both steric and electronic influences. When the electrophile is bulky (tri(*tert*-butyl)borane, **16**) F-strain becomes the most important factor and the order of reactivity of amines is

$$NH_3 > RNH_2 > R_2NH > R_3N,$$

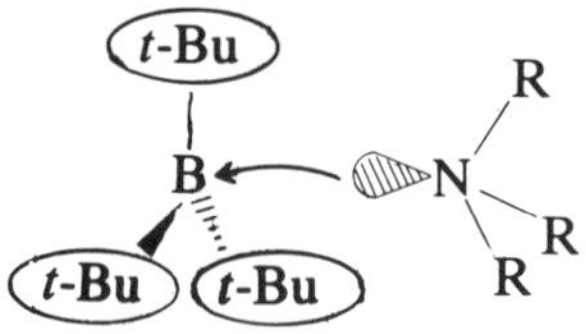

16

i.e. the smallest amine forms the complex most readily despite it being intrinsically the weakest base [17].

The Diels–Alder reaction

This reaction is subject to severe steric hindrance from substituents R_1 on the 2π-component (dienophile), **17**, although large groups, R_2, attached to the 4π-component (diene), have much less effect on rates [18,19]; from this one may deduce a transition state of the type **18**.

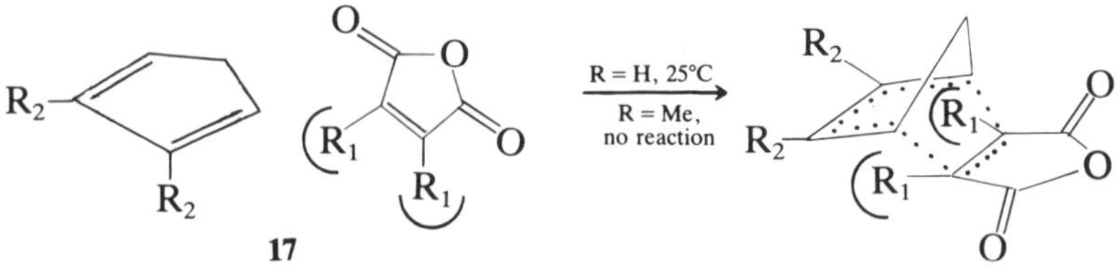

R = H, 25°C
R = Me, no reaction

17

18

Steric selection of the reaction site

Attack at a double bond, C=C or C=O, can, in principle, come from either side. If one side is sterically more accessible than the other by virtue of the surrounding structure, it will be favored [20]. A chiral group in the vicinity of a reaction center often provides a sterically asymmetric environment; for example, the menthyl ester of phenylglyoxylic acid, **19**, reacts with a Grignard reagent preferentially to give the *R*-alcohol. The asymmetric aluminohydride, **20**, can give stereoselection >90% [21]. The stereoselectivity of addition to the double bond of norbornene, **21**, is a similar,

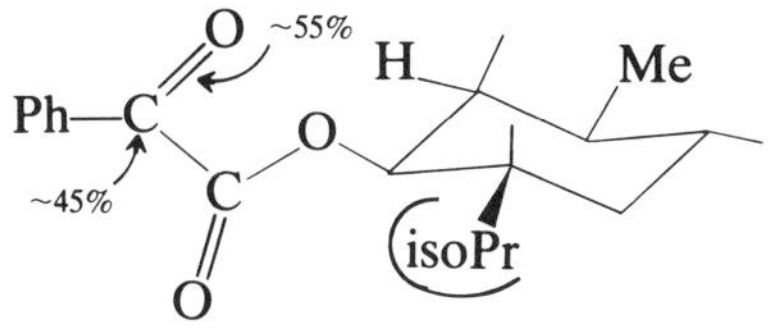

19

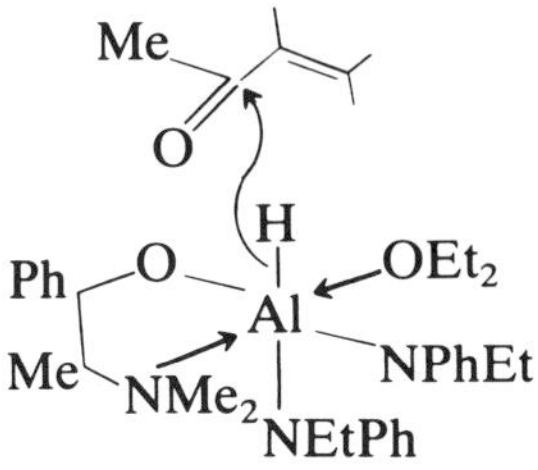

20

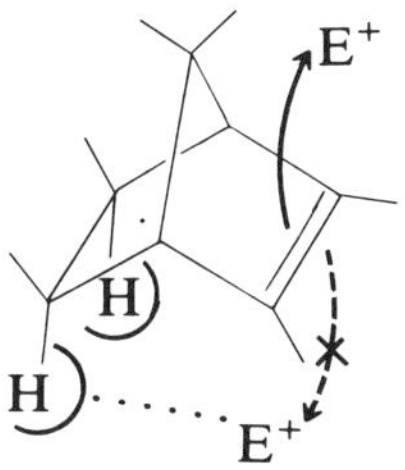

21

though non-chiral, example[22]. Electrophilic reagents tend to attack on the side of the one-carbon bridge (*exo*) in preference to the more hindered *endo* side.

8.2.3 Bond-angle strain

Three- and four-membered rings contain bond-angle strain (*I strain*)[23, 24] which contributes to their difficulty of formation and ease of opening (**22**), although other factors contribute to rates in a manner which is difficult to disentangle[25]. Eclipsing interactions are also present in three-, four- and five-membered rings **5**, **23**, **24** although four- and five-membered systems, which are not quite planar, are able to relieve some of this strain. The difficulty of synthesizing compounds, e.g. **25**, with a 'bridgehead' double bond (Bredt's rule[26–28]) is also is a reflection on the large amount of bond-angle strain they possess; see also Table 8.1c[29].

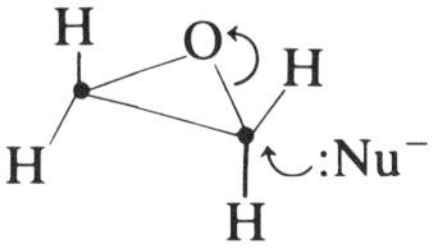

22

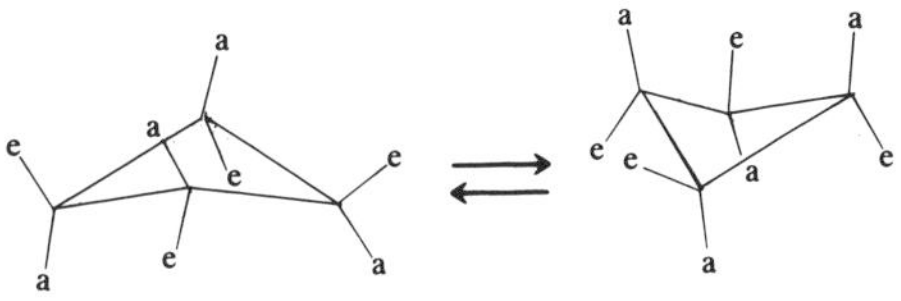

23

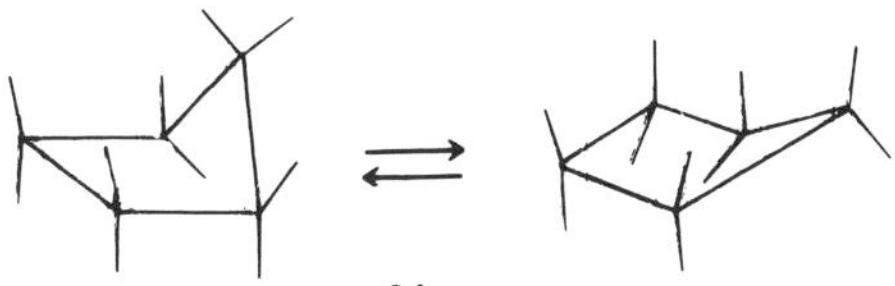

24

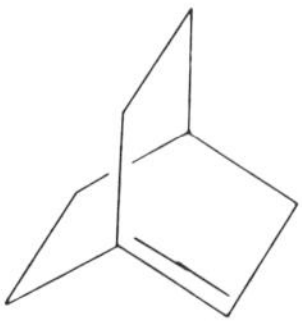

25

8.2.4 Steric inhibition of resonance

Nucleophilic displacement at an aromatic center is facilitated by the electronic effect of a nitro group in the *para* position (Section 10.4.7). If the nitro group is flanked by two alkyl groups it can no longer be coplanar with the ring and the conjugative effect is reduced: this effect makes **27** less reactive than **26**[30].

Biphenyls are twisted from the coplanar conformation on account of steric strain between the *ortho* substituents of adjacent phenyl rings. Resonance between the two rings is again greatly reduced, as shown by their almost independent reactivities (ratio of rates of nitration of biphenyl:4-nitrobiphenyl (at the 4'-position) are 30:1 whereas that of benzene:nitrobenzene is about 10^7:1). Suitably substituted biphenyls, **28**, are chiral and racemize by rotation through the coplanar transition state at rates which depend upon the size of *ortho* substituents[31,32].

X:	Me	NO$_2$	COOH	OMe
$10^3 k/\text{min}^{-1}$:	3·9	5·5	7·6	7·4

8.2.5 Steric acceleration

Release of strain should facilitate a reaction and probably contributes to rates of carbocation-forming ionization (S$_N$1 mechanism; e.g. **29**), though it

29 $\underset{(R^{\prime\prime\prime})}{\overset{(R}{\text{C--Br}}} \xrightarrow{k} \overset{R}{\underset{R}{\text{C}^{+}}} \text{Br}^{-}; \quad \dfrac{k\,(R = tert\text{-}Bu)}{k\,(R = Me)} = 1 \cdot 4 \times 10^{4}$

is often difficult to separate this from electronic effects on rates[33]. The inherent steric compression between three alkyl groups in the substrate is denoted *B* ('back') *strain*[34]. The large acceleration in rate of **30** (R = *tert*-Bu) compared with the rate for the methyl analogue is largely steric in origin[8,35,36]. Reduction of strained ketones affords relief of *I* strain and is consequently faster, other things being equal, than that of unstrained analogues. For example, equilibration **31** (by the Pondorff–Oppenauer catalyst) favors the products in which angle strain is relieved, eclipsing interactions remaining roughly equal on either side of the equation[37].

$\dfrac{k\,(R = tert\text{-}Bu)}{k\,(R = Me)} = 2 \cdot 2 \times 10^{5}$

30

31

$K = 33$

8.2.6   Steric enhancement of resonance

Steric enhancement of resonance is apparently unusual but possible. Substituents in the 3-position in 4-alkoxybenzoates, **32**, tend to hold the

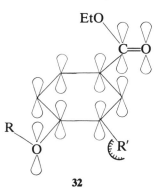

32

33

alkoxy group in the *trans* coplanar conformation in which resonance interactions with the carboxyl group are at a maximum. This reduces the reactivity of the ester function and causes deviations from LFER predictions of rate[38]. As an example of the drastic effects on reactivity which may be brought about by crowding, hexa-isopropenylbenzene, **33**, fails to react with bromine. Presumably each ethenic group is at right angles to the benzene ring and inhibits its neighbor from close enough approach by a bromine molecule[39].

8.2.7 Calculation of steric effects: the molecular mechanics method

The energy of a molecule relative to a strain-free system may be expressed as

$$E = E_r(r) + E_\theta(\theta) + E_\varphi(\varphi) + E_{nb}(d), \qquad (8.1)$$

in which the four terms are contributions to strain energy from, respectively, bond-length deformation (which may be divided into terms from σ- and π-bonds, E_r^σ and E_r^π), bond-angle strain, bond torsional strain and non-bonding interactions. If one knew how the energy varied with each of these types of deformation it would be possible to calculate E. It would also be possible to change the atomic coordinates systematically and locate energy minima. Calculations of this type were pioneered by Westheimer but it has required the advent of large computers to realize the full potential of this approach. The potential functions for each of the first three terms may be given in the form of equations for harmonic oscillators; together they generate the 'force field' of the molecule:

$$E_r = \tfrac{1}{2} \sum f_r (r - r_0)^2; \qquad (8.2)$$

$$E_\theta = \tfrac{1}{2} \sum f_\theta (\theta - \theta_0)^2; \qquad (8.3)$$

$$E_\varphi = \tfrac{1}{2} \sum f_\varphi (1 - \cos 3\varphi); \qquad (8.4)$$

in which f is the force constant (Section 1.3.3); r is the bond length and θ is the bond angle which have strainless 'standard values', r_0, θ_0. The non-bonding interactions may be described by

$$E_{nb} = \varepsilon[-c_1(r^*/r)^6 + c_2 \exp(-c_3 r/r^*)], \qquad (8.5)$$

in which c_1, c_2, c_3 are constants, ε an energy parameter, r the distance between a given pair of non-bonding but interacting atoms and r^* the sum of their van der Waals radii. There are many variants on the type of functions used and their parameterization but large computer programs are now generally available (e.g. 'MM2' of Allinger[40]) which will calculate E for a molecular assemblage of atoms whose coordinates are given (whether or not it actually exists or has been prepared) or will search for a minimum energy by changing bond lengths, angles and twist angles to predict favorable conformations. Quite large molecules can now be handled with some confidence but the programs are continually being refined and predictions improved. Some results of MM calculations are included in Table 8.3 and Fig. 8.1. Absolute energies of rotamers of phenylethane

Table 8.3 Relative energies/kJ mol^{-1} (kcal mol^{-1}) of principal rotamers of some 2-phenylethyl derivatives calculated by molecular mechanics[40]

(a) PhCHMeXYZR

X	Y	Z	R			
C	H	H	Me	0 [56]^a	2·68 [173] (0·568)	6·15 [302] (1·47)
			Et	0 [57]	2·85 [173] (0·681)	6·28 [303] (1·50)
			Pr^i	0 [53]	3·56 [178] (0·850)	11·17 [306] (2·67)
			Bu^t	0 [60]	8·33 [172] (1·99)	15·27 [308] (3·65)
C	OH	H	Me	0 [58]	0·50 [174] (0·12)	4·31 [300] (1·03)
			Et	0 [57]	0·96 [178] (0·23)	3·22 [301] (0·77)
			Pr^i	0 [54]	1·51 [180] (0·36)	8·24 [303] (1·97)
			Bu^t	0 [61]	3·01 [175] (0·72)	14·02 [308] (3·35)
C	H	OH	Me	0 [53]	5·94 [173] (1·42)	6·44 [307] (1·54)
			Et	0 [52]	5·86 [172] (1·40)	6·65 [304] (1·59)
			Pr^i	0 [48]	6·07 [175] (1·45)	11·09 [303] (2·65)
			Bu^t	0 [57]	8·12 [172] (1·94)	13·85 [308] (3·31)
C	=O		Me	0 [56]	5·61 [178] (1·34)	10·21 [314] (2·44)
			Et	0 [58]	6·15 [178] (1·47)	10·38 [314] (2·48)

^{a} The Ph–R dihedral angle φ is given in brackets.

Table 8.3 (*Continued*)

(a) PhCHMeXYZR

X	Y	Z	R	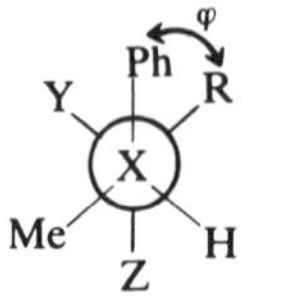	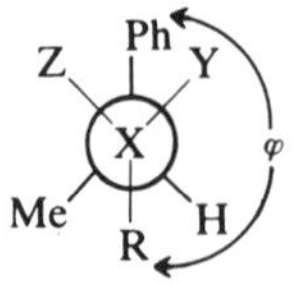	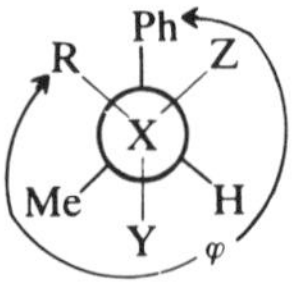
			Pr^i	0 [51]	6·86 [176] (1·64)	11·63 [283] (2·78)
			Bu^t	0 [79]	12·22 [178] (2·92)	16·53 [309] (3·95)
S	—	—	Me	0 [59]	3·93 [174] (0·94)	4·39 [309] (1·05)
			Et	0 [59]	4·31 [164] (1·03)	4·56 [309] (1·09)
			Pr^i	0 [59]	4·85 [180] (1·16)	5·52 [306] (1·32)
			Bu^t	0 [63]	8·03 [172] (1·92)	10·17 [307] (2·43)
S	—	O	Bu^t	0 [53]	10·54 [139] (2·52)	4·43 [309] (3·45)
S	O	—	Bu^t	0 [115]	13·14 [166] (3·14)	20·25 [308] (4·84)

(b) PhCH$_2$XYZR

X	Y	Z	R			
C	H	H	Me	0·29 [62] (0·07)	0 [180]	0·29 [298] (0·07)
			Et	0·29 [62] (0·07)	0 [180]	0·29 [298] (0·07)
			Pr^i	0·08 [62] (0·02)	0 [180]	0·08 [298] (0·02)
			Bu^t	3·31 [63] (0·79)	0 [180]	3·31 [297] (0·79)
C	H	OH	Me	0 [61]	1·72 [178] (0·41)	2·01 [301] (0·48)
			Et	0 [60]	1·63 [177] (0·39)	1·84 [300] (0·44)
			Pr^i	0 [60]	1·09 [179] (0·26)	1·21 [299] (0·29)
			Bu^t	1·38 [62] (0·33)	0 [171]	4·64 [299] (1·01)
C	=O		Me	0·54 [58] (0·13)	0 [180]	0·54 [302] (0·13)
			Et	0·54 [60] (0·13)	0 [180]	0·54 [300] (0·13)
			Pr^i	0 [53]	0·59 [173, 187] (0·14)	0 [307]
			Bu^t	0·75 [80] (0·18)	0 0 [180]	0·75 [280] (0·18)
S	—	O	Bu^t	0 [106]	1·34 [160] (0·32)	3·56 [294] (0·85)

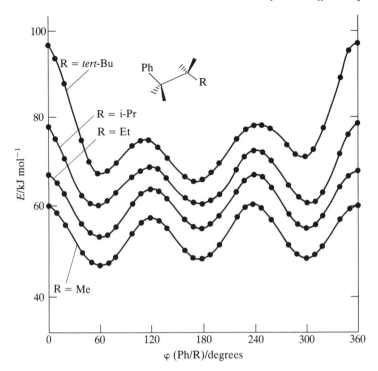

Fig. 8.1 Torsional energy as a function of dihedral angle Ph–R, φ, for 1-alkyl-2-phenylethanes, calculated by molecular mechanics.

derivatives including transition states can be obtained and show clearly the large steric energies associated with eclipsing of phenyl with *tert*-butyl groups[40,41]. Molecular mechanics has been used to define the steric parameter, $\mathscr{S}$. Since homolysis of the central C–C bond of a substituted ethane depends upon the intrinsic bond strength, D, offset by the relief of F strain, a measure of the latter would be obtained by taking the difference in bond dissociation energy between that of R—CH_3 and R—*tert*-Bu[42]:

$$\mathscr{S}(R)/\text{kJ mol}^{-1} = D_{\text{R—}\textit{tert}\text{-Bu}} - D_{\text{R—Me}} + 8 \cdot 87.$$

Values of $\mathscr{S}(R)$ (Table 8.4) provide an operational measure of strain release on converting the group from pyramidal to planar geometry and are applicable both in radical reactions (Section 15.4.2) as well as in polar processes.

8.3 MEASUREMENT OF STERIC EFFECTS UPON RATES

Whereas strain in a stable molecule can be estimated from the difference between the experimental heat of formation and that calculated from the sum of its constituent bond energies (Table 8.1), kinetic steric effects result

295

Table 8.4 Steric substituent constants $\mathscr{S}(R)/10^4 \, J \, mol^{-1}$ for acyclic and cyclic alkyl groups[42]

R	$\mathscr{S}$	R	$\mathscr{S}$
CH_3	$\equiv 0$	$C(CH_3)_2CH_2C(CH_3)_3$	4·40
C_2H_5	0·86	$C(CH_3)_2C_8H_{17}$	4·85
$CH_2CH_2CH_3$	0·89	$C(CH_3)(C_2H_5)_2$	5·26
$CH_2(CH_2)_2CH_3$	0·86	$C(C_2H_5)_3$	6·28
$CH_2(CH_2)_{3-5}CH_3$	0·85	$C(CH_3)_2C(CH_3)_3$	7·51
$CH_2(CH_2)_6CH_3$	0·84	Cyclo-C_4H_7	−0·21
$CH_2CH_2C(CH_3)_3$	0·85	C_8H_7 (Cubyl)	−0·04
CH_2-cyclo-C_6H_{11}	1·36	Cyclo-C_3H_5	1·33
$CH_2CH(CH_3)_2$	1·86	Cyclo-C_5H_9	1·81
$CH_2C(CH_3)_3$	2·29	Cyclo-C_6H_{11}	2·29
$CH_2C(CH_3)_2C_4H_9$	3·24	Cyclo-C_7H_{13}	2·94
$CH(CH_3)_2$	2·29	2-*exo*-**A**	1·83
$CH(CH_3)C_2H_5$	2·51	1-**A**	2·57
$CH(CH_3)CH_2C(CH_3)_3$	2·78	2-*endo*-**A**	2·71
$CH(CH_3)C_6H_{13}$	2·87	7-**A**	3·11
$CH(C_2H_5)_2$	3·29	2-**B**	2·37
$CH(CH_3)C(CH_3)_3$	4·93	1-**B**	3·49
$CH[CH(CH_3)_2]_2$	5·61	1-**C**	3·81
$CH(C_2H_5)C(CH_3)_3$	7·26	1-**D**	4·08
$CH(C(CH_3)_3)_2$	10·68	2-**D**	4·45
$CH(1\text{-adamantyl})_2$	12·65	1-**E**	4·92
$C(CH_3)_3$	3·82	2-*exo*-**F**	4·98
$C(CH_3)_2C_2H_5$	4·36	2-*endo*-**F**	6·20

from differences in strain energy between reagents and transition state (primary steric effects). The latter are inaccessible experimentally and when calculated (e.g. using molecular mechanics as discussed above) they are subject to much uncertainty since they contain so many quantities — precise geometry, energies of partial bonds and solvation, for example — that are unknown. Empirical approaches have long been developed for estimating the importance of steric factors upon reaction rates.

8.3.1 The Taft–Ingold hypothesis[3,43]

Hydrolytic rates of aliphatic esters under basic [B] and under acidic [A] conditions were shown in Section 4.8.2 to follow the LFER:

$$\log[k_X/k_{Me}]_B = 2 \cdot 48 \sigma_X^* + E_s \qquad (8.6)$$

$$\log[k_X/k_{Me}]_A = E_s \qquad (8.7)$$

where k_X, k_{Me} refer to substituent X and a reference substituent (Me). The quantity E_s is therefore a measure of the steric effect of the substituent, X, relative to the standard (Me was originally set at $E_s = 0$) derived from a reaction believed to respond to steric but not to electronic effects. Values of E_s, the Taft steric constant, may be used in LFER equations: they do appear to be related to the actual size of groups, as might be expected, although direct proportionality is not observed since solvation may influence effective size. It seems now to be generally agreed that the original E_s values are not purely steric parameters but contain an element of hyperconjugation[44–48], since the values appear to depend upon the number of α-hydrogens on the substituent (i.e. whether it is primary, secondary or tertiary alkyl). Corrections made according to Eq. (8.8) yield steric constants, E_s^c, known as Hancock constants[49], which are claimed to be more truly measures of steric properties (Table 8.5):

$$E_s^c = E_s + 0 \cdot 306(n - 3) \qquad (8.8)$$

(n = number of α-hydrogens).

The table of steric constants may be further extended by the observation that Eqs (8.9) and (8.10) hold quite well for substituents of the type —$CR_1R_2R_3$[50]:

$$E_s(CR_1R_2R_3) = -2 \cdot 467 + 9 \cdot 240 E_s(1) + 0 \cdot 774 E_s(2) + 0 \cdot 438 E_s(3) \qquad (8.9)$$

$$E_s^c(CR_1R_2R_3) = -2 \cdot 347 + 4 \cdot 589 E_s^c(1) + 0 \cdot 958 E_s^c(2) + 0 \cdot 630 E_s^c(3) \qquad (8.10)$$

in which the substituents R_1, R_2, R_3 on the α-carbon are taken in increasing order of size. For example, the steric constant for the 2-butyl group (CHMeEt) becomes

$$E_s^c = -2 \cdot 347 + (4 \cdot 589 \times 0 \cdot 32) + (0 \cdot 958 \times 0 \cdot 0) + (0 \cdot 630 \times -0 \cdot 38) = -1 \cdot 25.$$

$$(8.11)$$

8.3.2 Other steric parameters[54]

Rates of hydroboration of alkenes by disiamylborane (bis(1,2-dimethyl-propyl)borane) fit the general LFER (Eq. (8.12)) with a very low value of $\rho^* = 0 \cdot 01$:

$$\log k_X/k_{Me} = \rho^* \sigma^* + \delta E_s. \qquad (8.12)$$

This reaction is almost totally insensitive to electronic effects but the

297

Table 8.5   Steric constants[a]

	$-E_s$ (Refs 3, 48)	$-E_s'$ (Ref. 48)	$-E_s^c$ (Refs 50, 51)	v (Ref. 52)
H	−1·24	−1·12		0·00
Me	0·00	0·00	0·00	0·52
Et	0·07	0·08	0·38	0·56
n-Pr	0·36	0·31	0·67	0·68
n-Bu	0·39	0·31	0·70	0·68
isoPr	0·47	0·48	1·08	0·76
sec-Bu	1·13	0·48	1·08	0·76
isoBu	0·93	0·93	1·24	0·98
tert-Bu	1·54 (4.22)*	1·43		1·24
1-Pe		0·31		
3-Pe	1·98	2·00		1·51
isoPe		0·97		
neoPe	1·74 (1·84)*	1·63		1·34
tert-Pe	2·17 (4·74)*	2·28		
isoPrMe$_2$C				
tert-BuMe$_2$C	3·90 (6·00)*			
Et$_3$C	3·80			2·38
tert-BuisoPrMeC		7·56		
Ph	0·38	2·31		1·66
PhCH$_2$	0·38	0·39	0·69	0·70
PhCH$_2$CH$_2$		0·69	0·69	
Ph$_2$CH	1·76	1·50	2·59	1·25
Ph$_3$C		4·91		
o-Tolyl		2·82		
F		−0·57		0·27
Cl		0·02		0·55
Br		0·22		0·65
I		0·50		0·78
CH$_2$Cl	0·24	0·18		0·60
CH$_2$Br	0·27	0·24		0·64
CH$_2$I	0·37	0·30		0·64
CHCl$_2$	1·54	0·58		
CHBr$_2$	1·86	0·76		0·89
CF$_3$	1·16	0·78		
CCl$_3$	2·06	1·75		
CBr$_3$	2·42	2·14		
Cyclo-C$_3$H$_5$		1·09		
Cyclo-C$_4$H$_7$	0·06	0·03		
Cyclo-C$_5$H$_9$	0·51	0·41		
Cyclo-C$_6$H$_{11}$	0·71	0·69		

[a] Each of these scales forms a more or less consistent set of steric constants which correlates with rate data, usually without very high precision. A value of $E_s' = 1\cdot43$ (*tert*-Bu), for instance, means that in a reaction for which the susceptibility constant for steric interactions is 1·0, the effect on the rate of replacing methyl by *tert*-butyl would be a retardation by a factor of $\log^{-1} 1\cdot43 = 26$.

* Values of $-E_s^*$

moderate value of $\delta = 0.71$ confirms a considerable susceptibility to steric hindrance which makes it a suitable model system for evaluating steric constants. However, while values of E_s for primary and secondary alkyl groups agree with the Taft values, those for tertiary groups do not, and so a further scale of values, E_s^*, has been defined for this situation[52]. It is included in Table 8.6.

Charton has defined a steric parameter, v, from the intrinsic size of the substituent as defined by its average van der Waals radius, r_v[51,55]:

$$v_X = r_v(X) - r_v(H) = r_v(X) - 1.20. \tag{8.13}$$

Values (Table 8.5) correlate moderately well with E_s and have the advantage of being calculated without recourse to rate data. Statistical evidence has been put forward suggesting that v or r_v is a better measure of steric effects than E_s[56].

8.3.3 Examples of steric LFER

Equation (8.5) implies that the free energy of activation is affected in a systematic and consistent manner by a structural parameter related to the size of the substituent. If this equation is widely obeyed, it would enable LFER to be constucted in situations for which the Hammett equation and its extensions are not applicable by inclusion of a steric term, δE_s. Some examples follow.

Alkaline hydrolysis of aliphatic methyl esters[53,56]

$$XCOOMe + OH^- \rightarrow XCOO^- + MeOH;$$
$$\log k_X = 3.14 + 1.75\sigma_I + 3.75\sigma_R - 2.90v. \tag{8.14}$$

Inductive and resonance effects now reinforce each other and steric constraints are more severe than in the acid-catalyzed reaction (for which $\delta = 1$ by definition).

Halide displacement (S_N2) at a primary carbon[57]

$$XCH_2Br + LiCl(MeOAc, 20°C) \rightarrow XCH_2Cl + LiBr;$$
$$\log x_X = 20.8\sigma_I - 22.5\sigma_R - 6.3v + 2.83. \tag{8.15}$$

Again, cancelation of inductive and resonance effects makes this reaction rather insensitive to electronic perturbation. The very large value of δ indicating great sensitivity to steric effects points to a large increase in steric crowding on forming the transition state.

Table 8.6   LFER including *ortho* substituents[62]

$$\log k_X/k_H = \rho\sigma_o + \delta E_s + fF$$

Reaction	$\log k_X/k_H$	ρ	δ	f
1 Ionization of ArCOOH (K_A) (water, 25°C)	−4·179	0·9502	−0·392	1·469
2 Ionization of ArCH$_2$COOH (K_A) (butoxyethanol, 20°C)	−5·569	0·546	0·088	−0·2112
3 Ionization of ArOH (K_A) (water, 25°C)	−9·814	2·036	0·167	2·395
4 Alkaline hydrolysis of ArCOOMe (80% MeOH, 35°C)	−0·953	1·929	0·639	1·093
5 Acid hydrolysis of ArCOOMe (80% MeOH, 100°C)	0·979	−0·107	0·000	0·502
6 ArNMe$_2$ + MeI → ArNMe$_3$I (MeOH, 65°C)	0·041	−2·614	1·482	1·049

Reaction	LFER	Ref.
7 + Br$_2$ R$_1$, R$_2$ (R$_1$, R$_2$ = branched alkyl)	$\log k_X = -5\cdot43 \sum \sigma^* + 0\cdot90 \sum E_s + 7\cdot42$	63
8 —R + MeI	$\log k_X/k_{Me} \propto 1\cdot03 E_s$	64
9 O·COCH$_3$ (basic hydrolysis) R	$\log k_X = 1\cdot17\sigma^- + 0\cdot185 E_s + 0\cdot538 F - 9\cdot727$	65
10 CO·N(OH)Me (hydrolysis)	$\log k_X/k_{Me} = 0\cdot863\sigma^* + 0\cdot911 E_s$	66
11 + isoPe$_2$BH ⟶ [isoPe = isoPr—CHMe] isoPe isoPe	$\log k_X = 0\cdot01\sigma^* + 0\cdot71 E_s - 1\cdot727$	67
12 R—C(OMe)(OMe)—Me hydrolysis of	$\log k_X/k_{Me} = -4\cdot1 E_s + \text{constant}$	68
13 + MeI (relative to —R + MeI) (k) (k$_0$)	$\log k/k_0 = 2\cdot03 E_s$	69

Additions of bromine to alkenes[58]

$$R_1R_2C{=}CR_3R_4 + Br_2 \overset{k_X}{\rightarrow} R_1R_2CBrCBrR_3R_4.$$

The rates of these reactions correlate with each substituent contributing additively:

$$\log k_X = -5{\cdot}43 \sum_{}^{R} \sigma^* + 0{\cdot}96 \sum_{}^{R} E_s + 7{\cdot}42. \qquad (8.16)$$

Reactions at ortho substituted aromatic rings[59,60]

Ortho substitution normally causes failure of the Hammett equation whose constants, σ, only contain information pertaining to electronic effects. Inclusion of a steric term in the regression equation can lead to moderately successful correlations. For example, in the acid-catalyzed esterification of *ortho* substituted benzoic acids[56,61],

$$\log k_X = 1{\cdot}163 + 0{\cdot}188\mathscr{F} - 1{\cdot}277\mathscr{R} - 0{\cdot}814r_v \qquad (8.17)$$

where $\mathscr{F}$ and $\mathscr{R}$ are the Swain–Lupton field and resonance parameters (Section 4.8.2). The lack of sensitivity of this reaction to electronic effects is seen as a mutual cancelation of inductive and resonance contributions. It seems important to include the field effect because of the proximity of *ortho* groups to the reaction center. The three-parameter equation (8.18) which describes rate effects due to substituents located at *ortho*, *meta* or *para* positions has been applied to many reactions with considerable success[60];

$$\log k_X/k_H = \rho\sigma_{o,m,p}^{(0,+,-)} + \delta E_s + fF \qquad (8.18)$$

where the substituent constants σ are those appropriate to *ortho*, *meta* or *para* substitution and F is a field parameter (Table 8.6). The second and third terms are only applicable in the case of *ortho* substituents. Table 8.7 gives values of the regression constants, δ and f, for selected reactions.

8.4 CONFORMATIONAL BARRIERS TO BOND ROTATION

For a historical introduction to this subject, see Ref. 76.

While σ-bonds possess cylindrical symmetry, rotations about the bond axis are normally accompanied by small energy fluctuations, though these are usually too small to prevent rapid rotation. The term *conformer* refers to any specific geometry of a molecule obtained by rotation of single bonds and, in normal usage, is applied to the principal states, i.e. energy minima (or maxima). In ethane, for example, rotation of one methyl group relative to the other leads to an infinite number of conformations but attention naturally focuses on two, the (triply degenerate) maximum and minimum conformers, the 'eclipsed' (**34**) and 'staggered' (**35**), respectively, Fig. 8.2a.

Table 8.7   Regression constants for selected reactions obeying Eq. (8.18)[60]

Reaction	ρ	δ	f
Acid dissociation of ArCOOH	0·95	−0·392	1·469
Acid dissociation of ArOH	2·06 (ρ^-)	0·167	2·395
Esterification of ArCOOH with MeOH	−0·568	0·519	0·421
Esterification of ArCOOH with Ph_2CN_2	0·878	−0·193	0·718
Alkaline hydrolysis of ArCOOMe	1·93	0·64	1·1
Alkaline hydrolysis of $ArCH_2COOEt$	1·09	0·40	−0·32
Acetolysis of $ArCMe_2Cl$ (S_N1)	−4·54 (ρ^+)	0·609	−1·58
$ArNMe_2$ + MeI	−2·61	1·48	1·05
$ArC{\equiv}CCOOEt$ + tetracyclone (Diels–Alder)	0·716	0·314	0·295

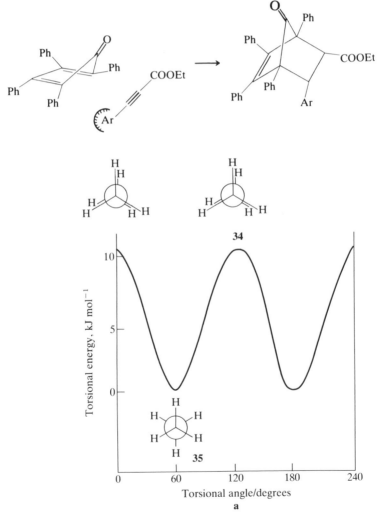

Fig. 8.2   Torsional energy as a function of rotational angle about the C–C bond: **a**, ethane; **b**, 1,2-dichloroethane; **c**, butane.

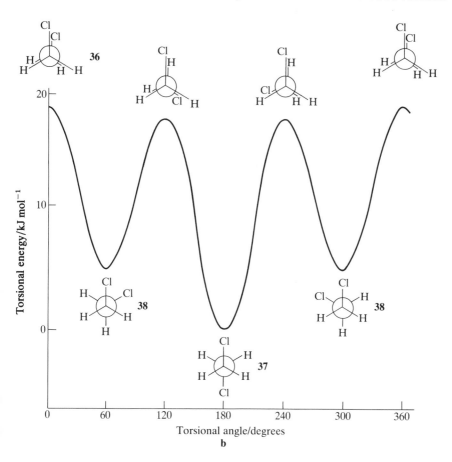

b

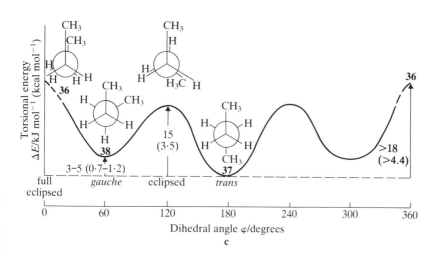

c

The study of energetics and origins of conformational preferences is known as *conformational analysis*[70-75]. Stable conformers (i.e. those lying in energy minima) are isomers and may be separately observable if methods of detection are sufficiently fast relative to the rate of interconversion. Low temperatures are therefore often used when spectroscopic detection of the separate species may be possible. Usually, bond rotations are so fast that all conformers are in dynamic equilibrium, the relative proportions being determined by their thermodynamic properties, and each with its own reactivity and activation parameters. This hints at a high degree of complexity which may be present but for which no reaction has yet been fully analyzed.

The conformational properties of two fundamental acyclic systems will be discussed, namely those of ethane and ethene and their analogues; also the special conformational properties of cyclic compounds, particularly cyclo-hexane, will be included. From these fundamental units larger molecules may be analyzed. However, preferred conformations and energy barriers to rotation are often highly dependent upon the state of the sample (gas, liquid, solid or solution).

8.4.1 Spectroscopic detection of individual conformers

Many of the following results to be discussed have been obtained from spectroscopic studies and it is appropriate at this point to examine the limitations of this group of methods. In general, a species in dynamic equilibrium may be directly observed by its characteristic absorption spectrum if the following conditions hold:

(a) The average lifetime of the species must be much greater than the (inverse) frequency of the radiation it is absorbing. For example:

	$1/v$/s	Minimum lifetime/s
NMR	10^{-8}	10^5
IR	10^{-12}	10^{-8}
UV	10^{-16}	10^{-12}

(b) Each conformer or species to be detected must possess different absorption characteristics and must be present within detection limits of concentration.

For the three spectroscopic methods indicated above, (b) is usually the limitation for IR and UV while condition (a) usually limits the usefulness of NMR but can often be overcome by working at sufficiently low tempera-tures. Some examples will illustrate the application of spectroscopy to conformational analysis.

Infrared

Furoate esters exist in conformations **A** and **B**:[77]

$$\nu_{CO} = 1732\ cm^{-1} \qquad \nu_{CO} = 1714\ cm^{-1}$$
$$\Delta G = 300\ J\ mol^{-1}\ (72\ cal\ mol^{-1})$$

Two carbonyl stretching bands are observed whose areas are a direct measure of K and its temperature dependence gives ΔG.

1,2-Dichloroethane shows two characteristic Raman bands assigned to the *gauche* and *trans* conformers both present in the gas phase:

$$\nu = 653\ cm^{-1} \qquad \nu = 753\ cm^{-1}$$
$$K = 1 \cdot 23\ (5°C)$$

Nuclear magnetic resonance

By far the majority of conformational studies are carried out by NMR, using ^{1}H, ^{13}C or other resonances. The full analyses of the spectra of interchanging conformers can be highly complex[78] but particularly simple interpretations may be made for the case in which conformational interchange is accompanied by the exchange of magnetic nuclei which are only slightly coupled, for example the axial and equatorial protons of cyclohexane or the two *N*-methyl groups in dimethylformamide, Section 8.5. Four regions of spectral type may be defined (Fig. 8.3):

(a) above coalescence temperature, T_c, a single peak is observed but broadened by exchange;
(b) at T_c two peaks are just discernible;
(c) below T_c two peaks are observed but broadened, the maxima separating as the temperature is lowered until
(d) two sharp peaks corresponding to infinitely slow exchange, as of a stable mixture of the two conformers, are obtained.

Rates of exchange can be obtained from the three regions by application of the appropriate equations indicated on the figure. Variable temperature studies then leads to ΔG for the conformational interchange.

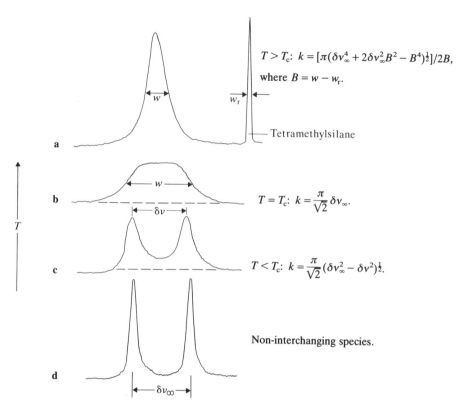

$$T > T_c: \ k = [\pi(\delta v_\infty^4 + 2\delta v_\infty^2 B^2 - B^4)^{\frac{1}{2}}]/2B,$$

where $B = w - w_r$.

Tetramethylsilane

$$T = T_c: \ k = \frac{\pi}{\sqrt{2}} \delta v_\infty.$$

$$T < T_c: \ k = \frac{\pi}{\sqrt{2}} (\delta v_\infty^2 - \delta v^2)^{\frac{1}{2}}.$$

Non-interchanging species.

Fig. 8.3 Characteristic NMR signals for a pair of interchanging non-coupled AB nuclei, e.g. methyl resonances of [structure]: **a**, above the coalescence temperature, T_c; **b**, at T_c; **c**, below T_c; **d**, non-interchanging species at very low temperature. Line widths, *w*, are measured at half height.

8.4.2 Acyclic compounds

Ethanes (rotations between two sp³-hybridized atoms)

Rotation of the two methyl groups in ethane about the central bond is accompanied by an alternation in potential energy for which the maxima correspond to the eclipsed conformation **34** and the minima to the staggered, **35**. The energy barrier amounts to $12 \cdot 3 \ kJ \ mol^{-1}$ ($2 \cdot 9 \ kcal \ mol^{-1}$) (obtained from spectroscopic information)[79]. This astonishingly high barrier may be thought of as originating from repulsions between hydrogen atoms on the adjacent carbons but in ethane itself these are spaced sufficiently far apart to account for only about $1 \ kJ \ mol^{-1}$ ($0 \cdot 24 \ kcal \ mol^{-1}$) of repulsive energy and the barrier appears to be inherent in MO properties. In 1,2-disubstituted ethanes, the threefold symmetry is removed and the resulting rotational potential-energy plot shows two different minima, each corresponding to a different staggered conformation (Fig. 8.2b, c); the most stable minimum corresponds to the *anti* (or *trans*) conformation, **37**, in which the substituents are at maximum separation, the other being the *gauche*, **38**, usually of higher energy due to non-bonding repulsions between substituent groups. Analysis of the conformational isomerism in more complicated acyclic systems may be made by considering each two-carbon component as a substituted ethane. The situation rapidly becomes complex; the number of staggered conformations possible for pentane, hexane and heptane are 3, 7 and 13. The most stable will be all-*anti*, **39**, each *gauche* interaction contributing some $3.4 \, \text{kJ mol}^{-1}$ $(0.81 \, \text{kcal mol}^{-1})$ of strain energy. Structures of the type **40**, though staggered, are of high energy because of the crowding at the positions marked * and may be excluded. In analyzing the proportions of each conformer in equilibrium, the entropies of each must also be taken into consideration. *Gauche* forms have higher symmetry numbers and there are more equivalent forms; this tends to favor such conformers statistically. It is the entropy content $(T\Delta S)$ which causes a change in the equilibrium proportions with temperature. The situation for pentane may be analyzed as follows:

Conformer	$\Delta\Delta H_f/$ kJ mol^{-1}	Symmetry number	Chiral forms	S	300 K		600 K	
					$\Delta\Delta G$	Equilibrium percentage	$\Delta\Delta G$	Equilibrium percentage
a,a	0	2	1	$-R \ln 2$	1·7	47	3·4	29
a,g	3·4	1	2	$+R \ln 2$	1·7	47	0	57
g,g	6·9	2	2	$R \ln 1 = 0$	6·9	6	6·9	14

In the solid state polymethylene compounds such as fatty acids (and their derivatives in biological membranes) and some stereoregular polymers often adopt the all-*anti* conformation as a result of favorable crystal packing. Substitution by large groups tends to increase rotational barriers as non-bonding repulsions between eclipsed atoms in the transition state become severe. The barrier to rotation of the *tert*-butyl group of **41** $(38 \, \text{kJ mol}^{-1} \, (9.08 \, \text{kcal mol}^{-1}))$ is sufficient to enable three separate methyl proton resonances to be observed. Oxygen and nitrogen analogues in which unshared pairs replace bonding pairs of electrons may be analyzed in a similar fashion to alkanes.

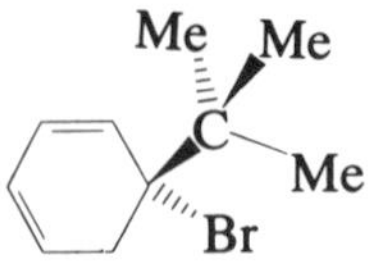

41 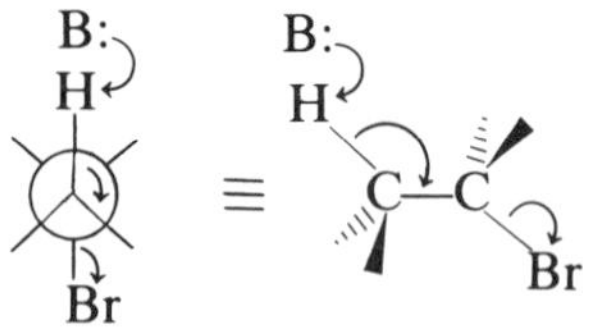   **42**

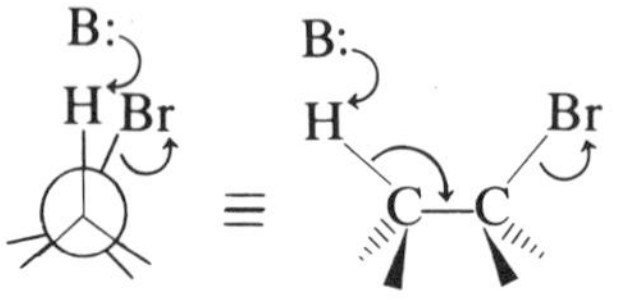

43   **44**

45a   **45b**

Ph D + MeCOOH   Ph Ph + MeCOOD

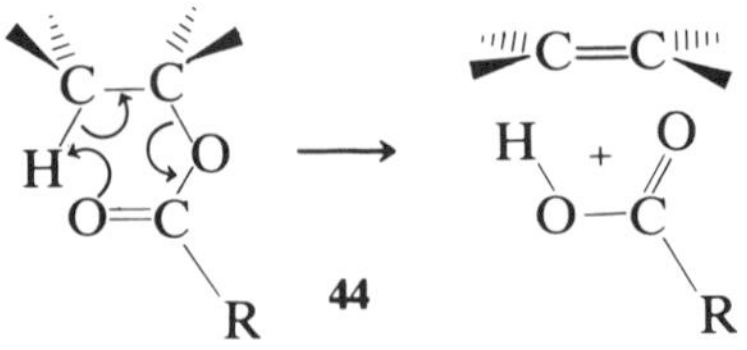

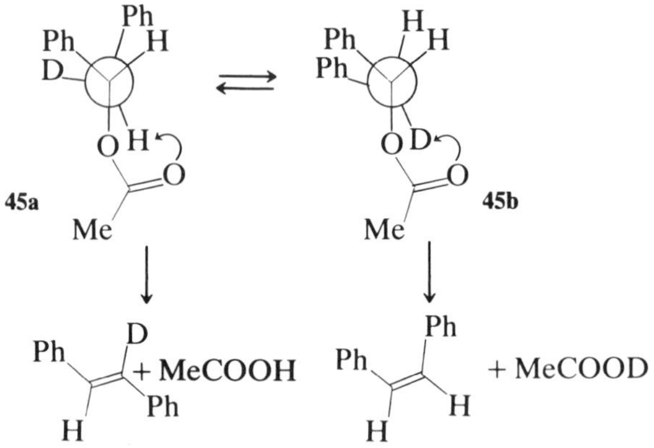

~100%   ~0%

Alkene-forming eliminations provide examples of reactions with strong conformational preferences. The base-promoted E2 reaction (Section 11.1) requires coplanarity of the β-H, the leaving group and the central carbons for molecular orbital reasons. The *anti-* geometry, **42**, is found to be energetically more favorable than the *syn-*, **43**, perhaps because it is staggered while the latter is eclipsed. Pyrolytic eliminations, for example of esters, must of necessity proceed via an eclipsed transition state, **44**, but conformational preferences may be discerned. For example, in the pyrolysis of **45**, two modes of *syn*-elimination are possible; that which avoids eclipsing of the two phenyl groups, **45a**, is energetically favored.

The propene model (rotations between sp² and sp³ centers)

The two-fold symmetry of the alkene carbon combined with the three-fold symmetry of the alkane gives rise to six conformers and it turns out that the favored one has the double bond eclipsed (**46**). 1,3-Dienes prefer the *anti*

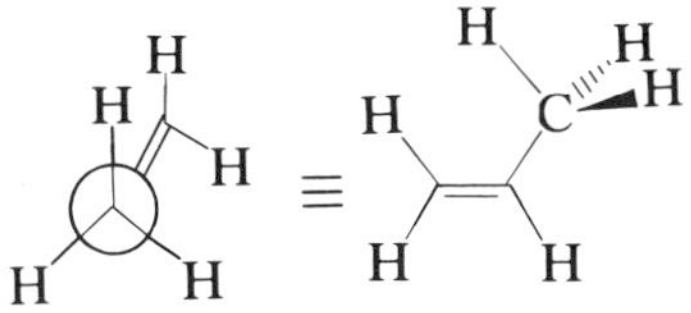

46

coplanar geometry, **47**, though a minor component believed to be nearly *syn* but slightly twisted appears to be present. Substituents at C2, C3 impart a preference for a nonplanar *gauche* conformation, **48**, and similarly for substituents at C1, C4 when *trans*. This is an unfavorable geometry for the Diels–Alder reaction, **47a**, in which the diene is required to approach a *syn* coplanar conformation.

Similar conformational isomerism occurs around C=O and C=N functions, examples being shown in structures **49a,b**. There is a preference for methyl to eclipse carbonyl oxygen and for two carbonyls to repel each other electrostatically.

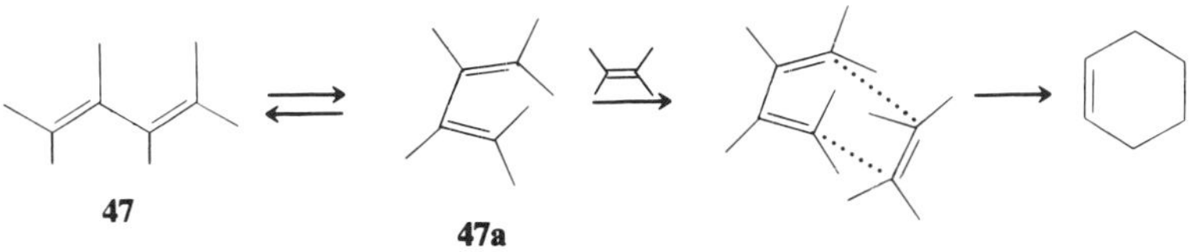

47   **47a**

47b

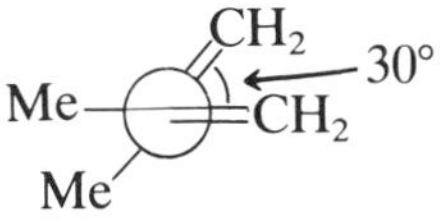

48 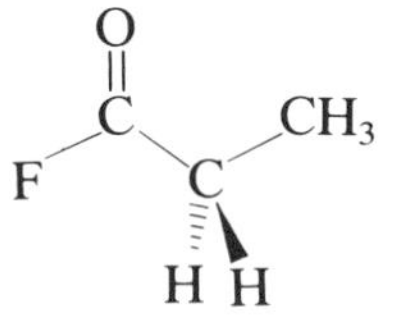   **49a** 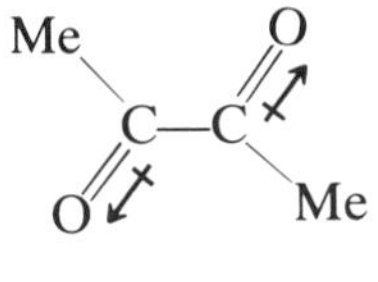   **49b**

8.4.3 Cyclic compounds

Cyclopropane necessarily possesses a rigid, planar ring with bond-angle strain (*I* strain) of about $100 \, \text{kJ mol}^{-1}$ ($24 \, \text{kcal mol}^{-1}$) and, in addition, eclipsing energy of $16 \, \text{kJ mol}^{-1}$ ($3 \cdot 8 \, \text{kcal mol}^{-1}$), making a total of $38 \cdot 5 \, \text{kJ mol}^{-1}$ ($9 \cdot 2 \, \text{kcal mol}^{-1}$) per CH_2, of strain energy. Cyclobutanes have perhaps less ring strain than three-membered rings ($110 \, \text{kJ mol}^{-1}$ ($26 \cdot 3 \, \text{kcal mol}^{-1}$) or $27 \cdot 6$ ($6 \cdot 6$) per CH_2) but possess some flexibility. This permits strain due to eclipsing of adjacent substituents to be relieved by bending of the ring by 30–33°, **23**, which differentiates the two positions at each carbon into an axial (a) and an equatorial (e) set. Substituents such as halogen prefer the equatorial positions which minimize 1,3-eclipsing inter-action. The barrier to inversion via the planar conformation is only $5\,\text{kJ mol}^{-1}$ ($1\cdot2\,\text{kcal mol}^{-1}$) for cyclobutane.

Cyclopentanes should have no bond-angle strain so the observed strain energy, $27\,\text{kJ mol}^{-1}$ ($6\cdot5\,\text{kcal mol}^{-1}$) ($5\cdot4$ ($1\cdot3$) per CH_2), must be due to eclipsing interactions, **8**. The ring is not completely planar; a conformation is adopted in which one atom lies out of the plane, the 'odd' atom rapidly alternating around the ring (*pseudo-rotation*; **24**). Ionization of halogen from a cyclopentyl halide reduces eclipsing strain: consequently these compounds react in solvolysis 20–100 times faster than cyclohexyl analogues for which no such driving force is present[80].

Cyclohexanes

The cyclohexane ring occupies a position of special importance in confor-mational studies, being a frequently encountered unit of structure among natural products and the smallest alicyclic system capable of existing in a non-planar strain-free conformation. Early workers showed that there were two possible non-planar arrangements of cyclohexane which satisfied the need for tetrahedral bond angles, and these became known as the *chair* (**2**, **50c**) and a *twist* form (**50g**), the former being more stable. Indeed, there are

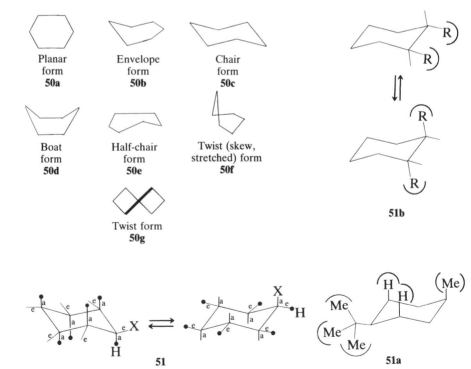

| Planar form **50a** | Envelope form **50b** | Chair form **50c** |

| Boat form **50d** | Half-chair form **50e** | Twist (skew, stretched) form **50f** |

Twist form **50g**

51b

51

51a

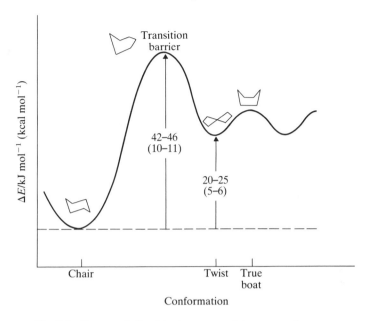

Fig. 8.4   Energy relationships between cyclohexane conformers.

two chair forms identical in the parent compound but chemically and physically distinct in substituted analogues (**51**) and a series of other conformers **50** through which the system passes during chair–chair interconversion (Fig. 8.4). The chair conformer contains two different sets of protons, also designated 'axial' and 'equatorial', each of which give separate NMR signals (when their rate of interchange is slowed down below −80°C).

Table 8.8   Conformational energies
(a) Energy barriers to rotation about methyl for

X	$\Delta G^{\ddagger}/kJ\,mol^{-1}\,(kcal\,mol^{-1})$
CH_3	12 (2·87)
CH_2CH_3	15 (3·58)
CMe_3	20 (4·78)
CH_2Cl	16 (3·82)
CH_2Br	15 (3·58)
CH_2I	13 (3·11)
CF_3	14 (3·35)
OH	5 (1·19)
SH	6 (1·43)
OMe	11 (2·62)
NH_2	8 (1·91)
NHMe	14 (3·34)
Ph	2 (0·48)

Table 8.8   (*Continued*)

(b) Free-energy differences (*A*-values), $-\Delta G = G(\text{eq}) - G(\text{ax})$, for mono-substituted cyclohexanes[a]

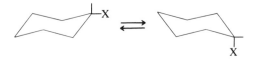

X	$-\Delta G^a/kJ\,mol^{-1}\,(kcal\,mol^{-1})$
Me	7·1 (1·69)
Et	7·5 (1·75)
isoPr	8·8 (2·10)
tert-Bu	23·8 (5·68)
Ph	12·9 (3·1)
COOR	4·6 (1·16)
CN	0·8 (0·19)
F	1·0 (0·24)
Cl	1·7 (0·41)
Br	2·1 (0·50)
I	1·7 (0·41)
OR	3.3 (0·78)
NH_2	7·5 (1·79)
HgBr	0 (0)

[a] A positive value in the Table means that the equatorial conformer is more stable than the axial.

(c) Free-energy differences, $G(\text{trans}) - G(\text{cis})$, for rotations about CO—N bonds of amides,

$$R-C\underset{N}{\overset{O}{\diagup}}\underset{R_2}{\overset{R_1}{\diagup}}$$

Amide			$\Delta G/kJ\,mol^{-1}\,(kcal\,mol^{-1})$
R	R_1	R_2	
H	H	Me	28 (6·7)
H	Me	Et	2·5 (0·6)
H	Me	*tert*-Bu	20 (4·8)
Me	H	Me	39 (9·3)
Me	H	Et	14 (3·3)
Me	Me	Et	2·5 (0·6)
Me	Et	Et	5·5 (1·3)

Energy barriers to rotation for amides

Amide	$\Delta G^{\ddagger}/kJ\,mol^{-1}\,(kcal\,mol^{-1})$
$HCONH_2$	105 (25)
$MeCONH_2$	97 (23)
$EtCONH_2$	85 (20)
$PhCONH_2$	77 (18)
$HCONEt_2$	98 (23)
$MeCONisoPr_2$	82 (19)
$HCSNMe_2$	127 (30)
$PhCSNMe_2$	88 (21)

A substituent on a cyclohexane ring can therefore occupy either an axial or an equatorial position. At normal temperatures the two conformers are rapidly interchanging, although a few examples of low-temperature isolation of the pure isomers have been reported[58]: at −150°C cyclohexyl chloride crystallizes as the axial isomer, with half-life 14 days for interconversion.

Normally, the axial conformer is less stable than the equatorial on account of steric repulsions with other axial substituents on the same side of the ring. This results in a free-energy difference (*A-value*), usually in the range 1–5 kJ mol^{-1} (0·2–1·2 kcal mol^{-1}); see Table 8.8. For *tert*-butylcyclohexane, however, the *A*-value is as large as 23, sufficient to maintain more than 99·9% of this compound in the equatorial form. Smaller groups in the same ring may then be forced to occupy an axial position; for instance, *cis*-1-methyl-4-*tert*-butylcyclohexane, **51a**, which must have one group axial, prefers it to be the smaller, methyl group. There is less conformational bias for compounds of the type **51b** since axial repulsions are compensated for by an additional *gauche* interaction. The conformational 'flip' which interchanges all axial with equatorial positions occurs via a transition state which is probably the *envelope* conformation, **50b**, leading to a series of *twist* (flexible) forms, each interchanging via the *boat* conformer, at an energy maximum (Fig. 8.5). The twist form may have been observed by very sudden cooling of cyclohexane vapor into a solid matrix[81]. The effects of conformation upon chemical reactivity can best be judged when the cyclohexane ring is so constructed that it is unable to invert. One method of achieving this is to place a *tert*-butyl group in the ring (conformational biasing); the other is to use fused-ring systems.

8.5 ROTATIONS ABOUT PARTIAL DOUBLE BONDS

Uncatalyzed rotation about olefinic double bonds does not normally occur at temperatures below about 500°C since the energy cost is the breaking of the π-bond, some 400 kJ mol^{-1} (96 kcal mol^{-1}). There are, however, compounds whose structure is best described in terms of a partial double bond for which the energy of activation for rotation is intermediate between that for formal single and double bonds[62,82,83]. Amides have been most extensively studied and can have barriers to rotation about the C–N bond of between 50 and 100 kJ mol^{-1} (12 to 25 kcal mol^{-1}) **52** (Table 8.8). This means that ^{1}H—NMR spectra of *N,N*-dialkyl amides will often show separate resonances for the two alkyl groups at room temperature (for example, Fig. 8.3) which will eventually coalesce at higher temperatures as a time-averaged spectrum is obtained. Smaller barriers to rotation exist in the analogous C–O bonds of esters which show a strong preference for the *trans* geometry, and also in carbocations, **53**, and carbanions, **54**, which have extended conjugation. In all cases energy minima correspond to those conformers which permit the conjugated system to be planar.

313

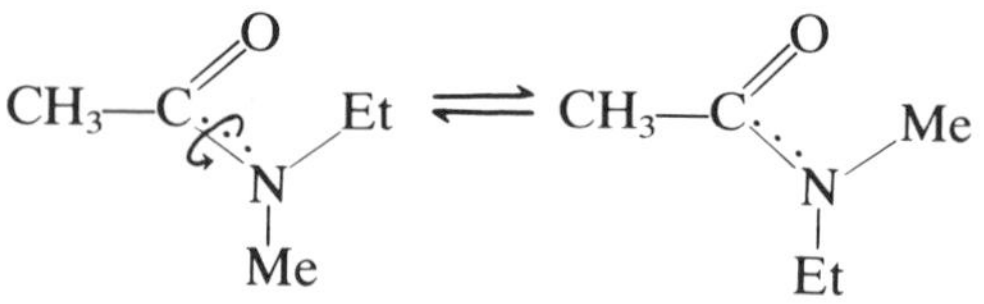

52

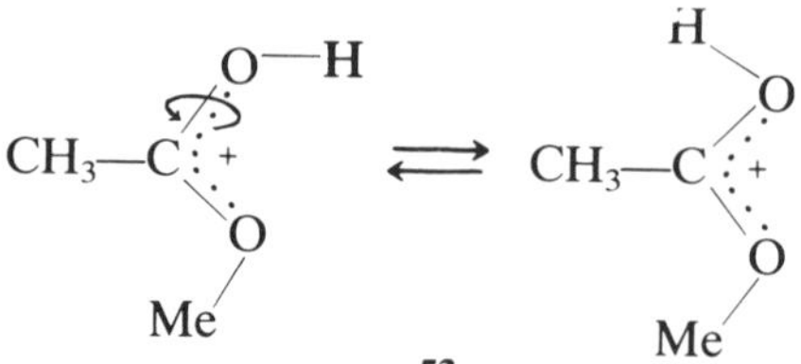

53

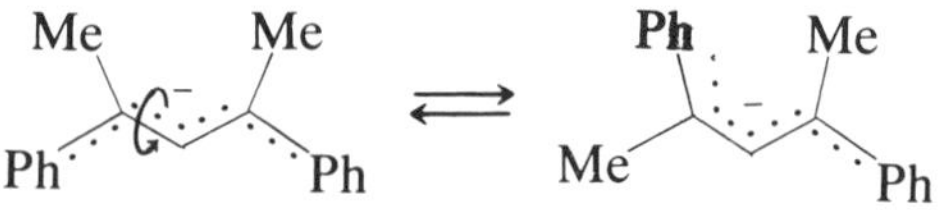

54

8.5.1 Inversion at Group V elements

A substituent on trivalent nitrogen is conformationally mobile and may invert via a planar transition state, **55**. In six-membered rings (piperidines), there is a slight preference for the favorable equatorial position to be occupied by hydrogen rather than the unshared pair and a larger preference for alkyl to be equatorial. Inversion at nitrogen is normally so much faster than chair–chair interconversion that the *N*-alkyl group remains equatorial. Inversion at nitrogen in aziridines, **56**, is much slower, with activation barriers around $30\,\text{kJ mol}^{-1}$ ($7\,\text{kcal mol}^{-1}$), perhaps a result of greater C—N—C bond-angle strain in the planar transition state[54]. Pyramidal molecules of other Group V elements have larger inversion barriers — phosphines, *ca* $125\,\text{kJ mol}^{-1}$ ($29\,\text{kcal mol}^{-1}$), arsines, *ca* 175 (42) — as also have sulfoxides, **57**, which are resolvable into stable enantiomers[85,86].

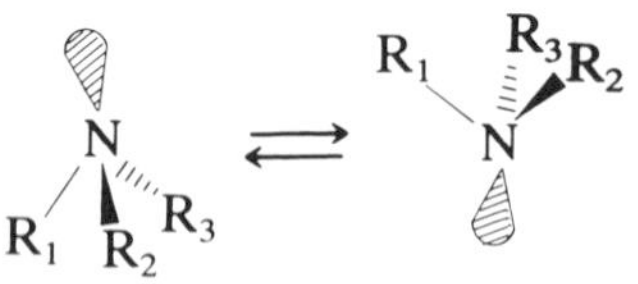

55

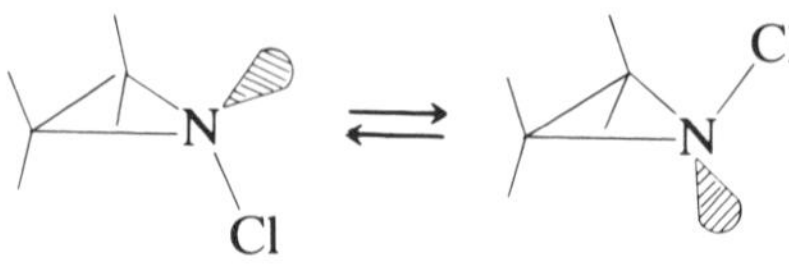

56

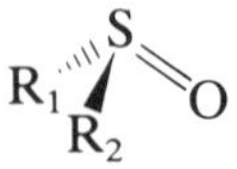

57

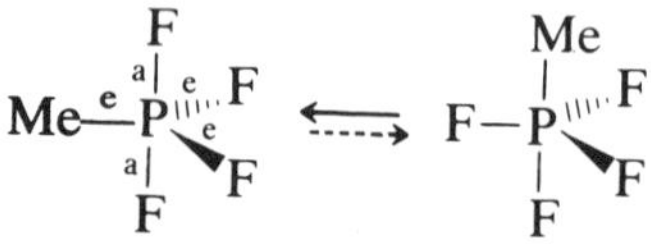

58

Pentavalent phosphorus compounds have two different bonding sites, the apical and the equatorial. Isomerism involving ligand interchange between the two (pseudorotation) is observable by NMR spectroscopy. In **58**, methyl preferentially occupies an equatorial site.

8.6 CHEMICAL CONSEQUENCES OF CONFORMATIONAL ISOMERISM—THE WINSTEIN–HOLNESS–CURTIN–HAMMETT PRINCIPLE

Each conformer of a compound is an individual chemical entity and must be expected to undergo reactions at its own characteristic rates. What is the relationship between these individual rates and that observed for the conformationally-mobile compound? Since conformational exchange is normally very fast, it is not possible to isolate each conformer and investigate its chemistry separately, and it would therefore be expected that observed rates are some weighted mean of those of the individual conformers. The formal analysis of this problem results in the Winstein–Holness equation[76,87,88], Fig. 8.5.

For a system:

$$P_1 \xleftarrow{\ k_1\ } C_1 \underset{k_{21}}{\overset{k_{12}}{\rightleftharpoons}} C_2 \xrightarrow{\ k_2\ } P_2, \qquad (8.19)$$

in which two conformers C_1 and C_2 in rapid equilibrium separately react at rates k_1 and k_2 to give products P_1 and P_2 (which might or might not be

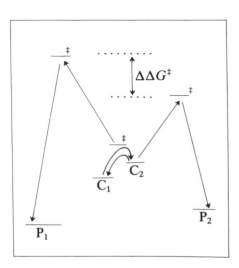

Fig. 8.5   Energy diagram showing a case in which the less stable conformer, C_2, leads to the major product, P_2. The pathway to P_1 requires additional activation energy of $\Delta\Delta G^{\ddagger}$.

identical), and assuming for the present k_1, k_2 to be unimolecular or pseudo-unimolecular, then

$$\frac{\partial[P_1]}{\partial t} + \frac{\partial[P_2]}{\partial t} = k_1[C_1] + k_2[C_2].$$ (8.20)

The Winstein–Holness equation defines the total rate of product formation,

$$\text{Rate} = \frac{\partial[P_1]}{\partial t} + \frac{\partial[P_2]}{\partial t} = k_{\text{obs}}\{[C_1] + [C_2]\}$$ (8.21)

or

$$k_1[C_1] + k_2[C_2] = k_{\text{obs}}\{[C_1] + [C_2]\}.$$

Hence

$$k_{\text{obs}} = k_1 \frac{[C_1]}{[C_1] + [C_2]} + k_2 \frac{[C_2]}{[C_1] + [C_2]}.$$ (8.22)

The concentration terms are respectively the mole fractions, x_1, x_2 of the two conformers, x_1, x_2; hence

$$k_{\text{obs}} = k_1 x_1 + k_2 x_2,$$ (8.23)

a result which can obviously be extended to a situation in which more than two conformers are present. Alternatively,

$$K = \frac{k_1 - k_{\text{obs}}}{k_{\text{obs}} - k_2},$$ (8.24)

a general expression which holds for any measured variables, e.g. for NMR chemical shifts or coupling constants on replacing k by δ or J. The measured rate constant, k_{obs}, is therefore a composite quantity, a parameter connecting a series of coupled reactions. From Eq. (8.24) it is clear that, in order to obtain the conformational equilibrium constant, $K = x_2/x_1$, from rate data, it is necessary to know the rates of reaction of the individual conformers. The best approximation which can be made to this intractable demand is to design compounds which are of fixed conformation and resemble the desired conformers as closely as possible. Rates of these models then serve as approximate values of k_1, k_2.

The original workers studied reactions at a cyclohexane ring in which the functional group X was in equilibrium between axial and equatorial locations (k_{obs})[66]. The *cis*- and *trans*-4-*tert*-Bu analogues were used as models for the X-axial and X-equatorial conformers since the bulky *tert*-butyl group was assumed to remain equatorial. Some results are shown in Table 8.9. The limitation of this kinetic method of conformational analysis is the degree to which the chemical behavior of the model compounds resembles that of the actual conformers which they model. There is evidence that the *tert*-butyl group can distort the ring and affect

Table 8.9　Rate data of conformationally mobile systems and their rigid analogues[87-89]

Reaction	k_{rel}		
	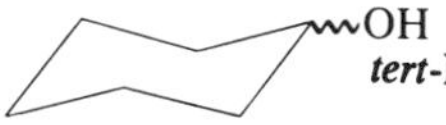	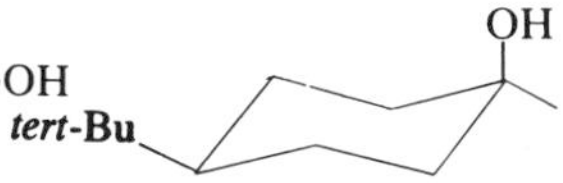 *cis*	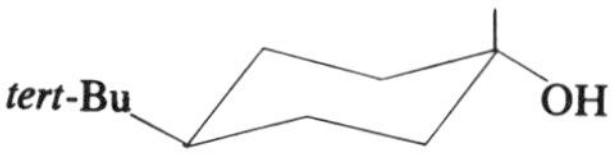 *trans*
Hydrolysis of acid phthalates, 50°C	1·02	0·11	1·00
Oxidation by CrO₃/AcOH, 25°C	1·24	2·97	1·00
Solvolysis of ROtos, HCOOH	1·23	3·58	1·00
Dehydration at 150°C	1·16	3·5	1·00
Acylation with Ac₂O/pyridine, 25°C	0·271	0·786	1·00

Acid-catalyzed esterification

Compound	$10^4 k/M\,s^{-1}$	*Compound*	$10^4 k/M\,s^{-1}$
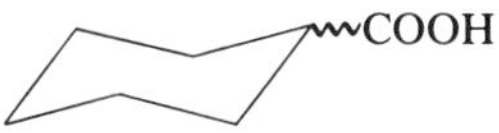	160		
	9·17[a]	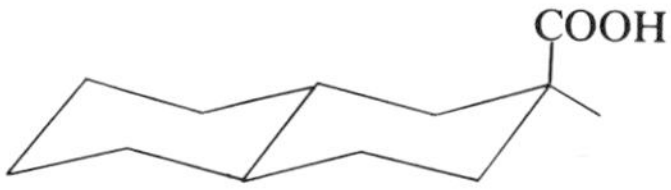	7·20
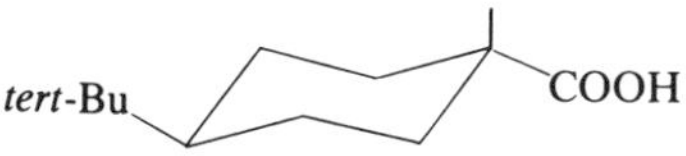	183	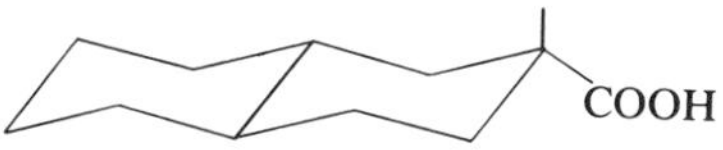	184

[a] From a comparison of this rate with that of the rigid 2-axial decalincarboxylic acid, it was concluded that the former is higher than that calculated for the pure axial conformer by 25%. This could be caused by the presence of just 1% of the more reactive conformer with an equatorial carboxylic acid group (and axial *tert*-butyl).

reactivity at other sites; it may also perturb solvation around the reaction site. Furthermore, large functional groups such as the solvated carboxylate ion may compete with the *tert*-butyl group for the equatorial location, rendering the primary assumption invalid[89-91]. Values of K obtained by this method should therefore be accepted with caution.

The Curtin–Hammett principle[76,92,93] focuses attention on the product

317

ratio in the kinetic scheme (8.19) considered above. Since

$$[P_1] = k_1[C_1]$$

and

$$[P_2] = k_2[C_2]$$

the product ratio may be expressed:

$$\frac{\partial[P_2]/\partial t}{\partial[P_1]/\partial t} = \frac{\partial[P_2]}{\partial[P_1]} = \frac{k_2[C_2]}{k_1[C_1]} \tag{8.25}$$

and

$$\int \partial[P_2] = \frac{k_2}{k_1} \int [C_2]/[C_1]\, \partial[P_1]. \tag{8.26}$$

Provided that the ratio of conformers, $[C_2]/[C_1]$, is constant (which will be the case if conformational equilibrium is much faster than the reaction rate),

$$\int \partial[P_2] = \frac{Kk_2}{k_1} \int \partial[P_1] \tag{8.27}$$

or

$$\frac{[P_2]}{[P_1]} = K\frac{k_2}{k_1}. \tag{8.28}$$

The Curtin–Hammett principle declares that the product ratio is dependent upon the conformational equilibrium constant and the rate ratio of the individual conformers. The point to note here is that the product ratio, an experimental quantity, will not in general be a reflexion of the conformer population of the reagents. It is not at all unlikely that products will be almost wholly derived from the minority conformer (see Fig. 8.5). The product ratio is controlled by the difference in Gibbs free energies of the two transition states, $\Delta\Delta G^{\ddagger}$. Examples of this principle follow.

Piperidines undergo pyramidal inversion at nitrogen, the unshared pair being either axial or equatorial. Oxidation gives the *N*-oxide, which may be *cis*- or *trans*- in relation to a 4-alkyl group.

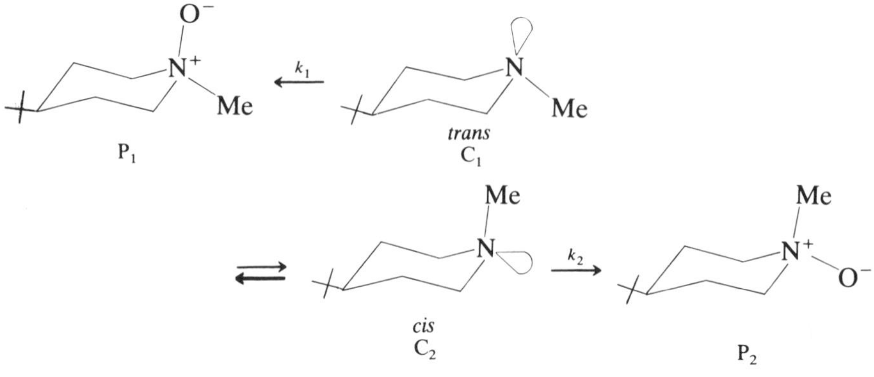

$$[P_1]/[P_2] = 19; \qquad K = [C_2]/[C_1] = 0.095.$$

Hence

$$k_2/k_1 = \frac{1}{19} \times \frac{1}{0 \cdot 095} = \frac{1}{2}$$

within the limits of experimental error. That is, for the experimental results to be consistent, the *cis* conformer with an equatorial unshared pair must be oxidized at a rate about half as fast as the *trans* conformer.

The methylation of **59** has been thoroughly studied by Seeman[98]. The *N*-methyl group exists *syn*- or *anti*- to the phenyl ring with a preference for the latter which increases with the size of R. Increasing the size of R forces more and more of the compound into the less reactive conformation (unshared pair *syn* to the phenyl ring) and the observed rate diminishes markedly. Actual rates of the more reactive conformer **59a** are not greatly affected by R (falling by a factor of 5, R = H to R = *tert*-Bu) while those of **59b** are much more sensitive (falling by a factor of 70). The product changes from being principally derived from **59b** when R = H to that from **59a** when R = *tert*-Bu (Table 8.10).

The main difficulty in applying the Curtin–Hammett relationship lies in obtaining a reliable value for the proportions of the conformers present under particular conditions, as they are known to be affected by both solvent and temperature[6,82,94,95].

Table 8.10 Conformational effects on the methylation of *N*-methyl-2-arylpyrrolidines

Compound	k_{obs} (*rel*)	K	Product ratio (59d/59c)	k_{syn} (59a)	k_{anti} (59b)	k_{syn}/k_{anti}
59: R = H	24	>17	1·7	200	20	10
R = Me	6·1	>30	1·4	98	4·6	21
R = Et	4·9	>30	1·3	81	3·5	23
R = isoPr	4·2	>30	1·3	69	3·0	23
R = *tert*-Bu	1	>40	0·28	40	0·28	143

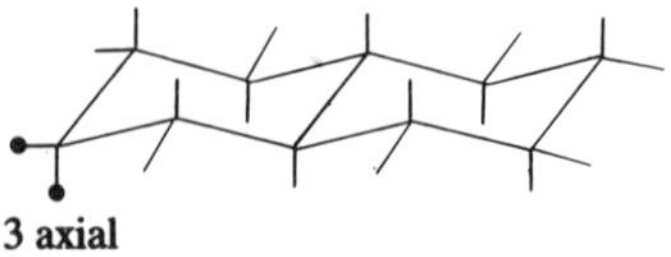

trans-**60** 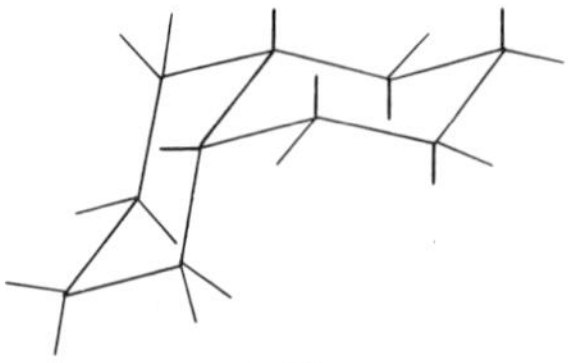   cis-**60**

Annelation of two cyclohexane rings can occur in two isomeric all-chair forms, for example *cis*- and *trans*-decalins, **60**. The *cis* isomer is conformationally rather mobile but the *trans* compound is rigid and neither ring is able to undergo a conformational 'flip'. Monosubstitution at peripheral positions may then be equatorial or axial, the two isomers being non-interchanging. This fixed stereochemistry makes *trans*-fused cyclohexane systems which includes steroids such as **61** — a very suitable method for investigating the conformational dependence of reactions, i.e. the interaction between geometrical requirements of a reaction and enforced geometry of the reagent[68].

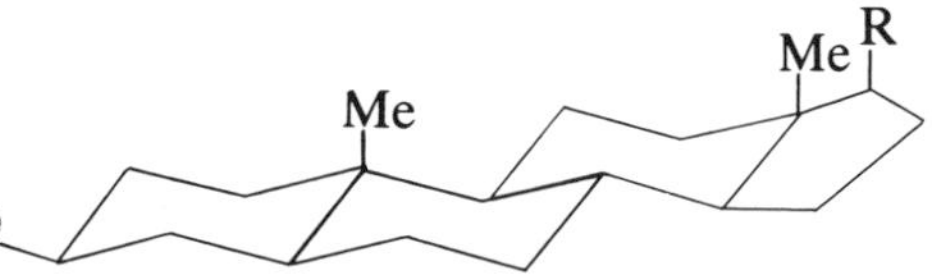

61 Cholestanol
R = isoPr(CH$_2$)$_3$CHMe—

Nucleophilic displacement at saturated carbon by the S$_N$2 route requires attack by the nucleophile at the side remote from the leaving group (Section 10.1). Consequently, an axial group is relatively easily displaced, **62a**, while under the same conditions the equatorial isomer **62b**, is inert. This results from inaccessibility of the rear face in the equatorial isomer due to steric hindrance within the structure of the secondary halide and also to axial hydrogens on the same side of the ring which impede the approach of the

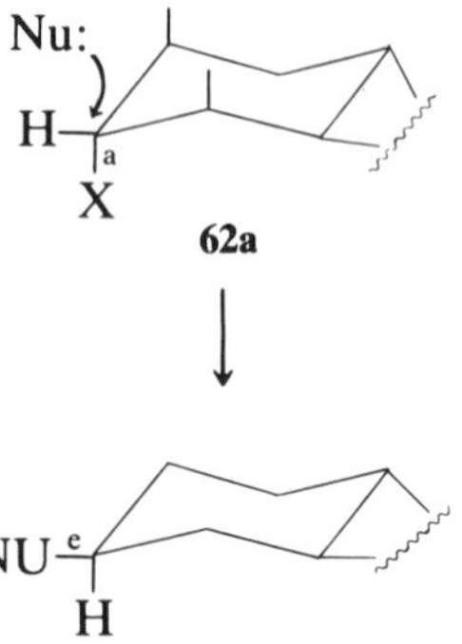

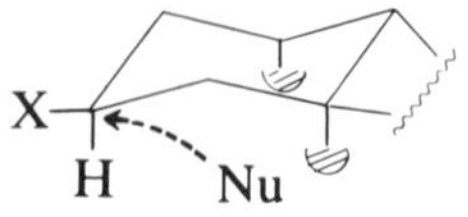

62a

62b

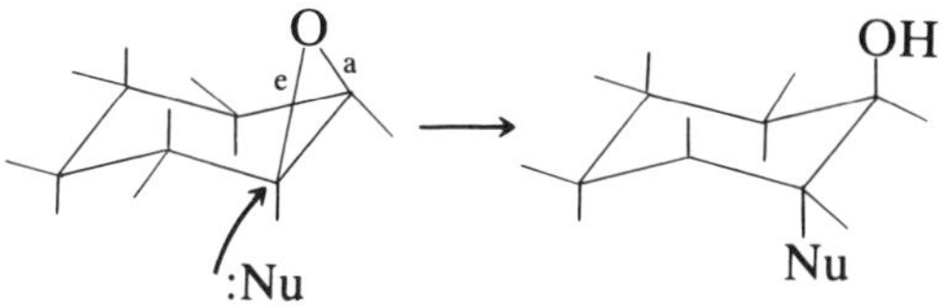

63

reagent. However, epoxides of cyclohexane systems (which have one C–O bond 'axial' and one 'equatorial'), **63**, undergo nucleophilic ring-opening such as to give the diaxial, rather than the diequatorial, product. This is because additional ring strain forces epoxycyclohexane (like cyclohexene) into a half-chair conformation from which displacement of the equatorial group is least hindered[67]. E2 eliminations occur readily only when both nucleophilic leaving group and β-hydrogen are axial since they then have the *anti* periplanar geometry required (see **42**).

Oxidations of axial secondary alcoholic groups by dichromate are faster than the corresponding equatorial isomers, **64**. The rate-determining step is the removal of the α-proton from an intermediate chromate ester by a base which can more readily approach an exposed equatorial position, than one screened by the other axial substituents[96] (Table 8.11).

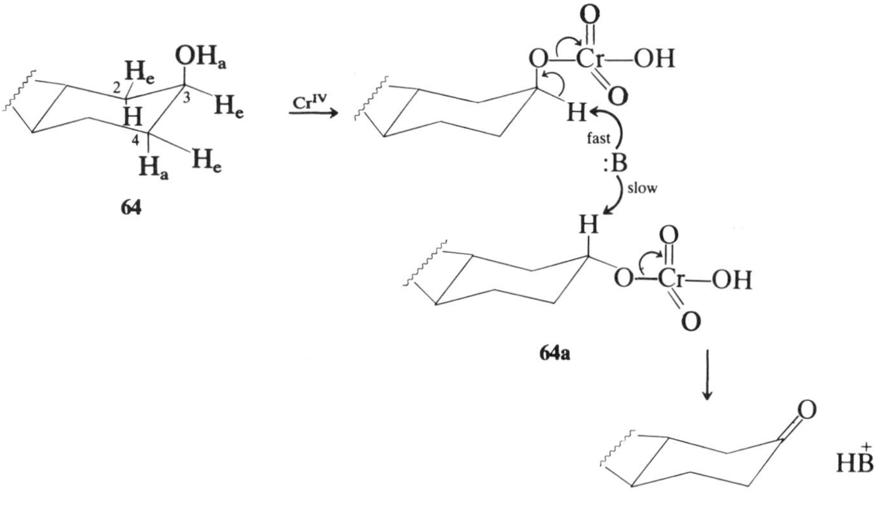

64

64a

Table 8.11   Relative oxidation rates of hydroxydecalins **64**

64: Equatorial position of OH	10^3k	**64a:** Axial position of OH	10^3k
trans-2	54	cis-2	250
cis-3	5	trans-2	30
trans-4	4	cis-4	13

The ease of formation of ketals of cyclohexanediols is in the order *cis*-1,2 (axial–equatorial), **65** > *cis*-1,3 (axial–axial), **66**, while *trans*-1,2 diols (axial–axial, **67**, or equatorial–equatorial, **68**) do not react. The reasons for this are steric; compared with **65a**, the reaction of **66** experiences some hindrance from the other axial substituent. The lack of reaction of **68** is obvious but is more subtle. The relative dispositions of hydroxyl groups in **65** and **68** are identical. However, on forming the five-membered ketal ring, the cyclohexane must distort such that the two C–O bonds become more nearly parallel. It is much easier to twist **65** in this sense, which tends towards a half-chair form, than to twist **68** in the required direction which tends towards a more puckered ring with axial substituents pointing more towards each other (**69**). These principles apply equally to heterocyclic rings[66], including piperidines[97] and the pyranose forms of carbohydrates; consequently, galactose readily forms a diacetonide, **70,** while glucose (all-equatorial sustituents) will only react with acetone after isomerization to the furanose form **71**, which possesses a *cis*-diol group in the side chain.

Asymmetric synthesis

When a prochiral group such as carbonyl is flanked by a chiral center (only one configuration of which is present), it is likely to exist in some preferred conformation with respect to the remainder of the molecule (e.g. **72a**; L, M and S are large, medium and small groups, respectively) and the attack of a Grignard reagent will also be preferred from one side, that which offers least steric hindrance to approach. Various rules have been devised to predict major products formed in particular reactions e.g. by Barton, Cram and Prelog[61]. The products of such reactions have an enantiomeric excess in the newly created chiral center which can range from the barely detectable to

323

near 100%. Needless to say, reactions in the latter category are highly valuable for the synthesis of natural products. In terms of the Curtin–Hammett principle, Eq. (8.28), a high enantiomeric excess will result from the appropriate precursors when K and k_1/k_2 have extreme values; at least one of these terms must be far from unity. The following are examples of 'asymmetric synthesis'.

The addition of methyl magnesium bromide to the chiral ketone **72b** leads to a 9:1 excess of one diastereoisomeric alcohol[100].

The addition of di-isocamphenylborane, **72c**, (made from addition of diborane to (R)-α-pinene—Section 12.2.1) to *cis*-but-2-ene occurs preferentially from one side with an enantiometric excess of 98%.

This leads to an oxidation (completing the hydroboration sequence) to almost pure (R)-butan-2-ol[100]. Allylic alcohols may be epoxidized by a combination of *tert*-butyl hydroperoxide with titanium isopropoxide and (+)-diethyl tartrate to give a 97% excess of one enantiomer (Sharpless' method), **72d**[101]. Only one enantiomer of this particular substrate, which additionally bears a chiral center, is attacked at all. The transition state of such a reaction is complicated but it seems clear that the reagents coordinate with the alcohol function and direct attack of the peroxide to one side of the double bond.

PROBLEMS

1 Biphenyls are chiral due to restricted rotation about the central bond (*atropoisomerism*) and may, when appropriate substituents are present, be resolved and their racemization kinetics examined. Compounds **1–4** are progressively more stable (Ref. *3*, p. 523):

	1	2; R = Me	3	4
$t_{\frac{1}{2}}$/min (120°C):	5	15	70	stable

(a) Sketch or model these molecules in their chiral conformations and confirm that enantiomers exist and are interchanged by rotation at the central bond.

(b) Using covalent bond lengths and atomic radii (Table 1.3) show that the degree of overlap of *ortho* groups in the planar conformations drawn above correlates with stability towards racemization.

(c) *Meta* groups, as in compound **2**, increase stability towards racemization ('buttressing effect'); give an explanation for this.

(d) Consider what relationship might exist between steric constants of the *ortho* groups and racemization rates. Should it be possible to use this approach to design optically-stable biphenyls?

2 Which of the structures **b**–**e** are conformers of **a**?

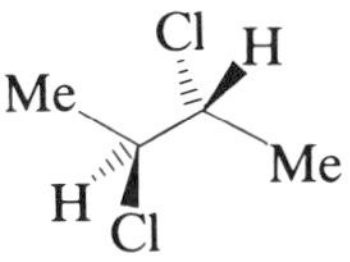

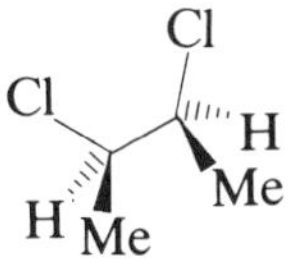

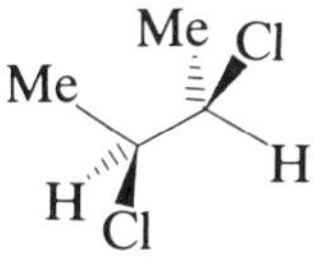

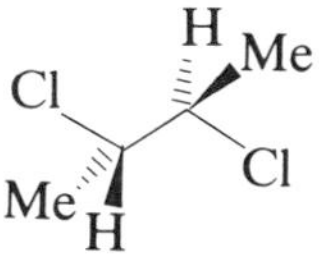

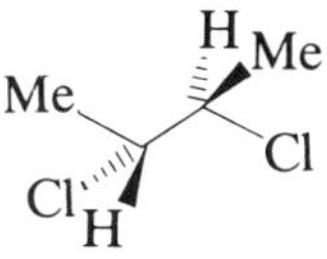

3 What stereochemical relationship exists between the following pairs of isomers?

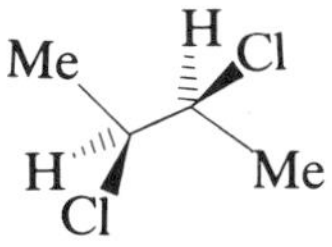

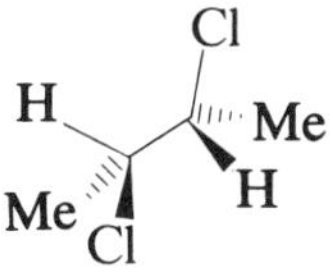

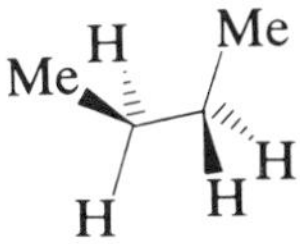

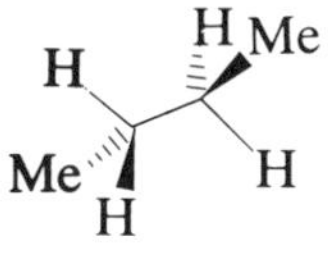

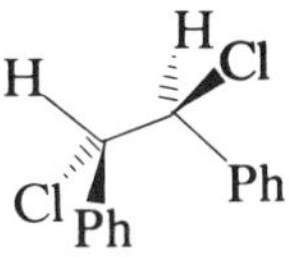

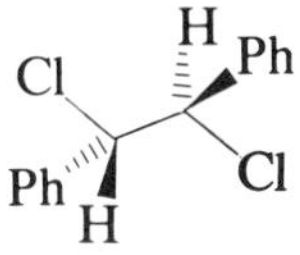

4 Draw the energy profile (schematically) for the rotation about the C2–C3 bond in (*S*)-2-iodobutane and sketch the principle conformers.

Steric and conformational properties

5 Predict the lowest energy conformers of the following;
(a) $ClCH_2CH_2I$;
(b) $MeCOOCH_2CH_2\overset{+}{S}Me_2$;
(c) $FCH_2CH(OH)Me$;
(d) $CH_3CO\cdot OCH_3$;
(e) $H\cdot COCH_2F$;
(f) *trans*-1,4-dibromocyclohexane.

6 (a) Of two stereoisomers of 3-bromocyclohexanecarboxylic acid, one eliminates Br^- readily when dissolved in aqueous ethanol while the other is inert. Identify the isomers.

(b) There are two stereoisomers of bicyclo[2,2,2]oct-2-ene-5-carboxylic acid, $C_9H_{12}O_2$. They give different products on treatment with bromine in water, thus:

$$\text{Isomer A: } C_9H_{12}O_2 + Br_2 \rightarrow C_9H_{12}O_2Br_2;$$
$$\text{Isomer B: } C_9H_{12}O_2 + Br_2 \rightarrow C_9H_{11}O_2Br + H^+, Br^-.$$

Identify the two isomers.

(c) There are eight stereoisomers of 1,2,3,4,5,6-hexachlorocyclohexane. Verify this statement and sketch the structures of the compounds. Of these, on treatment with methoxide/methanol, two isomers eliminate 3HCl readily; four eliminate 2HCl and one eliminates 1HCl. The remaining isomer is inert. Identify as far as possible the structures you have drawn with this reactivity pattern.

7 Interpret the accompanying ^{1}H-NMR spectra of lithium allyl,

$$Li^+H_2\overset{\cdots\cdots\overline{\cdots}\cdots}{C-CH-CH_2}$$

at 37° and −87°C.

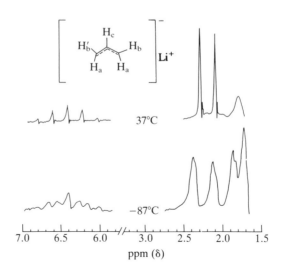

REFERENCES

1 E. L. Eliel, *Stereochemistry of Carbon Compounds*, McGraw-Hill, New York, 1962; A. Bassindale, *The Third Dimension in Organic Chemistry*, Wiley, Chichester, 1984.

2 E. Dreiding, *Angew. Chem. Int. Ed.*, **3**, 516 (1964).

3 M. S. Newman (Ed.), *Steric Effects in Organic Chemistry*, Wiley, New York, 1956.

4 D. F. DeTar and C. S. Tempas, *J. Org. Chem.*, **41**, 2009 (1976).

5 J. F. Liebmann and A. Greenberg, *Chem. Rev.*, **76**, 311 (1976).

6 A. Greenberg and J. F. Liebmann, *Strained Organic Molecules*, Academic Press, New York, 1978.

7 W. F. Maier and P. v. R. Schleyer, *J. Amer. Chem. Soc.*, **103**, 1891 (1981).

8 T. Tidwell, *Tetrahedron*, **34**, 1855 (1978).

9 A. Mugnoli and M. Simonetta, *J. Chem. Soc., Perkin II*, 1831 (1976).

10 P. T. Arnold, V. J. Wiehers and R. M. Dodson, *J. Amer. Chem. Soc.*, **74**, 368 (1953).

11 S. Singh and C. N. R. Rao, *J. Amer. Chem. Soc.*, **88**, 2142 (1966).

12 G. R. Stevenson, J. D. Kakosinki and Y. T. Chang, *J. Amer. Chem. Soc.*, **103**, 6558 (1981).

13 H. C. Brown, M. D. Taylor, M. Gerstein and H. Bartholomay, *J. Amer. Chem. Soc.*, **66**, 431 (1944).

14 H. C. Brown, M. Gerstein and H. Bartholomay, *J. Amer. Chem. Soc.*, **69**, 1332 (1947).

15 H. C. Brown, Science, **103**, 385 (1946).

16 W. Reeves, E. L. McCaffery and T. E. Kaiser, *J. Amer. Chem. Soc.*, **76**, 2280 (1954).

17 H. C. Brown, *J. Amer. Chem. Soc.*, **67**, 374, 378 (1945).

18 J. Sauer, *Angew. Chem., Int. Ed.*, **92**, 773 (1980).

19 J. G. Martin and R. K. Hill, *Chem. Rev.*, **61**, 537 (1961).

20 D. J. Cram and F. A. ElHafez, *J. Amer. Chem. Soc.*, **74**, 5828 (1952).

21 S. Terashima, N. Tanno and K. Koga, *Chem. Comm.* 1026 (1980).

22 H. C. Brown, J. H. Kawakomi and K. T. Liu, *J. Amer. Chem. Soc.*, **92**, 5536 (1970).

23 H. C. Brown, R. S. Fletcher and R. B. Johannesen, *J. Amer. Chem. Soc.*, **73**, 212 (1951).

24 A. Greenberg and J. F. Liebmann, *Chemistry of Strained Rings*, Academic Press, New York, 1978.

25 B. Capon, *Quart. Rev.*, **18**, 45 (1964).

26 J. Bredt, H. Thouet and J. Schmitz, *Ann.*, **431**, 1 (1924).

27 K. J. Shea, *Tetrahedron*, **36**, 1683 (1980).

28 K. B. Becker, *Tetrahedron*, **36**, 1717 (1980).

29 W. F. Maier and P. v. R. Schleyer, *J. Amer. Chem. Soc.*, **103**, 1891 (1981).

30 W. C. Spitzer and G. W. Wheland, *J. Amer. Chem. Soc.*, **62**, 2995 (1940).

31 R. Adams and H. C. Yuan, *Chem. Rev.*, **12**, 261 (1933).

32 F. H. Westheimer and J. E. Meyer, *J. Chem. Phys.*, **14**, 733 (1946).

33 P. D. Bartlett and M. Stiles, *J. Amer. Chem. Soc.*, **77**, 2806 (1955).

34 H. C. Brown and R. S. Fletcher, *J. Amer. Chem. Soc.*, **71**, 1845 (1949).

35 J. S. Lomas, P. K. Luong and J.-E. Dubois, *J. Amer. Chem. Soc.*, **99**, 5478 (1977).

36 R. C. Badgerand and J. L. Fry, *J. Amer. Chem. Soc.*, **101**, 1680 (1979); J. S. Lomas, P. K. Luong and J.-E. Dubois, *J. Org. Chem.*, **44**, 1647 (1979).

37 P. Mueller and J. Blanc, *Tetrahedron Lett.*, **22**, 715 (1981).

38 V. Baliah and V. M. Kanagasathapathy, *Tetrahedron*, **34**, 3611 (1978).

39 E. M. Arnett, J. M. Ballinger and J. C. Sanda, *J. Amer. Chem. Soc.*, **87**, 2050 (1965).

40 M. Hirota, T. Sekiya, K. Abe, H. Tashiro and M. Karatsu, *Tetrahedron*, **39**, 3091 (1983); M. Simonetta, *Acc. Chem. Res.*, **7**, 345 (1974).

41 S. Patai and Z. Rappoport (Eds), *Chemistry of the Halides, Pseudo-halides and Azides*, Vol. 1, Wiley, Chichester, 1983, p. 1.

42 H. D. Beckhaus, *Angew. Chem., Int. Ed.*, **17**, 592 (1978).

43 R. W. Taft, *J. Amer. Chem. Soc.*, **74**, 3120 (1952).

Steric and conformational properties

44 J. A. McPhee, A. Panaye and J.-E. Dubois, *Tetrahedron*, **34**, 3553 (1978).
45 A. Panaye, J. A. McPhee and J.-E. Dubois, *Tetrahedron*, **36**, 759 (1980); *Tetrahedron Lett.*, 3293, 3297 (1978); 3485 (1980).
46 J.-E. Dubois, J. A. McPhee and A. Panaye, *Tetrahedron*, **36**, 919 (1980).
47 P. P. Mager, *Sci. Pharm.*, **49**, 1 (1981).
48 H. Mager and A. Barth, *Experientia*, **38**, 423 (1982).
49 K. Hancock, E. A. Meyers and R. J. Yater, *J. Amer. Chem. Soc.*, **83**, 4211 (1961).
50 T. Fujita, C. Takayama and M. Nakayima, *J. Org. Chem.*, **38**, 1623 (1976).
51 M. Charton, *J. Org. Chem.*, **43**, 3995 (1978).
52 R. Fellous and R. Luft, *J. Amer. Chem. Soc.*, **95**, 5593 (1973).
53 M. Charton, *J. Amer. Chem. Soc.*, **97**, 3691 (1975).
54 R. J. Gallo, *Prog. Phys. Org. Chem.*, **14**, 115 (1983).
55 M. Charton, *J. Amer. Chem. Soc.*, **97**, 1554 (1975).
56 H. Mager, *Tetrahedron*, **37**, 509, 523 (1981).
57 M. Charton, *J. Amer. Chem. Soc.*, **97**, 3694 (1975).
58 J.-E. Dubois and E. Bienvenue-Goetz, *Bull. Soc. Chim. France*, 2094 (1968).
59 S. H. Ungar and C. Hantzsch, *Prog. Phys. Org. Chem.*, **12**, 91 (1976).
60 T. Fujita and T. Nishioka, *Prog. Phys. Org. Chem.*, **12**, 49 (1976).
61 J. P. Lowe, *Prog. Phys. Org. Chem.*, **6**, 1 (1968).
62 J. Hauer, G. Volkel and H.-D. Ludemann, *J. Chem. Res. (S)*, 16 (1980); *(M)* 426 (1980).
63 J. E. Dubois and G. Mouvier, *Tetrahedron Lett.*, 1325 (1963).
64 U. Berg, R. Gallo, J. Metzer and M. Channon, *J. Amer. Chem. Soc.*, **98**, 1260 (1976).
65 G. B. Barlin and D. D. Perrin, *Quart. Rev.*, **20**, 75 (1966); J. Clark and D. D. Perrin, *Quart. Rev.*, **18**, 295 (1964); P. J. Pearce and R. J. J. Simkins, *Can. J. Chem.*, **46**, 241 (1968).
66 D. C. Berndt and I. E. Ward, *J. Org. Chem.*, **43**, 13 (1978).
67 R. Fellous and R. Luft, *Tetrahedron Lett.*, 1505 (1970).
68 K. B. Wiberg and R. R. Squires, *J. Amer. Chem. Soc.*, **101**, 5512 (1979).
69 R. Gallo, M. Channon, H. Lund and J. Metzger, *Tetrahedron Lett.*, 3857 (1972).
70 O. Hassel, *Quart. Rev.*, **7**, 221 (1953); J. Bassindale, *The Third Dimension in Chemistry*, John Wiley, London, 1984.
71 D. H. R. Barton and R. C. Cookson, *Quart. Rev.*, **10**, 44 (1956).
72 E. L. Eliel, N. L. Allinger, S. J. Angyal and G. A. Morrison, *Conformational Analysis*, Wiley–Interscience, New York, 1965.
73 M. Hanack, *Conformational Theory*, Academic Press, New York, 1965.
74 G. Chuiradoglu, *Conformational Analysis*, Academic Press, New York, 1971.
75 J. Dale, *Stereochemistry and Conformational Analysis*, Verlag Chemie, Weinheim, 1978.
76 J. I. Seeman, *Chem. Rev.*, **83**, 83 (1983).
77 D. J. Chadwick, J. Chambers, G. D. Meakins and R. L. Snowdon, *Chem. Comm.*, 625 (1971).
78 C. S. Johnson, *Adv. Mag. Reson.*, 1, 33, 1965; L. H. Piette and W. A. Anderson, *J. Chem. Phys.*, **25**, 1228 (1956); J. A. Pople, W. G. Schneider and H. J. Bernstein, *High Resolution Nuclear Magnetic Resonance*, McGraw-Hill, New York, 1959.
79 A. G. Robbiette, Specialist Periodicals Report No. 20, *Molecular Structure by Diffraction Methods* (L. E. Sutton and G. A. Sims, Eds), Vol. 3, The Chemical Society, London, 1975.
80 A. Streitweiser, *Solvolytic Displacement Reactions*, McGraw-Hall, New York, 1962.
81 J. Offenbach, L. Fredin and H. L. Strauss, *J. Amer. Chem. Soc.*, **103**, 1001 (1981).
82 A. H. Lewin and M. Frucht, *Org. Mag. Reson.*, **7**, 206 (1975).
83 W. E. Stewart and T. H. Siddall, *Chem. Rev.*, **70**, 517 (1970).
84 R. G. Kastyanovsky, Z. E. Samijlora and I. I. Tschervin, *Tetrahedron Lett.*, 719 (1969).
85 G. H. Senkler and K. Mislow, *J. Amer. Chem. Soc.*, **94**, 291 (1972).
86 W. Egan, R. Tang, G. Zon and K. Mislow, *J. Amer. Chem. Soc.*, **92**, 1442 (1970).

87   S. Winstein and N. J. Holness, *J. Amer. Chem. Soc.*, **77,** 5562 (1955).
88   E. L. Eliel and C. A. Lukach, *J. Amer. Chem. Soc.*, **79,** 5986 (1957).
89   N. B. Chapman, A. Ehsan, J. Shorter and K. J. Toyne, *J. Chem. Soc. (B)*, 570 (1967).
90   D. R. Brown, P. G. Levinston, J. McKenna, J. M. McKenna, R. A. Melia, J. C. Ratt and B. G. Hutley, *J. Chem. Soc., Perkin II*, 838 (1976).
91   J. McKenna, *Tetrahedron*, **30,** 1555 (1974).
92   D. Y. Curtin, *Rec. Chem. Prog.*, **15,** 111 (1954).
93   W. G. Dauben and K. S. Pitzer, *Steric Effects in Organic Chemistry* (M. S. Newman, Ed), Wiley, New York, 1956, Chapter 1.
94   E. L. Eliel, L. A. Plato and J. C. Richer, *Chem. Ind.*, 2007 (1961).
95   F. Riddell, *Conformational Analysis of Heterocyclic Compounds*, Academic Press, New York, 1980.
96   L. F. Fieser and M. Fieser, *Steroids*, Reinhold, New York, 1959.
97   P. J. Crowley, M. J. T. Robinson and M. G. Ward, *Tetrahedron*, **33,** 915 (1977).
98   J. I. Seeman, H. J. Secor, H. Hartung and R. Galzeroni, *J. Amer. Chem. Soc.*, **102,** 7741 (1980).
99   D. H. R. Barton, *J. Chem. Soc.*, 1027 (1953); D. J. Cram and F. A. Abd. Al Hafez, *J. Amer. Chem. Soc.*, **74,** 5828 (1952).
100 J. D. Morrison, Ed., *Asymmetric Synthesis*, Academic Press, New York, 1983.
101 T. Katsuki and K. B. Sharpless, *J. Amer. Chem. Soc.*, **102,** 5974, (1980); K. B. Sharpless, *Chem. in Brit.*, **22,** 38 (1986).
102 O. Arjona, R. Perez-Ossario, A. Perez-Rubalcaba and M. L. Quiroga, *J. Chem. Soc., Perkin II*, 597 (1981).

9 Homogeneous catalysis

The rates of many organic reactions are enhanced by the presence of catalysts, species which do not themselves appear in the products but which in various ways assist the progress of the reaction. Homogeneous catalysts[1-4] (i.e. those which act in solution rather than, say, at solid surfaces) clearly must participate in the chemistry of the reaction and be regenerated at the formation of the product. They are frequently nucleophiles or electrophiles and their role is to provide an alternative channel between reagents and products with a lower activation barrier (Fig. 9.1). In this sense the solvent itself is a catalyst (Chapter 5), but the present discussion will center on specific additives which enhance the rates of reaction.

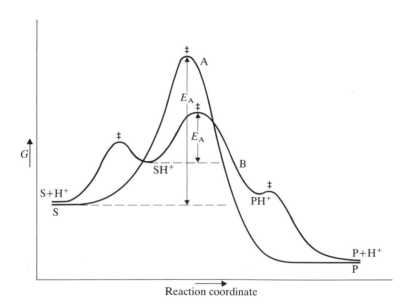

Fig. 9.1 Schematic energy profile for uncatalyzed (A) and acid-catalyzed (B) reactions in which a substrate, S, is converted to a product, P: S, substrate; P, product; E_A, principal energy barrier.

9.1 ACID AND BASE CATALYSIS[5-10]

Many reactions will only take place at reasonably observable rates when under the influence of added acids and bases; the hydrolyses of esters and amides to the parent acids, for instance, occur at almost undetectable rates in neutral water but readily on the addition of either acid or alkali. Acid catalysis implies an initial protonation of the substrate, S, which, because of the rapidity of proton transfers, frequently but not necessarily reaches equilibrium (pre-equilibrium protonation), **1**. An alternative possibility is proton transfer concerted with bond fission, **2**, an electrophilic substitution at carbon denoted A–S_E2. Catalysis by acid occurs since the substrate is made more reactive on protonation. Base catalysis may be of two types; the base, such as OH⁻, may take part in a hydrolytic reaction much faster than water since it is by far the more powerful nucleophile (**3b**). Alternatively, catalysis may be achieved by removal of a proton from the substrate, **3a**, whose conjugate base is more reactive; this occurs in the bromination of a ketone (**16b**).

1 Acid catalysis   **3** Base catalysis

2 A–S_E2 mechanism

9.1.1 Specific and general catalysis

The kinetics of an acid- or base-catalyzed reaction are frequently complex, the empirical rate equation containing several terms. The mutarotation of glucose (G), **4**, i.e. the conversion of the β-anomer to the $(\alpha \rightleftarrows \beta)$ equilibrium mixture, in aqueous acetate buffer solution follows the rate law:

$$\text{Rate} = [\text{G}](k_{\text{H}_2\text{O}} + k_{\text{H}_3\text{O}^+}[\text{H}_3\text{O}^+] + k_{\text{AcOH}}[\text{AcOH}]). \qquad (9.1)$$

Three parallel routes to product are evidently available, mediated respectively by water, hydronium ion and molecular acetic acid, all of which are capable of transferring a proton to the glucose though at very different rates. This is termed *general acid catalysis* and can be inferred from a plot of rate against buffer concentration; as the latter changes, constancy of the hydronium ion concentration is ensured but [AcOH] varies (Fig. 9.2). The

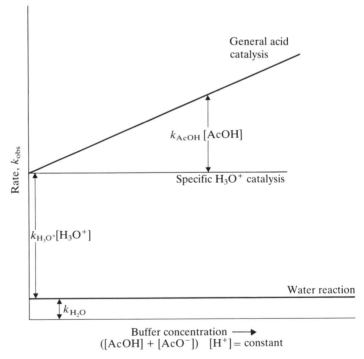

rate law for general acid catalysis will be of the form

$$\text{Rate} = \overset{\text{acids}}{\sum} k_{\text{cat}}[\text{S}][\text{HA}],$$

in which the summation is over all acidic species, each with their specific rate constant, k_{cat}, the catalytic constant. Similarly, a reaction may be

Fig. 9.2 Effect of buffer concentration of rates of specific H_3O^+-catalyzed and general acid-catalyzed reactions.

catalyzed by all bases present (e.g. H_2O, OH^-, A^-) — *general base catalysis* — and it is also possible to have a reaction which is catalyzed by both acids and bases. The bromination of a ketone **16a,b** is an example.

$$\text{Rate} = \overset{\text{acids}}{\sum} k_{cat}[S][HA] + \overset{\text{bases}}{\sum} k_{cat}[S][B]. \tag{9.2}$$

On the other hand, some reactions respond to catalysis only by hydronium or hydroxide ions (in general, lyonium and lyate ions) and are then deemed to show specific acid or base catalysis. The hydrolysis of oxiranes, **5**, is an example. In such cases, catalysis by the weaker acids and bases present is negligible (Table 9.1). Specific acid catalysis implies a rapid pre-equilibrium protonation of the substrate. General acid catalysis implies that proton transfer occurs in a rate-determining step and may take place through a molecule of water or other protic solvent.

Table 9.1   Examples of acid- and base-catalyzed reactions

Reaction	Type[a]
Esterification and hydrolysis of carboxylic and phosphoric esters and amides	A, B
Hydrolysis of acetals, ketals and orthoesters	(sp) A
Hydrolysis and alcoholysis of anhydrides	A
Carbonyl condensation reactions with X—NH_2	A, B
Enolization and electrophilic substitution α- to a carbonyl group	A, B
Hydration of alkenes and dehydration of alcohols	A
Additions of CN^-, $HSO_3^- H_2O$, etc., to $\rangle C{=}O$	A, B
Cleavage of C–Hg, C–Sn bonds	A
Decomposition of NH_2NO_2 to N_2O and H_2O	B
Aldol and Michael reactions	B
Aromatic electrophilic proton exchange	A
Beckmann rearrangement	A
Hydrolysis of epoxides	(sp) A, B

[a] A, acid-catalyzed; B, base-catalyzed; sp, specific.

The derivations of the rate laws can be made as follows. For the reaction scheme with substrate, S:

$$S + H^+ \rightleftarrows SH^+;$$

$$SH^+ \ (+ \ \text{solvent}) \xrightarrow[\text{slow}]{k} \text{products},$$

$$\text{Rate} = k[SH^+]$$

or, since the acidity of the substrate, $K_A^S = [SH^+]/[S][H^+]$,

$$\text{Rate} = kK_A^S[S][H^+],$$

an expression of specific hydronium ion catalysis. However, if the anion of the catalyzing acid is involved, there will be a compensating situation in that increasing acid strength displaces the pre-equilibrium to the right but also slows down the rate-limiting step. Tautomerization is an example of this.

$$\text{Rate} = k[SH^+][A^-],$$

but $K_A^S = [SH^+]/[S][H^+]$ and $[A^-] = K_A^C[HA]/[H^+]$. Hence

$$\text{Rate} = kK_A^S K_A^C[S][HA],$$

an expression of general acid catalysis by a weak acid, dissociation constant K_A^C. There will be a sum of such terms if more than one acid is present.

Specific hydronium ion catalysis will be faster in D_2O than in H_2O since D_3O^+ is a stronger acid than H_3O^+ by a factor of about 3. Hence, $k_{H_2O}/k_{D_2O} \sim \frac{1}{3}$. However, general acid catalysis requires a proton-transfer equilibrium with solvent to be followed by a rate-determining removal of a proton, a primary isotope effect. Hence $k_{H_2O}/k_{D_2O} \sim \frac{1}{3} \times 7$.

9.1.2 Mechanisms of acid catalysis

Solvolytic reactions, whether of alkyl halides, esters, acetals, or epoxides, are nucleophilic displacements in which water or another protic solvent (SH) supplies a lyate group (—S) to replace a nucleophilic leaving group. Acid-catalyzed solvolyses take place from the protonated solute and, usually, it is the leaving group which becomes protonated and its leaving

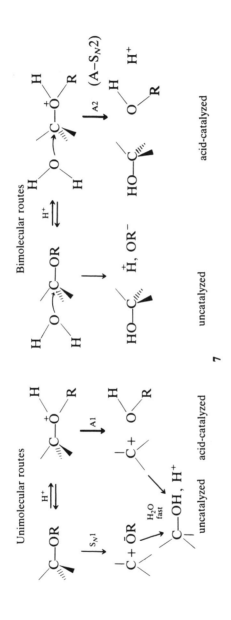

ability enhanced; this is the reason for the catalysis. It is much easier to displace —$\overset{\frown}{O}H_2^+$ as H_2O than —$\overset{\frown}{O}H$ as OH^-.

The fission of the leaving group from protonated reagent may be one of two mechanistic types; either it occurs spontaneously, the A1 process (i.e. unimolecular), or it occurs under the action of solvent attack, a bimolecular reaction denoted A2 ($A-S_N2$). This dichotomy corresponds to that as-

Table 9.2   Examples of acid-catalyzed reactions

Reaction	*Reference*
A1 mechanism	
1 Esterification	11
$ROH + R'COOH \xrightarrow{H_2SO_4} R'CO \cdot OR$	
2 Hydrolyses of tertiary alkyl esters ($A_{Al}1$)	12, 13
$RCO \cdot O t Bu + H^+, H_2O \rightarrow RCOOH + t BuOH$	
3 Hydrolyses of β-lactones	14
4 Hydrolyses of anhydrides	15
$RCO \cdot O \cdot COR + H_2O, H^+ \rightarrow 2RCOOH$	
5 Cleavage of tertiary ethers	16
$R_3C—OR' \xrightarrow{H^+} R_3C—OH + R'OH$	
6 Hydrolyses of acetals and ketals	17
$R_2C(OR')_2 \xrightarrow{H^+} R_2C{=}O + 2R'OH$	
7 Pinacol rearrangement	18
$R_2C—CR_2 \xrightarrow{H^+} R_3C—CO \cdot R$ with OH OH	
8 Beckmann rearrangement	19
$Me—C—Ar$ (with $N—OH$) $\xrightarrow{H^+}$ ArCO·NHMe	
9 Epoxide ring-opening	20
$RCH—CH_2 + H^+, H_2O \longrightarrow RCH(OH)—CH_2OH$ (with O)	
A2 Mechanism	
10 Cleavage of primary ethers	21
$Et—O—Et + H^+, H_2O \rightarrow 2EtOH$	
11 Ester hydrolysis ($A_{Ac}2$)	22
$R \cdot CO \cdot OR' + H^+, H_2O \rightarrow R \cdot COOH + R'OH$	
12 Amide hydrolysis ($A_{Ac}2$)	23, 24
$R \cdot CONR'_2 + H +, H_2O \rightarrow R \cdot COOH + R_2NH$	
13 Hydration and dehydration of aldehydes	25
$RCHO + H^+, H_2O \rightarrow RCH(OH)_2$	
14 Hydration of alkenes	26
$RCH{=}CHR' + H^+, H_2O \rightarrow RCH_2—CH(OH)R'$	

For reaction 3 (β-lactones):

$$\text{—O + H}^+, \text{H}_2\text{O}, \text{—CO} \longrightarrow \text{—OH}, \text{—COOH}$$

sociated with uncatalyzed solvolytic reactions designated S_N1 (unimolecular) and S_N2 (bimolecular) (Section 10.1). The relationships are summarized in the scheme **7** and some examples are given in Table 9.2.

9.1.3 Methods of distinguishing between A1 and A2 reactions

Reaction kinetics cannot usually distinguish the involvement of solvent since it is present in large and constant excess.

The Hammett–Zucker postulate[5,6]

The A1 process requires the substrate to be protonated by the acidic species of the medium whose ability to do so should be parallel to protonation of the bases which define the Hammett acidity function, H_0 (Section 6.11.1). Catalytic rates should therefore follow H_0 rather than $[H_3O^+]$ and a plot of $\log k$ against H_0 should have unit slope. On the other hand, for the A2 process,

$$\text{Rate} = k_2[SH^+][H_2O] = k_2/K[S][H_3O^+] \tag{9.3}$$

and it was suggested that rates should follow hydronium ion concentration. It was subsequently found that this clear distinction of rate dependence could seldom be made in the type of media for which H_0 and $[H_3O^+]$ differ significantly, of necessity highly acidic solutions (Section 6.6).

Bunnett's w-values[2,7,27]

The failure of many $\log k$ versus H_0 plots to show linearity can be made quantitative by plotting $(\ln k - H_0)$ against the activity of water, a_{H_2O} (Table 9.3). The slope, w, varying over the range -8 to $+7$ (Table 9.4), is a measure of this deviation, which has mechanistic significance. By examination of many systems, Bunnett ascribed the following interpretation to various ranges of w assuming protonation is occurring at nitrogen or oxygen.

w	Function of water in rate-determining step
-2.5 to 0	Not involved; A1 mechanism,
$+1.2$ to $+3.3$	Acts as nucleophile; A–S_N2 reaction.
$>+3.3$	Proton transfer agent; A2 reaction.

A difficulty with both approaches is that the substrates which are protonated are not necessarily H_0 bases which, it will be recalled, were aromatic amines. More valid relationships of this type might result from a replace-

Table 9.3 Activities of water in some strong acid solutions

Molarity (M)	$\log a_{H_2O}$		
	HCl	H_2SO_4	$HClO_4$
0·5	−0·008	−0·008	−0·008
1·0	−0·017	−0·018	−0·018
1·5	−0·027	−0·030	−0·030
2·0	−0·039	−0·043	−0·043
2·5	−0·053	−0·063	−0·060
3·0	−0·070	−0·085	−0·081
3·5	−0·087	−0·111	−0·106
4·0	−0·107	−0·142	−0·135
4·5	−0·130	−0·176	−0·172
5·0	−0·155	−0·219	−0·215
5·5	−0·181	−0·267	−0·271
6·0	−0·211	−0·320	−0·330
6·5	−0·244	−0·377	−0·411
7·0	−0·279	−0·439	−0·496
7·5	−0·318	−0·510	−0·602
8·0	−0·358	−0·587	−0·714
8·5	−0·399	−0·670	−0·842
9·0	−0·444	−0·761	−0·983
9·5	−0·490	−0·859	−1·150
10·0	−0·539	−0·968	—
10·5	−0·591	−1·082	—

Table 9.4 Examples of Bunnett's w- and ϕ-parameters[2, 7, 27]

Reaction	w	ϕ
A1 reactions: $w < 0$, $\phi < 0$		
Hydrolysis of MeCH(OEt)$_2$, HCl, 0°C	−7·86	−1·02
Paraldehyde → CH$_3$CHO, 25°C	−4·32	−0·58
Hydrolysis of CH$_3$COOCH$_2$OEt, 25°C	−3·57	−0·44
Inversion of sucrose (→ glucose + fructose)	−0·43	−0·04
H–D exchange in anisole, para-position, H$_2$SO$_4$		−0·24
A–S$_N$2 reactions: $w = 1·2$ to $3·3$		
tert-Bu^{18}OH $\xrightarrow{H_2S^{16}O_4}$ tert-Bu^{16}OH	+0·82	−0·29
Racemization of (+)-CH$_3$CHOH·Et, HClO$_4$, 100°C	+1·28	+0·17
Hydrolysis of CH$_3$CONH$_2$, HCl, 50°C	+2·65	
Hydrolysis of CH$_3$CONHNH$_2$, 60°C		+1·07
Hydrolysis of (MeO)$_2$S=O, HClO$_4$, 0°C	+2·36	
Proton transfer, $w > 3·3$		
Hydrolysis of HCOOMe, HCl, 25°C	+4·21	+0·74
Hydrolysis of CH$_3$COOEt, H$_2$SO$_4$, 25°C	+4·5	+0·84
Hydrolysis of γ-butyrolactone, HClO$_4$	+8·5	+1·04
Pinacol rearrangement	+4·99	
Enolization (iodination) of acetone		+0·83
Carbon-centered reactions		
C$_6$D$_6$ $\xrightarrow{H_2SO_4}$ C$_6$H$_6$	−0·58	
PhC^{18}O·OH $\xrightarrow{H_3^{16}O^+}$ PhC^{16}O·OH		+0·94
Cis–trans isomerization of PhCOCH=CHPh		+0·44

ment of H_0 by the appropriate acidity function, H_X in each case, interpreting r in the same way as w above[8].

$$\log k_{obs} + H_X = r \log a_{H_2O} + \text{constant.} \tag{9.4}$$

The Bunnett and Olsen Parameter, ϕ[11,12,39]

From the definition of the acidity function, $H_0 = pK_A + \log([B]/[BH^+])$ (Eq. 6.32), for an aromatic amine of $pK_A = 0$, one may write

$$\log[BH^+]/[B] = \log I = -H_0.$$

Hence

$$(\log I - \log[H^+]) = (1 - \phi)(-H_0 - [H^+]) + \text{constant,}$$

in which the constant of proportionality $(1 - \phi)$ is the slope of a plot of the left-hand expression against the right. This equation may be written

$$\log I + H_0 = \phi(H_0 + \log[H^+]) + pK_A. \tag{9.5}$$

Linear plots are obtained for many weak bases in moderately strongly acidic media and the slopes, ϕ, express the response of the protonation equilibrium of each to changing acid concentration. A positive ϕ means that $\log I$ for the base increases less rapidly than does H_0; the converse is true for a negative value (Table 9.4): for each base ϕ expresses the difference in protonation behavior upon which various acidity functions are based. An analogous expression (Eq. (9.6)) may be written for acid-catalyzed rate processes of weakly basic substrates:

$$\log k + H_0 = \phi(H_0 + \log[H^+]) + \text{constant.} \tag{9.6}$$

The slope, ϕ, characterizes the kinetic effect of changing the acidity of the reaction medium brought about by changes in the activity coefficients of reagents and transition state, which in turn are caused by solvation changes: the intercept, C, is $\log k$ in water. Linear relationships expressed by Eq. (9.6) are found for hundreds of acid-catalyzed reactions. Since changes in activity coefficients in these moderately acid media are caused by changes in hydration, ϕ measures the relationship between the hydration change on protonation of the substrate compared with that on protonation of a primary aromatic amine, an H_0 base, and may be interpreted similarly to the w values:

ϕ	*Function of water in rate-determining step*
<0	Not involved.
+0·22 to +0·56	Acts as nucleophile.
>+0·58	Acts as proton-transfer agent.

Table 9.5 Entropies and volumes of activation for hydrolytic reactions[28-31]

Substrate	Mechanism	$\Delta S^{\ddagger}/J\,K^{-1}\,mol^{-1}$ $(cal\,K^{-1}\,mol^{-1})$	$\Delta V^{\ddagger}/cm^3\,mol^{-1}$
Methyl acetate	$A_{Ac}2$	−88 (−21)	−8·3
Ethyl acetate	$A_{Ac}2$	−96 (−23)	−8·4
Ethylene oxide	A2	−25 (−6)	−9
Acetamide	A2	−154 (−38)	−11
γ-Butyrolactone	A2	−108 (−25)	−8·4
Ethylene sulfite	A2	−66 (−16)	
Ethyleneimine	A2	−34 (−9)	
$PhC^{18}O\cdot OH$ (oxygen exchange)	A2	−125 (−30)	
tert-Butyl mesitoate	$A_{Al}1$	+41 (+10)	
tert-Butyl acetate	$A_{Al}1$	+59 (+14)	
Mesitoic acid–$C^{18}O\cdot OH$ (oxygen exchange)	A1	+37 (+9)	
Methyl α-D-glucopyranoside	A1	+17 (+)	+6·2
Methyl β-D-glucofuranoside	A2 (ring opening)	−10 (−2·4)	−3·6
Sucrose inversion	A1		+4·3

Volumes and entropies of activation

The A1 mechanism requires a proton transfer which releases a water molecule in the slow step and should therefore have a positive entropy and a positive volume of activation. On the other hand, the A2 reaction requires the formation of a bond between protonated substrate and water which should be accompanied by a loss of entropy and a decrease in volume. Numerous examples support this criterion (Table 9.5).

9.1.4 Linear free-energy relationships; the Brønsted Catalysis Law[2,8-10,32]

Catalytic rate constants might be expected to increase with increasing strength of the acid or base, and indeed this is so. It is perhaps more surprising to find a very good linear relationship between k_{cat} and K_A expressed by Eqs (9.7) and (9.8) and known as the Brønsted Catalysis Law, the first example of a linear free-energy relationship to be revealed[33].

$$k_A = G_A K_A^{\alpha} \quad \text{or} \quad \log k_A = -\alpha pK_A + \text{constant}; \qquad (9.7)$$

$$k_B = G_B(1/K_A)^{\beta} \quad \text{or} \quad \log k_B = \beta pK_A + \text{constant}; \qquad (9.8)$$

where k_A, k_B are catalytic constants for individual acids and bases of acid dissociation constants K_A, respectively. A good linear relationship has been found for numerous reactions to fit a wide range of catalyst types (Fig. 9.3). This relationship would be expected if a catalytic mechanism involved pre-equilibrium protonation, $k_{obs} = Kk_2$ when k_2 would be constant for all catalysts and K, the extent of protonation, would be determined by K_A of the catalyst.

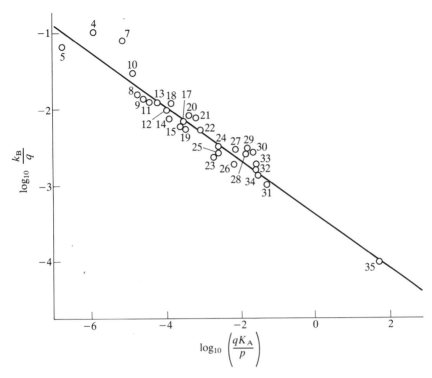

Fig. 9.3 A typical well-behaved Brønsted plot, for mutarotation of glucose, slope = β = 0·94; p and q are statistical factors, the number of dissociable protons and equivalent sites for protonation of the catalyst, respectively[10]. The catalysts were, respectively: 5, phenoxide ion; 4, histidine; 7, pyridine; 8, trimethylacetate ion; 9, propionate ion; 10, quinoline; 11, acetate ion; 12, phenylacetate ion; 13, glutamate ion; 14, benzoate ion; 15, *o*-toluate ion; 17, glycolate ion; 18, aspartate ion; 19, hippurate ion; 20, formate ion; 21, α-alanine; 22, mandelate ion; 23, salicylate ion; 24, *o*-chlorobenzoate ion; 25, chloroacetate ion; 26, cyanacetate ion; 27, *p*-benz-betaine; 28, sarcosine; 29, lysine-HCl; 30, arginine-HCl; 31, sulfate ion; 32, proline; 33, dimethylglycine; 34, betaine; 35, water.

Proton transfers between oxygen and nitrogen bases and acids with hydronium or hydroxide ions are normally diffusion-controlled, but rates of proton exchange with water are proportional to their respective K_A values. This can be expressed in the intersecting potential energy curves for the two B–H bonds involved in a proton transfer. The replacement of one base by a stronger one (lower energy) lowers the energy of activation for the proton transfer, which varies linearly with pK_A and with unit slope until, at pK_A = 0 a constant maximum rate is attained (Figs. 9.4, 9.5). It is within the linear portion of this plot that the Brønsted relationship is normally tested, though it has been inferred by Eigen that a more exact relationship would be a non-linear curve observable if catalysts with a sufficiently large span of pK_A were examined.[34]. In fact, Brønsted plots for proton transfers at carbon acids and bases are frequently curved (Fig. 9.5c) as a result of the

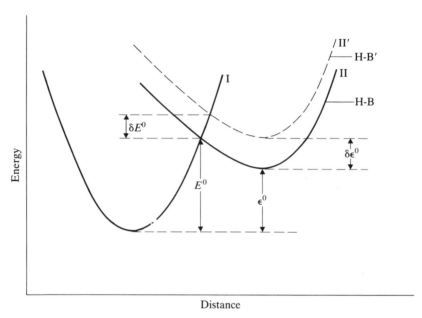

Fig. 9.4 Potential curves (energy versus distance) for the systems A—H$\cdots$B (curve I) and A$^-\cdots$H—B$^+$ (curve II). BH$^+$ is a weaker acid than AH by an amount proportional to ε^0. The proton transfer proceeds by stretching of the A–H bond until the energy is such that the two curves intersect. Continuation of the motion results in product formation with an energy barrier E^0. H—B' is a still weaker acid and the additional activation energy required for proton transfer from HA is δE^0, proportional to its difference in pK_A, $\delta\varepsilon$. The Brønsted relationship shows linearity as long as curves II, II' are parallel.

greater complexity of carbon-acid behavior due to the need for rehybridization and changes in the geometry of the acid or base upon proton transfer.

9.1.5 Interpretation of the Brønsted coefficients

The coefficients α and β have the significance of LFER susceptibility parameters and are a measure of the sensitivity of the catalyzed reaction to the strength in water of its catalysts as acids and as bases respectively. While the simple model suggests values of unity should be found, this is the case only for equilibrium protonation or deprotonation. Reactions in which the proton transfer is a part of the activation step will show smaller values of α, β over the range 0–1. Assuming the thermodynamic properties of the transition state for a proton transfer to be intermediate between those for reagents and products, a structural change will affect the free energy of the transition state by the amount $\delta G^{\ddagger}$ such that

$$\delta G^{\ddagger} = \alpha\delta G_p + (1 - \alpha)\delta G_r, \qquad (9.9)$$

where G_p, G_r are free energies of reagents and products. Therefore, α is a

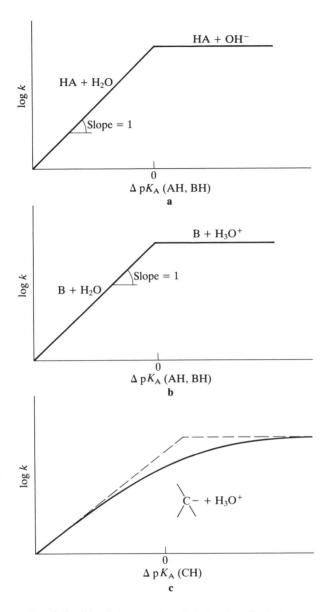

Fig. 9.5 Schematic relationships between rates of proton transfer between two bases, and their difference in pK_A[15]. As long as pK_A of the proton acceptor is greater than that of the donor, the reaction is diffusion-controlled and the rate is large and constant, independent of the donor strength. At $\Delta pK \sim 0$ there is a change in slope giving a normal Brønsted plot, $\alpha, \beta \sim 1$, as the reverse reaction becomes diffusion-controlled. Carbon acids, c, tend to transfer protons at less than diffusion-controlled rates and may give curved Brønsted plots.

Table 9.6 Examples of Brønsted coefficients for some acid- and base-catalyzed reactions

Reaction	α^a	β	Reference
$CH_3CH(OH)_2 \xrightarrow{H^+} CH_3CHO + H_2O$	0·54		35
α-D-Glucose $\xrightarrow{H^+}$ $(\alpha \rightleftarrows \beta)$-D-glucose	0·30		36
Mutarotation $\xrightarrow{B}$		0·40	
$CH_2(OH)_2 \overset{H^+}{\underset{B}{\xrightarrow{\quad\quad}}} HCHO + H_2O$	0·10	0·40	37
$(EtO)_4C \xrightarrow{H^+} 4EtOH + CO_2$	0·69		17, 38
$(EtO)_3CH \xrightarrow{\underset{aq.}{H^+}} 2EtOH + HCOOH$	0·77		
$\xrightarrow{aq.}$ dioxane	1·00		
$(EtO)_2CPh_2 \xrightarrow{H^+} 2EtOH + Ph_2CO$	0·78		

^{a} $\alpha \sim 1$ indicates specific H_3O^+ catalysis with pre-equilibrium protonation of the substrate; $\alpha < 1$ indicates general acid-catalysis and is
a measure of the extent of protonation in the transition state.

parameter measuring the position of the transition state on the reaction
coordinate. For $\alpha = 0$ the transition state is close to reactants, while when
$\alpha = 1$ it resembles closely the products. The Brønsted coefficients measure
the degree of proton transfer at the transition state, though it might be
argued that the position of the transition state along the reaction coordinate
will not be constant when the pK_A of the catalyst changes over 10–15 units,
but will vary according to the Hammond postulate. This should introduce
some curvature into the plot. Values of Brønsted constants for various
reactions (Table 9.6) are of diagnostic value in mechanistic studies. For
example, the dehydration of the carbinolamine, **8**, a common type of
intermediate in ketone condensation reactions, is general base-catalyzed
and could in principle occur by proton abstraction from either carbon or
nitrogen. Since ΔpK_A between base and the C–H bond is very large,
dissociation at this center will be very base-selective and $\beta \sim 1$. Dissociation
at nitrogen, $\Delta pK_A \sim 0$, will show much less selectivity and $\beta \ll 1$, which is
supported by the experimental value $\beta = 0·4$. The value of the Brønsted α
and β coefficients is therefore a measure of the discriminating capacity of
the system towards catalysts of various strength; $\alpha, \beta = 1$ implies high
selectivity and, indeed specific catalysis by H_3O^+ or OH^-; a value of
$\alpha, \beta = 0$ on the other hand implies complete indifference to the nature of the
catalyst so that, in a general catalyzed reaction, the proportion of product
formed by each catalytic species will be in proportion to their concentrations. Table 9.7 illustrates this principle[10].

Table 9.7   Percentage of product formed by acidic species present in 0·1M acetate buffer as a function of Brønsted α

α	H_3O^+	$AcOH$	H_2O
1	99·8	0·2	5×10^{-12}
0·5	3·6	96·4	0·01
0·1	0·002	2	98

9.1.6 Nucleophilic catalysis

True base catalysis requires a nucleophilic attack by the catalyst on a proton. Nucleophilic catalysis, however, occurs by attack on some other center and must be distinguished from base catalysis by the non-involvement of a proton transfer in the catalytic step. For example, the hydrolysis of *p*-nitrophenyl acetate is catalyzed by imidazole, **9**, by direct interaction of the catalyst at the carbonyl carbon to produce an intermediate acylimidazole, **9a**. This is very rapidly hydrolyzed by water. Catalysis occurs since all steps in this multistage reaction are very much faster than that of the simpler hydrolysis by water alone. General base catalysis also occurs and the imidazole acts as a proton sink, increasing the basicity of water, **10**. The following characteristics enable general base and nucleophilic catalysis to be distinguished.

(a) Catalysis by the leaving group (i.e. a common-ion effect) may occur. This can only be due to general base action, **10**, since attack by the leaving group directly on the substrate could only lead to regeneration of substrate and no catalysis[1,2,36,39,40] Negative catalysis, i.e. rate retardation by the leaving group, may also occur but this suggests nucleophilic catalysis. Acetate ion, for example, retards the hydrolysis of *p*-nitrophenyl acetate catalyzed by pyridine[41]. This is due to the reversible formation of the intermediate, acylpyridinium, and lowering of the available catalyst concentration.

(b) Reactions which are second order with respect to the catalyst also suggest general base behavior, one molecule activating the other just as, in **6**, a water molecule was activated[38,42].

(c) Base strength and nucleophilic activity at carbon do not in general run parallel (Section 6.12) so that a Brønsted plot with much scatter (Fig. 9.6) suggests nucleophilic catalysis. In particular, separate lines may be drawn for different families of reagents. Sterically hindered nucleophiles, e.g. 2-substituted pyridines, should also depart from linearity and be less

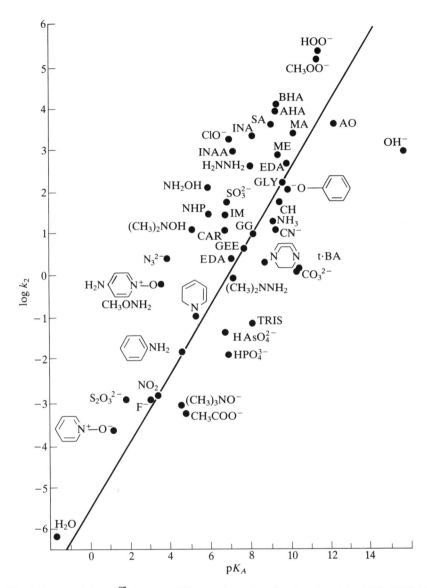

Fig. 9.6 Brønsted plot[87] of nucleophilic reactions of *p*-nitrophenyl acetate, H₂O, 25°C. Key: AHA, acetohydroxamic acid; AO, acetoxime; BHA, *n*-butyrylhydroxamic acid; CAR, carnosine; CH, chloral hydrate at 30°C; EDA, ethylenediamine; GEE, glycine ethylisonitroso-acetylacetone; GG, glycylglycine; GLY, glycine; IM, imidazole; INAA, indoleacetic acid; MA, sodium mercaptoacetate; ME, mercaptoethanol; NHP, *N*-hydroxyphthalimide; SA, salicylal-doxime; t·BA, *tert*-butylamine; TRIS, tri(hydroxy-methyl)aminomethane.

reactive than expected. This is because proton transfer is rather insensitive to steric factors (since the proton is so small), while nucleophilic attack is very sensitive to the size of both reactant and catalyst. Hence, while pyridine is an efficient catalyst for the hydrolysis of acetic anhydride, 2,6-dimethylpyridine, a stronger base, is not a catalyst.

(d) Occasionally, a reactive intermediate can be isolated, confirming the role of the catalyst as a nucleophile rather than as a general base. Acetylimidazole, **9a**, and the acylpyridinium ion[42] formed in the pyridine-catalyzed hydrolyses of anhydrides are examples.

9.1.7 Potential-energy surfaces for proton transfers[35,43,44]

To a good approximation, the potential energy of a proton-transfer process may be held to depend upon two geometrical variables, the distances $A \cdots H$ and $B \cdots H$. The system may therefore be represented by a three-dimensional graph, which for convenience is usually expressed as a two-dimensional plot of the distance variables with energy contour lines. These plots (e.g. Fig. 9.7) are often referred to as More O'Ferrall–Albery–Jencks plots. Reagents and products are located at corners **A** and **C** respectively, while points near **B**, **D** represent intermediates located at potential-energy minima which may be of high or low energy with respect to the reagents. The most favorable pathway will be that which permits passage from **A** to **C** surmounting the least energy barrier. A two-step process will progress from **A** to **B** or **D** and thence to **C**. A concerted process will be expressed as direct passage diagonally from **A** to **C** with a possible inclination towards one or other corner corresponding to asymmetry of bond making and bond breaking. Figure 9.7a represents the hydrolysis of an acetal proceeding in two steps by a pre-equilibrium protonation corresponding to specific hydronium ion catalysis and Brønsted $\alpha \sim 1$[17]. Note the high energy barrier for the uncatalyzed S_N1 route, $A \rightarrow D \rightarrow C$. Figure 9.7b is the situation when both intermediates B and D are of high energy, when the concerted process becomes more favorable as for example in the hydrolysis of aryl acetals and orthoesters ($\alpha \sim 0.5$).

The same concept is applicable to nucleophilic additions to the carbonyl group. The axes in Fig. 9.7c–e represent $Nu \cdots C{=}O$ and $C{=}O \cdots H \cdots A$ bond distances of the system:

$$Nu: \overset{\frown}{} \overset{\backslash}{C}{=}O \overset{\frown}{} H{-}A \longrightarrow Nu^+{-}\overset{\backslash}{C}{-}O{-}H \quad A^-$$

The route $A \rightarrow B \rightarrow C$ (d) represents the uncatalyzed addition of nucleophile to form a rather stable intermediate which is then protonated; $A \rightarrow D \rightarrow C$ (c) represents the stepwise pre-equilibrium protonation followed by nucleophilic attack. With increased stability of the carbonyl compound, a weaker nucleophile or a weaker acid, the situation tends to change into e,

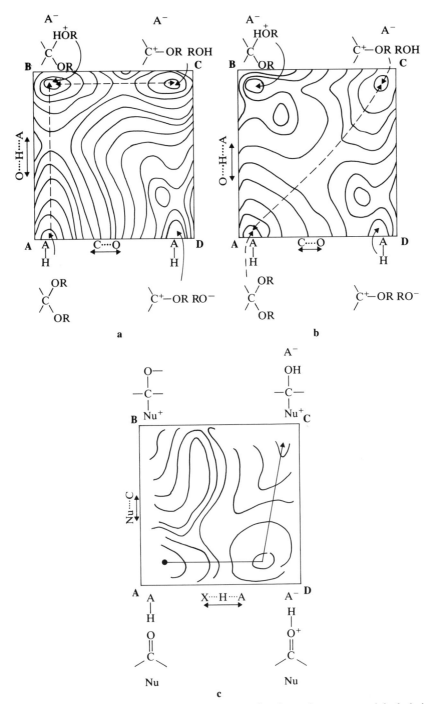

Fig. 9.7 More O'Ferrall–Jencks plots for some catalyzed reactions: **a**, acetal hydrolysis, specific acid-catalyzed by two-step path; **b**, acetal hydrolysis by general acid-catalyzed (A–S_E2) route; **c**, nucleophilic addition to carbonyl, catalysed by a strong acid; **d**, as for c, catalyzed by a weak acid; **e**, as for c, concerted reaction catalyzed by acid of intermediate strength.

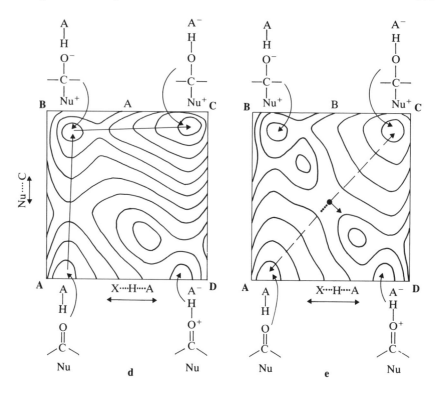

Fig. 9.7 (*Continued*)

the concerted route to products. The guiding principle has been summarized as follows: concerted acid–base attack is preferred only at sites which undergo a large change in pK during the course of the reaction ($\supset$C$=$O to $\supseteq$C—O$^-$, for example) and when the pK of the catalyst is intermediate between those of reagent and product sites. In this example, the pK_B values of $\supset$C$=$O and $\supseteq$C—O$^-$ groups are of the order of 16 and 0 respectively so a choice of acid catalyst for the concerted reaction would have p$K_B \sim 8$, p$K_A \sim 6$: for example, acetic acid. This would be a favorable situation since the catalyst is too weak to protonate $\supset$C$=$O but amply strong enough to protonate $\supseteq$C—O$^-$. H$_3$O$^+$, however, would be strong enough to protonate sufficient of the $\supset$C$=$O for the purposes of reaction and so would catalyze the stepwise route, **A**→**D**→**C**, specifically. If the catalyst is a weaker acid than $\supseteq$C—OH, the preferred route would be the two-step uncatalyzed addition, **A**→**B**→**C**.

9.1.8 Solvent isotope effects[32,45,46]

The proton transfers and participation of water typical of acid–base catalytic reactions will be accompanied by solvent deuterium isotope effects (Table

Table 9.8   Some solvent isotope effects in acid-catalyzed hydrolyses

Substrate	k_{H_2O}/k_{D_2O}
Ethyl orthocarbonate, $(EtO)_4C$	0·714
Ethyl orthoacetate, $MeC(OEt)_3$	0·534
Ethyl orthoformate, $HC(OEt)_3$	0·335
Acetaldehyde diethylacetal, $MeCH(OEt)_2$	0·379
1,3-Dioxane,	0·36
2-Methoxytetrahydrofuran, —OMe	0·34
Sucrose ($\rightarrow$ glucose + fructose)	0·568
Ionization of $PhCOCH^*$	(SKIE)1·24
at H^* (L = H, D)	

9.8) determined by a combination of fractionation factors and zero-point energy effects (Chapter 7). Acid-catalyzed reactions are faster in D_2O but A1 and A2 mechanisms cannot be distinguished from the magnitudes of the isotope effects. In general, values of k_{H_2O}/k_{D_2O} are expected to lie in the range 0·3–0·43 for pre-equilibrium protonation but will be closer to unity, usually >0·45, for reactions involving concerted proton transfer and bond breaking. The hydrolyses of acetaldehyde acetals (Table 9.9**b**, R = Me), indicated by other criteria to be A1, have values for k_{H_2O}/k_{D_2O} around 0·37 while ethyl orthocarbonate, (Table 9.9**p**, R = Et), reacting by A–S_E2, has $k_{H_2O}/k_{D_2O} = 0·7$. When proton transfer becomes rate-determining, as is often the case for ionization of carbon acids, a full PKIE may be observed. The halogenation of nitroacetate is an example (B: = acetate):

$$O_2N—CH_2COOEt + B: \xrightarrow[slow]{} O_2N—\overset{-}{C}H—COOEt \underset{+BH^+}{} \xrightarrow[fast]{I_2}$$

$$O_2N—CHICOOEt; \qquad k_H/k_D = 7·7.$$

9.1.9 Electrophilic catalysis

Electrophiles beside H^+ are able to bring about catalysis by a similar mechanism. Metal cations (Ag^+, Hg^{2+}) may participate in aqueous solution, while in non-protic media strong electrophiles such as $AlCl_3$, BF_3 and similar species can exert catalytic effects. Just as in proton catalysis, the action of these species is to increase the electrophilicity of one of the reagents. The electrophilic catalyst may release a proton from water (**11**); it

Table 9.9 The ketal family

R₂C(OR)(OR) Ketal **a**	H,R–C(OR)(OR) Acetal **b**	R₂C(OH)(OR) Hemiketal **c**	H,H–C(OH)(OR) Hemiacetal **d**	

$$
\begin{array}{ccccc}
R_2C\diagup_{SR}^{SR} & \begin{array}{c}H\\R\end{array}\diagup_{SR}^{SR} & R_2C\diagup_{SR}^{OH} & \begin{array}{c}H\\R\end{array}\diagup_{SR}^{OH} & R_2C\diagup_{SR}^{SH} \\
\text{Thioketal} & \text{Thioacetal} & \text{Thiohemiketal} & \text{Thiohemiacetal} & \text{Dithiohemiketal} \\
\mathbf{e} & \mathbf{f} & \mathbf{g} & \mathbf{h} &
\end{array}
$$

$$
\begin{array}{ccc}
R{-}\underset{OR}{\overset{OR}{C}}{-}OR & R{-}\underset{OR}{\overset{OH}{C}}{-}OR & R{-}\underset{SR}{\overset{OH}{C}}{-}SR \\
\text{Orthocarboxylate} & \text{Hemiorthocarboxylate} & \text{Hemithioorthocarboxylate} \\
\mathbf{i} & \mathbf{j} & \mathbf{k}
\end{array}
$$

$$
\begin{array}{ccccc}
R_2C\diagup_{SR}^{OR} & R_2C\diagup_{NR_2}^{SR} & R_2C\diagup_{NR_2}^{OR} & R_2C\diagup_{OH}^{OH} & \begin{array}{c}H\\R\end{array}\diagup_{OH}^{OH} \\
O,S\text{-Ketal} & N,S\text{-Ketal} & O,N\text{-Ketal} & \text{Ketone hydrate} & \text{Aldehyde hydrate} \\
\mathbf{l} & \mathbf{m} & \mathbf{n} & &
\end{array}
$$

C(OR)₄

Tetraalkyl orthocarbonate
p

Glycopyranoside
q

Glycofuranoside
r

---

may coordinate with an electrophilic reagent and increase its electrophilicity (for example the carbonyl group, **12**) or it may coordinate to a nucleophilic

$$
\begin{array}{c}
\diagup_{\diagdown}C{=}O \rightarrow H \quad \nearrow E^+ \longrightarrow \quad \diagup_{\diagdown}C{=}\overset{+}{O} \qquad O{-}E \\
\mathbf{11} \qquad \underset{H}{\overset{|}{O}} \qquad\qquad H \quad H
\end{array}
$$

$$
\updownarrow
$$

$$
\diagup_{\diagdown}\overset{+}{C}{-}O \\
 H
$$

$$
\diagup_{\diagdown}\overset{\delta+}{C}{=}\overset{\delta-}{O} \; Zn^{2+} \longrightarrow \diagup_{\diagdown}\overset{+}{C}{-}O \\
\mathbf{12} \qquad\qquad\qquad\qquad Zn^+
$$

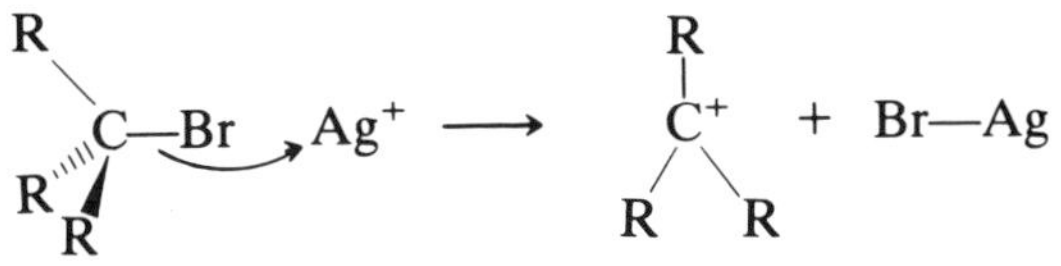

13

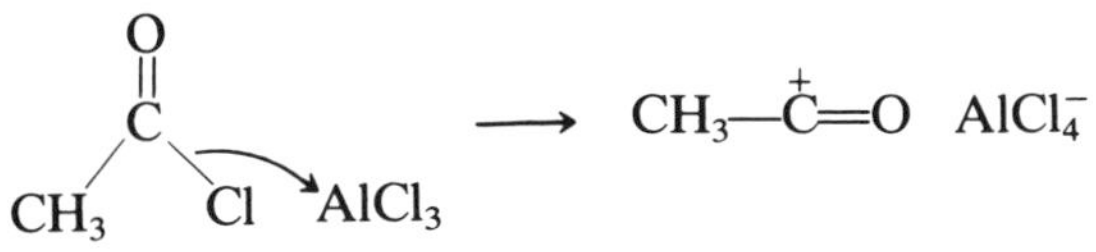

$$\longrightarrow CH_3 - \overset{+}{C} = O \quad AlCl_4^-$$

14

leaving group and increase its nucleofugacity (**13, 14**). In MO terms, the catalyst may increase positive charge (hard electrophilicity) or lower the energy of the LUMO (soft electrophilicity) at the reaction center. The hard/soft properties of catalyst and reaction site should match in order to obtain strong bonding and efficient catalysis. It is natural, therefore, that the Ag^+ and Hg^{2+} cations should catalyze the fission of halide ion in solvolyses and that $AlCl_3$ or BF_3 should catalyze the Friedel–Crafts reaction by generating the strongly electrophilic cation CH_3CO^+ from CH_3COCl[47].

9.2 THE MECHANISMS OF SOME CATALYZED REACTIONS

9.2.1 Substitutions α- to a carbonyl group

Aldehydes and ketones and, indeed, all compounds with a proton on a carbon atom adjacent to a Z substituent, **15**, (—CH—Z where Z = —C(R)=O, —C≡N, —NO$_2$, —SO$_2$R) undergo electrophilic substitutions by halogen, alkylating agents ('virtual' carbocations) or hydrogen-ion exchange which may be catalyzed either by acids or by bases. The bromination of acetone, **16**, is a typical example. In an acetate buffer, the empirical rate law is[48–51]:

$$\text{Rate} = 5 \times 10^{-4}[H_2O] + 1600[H^+] + 1.5 \times 10^7[OH^-] + 5.0[AcOH]$$
$$+ 15[OAc^-] + 20[AcOH][OAc^-],$$

and, as a consequence, the reaction is general acid- and general base-catalyzed. The final term suggests a contribution from a concerted acid +

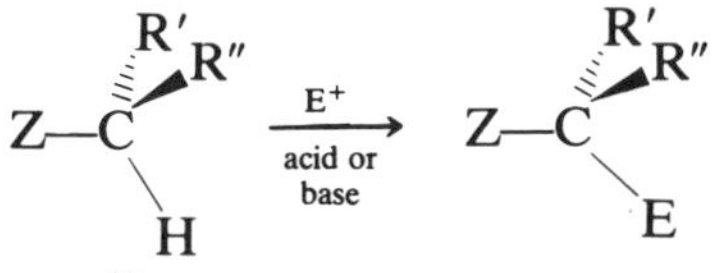

$Z = $ —CO·R, —CN, —SO$_2$R, —NO$_2$
$E^+ = $ Hal$^+$, D$^+$, R$^+$

15

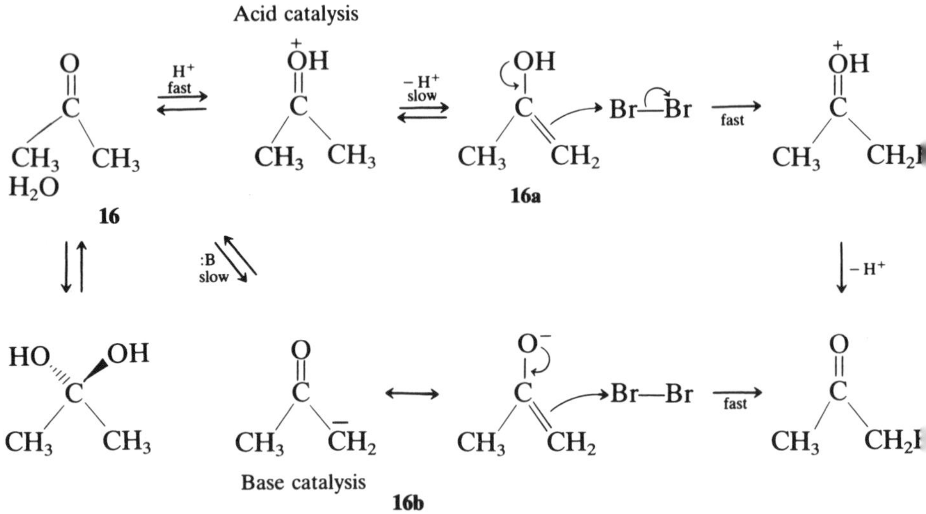

Acid catalysis

16 **16a**

Base catalysis
16b

base ('push–pull') process. The absence of any rate dependence on the bromine concentration and the observation (by Lapworth[52] in 1904) that substitutions by chlorine, bromine and iodine occurred at identical rates, means that the rate-determining steps in all these reactions lead to the same reactive intermediate which is rapidly brominated. The intermediate can be identified with the enol, **16a**, formed under acid catalysis or the enolate, **16b**, formed by base. Both are highly reactive towards electrophilic attack.

The Brønsted parameters, $\alpha = 0.88$, $\beta = 0.55$, are interpreted as showing a stepwise protonation or deprotonation rather than a concerted process for which $(\alpha + \beta)$ should be unity. Hammett reaction constants, ρ, are small and correlate with σ:

$$ArCOCH_3 + Br_2(+catalyst) \rightarrow ArCOCH_2Br$$

Catalyst	ρ
H_2O	0.42
aq. AcOH	−0.64
aq. AcOH + HCl	−0.42

Both protonation and deprotonation separately should be subject to larger substituent effects than these, so there is evidently a canceling of the two and the rate of formation of the enol must depend upon both steps (**17**).

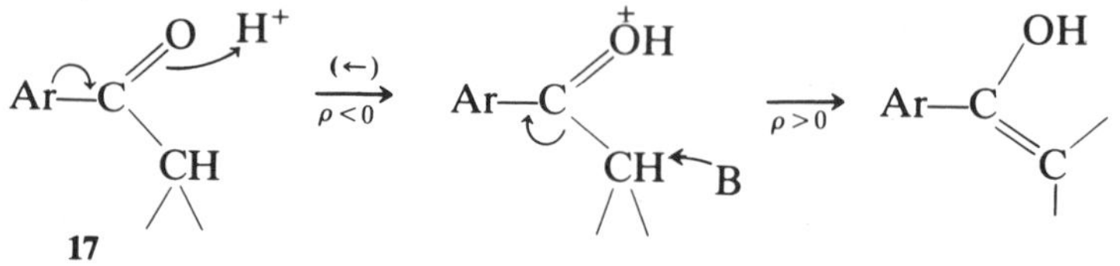

17

The Bunnett w and ϕ values are large and positive (3·8 and 0·8, respectively), showing the involvement of water in the transition state as a proton-transfer agent. Catalytic constants correlate with entropies of activation; this also suggests the importance of solvation changes in the slow step. These observations accord with the mechanism **16**. Other reactions of enolates, such as the family of aldol reactions (Section 12.3), and of their nitrogen analogues, the enamines, **18**, must be of the same type. The greater $+R$ effect of nitrogen makes these species more reactive than enols. Pyrrolidine enamines (**18**) are isolable and readily alkylated (Stork reaction), a convenient method for α-alkylation of a ketone[53,54].

9.2.2 Keto–enol equilibria[55,56]

This is the best-known example of a prototropic change (i.e. isomerization by H^+ migration) and all carbonyl species with an α-H can be considered to reach an equilibrium between the two tautomers. The equilibria lie far in favor of the carbonyl form in simple aldehydes and ketones (Table 9.10) but even as such minority species enols are the reactive components in reactions at the α-carbon[53]. Enols of β-dicarbonyl compounds are far more stable as they exist as the cyclic hydrogen-bonded structure, **19**. The enols of α-diketones are not especially stable: an exception is the enol of cyclopentane-1,2-dione, **20**, in which dipole–dipole repulsions in the keto form are reduced by enolization. Equilibrium constants differ somewhat in

Table 9.10 Acid–base properties of carbonyl compounds

Compound	$K_T = \dfrac{[enol]}{[keto]}$	$10^5 k$ (enolization, H_2O)	$\log K_B$ (protonation)
MeCHO		1·6	−5·3
MeCOMe	$2\cdot5 \times 10^{-4}$	2·76	−7·2
MeCOEt			−8·3
EtCOEt			−7·8
PhCOMe		1·26	−7·8
	$4\cdot8 \times 10^{-3}$	4·4	−7·2
	$2\cdot0 \times 10^{-2}$	23	−5·7
MeCOCH$_2$COMe	11		
MeCOCH$_2$COOEt	0·062		

the gas phase but lifetimes are often longer so that it is possible to obtain a microwave spectrum of CH$_2$=CHOH and characterize its structure. Pure enolic compounds can be isolated when the corresponding keto tautomers are destabilized by $-R$ or very bulky substituents, as for example in **21** and **22**[57–59].

9.2.3 Hydrolyses of acetals, ketals, orthoesters and related compounds[17]

Compounds in which a saturated carbon has two or more alkoxy or related (i.e. $+R$) groups attached (Table 9.9) are susceptible to acid-catalyzed hydrolysis, to form the carbonyl compound, though most are stable towards base. Under acidic conditions equilibrium is set up, the position depending upon the concentration of water (**23**). All reactions involve the cleavage of

Table 9.11   LFER constants for acetal and ketal hydrolysis

Compound hydrolysed	ρ	w	ϕ	k_{D_2O}/k_{H_2O}
ArCH(OEt)$_2$	$-3\cdot35$ to $-4\cdot1$		$-1\cdot02$	
RR'C(OR'')$_2$	$-3\cdot9$	$0\cdot6$		$2\cdot6$
CH$_2$(OR)$_2$	$-4\cdot3$			
ArC(OMe)$_3$	$-2\cdot0$	$0\cdot7$		$2\cdot2$
(Glucose)—OAr (acid catalyzed)	$-0\cdot06$			
(base catalyzed)	$+2\cdot48$			

the C—OR bond, rather than O—R, as shown by the retention of configuration of a chiral group R. The formation of free R$^+$ would, of course, be accompanied by loss of chirality while a nucleophilic displacement at R (S$_N$2) would give inversion. Rates are quite strongly accelerated by electron-donating substituents in either the aldehyde or the alcohol moiety with values of ρ mainly in the range -2 to -4 although very small for glycoside hydrolyses (Table 9.11).

Secondary deuterium isotope effects α to the central carbon are in the range 10–15%: this points to a change in hybridization in the slow step at the central carbon and the formation of an oxocarbocation. Rates of hydrolysis of aliphatic acetals are specific acid-catalyzed (A1). Stereoelectronic factors require the geometry of decomposition for the protonated acetal as indicated, the unshared pair on oxygen coplanar with the departing ligand[60]. Orthoesters of aliphatic alcohols and phenolic acetals react by a general (A–S$_E$2) acid-catalyzed route.

23

9.2.4 Dehydration of aldehyde hydrates and related compounds

The structural unit $>$C(OH)$_2$ is found in the hydrates of aldehydes and ketones, in hemiacetals and in the hemiorthoesters, which are intermediates in the hydrolyses of carboxylic esters (Section 10.6) and which rapidly revert to the carbonyl form. Aliphatic aldehydes are largely hydrated in aqueous solution (formaldehyde to about 99·95%). Both hydration of a carbonyl group and the reverse, dehydration, are general acid- and base-catalyzed. A cyclic concerted process has been proposed for this reaction, **24**, water acting as general acid and base.

24

9.2.5 The formation of oximes, semicarbazones and hydrazones

Initial nucleophilic addition of many species X—NH$_2$ to aldehydes and ketones, **25**, leads to the carbinolamine **25a** (an *N,O*-acetal) which rapidly undergoes acid- and base-catalyzed dehydration. Since the rate-determining step of the reaction is the initial nucleophilic attack on the ketone, as shown by secondary ^{14}C-isotope effects at C=O and a Brønsted $\alpha = 0$ (overall rate unaffected by strength of catalytic acid), the kinetics cannot reveal details of the fast dehydration step but it is probably similar to that deduced for the aldehyde hydrates, **24**.

9.2.6 Decarboxylation

Carboxylic acids lose CO$_2$ with a facility which reflects the stability of the product, a carbanion. Among the easiest to decarboxylate are α-nitro, α-acetylenic and β-keto acids, the latter giving an enolate, **26a**. Primary and secondary, but not tertiary, amines catalyze this reaction as also does CN$^-$. It is postulated that initial imine formation occurs in a rate-determining step, followed by rapid fragmentation, **26**. This is supported by measurement of the rates of imine formation (which is observable) by aniline (**26**; R$'$ = Ph) and of aniline-catalyzed decarboxylation: they are identical as required by this scheme. Simple aliphatic acids such as acetic acid require the forcing conditions of soda-lime at a high temperature to decarboxylate to methane.

$$\text{Na}^+\text{OH}_\text{s}^- + \text{CH}_3\text{COO}^- \rightarrow [\text{Na}^+\text{CH}_2^-\text{COO}^-] + [\text{Na}^+\text{CH}_2^-] + \text{CO}_2^+ \xrightarrow{(\text{H}^+)} \text{CH}_4$$

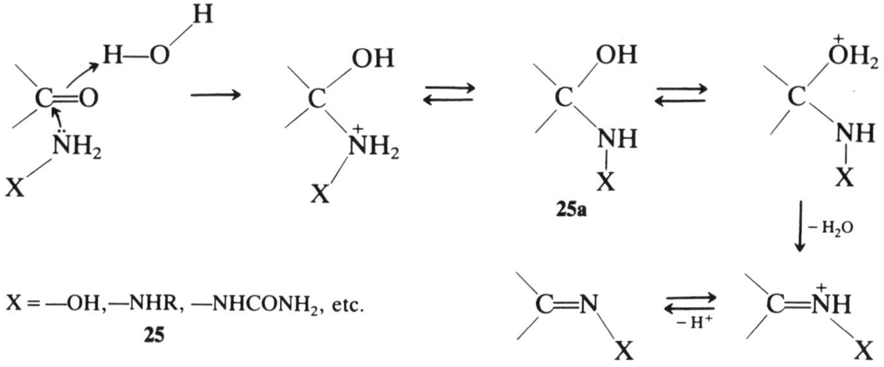

X = —OH, —NHR, —NHCONH$_2$, etc.

25

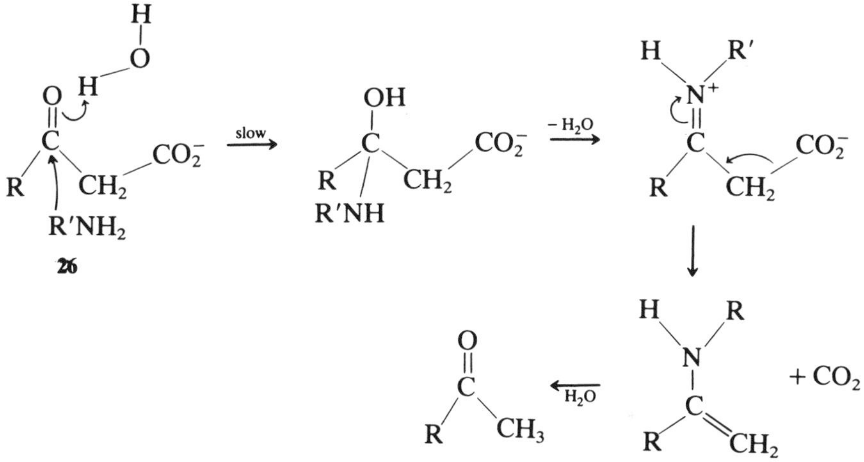

25a

26

slow

$-H_2O$

$-H^+$

$+CO_2$

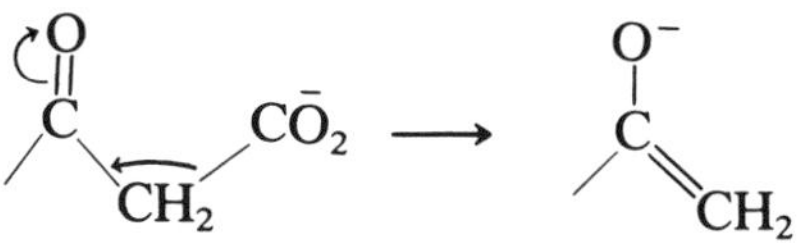

26a

The molecular properties and reactivities of enolates, ambident nucleophiles, are discussed in Section 6.12.6.

9.2.7 Acid-catalyzed alkene–alcohol interchange[61–66]

Alkenes become reversibly hydrated to alcohols under strongly acidic conditions. Both steps must occur by the same mechanism and the dehydration aspect is considered with other eliminations in Chapter 11. Hydration of alkenes in predominantly aqueous solution was originally thought to show specific catalysis and the A1 scheme **27**, via an intermediate π-complex, **27a**, was proposed. However, the solvent KIE, $k_{D_2O}/k_{H_2O} < 1$, precluded pre-equilibrium protonation and, since several examples of general acid catalysis are now established and there seem to be no valid reasons for differentiating any simple alkenes mechanistically, it is concluded that the mechanism $A–S_E2$ operates. A rate-determining protonation to form an intermediate carbocation (probably solvated — an 'encumbered' species) is followed by its rapid hydrolysis. Rates follow the acidity function H_R better than they do H_0, and large negative LFER coefficients indicating electron demand by the reagent agree with this mechanism; the rates of hydration of substituted styrenes, **28**, correlate with σ^+, and those of alkylated ethenes, **29**, correlate with σ_p^+ [61,67,68].

$$28 \quad ArCH{=}CH_2$$

$$\rho^+ = -3\cdot6$$

$$29 \quad R_1R_2C{=}CR_3R_4$$

$$\log k = -8\cdot96 + 10\cdot7 \sum^R \sigma_p^+$$

Steric effects, if present, would be revealed particularly in the latter series by deviations from linearity of alkenes with bulky substituents. These deviations do not, however, appear to be important: this is in accordance with the transfer of a proton, a species with small steric requirements, in the rate-determining step. Regioselectivity is high and results from the formation of the most highly substituted and hence most stable carbocation. Alkyl, aryl and particularly alkoxy substituents aid an alkene carbon to take

30

positive charge. Regiospecificity is very high and follows the Markownikoff Rule (Section 12.1) of addition via the more stable (more highly substituted) carbocation.

Alkenes may be hydrated by dissolving them in concentrated sulfuric acid and diluting with water. In this case the sulfate ester is an intermediate which rapidly hydrolyzes in aqueous acid, **30**. In the presence of ethanol at 140°C, an ether is formed.

9.2.8 Some acid-catalyzed rearrangements[47,69,70]

Skeletal rearrangements promoted by strong acids are plentiful: some important examples are set out in Table 9.12. The catalysts which are most effective may be strong proton acids such as sulfuric acid or powerful Lewis acids such as $AlCl_3$, but their purpose is to loosen a potential nucleophilic leaving group. On its departure, either sequentially or synchronously, this permits a molecular rearrangement to take place, frequently a 1,2-shift of an alkyl or aryl group (the Wagner–Meerwein rearrangement; Section 14.8), which occurs for example on dehydration of alcohols with a neopentyl structure, **31**. The pinacol rearrangement, **33**, is of this type while the Beckmann, **32**, and Schmidt, **34**, rearrangements are analogous but with a carbon–nitrogen alkyl migration. It is probable that rearrangement and bond fission are concerted in these reactions. This accounts for the strictly *anti* geometry of the migrating and leaving groups in the rearrangements of oximes. The Schmidt, **34**, and related Curtius, **35**, and Lossen, **36**, rearrangements proceed by way of an intermediate isocyanate which can be observed but is normally hydrolyzed with decarboxylation of the intermediate carbamic acid. Rearrangements of alkylbenzenes under acidic conditions (Jacobsen rearrangement, for example **37 → 38**) are now understood to take place within the benzenium ion by a series of 1,2-alkyl shifts. The transformations of one isomer to another in superacid solution may be observed by NMR and rates may be measured. These rearrangements tend to proceed until the most stable benzenium ion is formed, i.e. that with the maximum number of $+R$ groups in 2-, 4- or 6-positions stabilizing the positive charge. It may be necessary for relatively high-energy intermediates to be formed en route[71,72] to the final product.

Another major type of rearrangement is the migration of a substituent group from an aryl sidechain to the nucleus, normally at the *para* position as in **39**. In most cases 'crossover' experiments, **40**, indicate the reactions to be

31

Table 9.12 Some catalyzed rearrangements

Name	Reaction

Beckmann, 32

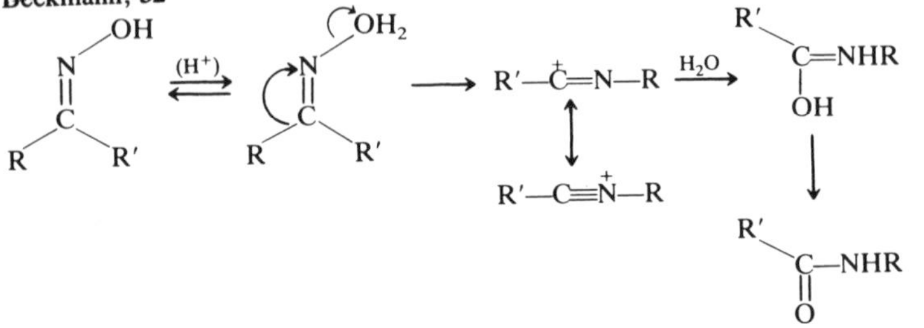

Pinacol, 33

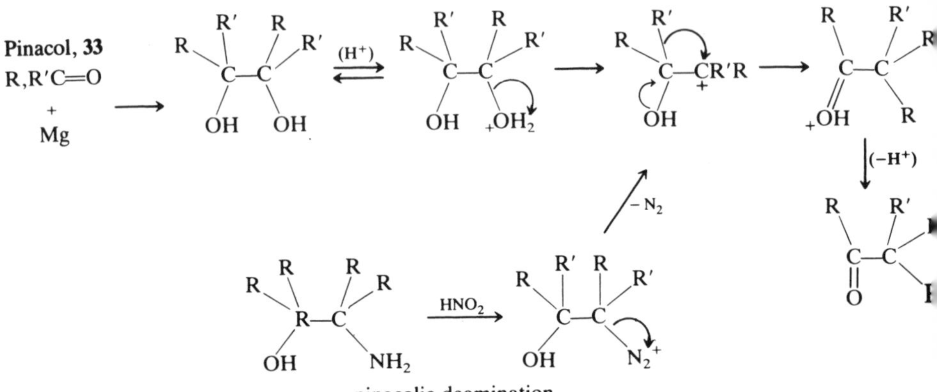

R,R′C=O + Mg

pinacolic deamination

Schmidt, 34

$RNH_2 \xleftarrow{-CO_2} RNHCOOH \xleftarrow{H_2O} R-N=C=O$

Lossen, 36

$\longrightarrow R-N=C=O + OAc^-$

Curtius, 35

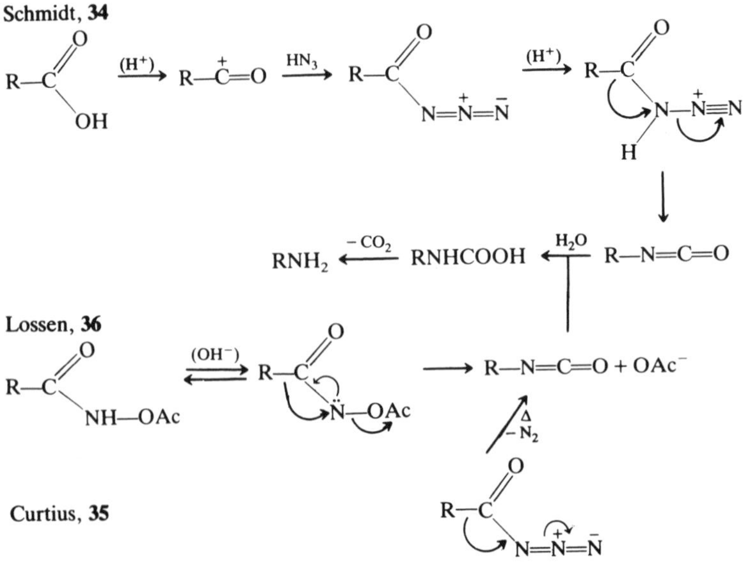

Table 9.12   (*Continued*)

Name	Reaction

Jacobsen, 37

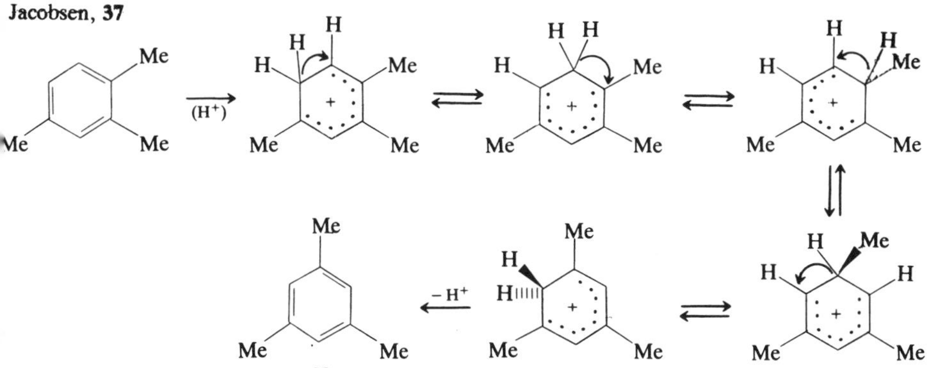

intermolecular; that is, the migrating entity may be transferred to another molecule, usually a second reactive aromatic species suggesting, although not proving, that it has passed through a free stage[73–75]. The formation of a high proportion of *para* products is also suggestive of intermolecular reaction although *ortho/para* ratios are often not those obtained in substitution reactions leading to the same products. The Fries rearrangement of phenyl acetate gives *ortho*- and *para*-acetylphenols in the ratio 12:1 when catalyzed by $AlCl_3$, and 1·04:1 when catalyzed by sulfuric acid compared with 0·54:1 for the acetylation of phenol. It seems likely that both inter- and intra-molecular pathways are available and competitive, their proportions changing subtly with conditions. Thus, the related rearrangement of (+)-2-butyl phenyl ether to *p*-2-butylphenol occurs with retention of configuration of the migrating 2-butyl group although racemization accompanies rearrangement of the mesityl analog[76]. One example which

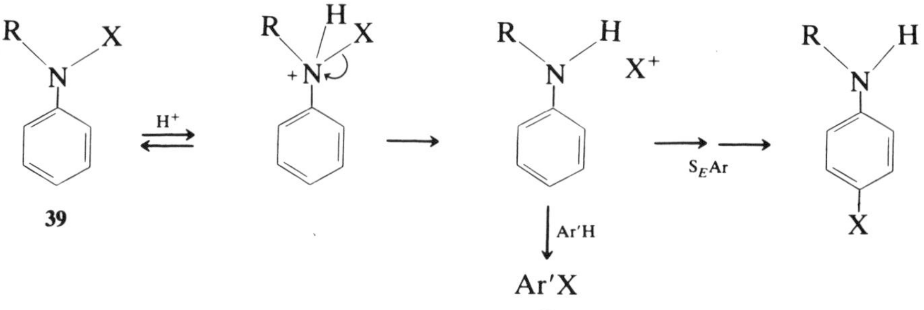

X = —NO (Fischer–Hepp), —Hal (Orton), —N≡NAr, —NO₂

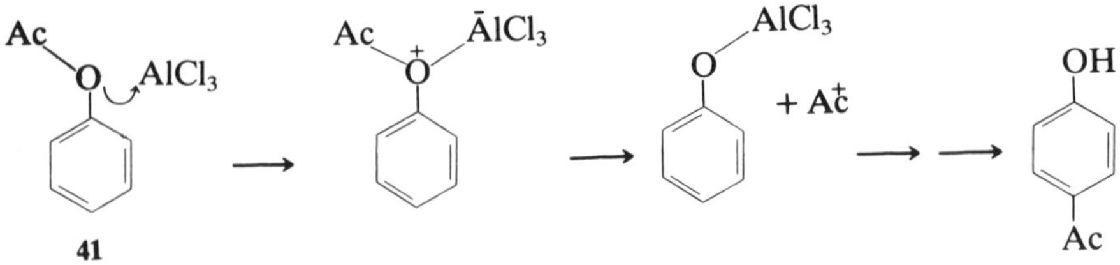

41

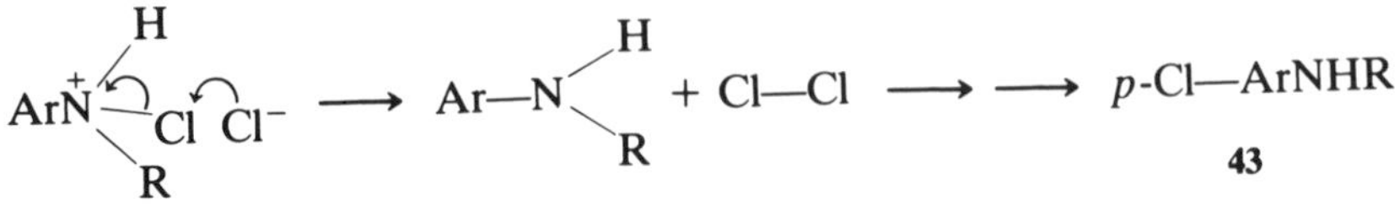

42   **43**

must be largely intermolecular, however, is the rearrangement of *N*-nitroanilines[77] (**39**; X = —NO₂) to give predominantly *ortho*-nitroanilines, since the nitronium ion is such a powerful electrophile that it would be energetically difficult to displace as the free species. The Orton rearrangement of *N*-haloanilines, **42**, is specifically catalyzed by HCl. The acid protonates the nitrogen but chloride ion is required as a nucleophilic catalyst to remove Cl⁺ as Cl₂ which then chlorinates the aromatic ring in the usual way (Section 10.4).

The benzidine rearrangement is more curious[47]. It is kinetically second order in H⁺ (though first, second or fractional order for homologues) and is intramolecular. It appears that the doubly protonated hydrazobenzene, **44**,

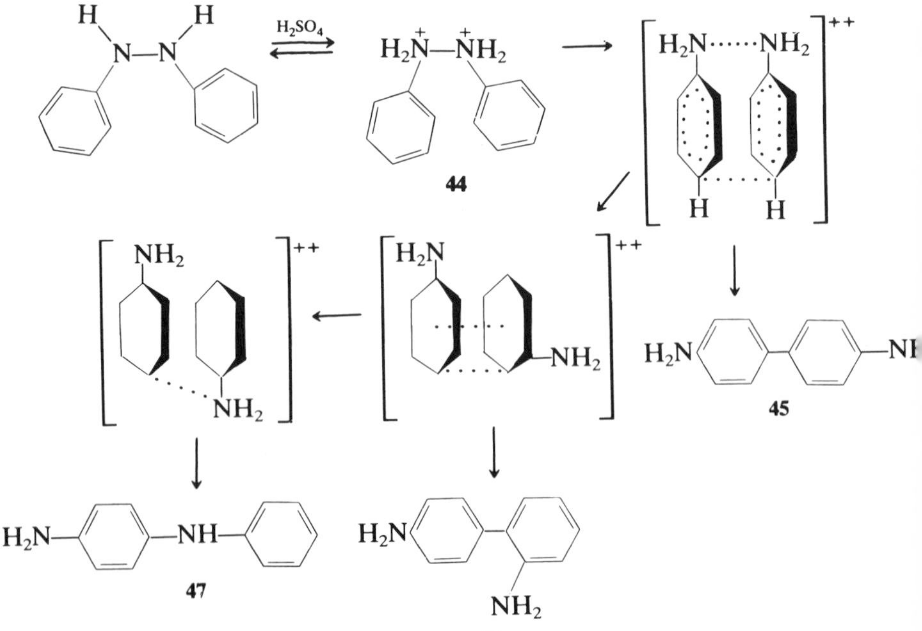

44

45

47

46

must fold over, the two rings being held together by weak forces while bonding develops at the *para* positions leading to benzidine, **45**. If the *para* positions are blocked by substitution, coupling at the *ortho* positions may occur or rotation of one ring with respect to the other may lead to 2,4-, 2,*N*- or 4,*N*-bonding and formation of diphenylines, **46**, and semidines, **47**, as alternative products.

9.2.9 Rate-limiting proton transfers

Proton transfers from carbon acids and to carbon bases are normally far slower than those involving nitrogen or oxygen acids and bases. Reactions of this kind may become rate-limiting in an acid-catalyzed reaction. Notable examples of this behavior are the acid decompositions of diazoalkanes, **48** ($R_2C\overset{+}{=}\overset{-}{N}=N$) which undergo esterification with carboxylic acids[78–82]:

$$R_2CN_2 + R'COOH \rightarrow R'CO \cdot OCHR_2 + N_2.$$

Table 9.13   Deuterium kinetic isotope effects on esterifications by diphenyldiazomethane

Reaction	Solvent	k_H/k_D for	
		k_0	k'
DDM + PhCOOH (D)	Ethyl acetate	1·12	7·3
DDM + PhCOOH (D)	Acetone	1·78	4·2
DDM + MeCOOH (D)	Ethyl acetate	0·97	4·8
DDM + MeCOOH (D)	Acetone	0·63	5·9

Diazomethane, CH_2N_2, provides a clean and quantitative route to methyl esters in non-protic solvents. Most studies of the reaction have been performed with diphenyldiazomethane (DDM), Ph_2CN_2, which is more stable. That proton transfer is rate-determining is shown by the primary isotope effect on reactions using RCOOH(D). However, simple second-order kinetics are not observed but the reaction obeys the relationship:

$$\text{Rate} = [\text{DDM}]([k_0[\text{RCOOH}] + k'[\text{RCOOH}]^2). \qquad (9.10)$$

The kinetic isotope effect is on k', not on k_0, which even shows an inverse effect (Table 9.13). This has been interpreted as indicating the importance of two acidic species, one the monomeric acid and the other a linear hydrogen-bonded dimer. An alternative possibility seems to be that the diazo compound has two sites for protonation, at carbon (**48a**) or at the terminal nitrogen (**48b**). The former leads immediately to reaction by the expulsion of N_2 and so shows a large PKIE. The latter cannot lead directly to reaction but instead establishes an equilibrium (for which either K_H or K_D may be greater than 1, depending upon the solvent). There is then a slow internal proton transfer to the same diazonium ion-pair as before. 1,3-Proton transfers of this kind are analogous to keto–enol tautomerism and to that of triazenes, **49**. The slow step involves a moderate increase in dipolar character as shown by the solvent effect; $k(\text{MeCN})/k(\text{dioxane}) = 65$, for which the degree of proton transfer at the transition state has been inferred to be about 30%. Solvent effects on rates of the analogous reactions of triazenes are found to correlate with a combination of solvent polarity and donor character, pointing to the importance of specific solvation (Eq. (9.11))[75,78,80,83]

$$\log k = -4.297 - 0.119\,\text{DN} + 0.105E_T. \tag{9.11}$$

In alcoholic solvents, Eq. (9.12) holds:

$$\log k = 3.341 - 0.0056B - 1.752f_\varepsilon - 0.0351E \tag{9.12}$$

in which B and E are basicity and electrophilicity parameters (Section 5.8.1), and f_ε is the Kirkwood function $(\varepsilon - 1)/(2\varepsilon + 1)$ where $\varepsilon =$ dielectric constant. In protic solvents products contain both the ester and the benzhydryl ether derived from the solvent; the ratio of these is independent of the strength of the acid. Product formation therefore appears to depend upon the rate at which the benzhydryl cation can diffuse out of the solvent cage, in competition with reaction with the carboxylate ion within the cage.

9.3 CATALYSIS BY NON-COVALENT BINDING

Rates of reaction of a homologous series usually do not change greatly after the first few members, as the chain length is increased. There are, however, some substantial rate effects sometimes evident when both reagents in a bimolecular process have long alkane chains. For example, the aminolysis of

50

nitrophenyl esters, **50**, depends upon chain length and is faster for decyl compounds than for methyl by a factor of 400. The same is found for the hydrolysis of alkylsulfonates[84]:

$$\text{R—OSO}_2\text{O}^- + \text{H}_3\text{O}^+ \rightarrow \text{ROH} + \text{H}_2\text{SO}_4;$$

R	β_{rel}
$C_1–C_5$	1
C_{10}	8
C_{14}	56
C_{18}	100

Protein is hydrolyzed 100 times faster by long-chain sulfonic acids than by low molecular weight homologs of similar strength and concentration. It is known that hydrocarbon chains tend to associate in water not because of intrinsic attractive forces between them but because when they do so, the water is enabled to maintain the maximum hydrogen bonding to itself. This 'hydrophobic bonding' must promote association between the aliphatic

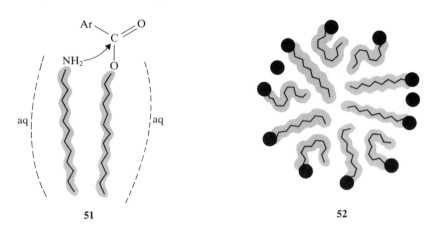

51   52

chains of the reagents so that the functional groups are more likely to interact (**51**). This brings advantages similar to those of an intramolecular reaction (Chapter 14).

Solutes with long aliphatic chains terminated by an ionic group are stabilized in aqueous solution by aggregation such that the hydrocarbon chains are in close proximity (**51**) and exclude water while the ionic groups protrude into the aqueous phase. The result is a semi-ordered structure known as a micelle, **52**. Micelles are typically 150–500 pm diameter and approximately spherical, and they contain 50–100 molecules of the monomer. They may be cationic, of the type R—NMe$_3^+$, or anionic, e.g. R—SO$_2$O$^-$ (R = C$_{10}$–C$_{20}$).

Compounds of this type are of commercial importance since they act as detergents and are capable of dispersing and solubilizing hydrophobic molecules which enter into the micelle. Micelles are themselves capable of catalyzing a wide variety of reactions[85,86]: three modes of action which affect reaction rates in the presence of detergents may be recognized.

(a) Concentration effects occur if bimolecular reactions are accelerated through incorporation of the reagents into the micelle, in which their concentration will be greatly enhanced compared with the bulk solution. This effect will be expected to operate when both reagents have substantial hydrocarbon components.

(b) There may be electrostatic effects between the high concentration of charges at the surface of the micelle and the reagents either stabilizing or destabilizing the transition state. These should be most prominent when one reagent is incorporated partly into the micelle, its reactive site protruding near the ionic surface towards which the other reagent must approach. In general a cationic micelle will attract anions but repel cations. Rates of reaction between a neutral species in the micelle and an anionic reagent will be accelerated by a cationic micelle, and conversely an anionic micelle accelerates the action of a

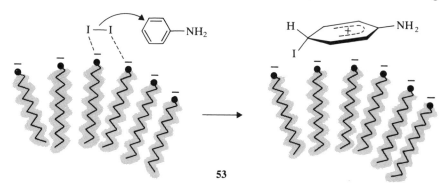

53

cationic agent on a neutral species in the micelle. The effect is similar in origin to a Brønsted salt effect (Chapter 5). An anionic micelle will stabilize the benzenium ion intermediate formed in the iodination of aniline (**53**)[96].

(c) A third effect for reactions which take place within the micelle will be

Table 9.14   Some reactions catalyzed by micelles

Reaction	Ionic type of surfactant	k_{cat}/k_0
O_2N⟨⟩O—COC_8H_{17} + OH^-	−	30
O_2N⟨⟩O—$COC_{10}H_{21}$ + OH^-	−	21
H_2N⟨⟩CO—OEt + OH^-	$\{+ \atop -}$	$1 \atop 18$
$Na^+\bar{O}SO_2O$—$C_{10}H_{21}$ + H_3O^+	−	36
O_2N⟨⟩O—$PO\cdot(OH)_2$, H^+ NO_2	+	22, 50
O_2N⟨⟩O—$PO\cdot(OPh)_2$, H^+ NO_2	$\{+ \atop -}$	$50 \atop 70$
Cl⟨⟩CH=$N\cdot tBu$ + H_2O	$\{+ \atop -}$	$1 \atop 290$

that due to transfer from an aqueous to a hydrocarbon environment, a medium of much lower polarity. Rates will be affected according to the principles discussed in Chapter 5. It is often difficult to separate these contributions completely. In Table 9.14 are set out examples of reactions catalyzed by micelles.

PROBLEMS

1 True acid catalysis (in which the catalyzing acid is not consumed) must also require the availability of a base. Consider the logic of this statement.
2 A general scheme for acid catalysis is the following:

$$S + HA \underset{k_{-1}}{\overset{k_1}{\rightleftharpoons}} SH^+ + A^-;$$

$$SH^+ \underset{k_{-2}}{\overset{k_2}{\rightleftharpoons}} PH^+;$$

$$PH^+ + A^- \overset{k_3}{\longrightarrow} P + HA.$$

Derive kinetic expressions for catalyzed reactions when
(a) $k_{-1} \gg k_2$ and $k_3 \gg k_{-2}$,
(b) $k_2 \gg k_{-1}$.
3 Strong bases promote isomerization of alkenes with a tendency to move double bonds into conjugation. The following isomerizations were carried out in a deuterated solvent and deuterium incorporation was determined:

$$PhCH_2CH{=}CH_2 \xrightarrow{\textit{tert-BuO}^-/\textit{tert-BuOD}} PhCH{=}CHCH_3 + PhCH{=}CHCH_2D$$
$$59\% \qquad\qquad 41\%$$

How may these results be accounted for?
4 Sketch out the steps needed to convert ethyl orthocarbonate into CO_2:

$$C(OEt)_4 \xrightarrow{H^+,H_2O} \to CO_2 + 4EtOH.$$

5 Suggest mechanisms for the following catalyzed reactions.
(a) Amide dehydration:

(b) The 'abnormal' Beckmann rearrangement:

$$\xrightarrow[\text{(H}_2\text{O)}]{\text{H}^+} \quad \text{MeC}\equiv\text{N} + \text{Ph}_2\text{CHOH}.$$

(c) The Stieglitz rearrangement:

$$\xrightarrow[\text{MeOH}]{\text{Ag}^+}$$

(d) The Sommelet–Hauser rearrangement:

$$\xrightarrow{\text{K}^+\text{NH}_2^-}$$

(e) The Bamberger rearrangement:

$$\xrightarrow{\text{H}^+}$$

(f) Wagner–Meerwein rearrangements in the terpene series:

$$\xrightarrow{\text{HONO}}$$

Homogeneous catalysts

(g) The S_N2' reaction:

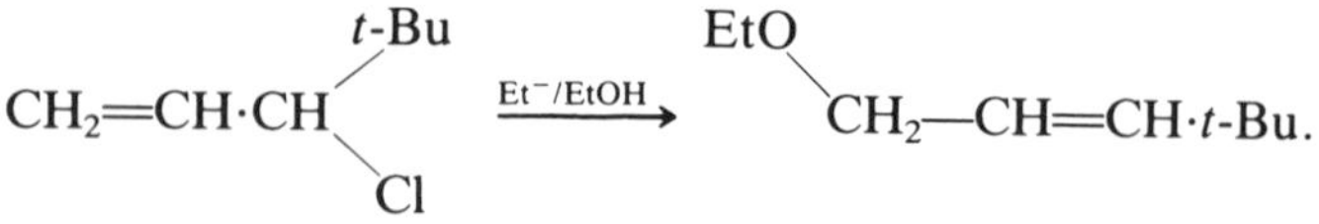

$$CH_2{=}CH{\cdot}CH\Big\langle\begin{matrix}t\text{-Bu}\\\\Cl\end{matrix}\xrightarrow{\ Et^-/EtOH\ }\begin{matrix}EtO\\\\CH_2{-}CH{=}CH{\cdot}t\text{-Bu.}\end{matrix}$$

(h) The Ritter reaction:

$$Me_2CHOH + PhCN \xrightarrow{\ H_2SO_4\ } PhC\Big\langle\begin{matrix}O\\\\NHCHMe_2.\end{matrix}$$

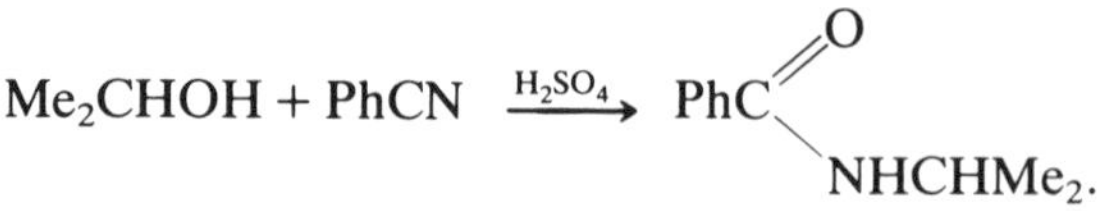

6 Suggest a mechanism for the Baeyer–Villiger oxidation of a ketone to an ester consistent with the isotopic labeling shown and with the presence of a ^{13}C-PKIE at the carbon indicated:

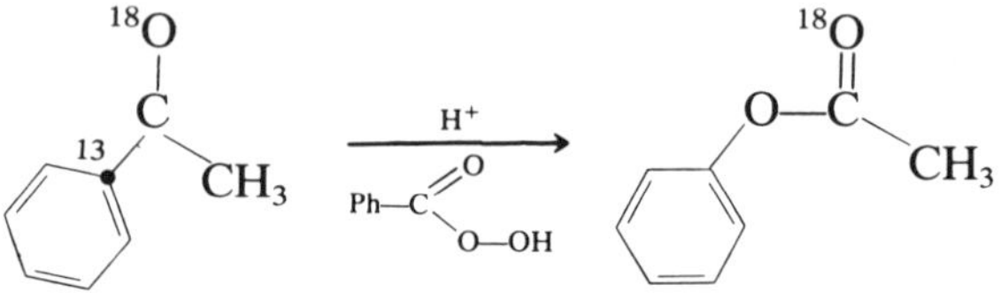

Substitution in the aromatic ring of the ketone leads to a Hammett plot, $\rho^+ = -1{\cdot}45$. Does this accord with your mechanism?

REFERENCES

1 W. Jencks, *Catalysis in Chemistry and Enzymology*, McGraw–Hill, New York, 1969.
2 M. L. Bender, *Mechanisms of Homogeneous Catalysis from Protons to Proteins*, Wiley–Interscience, New York, 1971.
3 P. G. Ashmore, *Catalysis and Inhibition of Chemical Reactions*, Butterworth, London, 1963.
4 C. H. Bamford and C. F. H. Tipper (Eds), *Comprehensive Chemical Kinetics*, Vol. 20, Elsevier, Amsterdam, 1978.
5 L. Zucker and L. P. Hammett, *J. Amer. Chem. Soc.*, **61**, 2791 (1939); R. P. Bell, *Acid–Base Catalysis*, Oxford University Press, 1941.
6 F. A. Long and M. A. Paul, *Chem. Rev.*, **57**, 938 (1957).
7 J. F. Bunnett, *J. Amer. Chem. Soc.*, **83**, 4956, 4968, 4973, 4978 (1961).
8 H. Paul and F. A. Long, *Chem. Rev.*, **57**, 1 (1957).
9 M. Bender, *Chem. Rev.*, **60**, 53 (1960).
10 R. P. Bell, *Acid–Base Catalysis*, Oxford U.P. 1941; E. Caldin and V. Gold (Eds), *Proton Transfer Reactions*, Chapman and Hall, London, 1975.
11 G. Williams and D. J. Clark, *J. Chem. Soc.*, 1304 (1956).
12 C. A. Bunton and J. L. Wood, *J. Chem. Soc.*, 1522 (1955).
13 P. Salomaa, *Suomi Kim.*, **145**, 328 (1959).
14 F. A. Long and M. Purchase, *J. Amer. Chem. Soc.*, **72**, 3267 (1950).
15 C. A. Bunton, J. H. Fendler, A. Fuller, S. Perry and J. Rocek, *J. Chem. Soc.*, 6174 (1965).

16   R. L. Burwell, *Chem. Rev.*, **54,** 615 (1954).

17   E. H. Cordes and H. G. Bull, *Chem. Rev.*, **74,** 5812 (1974).

18   C. A. Bunton, T. Hudwick, D. R. Llewellyn and Y. Pocker, *J. Chem. Soc.*, 402 (1958).

19   L. P. Hammett and A. J. Deyrup, *J. Amer. Chem. Soc.*, **82,** 6104 (1960).

20   J. G. Pritchard and F. A. Long, *J. Amer. Chem. Soc.*, **78,** 2667 (1956).

21   C. A. Lane, *J. Amer. Chem. Soc.*, **86,** 2521 (1964).

22   R. P. Bell, A. L. Dowding and J. M. Noble, *J. Chem. Soc.*, 3106 (1955).

23   K. Yates and J. B. Stevens, *Can. J. Chem.*, **43,** 529 (1965).

24   J. T. Edward and S. C. R. Meacock, *J. Chem. Soc.*, 2000 (1957).

25   Y. Pocker, J. E. Meary and B. J. Nist, *J. Phys. Chem.*, **71,** 4509 (1967).

26   N. C. Deno, F. A. Kish and H. J. Peterson, *J. Amer. Chem. Soc.*, **87,** 2157 (1965).

27   K. Yates and J. B. Stevens, *Can. J. Chem.*, **43,** 529 (1965).

28   L. L. Schaleger and F. A. Long, *Adv. Phys. Org. Chem.*, **1,** 1 (1963).

29   E. Whalley, *Prog. Phys. Org. Chem.*, **2,** 93 (1964).

30   T. Asano and W. J. LeNoble, *Chem. Rev.*, **78,** 407 (1978).

31   N. S. Isaacs, *Liquid Phase High Pressure Chemistry*, Wiley, Chichester, 1981.

32   R. P. Bell, *The Proton in Chemistry*, Cornell U.P., Ithaca, N.Y., 1959.

33   J. N. Brønsted and K. J. Pedersen, *Z. Phys. Chem.*, **108,** 185 (1924); J. N. Brønsted and W. F. K. Wynne-Jones, *Trans. Farad. Soc.*, **25,** 59 (1929).

34   M. Eigen, *Angew. Chem. Int. Ed.*, **3,** 1 (1964).

35   R. A. More O'Ferrall, *J. Chem. Soc., B*, 274 (1970).

36   A. R. Butler and V. Gold, *J. Chem. Soc.*, 2305 (1961).

37   A. R. Butler and V. Gold, *J. Chem. Soc.*, 1334 (1962).

38   J. F. Kirsch and W. P. Jencks, *J. Amer. Chem. Soc.*, **86,** 833 (1964).

39   J. F. Bunnett and F. P. Olsen, *Can. J. Chem.*, **44,** 1899, 1917 (1966).

40   M. Kilpatrick, *J. Amer. Chem. Soc.*, **50,** 2891 (1928).

41   S. L. Johnson, *J. Phys. Chem.*, **67,** 495 (1963).

42   T. C. Bruice and S. J. Benkovic, *J. Amer. Chem. Soc.*, **86,** 418 (1964); M. L. Bender and B. W. Turnquist, *J. Amer. Chem. Soc.*, **79,** 1652 (1957).

43   W. P. Jencks, *J. Amer. Chem. Soc.*, **94,** 4731 (1972).

44   W. P. Jencks, *Chem. Rev.*, **72,** 702 (1972).

45   R. L. Schowen, *Prog. Phys. Org. Chem.*, **9,** 275 (1972).

46   V. Gold, *Adv. Phys. Org. Chem.*, **7,** 259 (1969).

47   C. K. Ingold, *Structure and Mechanism in Organic Chemistry*, Bell, London, 1969.

48   H. M. Dawson and E. Spivey, *J. Chem. Soc.*, 2180 (1930).

49   R. P. Bell and P. Jones, *J. Chem. Soc.*, 88 (1953).

50   K. J. Pedersen, *J. Phys. Chem.*, **38,** 590 (1934).

51   H. Hart, *Chem. Rev.*, **79,** 515 (1979).

52   A. Lapworth, *J. Chem. Soc.*, **85,** 30 (1904).

53   I. Fleming, *Chimia*, **34,** 265 (1980).

54   J. Szmuczkovic, *Adv. Org. Chem.*, **4,** 1 (1963).

55   H. Hart, *Chem. Rev.*, **79,** 515 (1979); R. C. Fuson, J. Case and C. H. McKeever, *J. Amer. Chem. Soc.*, **62,** 3250 (1940).

56   J. P. Guthrie and P. A. Cullimore, *Can. J. Chem.*, **57,** 240 (1979).

57   C. H. DePuy, R. N. Greene and T. E. Schroer, *Chem. Comm.*, 1225 (1968).

58   M. G. Voronkov, N. A. Keiko, T. A. Kuznetsova, V. A. Pestunovich and I. D. Kalekhman, *Dokl. Akad. Nauk. SSSR*, **247,** 110 (1979).

59   G. S. Hammond, *Steric Effects in Organic Chemistry*, M. S. Newman (Ed.), Wiley, New York, 1956.

60   P. Deslongchamps, *Stereoelectronic Effects in Organic Chemistry*, Pergamon, London, 1983.

61   V. J. Nowlan and T. T. Tidwell, *Acc. Chem. Res.*, **10,** 252 (1977).

62   G. R. Lucas and L. P. Hammett, *J. Amer. Chem. Soc.* **64,** 1938 (1942)

63   F. G. Ciapetta and M. Kilpatrick, *J. Amer. Chem. Soc.* **70,** 639 (1948).

64 P. B. D. de la Mare and R. Bolton, *Electrophilic Additions to Unsaturated Systems*, Elsevier, Amsterdam, 1982.
65 P. D. Bartlett and G. D. Sargent, *J. Amer. Chem. Soc.*, **87**, 1297 (1965).
66 R. H. Boyd, R. W. Taft, A. P. Wolf and D. R. Christmas, *J. Amer. Chem. Soc.*, **82**, 4729 (1960).
67 H. C. Brown and E. N. Peters, *J. Amer. Chem. Soc.*, **95**, 2400 (1973).
68 K. Oyama and T. T. Tidwell, *J. Amer. Chem. Soc.*, **98**, 947 (1976).
69 M. J. Dewar, *Molecular Rearrangements* (P. De Mayo, Ed.), Wiley, New York, 1963.
70 P. A. S. Smith, *Molecular Rearrangements,* (P. De Mayo, Ed.), Wiley, New York, 1963, Chapter 8.
71 D. M. Brouwer, E. L. Mackor and C. Maclean, Ref. *72*.
72 G. A. Olah and P. v. R. Schleyer (Eds), *Carbonium Ions,* Wiley, New York, 1970.
73 Y. Yukawa, *J. Chem. Soc. Japan,* **71**, 547, 603 (1950).
74 R. A. Becker, G. G. Melikyan, B. L. Dyatkin and I. L. Knunyants, *Zh. Org. Khim.*, **11**, 1370 (1975).
75 K. J. P. Orton, F. G. Soper and G. Williams, *J. Chem. Soc.*, 998 (1928).
76 M. M. Sprung and E. E. Wallis, *J. Amer. Chem. Soc.*, **56**, 1715 (1934).
77 E. D. Hughes and G. T. Jones, *J. Chem. Soc.*, 2678 (1950).
78 R. A. More O'Ferrall, W. K. Kwok and S. I. Miller, *J. Amer. Chem. Soc.*, **86**, 5553 (1964).
79 N. B. Chapman, M. J. R. Dack, D. J. Newman, J. Shorter and R. Wilkinson, *J. Chem. Soc., Perkin II,* 962, 971 (1974).
80 M. H. Aslam, N. B. Chapman, M. Charton and J. Shorter, *J. Chem. Res.*, (*S*), 188; (*M*), 2301 (1981).
81 R. A. More O'Ferrall, *Adv. Phys. Org. Chem.*, **5**, 331 (1967).
82 N. B. Chapman, A. Ehsan, J. Shorter and K. J. Toyne, *Tetrahedron Lett.,* 1049 (1968).
83 R. A. More O'Ferrall, *Adv. Phys. Org. Chem.*, **5**, 331 (1967).
84 J. L. Kurtz, *J. Phys. Chem.*, **66**, 2239 (1962).
85 E. J. Fendler and J. H. Fendler, *Adv. Phys. Org. Chem.*, **8**, 271 (1970).
86 J. H. Fendler, *Acc. Chem. Res.*, **13**, 7 (1980).
87 W. P. Jencks and J. Carriulo, *J. Amer. Chem. Soc.*, **82**, 1779 (1960).

10 Substitution reactions at carbon

There is a large group of organic reactions which involve the replacement of one covalently bound entity by another at a carbon atom. The incoming reagent and the leaving group may both be nucleophiles or electrophiles (or radicals; Chapter 15), and the overall reactions may be expressed:

Nucleophilic: $\text{Nu:}^- + \text{C—X} \longrightarrow \text{C—Nu} + :\text{X}^-.$

Electrophilic: $\text{E}^+ + \text{C—X} \longrightarrow \text{C—E} + \text{X}^+.$

(The charges indicated are relative.)

Reactions of these types will occur at any variety of the central carbon atom — saturated (sp^3-hybridized), or unsaturated (sp^2- or sp-hybridized), whether of aromatic, alkene, alkyne or carbonyl type. The characteristics of each differ widely, however, and will be considered separately.

10.1 SUBSTITUTIONS AT SATURATED CARBON

10.1.1 Nucleophilic substitution

In Table 10.1 a selection of common reactions which fall into this category are set out, clearly a large and important part of organic chemistry[1]. These may all be discussed in the context of the mechanistic scheme (shown in Fig. 10.1) which is due to Winstein[2] and is an elaboration of older and simpler schemes shaped by Hughes, Ingold and others[3][4] from 1935 onwards. Figure 10.1 indicates that a range of mechanisms may operate depending upon the nature of the reactants and solvents. There is a fundamental division of these pathways according to whether the substitution is concerted (designated S_N2) or a two-step bond fission–bond formation sequence (designated S_N1). Further refinement in the latter class distinguishes between various types of intermediate — intimate or solvent-separated ion-pairs or free carbocations. The relevant free-energy profiles of these basic pathways are shown in Fig. 10.2. The one which is followed by any given system will be

Table 10.1 Nucleophilic substitutions at saturated carbon[1]

Reaction	Nu:	Product, Nu—R
Displacements of halide, Nu: + R—Br → Nu—R + :Br⁻		
1 Hydrolysis	H_2O, OH^-	R—OH
2 Alcoholysis (Williamson synthesis)	OEt^-	R—OEt
3 Aminolysis	$EtNH_2$,	R—NHEt R_2NEt
4 Menshutkin reaction[5]	Et_3N	R—$\overset{+}{N}Et_3hal^-$
	Et_2S	R—$\overset{+}{S}Et_2hal^-$
5 Halogen exchange	*Br	R—*Br
Finkelstein reaction	I^-	R—I
6 Formation of cyanides	CN^-	R—CN
Formation of azides	N_3^-	R—N_3
Formation of thiocyanates	SCN^-	R—S—CN
Formation of nitroalkanes	O—N—O⁻	R—NO_2
7 Acetoacetate and malonate alkylation	$\bar{C}H(COOEt)_2$	R—$CH(COOEt)_2$
8 Reduction by complex hydride[6]	AlH_4^-	R—H
Displacements of oxygen ligands		
9 Acid-catalyzed cleavage of alcohols, R—OH ⇌ R—$\overset{+}{O}H_2$	Cl^-	R—Cl
10 Acid-catalyzed cleavage of ethers, $R_2\overset{+}{O}$—H	I^-	R—I, ROH
11 Methylation with dimethyl sulfate CH_3—OSO_2OMe (similarly with methyl triflate, CH_3OCOCF_3, and Meerwein reagent[7], CH_3—$\overset{+}{O}Me_2$)	PhO^-	PhO—CH_3
12 Ring opening of epoxides[9],	Nu:	
13 Displacements of arylsulfonates, R—OSO_2Ar	Nu:	R—Nu
Displacements of nitrogen ligands		
14 Decomposition of diazonium ions, R–N_2^+	H_2O	R—OH
15 Hofmann degradation of R–N⁺R_3[8]	OH^-	R—OH, NR_3

that with the lowest activation energy. To anticipate later discussion slightly, the tendency to react by prior ionization, i.e. by the S_N1 route rather than the synchronous S_N2, is increased by:

(a) substitution at the carbon center, i.e. primary < secondary < tertiary;
(b) the presence of electron-donating (+R)substituents (Ar > R > H);
(c) increasing solvent polarity, depending upon charge type;
(d) easy leaving groups (high nucleofugacity of X^-).

10.1.2 The bimolecular mechanism, S_N2

Reactions which normally proceed by this route include most of the examples of Table 10.1 when the reaction center is methyl, a primary group,

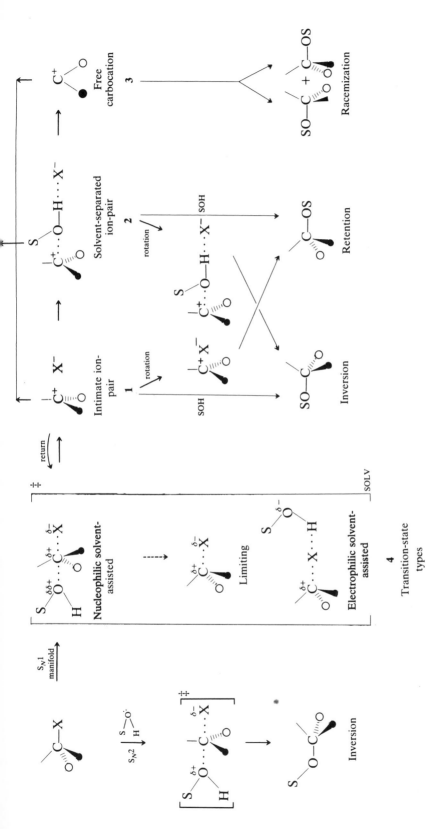

Fig. 10.1 Mechanistic relationships in S_N reactions at saturated carbon.

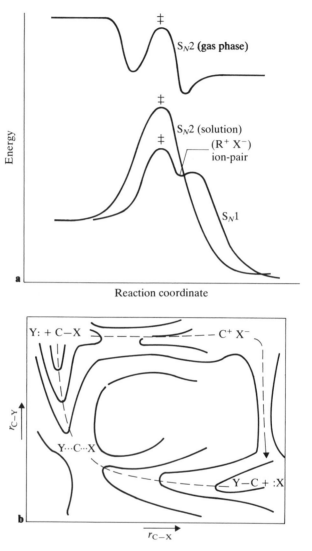

Fig. 10.2 Energies of S_N1 and S_N2 reactions: (**a**) schematic energy profiles; (**b**) schematic energy contours.

RCH_2—, or possibly also a secondary group, $RR'CH$—, particularly when the solvent is one of relatively low polarity.

Kinetics and activation parameters

Since the reaction is initiated by collision between substrate and nucleophile, a bimolecular process, second-order kinetics would be ex-

Table 10.2 Dependence of entropies and volumes of activation of S$_N$2 reactions on charge type

Charge type	Typical $\Delta S^\ddagger$, $\Delta V^\ddagger$	$\Delta S^\ddagger / J\,K^{-1}\,mol$ $(cal\,K^{-1}\,mol^{-1})$	$\Delta V^\ddagger / cm^3\,mol^{-1}$
1. Charge separation[5]	Large, negative	−90 to −120 (−20 to −30)	−30 to −50
2. Charge neutralization[6]	Positive	+50 to +70 (12 to 16)	+25 to +45
3. Charge delocalization	Moderate negative	−20 to −40 (−5 to −10)	−5 to −20

pected and are observed, for example,

$$PhCH_2Cl + N_3^- \rightarrow PhCH_2N_3 + Cl^-;$$
$$Rate = k_2[PhCH_2Cl][N_3^-].$$

This in itself is a necessary but insufficient criterion by which to assign the S$_N$2 mechanism, as will be seen below. Rates and activation energies will, of course, vary widely according to the nature of Nu:, —X and the alkyl moiety, but entropies of activation better characterize the activation process. Since this is associative, $\Delta S^\ddagger$ should be negative but, as solvation plays an important part, the range of values will depend upon the charge type under consideration (Table 10.2) and upon the solvent (Section 5.5).

For similar reasons, volumes of activation are generally negative, highly so when increased solvation of the transition state is occurring (type 1), but they become positive when reduced solvation accompanies activation (type 2)[7].

(1) $Et_3N: + \overset{Me}{\underset{|}{CH_2}}{-}I \rightarrow [Et_3\overset{\delta+}{N}\cdots\overset{Me}{\underset{|}{CH_2}}\cdots\overset{\delta-}{I}] \rightarrow Et_3\overset{+}{N}{-}\overset{Me}{\underset{|}{CH_2}} + I^-;$
   $\Delta S^\ddagger = -115\,J\,K^{-1}\,mol^{-1}\,(-27\,cal\,K^{-1}\,mol^{-1});\quad \Delta V^\ddagger = -38\,cm^3\,mol^{-1}.$

(2) $Br^- + \overset{Me}{\underset{|}{CH_2}}{-}\overset{+}{S}Et_2 \rightarrow [\overset{\delta-}{Br}\cdots\overset{Me}{\underset{|}{CH_2}}\cdots\overset{\delta+}{S}Et_2] \rightarrow Br{-}\overset{Me}{\underset{|}{CH_2}} + SEt_2;$
   $\Delta S^\ddagger = +76\,J\,K^{-1}\,mol\,(+18\,cal\,K^{-1}\,mol^{-1});\quad \Delta V^\ddagger = +32\,cm^3\,mol^{-1}.$

(3) $EtO^- + CH_3{-}Br \rightarrow [\overset{\delta-}{EtO}\cdots\overset{\overset{\displaystyle H}{|}}{C}\cdots\overset{\delta-}{Br}] \rightarrow EtOCH_3 + Br^-;$
   $\Delta S^\ddagger = -26\,J\,K^{-1}\,mol^{-1}\,(-6\,cal\,K^{-1}\,mol^{-1});\quad \Delta V^\ddagger = -2\cdot7\,cm^3\,mol^{-1}.$

Structural effects[4,8]

S_N2 reactions are rather insensitive to electronic effects of substituents at carbon since the requirements of bond formation and bond breaking, synchronously occurring, are opposite. A substituent which aids the attachment of the nucleophile retards the departure of the leaving group and conversely, one which retards attachment will aid departure, producing a cancelation of effects. Reactions generally correlate with σ^0 indicating a lack of conjugative interactions and values of ρ are small, $+1$ to -1 depending upon the relative importance of the bond-making and -breaking processes in reaching the transition state, Table 10.3. A more detailed analysis (Eq. (4.7); entry 6 in Table 4.10) of reactivities reveals a positive ρ_I and negative ρ_R. Evidently inductive effects aid the early stages of nucleophilic attachment but resonance donation towards some developing cationic character also facilitates reaction. On the other hand, steric effects are of the utmost importance (Section 8.2). The approach of the nucleophile towards the reaction center is required to occur from the rear face, so that any structure on this side will impede the reaction. All substituents attached to either the α- or the β-carbon will cause steric hindrance (Table 10.4). Introduction of the third alkyl group in either series makes the largest rate reduction, particularly in the β-series for which it is possible to turn a hydrogen towards the incoming nucleophile until the neopentyl structure is reached, in which case reaction is effectively prevented. Tertiary compounds also are inert to S_N2 attack. That rapid reactions of these compounds are observed is due to the incursion of the ionization mechanism (S_N1), which becomes far more favorable. Secondary compounds also may go over to this route when encouraged by polar solvents. The presence of unsaturated and aromatic groups at the reaction center is accompanied by a considerable increase in reactivity due to their ability to stabilize partial positive charge in the transition state; allyl compounds may react at both the α- and γ-carbons, **5**, the latter reaction being denoted S_N2':

Reaction center:	CH_3CH_2—Br	$PhCH_2$—Br	CH_2=CH—CH_2—Br
Av . k_{rel}:	1	120	40

S_N2' S_N2

5

Table 10.3   Hammett reaction constants for S_N2 substitutions at saturated carbon[30,31], for $X—C_6H_4CH_2Br + Nu:$

Nu:	ρ
OH^-/H_2O	−0·33
I^-	0·81
Me_3N	0·37
Pyridine	−0·26
$X—C_6H_4COCH_2—Cl$ + pyridine	−1·98[32]
$EtO^- + CH_3OSO_2Ar$	1·0
$X—C_6H_4NMe_2 + CH_3—OClO_3$	−3·0

Halogen tends to reduce reactivity but oxygen substituents exert a strong activating effect at the α-carbon and a mild retarding one at the β-carbon. These results show the complex interaction of a combination of factors: steric hindrance (Br > Cl), the balance of inductive and resonance effects,

Table 10.4   Steric substituent effects on S_N2 reactions

α-Methylation

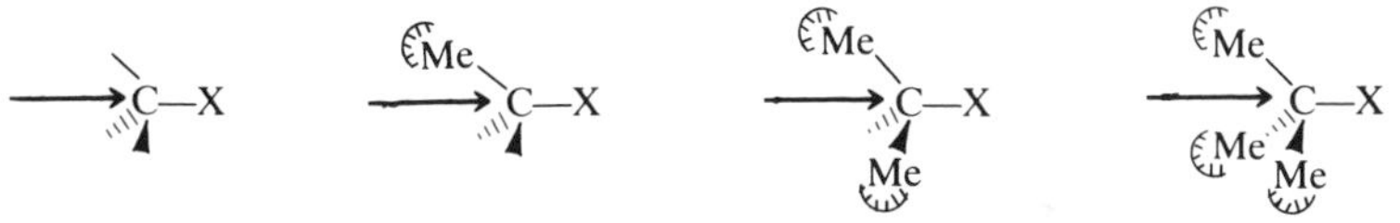

	$CH_3—$ Methyl	$MeCH_2—$ Ethyl	$Me_2CH—$ Isopropyl	$Me_3C—$ *tert*-Butyl
Av. k_{rel}:	1	0·05	0·01	(10^{-5})
$R—Br + OH^- \rightarrow R—OH + Br^-$; $EtOH-H_2O$				
k_{rel}:	1	0·08	0·01	$(10^{-5}?)$
$R—Br + EtOH \rightarrow R—OEt + HBr$				
k_{rel}:	1	0·06	0·016	$2·4 \times 10^{-7}$

β-Methylation

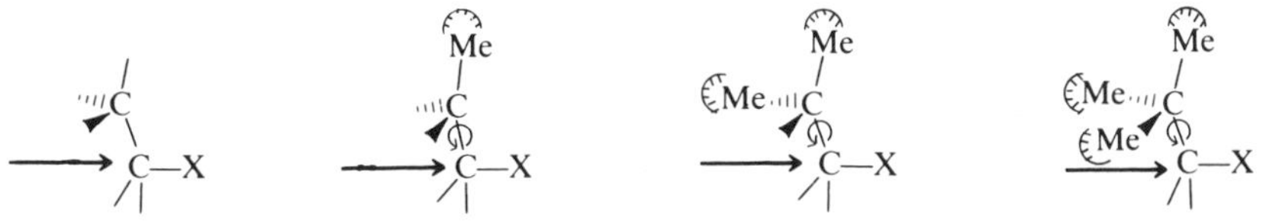

	$CH_3CH_2—$ Ethyl	$MeCH_2CH_2—$ *n*-Propyl	$Me_2CHCH_2—$ Isobutyl	$Me_3CCH_2—$ Neopentyl
Av. k_{rel}:	1	0·4	0·03	10^{-5}
$R—Br + Br^{-*} \rightarrow R—Br^* + Br^-$; MeOH				
k_{rel}:	1	0·65	0·033	$1·5 \times 10^{-6}$

and electrostatic repulsion of the approaching nucleophile[9-11]:

$$RCl + Et_2NH \rightarrow R\text{---}NEt_2$$

RCl:	MeOCH$_2$Cl	CH$_3$Cl	CH$_2$Cl$_2$	CHCl$_3$	CCl$_4$
k_{rel}:	(*ca* 600)	1	0·046	0·110	0

$$XCH_2CH_2Br + I^- \rightarrow XCH_2CH_2I$$

X:	Et	F	Cl	Br	OMe
k_{rel}	1	0·18	0·21	0·19	0·13

Reactivities of alicyclic compounds vary with ring size in a manner which shows that more than one influence is felt[12]:

$$(\overline{CH_2)_n}CHBr + I^-/acetone$$

				$(\widehat{CH_2)_n}CHBr$				
Bromide:	Me$_2$CHBr	$n = 2$	$n = 3$	$n = 4$	$n = 5$	$n = 6$	$n = 7$
k_{rel}:	1	10^{-5}	0·0075	1·6	0·01	0·97	0·22

The inertness of cyclopropyl and cyclobutyl halides is due to a combination of steric hindrance and increased strain in an S$_N$2 transition state and, in the former case, the π-character of the ring, making cyclopropyl compounds — in which they are analogous to vinyl, also notably inert. Some driving force to reaction of planar ring compounds ($n = 2$–4) is created by the reduction of eclipsing interactions at the transition state. The low reactivity of cyclohexyl halides seems to be due to the preponderance of the unreactive equatorial conformer[13] present and to an increase in eclipsing at the transition state, absent in the ground state. The reactivities of larger ring compounds are presumably controlled by their complex conformational properties.

Isotope effects[14-23]

Isotopic substitution at either the central carbon or the leaving group will result in a heavy-atom PKIE (Table 10.5). Secondary deuterium isotope effects have also been measured for α, β and γ positions but these in general are very small; α-effects are frequently inverse (<1) and more remote effects negligible. Such values are, however, characteristic of this mechanism and contrast with the much larger ones associated with the S$_N$1 route and the development of carbocation character.

Table 10.5   Kinetic isotope effects associated with S_N2 reactions

Reaction	PKIE	Ref.
Heavy atom PKIEs (isotopic atom denoted *)		
⟨—⟩N + CH$_3$Cl*	$^{35}k/^{37}k = 1 \cdot 00355$	19
H$_2$O + PhCH$_2$Cl*	$^{35}k/^{37}k = 1 \cdot 0091$	19
PhNMe$_2$ + PhC*H$_2$Otos	$^{12}k/^{14}k = 1 \cdot 135$	20
H$_2$O + PhCH$_2$S*Me$_2^+$	$^{32}k/^{34}k = 1 \cdot 0098$	21
EtO$^-$ + Me$_3$S^{+*}	$^{32}k/^{34}k = 1 \cdot 0085$	22
MeO$^-$ + PhC*HMeBr	$^{12}k/^{13}k = 1 \cdot 0032$	23
		24
Secondary deuterium isotope effects (L = H, D)		
CL$_3$I + S$_2$O$_3^{2-}$	$k_H/k_D = 0 \cdot 970$	25
CL$_3$I + N$_3^-$	$k_H/k_D = 0 \cdot 85$	26
CL$_3$OClO$_3$ + R$_2$NH	$k_H/k_D = 0 \cdot 94$	27

Stereochemistry

Perhaps the most characteristic feature of the S_N2 mechanism is the overwhelming preference for attack by the incoming nucleophile at the rear face with respect to the leaving group (6). This may be explained as the requirement to afford maximum overlap between interacting orbitals, that in the substrate being the vacant antibonding σ^* orbital complementary to the bond being broken. The substituents remaining at the reaction center therefore undergo an 'umbrella' inversion so that the product has the opposite configuration (the Walden Inversion). This is easily demonstrated

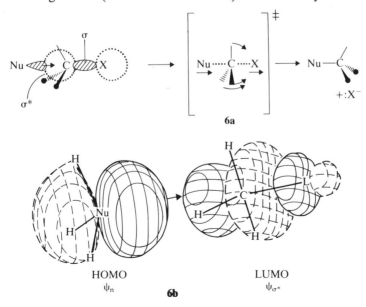

6a

HOMO
ψ_n

LUMO
ψ_{σ^*}

6b

trans
2-Bromo-1-methylcyclopentane
7

cis
2-Methylcyclopentanol

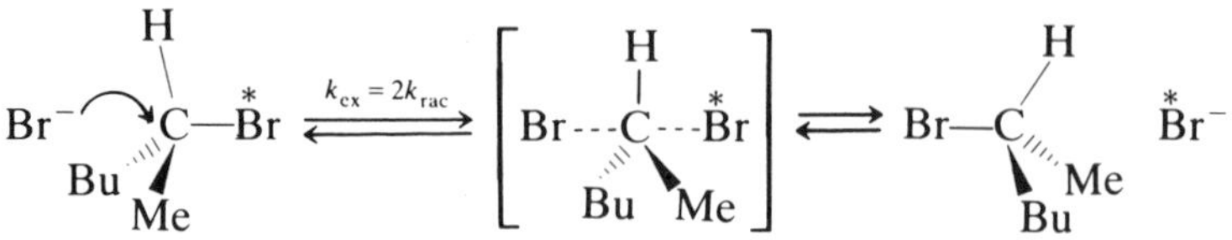

8   **8a**

when the substrate is chiral[24] or cyclic (**7**). A further elegant demonstration using symmetrical reaction, **8**, showed that the rate of racemization was exactly twice that of isotopic iodide exchange (the factor 2 arises since only half the starting material has to react in order to convert the *R* enantiomer to the *RS* racemic mixture). This relationship confirms that the transition state has become achiral[25,26] and may best be depicted as in **8a**.[17,27] High-level MO theory has been used to justify the stereochemical preference of S_N2 reactions and also to show a much reduced preference of inversion over retention for electrophilic substitutions at saturated carbon (S_E2), Section 10.2[28]. It is even possible to include a large assemblage of solvent molecules into the calculation[29].

Medium effects

Reactions of charge type 1 (charge separation) are powerfully accelerated and those of charge type 2 (charge neutralization) strongly retarded by transfer to a more polar solvent (Chapter 5); neutral reactions are mildly retarded. Rates of reactions of anionic nucleophiles are very greatly increased by transfer from hydrogen-bonding protic solvents to dipolar aprotic ones (Section 6.12)[32,33]. This is due partly to changes in solvation of the reagents and partly to changes in solvation of the transition state as shown by the following heats of transfer[34,35]:

	Reagents	Transition state	Total
ΔH_{tr}(MeOH→DMF)/kJ mol^{-1}:	−17	−18	−35
(k cal mol^{-1}):	(−4)	(−4·3)	(−8·3)

Salt effects

The rates of S_N2 reactions are only affected by addition of ionic solutes insofar as this effectively increases the polarity of the medium. If an added ion also acts as a nucleophile it will simply undergo an additional parallel reaction, with

$$\text{Rate} = k'[\text{RX}][\text{Nu}_1^-] + k''[\text{RX}][\text{Nu}_2^-], \quad \text{etc.}$$

Variation of the nucleophile

Rates are sensitive to the nature of the nucleophile (in particular the energy of the HOMO and the size of its coefficient at the nucleophilic center) as shown by the large negative value of ρ for $ArNMe_2$ in Table 10.3. The Swain–Scott and Edwards parameters for nucleophilicity (Section 6.12.3) are based on reactivities in the S_N2 reaction and this order will in general apply. Relative rates for a given set of nucleophiles will, however, depend upon the substrate used for their evaluation according to the reactivity-selectivity principle, which determines the coefficients in Eqs (6.37), (6.38).

Variation of the leaving group

Rates will depend strongly upon the nature of the leaving group and the properties of the C–X bond, its dissociation energy and particularly its polarizability. Relative rates for a given leaving group will depend also upon the strength of the nucleophile used. In general, highly reactive nucleophiles will tend to level rates for a range of leaving groups while weaker, more selective reagents will differentiate them[36] (Table 10.7g).

The overall heat of reaction is roughly the difference in bond dissociation energies for the C–X and C–Nu bonds and this will be reflected in the activation energy. Whether or not a displacement reaction proceeds spontaneously depends on it being exothermic, or at least not highly endothermic, since the entropy of activation then tends to be mildly unfavorable. It is for this reason that it is not possible to displace OH^- or NH_2^- as such. The reaction

$$R\text{—}OH + Cl^- \rightarrow RCl + OH^-$$

Table 10.6 Relative rates of different leaving groups in S_N2 reactions[37,38]

$$N_2^+ \gg CF_3SO_2 > tos \sim I > Br > Cl > \overset{+}{S}R_2 > F \gg OH$$

In $XC_6H_4SO_2OAr$,

$$X = NO_2 > Cl > H > Me > OMe$$

(for $CH_3OSO_2Ar + EtOH$ at 70°C, $\rho = 1.32$)

Leaving group, X:	tos	$PhSO_2O$	I	Br	Cl	F	$\overset{+}{S}Me_2$
Av. k_{rel}:	500	10	3	1	0.02	10^{-4}	0.5

has $\Delta H = +80 \, \text{kJ mol}^{-1}$ ($19 \, \text{k cal mol}^{-1}$) and, in consequence, the reverse reaction is favored. Protonation of leaving groups such as —OH ($\rightarrow$ —OH$_2^+$) will convert them into highly displaceable ligands so the conversion of alcohols into alkyl halides takes place in concentrated hydrogen halide solutions or in the presence of electrophilic catalysts, such as Zn^{2+} (Lucas' reagent) and Ag$^+$, which is particularly effective in catalyzing the removal of halide ion to which it binds by a strong soft–soft interaction[39]. Even neopentyl iodide reacts releasing halide faster than does methyl. There is evidence for a pre-association of the reagents followed by release of a carbocation which undergoes Wagner–Meerwein rearrangement.

$$\text{Me}_3\text{C}\!-\!\text{CH}_2\text{I} + \text{Ag}^+ \rightarrow [\text{Me}_3\text{C}\!-\!\text{CH}_2\text{I}\cdot\text{Ag}]^+ \rightarrow \text{Me}_2\overset{+}{\text{C}}\!-\!\text{CH}_2\text{Me} + \text{AgI}.$$
$$\downarrow$$
$$\text{Products}$$

10.1.3 Solvolytic reactions — the S$_N$1 spectrum[3,4–13,40–43]

The hydrolyses of the 'α-methylated' series of alkyl halides, methyl, ethyl, *n*-propyl, *tert*-butyl show a gradual reduction in rates for the first three members but a sudden dramatic increase for the last (Fig. 10.3). This evidently signals a change in mechanism. The rate of hydrolysis of the tertiary compound is independent of [OH$^-$], unlike the others; therefore a

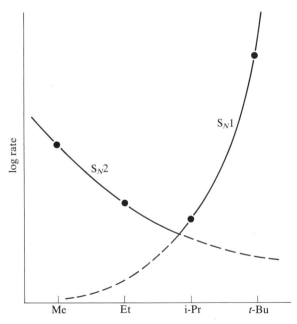

Fig. 10.3 Mechanistic change revealed by the effects of α-substitution on rates of S$_N$ reactions (after Ingold[4]).

386

$$\left(\begin{array}{c} S \\ \diagdown_{\delta+} \quad | \quad _{\delta-} \\ O \cdots C \cdots X \\ H \end{array} \right) \quad \left(\begin{array}{c} S \\ \diagdown_{\delta\delta+} \quad |_{\delta+} \quad _{\delta-} \\ O \cdots C \cdots X \\ H \end{array} \right) \quad \left(\begin{array}{c} \diagdown_{\delta+} \quad _{\delta-} \\ C \cdots X \\ \diagup \end{array} \right)_{Solv} \quad \left(\begin{array}{c} \diagdown \\ C \cdots X \cdots E \\ \diagup \end{array} \right)$$

S_N2	solvent-assisted	limiting	electrophilic
9	(nucleophilic)	S_N1	catalysis

unimolecular slow step is indicated, the reaction proceeding via an intermediate. This was originally identified as a carbocation formed by rate-determining ionization of the substrate and the process denoted S_N1. While ionization is undoubtedly the initial step in displacements at tertiary centers, the nature of the intermediate must be reconsidered and the 'S_N1' mechanism modified to accommodate a range of possibilities. It is better described as a 'spectrum' or 'manifold' of mechanisms (**9**), the details depending upon the nature of the substrate and on the solvent. It is for this reason that S_N1 reactions are intimately associated with solvolysis, which is nucleophilic substitution by a hydroxylic solvent, SOH (an alcohol, water or a carboxylic or sulfonic acid) which acts both as ionizing medium and as nucleophile (Fig. 10.1):

$$R—X + S\ddot{O}H \rightarrow R—OS + X^- + H^+.$$

Mechanisms within the scope of solvolytic reactions may be characterized by the following variables.

(a) The degree to which the solvent participitates as nucleophile at the rearface of the substrate may have any value between the extremes: no participation at all is referred to as 'limiting' behavior, and full nucleophilic attack by the solvent constitutes an S_N2 reaction. This implies a continuum of mechanistic change between extreme S_N1 and S_N2, one popular current view[44], the other being of concurrent reactions of the two types[45, 46].

(b) Nucleophilic participation by an internal electron pair may occur ranging from 'assistance' to full neighboring-group participation (Chapter 13).

(c) There appear to be a range of intermediate structures from which intimate, **1**, and solvent-separated, **2**, ion-pairs and free (solvated) cations, **3**, may be sequentially formed and products may be derived from any stage. Anion exchange in the ion-pairs may also occur.

10.1.4   Measurement of solvent participation

The nature of the solvent is crucial in deciding which route will be energetically most favorable for a given substrate (Table 10.8). Solvent ionizing power (the ability to stabilize ions) and nucleophilicity (which tend to have an inverse relationship), measured by the parameters Y and N respectively (or, better, Y_{tos} and N_{tos}) of the Grunwald–Winstein equation

(Eq. 5.15) are of particular importance:

Solvent:	EtOH	MeOH	AcOH	CH$_3$CH$_2$OH	HCOOH	H$_2$O	CF$_3$COOH
Y_{tos}:	−1·75	−0·96	−0·61	1·80	3·04	4·0	4·57
N_{tos}:	0·0	−0·04	−2·35	−3·0	−2·35	−0·41	−5·56

Solvolyses which depend upon some degree of nucleophilic solvent assistance will occur faster than the theoretical limiting rates. In order to judge the magnitude of the discrepancy, a model substrate is required which exhibits limiting behavior in all solvents and a solvent which induces limiting behavior in all substrates (i.e. one which is non-nucleophilic). Models fitting these requirements have been proposed by several workers; the following is due to Schleyer[47,55]. As a limiting substrate, the 2-adamantyl (2-Ad)system, **10**, was chosen since rearface attack is impossible, an assumption supported by the following evidence.

(a) Limiting ionization will be very sensitive to the ionizing power of the solvent measured by the Grunwald–Winstein solvent sensitivity constant m_s. Solvolyses of 2-Ad halides and sulfonates show higher values than any other secondary systems; $m_s > 1$, comparable with the reactions of tertiary compounds (Table 10.7). Indeed, rates of solvolysis of 2-adamantyl compounds correlate excellently with those of the tertiary 1-adamantyl (**11a**) or bicyclo-octyl (**11b**) analogues, which are unable for obvious reasons to receive nucleophilic assistance from the rear.

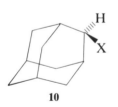
10

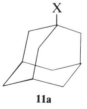
11a

11b

(b) Ethanol–water (EW) mixtures about 98:2(v/v) and acetic acid have identical Y-values but the former are more nucleophilic. The rate ratio for limiting behavior should be $k_{EW}/k_{AcOH} = 1$; this is approximately so for 2-adamantyl compounds, whereas it becomes increasingly large for less hindered compounds (Table 10.8).
(c) 2-Adamantyl compounds solvolyze with the largest secondary isotope effects known (see below).
(d) Conventional S$_N$2 reactions of 2-adamantyl halides, e.g. by azide ion in ethanol, are exceedingly slow or fail altogether.

The solvent chosen as possessing the highest ionizing capacity with negligible nucleophilicity was trifluoroacetic acid: $Y = 4·57$, $N = −5·56$. Now, a quantitative measure of nucleophilic solvent assistance can be made

Table 10.7 Linear free-energy correlations of solvolysis reactions [13,41,44,56]

(a) Taft correlations of aliphatic substrates, $\log k_X/k_H = \rho^*\sigma^*$

System	Solvent	ρ^*
RCH$_2$Otos	EtOH	-0.74
RCH$_2$Otos	AcOH	0
RR'CHOtos	AcOH	-2.6
RR'CHOtos	HCOOH	-3.5
RR'CHOtos	CF$_3$COOH	-7.1
RR'R''C—Otos	EtOH	-3.3

(b) Hammett–Brown correlations of aralkyl systems, $\log k_X/k_H = \rho^+\sigma^+$.

System	Solvent	ρ^+
ArCMe$_2$Cl	EtOH	-4.67
ArCPh$_2$Cl	SO$_2$	-3.73

(c) Grunwald–Winstein parameters[a,b],
$$\log k/k_0 = lN_{tos} + m_s y_{tos}$$

Group	l	m_s
Me	1·0	0·3
Et	0·83	0·41
PhCH$_2$	0·75	0·64
isoPr	0·40	0·58
3-Pentyl	0·26	0·72
Cyclopentyl	0·26	0·71
Cyclohexyl	0·23	0·75
2-Ad	0·0	1·0
Bicyclo[2,2,2]octyl	0·02	1·05
tert-Bu	0·0	1·0

[a] l = Sensitivity to solvent nucleophilicity.
[b] m_s = Sensitivity to solvent ionizing power.

(d) Carbocation stabilities measured by pK_{R^+}

$$ROH \underset{K_{R^+}}{\overset{conc. H_2SO_4}{\rightleftharpoons}} R^+$$

Triarylmethanols with substituents in three rings

X_1	X_2	X_3	pK_{R^+}
4-OH	4-OH	4-OH	$+1.97$
4-OMe	4-OMe	4-OMe	$+0.82$
4-OMe	4-OMe	—	-1.24
4-OMe	—	—	-3.40
2-Me	2-Me	2-Me	-3.4

Table 10.7   (*Continued*)

Triarylmethanols with substituents in three rings

X_1	X_2		pK_{R^+}
4-Me	4-Me	4-Me	−3·56
4-Me	4-Me	—	—
4-Me	—	—	−5·41
4-CD$_3$	—	—	−5·43
3-Me	3-Me	3-Me	−6·35
4-*tert*-Bu	—	—	−6·1
4-*tert*-Bu	4-*tert*-Bu	—	−6·6
4-*tert*-Bu	4-*tert*-Bu	4-*tert*-Bu	−6·5
4-isoPr	4-isoPr	4-isoPr	−6·54
—	—	—	−6·63
4-F	4-F	4-F	−6·05
4-Cl	4-Cl	4-Cl	−7·74
4-NO$_2$	—	—	−9·15
3-Cl	3-Cl	3-Cl	−11·03
4-NO$_2$	4-NO$_2$	—	−12·90
4-NO$_2$	4-NO$_2$	4-NO$_2$	−16·27
F$_5$	F$_5$	F$_5$	−17·5

Diarylmethanols with substituents in two rings

X_1	X_2	pK_{R^+}
4-OMe	4-OMe	−5·71
2,4,6-Me$_3$	2,4,6-Me$_3$	−6·6
4-OPh	4-OPh	−9·85
4-Me	4-Me	−10·4
2-Me	2-Me	−12·45
4-F	4-F	−13·03
4-*tert*-Bu	4-*tert*-Bu	−13·2
—	—	−13·3
4-Cl	4-Cl	−13·96
4-Br	4-Br	−14·16
4-I	4-I	−14·26

Table 10.7   (*Continued*)

Others

Compound	pK_{R^+}
Xanthydrol	-0.84
$\alpha,\alpha,2,4,6$-Pentamethylbenzyl alcohol	-12.2
$\alpha,\alpha,2,3,4,5,6$-Heptamethylbenzyl alcohol	-12.4
9-Methylfluorenol	-16.60

(e) Carbocation stabilities measured by enthalpies of ionization in superacid;[a]

$$R—ClR^+, Cl^-: \Delta H_{RCl}$$

R	$\Delta H_{RCl}/kJ\,mol^{-1}\,(kcal\,mol^{-1})$
2-Propyl	$-64\,(-15.3)$
2-Butyl	$-66\,(-15.9)$
tert-Butyl	$-104\,(-24.8)$
tert-Pentyl	$-113\,(-27.1)$
tert-Hexyl	$-116\,(-27.9)$
Cyclopentyl	$-72\,(-17.3)$
2-Norbornyl	$-99\,(-23.6)$
1-Methylcyclopentyl	$-113\,(-27.1)$
Cumyl	$-137\,(-30.3)$

[a] E. M. Arnett and T. C. Hoferich, *J. Amer. Chem. Soc.*, **105**, 2889 (1983).

(f) Gas-phase enthalpies of heterolysis, $D(R^+H^-)$[b]

R	$D(R^+H^-)/kJ\,mol^{-1}\,(kcal\,mol^{-1})$
Methyl	1309 (313)
Ethyl	1134 (271)
1-Propyl	1121 (268)
2-Propyl	1054 (252)
tert-Butyl	975 (233)
Allyl	1071 (256)
Propargyl	1134 (271)
Phenyl	1230 (294)
Benzyl	996 (238)
Acetyl	962 (230)
—CH_2CN	1276 (305)
—CH_2NH_2	912 (218)
—CH_2OH	1054 (252)

[b] F. P. Lossing and J. L. Holmes, *J. Amer. Chem. Soc.*, **106**, 6917 (1984).

391

Table 10.7 (*Continued*)

(g) S_N2 reactions in the gas phase

$$R\text{-Hal} + Nu:^- \rightarrow R\text{-Nu} + Hal^-$$

R-Hal	Nu:⁻	$k/10^{11}\,M^{-1}\,s^{-1}$	Efficiency	$\Delta H\,kJ\,mol^{-1}$ $(kcal\,mol^{-1})$
CH₃Cl	OH⁻	3·51	0·68	−209 (−50)
	F⁻	4·8	0·35	−130 (−31)
	MeO⁻	3·6	0·25	−255 (−61)
	MeS⁻	0·46	0·045	−134 (−61)
	CN⁻		0·0005	−126 (−30)
CH₃Br	OH⁻	11·4	0·84	−239 (−57)
	F⁻	3·6	0·28	−160 (−38)
	MeO⁻	4·38	0·40	−285 (−68)
	MeS⁻	0·84	0·091	−163 (−39)
	Cl⁻	0·48	0·070	−29 (−7)
	CN⁻		0·01	−155 (−37)

(data from J. M. Riveros, S. M. Jose and K. Takashima, *Adv. Phys. Org. Chem.*, **21**, 197 (1985).

Table 10.8   Estimates of nucleophilic assistance in solvolyses of ROtos

	SOH						
R	EtOH	50% aq. EtOH	AcOH	H₂O	HCOOH	CF₃CH₂OH	CF₃COOH
2-Ad	1	1	1	1	1	1	1
Cyclohexyl	360	61	27	10	2	0·6	1
Cyclopentyl	2315	160	104	15	10	1	1
3-Pentyl	1835	103	47	12	6	0·6	1
isoPr	32,500	1130	464	57	3	0·6	1

Sensitivities[a] to nucleophilic power of solvent[44]

Substrate	k_{EW}/k_{AcOH}	k_{EW}/k_{HCOOH}
MeBr		200
MeOtos	97	55
EtBr		80
EtOtos	80	18
isoPrBr		20
isoPrOBros	6	2
2-Pentyl-Otos	4	0·4
tert-BuCl	1	1
Bicyclo[2,2,2]octyl bromide	0·7	0·5
Benzyl chloride	58	4
Benzyl tosylate	30	
Allyl chloride		40
2-Ad—Otos	0·2	
Ph₂CHBr	90	
tert-BuBr	3·3	
tert-BuCl	1·0	
PhCHMeBr	20	

[a] Rate in ethanol–water relative to rate in acetic or formic acids at constant ionizing power (*Y*).

Table 10.9   Variation of solvolysis rates of alkyl tosylates, R—Otos, with solvent

Primary substrates

Solvent	Substrate				
	Me	Et	n-Pr	isoBu	neoPe
AcOH	1·0	0·92	0·731	0·28	0·1
CF₃COOH	1·0	12·5	93	3060	6000
FSO₃H	1·0	118	33 000	$5·4 \times 10^5$	$1·1 \times 10^6$

Secondary substrates

Solvent	Substrate						
	2-Ad	2-Norbornyl		Cyclohexyl	Cyclopentyl	3-Pentyl	isoPr
		exo	endo				
EtOH	1	10 400	34	108	6250	1560	910
AcOH	1	4 140	14	8·3	280	40	13
HCOOH	1	1 940	1·1	1·5	27	5·3	0·9
(CF₃)₂CHOH	1	1 300	0·8	0·2	3·1	0·5	0·016
CF₃COOH	1	520	0·46	0·3	2·7	0·85	0·03

Percentage rearrangement in primary alkyl tosylates

Solvent	Et	n-Pr	iso-Bu	neoPe
EtOH	0	0	5	92
AcOH	0	0·2	79	100
HCO₂H	0	0·8	97	100
CF₃CO₂H	0	87	100	100
96% H₂SO₄	0	>95	100	100
FSO₃H	30–40	100	100	100

for any substrate RX solvolyzing in any solvent, SOH, using Eq. (10.1):

$$\text{Solvent assistance} = \frac{[k_{RX}/k_{2AdX}]_{SOH}}{[k_{RX}/k_{2AdX}]_{CF_3COOH}}. \tag{10.1}$$

Taking relative rates from Table 10.9, it may be shown that cyclohexyl tosylate in ethanol has

$$\text{Solvent assistance} = (108/1)/(0·3/1) = 360,$$

i.e. the observed rate is enhanced by this factor over the limiting ionization rate. Values of solvent assistance increase the more nucleophilic the solvent and the less stable the carbocation, R^+ (Table 10.8). Tertiary substrates will tend to undergo limiting ionization in all but the most nucleophilic solvents while primary substrates, with the possible exception of benzyl compounds[57], will receive maximum assistance reacting essentially by the S_N2 mechanism. The test of electron supply can be applied; the more limiting the solvent, the greater the electron demand from the structure and

393

hence the larger (negative) the value of the reaction constant. It is notable that the value of $\rho^* = -7.2$ for solvolyses of secondary alkyl tosylates in trifluoroacetic acid is exceeded in the solvent hexafluoro-isopropanol ($\rho^* = -9.3$), which therefore appears to be even less nucleophilic[45].

Rates, of course, measure $\Delta G^{\ddagger}$ and it might be expected that this function changes smoothly with composition in a mixed solvent system such as water–ethanol and this is found to be the case. If $\Delta G^{\ddagger}$ is dissected into the two components, $\Delta H^{\ddagger}$ and $T\,\Delta S^{\ddagger}$, both show most irregular behavior, Fig. 10.6b[58]. Evidently, changes in solvation which occur as solvent composition is changed, are complex and not readily interpreted. That these effects are real, however, is confirmed by very similar curves for the free energies, enthalpies and entropies of transfer of glycine, $H_3\overset{+}{N}$—CH_2—COO^-, which acts as a model for the solvolysis transition state, $R^{\delta+} \cdots X^{\delta-}$.

10.1.5 Kinetic isotope effects[41,48–54]

Solvolytic reactions will show heavy-atom PKIEs at both carbon and the leaving-group atom, confirming that bond breaking is the slow step. More characteristic of the detailed reaction pathway are secondary deuterium isotope effects at α, β, or even γ positions. An SKIE at the α position, $k_H/k_D > 1.2$, is characteristic of a carbon undergoing hybridization change from sp^3 to sp^2 and is expected for limiting solvolysis. Theoretical estimates suggest 1.4 as the upper limit. Nucleophilic participation tends to reduce this value towards that characteristic of S_N2 reactions, 0.95–1.05. The resultant SKIE (Table 10.10) correlates with the degree of solvent participation determined as described above (Table 10.9; Fig. 10.4). Primary compounds tend to show very low values, consistent with reaction by the S_N2 mechanism, while tertiary compounds tend to show maximum values, insensitive to solvent, characteristic of limiting ionization. Consequently this provides an index of limiting behavior. Secondary deuterium isotope effects are also observed at more remote positions. β-Effects are slightly greater than α-effects and are attributable to stabilization of the incipient cation by hyperconjugation from a periplanar C–H bond (Section 7.4).

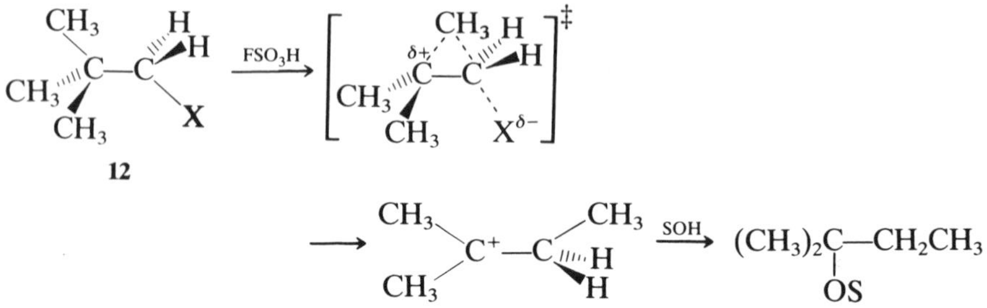

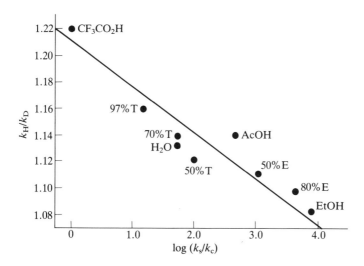

Fig. 10.4   The ratio of solvent-assisted (k_s) to unassisted (k_c) rates of solvolysis (as shown by α-SKIE) as a function of the solvent. The more nucleophilic the solvent (bottom, right), the larger is k_s/k_c, indicated by a very small SKIE. 'Limiting' ionization increases towards the top left. Solvents are: E, ethanol; T, trifluoroacetic acid; $x\%$ indicates the proportion of cosolvent with $(100-x)\%$ water; substrate is 2-propyl tosylate.

γ-Effects may be pronounced if there is participation by a neighboring group. A typical example is the neopentyl system, **12** (see Section 10.1.14).

10.1.6   The structures of intermediates in S$_N$1 reactions

Species with at least some degree of positive charge localized on carbon are referred to as *carbocations* (systematic name *carbenium ions,* older name 'carbonium ions') and, specifically, as 'methyl cation, CH_3^+' for example. Simple alkyl cations are planar and in consequence achiral. They possess a very low-lying LUMO (Fig. 10.5) which engenders the most powerful electrophilic character that dominates their chemistry. It is probably true to say that a species as reactive as CH_3^+ is never formed except possibly in the following nuclear reaction leading to the most nucleofugic atom (He) possible:

$$CH_3T \xrightarrow{\text{β-decay}} CH_3\text{—}^3He^+ + e^- \rightarrow CH_3^+ + {}^3He.$$

Substituent groups, alkyl and especially aryl, interact with the cationic center and raise the LUMO energy, thereby increasing the stability. Much

Table 10.10   Secondary kinetic isotope effects in solvolyses (L = H, D) [41,48-54]

Substrate	Solvent	k_H/k_D
α-Effects		
Primary		
CH_3Otos	H_2O	0·985
CH_3CL_2Otos	H_2O	1·020
CH_3CL_2Br	H_2O	0·98
$CH_3CH_2CL_2OMes$	H_2O	1·02
$PhCL_2OBros$	90% EtOH–H_2O	1·06
$PhCL_2OBros$	97% CF_3CH_2OH–H_2O	1·17
Secondary		
$(CH_3)_2CLBr$	EtOH–H_2O	1·11
$(CH_3)_2CLBr$	CF_3COOH	1·22
$CH_3CH_2C(Me)LOBros$	AcOH	1·17
Cyclopentyl—Otos(1-L)	AcOH	1·16
Cyclohexyl—(1-L)	AcOH	1·22
2-Ad—Otos		1·23
β-Effects		
Primary		
CL_3CH_2Otos	H_2O	1·018
CL_3CH_2Otos	AcOH	1·01
CL_3CH_2Otos	H_2SO_4	1·2
CL_3CH_2Otos	HSO_3F	1·58
$Me_2CHLOtos$	HSO_3F	1·9
Secondary		
$(CL_3)_2CH_2OBros$	CF_3COOH	2·12
CH_3CL_2CHEt—OBros	EtOH–H_2O	1·32
Cyclopentyl—OBros(*cis*-2-L)	AcOH	1·15
Cyclopentyl—OBros(*cis*-2-L)	AcOH	1·18
Cyclohexyl—OBros(*cis*-2-L)	AcOH	1·25
Cyclopentyl—OBros(*trans*-2-L)	AcOH	1·30
$EtCL_2CHEt$—OBros	EtOH	1·31
2-Ad—Otos(1-L)	EtOH–H_2O	1·23
2-Ad—Otos(1-L)	CF_3COOH	1·23
Tertiary		
$(CL_3)_3Cl$	60% EtOH–H_2O	2·416 (1·1 per D)
$(CL_3)_3Cl$	CF_3CH_2OH	1·38
$(CL_3)_3Cl$	AcOH–H_2O	1·34
γ-Effects		
$CL_3CH_2CH_2$—Br	H_2O	0·92
$(CL_3)_2CHCH_2OMes$	H_2O	0·97
Cyclopentyl—OBros(3,3,4,4-L_4)	AcOH	0·95
$CL_3CH_2CMe_2Cl$	EtOH–H_2O	0·97
$(CL_3)CCH_2$—OMes	H_2O	1·017
$(CL_3)CCH_2$—ODNB	CF_3COOH	1·027

of the positive charge is taken by β-hydrogens in the relatively stable
tert-butyl cation, Me_3C^+, a result of hyperconjugation. This results in a
shortening of the C–C bond lengths and a lengthening of C–H bonds which
lie out of the plane, by 2·8 and 0·7 pm, respectively. This is a result of
electron density transferring from the σ-bonds to the positive center leaving
approximately 0·05 units of positive charge on each hydrogen.

Many carbocations can be generated and observed as stable species if
removed from a nucleophilic environment, which in the case of alkyl cations
means superacid solution at low temperature. This technique is limited to
those species which do not undergo unimolecular rearrangement including
tertiary alkyl or aryl, benzenium, and various heteroatom-substituted
species, for example:

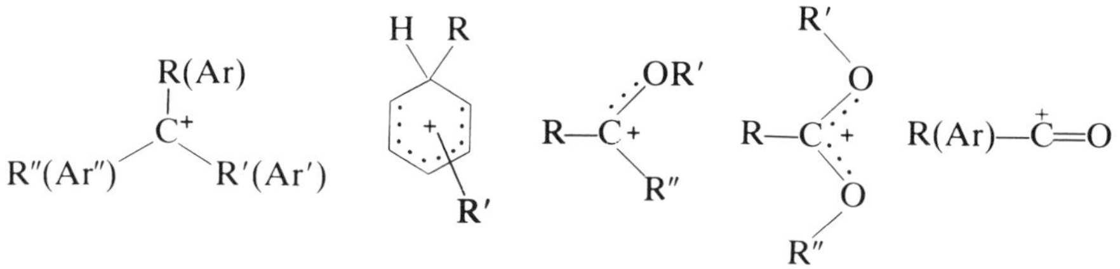

Secondary species which are observable include isopropyl and 1-
phenylethyl, but others will rapidly isomerize to a more stable species by a
series of 1,2-shifts; all isomeric butyl cations, for example, revert to
tert-butyl.

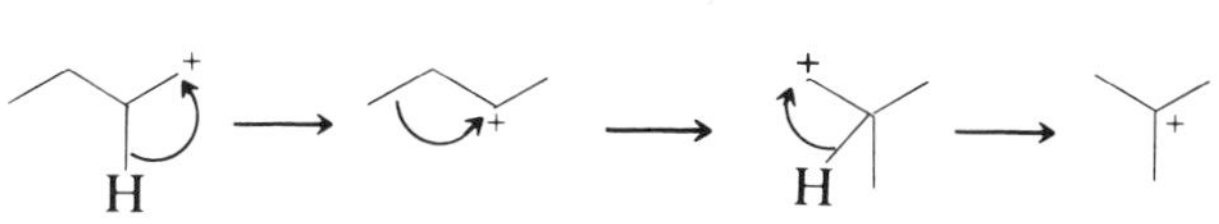

In more nucleophilic solvents — water, alcohols or carboxylic acids —
carbocations are normally very transient and react with the solvent at
diffusion-controlled rates. Nevertheless, such species were originally postulated as the reactive intermediates in the solvolyses of tertiary halides for
which the rate-controlling step was demonstrated to be independent of
nucleophilic species present, 3. This is the classical S_N1 mechanism now
recognized as an extreme situation, though one which can occur when a
particularly stabilized carbocation such as Ph_3C^+ is involved. Far more usual
is a reactive intermediate which is composed of carbocation, leaving-group
anion and possibly solvent, electrostatically bound together, i.e. a species of
ion-pair. The following evidence is compelling in pointing to the existence
of several different types of ion-pair intermediates in solvolytic reactions
(Fig. 10.1).

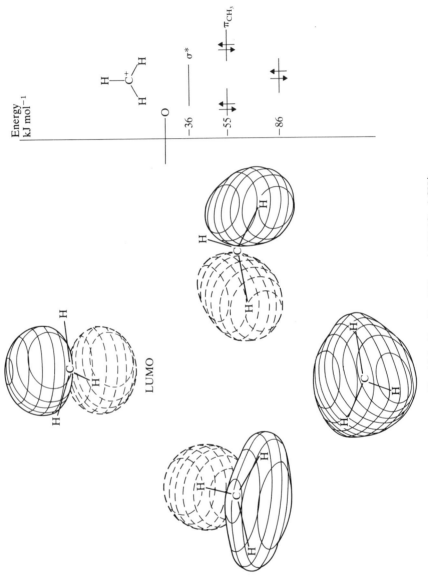

Fig. 10.5 Bonding MOs and LUMO of CH_3^+.

10.1.7 The phenomenon of 'return'

This term is used to signify the reverse process to ionization, i.e. the recombination of carbocation and leaving group (k_{-1}). If these are the identical pair which arose from ionization, the return is *internal*; if they result from exchange with other species, *external*.

Kinetic criteria for return

External return is revealed by a detailed study of the kinetics.

$$R—X \underset{k_{-1}}{\overset{k_1}{\rightleftharpoons}} R^+, X^- \xrightarrow[k_2]{SOH} R—OS;$$

$$Rate = d[product]/dt = k_2[R^+] \qquad ([SOH] = const).$$

Applying the steady state assumption, $[R^+]$ is very small and constant:

$$k_1[RX] = k_{-1}[R^+][X^-] + k_2[R^+].$$

Hence

$$[R^+] = \frac{k_1[RX]}{k_{-1}[X^-] + k_2} \quad and \quad Rate = \frac{k_1 k_2[RX]}{k_{-1}[X^-] + k_2}. \qquad (10.2)$$

(At $k_{-1} \ll k_2$, rate $= k'[RX]$, the first-order limit.)

If the intermediate is a free carbocation, R^+, in kinetic equilibrium with X^-, the rate will be depressed by the addition of X^- to the system (common ion effect). Also, k_{obs} will fall with time as product X^- accumulates, as shown for example in Table 10.11. Ionization of a neutral substrate would normally be enhanced by addition of an inert salt (LiBr) which increases the polarity of the medium. This effect is present in LiCl but is overwhelmed by the common ion effect. In this case, the product-determining intermediate

Table 10.11 Relative rates of reaction for hydrolysis of $(p\text{-MeC}_6\text{H}_4)_2\text{CHCl}$ in H_2O–acetone

Salt added	Extent of reaction				
	10%	30%	50%	70%	80%
None	1	0·95	0·83	0·79	0·76
0·1M LiBr	1·16	(ionic strength effect)			
0·1M LiCl	0·87	(common ion + ionic strength effects)			

Table 10.12 Estimates of proportions of return in solvolyses

Substrate	Solvent	Return (%)
Internal return		
Ph$_2$CH—Opnb	acetone–H$_2$O, 2:8 v/v	75
Ph·tol·CH—Opnb	acetone–H$_2$O, 2:8 v/v	72
PhCH$_2$CH$_2$—Otos	AcOH	78
PhCH$_2$CH$_2$—Otos	EtOH	51
PhCH$_2$CH$_2$—Otos	HCOOH	15
CH$_3$CH·p-AnCHMe—Otos	AcOH	37 (+ 80% external)
Cyclo-C$_4$H$_7$—Cl	AcOH	43
Cyclo-C$_4$H$_7$—Otos	AcOH	8
Cyclo-C$_4$H$_7$—OSO$_2$Me	AcOH	20
Cyclo-C$_4$H$_7$—OSO$_2$Me	EtOH	20
Cyclo-C$_4$H$_7$—OSO$_2$Me	AcOH	68
External return		
tert-BuCl	acetone–H$_2$O, 8:2 v/v	0
Ph$_2$CHCl	acetone–H$_2$O, 8:2 v/v	10
tol$_2$CHCl	acetone–H$_2$O, 8:2 v/v	70

Common ion mass law coefficients for R—Hal in acetone–H$_2$O

R—Hal	k_{-1}/k_2
tert-BuCl	0·25
Ph$_2$CHCl	12
Ph$_2$CHBr	60
4-MeC$_6$H$_4$CHPhCl	30
(4-MeC$_6$H$_4$)$_2$CHCl	90

must be exchanging with added Cl$^-$, normally only found when the carbocation is especially stable and long-lived. Had the isotopic ^{36}Cl$^-$ been added it would have been found incorporated into unreacted RCl. Isotopic exchange, however, can occur without there being any kinetic common ion effect and must then originate from a solvent-separated ion-pair which will not affect the proportion of return (Table 10.12).

10.1.8 Rearrangement criteria for return

Return from intimate ion-pairs occurs even more frequently. This may be revealed if either the carbocation or the anion is ambident, i.e. has more than one reactive site, when return may be accompanied by rearrangement. Allyl cations on the one hand, and carboxylate ions on the other, both satisfy this requirement. Solvolysis of α, α-dimethylallyl halides, **14**, leads to the formation of the isomer **14a** (which, being the less reactive primary compound, accumulates) together with solvolysis products **15**[59].

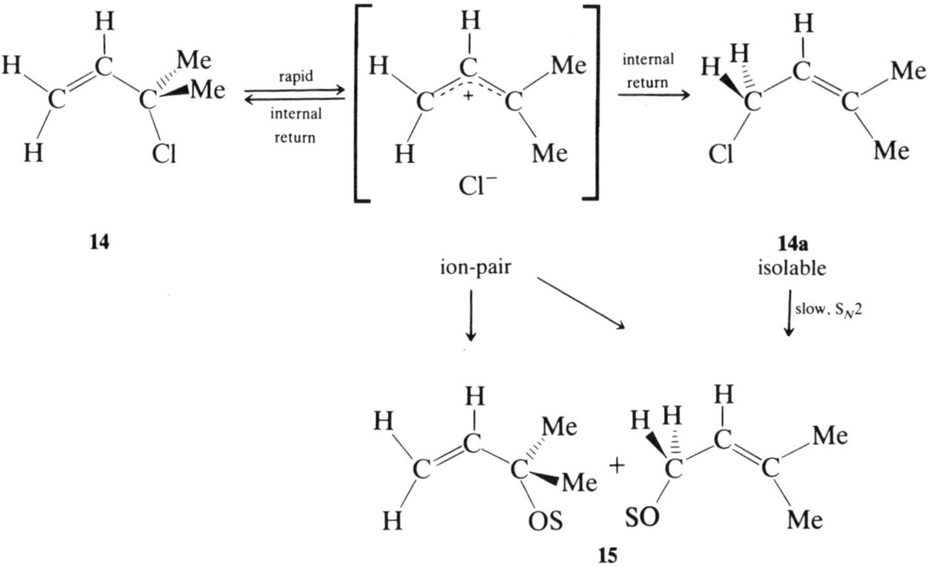
rapid
internal return
internal return
14
ion-pair
14a
isolable
slow, S_N2
15

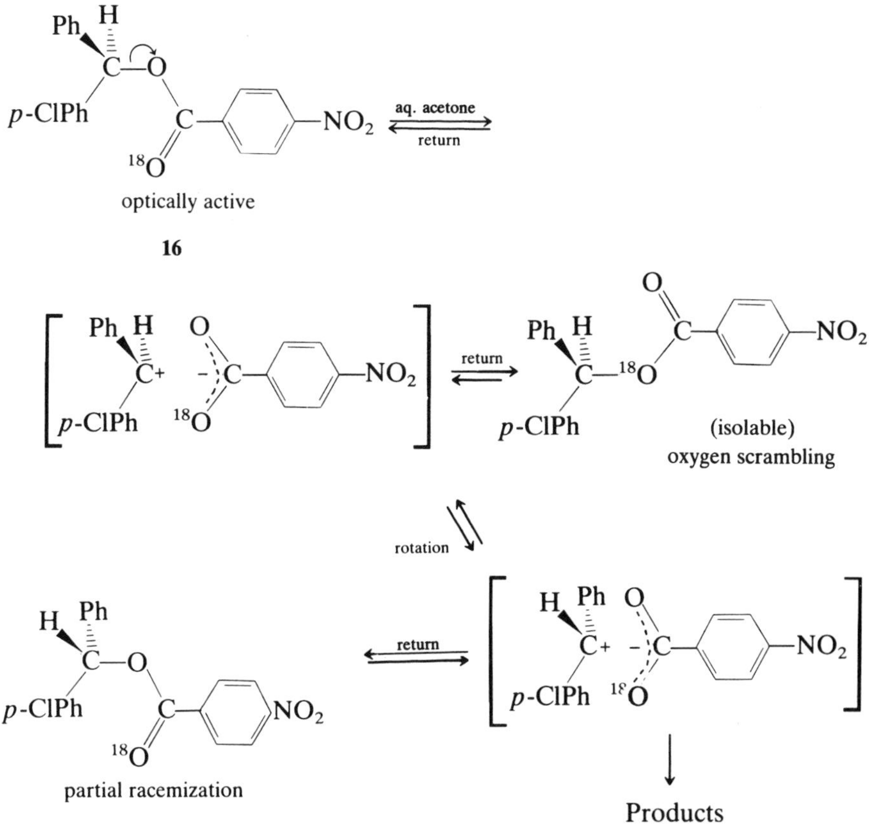
aq. acetone
return
optically active
16
return
(isolable)
oxygen scrambling
rotation
return
partial racemization
Products

Carboxylate esters with an ^{18}O label may solvolyze with prior scrambling of the label due to internal return from an intimate ion-pair, **16**. Partial racemization of the starting material is also observed[60,61]. Presumably the cation is able to rotate before recombination occurs to the opposite configuration. For this reason, rates determined by change of optical rotation (k_α) are always greater than those determined by conductivity, k_{titr}. The ratio k_α/k_{titr} provides an index of internal return. The amount of return detected by these methods represents a lower limit since 'hidden' return (to the identical starting material) presumably will occur in many cases but will be undetectable. The measured rate of ionization will be less than the true value.

Some enantiomeric preference is observed when solvolysis of a chiral leaving group occurs in a chiral solvent. There will then be formed two diastereoisomeric ion-pairs which pass to product at different rates[62].

$$(R)\text{-R—Ocas} \xrightarrow{\text{(+)-ethyl lactate}} [R^+((R) - Ocas^-)] \xrightarrow{k_R}$$

$$(S)\text{-R—Ocas} \xrightarrow{\hspace{2cm}} [R^+((S)\text{-Ocas}^-)] \xrightarrow{k_S}$$

$$\left.\begin{array}{r}\\\\\end{array}\right\} \text{products} \left\{ k_R \neq k_S. \right.$$

[cas = camphor-10-sulfonate]

10.1.9 The 'special' salt effect: an ion exchange in an ion-pair

This is the third distinct effect which may be produced by added ionic substances and confirms the importance of return from the solvent-separated ion-pair. When an inert salt such as $NaClO_4$ is added to solutions of certain substrates undergoing solvolysis, at very low salt concentration a very large rate increase occurs which eventually falls off to become a normal ionic strength effect (Fig. 10.6a). This was observed by Winstein during the acetolysis of **17** and named the *special salt effect*. The explanation appears to be as follows; in the absence of added salt solvolysis will proceed to the intimate ion-pair stage and considerable return to starting material will occur. The perchlorate ion is able to exchange with leaving-group ion and thereby prevent return since solvent attack proceeds far more rapidly than *external* return of the leaving-group ion. The net rate is then increased since a larger proportion of ionizations will proceed to product, as an additional channel for reaction is available[63,64].

Another interesting result comes from the solvolysis of **18**, from which Ad—Otos was isolated among the products. This must have arisen from an ion-pair which, after formation, expelled N_2O with some subsequent recombination of the proximate ions[65]. That anion exchange can occur is shown by the incorporation of ^{13}C-labeled tosylate ion into the substrate during solvolyses of ArCHR·Otos; the exchange process is presumed to occur at the ion-pair stage[66].

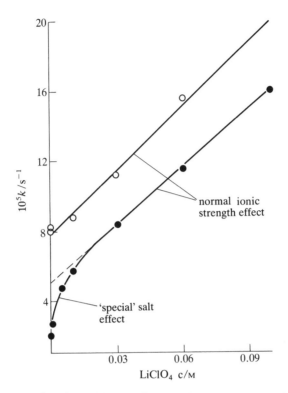

Fig. 10.6a Ionic strength and 'special' salt effects on the rates of acetolysis of *threo*-3-(*p*-anisyl)-2-butyl *p*-bromobenzenesulfonate in the presence of LiClO$_4$:

10.1.10 Structural effects upon ionization

Rates of limiting ionization with a common leaving group and solvent, if they can be reliably measured, give a direct measure of the relative stabilities of the carbocations produced. If solvolytic reactions are to be used as a probe into carbocation stability one must ensure that limiting ionization with negligible return occurs. A highly ionizing and non-nucleophilic solvent is clearly necessary so that rates of primary or secondary substrates are not boosted by nucleophilic solvent participation. Using a solvent such as CF$_3$COOH or, even better, FSO$_3$H, a very large span of reactivities may then be observed, much reduced in a nucleophilic solvent such as ethanol. Stabilities of carbocations may also be measured by

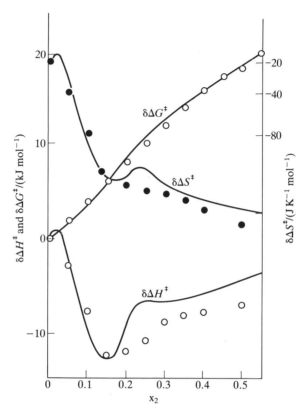

Fig. 10.6b Solvolysis of *tert*-butyl chloride: $\Delta G^{\ddagger}$, $\Delta H^{\ddagger}$ and $\Delta S^{\ddagger}$ as a function of solvent composition (ethanol–water). ($\bigcirc$ and $\bullet$ show parallel trends in transfer functions for glycine, H_3N—CH_2—COO^-, e.g. ΔG_{tr}(gas→soln.).

comparison of the enthalpies of a reaction such as that detailed in Table 10.13.

It is interesting to note that differences in gas-phase ionization enthalpies for secondary and tertiary halides are similar to those in FSO_3H, supporting the lack of nucleophilic solvation here. Quantitatively these trends are expressed by the Grunwald–Winstein coefficients (Table 10.7c), for which limiting ionization is shown by $l \to 0$, $m_s \to 1$, and a low $k(EtOH)/k(AcOH)$ (Table 10.12).

10.1.11 Leaving-group effects

The order of leaving-group ability is essentially that described for S_N2 reactions. However, the sensitivity of the rate to changes in the leaving group is much greater for the ionization mechanism and increases with limiting character of the reaction reflecting a greater degree of C–X bond

$$\text{Ad--}^+\overset{\bar{\text{O}}}{\text{N}}=\text{N--Otos} \rightleftharpoons \left[\text{Ad--}^+\overset{\text{O}^-}{\text{N}}\underset{\overset{+}{\text{N}}}{}\bar{\text{O}}\text{tos}\right] \rightleftharpoons \left[\text{Ad}^+\text{Otos}^-\right] + \text{N}_2\text{O}$$

18

$$\text{SOH}\diagup \qquad \diagdown$$

$$\text{AdOS} \qquad \text{Ad--Otos}$$

breaking in the transition state. The ratio $k_{\text{tos}}/k_{\text{Br}}$ is often taken as a guide to transition-state character (Table 10.14).

Electrophilic assistance at the leaving group

The role of the solvent in bringing about ionization is multifarious and has been shown to include nucleophilic attack and electrostatic stabilization of ionic products. The third aspect of solvent behavior which may play a prominent part in ionization is electrophilic catalysis. Protic solvents are hydrogen-bond acceptors (the stronger as acids the better), while leaving groups, halide or oxide ions, are donors. It seems likely that one role of the more acidic solvents in promoting ionization is their ability to solvate the leaving group, a form of acid catalysis less extreme than a full proton transfer. Leaving groups are usually either halide or oxide (e.g. sulfonate) types: halide may form two hydrogen bonds to a protic solvent while oxygen only seems to form one. Nonetheless, oxygen, being far more basic, is capable of the stronger H-bond. In the extreme this develops into full acid catalysis; S_N reactions of alcohols may be acid-catalyzed, those of halides

Table 10.13 Stabilities of carbocations

(a) Rates and activation enthalpies for gas-phase heterolysis of alkanes compared with solvolytic rates of alkyl chlorides[67]

$$R\text{—}H \rightarrow R^+ + H^-: \quad \Delta H = \Delta H_f(R^+) + \Delta H_f(H^-)^a - \Delta H_f(RH)$$

R	$\Delta H/kJ\,mol^{-1}\,(kcal\,mol^{-1})$	k_{rel}	Limiting (averaged) $k_{rel}(RCl/EtOH)$	Mechanism
Me	1308 (312)	1	1	
Et	1139 (272)	2	0·08	S_N2
isoPr	1403 (249)	50	0·02	
$PhCH_2$	999 (238)	10^6	40	Borderline
tert-Bu	981 (234)	20^8	45 000	
Ph_2CH		10^9	75 000	S_N1
Ph_3C		10^{13}	10^{12}	

$^a\ \Delta H_f(H^-) = +146\,kJ\,mol^{-1}\,(+34\cdot9\,kcal\,mol^{-1})$.

(b) Enthalpies of formation of gaseous carbocations, R^+, and hydrocarbons, RH ($\Delta H_f/kJ\,mol^{-1}\,(kcal\,mol^{-1})$)

R	$\Delta H_f(RH)$	$\Delta H_f(R^+)$	R	$\Delta H_f(RH)$	$\Delta H_f(R^+)$
Me	−74·5 (−17·8)	+1088 (+260)	$CH_2{=}CH$	+52·3 (+12·5)	+1125 (+269)
Et	−84·7 (−20·0)	+916 (+219)	$CH_2{=}CH\text{—}CH_2$	+20·4 (+4·88)	+903 (+216)
1-Pr	−104·7 (−25·0)	+874 (+209)	△	+53·3 (+12·7)	+1000 (+239)
2-Pr	−104·7 (−25·0)	+795 (+190)	▢	+26·6 (+6·4)	+891 (213)
1-Bu	−125·6 (30·0)	+912 (218)	⬠	−77·2 (−18·5)	812 (194)
2-Bu	(30·0)	+803 (+192)	Ph	+82·9 (+19·8)	+1192 (+285)
isoBu	−134·5 (−32·1)	+857 (+205)	$PhCH_2$	+50·0 (+11·9)	+904 (+216)
tert-Bu	(−32·1)	+736 (+176)	$m\text{-}MeC_6H_4CH_2$	+17·2 (+4·1)	+861 (+206)
			$p\text{-}MeC_6H_4CH_2$	+17·9 (+4·3)	+845 (+202)

Table 10.14. Sensitivity of solvolyses to leaving groups, k_{tos}/k_{Br}[68, 69]

Solvent	R							
	Me	Et	iso-Pr	Cyclohexyl	2-Ad	tert-Bu	Bicyclo-octyl	1-Ad
80% aq.EtOH	11	10	40		230	>4000	5000	9750
CF_3COOH			470		16,000		90000	200,000

are not. It may be seen from Table 10.14 that while the leaving-group ability, measured by k_{tos}/k_{Br}, is large for limiting ionization and very small for S_N2 displacements, its value continues to increase the more acidic the solvent even after fully limiting behavior is reached. This suggests that as nucleophilic solvent participation at the rearface diminishes, so electrophilic participation at the leaving group increases. The result is a 'push–pull' effect by these solvents capable of amphiphilic behavior.

α-Methyl effects [70]

Replacing an α-H by α-Me transforms a secondary substrate into a tertiary one. Rate changes observed depend upon the mechanisms of solvolysis of each, as summarized below:

$$S_N2 \longleftarrow ---- S_N2 ------\longrightarrow \text{ solvent assisted } ------\longrightarrow \text{ limiting } S_N1$$
(tertiary)   (secondary)

$$\xleftarrow{\hspace{2cm} \text{decreasing} \hspace{2cm}} k_{Me}/k_H \xrightarrow{\hspace{3cm} \text{increasing} \hspace{3cm}}$$
                                       [moderate]   [large]

The large rate increase (*ca* 10^8) on introducing an α-methyl group is a criterion that indicates the secondary compound is already reacting by a limiting ionization mechanism. A smaller increase would mean that the rate of the secondary compound was higher than the limiting rate due to nucleophilic assistance. For example, the rate changes for α-methyl substitution,

$$\text{2-Ad:} \quad k_{Me}/k_H = 10^{7.5};$$
$$\text{isoPr:} \quad k_{Me}/k_H = 10^{3.7},$$

indicate limiting solvent assistance in the secondary compound.

10.1.12   Bridgehead systems [71–74]

Tertiary substrates which have the leaving group at the junction of two rings ('bridgehead' positions) tend to be very inert in solvolysis although bridgehead carbocations will form in superacid solution [73]. The principle cause is the additional strain energy accompanying ionization due to deformation of the carbocation from the preferred planar conformation analogous to that in bridgehead alkenes. Rates of ionization therefore correlate with the amount of strain so produced (Table 10.15).

10.1.13   Linear free-energy relationships [13,41,44]

Whereas Hammett or Taft reaction constants, ρ and ρ^*, for S_N2 reactions tend to be small, those for S_N1 processes are usually very large and negative

Table 10.15   Reactivities of bridgehead tosylates in acetolysis

Strain energy/kJ mol^{-1}:	0	51	68	84	96	
(k cal mol^{-1}):	(0)	(12)	(16)	(20)	(23)	
k_{rel}:	1	2×10^{-4}	5×10^{-8}	10^{-11}	10^{-14}	10^{-19}

indicating the high electronic demands at the central carbon in the transition state. Rates correlate with σ^+ rather than σ^0 which indicates a strong conjugative interaction between substituent and reaction center. This is in accordance with carbocation character which is developing in the course of reaction. At the same time, nucleophilic solvent participation reduces the value of σ^+, which may be used as an index of such assistance (Tables 4.4, 10.7). This is because the more limiting the solvolysis, the greater the stabilization needed from the surrounding structure. Values of reaction constants in excess of -4 are common: i.e. solvolytic rates for $XC_6H_4CR_2Cl$ can increase by a factor of $10^{(4+1\cdot6)} = 2\cdot5 \times 10^6$ on replacement of p-NO$_2$ by p-OMe ($\Delta\sigma^+ = 1\cdot6$) at X. Stabilities of carbocations, defined in general as differences in energy between the carbocation (or transition state leading to it) and a covalent precursor, may be estimated in several different ways:

(a) rates of limiting ionization, $\ln k \equiv \Delta G^{\ddagger}$

$$R\!-\!X \xrightarrow{\ k\ } R^+, X^-$$

For a series of similar compounds, e.g. substituted cumyl chlorides, values of σ^+ for the substituents serve as a measure of their carbocation-stabilizing ability.

(b) Equilibrium ionization of alcohols in concentrated sulfuric acid may be used ($pK_{R^+} = -\log K_{R^+}$) provided the carbocations are rather stable: Table 10.7d.

$$R\!-\!OH + 2H_2SO_4 \overset{K_{R+}}{\rightleftharpoons} R^+ + 2HSO_4^- + H_3O^+.$$

(c) Enthalpies of ionization of halides in superacid solution become increasingly negative, the more stable the carbocation produced: Table 10.7e.

$$R\!-\!Cl \xrightarrow{\text{SbF}_5/\text{HSO}_3\text{F}/\text{SO}_3\text{ClF}, \ -40°C} R^+, Cl^- : \Delta H_{RCl}.$$

(d) Heterocyclic bond dissociation energies, $D(R^+H^-)$ may be measured by the thermal cycle; by means of:

$$R—H \xrightarrow{\Delta H} R^+ \quad H^-$$

$$D(R—H)\downarrow \qquad\qquad \uparrow E.A.(H\cdot)$$

$$R\cdot \quad H\cdot \xrightarrow[I.P.(R\cdot)]{} R^+ \quad H\cdot$$

$$D(R^+, H^-) = \Delta H = D(R—H) + I.P.(R^\cdot) + E.A.(H^\cdot).$$

Despite $D(R—H)$ being measured in the gas phase and the other scales in solution there is broad agreement between these measures. The following correlation equations are found to hold:

$$\Delta H_{RCl}/kJ\, mol^{-1} = -6.9pK_{R^+} - 250 = 36\cdot3\sigma^+ - 176\cdot5.$$

The factor which appears most important in stabilizing carbocations is the ability to delocalize positive charge: requirements for this include $+R$ substituents in conjugation, a planar center and (in the gas phase at least) a large number of atoms in the ion.

Electronic demands at the incipient carbocation may be met both by nucleophilic solvent assistance and by structural features. If solvolysis is truly limiting, as it must be in the most highly ionizing and non-nucleophilic solvents, reaction constants reflect intrinsic structural features of the carbocation which may be either stabilizing or destabilizing, for which ρ^+ will tend to be less or more negative, respectively[75].

A planar, trigonal geometry is preferred by a carbocation and deviations from this are accompanied by an increase in its energy. Bridgehead cations are necessarily non-planar and small rings impart angle strain; for example:

$$R—Cl \longrightarrow R^+ + Cl^-$$

k_{rel}	1	5×10^{-8}	0·002
ρ^+:	−4·7	−7·1	−5·6

Anti-aromatic structures are highly energy-prohibitive. The most common example is the cyclopentadienyl cation, for which the following comparisons may be made:

$\Delta H/kJ\, mol^{-1}$ (kJ mol^{-1}) for $R^+ + H^- \rightarrow RH$:	−1078(−258)	−940(−225)	—
pK_{R^+}:	−40	—	−20
k for $RI \rightarrow R^+I^-$:	10^{-5}	—	1

Electron-withdrawing substituents will also raise the electronic demands on the surrounding structure. Gas-phase stabilization energies for cations $Y-CH_2^+$ increases as σ^+ becomes more negative and the electronic charge transferred, Δq, increases:

Y:		NO$_2$	CN	H	F	Cl	Me	Br	Ph	OH	OMe	NMe$_2$
E_s:			-10	0	26	32	35	51	55	60	69	95
$\Delta q/e$:		-76	-262	0	-353		-113			-486		-566

The CF$_3$ group, a strong σ-acceptor, will destabilize a methyl cation by 55 kJ mol^{-1} (13 kcal mol^{-1}) according to the equation

$$CH_4 + CF_3CH_2^+ \rightarrow CH_3^+ + CF_3CH_2^+;$$

$$\Delta H = -55 \text{ kJ mol}^{-1} (-13 \text{ kcal mol}^{-1}),$$

and retard ionization rates, for example:

$$k_{HCMe_2OTf}/k_{CF_3CMe_2OTf} \text{ (in EtOH)} = 6700.$$

Carbonyl is a π-acceptor and will have a similar effect at the β position, but little at the α at which it can share the charge:

k:	8:1		10^5:1

$$\rho = -7.1$$

α-Cyano, on the other hand, has a strong deactivating effect:

$\dfrac{k (R = H)}{k (R = CN)}$:	3×10^3	10^2	10^6	10^7

10.1.14 Intramolecular assistance in ionization

In a highly ionizing medium such as fluorosulfonic acid, FSO_3H, rates of solvolysis of primary compounds seem actually to increase in a dramatic way with increasing substitution at the α-carbon (Table 10.8). Neopentyl compounds, **12**, in this medium are far more reactive than methyl, although the reverse is true for solvent-assisted processes. It is unlikely that the neopentyl cation is much more stable than any other primary cation since electronic effects in an alkane will not be large. Moreover, there is a substantial α-SKIE in neopentyl solvolysis, not present in propyl (Table 10.10), and products are entirely the isoamyl derivatives resulting from a Wagner–Meerwein rearrangement. It seems likely that both rearrangement and the accelerated rate are related and that assistance to ionization by the σ-electrons which bind the methyl group is responsible. Similarly, rearrangement of 1- to 2-propyl occurs extensively when the solvolysis is carried out in highly ionizing media but not when in a nucleophilic solvent. These are examples of 'neighboring-group participation' (Chapter 13). Winstein analyzed the kinetics in terms of two competing pathways, k_s (solvent assisted) and k_Δ (neighboring-group assisted), both rate-determining, so that[76]

$$k_{obs} = k_s + Fk_\Delta, \tag{10-3}$$

	Retention or inversion	Rearrangement?
21	Inv.	No
22	Ret.	Yes
23	Ret.	No

where F is the fraction which proceeds to products, $(1 - F)$ returning to reagents. The 3-aryl-2-butyl system provides a suitable example[77-79]. Ionization may be assisted by aryl participation, **19** representing a transition state or perhaps an intermediate **20**. The symmetry of **20** permits it to form products by solvent attack at C2 or C3 and therefore an isotopic label is needed to distinguish these. Both of these centers are chiral and the stereochemistry at each in the products can be determined. The importance of each route may then be ascertained by analysis of the products:

$$k_s/k_\Delta = [21]/[22] + [23]. \qquad (10.4)$$

A second approach was developed by Schleyer[76], who constructed a Taft plot of rate against σ^* for a series of secondary substrates which were not capable of internal participation. A good linear relationship resulted (Fig. 10.7). Points for β-arylalkyl substrates fell above the line, the difference between the predicted value on the line (k_s) and the experimental point being attributed to k_Δ. On the whole, reasonable agreement as to the proportion of assistance is obtained between the two methods (Table 10.16). As would be expected, aryl participation increases with the presence of $+R$ substituents in the ring and conversely it decreases with $-R$. Taking the view that rates of unassisted solvolysis should be predictable from the steric strain introduced into a molecule when it ionizes, Foote and Schleyer proposed Eq. (10.5)[80] relating this to solvolytic rates: ν_{CO} is the stretching frequency of the corresponding ketone, φ the twist angle in R^+:

$$\log k_{rel} = 0 \cdot 125(1715 - \nu_{CO}) + 1 \cdot 32(1 - \cos 3\varphi) + nbi + I. \qquad (10.5)$$

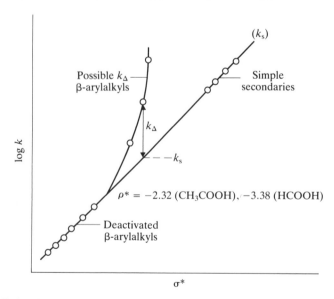

Fig. 10.7 Estimation of the extent of internal nucleophilic assistance to ionization (k_Δ) from the deviation of observed rates from the LFER.

Table 10.16   Observed and estimated rates of acetolysis of secondary substrates[80]

(a) Unassisted   (b) Assisted

	$\log k_s$		$\log k_\Delta{}^b$		$\log k_s$		$\log k_\Delta{}^b$
	obs.	calc.^{a}			obs.	calc.^{a}	
7-Norbornyl	−7·0	−7·00	0·0	*anti*-Dicyclopenta-dien-8-yl	−8·8	4·33	13·1
endo-8-Bicyclo[3,2,1]-octyl	−4·2	−4·11	0·1	*anti*-7-Norbornenyl	−8·8	4·11	12·9
endo-2-Benznorborn-enyl	−2·4	−2·22	0·2	7-Dibenzonorborna-dienyl	−11·3	−0·79	10·5
endo-2-Norbornenyl	−1·0	−1·48	−0·4	*anti*-7-Benzonor-bornenyl	−10·3	−1·22	9·1
Adamantyl	−1·1	−1·18	0·1	7-Quadricyclyl	−4·9	3·31	8·2
Cyclohexyl	−0·1	0·00	0·1	3-Nortricyclyl	−6·2	1·82	8·0
Cyclotetradecyl	+0·1	+0·08	0·0	*anti*-Bicyclo[3,2,1]oct-2-en-8-yl	−6·1	−0·13	6·0
Isopropyl	−0·4	+0·15	0·6	7-*syn*-Norbornenyl	−8·9	−3·28	5·6
endo-2-Norbornyl	−0·2	+0·18	0·0	Cyclobutyl	−4·2	0·99	5·2
Cyclopentadecyl	0·0	+0·42	0·4	*exo*-2-Benzonorbornenyl	−2·8	1·63	4·4
cis-4-*tert*-Butylcyclo-hexyl	+0·4	+0·42	0·0	*exo*-Bicyclo[3,2,1]oct-8-yl	−4·4	−0·21	4·2
Cyclododecyl	+0·3	+0·50	0·2	*exo*-2-Norbornenyl	−1·4	2·42	3·8
2-Butyl	−0·3	+0·53	0·8	Cholesteryl	−1·7	2·01	3·7
3,3-Dimethyl-2-butyl	+1·5	+0·62	−0·9	*exo*-Bicyclo[2,2,2]oct-5-en-2-yl	0·5	4·10	3·6
Cyclotridecyl	+0·3	+0·66	0·3	*exo*-2-Norbornyl	−0·6	2·71	3·3
3-Methyl-2-butyl	+0·6	+0·93	0·3	*endo*-Bicyclo[2,2,2]oct-5-en-2-yl	−0·7	2·49	3·2
Cyclopentyl	+1·5	+1·51	0·0	Bicyclo[2,1,1]hex-2-yl	−3·3	−0·37	2·9
Cycloheptyl	+2·0	+1·78	−0·2	*epi*-Cholesteryl	−1·4	1·40	2·8
trans-2-*tert*-Butylcyclo-hexyl	+2·2	+2·20	0·0	Cyclopropyl	−7·2	−5·32	1·9
cis-2-*tert*-Butylcyclohexyl	+2·6	+2·61	0·0	Bicyclo[3,3,1]non-9-yl	−1·0	0·48	1·5
1,4-α-5,8-β-Dimethano-perhydro-9-anthracyl	+2·8	+2·67	−0·1	*exo*-Trimethylene-norborn-*exo*-2-yl	−0·6	0·84	1·4
Cyclo-octyl	2·8	2·76	0·0	*cis*-Bicyclo[3,1,0]hex-3-yl	−0·2	1·14	1·3
Cyclononyl	2·8	2·70	−0·1	a-Bicyclo[3,2,1]oct-2-yl	0·4	1·62	1·2
Cyclodecyl	2·7	2·98	0·3	Bicyclo[2,2,2]oct-2-yl	0·9	1·85	0·9
Cycloundecyl	2·1	2·05	0·0	e-Bicyclo[3,2,1]oct-2-yl	0·1	0·47	0·4
				a-Bicyclo[3,1,0]hex-3-yl	0·2	0·17	0·0
				syn-Bicyclo[3,2,1]oct-2-en-8-yl	−5·5	−5·54	0·0

^{a} Calculated by Eq. (10.5).
^{b} k_Δ is the difference, $\log k_s$ (obs.) $- \log k_s$ (calc.), and is a measure of internal assistance to ionization; see Chapter 13.

The first two terms measure internal bond angle strain and torsional strain in the trigonal carbocation; nbi and I are small additional terms for non-bonded interactions and inductive effects if present. This equation predicts solvolytic rates with remarkable accuracy for many simple secondary systems which solvolyze at limiting rates with no nucleophilic assistance either from solvent or internal nucleophile (Table 10.16). However, many systems are known whose experimental rates are far in excess of predicted values and these are inevitably systems for which the excess rate can be plausibly explained by internal nucleophilic assistance. In some this assistance comes from π-electrons, and in others from σ-electrons made available at the rear of the leaving group. The value of $\log(k_{obs}/k_{calc})$ is a measure of this assistance.

10.1.15 Activation parameters[81-84]

Entropies of activation have been measured for many solvolyses but do not provide an unambiguous guide to mechanism. In general, primary halides and sulfonates solvolyze with a negative value of $\Delta S^{\ddagger}$ since the nucleophilic attack of solvent is an associative process. The value depends upon the degree of nucleophilic solvent assistance and diminishes with neighboring-group participation; for example, for trifluoroacetolysis of R—CH_2Otos,

R:	Me	Et	n-Pr	Neopentyl	
$\Delta S^{\ddagger}/J\,K^{-1}\,mol^{-1}$:	-130	-111	-72	-37	(-4^a)
$(cal\,K^{-1}\,mol^{-1})$:	(-31)	(-26)	(-17)	(-9)	(-1^a)

^{a} For acetolysis.

Values of $\Delta S^{\ddagger}$ for limiting solvolysis of tertiary compounds tend to be positive but are still highly variable; e.g. for hydrolysis of *tert*-BuCl $\Delta S^{\ddagger} = +42\,J\,K^{-1}\,mol^{-1}$ ($10\,cal\,K^{-1}\,mol^{-1}$). Volumes of activation tend to follow entropies and are also not very indicative of mechanism. Despite the dissociation of the substrate which would tend to increase the volume, electrostriction of the solvent around the developing ions has the opposite effect, which is usually dominant. As a result, $\Delta V^{\ddagger}$ for solvolyses of halides and sulfonates, primary or tertiary, tend to fall in the range -10 to $-20\,cm^3\,mol^{-1}$, becoming more negative in aqueous organic solvent mixtures as the dielectric constant of the medium diminishes[85] (Table 10.17).

Solvolytic rate constants may be measured by conductimetry to very high precision sufficient to show up deviations from linearity of the Arrhenius plots of $\ln k$ against $1/T$. This enables $\Delta C_p^{\ddagger}$, the heat capacity of activation, to be measured[86-89]. The values in water, differences in heat capacity between reagents and transition state, reflect changes in the pattern of solvation and depend upon the anion, for the reaction $MeX + H_2O$,

X:	Cl	Br	I	$PhSO_2O$	$MeSO_2O$
$\Delta C_p^{\ddagger}/J\,mol^{-1}\,K^{-1}$:	217	194	236	140	157
$(cal\,mol^{-1}\,K^{-1})$:	(52)	(46)	(56)	(33)	(37)

Table 10.17   Some activation parameters for solvolytic reactions

Substrate	Solvent	$\Delta S^{\ddagger}/$ $J\,K^{-1}\,mol^{-1}$ $(cal\,K^{-1}\,mol^{-1})$	$\Delta C_p^{\ddagger}/$ $J\,K^{-1}\,mol^{-1}$ $(cal\,K^{-1}\,mol^{-1})$	$\Delta V^{\ddagger}/cm^3\,mol^{-1}$
MeF	H_2O	−109 (−26)	−275 (−65)	
MeCl	H_2O	−51 (−12)	−204 (−49)	
MeBr	H_2O	−42 (−10)	−192 (−46)	−17
MeI	H_2O	−34 (−8)	−234 (−56)	
MeOCOCF$_3$	H_2O	160 (38)	−85 (−20)	
EtBr	H_2O	−25 (−6)	−204 (−49)	−11
n-PrBr	H_2O	−60 (−14)	−180 (−43)	−15
isoPrBr	H_2O	−34 (−8)	−158 (−38)	−9 (−15, HCOOH)
tert-BuBr	H_2O	+42 (10)	+346 (82)	−24 (acetone–H_2O, 9:1v/v) −42 (acetone–H_2O, 9:1 v/v)
PhCH$_2$Cl	H_2O	+46 (11)	−154 (−37)	−7
CycloC$_6$H$_{11}$Br	H_2O	+4·5 (1)	−246 (−59)	−12
ArCMe$_2$Cl	H_2O			
Neopentyl—Otos	AcOH	−4 (−1)		
EtOtos	CF$_3$COOH	−130 (−31)		
Neopentyl—Otos	CF$_3$COOH	−37 (−9)		

These values were once thought to reflect the degree of hydration of the departing anion and hence the position of the transition state. More recently, a re-interpretation in terms of a two-step mechanism (association of the solvent followed by product formation) has been considered[90–93]. This seems reminiscent of the S_N2 intermediate mechanism[94], in which what is normally considered a transition state is interpreted as an intermediate. These matters are still speculative.

10.1.16 The S$_N$i reactions[95]

An unusual stereochemical result may accompany substitution of a chiral secondary alcohol by thionyl chloride to form the alkyl chloride. Lewis recorded examples in which almost complete retention of configuration occurred although the amount is very solvent-sensitive, being highest in dioxane. It has been established that the chlorosulfite, **24**, is an intermediate. Although various cyclic transition states have been suggested, it seems

24

unnecessary to postulate any unique mechanism. This intermediate may therefore undergo rate-determining heterolysis or fragmentation with the expulsion of SO_2 and the resulting carbocation and chloride ion recombine within the solvent cage. Possibly the nucleophilic solvent is involved in the configuration-holding which, in the extreme, becomes a double inversion.

10.1.17 Aliphatic S_N2 reactions in the gas phase[96-99]

Nucleophilic anions such as OH^- can be sustained in orbit in a cyclotron under high vacuum and neutral species such as alkyl halides introduced. Under these conditions displacement reactions occur analogous to those in solution but in the absence of solvent molecules. It is now possible to study the energetics of a wide variety of S_N2 reactions in this way and estimate the intrinsic effects on reactivity. Table 10.7g gives some examples of rate constants, heats of reaction and efficiencies. Reaction efficiency is the fraction of collisions which result in reaction and can be as high as 0·8 (80%) although they vary with structure by a factor of at least 1000. Rates themselves are very high, as much as 10^{20} times faster than similar reactions in solution owing to more favorable A-factors since there is no longer any need for solvent reorganization in the activation step. The scope of observable reactions is somewhat increased; for example, displacements of OR^- and SH^- are found to occur. Careful analysis of the rates as a function of pressure yields a more detailed knowledge of the potential energy profile of the reaction and indicates a double potential minimum as in Fig. 10.2a. Evidently, the approach between nucleophilic anion and alkyl halide is first facilitated by an attraction due to long-range forces and leads to a minimum corresponding to a complex, $[Nu^- \cdot RX]$ from which the alkyl cation exchange occurs. The stereochemistry of the reaction can be studied under these unusual conditions and inversion still seems to be the rule[100, 101] so the transition state structure is evidently similar to that inferred for reactions in solution. Structure–reactivity trends also tend to confirm this inference. Variation of the alkyl group affects rates in much the same way as in solution with methyl being by far the most reactive of the alkyl halides:

$$Cl^- + R\!\!-\!\!Br \rightarrow Cl\!\!-\!\!R + Br^-$$

R:	Me	Et	nBu	isoPr	isoBu
$E_A/\text{kJ mol}^{-1}$:	$-10\cdot5$	$2\cdot1$	$-2\cdot1$	$21\cdot3$	$23\cdot8$
(kcal mol^{-1}):	$(-2\cdot5)$	$(0\cdot5)$	$(-0\cdot5)$	$(+5\cdot1)$	$(+5\cdot7)$

This evidently means that the invocation of steric factors rather than solvation to explain this sequence is correct.

The reactivity sequence for nucleophiles (e.g. reacting with MeCl) falls in the following order which may be denoted 'intrinsic nucleophilicity';

$$OH^- > F^- \sim OMe^- > SMe^- \gg Cl^- > CN^- > Br^-.$$

The behavior of fluoride and cyanide, in particular, is surprising. The former is now highly reactive although in protic solvents it is so solvated as to be very low on the nucleophilicity scale. Cyanide is a soft nucleophile and it would appear that this quality (i.e. the availability of a high level HOMO, polarizability), is no longer of such importance as in solution chemistry. Rates are to a greater extent controlled by charge interactions which presumably make up the attractive forces present in the early stages of reaction. This suggests that two reactants 'find' each other in liquid milieu by attractions relayed through the intervening solvent molecules via polarizability forces. In a vacuum, with the lowest possible dielectric constant, electric fields are experienced at longer distances and electrostatic forces become increasingly the medium of intermolecular 'communication'.

It is found that rates of gas-phase S_N2 reactions are quite well correlated with 'methyl cation affinities', defined as ΔH for the process;

$$CH_3\!\!-\!\!X \rightarrow CH_3^+ + X^-: \qquad \Delta H = \text{methyl affinity of } X^-.$$

This fact suggests that, in the transition state the central carbon bears a considerable degree of positive charge,

$$[Nu^- \cdots \overset{|}{\underset{\diagdown}{C}}^+ \cdots Nu^-].$$

It is also possible to admit solvent vapor into the experimental chamber and to observe solvated ions with $1, 2, 3, \ldots$ solvent molecules coordinated. The rates of reaction of these species with the alkyl halides can also be measured and so the gradual deactivation of nucleophilic power by solvent demonstrated beyond doubt, thus[102]:

$$OH(H_2O)_n^- + MeBr$$

n:	0	1	2	3
k_{rel}:	1·0	0·63	0·002	0·0002

10.2 ELECTROPHILIC SUBSTITUTIONS AT SATURATED CARBON[103]

A set of mechanisms parallel to the S_N types may be written for displacements of one electrophile for another at a saturated carbon. Such reactions can be realized provided that there is a leaving group present which will easily depart as an electrophile i.e. leaving its bonding electrons behind. Electrophilic leaving groups may include the proton (although such reactions are usually considered carbon acid–base processes), stable carbocations and highly electropositive elements, especially metals such as tin, mercury, lead and metalloids silicon, germanium and boron.

This topic therefore includes reactions at carbon ligands of organometallic compounds and two basic mechanistic types emerge, S_E1 and S_E2, analogous to their nucleophilic counterparts.

$$S_E1: \quad \overset{\backslash}{\underset{/}{C}}-M \longrightarrow \overset{\backslash}{\underset{/}{C}}{}^- \;\; M^+ \overset{E^+}{\longrightarrow} \overset{\backslash}{\underset{/}{C}}-E + M^+.$$

$$S_E2: \quad E^+ \;\; + \;\; \overset{\backslash}{\underset{/}{C}}-M \longrightarrow E-\overset{/}{\underset{\backslash}{C}} + M^+.$$

There is, however, a range of behavior within the latter category which has no parallel in nucleophilic displacements; frequently electrophilic displacements are too fast for convenient rate measurements so less kinetic information is available than for nucleophilic displacements. Reactions sometimes show complexities which have not yet been unraveled.

10.2.1 The S_E1 mechanism[104,105]

Ionization of an electrophilic leaving group generates a carbanion, so this type of behavior is to be found when the reaction center is flanked by $-R$ substituents which can stabilize negative charge. The first to be identified was the mercury isotopic exchange reaction,

$$R-Hg-Br + Hg^*Br_2 \xrightarrow{\text{DMSO}} R-Hg^*Br + HgBr_2,$$

25

in which R = PhCHCOOEt, which may be attached to mercury in optically active form. The isotopic exchange is of first order with respect to RHgBr and independent of $[Hg^*Br_2]$. Rates of exchange are equal to rates of racemization, as expected for a reaction forming a readily inverted carbanion in the slow step. The mechanism may therefore be written:

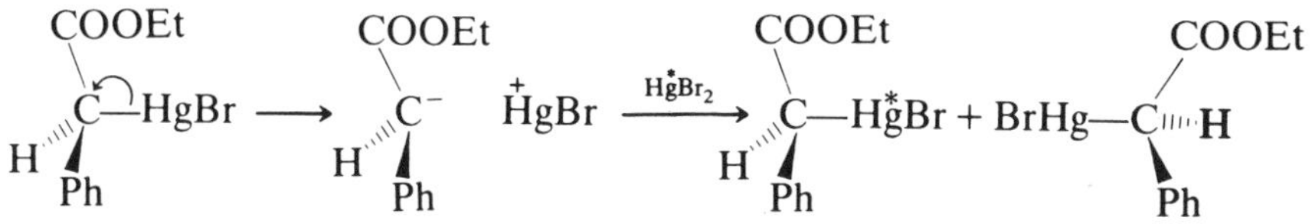

26

Catalysis by bromide ion is observed with the kinetic form

$$k_{obs} = k_1 + k_c[Br^-]^2.$$

It is supposed that this indicates attachment of two bromide ions forming the more easily displaced —HgBr$_3^-$ group. A similar mechanism operates in the cleavage of the pyridiniummethyl mercury, **27**, in which C–Hg bond dissociation is aided both by nucleophilic catalysis by Cl$^-$ and by the neutralization of charge.

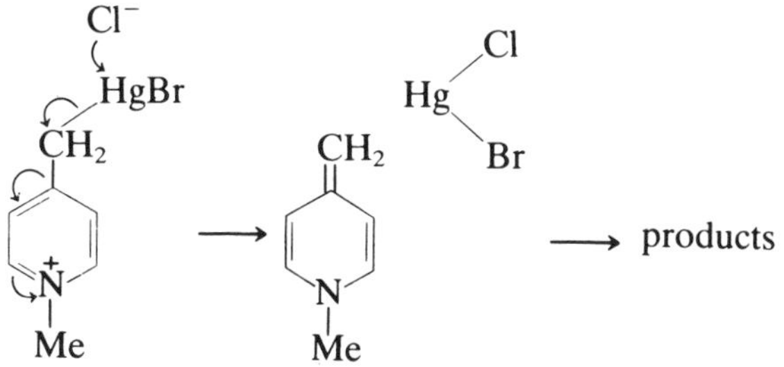

27

10.2.2 The S$_E$2 mechanism

The halogenolysis of many organometallic compounds provides examples of a range of behavior within the scope of the bimolecular displacement.

Cleavage of alkylmercuric halides,

$$R—HgBr + Br_2 \longrightarrow R—Br + HgBr_2,$$

is a reaction whose stereochemistry can be examined by the use of cyclic or chiral ligands, R (Table 10.18). It is clear that the results are not as clearcut as is the case with S$_N$2 reactions. Products resulting from complete inversion to retention of configuration can be obtained depending upon solvent and the availability of air. The latter influence suggests a competing radical pathway which is suppressed by oxygen. Even so, the stereochemistry seems to vary in a capricious manner with solvent. Better behaved are the

Table 10.18 Rate and activation data for some electrophilic substitutions at saturated carbon

(a) Effect of structure on mechanism of S_E2 processes (k_{rel})

System	R					Mechanism
	Me	*Et*	*n-Pr*	*isoPr*	*tert-Bu*	
$R_4Sn + HgCl_2/MeOH$	100	0·21	0·04	0·04	10^{-7}	Open
$R_4Sn + I_2/AcOH$	100	33	3·9	0·03		Open
$R_4Sn + Br_2/AcOH$	100	83	12	2·5		Open
$R_4Sn + I_2/PhCl$	100	600	78	560		Cyclic
$R_4Sn + HCl/PhH$	100	750	300	300		Cyclic
$R_4Sn + Br_2/PhCl$	100	1230	470	310		Cyclic
$R_4Sn + Br_2/CCl_4$	100	9500	4400	86 000		Cyclic
$RB(OH)_2CrO_3$	100	3×10^5			3×10^7	Coordinated

(b) Effect of leaving group for Me—SnR_3 + I_2/MeOH system

Me—SnR_3	k_{rel}
—$SnMe_3$	1·00
—$SnMe_2Et$	1·32
—$SnMe_2Pr$	0·85
—$SnMe_2t$-Bu	$5·6 \times 10^{-3}$

(c) Effect of metal, k_{rel}

Electrophile	System		
	AcOH	*I_2/CCl_4*	*$HgCl_2/MeOH$*
SnR_4	1	3·3	10^6
PbR_4	18·3	10^6	
HgR_2	4·5	0·036	3×10^{-5}

(d) Stereochemistry of the reaction

$$R_3Sn\text{–}2\text{-Bu} + Br_2 \rightarrow R_3SnBr + 2BuBr$$

R	Solvent	Predominant stereochemistry	Excess	Ref.
isoPr	MeOH/cyclohexane	ret.	34	239
2-Bu	MeOH/cyclohexane	ret.	7·5	
3-Pentyl	MeOH/cyclohexane	ret.	2	
Neopentyl	MeOH/cyclohexane	inv.	7	
Neopentyl	PhCl	inv.	0·4	240
isoPr	CCl_4	ret.	70	
isoPr	MeOH	ret.	22	
isoPr	MeCN	inv.	9	
isoPr(neoPe)$_2$	CCl_4	ret.	76	
isoPr(neoPe)$_2$	MeCN	inv.	100	

Table 10.18 (*Continued*)

(e) Effect of solvent[240]

Solvent	Observed stereochemistry/%	
	iso-Pr₃Sn-2-Bu	*iso-Pr₂neoPeSn-2-Bu*
Methanol	9·3 ret.	3·5 ret.
Ethanol	32·8 ret.	4·8 ret.
n-Propanol	25·1 ret.	8 ret.
Acetonitrile	14·5 inv.	65·7 inv.
Dimethylformamide	3·8 inv.	48·6 inv.
Benzonitrile	12·2 ret.	4·4 ret.
1,2-Dichloroethane	7·7 ret.	24·4 ret.
Chlorobenzene	27·6 ret.	33·1 ret.
Tetrahydrofuran	23·7 ret.	51·7 ret.
Carbon tetrachloride	56·6 ret.	73·8 ret.

(f) Effect of electrophile[240]

Solvent	Electrophile	Observed stereochemistry/%	
		iso-Pr₃Sn2-Bu	*iso-Pr₂neoPeSn2-Bu*
Methanol	I₂	15·2 ret.	11 ret.
Methanol	IBr	30·7 ret.	73 ret.
Methanol	ICl	43·1 ret.	40·6 ret.
Methanol	Br₂	9·2 ret.	3·5 inv.
Acetonitrile	I₂	5 inv.	13·2 inv.
Acetonitrile	IBr	40·7 inv.	100 inv.
Acetonitrile	ICl	62·5 inv.	100 inv.
Acetonitrile	Br₂	14·5 inv.	65·7 inv.

halogenolyses of tetra-alkyltins[106]:

$$R_4Sn + Hal_2 \rightarrow R_3SnHal + R\text{---}Hal.$$

The rates are first order in each reagent and are very sensitive to solvent polarity, evidently indicating the formation of a highly polar transition state. The bromination of tetramethyltin has been mentioned as the basis of the X scale of solvent polarity (Eq. (5.17)), which is suitable since rates vary by a factor of 63 000 between solvents acetic acid and carbon tetrachloride. A dipolar transition state leading to non-polar products is also supported by the volume profile, which shows a negative volume of activation but positive volume of reaction (-60 and $+30\,cm^3\,mol^{-1}$, respectively). Chiral alkyl ligands may also be attached to tin and the reactions are shown to proceed with a mixture of inversion and retention (Table 10.18). Retention seems to be predominant in simple cases, though when neopentyl ligands are present

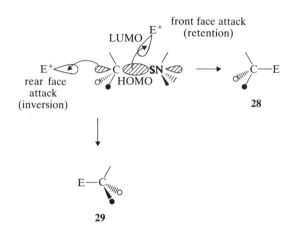

28

29

a 2-butyl group is displaced with excess inversion. Evidently, two mechanistic pathways are operating with very little difference in their energies of activation so that the proportions of retained relative to inverted products vary widely with changes in the conditions — structure, solvent and concentration of bromide ion added. The frontier orbitals will be the antibonding σ^* LUMO of bromine which must interact with the C–M σ-bond of the organometal, accessible from either front or rear faces (**28, 29**). The effect of varying the alkyl groups upon rates of halogenolysis of R_4Sn and upon the stereochemistry of the reaction is complex. At least two directing influences are operative (Table 10.18); they may follow a steric (E_s) or an inductive (σ^*) order, characterized by sequences **a** and **b**:

a	Me	>	Et	>	*n*-Pr	>	isoPr	>	*tert*-Bu
cf. E_s:	0		−0·07		−0·36		−0·47		−1·54

b	Me	<	*n*-Pr	<	isoPr	<	*tert*-Bu
cf. σ^*:	0		−0·1		−0·19		−0·3

The most coherent explanation of these effects has been offered by Gielen and by Abraham [103] as follows. The electrophile, Hal_2, is not a simple cation but must be regarded as an electrophile–nucleophile couple (Br^+Br^-), and there are two pairs of interacting centers $C\cdots Br^+$ and $M\cdots Br^-$. There is a spectrum of transition states over which the emphasis on initial bonding ranges, **30**, going from those described as S_E2(open) via S_E2(cyclic) to

inversion
30

open

cyclic

coordinated

422

S_E2(coordinated). The open type responds to changing alkyl groups according to their steric size (E_s) since the coordination number of carbon is temporarily increased. The coordinated type of transition state responds to polar effects of the alkyl groups (σ^*) — the more inductively electron donating they are the greater the rate of reaction. The cyclic type of transition state is perhaps the commonest, with some degree of bonding at each center occurring simultaneously. The reaction pathway results from a complex interaction of structural and environmental factors (Table 10.18). As additional complicating factors, the rates depend on the electrophilic leaving group; this means that variation of R in a series R_4Sn includes both structural and leaving-group effects in rate changes. There is also some evidence that the presence of products of the type R_3SnHal reduces rates, possibly by coordinating with the halogen. Varying the electrophile from the relatively hard acetic acid (H^+) to the soft $HgCl_2$ results in a very large increase in rate constants for reactions at tin and lead alkyls but a large reduction in rates for mercury compounds. The leaving-group reactivity order $Pb > Hg > Sn$ holds for attack by acetic acid, $Pb \gg Sn > Hg$ for attack by iodine and $Sn \ll Hg$ for attack by $HgCl_2$. Rates seem to be under the control of several factors and it cannot be claimed that these reactions are well understood.

10.2.3   Electrophilic substitution via enolization

This mechanism has already been discussed in Sections 9.2.1 and 9.2.2. It is essentially of the S_E family and requires prior removal of the leaving electrophile (normally H^+) to give a stabilized carbanion (enolate) or its *O*-protonated form (enol) followed by rapid reaction with an electrophilic reagent. The structural requirement is at least one $-R$ group ($-C=O$, $-C\equiv N$, $-NO_2$) on the adjacent carbon.

The reaction encompasses the halogenation, D-exchange and racemization of ketones and related compounds and all aspects of carbanion chemistry (Section 6.6).

10.3 NUCLEOPHILIC DISPLACEMENTS AT VINYL CARBON

Compared with saturated analogues, vinyl halides and sulfonates are very unreactive towards nucleophilic displacement. Rates of solvolysis can be less than those of tertiary alkyl by a factor of 10^6 and reflect the lack of stabilization of a vinyl cation, in which the vacant n- and π-orbitals are orthogonal (31); this does not facilitate an S_N1 reaction and the π-electrons

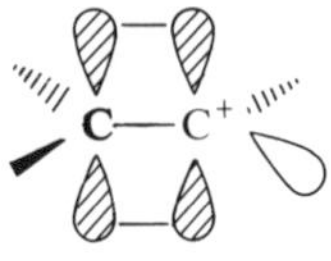

31

hinder approach of a nucleophile for S_N2. Nonetheless, these reactions do occur under forcing conditions or when promoted by electron-withdrawing substituents in the vinyl compound[107]. Examples are listed in Table 10.19.

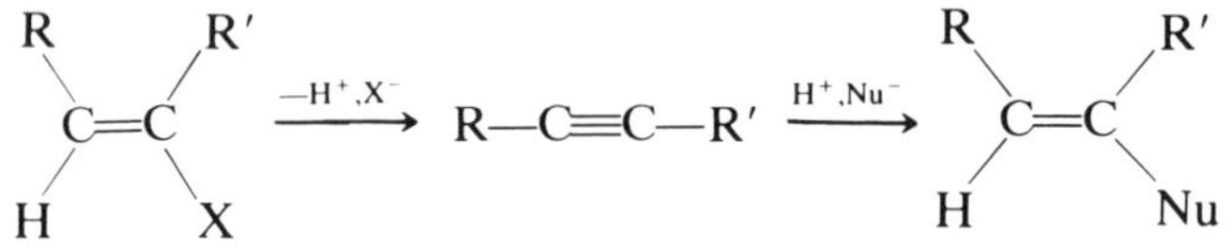

32

There are many mechanisms available to vinylic systems; besides the two mentioned, nucleophilic addition–elimination, elimination–addition via carbenes, alkynes or allenes, 32, or radical processes are possible but the following examples may be discussed in terms of a variable transition state. There appears to be a strong preference for retention of configuration in a nucleophilic substitution at alkene carbon, irrespective of its geometry; the

Table 10.19   Rates and activation parameters for some vinyl solvolyses[107]

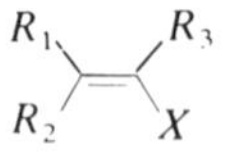 R_1	R_2	R_3	X	Conditions of reaction	$10^4 k/s^{-1}$ [a]	$\Delta H^{\ddagger}/kJ\,mol^{-1}$ ($kcal\,mol^{-1}$)	$\Delta S^{\ddagger}/J\,K^{-1}\,mol$ ($cal\,K^{-1}\,mol^{-1}$)
H	H	H	OTf	Aq. EtOH/76°C	0·49	103 (25)	−32 (−8)
Me	Me	Me	TNBS	Aq. EtOH/100°C	0·55	105 (25)	−50 (−12)
Ph	Ph	Ph	OTf	AcOH/25°C	0·0014	100 (24)	−83 (−20)

[a] Compare *tert*-BuCl, H_2O, 100°C, $k = 115\,s^{-1}$.

424

following reactions are >98% stereospecific for both *E*- and *Z*-isomers[108]:

$$PhCH{=}CHBr + Ph_3P \rightarrow PhCH{=}CH{-}\overset{+}{P}Ph_3 + Br^-;$$

$$NCCH{=}CHCl + EtS^- \rightarrow NCCH{=}CH{-}SEt + Cl^-;$$

$$EtO{\cdot}COCH{=}CHCl + R_2NH \rightarrow EtO{\cdot}COCH{=}CH{-}NR_2.$$

However, there are exceptions:

$$(E){-}CF_3CCl{=}C(CF_3)Cl + RO^- \rightarrow CF_3CCl{=}C(CF_3){\cdot}OR + Cl^-;$$

$$(E)/(Z) = 70/30.$$

The nucleophile must enter perpendicular to the plane of the alkene to produce an intermediate carbanion (**33a,b**), with the bisected geometry which it prefers, or a transition state (**33c,d**). In order to reach the observed product, a rotation of 60° must be performed (**33a→e; b→f**), while to reach products with the opposite geometry, the rotation must be 120°. It appears that the former is strongly preferred and seems a natural consequence of a concerted reaction and possibly an example of the 'Principle of least motion'[109], a somewhat vague precept which indicates a preference,

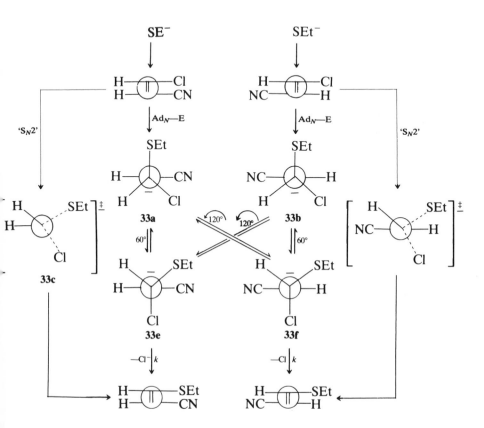

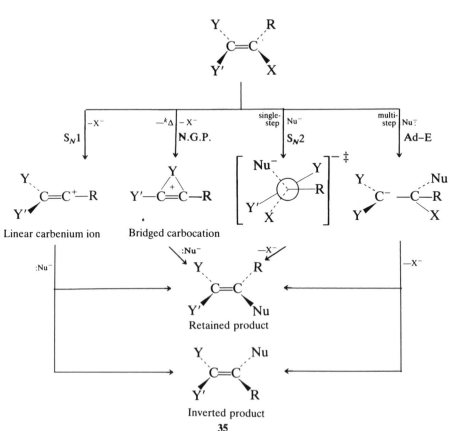

34

35

N.G.P., neighbouring-group participation

other considerations being equal, for products to form having geometry closest to their precursors. If a carbanion **33a,b** were formed as an intermediate it would seem to have a good probability of rotation in either direction to give a mixture of stereoisomers and, if sufficiently long-lived, of forming the same mixture from both *E*- and *Z*-reagents ('stereoconvergence'); this is observed for **34**, in which two −*R* groups are present.

On the other hand, leaving-group effects can indicate a rate-determining step in which the C–X bond is not broken: fluorides often react faster than chlorides and, since the C–F bond is stronger than the C–Cl bond, it may be argued, as in the interpretation of a PKIE, that this bond is not broken in the slow step. Rappoport has inferred a variable transition state in these reactions that depends upon the substitution pattern and ranges from a full vinyl cation to an intermediate carbanion[108], **35**. A scheme such as this seems to account for the observed rates and stereochemistry of most, if not all, vinylic substitutions.

Vinyl cations are undoubtedly formed under the right conditions, i.e. when stabilized by *α*-aryl groups and when generated from vinyl precursors with very good leaving groups[111] and in highly ionizing media. Because of the variety of possible pathways which a solvolysis might follow, it is not easy to be sure that vinyl cations are indeed formed intermediately. One indication is the facile rearrangements which can occur as in saturated analogues; for example.

$$EtOC(CH_3)_2C(CH_3){=}CH_2$$

Another is independence of base concentration; the rate of hydrolysis of 1-bromo-1-*p*-anisylethene is constant over the pH range 3–13. Both addition–elimination and elimination–addition pathways would be expected to increase in rate with pH:

Table 10.20 Additional criteria by which Ad_E–E and S_N1 mechanisms of vinyl solvolyses may be distinguished

		Ad_E–E	S_N1
a	Reaction in ROH or aprotic solvent	Very slow	Reaction occurs
b	Effect of increasing acidity	Rate increase	No effect
c	Effect of added Ag^+	No effect	Rate increase for R—Hal
d	Effect of β-OMe or similar $+R$ group	Large rate increase	Little effect
e	Solvent isotope effect	Normal inverse	None
f	Solvolysis in RCOOD	Incorporation of D in product	No incorporation of D in product
g	Common ion effect	Not expected	Possible
h	Scrambling of α and β carbons	Not expected	Likely

The additional criteria listed in Table 10.20 may be used to distinguish between addition–elimination and ionization via vinyl cations in solvolyses of vinyl substrates[112].

10.4 ELECTROPHILIC DISPLACEMENTS AT AN AROMATIC CARBON

Substitutions at an aromatic center which retain the integrity of the aromatic ring are among the most familiar and thoroughly studied of all organic processes[4,110–123]. The scope of this reaction is summarized in Table 10.21.

It is significant that most of the reactions take place in acidic — sometimes highly acidic — solution and none under basic conditions. Acidic media would be suitable environments for electrophilic species. Furthermore, the most common leaving group is hydrogen, which clearly must depart as its electrophilic form, H^+, since the other alternatives, $H:^-$ or $H^{\cdot}$, would lead to the evolution of H_2, which does not occur. Other leaving groups which are known (HgX, SiR_3, SO_2, CO_2) are also typically electrophilic[124]. Therefore it may be inferred from the stoichiometry that all these reactions are by nature electrophilic displacements.

10.4.1 Timing of bond breaking and making

Since sp^2-hybridized carbon can increase its coordination number, three possibilities may be considered which can be differentiated by their kinetic isotope effects.

Table 10.21   Summary of aromatic electrophilic substitution reactions

Reaction	Effective electrophile	Typical conditions
Leaving group H⁺		
1 Nitration	NO_2^+	HNO_3, H_2SO_4
2 Sulfonation	SO_3	H_2SO_4/SO_3
3 Halogenation	'Hal^{+}'^{a}	Hal_2, LA^b
4 Acylation	$RC{=}O^+$	R·COCl, $(RCO)_2O$, LA
5 Alkylation	R^+	R—Hal, LA
6 Diazo coupling	ArN≡N⁺	ArN_2^+, phenols, amines
7 Nitrosation	NO^+	HNO_2, H^+
8 Mercuration	'HgOAc'^{a}	$Hg(OAc)_2$
9 Hydroxylation	'OH$^+$'	$CF_3CO·OOH$
10 Chloromethylation	'H—CO^{+}'^{a}	HCl + CO + LA
11 Thalliation	'Tl(OAc)$_2$'^{a}	$Tl(OAc)_3$
12 Hydrogen exchange	'H⁺(D⁺, T⁺)'^{a}	CF_3COOH/H_2SO_4
13 Amination	'NH$_2^+$'^{a}	NH_2OSO_3H, LA

^{a} In these cases it is unlikely that the electrophile becomes free as such; it is more probably transferred.
^{b} LA, Lewis Acid (electrophile) such as $AlHal_3$, $FeHal_3$, $GaHal_3$, BF_3, etc.

Electrophile = H^+, leaving groups other than hydrogen

Reaction	Leaving group	Typical conditions
13 Protodemercuration	—HgR	AcOH
14 Protodemagnesation	—MgR	H_2O (hydrolysis
15 Protodelithiation	—Li	of Grignard reagent, etc.)
16 Protodecarbonylation	—CO·R	Hot H_3PO_4
17 Protodecarboxylation	—COOH	H_2SO_4
18 Protodesilylation	—SiR$_3$	$AcOH/H_2SO_4$

Two-step processes

(1a) S_E1: C–H bond breaking followed by addition of electrophile:

$$Ar—H\ (+B) \xrightarrow{\text{slow}} [Ar \cdots H \cdots B]^{\ddagger} \rightarrow Ar^-(+BH^+)$$

$$Ar^- + E^+ \xrightarrow{\text{fast}} Ar—E.$$

Full PKIE (requires ArH to act as a carbon acid).

(1b) S_E2–Ar: addition of electrophile followed by loss of H^+:

$$Ar \underset{H}{\overset{E}{<}} + (+B:) \xrightarrow{\text{fast}} ArE + BH^+$$

No PKIE but perhaps a small SKIE.

Single-step reaction

(2) S_E3 (a base is required also):

$$Ar - H + E^+ \rightarrow \left[\underset{+ : B}{Ar.} \overset{E}{\underset{H \cdots B}{\cdot \cdot}} \right]^{\ddagger} \rightarrow Ar—E + BH^+$$

Full PKIE.

It was shown by Melander and others[125, 126] that the nitrations of benzene and benzene-d_6 proceed at the same rate, i.e. there is no PKIE and the value of $k_H/k_D = 1\cdot00$. This appears to rule out any mechanism which requires the breaking of the C–H bond in the slow step of the reaction (i.e. 1a and 2) while supporting 1b (the secondary effect here would probably not exceed 1%). Since C–H bond fission does occur but is not rate-determining, the reaction must take place in more than one discrete step. However, not all aromatic substitutions show a complete absence of a kinetic isotope effect, though values of k_H/k_D are usually small (Table 10.22); this must also be permitted by any proposed mechanism.

10.4.2 The general mechanism for electrophilic aromatic substitution

All reactions of this category may be discussed within the context of the mechanism **36** and Fig. 10.8. Minor differences arise, in the main, from changes in the relative rate constants for the various steps and from rate or

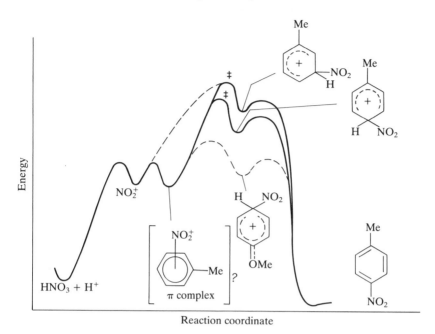

Fig. 10.8   Energy profile for the nitration of toluene in pure nitric acid.

equilibrium processes leading to the formation of the active electrophile. The aromatic substrate acts as a nucleophile supplying an electron pair from a high-lying π-orbital to the electrophilic reagent, possibly via a π-complex but in any case leading to an intermediate σ-complex, a cyclohexadiene cation or benzenium (arenium) ion, **36a**[127, 128]. This is also a protonated aromatic and it subsequently releases a proton to a base in a second step to complete the reaction, the product regaining the aromatic structure. This reaction is designated S_E2–Ar and applies to reactions of substituted benzenes, polycyclic aromatics and many heterocyclics.

10.4.3   The nature of the electrophilic reagents

Many of the electrophiles which participate in the S_E2–Ar reaction are stable species whose identity is not in doubt, for example, diazonium ions, $Ar\overset{+}{N}_2$, mercury(II) salts and bromine. Others may be highly reactive and transient, generated in the course of the reaction.

Nitration

There is no doubt that the species NO_2^+, the nitronium ion, is a powerful nitrating agent and may be used in the form of its stable salts, for example $NO_2^+BF_4^-$, $NO_2^+PF_6^-$ and $NO_2^+NO_3^-$ (N_2O_5), which are capable of nitrating the most unreactive of substrates[113]. The more usual conditions for laboratory nitration (concentrated nitric acid and sulfuric acid) also generate the nitronium ion in equilibrium, complete in concentrated acid:

$$HO-NO_2 + (H^+) \rightleftarrows H_2\overset{+}{O}-NO_2;$$
$$H_2\overset{+}{O}-NO_2 + (H^+) \rightleftarrows H_3O^+ + NO_2^+.$$

In solutions of nitric acid in concentrated sulfuric acid NO_2^+ may be detected by its IR spectrum: there is also NO_2^+ present in pure nitric acid:

$$2HO-NO_2 \rightleftarrows H_2\overset{+}{O}-NO_2 + NO_3^- \rightleftarrows H_2O + NO_2^+ + NO_3^-.$$

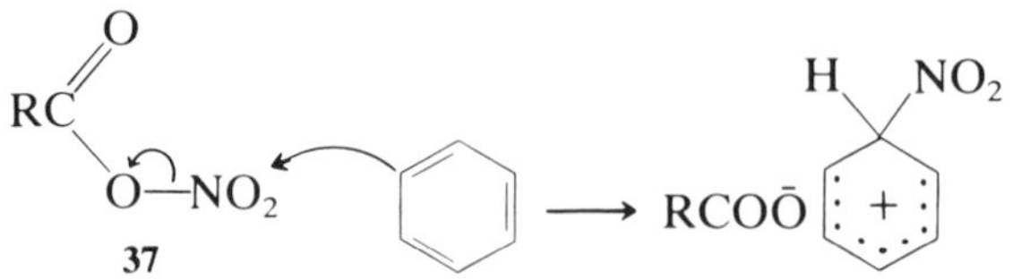

Other nitrating agents may be termed 'transfer' agents since they nitrate by transferring NO_2^+ which does not become free. Acetyl nitrate [129], $CH_3CO-O-NO_2$, and benzoyl nitrate, $PhCO-O-NO_2$, are examples (37); another is the nitric acidium ion, $H_2\overset{+}{O}-NO_2$, formed in the equilibrium depicted above. All such nitrating agents contain the nitronium ion covalently bound to a good nucleophilic leaving group, $Nu-NO_2$. Nitrations, however, are carried out under the most diverse conditions and it is often difficult to be sure of the exact nature of the reagent. Very reactive substrates such as phenols may be nitrated in dilute aqueous nitric acid. It is hardly possible for NO_2^+ to exist in such a basic medium as water and it is likely that nitration occurs via the nitroso compounds (produced by attack of the weak electrophile, NO^+) which are then oxidized by HNO_3:

$$ArH + NO^+ \rightarrow Ar\overset{H}{\underset{NO}{\diagdown}} \rightarrow Ar-NO \xrightarrow{\text{oxidation}} Ar-NO_2$$

Acylation and alkylation

Friedel–Crafts reactions are carried out by acyl halides or anhydrides (for acylation) or alkyl halides (for alkylation) in the presence of an electrophilic catalyst such as aluminum chloride or boron trifluoride, whose function is to remove halide ion and to generate an acyl or alkyl cation, the active electrophile, 38a[111]. Acyl cations are reasonably stable but highly reactive electrophiles isolable as salts such as $CH_3\overset{+}{C}O\ \bar{B}F_4$ which will acetylate aromatic compounds very rapidly and with no further need for catalysts. Alkyl cations are less stable but many (tertiary carbocations especially) can be observed in 'superacid' solution and are then capable of alkylating aromatic molecules. Furthermore, primary carbocations are known to

38 $R-\underset{\underset{\overset{|}{Cl}\;\;AlCl_3}{}}{\overset{\overset{O}{\parallel}}{C}} \longrightarrow R-\overset{+}{C}=O \;\; AlCl_4^-$
    35a

39 $CH_3CH_2CH_2-Cl + AlCl_3 \longrightarrow CH_3CH_2\overset{+}{C}H_2 \; AlCl_4^-$

$\longrightarrow CH_3\overset{+}{C}HCH_3 \xrightarrow{ArH} Ar\cdot CH\overset{\displaystyle CH_3}{\underset{\displaystyle CH_3}{\big\langle}}$

isomerize to secondary or tertiary if possible and this occurs during Friedel–Crafts alkylation, **39**. Analogously, chlorosulfonation with SO_2Cl_2 in the presence of a Lewis acid is presumed to occur by attack of $ClSO_2^+$.

Sulfonation

Sulfur trioxide, SO_3, is known to be a very powerful sulfonating agent and is presumably the reactive electrophile when sulfonations are carried out in oleum (H_2SO_4/SO_3). It seems also to be the active species in sulfuric acid and may result from the equilibrium **40**:

40 $2H_2SO_4 \rightleftarrows SO_3 + H_3O^+ + HSO_4^-.$

The evidence for this comes from the effect of added water on rates of sulfonation in sulfuric acid. Water is completely dissociated in sulfuric acid:

41 $H_2SO_4 + H_2O \rightleftarrows H_3O^+ + HSO_4^-.$

The product ions will suppress the concentration of SO_3, since they are both common to equilibrium **40**. In fact, if the rate of sulfonation is proportional to $[SO_3]$ it would be predicted that

$$\text{Rate} \propto [SO_3] \propto 1/[H_2O]^n; \qquad n = 2 \qquad (10.6)$$

since each water molecule is converted to two common ions of equilibrium **41**. This is observed.

Consideration of other possible contenders as the active electrophiles leads to different predictions. For example,

$$2H_2SO_4 \rightleftarrows H_3SO_4^+ + HSO_4^-; \qquad n = 1,$$
$$3H_2SO_4 \rightleftarrows HSO_3^+ + H_3O^+ + 2HSO_4^-; \qquad n = 3,$$
$$4H_2SO_4 \rightleftarrows S_2O_6 + 2H_3O^+ + 2HSO_4^-; \qquad n = 4.$$

Halogenation[112]

The halogens Cl_2, Br_2 and I_2 are a set of progressively weaker electrophiles which will act as mild aromatic halogenating agents, by *molecular halogena-*

42      $\xrightarrow{Br_2}$ 42a $\longrightarrow$ $AlCl_3Br^-$

tion. More potent electrophilic halogen is obtained when these are used in combination with an electrophilic catalyst, because, as in the Friedel–Crafts reaction, it facilitates the departure of Hal^-. This reaction is often referred to as *positive halogenation,* though it seems unlikely that free Hal^+ is formed. Probably, the catalyst removes Hal^- from a complex between ArH and Hal_2 permitting the formation of the benzenium ion, a form of transfer halogenation (**42**). Compounds with N–Hal and O–Hal bonds (i.e. Nu–Hal) also transfer Hal^+ and act as mild halogenating agents[130]. Acid catalysis is common and is due to protonation of the leaving group ($\rightarrow HNu^+-Hal$) making Hal^+ more readily available. Iodine in the presence of oxidizing agents such as HNO_3, IO_3^- and peroxides forms a potent iodinating species of uncertain structure, possibly I_3^+ or even I^+ since iodine has the lowest oxidation potential of all the halogens.

Proton exchange

This symmetrical reaction commences with the coordination of a proton as electrophile to an aromatic ring, an acid–base reaction involving a very weak base. The basicity of aromatic molecules may be measured in superacids, and the tendency to protonate follows the acidity function H_0[131]. Benzene is so weak a proton acceptor that it requires a medium of $H_0 \sim 22$ to achieve half-protonation. However, exchange reactions will take place in aqueous perchloric or sulfuric acids. Alkylated benzenes are more basic; hexamethylbenzene is half-protonated at $H_0 = 8.45$ (87% sulfuric acid) and proton exchange reactions in alkylbenzenes are proportionately faster as the number of alkyl groups *ortho* and *para* to the site of reaction increases.

10.4.4 Kinetic isotope effects

The absence of a kinetic isotope effect that is observed in the nitration of benzene and benzene-d_6 is by no means general, although numerous similar examples are known[125]. Isotope effects ranging from $k_H/k_D \sim 1$ to the maximum value ~ 7 are known (Table 10.22), although values are usually small[128]. It is clear that the two-step mechanism S_E2–Ar can permit a PKIE greater than 1. This would necessarily be the case if k_2 were to become rate-determining either because k_1 becomes very high or is highly reversible, $k_1 \ll k_{-1}$ (see mechanism **36**). Accordingly, large isotope effects are often to be found for reactions by weak electrophiles such as diazonium and

Table 10.22   Kinetic isotope effects in electrophilic aromatic substitution [132-134]

Substrate (position)	Reaction	k_H/k_D
(a) Primary		
Nitration		
1 Benzene	HNO_3	1·0
2 Nitrobenzene (3)	HNO_3	1·0
3 Naphthalene (1, 4)	$NO_2^+ BF_4^-$	1·15
4 1,3,5-Tri(*tert*-butyl)-2-nitrobenzene (4, 6)	HNO_3–H_2SO_4	3·0
5 Anthracene (9)	$NO_2^+ BF_4^-$ in MeCN	6·1
Halogenation		
6 *N,N*-Dimethylaniline (2)	Br_2	2·6
N,N-Dimethylaniline (4)	Br_2	1·0
7 1,2,3-Trimethyl-5-*tert*-butylbenzene	Br_2	2·7
8 Anisole (2 + 4)	ICl	3·8
9 4-Nitrophenol (2)	I_2 ($[I^-] = 0·00001$M)	2·3
4-Nitrophenol (2)	I_2($[I^-] = 0·006$M)	5·5
10 1,3,5-Tri(*tert*-butyl)-benzene-2,4,6-t_3 (2)	Br_2(k_H/k_T)	10·0
11 Benzoic acid (3)	$I_2 + IO_3^-$	2·1
12 Nitrobenzene (3)	$I_2 + IO_3^-$	3·4
13 2,3,5,6-Tetramethyl-bromobenzene (4)	Cl_2	1·00
14 Imidazole (4)	I_2	4·5
Diazo coupling		
15 2-Naphthol-6,8-disulfonic acid (1)	4-chlorobenzenediazonium	5·5
16 2-Naphthol-8-sulfonic acid	4-chlorobenzenediazonium	6·2
17 1-Naphthol-4-sulfonic acid	4-chlorobenzenediazonium	1·04
Sulfonation		
18 Bromobenzene (4)	oleum	1·5
19 Nitrobenzene (3)	oleum	1·6
Bromodecarboxylation ($ArCO_2^- + Br_2 \rightarrow ArBr + CO_2 + Br^-$)		
20 Benzoate	Br_2	1·002[a]
21 Benzoate	Br_2 + added Br^-	1·045[a]
Nitrosation		
22 *N,N*-dimethylaniline	NO^+	1·7
23 Phenol	NO^+	4·1
Metallation		
24 Benzene	$Hg(OAc)_2$	6·0
(b) Secondary		
25 Benzene-d_6	$NO_2^+ \cdot BF_4^-$	0·89
26 Toluene-d_8	$NO_2^+ \cdot BF_4^-$	0·85
27 Toluene-d_8	Br_2	1·15
28 1,3,5-Trimethoxy-benzene-2,4,6-d_3	4-Chlorobenenediazonium	0·88

Table 10.22   (*continued*)

Substrate (*position*)	Reaction	k_H/k_D
(c) Solvent isotope effects		
29 Phenol	Br_2 in AcOL	1·9
30 2,4,6-Tri-isopropyl-benzaldehyde	acid cleavage in L_2SO_4 of formyl group	2·0

[a] Carbon isotope effect, k_{12}/k_{13}.

nitrosonium ions, iodine and metals such as Hg^+, Tl^+ and for sulfonation, known to be reversible. Moreover, when an isotope effect is present, its magnitude increases with the concentration of base present showing the kinetic involvement of base (in the k_2 reaction). A further cause of large isotope effects is steric interactions. If the departing proton is flanked by bulky *ortho* groups, k_2 may be reduced both on account of hindrance to the approach of a base and also since the reagent is forced into the plane of the ring where it increases steric compression. This is presumably the case in the nitration of anthracene and of tri(*tert*-butyl)benzene. Halogenation becomes reversible by the addition of halide ion, **42**, which increases the PKIE. While the magnitudes of isotope effects vary widely with small changes in structure or conditions, which evidently perturb the balance between k_1, k_{-1} and k_2, the sensitivity of rates towards electronic perturbation (i.e. ρ) is no different in kind for all these reactions (though it may be in magnitude). No kinetic involvement of a third species such as a base is observed so that there appears no reason for proposing any basic differences of mechanism for reactions with and without a PKIE.

10.4.5   Kinetics of S_E2–Ar reactions[135]

The coordination of an electrophile to an aromatic molecule is a bimolecular process which would be expected to lead to second-order kinetics when rate-determining. This is often the case, especially when the electrophile is a stable species of low reactivity such as $Hg(OAc)_2$ or $Tl(OCOCF_3)_3$. When the electrophile is a highly reactive transient species, e.g. NO_2^+, which needs to be generated from a precursor, the empirical rate equation may become much more complex, and may reveal important mechanistic detail.

Nitration

It is not possible to generalize in a discussion of this topic since the observed kinetics depend very much upon the specific conditions: the nature of the nitrating species, concentrations, solvent and substrate all have influence upon rates and even upon the reaction orders which are observed.

Using nitric acid in sulfuric acid, an aromatic substrate, ArH, of rather low activity (such as chlorobenzene) nitrates according to the second-order rate law:

$$HNO_3 \underset{k_{-1}}{\overset{k_1}{\rightleftharpoons}} NO_2^+ : ArH + NO_2^+ \underset{k_{-2}}{\overset{k_2}{\rightleftharpoons}} ArHNO_2^+ \overset{k_3}{\underset{:B}{\longrightarrow}} ArNO_2 + BM^+$$

$$Rate = k[ArH][HNO_3].$$

This indicates attack of nitronium ion on the substrate to be rate-determining, i.e. $k_2 \ll k_1$, k_{-1} and k_3. When the nitric acid is in large excess, $[HNO_3]$ becomes constant and pseudo first-order kinetics are observed:

$$Rate = k'[ArH].$$

If, on the other hand, successively more reactive substrates are nitrated under similar conditions, rates become progressively more independent of [ArH] and tend towards a constant upper limiting value (Figs. 10.9[136], 10.10a). A substrate such as anisole therefore nitrates with first-order kinetics (zero-order if nitric acid is in large excess) at a rate determined by the production of nitronium ion from its precursors. Benzene falls in between these extremes and shows a fractional order rate dependence of substrate concentration. In the latter cases, k_a, the rate of production of NO_2^+, is rate-determining since k_1 has become so great (approaching the encounter rate) and is independent of [ArH]. The onset of diffusion control in the reaction between NO_2^+ and ArH means that reactivities tend to be leveled and one can learn nothing of strucural effects on reactivity from the measured rates, which vary with the viscosity of the medium. The situation

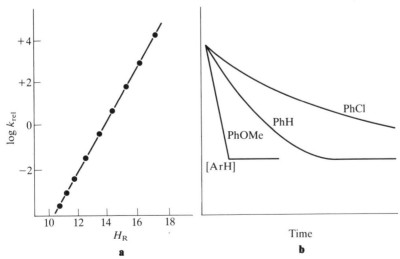

Fig. 10.9 **a**, Rate of nitration of benzene in HNO_3–H_2SO_4 as a function of H_R. **b**, Progress of reaction plots in the nitrations of chlorobenzene (first order), benzene (order between 0 and 1) and anisole (zero order).

arises because of the extremely high reactivity of the nitronium ion. As the water content of the sulfuric acid is reduced, its activity increases and so do rates of nitration, due to the higher equilibrium concentration of NO_2^+. Rates follow the acidity function H_R, showing a similar solvation change between the generation of NO_2^+ from HNO_3 and of a carbocation R^+ from ROH (Fig. 10.9a).

Nitration by nitric acid in an organic solvent, such as acetonitrile, obeys the rate law expressed by Eq. (10.7):

$$\text{Rate} = k_3[\text{ArH}][\text{NO}_2^+] = \frac{k_3 K_1 K_2 [\text{ArH}][\text{HNO}_3]^2}{[\text{NO}_3^-][\text{H}_2\text{O}]}, \qquad (10.7)$$

$$K_1 = k_1/k_{-1} \qquad K_2 = k_2/k_{-2}$$

The second power of nitric acid concentration is involved since two molecules are required to generate NO_2^+ and inhibition by both nitrate ion and by water are mass law effects as required by Eq. (10.7). Complexity arises now on account of the many reactions involved in the production of NO_2^+. Other difficulties may arise if, in order to circumvent this problem,

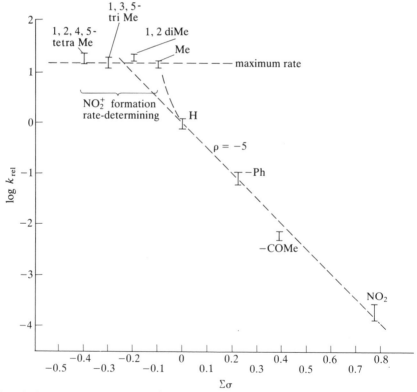

Fig. 10.10a Plot of rate against σ^+ for nitration by HNO_3–H_2SO_4 of polymethylbenzenes, showing eventual maximum rate due to the rate-controlling step becoming the formation of NO_2^+.

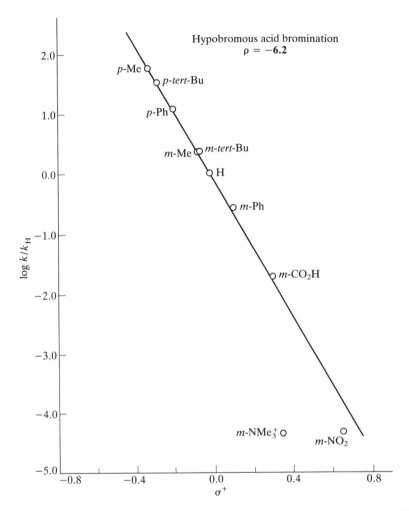

Fig. 10.10b   Brown–Okamoto plot for bromination of monosubstituted benzenes by HOBr ($\rho^+ = -6 \cdot 2$).

preformed nitronium ion is used as nitrating agent. Reactions may then be so fast that product ratios are governed by the rate of mixing with the substrate. This was elegantly demonstrated by nitration of an excess of 1,2-diphenylethane using $NO_2^+BF_4^-$ [137]. The two phenyl groups in each molecule react independently, unaffected by nitration of the other. The statistical ratio of mononitro (2- and 4-) to dinitro (2,2'- + 2,4'- + 4,4'-) products from a homogeneous reaction is calculable (7:1). This ratio was observed when nitration by nitric acid was carried out. However, the nitronium ion produced far more dinitro compound (mononitro:dinitro = 0·28:1) and also more unreacted diphenylethane than expected; the result must clearly be due to incomplete mixing before much reaction had occurred: substrate finding itself in a region with a high concentration of

439

NO_2^+ would tend to be dinitrated, while in regions of low concentration no nitration at all would be probable.

Early workers found further difficulties in obtaining reproducible rate constants for the nitrations of reactive substrates such as phenol and anisole. This problem was traced to the presence in samples of nitric acid of nitrous acid (or N_2O_4), which exerts strong catalysis due to its conversion to the nitrosonium ion, NO^+. Nitrosation of the reactive substrate occurs by the normal S_E2–Ar route and the nitroso products are then oxidized by nitric acid to nitro compounds and nitrite is regenerated[120,138]. This opens up a new pathway to products.

Sulfonation

Kinetics of sulfonation are not easy to examine since sulfuric acid is normally the medium. The rates in oleum (H_2SO_4/SO_3) are given by:

$$Rate = kH_0 + k'[SO_3],$$

where H_0 is the Hammett acidity function; this may mean that the medium is supplying a proton and an SO_3 molecule in the transition state. The reaction is complicated by the fact that it is reversible, protodesulfonation being more favored the lower the acidity.

Halogenation

Halogenation in the absence of Lewis acids is usually well-behaved second-order, but in non-polar solvents second-order terms in bromine may be observed. The bromide ion produced as product will complex with reagent bromine to form Br_3^- which has a different reactivity and different absorbance from Br_2.

Difficulties in obtaining absolute rate constants for aromatic substitutions are usually by-passed by the use of competition methods which yield relative rates for a series of substrates; this normally provides sufficient information for mechanistic studies.

10.4.6 Structural effects on rates

Substituents on the aromatic ring have an enormous influence upon its reactivity, reflected in correlations of rates with the enhanced resonance parameter, σ^+ (Section 4.5; Fig. 10.10.b). Values of ρ^+ are invariably large and negative ranging between -3 and -13 (Table 10.23). In the extreme case, for example, relative rates of iodination of toluene and nitrobenzene ($\delta\sigma^+ \sim 1$) are $1:10^{13}$, a difference which is too great for practical measurement. The LFER parameters indicate a transition state powerfully stabilized by donor resonance ($+R$) interactions, suggesting a structure close to a

Table 10.23   Reaction constants for aromatic substitutions

Reaction	ρ^+
1 Nitration (HNO$_3$, MeCN or Ac$_2$O)	$-6\cdot38$
Nitration (HNO$_3$, H$_2$SO$_4$)	$-9\cdot7$
2 Chlorination (Cl$_2$, AcOH)	-10
Chlorination (SO$_2$Cl$_2$)	-4
3 Bromination (Br$_2$, MeNO$_2$)	$-8\cdot6$
Bromination (Br$_2$, AcOH)	-12
Bromination (HOBr, H$_2$O)	$-6\cdot2$
4 Iodination (I$_2$, AcOH)	-13
Iodination (I$_2$, IO$_3$–'I$^+$')	$-6\cdot4$
5 Protodetritiation (aq. HClO$_4$)	-8
6 Acetylation (CH$_3$COCl, AlCl$_3$, C$_2$H$_4$Cl$_2$)	$-9\cdot1$
7 Mercuration (HgOAc$_2$, AcOH)	$-4\cdot0$
8 Alkylation (EtBr, GeBr$_3$)	$-4\cdot0$
9 Protodesilylation (aq. HClO$_4$)	$-4\cdot6$

Yukawa–Tsuno constants,

$$r \log k_X/k_H = \rho[\sigma^0 + r(\sigma^+ - \sigma^0)] \text{ (Eq (4.5))}.$$

Reaction	r
1 'Molecular halogenation'	$1\cdot66$
2 'Positive' halogenation	$1\cdot15$
3 Nitration	$0\cdot9$
4 Protodesilylation	$0\cdot7$
5 Protodestannylation	$0\cdot4$

benzenium ion (**43**), the proposed intermediate. A late transition state also accords with the Hammond postulate: coordination of the electrophile is an endothermic process due to the loss of aromatic resonance energy so that its transition state will be product-like. Consequently, structures and energies of benzenium ions can be taken as representing those of the transition states which govern the rates of reaction. In the transition state, as in the benzenium ions themselves, aromatic character has been largely lost and the positive charge, accumulating on the *ortho* and *para* positions, is available to be shared by $+R$ substituents. $-R$ and $-I$ substituents have an equally powerful destabilizing effect since they juxtapose an atom bearing a partial positive charge with the positively charged ring, as discussed in Section 4.4.

43

43a   43b   43c   43d   43e

$X = +R$ groups, $Z = -R$ groups (Section 4.4).   Note: for $X = $ Hal stability order is $\mathbf{c} > \mathbf{a} > \mathbf{b}$

Both resonance effects are markedly less from a *meta* position, **43b,d**. This analysis explains the familiar regiospecificity pattern of aromatic substitution that 'activating groups' (+R groups with unshared pairs, e.g. —NR$_2$, —OR, —SR, —R and halogen) orient further substitution at *ortho* and *para* positions while 'deactivating groups' (−R types with a 'positive' atom, e.g. —C=O, —C≡N, —NO$_2$; also −I types, e.g. −$\overset{+}{N}$Me$_3$) orient *meta*. In all cases the most powerful influence on orientation of substitution is the relative stability of the isomeric benzenium ions (as models of the transition states). The fastest-formed product is that derived from the most stable intermediate, provided that the overall reaction is not reversible and subject to thermodynamic control. An additional factor which may play a part in determining rates and orientation of substitution is the partial charge upon each site in the aromatic substrate (initial-state effect). Certainly the *ortho* and *para* positions of a mono-substituted benzene bear a high electron density when the substituent exerts a +R effect and a low electron density when the effect is −R. It is probable that, in the early stages of reaction at least, attractive forces between a cationic electrophile and the sites of high electron density on the aromatic ring might contribute to the fruitful collision leading to reaction. Charge densities may be measured by ^{13}C-NMR or may be calculated by MO theory (Fig. 10.11) [139]. However, this type of interaction (charge control) is not of major importance in determining rates or products, as shown by two lines of evidence. Firstly, cationic and neutral electrophiles give similar product distributions when the latter should not be affected by electrostatic forces. Secondly, there is no

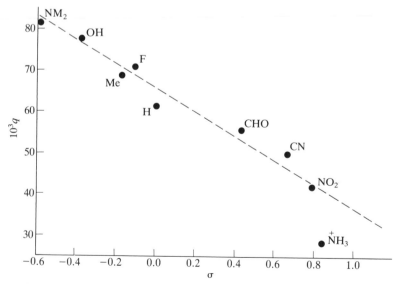

Fig. 10.11   Calculated charge density at the *para* position of substituted benzenes as function of σ (σ$^+$).

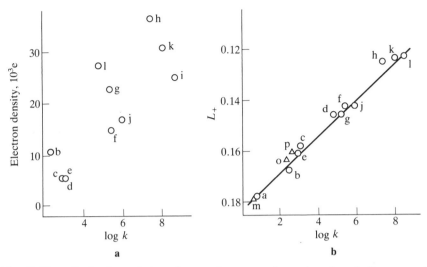

Fig. 10.12 **a**, Lack of correlation of rates of proton exchange in polymethylbenzenes with initial state charge densities. **b**, Correlation of the same rates with stabilities (localization energies L_+) of the benzenium ions. Substitution patterns are as follows: a, no substitution; b, 1; c, 1, 2; d, 1, 3; e, 1, 4; f, 1, 2, 3; g, 1, 2, 4; h, 1, 3, 5; j, 1, 2, 4, 5; k, 1, 2, 3, 5; l, 1, 2, 3, 4, 5.

correlation between reaction rates and partial charges at the reaction site, whereas there is a good correlation between reaction rates and stabilities of the corresponding benzenium ion intermediates (Fig. 10.12)[140]. The MO interpretation of aromatic reactivity is developed in Section 10.4.10.

Total reaction rates of individual substrates comprise the sum of rates at each reactive position, for separate and competitive reactions:

$$k_{\text{total}} = 2k_o + 2k_m + k_p.$$

Table 10.24 Values of some partial rate factors for aromatic substitution (For example, $f_o = 50$ means that in nitration under the stated conditions, a single *ortho* position of the compound reacts 50 times faster than a single position of benzene.)

(a) Reactivity of toluene towards various electrophiles

Reaction and conditions	f_o	f_m	f_p
Nitration, HNO_3/AcOH, 45°C	42	2·5	58
Nitration, HNO_3/$MeNO_2$, 30°C	37	2·8	47
Sulfonation, conc. H_2SO_4, 25°C	63	5·7	258
Acetylation, AcCl, $AlCl_3$, 25°C	4·5	4·8	749
Chlorination, Cl_2, AcOH, 25°C	617	5	820
Chlorination, 'Cl^+'	134	40	82
Bromination, Br_2, AcOH, 25°C	600	5·5	2420
tert-Butylation, *t*-BuBr, $SnCl_4$, 25°C	0	3·2	93
Methylation, MeBr, $GaBr_3$	9·5	1·7	11·8
Benzoylation, PhCOCl, $AlCl_3$	33	5	830
Mercuration	4·6	2·0	16·8
Protodesilylation, $HClO_4$, 50°C	18	2·1	14

Table 10.24 (*continued*)

(b) Nitration (HNO$_3$, Ac$_2$O, 0°C) of Ph—R

R	f_o	f_m	f_p
Me	50	1·3	60
Et	31	2·3	69
isoPr	15	2·4	72
tert-Bu	4·5	3·0	75

(c) Nitration (AcONO$_2$, MeNO$_2$) of Ph—Hal

Hal	f_o	f_m	f_p
F	0·056	0·0009	0·79
Cl	0·029	0·0009	0·137
Br	0·033	0·0011	0·122
I	0·252	0·012	0·78

(d) Nitration of some deactivated benzenes, Ph—X

X	Reaction conditions	f_o	f_m	f_p
NO$_2$	HNO$_3$, H$_2$SO$_4$	$1·08 \times 10^{-8}$	$16·2 \times 10^{-8}$	$7·3 \times 10^{-8}$
SO$_2$Et	HNO$_3$, Ac$_2$O	$0·9 \times 10^{-3}$	$9·3 \times 10^{-3}$	$0·7 \times 10^{-3}$
COOEt	HNO$_3$, Ac$_2$O	$2·6 \times 10^{-3}$	$7·9 \times 10^{-3}$	$0·9 \times 10^{-3}$
CO·Me	HNO$_3$, H$_2$SO$_4$	$10·2 \times 10^{-6}$	$27·7 \times 10^{-6}$	$1·55 \times 10^{-6}$

(e) Nitration of 4-haloanisoles[124] →

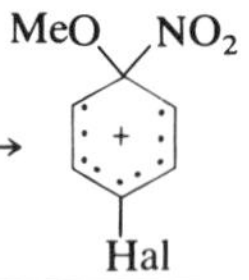

with *ipso* attack at C1

Hal	f_{ipso}
Cl	0·06
Br	0·07
I	0·18

(f) Nitration of halotoluenes

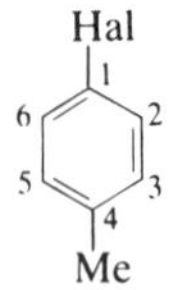

	Attack (%) at:			
Hal	1	2,6	3,5	4
Br	17	8	22	15
Cl	30		15	17

Meaningful reactivity relationships can best be established on the basis of single position rates and so relative reactivities are usually quoted in the form of partial rate factors, f, defined as follows and listed in Table 10.24. Partial rate factor for position x of a substituted benzene

$$= f_x = \frac{\text{rate of reaction at position } x}{\text{rate of reaction at a single benzene site}}.$$

Also,

$$\log f_x = \rho^+ \sigma_x^+.$$

Partial rate factors for each reactive position of an aromatic substrate are measured by competitive reactions of a mixture of the compound with benzene both in large excess over the reagent. Gas chromatographic analysis gives the product ratio from which the rate ratio for each may be derived. Values of f may be used in a Hammett plot. Since, by definition, f(benzene) = 1, deactivation is shown by $f < 1$ and activation by $f > 1$, and the more divergent the value of f from unity, the more selective the reagent and the more negative the value of ρ. The figures in Table 10.24 show that $+R$ substituents strongly activate *ortho* and *para* positions but also weakly activate *meta*. $-R$ substituents strongly deactivate *ortho* and *para* but less strongly deactivate *meta*. The reaction constants, ρ^+, are measures of the selectivities of the reagents. In general, electrophiles of low reactivity, and consequently, high selectivity, will show a large negative value and conversely those of high reactivity have a large positive value. This explains the trend in the reactivity sequence of the halogens: in reactivities, $Cl_2 > Br_2 > I_2$, for which $\rho^+ = -10, -12, -13$ respectively, and the value for molecular iodination, -13, may be compared with the value for the far more reactive 'positive' iodine, $-6 \cdot 4$. However, one cannot use the reaction constants as an index of electrophilicity. The position of the transition state may vary from reaction to reaction so that the earlier the occurrence of the transition state and the less developed the bonding of the electrophile, the less differentiated are their reactivities. The degree of reversibility of the addition, k_1/k_{-1}, will also affect the apparent reactivity of the electrophile.

Aromatic heterocycles undergo the reactions of Table 10.21. Those in which the hetero atom contributes a lone pair to the 6π-system (pyrrole, furan and thiophene) are highly activated[141]:

k_{rel}:   1 10^3 150×10^3 10^{11-13}

Orientation is determined by the stabilities of isomeric intermediates, as usual, and may be summarized (X: is a $+R$ and Z a $-R$ group):

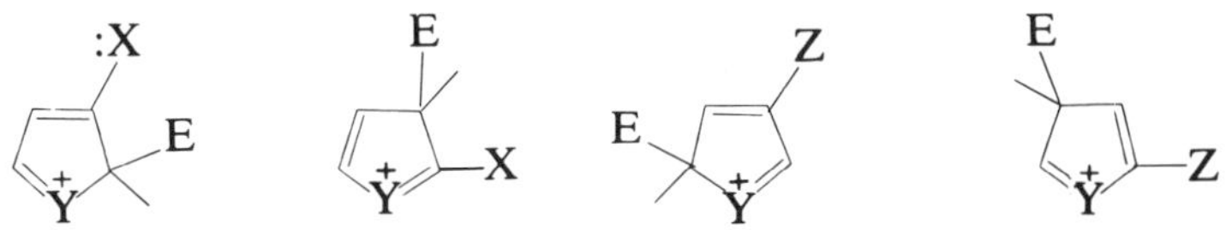

Heterocycles whose lone pairs do not take part in the aromatic π-system (pyridine, for example) are highly deactivated towards S_E2–Ar reactions.

445

10.4.7 The *ortho–para* selectivity ratio, $S_{o,p} = (2f_o/f_p)$

The resonance effect of a $+R$ substituent should be experienced equally at the *ortho* and *para* positions, in the Hückel approximation, and lead to the statistical ratio of 2 for relative yields of *ortho* and *para* products. In practice, this ratio may deviate in either direction as a result of several perturbing features. Firstly, 2-substituted benzenium ions have slightly greater stability than 4-substituted ones since they have a linear rather than a branched conjugated system and this favors *ortho* substitution. On the other hand, the *ortho* position is subject to steric hindrance from a substituent and is therefore disfavored on this count. Steric effects are clearly dominating the values of f_o for the nitrations of Ph—R (R = Me, Et, isoPr, *tert*-Bu) in order of increasing size of substituent R (Table 10.24). Within this series, the value of f_p is fairly constant, as is f_m. In a similar way, the size of the electrophile will discriminate between *ortho* and *para* attack. Friedel–Crafts alkylation of toluene by MeBr is characterized by a selectivity $S_{o,p} = 1·6$, but for alkylation by *tert*-BuBr $S_{o,p} \sim 0$. The reverse of this effect, steric acceleration, may sometimes be diagnosed to explain an abnormally high S-value. If the leaving group is bulky, steric compression with an *ortho* substituent may be relieved on conversion to the benzenium ion, especially if the electrophile is small, for example a proton. Another possibility is the initial coordination of the electrophile to a substituent group; this is possible if it bears a lone pair, as does OR (**44**). Intramolecular rearrangement of this initially-formed complex will then lead to attack at the nearest (i.e. *ortho*) position in preference to *para*[114,142,143]. This explanation has been advanced to explain the high proportion of *ortho* product obtained in the nitration of anisole. As discussed below, frontier orbital control of rates is important and the ratios of products are probably affected by the coefficients of the HOMO at the *ortho* and *para* carbons. The assessment of these diverse factors and their relative importance is a matter of some difficulty. As a consequence of the Hammond postulate, selectivity will be expected to diminish as reactivity increases, provided that the series of reactions studied adhere to the same mechanism.

One of the best tested selectivity relationship is that demonstrated by Stock and Brown[117]. The partial rate factor for *para* attack at toluene, f_p^{Me}, can be taken as a measure of reactivity of a given electrophile. Values will

44   OAc⁻   75%
(+25% *para*)

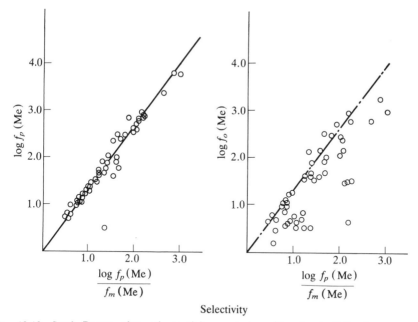

Fig. 10.13 Stock–Brown plots of reactivity against positional selectivity for aromatic substitutions. **a**, *p–m* selectivity; **b**, *p–o* selectivity; for identification of points, see Ref. *100*.

be very high for weak, selective reagents and tend towards unity for those reacting at diffusion-controlled rates. Similarly, the ratio of partial rate factors at *para* and *meta* positions of toluene, f_p^{Me}/f_m^{Me}, will be a measure of the positional selectivity of the reagent. The logarithms of these quantities should measure $\Delta\Delta G^{\ddagger}$ and should be linearly related. That this is the case is seen from the Stock–Brown plot, Fig. 10.13a. The regression equation,

$$\log f_p^{Me} = -0.17 + 1.38 \log(f_p^{Me}/f_m^{Me}) \qquad (10.8)$$

fits tolerably well over 100 reaction series, including nitrations, Friedel–Crafts reactions and halogenations under a variety of conditions, including the gas phase. Such a line argues powerfully that the basic mechanisms of all these reactions are similar. One would not expect such a linear relationship for *ortho–para* selectivity, Fig. 10.13b, because of other directing factors which supervene. The plot is interesting in that most of the points are found to lie on one side of the previous line expressed by Eq. (10.8), indicating the *ortho* positions to be more reactive than predicted. This suggests that deviations from LFER by *ortho* substituents are caused more by field and other electronic effects than by steric factors. Steric factors show up clearly only when size is varied while electronic effects are constant, as in the series of alkylbenzene reactions.

10.4.8 The nature of the intermediate

The most direct evidence for the structure of the intermediates in S_E2–Ar reactions, inferred from kinetic measurements, comes from direct observation under conditions in which they may be made stable. Benzenium ions* which possess at least one hydrogen at the tetrahedral center may be regarded as protonated benzenes, i.e. the conjugate acids of aromatic species, **45**. Since the basicity of aromatic rings in general is so low, their conjugate acids are exceedingly strong. Consequently benzenium ions will only be stable in highly acidic media in association with non-nucleophilic counterions such as BF_4^- and SbF_6^- [144]; they can be observed in the gas phase in the absence of a counter ion [130]. The first arenium ion to be isolated with no acidic proton was prepared by Doering [145] by methylation of hexamethylbenzene, **46**, and numerous analogues were characterized by Olah [146], for example ones in which the pentadienyl cation moiety is stabilized by alkyl groups at C2, C4 and C6. Arenes of basicity no less than that of benzene protonate in superacid solutions such as SbF_5–HF, BF_3–HF and $AlCl_3$–HCl, to give arenium ions, for example, **47**[147,148], whose identity may be confirmed by their NMR spectra. A study of these spectra reveals that not only the proton but also methyl groups are mobile and may migrate by 1,2-shifts to yield the most stable isomeric structure **48**. This behavior has long been known to occur in electrophilic substitutions of polyalkylbenzenes (the Jacobsen rearrangement, **49**, also known as 'cine'-substitution; see also

45

46

^{1}H chemical shifts/ppm

47

---

* Older texts refer to these species as 'Wheland intermediates' commemorating the contributions in this field of G. W. Wheland [127]. Ingold [115] states that prior recognition is due to Pfeiffer [128].

48

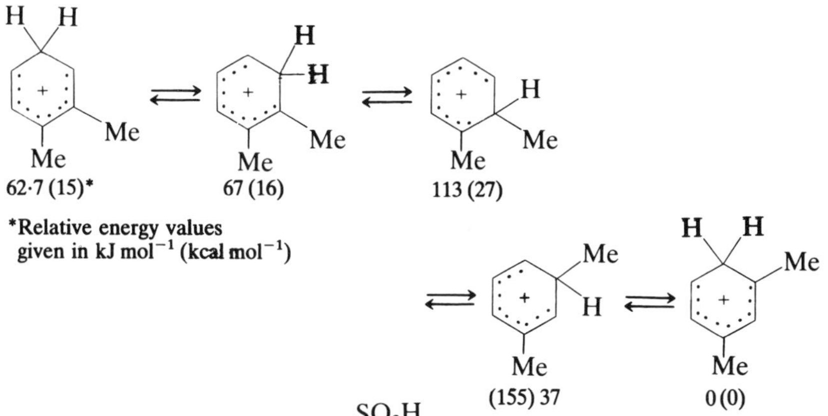

62·7 (15)*   67 (16)   113 (27)

*Relative energy values
given in kJ mol^{-1} (kcal mol^{-1})

(155) 37   0 (0)

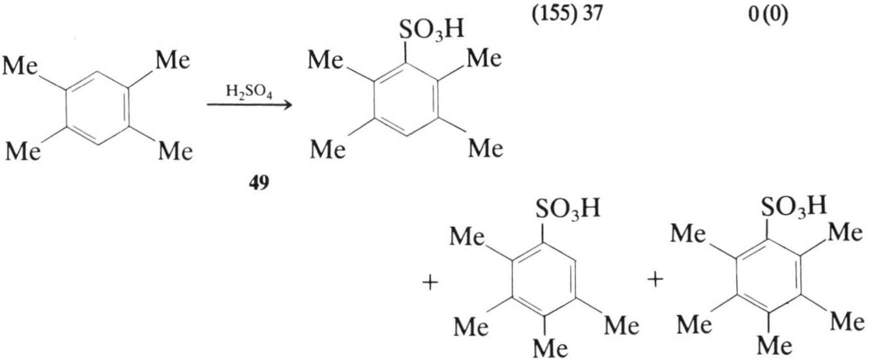

49

SO$_3$H

+   SO$_3$H   +   SO$_3$H

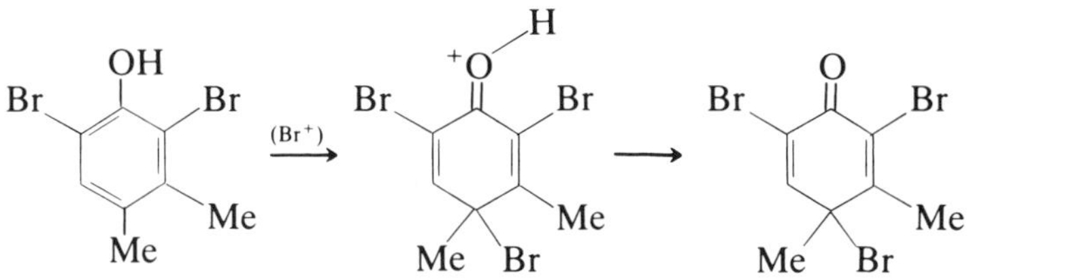

50   **50a**   **50b**

Section 10.5.3) which lends further support to the formation of these species as intermediates. The spectra show also that the two groups at the tetrahedral carbon of the arenium ion are non-identical so that the ions must be non-planar (**45**). Phenolic groups *ortho* or *para* to the tetrahedral carbon of the arenium ion may lose a proton in preference to carbon, leading to a dienone, **50b**, which may on occasion be isolated and from which may arise a complicated set of non-aromatic products[122].

π-Complexes

It is well known that arenes and certain electrophiles will form well-defined 1:1 complexes with characteristic spectra and other properties (**42a**). It is certainly plausible that an aromatic substitution may commence by forming such a π-complex in which the electrophile is not attached to a particular ring site but which would rapidly progress to one or other of the possible σ-complexes (arenium ions) available. However, despite the fact that this idea has been mooted for the past 40 years, there is no evidence that a π-complex lies on the reaction path of any such reaction. The selectivity plot, Fig. 10.13a, is evidence against any of the reactions involving a change of mechanism via a π-complex.

10.4.9 *Ipso* attack[120,121]

Coordination of the electrophile at a site already occupied by a substituent, especially one incapable of displacement, for example an alkyl group, is called *ipso* attack. In such cases, migration of either E or R must occur before a stable aromatic product can be formed. If the electrophile migrates, then no evidence of the intermediacy of the *ipso* benzenium ion will be discernible, while if R migrates, a cine substitution will result. Since intramolecular rearrangement may be slow, the intermediate can add a nucleophile to give a cyclohexadiene derivative, an isolable product (**51**). There is now substantial evidence that *ipso* attack is important in nitration[124], halogenation and other related reactions, and may be a factor determining product ratios, particularly at *ortho* positions[149]. Some partial rate factors are included in Table 10.24.

51

10.4.10 The MO interpretation of aromatic reactivity

Hückel MOs can be used very simply to illustrate the effects of $-R$ and $+R$ substituents upon the reactivity of the benzene ring, from the viewpoints of both product (i.e. benzenium ion) stability and also initial-state interactions via frontier orbitals.

Benzenium ion stability

As models for $-R$ and $+R$ substituents (X: and Z) it is convenient to use —CH$_2^-$ and —CH$_2^+$ respectively. Though providing extreme examples of conjugative effects, these groups exert no inductive effects, while the calculations are simplified by avoidance of heteroatoms. Hückel solutions for the benzyl anion and cation together with those for the three isomeric benzenium ions derived from each are given in Table 1.1, entries 21, 10–12. The total π-energy of each is given by $\sum^{occ_i} n\varepsilon_i$, the energy of each occupied orbital (electronic occupancy c) being summed.

$\ddot{X}$—C_6H_5 ($\ddot{X} = -\ddot{C}H_2^-$)

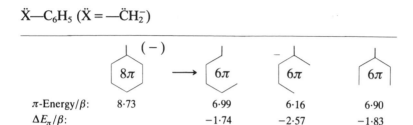

π-Energy/β:	8·73	6·99	6·16	6·90
$\Delta E_\pi/\beta$:		−1·74	−2·57	−1·83

H—C_6H_5

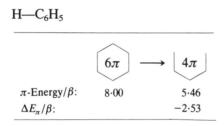

π-Energy/β:	8·00	5·46
$\Delta E_\pi/\beta$:		−2·53

Z—C_6H_5 (Z $= -CH_2^+$)

π-Energy/β:	8·73	6.10	6·16	5·86
$\Delta E_\pi/\beta$:		−2·62	−2·56	−2·86

The stability of each isomeric benzenium ion is assessed by the loss of π-energy (ΔE_π) in units of energy, β, on formation from the aromatic system. Since this involves the notional removal of one atom of the ring from conjugation, it is known as *localization energy*. The smaller the localization energy, the greater the stability and hence reactivity at the position considered. For the benzyl anion this clearly points to *ortho* and *para* positions while for the benzyl cation, the *meta* position emerges as that costing least π-energy when removed from the conjugated system. Reactivity in the series above is therefore predicted to be in the order:

$$X(o) > X(p) > H > X(m) > Z(m) > Z(p) > Z(o).$$

$\underbrace{\qquad\qquad\qquad}_{\text{activated}} \qquad \underbrace{\qquad\qquad\qquad\qquad\qquad\qquad}_{\text{deactivated}}$

451

This is essentially the order observed for reactions of $\ddot{X}$- and Z-substituted benzenes.

The equal π-energies of benzyl anion and cation (odd AH systems) are due to the additional electron-pair in the former occupying an NBMO, but in the benzenium ions all electrons occupy bonding orbitals. Those derived from the anion possess an additional occupied orbital compared with those from the cation and are always, therefore, intrinsically more stable.

Frontier orbitals

A substituted benzene is an ambident nucleophile with several sites for interaction with a vacant orbital of an electrophile. Although reactivity seems to be controlled mainly by benzenium ion stability, consistent with a late transition state, initial attractive interactions between the aromatic species and the electrophile will be frontier-orbital controlled and may assist in bond formation. The benzyl anion and cation models may again be used to examine the perturbations produced on the HOMO by $-R$ and $+R$ substituents, respectively. The required information is the coefficients for ψ_4 and for ψ_3 of the benzyl system, which are HOMOs of the anion and cation respectively (Table 10.25).

For the benzyl anion, a model for $\ddot{X}$—C_6H_5 and an odd AH, the HOMO is its NBMO (ψ_4) and consequently the coefficients are zero at the 'unstarred' positions, *ipso* and *meta*. The *ortho* and *para* atoms are the sites at which frontier orbital interactions are greatest preparatory to formation of a benzenium ion and must reinforce this process:

$$c^2(\psi_4): \quad ortho, \ 0\cdot143; \quad meta, 0; \quad para, \ 0\cdot143.$$

Table 10.25   Hückel MO solution for benzyl

x/β:		ψ_1 2·101	ψ_2 1·259	ψ_3 1·000	ψ_4 0·000	ψ_5 −0·999	ψ_6 −1·250	ψ_7 −2·100
(α)	c_1	0·238	−0·397	0·000	−0·755	0·000	0·397	0·238
	c_2	0·500	−0·500	0·000	0·000	0·000	−0·500	−0·500
(o)	c_3	0·406	−0·118	−0·500	0·378	−0·500	0·116	0·406
	c_4	0·354	0·354	−0·499	0·000	0·500	0·354	−0·353
(m)	c_5	0·337	0·562	0·000	−0·378	0·000	−0·561	0·337
(p)	c_6	0·354	0·354	0·500	0·000	−0·500	0·354	−0·353
	c_7	0·406	−0·116	0·500	0·378	0·500	0·116	0·406

(structure of benzyl anion with orbitals)   ↑↓   ↑↓   ↑↓   ↑↓

(structure of benzyl cation with orbitals)   ↑↓   ↑↓   ↑↓

The benzyl cation is more of a problem. Its HOMO is now ψ_3, for which the coefficients are equal at the *ortho* and *meta* positions, though zero at the *para* position which will therefore contribute nothing. There is a further complication in that ψ_2 is rather near in energy to ψ_3 and will also contribute to a bonding interaction. Since the overlap term is $\sum^{occ} c^2/(E - E_{occ})$, the contributions from these two orbitals may be included; assuming for convenience that the energy of the LUMO of E^+ in the reactant electrophile is -1, then these contributions are given by:

	ortho	*meta*	*para*
ψ_2/β	$0{\cdot}13/(1 + 1{\cdot}259)$	$0{\cdot}25/(1 + 1{\cdot}259)$	$0{\cdot}315/(1 + 1{\cdot}259)$
ψ_3/β	$0{\cdot}25/(1 + 1)$	$0{\cdot}25/(1 + 1)$	
Sum	$0{\cdot}13$	$0{\cdot}18$	$0{\cdot}14$

The inclusion of ψ_2 then predicts that there will be greater initial-state interactions at the *meta* position than at *ortho* or *para* but the difference is not great. It thus appears that there is no conflict between rationalizations of aromatic reactivity based on either initial-state or product-state properties.

10.5 NUCLEOPHILIC SUBSTITUTION AT AN AROMATIC CENTER

10.5.1 The addition–elimination pathway (S$_N$Ar–Ad,E)[151–157]

Displacement of such typical nucleophilic leaving groups as halide or arylsulfonate from an aromatic carbon is far more difficult than from saturated carbon, the most powerful nucleophiles (MeO$^-$ in HMPT at 90°C) being required[158]. The S$_N$2 mechanism, **52**, cannot operate because of the steric impossibility of rearface attack: S$_N$1 ionization is very unfavorable since the aryl cation, **53**, is not resonance-stabilized as the vacant *p*-orbital is orthogonal to the π-system. As will emerge, other mechanisms come into play when reactions of this type occur. It has long been known that nitro groups in the ring, *ortho* or *para* to a halogen leaving group, activate it towards displacement by amines or alkoxide ion as typical nucleophiles. Two or three nitro groups exert a very powerful effect so that, while chlorobenzene requires temperatures around 300°C for hydrolysis (an

52   **53**

$$k_{obs} = \frac{k_1 k_2 + k_1 k_3 [B]}{k_{-1} + k_2 + k_3 [B]}$$

obsolete industrial process for phenol), 2,4,6-trinitrochlorobenzene (picryl chloride) is comparable in hydrolytic reactivity with acyl chlorides. Rates of reaction of a series of *para*-substituted 2-nitrochlorobenzenes, **54**, with amines are essentially of second order and correlate with σ^-; $\rho^- = +4$[159,181], which indicates a transition state with anionic charge developing in the ring capable of conjugation with a $-R$ substituent in the *ortho* or *para* position. The effects of *para*-substituents on rates of displacement of chloride by OMe^- is illustrated by the following 'substitution rate factors'.

Z:	$-H$	$-COMe$	$-SO_2Me$	$-CN$	$-NO_2$	$-\overset{+}{N}Me_3$	$-CF_3$
k_{rel}:	1	2000	12,000	8000	114,000	21,000	800

A wide variety of nucleofugic groups, not only halogen, may take part in this reaction and the effect of leaving group on rate is very different from that found with S_N1 or S_N2 reactions (Table 10.26). Among the halogens the order is $F \gg Cl, Br > I$, quite the reverse of that for solvolytic

454

Table 10.26   Rates of nucleophilic aromatic substitutions [162,163]

(a) Substituent effects [180,181]

X	$10^4 k/\text{M}^{-1}\,min^{-1}$	X	$10^4 k/\text{M}^{-1}\,min^{-1}$
NH_2	0·0036	F	7·55
OH	0·0169	H	29
NMe_2	0·035	COO^-	73
OEt	0·44	I	150
OMe	0·52	Cl	162
Me	4·2	Br	227
tert-Bu	4·9	NO_2	v. fast

(b) Variation of the nucleophile

Nu:	$10^4 k/\text{M}^{-1}\,min^{-1}$
$PhNH_2$	<0·05
OH^-	0·0662
OPh^-	0·718
OMe	2·2
$(CH_2)_5NH$	4·48
PhS^-	1300

(c) Variation of the leaving group

	k_{rel}	
Leaving group, X	$Nu = MeO^-$	$Nu = MeS^-$
I	1	1
Cl	5·2	0·7
Br	3·4	1·5
F	3100	27
OPh	0·88	0·01
Opnp	13·3	0·3
Spnp	0·45	0·28
NO_2	2590	1345

Table 10.26   (*continued*)

(d) Decomposition of phenyldiazonium ions

$p\text{-}X\text{—}C_6H_4\text{—}N_2^+ \;\; \rightarrow \;\; p\text{-}X\text{—}C_6H_4^+ + N_2$

X	k/min^{-1}
OH	0·93
OMe	0·11
Ph	37
Me	91
H	740
COOH	91
SO_3^-	42
Cl	1·4
NO_2	3.1

reactions[160,161]. The most easily displaced groups include nitro, as NO_2^-, and PhSO$^-$, which are not normally considered to be nucleofuges at aliphatic carbon. These observations clearly require a mechanism in which loss of the leaving group does not occur in the rate-determining step. In conjunction with the substituent effects the addition–elimination scheme 54 best fits the evidence. The rate-determining step is the coordination of the nucleophile to the aromatic ring at the site bearing the leaving group. The nature of this group will affect the ease of bond formation and hence control the rate. Electron-withdrawing groups such as —F, —NO_2 and N_2^+ will facilitate bond formation by their inductive or resonance interactions. The situation is complex, however, as shown by the large dependence of relative rates upon the nature of the nucleophile (Table 10.26). What seems to emerge from an analysis of nucleophile–nucleofuge effects is that rates increase with the polarizabilities of either. If PhS$^-$ and PhO$^-$ are taken as examples of polarizable and non-polarizable (or soft and hard) nucleophiles of similar steric properties, then the rate ratio k_{PhS^-}/k_{PhO^-} increases with the polarizability of the leaving group, in the order $F < Cl < Br < I$, and will fit the linear relationship[162]:

$$\log k_{PhS^-}/k_{PhO^-} = A + B \log \mathbf{b}_{C\text{-}X} \tag{10.8}$$

in which $\mathbf{b}_{C\text{-}X}$ is the bond polarizability of the C-leaving group bond, Section 1.3.5. The slow step therefore is the coordination of the nucleophile to carbon, assisted in the case of a soft nucleophile by a polarizable group at the reaction center and for hard nucleophiles by the inductive polarization of, for example, the C–F bond. The second step, the expulsion of the leaving group, no doubt takes place at the 'normal' rate order $(I > Br > Cl > F)$ but as it occurs after the slow step, this cannot be ascertained. Catalysis by amines including substrate and by alkoxide or carboxylate ion occurs. It is likely that this brings about proton removal from the substrate, increasing its nucleophilicity. However, a number of examples are found to be general acid- and specific base (OH$^-$ or OR$^-$)-catalyzed[163–167], and this can be accounted for by a reversible addition of the nucleophile followed by rate-controlling loss of the nucleofuge at a rate k_2. This model also leads to an explanation of the small primary isotope effects found when secondary amines and their deuterated analogues react ($k_H/k_D \sim$ 1·25)[168,169].

The intermediate complex

Further support for the mechanism of the bimolecular S_NAr reaction comes from direct observation of the proposed intermediates, which can be rather stable. It has long been known that nitrophenyl ethers will form colored complexes in solution upon addition of alkoxide; these are known as *Meisenheimer complexes*, e.g. **54a**[170]. Examples such as **55** may be crystallized and their molecular structures determined by X-ray crystallography[171,172] or their kinetics of reaction studied[167,173].

55

MO calculations

There seems little doubt that the transition states of these reactions come late and approximate to the intermediate Meisenheimer complexes in energy. One can use arguments similar to those put forward to explain benzenium ion stability in this case also. The reactions to be considered are **56** and **57**, in which a Z and an X substituent (NO$_2$ and OMe, for example) are represented by CH$_2^+$ and CH$_2^-$ respectively. It is easy to see that the resulting localization energy is more favorable for the former, which has six π-electrons to fill three bonding orbitals, while for the latter the extra

π-energy/β: 8·72   6·9   8·72   5·9

$L_- = 1·82\beta$ $L_- = 2·82\beta$

electron-pair must be accommodated in an antibonding MO. The difference in localization energy is 1β, a considerable effect. Similarly, activation by *ortho*-CH_2^+ and deactivation by *ortho*-CH_2^- can be shown to be predicted.

The availability of the frontier orbitals leads to the same prediction. The aromatic molecule provides the LUMO with which the HOMO of the nucleophile interacts. For $ArCH_2^+$ the LUMO is ψ_4, for which the coefficient at the reaction site is $0{\cdot}3779$ ($c^2 = 0{\cdot}143$) whereas for $ArCH_2^-$ the LUMO is ψ_5, the coefficient at the reaction site being zero. Hence a Z-substituent, while capable of accepting charge, also provides for good overlap at a *para*-carbon with a nucleophilic reagent.

Solvent effects

Rates of reaction of a secondary amine such as piperidine with 2,4-dinitro-fluorobenzene are very sensitive to the solvent and are greatly accelerated by polar media,

Solvent:	benzene	ethyl acetate	$CHCl_3$	CH_2Cl_2	CH_3CN	CH_3NO_2
k_{rel}:	1	4	8	17	42	73

This is consistent with a highly dipolar transition state as is also the very large, negative volume of activation (-30 to $-40\,cm^3\,mol^{-1}$ dependent upon the solvent). The relative ease of displacement of fluoride and chloride is solvent-dependent and may be moderate, e.g. $k_F/k_{Cl} = 30$ in acetone or very large, e.g. 375 in acetonitrile. It seems likely that in the latter case the formation of the intermediate, (k_1), is rate-determining while in the former, expulsion of halide ion is acquiring some importance.

10.5.2   The unimolecular mechanism [174]

Since little driving force to the expulsion of the leaving group is provided by the stability of the phenyl cation, the S_N1 route is only to be expected in the case of an exceptionally labile leaving group. This, in practice, is limited to the diazonium ion, which can undergo heterolysis with the expulsion of N_2:

$$Ar\text{–}N_2^+ \rightarrow Ar^+ + N_2.$$

The reaction is complicated by competing homolytic decomposition, especially in basic or non-aqueous solution or in the presence of oxidizing agents, notably Cu^{2+}.

In aqueous, acidic solution, diazonium salts decompose at rates independent of the concentrations of nucleophiles which may become incorporated in the aromatic ring. Unexpectedly, rates of decomposition of a series of *para*-substituted phenyldiazonium ions show no simple Hammett plot but a maximum rate for the parent compound (*para*-H). Rate retardation by $-R$ and $-I$ groups is to be expected since they would destabilize a cation but retardation by $+R$ groups may indicate a change in mechanism or, plausibly, the additional stabilization of the diazonium ion expressed by such structures as **58**. In reactions of this type, no cine substitution is observed (see Section 10.5.3), which rules out an intermediate aryne, while ^{15}N-labeled N_2 can be incorporated to some extent into a diazonium ion, indicating reversible decomposition.

Reactions of diazonium ions with hydroxide or alkoxide ion probably occur by attack on nitrogen followed by homolysis. Displacement by iodide occurs by initial electron transfer.

10.5.3   The aryne mechanism (E–Ad)[159,175–178]

Frequently, on treating unactivated aryl halides with powerful bases, substitution of halide occurs accompanied by some rearrangement to the adjacent position (cine substitution), **59**. No reaction will occur if the halide is flanked by two alkyl groups (although it is reasonable to argue this as a steric effect). The incorporation of an isotopic label, **60**, removes this complication and the products are now two isomers in almost equal amounts[176–178]. Leaving-group effects are quite different from those of the

59

bimolecular S_NAr mechanism; for displacements by NH_2^- in liquid NH_3 the order $Br > I > Cl \gg F$ is observed indicating, evidently, loss of halide in the slow step. These facts are accommodated by the mechanism **60**. Base-promoted elimination of H^+ and X^- at the aromatic ring, either synchronously (E2–Ar) or sequentially (E1cb–Ar) leads to a species (**59a,b; 60a,b**) known as an aryne (specifically benzyne) and written with a triple bond although the geometry must preclude any resemblance to an alkyne. Arynes are extremely reactive and will rapidly undergo nucleophilic addition to regenerate the aromatic structure. In the absence of a nucleophile, arynes will dimerize (**61a**), or will undergo Diels–Alder cycloadditions with great facility (e.g. **61b**). Other routes which allow the convenient generation of benzyne in inert media in which these alternative pathways may be studied are **62** and **63**.

10.5.4   Nucleophilic substitution via ring opening; the S_N(ANRORC) route[179]

Nucleophilic substitution of halogen in pyridine rings may occur by the Ad–E or E–Ad mechanisms previously encountered in benzene systems. Thus, amination of 2-halopyridines occurs by the Ad–E route to give 2-aminopyridine, **64**, while a similar reaction on 3-halopyridines prefers the E–Ad mechanism via 3,4-pyridyne, **65**, and leads to a mixture of 3- and 4-aminopyridines. 2,3-Pyridyne seems not to form. However, a different mechanism is available for the amination of halo-pyrimidines and -pyrazines. Treatment of the ^{15}N-labeled compound, **66**, with potassium amide leads to substitution with scrambling of the label on to the amino nitrogen. Furthermore, when the pyrimidine, **67**, is treated with lithium piperide, the product is the acyclic compound, **68**. It appears that the

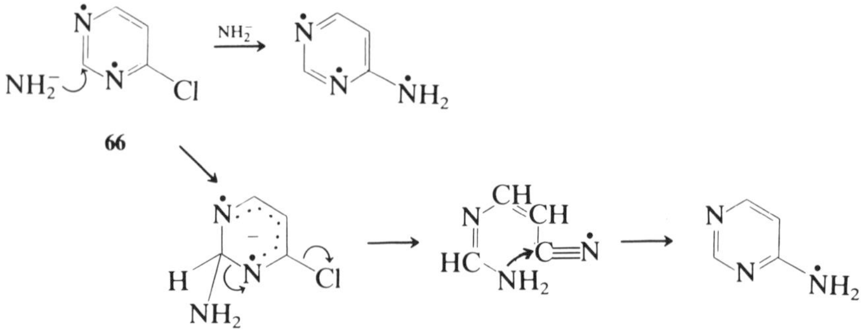

66

69

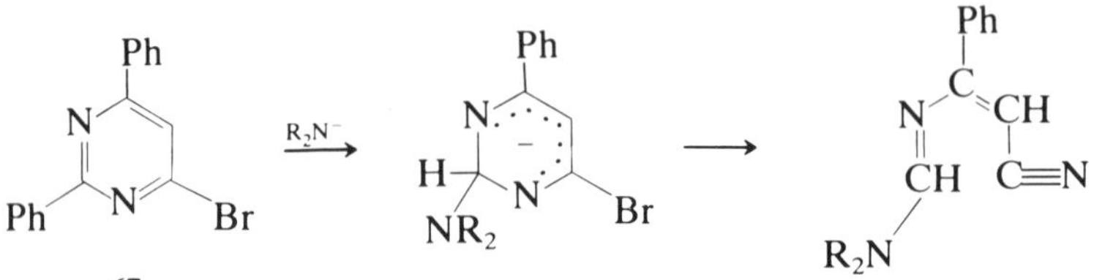

67

68

substitution reaction of **66** takes place via an analogous intermediate which recyclizes, the amide ion becoming incorporated into the heterocyclic system **69**. This sequence has been denoted ANRORC (Addition of Nucleophile, Ring Opening, Ring Closure).

10.6 NUCLEOPHILIC SUBSTITUTIONS AT CARBONYL CARBON[182]

The displacement of one nucleophile by another at a carbonyl center, an acyl transfer, is of wide occurrence (Table 10.27):

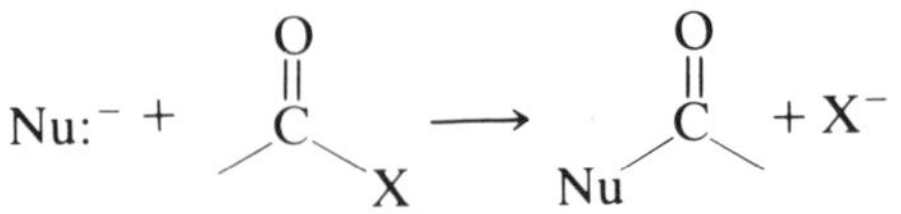

A nucleophilic leaving group is required, the order of reactivity normally observed being similar to that for displacements at saturated carbon:

Acyl halides > anhydrides > esters > amides

(X = Hal > OCOR > OR > NR$_2$).

Acid catalysis is often observed in which the leaving group is converted into a protonated form (OHR$^+$ > OR and NHR$_2^+$ > NR$_2$), which weakens the

$\overset{\text{O}}{\underset{\|}{\text{C}}}$—X bond. The nucleophilic reagents may be many and varied. Carbonyl carbon differs from alkene carbon in bearing positive charge resulting from

462

Table 10.27 Substitutions at carbonyl carbon

Scope of the reaction

$$\text{Nu:} + \text{R—C=O} \rightarrow \text{R—C=O} + \text{X:}$$
$$\qquad\quad |\qquad\qquad\; |$$
$$\qquad\quad X\qquad\quad\;\; Nu$$

	X					
Nu:	*OR* esters	*NR$_2$* Amides	*O·COR* Anhydrides	*Hal* Acyl halides	*SR* Thiol esters	*OC$_6$H$_4$NO$_2$(Opnp)* p-Nitrophenyl esters
H$_2$O(OH$^-$, H$_3$O$^+$ hydrolysis)	✓	✓	✓	✓	✓	✓
ROH (alcoholysis)	✓		✓	✓	✓	✓
R$_2$NH (aminolysis)	✓		✓	✓	✓	✓
RSH	✓		✓	✓		
H$^-$(AlH$_4^-$/reduction)	✓	✓	✓	✓	✓	✓
\C⁻ /|			✓	✓		
N$_3^-$, S$_2$O$_3^{2-}$, ROO$^-$, OCl$^-$, etc.	✓		✓	✓		✓

polarization of the double bond,

$$\text{\Large >C=O} \longleftrightarrow \text{\Large >C}^+\!\!-\!\text{O}^-$$

which gives it electrophilic character due to a low-lying LUMO centered principally on carbon (Fig. 10.14). The typical reaction is therefore the addition of a nucleophile. What happens subsequently depends upon the nature of the 4-coordinate species produced; if a leaving group is available it may be displaced (substitution) or, if not, a proton may complete an addition reaction. If the nucleophile possesses a labile hydrogen, water may be eliminated and a condensation reaction is observed (**70**).

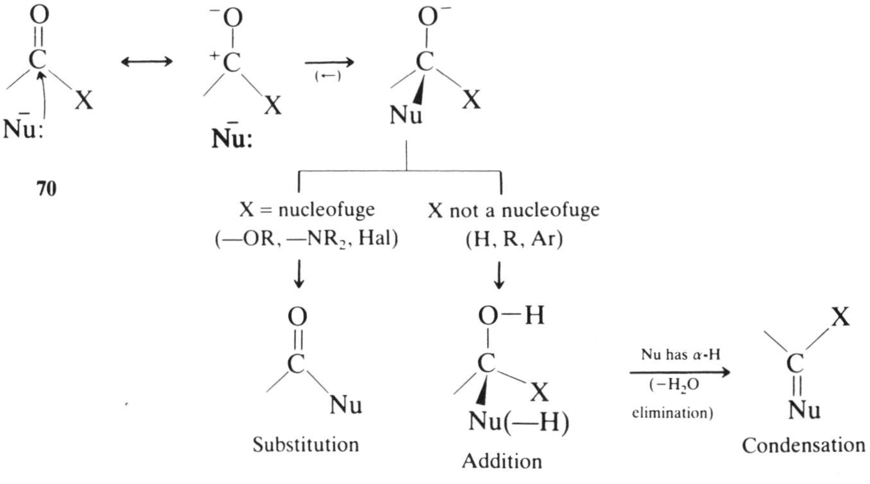

70

X = nucleofuge (—OR, —NR$_2$, Hal)	X not a nucleofuge (H, R, Ar)	
Substitution	Addition	Condensation

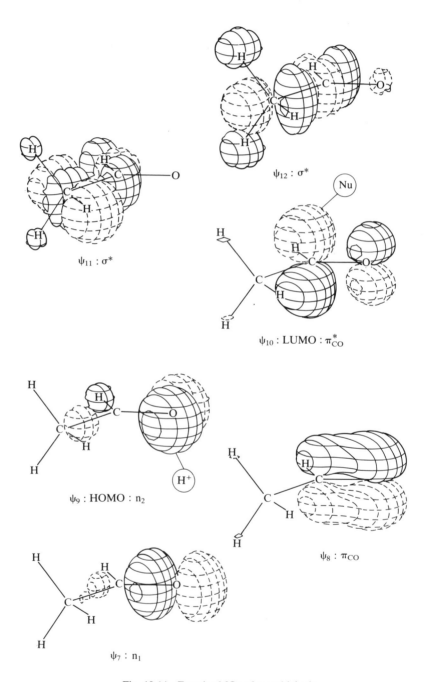

$\psi_{12} : \sigma^*$

$\psi_{11} : \sigma^*$

$\psi_{10} : \text{LUMO} : \pi^*_{CO}$

$\psi_9 : \text{HOMO} : n_2$

$\psi_8 : \pi_{CO}$

$\psi_7 : n_1$

Fig. 10.14 Frontier MOs of acetaldehyde.

Nucleophilic reactivity at the carbonyl carbon differs from that described previously for attack at a saturated carbon since it is a harder electrophilic center and tends to prefer hard nucleophiles which react under predominantly electrostatic influence. As a result, rates of attack of a series of nucleophiles on an ester do not correlate well with *n*-values (based on attack at MeI; Section 6.12.3) or with pK_A (attack at H). Figure 10.15 emphasizes the need for different scales of nucleophilicity for each type of electrophile. The N_+ scale (based on attack at R$^+$) is the most appropriate for carbonyl reactions (Section 5.6.2).

P=O and S=O are other reaction centers which are analogous to carbonyl and whose derivatives may react by similar pathways.

Carboxylic esters undergo pH-dependent hydrolysis to the corresponding acids and alcohols. The effectiveness of acid and basic catalysis is shown by

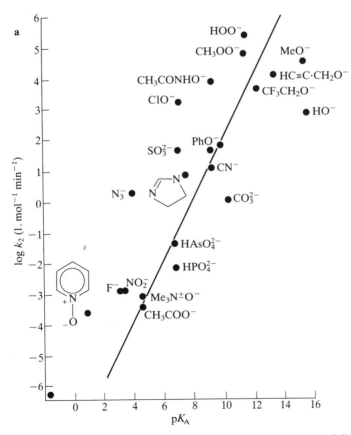

Fig. 10.15 Brønsted plots for displacements from carbonyl carbon. **a**, Rates of displacement of *p*-nitrophenol from *p*-nitrophenyl acetate by nucleophiles as a function of their pK_A. There is only a moderate correlation and notable deviations by 'α' nucleophiles. **b**, Lack of correlation between rate of nucleophilic displacements at carbonyl in *p*-nitrophenyl acetate and at saturated carbon of methyl iodide (*n*).

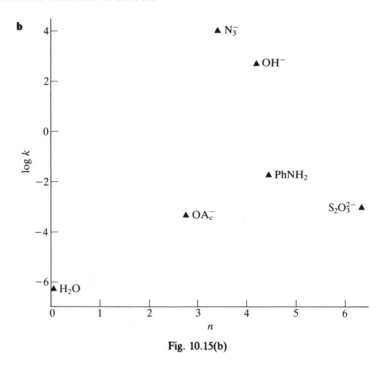

Fig. 10.15(b)

71

72

73

Table 10.28   Rates of hydrolysis of ethyl acetate

$$CH_3CO \cdot OEt + H_2O \rightarrow CH_3CO \cdot OH + EtOH$$

	Acidic	Neutral	Alkaline
$k_2/M^{-1}s^{-1}$	1.03×10^{-4}		1.52×10^{-1}
k'/s^{-1}		3.16×10^{-10}	
$\Delta G^{\ddagger}/kJ\,mol^{-1}$ (kcal mol^{-1})	96 (23)	120 (28.6)	64 (15)

the rate comparison for ethyl acetate summarized in Table 10.28. Energy profiles for the acid, base and neutral rates are shown in Fig. 10.16.

10.6.1   Basic hydrolysis of carboxylic esters[183-188]

Treatment of a carboxylic ester with aqueous alkali, **74**, releases carboxylic acid and alcohol quantitatively with a rate law:

$$\text{Rate} = k_2[\text{ester}][\text{OH}^-].$$

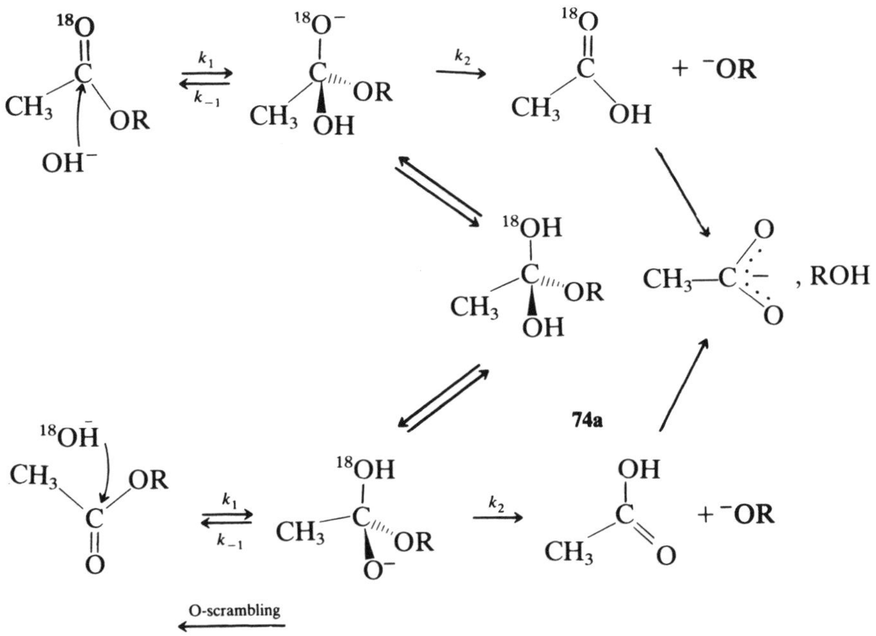

Entropies of activation are large and negative, indicative of association in the activation step (Table 10.29). By analogy with similar reactions at saturated carbon, a concerted displacement **71**, an S$_N$2 reaction, might be considered. However, the following evidence shows this to be untenable.

467

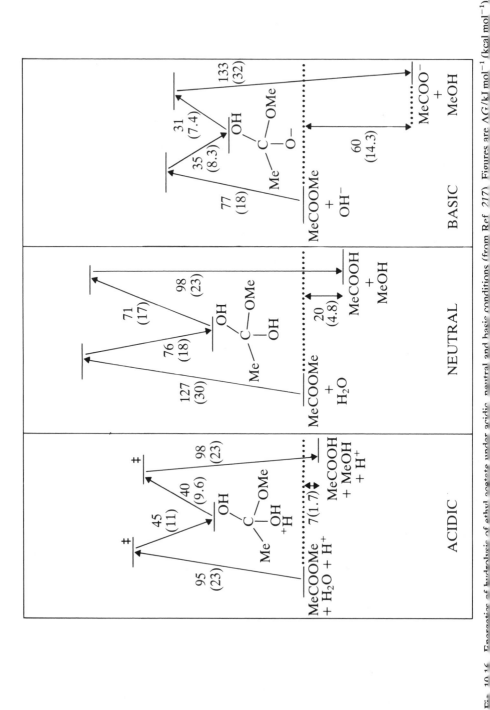

Fig. 10.15 Energetics of hydrolysis of ethyl acetate under acidic, neutral and basic conditions (from Ref. 217). Figures are $\Delta G/\text{kJ mol}^{-1}$ (kcal mol^{-1}).

Table 10.29 Activation parameters for ester hydrolysis and related reactions

Substrate	$-\Delta S^{\ddagger}/J\,mol^{-1}\,K^{-1}\,(cal\,K^{-1}\,mol^{-1})$
(a) Base-promoted reactions, OH^{-}, aqueous acetone, 25°C	
HCOOEt	41 (98)
CH$_3$COOEt	136 (32)
CH$_3$COOt-Bu	132 (31)
CH$_3$COOPh	92 (22)
Me$_3$CCOOEt	138 (33)
Cl$_3$CCOOEt	159 (38)
CH$_2$=CHCOOEt	61 (14)
PhCOOEt	109 (26)
p-NO$_2$C$_6$H$_4$COOEt	84 (20)
PhCOOPh	63 (15)
PhCOOMe + Me*O^{-}	96 (23)
PhCOOEt (^{18}O exchange)	106 (25)
(b) Hydrolyses by neutral water	
CF$_3$COOPh	192 (45)
HCOOCH$_2$Cl	163 (39)
ClCH$_2$COOCH$_2$Cl	154 (37)
CF$_3$COOH (^{18}O exchange)	146 (35)
Ac$_2$O	179 (42)
AcCl	58 (14)
PhCOCl	30 (7)
(c) Acid-catalyzed hydrolyses	
R·COOEt	
R = H	85 (20)
R = Me	109 (26)
R = Et	110 (26)
R = isoPr	117 (28)
R = $tert$-Bu	98 (23)
p-X—C$_6$H$_4$COOEt	103 (25)
X = OMe	110 (26)
X = H	133 (32)
X = NO$_2$	121 (29)
ClCH$_2$COOEt	142 (34)
Cl$_2$CHCOOEt	158 (38)
Cl$_3$CCOOEt	
(d) Acid-catalyzed esterification (RCOOH + MeOH)	
R = H	115 (27)
R = Me	136 (32)
R = Et	138 (33)
R = isoPr	148 (35)
R = $tert$-Bu	145 (35)
R = Ph	101 (24)

(a) The position of bond fission is inappropriate. An ^{18}O label in the ester, as in **72**, appears entirely in the alcohol product[189–191]. If the labeled oxygen is present in the water, some (but very little) appears in the carboxylic acid but not in the alcohol[192].
(b) The stereochemistry at the alcohol carbon is retention, i.e. a chiral ester

73 hydrolyzes to the chiral alcohol with the same configuration[193]. An S_N2 process would be accompanied by inversion.

(c) There is no special difficulty in the hydrolysis of esters of cyclic or neopentyl alcohols, for which rearface attack would be sterically hindered. LFERs show only moderate steric hindrance to be present.

(d) No Wagner–Meerwein rearrangement takes place in the alcohol structure; such rearrangement would be observed if ionization to a carbocation (S_N1) had occurred[194]. For example, neopentyl esters hydrolyze to neopentyl alcohol.

All of these observations accord with the occurrence of acyl–oxygen fission rather than alkyl–oxygen fission, but do not distinguish between one- and two-step mechanisms.

The case for a two-step reaction may be summarized as follows.

(a) LFER. S_N2 reactions typically are very insensitive to substituent effects since there is a cancelation of electronic demands for the bond-making and bond-breaking components (Table 10.3). Ester hydrolysis in alkaline medium, on the other hand, is very sensitive to structure and rates are strongly accelerated by electron-withdrawing groups. For example, for the alkaline hydrolysis of R·COOEt:

R:	—CH$_3$	—CH$_2$Cl	—CHCl$_2$	—CCl$_3$
k_{rel}:	1	760	16 000	10^5

This suggests that the addition of the nucleophile is kinetically more important than fission of the leaving group. Values of ρ for hydrolyses of benzoate esters and ρ^* for aliphatic esters are around 2·5 (Table 10.30) and confirm this view[184, 195].

(b) The observation of a break in the pH–rate profile (Fig. 10.17a, curve B) or even more irregular kinetic behavior (Fig. 10.17a, curve A). Also, a reaction order in nucleophile varying with concentration (Fig. 10.17b) points to reaction complexity which can be plausibly explained in terms of an adduct of the ester and the nucleophile as a reactive intermediate[196].

(c) Isotopic oxygen scrambling. When a carbonyl-^{18}O-labeled ester is hydrolyzed, a part of the label is often exchanged with solvent water[191].

Table 10.30 Reaction constants for hydrolysis of esters

Hydrolysis	Conditions	ρ
ArCOOMe + OH⁻	MeOH–H$_2$O, 25°C	2·23
ArCOOEt + OH⁻	MeOH–H$_2$O, 25°C	2·19
ArCOOEt + OH⁻	acetone–H$_2$O	2·26
ArCOOisoPr + OH⁻	acetone–H$_2$O	2·14
RCOOEt + OH⁻	acetone–H$_2$O	2·42 (ρ^*)
ArCOOEt + H$_3$O⁺	EtOH–H$_2$O, 100°C	0·14

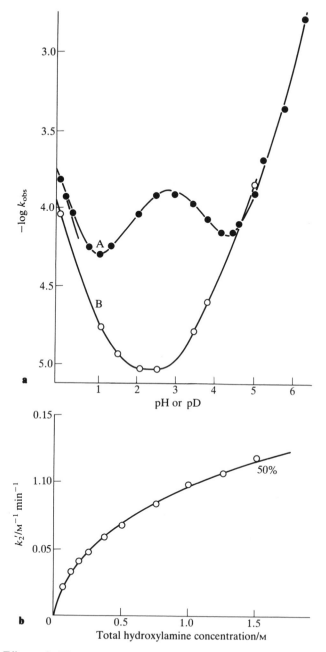

Fig. 10.17 Effects of pH on rates of ester hydrolysis. **a**, pH–rate profiles for hydrolyses (water, 100°C) of

(A) [structure with CO/NH/CO ring and COOH substituent]   (B) [structure with CO/NH/CO ring]

(from Ref. *176*). **b**, Variation of rate of reaction of formaldehyde with hydroxylamine as a function of [NH$_2$OH], indicating complex rate law.

Table 10.31 Relative rates of hydrolysis, k_h, and oxygen exchange, k_{ex}, in the hydrolysis of some carboxylic acid derivatives

Substrate	Conditions	k_h/k_{ex}
(a) Basic hydrolysis		
PhCOOEt	H_2O, 100°C	5·2
PhCOOMe	dioxane–H_2O, 25°C	5·2
PhCOOEt	dioxane–H_2O, 25°C	10·6
PhCOOPh	dioxane–H_1O, 25°C	>100
PhCONH$_2$	MeCN–H_2O	0·21
CF$_3$COOEt	H_2O, 25°C	2·0
CF$_3$COOMe	H_2O, 25°C	12·6
(b) Neutral hydrolysis		
Ac$_2$O	dioxane–H_2O	300
(PhCO)$_2$O	dioxane–H_2O	200
PhCOCl	dioxane–H_2O	18
(c) Acidic hydrolysis		
Ac$_2$O	H_2O, 0°C	V. large
(PhCO)$_2$O	H_2O, 0°C	V. large
PhCONH$_2$	H_2O, 100°C	>350
PhCOOEt	H_2O, 100°C	5·2

Furthermore, if the ester is re-isolated before completion of the reaction it may be found to have lost a substantial fraction of the label to solvent. Rates of hydrolysis, k_h, and of exchange, k_{ex}, may be determined separately and the ratio calculated as a measure of this effect (Table 10.31). An intermediate is therefore required in which the carbonyl and the hydroxyl oxygens are equivalent at some stage. Its formation must also be reversible. Points (a) and (b) require this intermediate to be on the reaction path, which is not mandatory for (c) although it is consistent with it. Altogether, these requirements can only be met by the mechanistic scheme **74**. The intermediate is a hemiorthocarboxylate, **74a**, reversibly formed by coordination of OH$^-$ to the carbonyl group[197]. The kinetics of the reaction therefore take the form:

$$k_{obs} = \frac{k_1 k_2}{k_1 + k_2} = \frac{k_1}{(k_1/k_2 + 1)},$$

assuming $k_2, k_3 \gg k_1$ (cf. Eq. (10.2)). Consequently, $k_{ex}/k_h = k_{-1}/2k_2$, the factor $\frac{1}{2}$ arising since only half the 'regression' steps will lead to exchange. The two hydroxyl functions are interchanged by proton transfer, expected to be very rapid and diffusion-controlled, and the expulsion of either reforms the ester. Expulsion of OR$^-$ leads to products. This mechanism is designated $B_{Ac}2$ (base-promoted, acyl–oxygen fission, bimolecular).

The tetrahedral intermediates in ester hydrolysis normally undergo decomposition too rapidly to be detected, much less isolated. Rates of breakdown of these species have been measured and found to occur at

75   **76a**   **76b**

almost diffusion control, $k \sim 10^8 - 10^9\,\text{s}^{-1}$, generally catalyzed both by acids and bases[198, 199]. Examples are known which are especially stable due to one of the following features[200].

(a) Strongly electron-withdrawing groups in the acid moiety, as in **75**.

(b) Destabilization of ester or amide form by reducing the resonance interaction with the carbonyl group as in the aromatic pyrrole amide, **76a** or the amidine, **76b**[201, 202].

(c) Incorporation of the function within a stable cyclic structure such as the adamantane-like tetrodotoxin, **77**.

Although isolable, these compounds break down under acidic or alkaline conditions and in general behave as stabilized examples of the tetrahedral intermediates inferred in carbonyl reactions.

77

Solvation effects

The large, negative values of $\Delta S^{\ddagger}$ are not explained by the structure of the intermediate alone. They suggest in addition considerable immobilization of solvent molecules, presumably by hydrogen bonding of water to the carbonyl oxygen. Solvent kinetic isotope effects ($k_{\text{H}_2\text{O}}/k_{\text{D}_2\text{O}}$) for neutral hydrolyses of esters, anhydrides and acyl chlorides usually fall in the range 1·5–3 and are taken to indicate involvement of water molecules in the transition state[203].

10.6.2 Acidic hydrolysis of esters

Aqueous acid will also bring about hydrolysis of an ester, with the rate law again of second order, and specific hydronium ion catalysis is observed:

$$\text{Rate} = k_2[\text{ester}][\text{H}_3\text{O}^+].$$

A molecule of ester and a proton are required for the transition state, presumably in addition to the requirement for a water molecule which would not be apparent from the kinetic information. As with basic hydrolysis there are large negative entropies of activation typical of an associative activation process (Table 10.29). The overall pH–rate profile therefore takes the form of curve B in Fig. 10.17a[204], but anomalies such as curve A require there to be a two-step reaction, one step being acid-catalyzed. The retention of the ether oxygen in the alcohol product was again established by [18]O labeling:

$$\text{EtCO·}^{18}\text{OMe} \xrightarrow{\text{H}^+,\text{H}_2\text{O}} \text{EtCOOH} + \text{Me}^{18}\text{OH},$$

and by retention of chirality of the alcohol structure. Oxygen scrambling between the carbonyl group and solvent again takes place and the isotopic enrichment of an ester such as PhC[18]O·OEt diminishes as the reaction proceeds. All these characteristics are so similar to those of the basic hydrolysis that the two mechanisms must be closely related. They differ, in fact, only by the addition of two protons to the transition state[205].

Acidic hydrolysis commences with pre-equilibrium protonation of the ester. Esters are known to be weak bases, $pK_A \sim -7$, and may be converted to their conjugate acids in superacid media. Their NMR spectra show that protonation occurs at the carbonyl oxygen rather than at —OR, (**78**)[206]. The carbonyl center is therefore made more receptive to nucleophilic attack, in this case by a water molecule. The result is the same tetrahedral intermediate, **79a**, a hemiorthocarboxylate which decomposes to products or reverts to reagents both with and without oxygen exchange as before, but with acid catalysis of all these processes. This stage of the reaction is analogous to the acid-catalyzed hydrolyses of acetals and related species (Section 9.2.3). All steps are reversible so that acid-catalyzed esterification (in solvent alcohol) follows exactly the same pathway in reverse by the

78

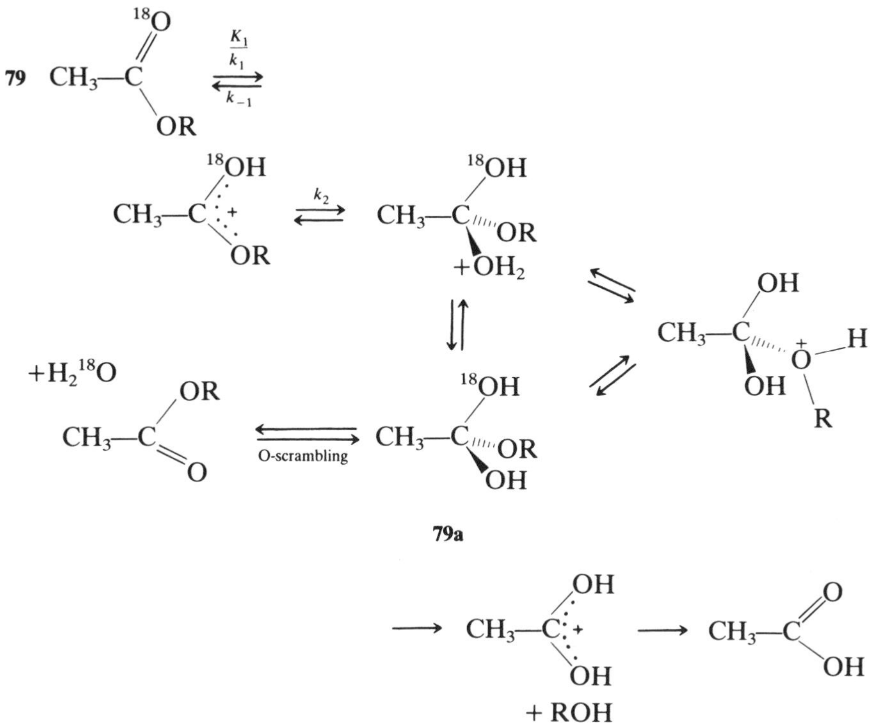

79 CH_3-C $\overset{^{18}O}{\underset{OR}{\diagup\diagdown}}$ $\overset{K_1}{\underset{k_{-1}}{\overset{k_1}{\rightleftharpoons}}}$

$CH_3-C\overset{^{18}OH}{\underset{OR}{\diagup\diagdown}}{:}^+$ $\overset{k_2}{\rightleftharpoons}$ $CH_3-C\overset{^{18}OH}{\underset{+OH_2}{\diagup}}\cdots OR$

$\Updownarrow$ $\rightleftharpoons$ $CH_3-C\overset{OH}{\cdots}\underset{OH}{\overset{}{}}\overset{+}{O}\diagdown^H_R$

$+H_2{}^{18}O$ $CH_3-C\overset{OR}{\underset{O}{\diagup\diagdown}}$ $\underset{\text{O-scrambling}}{\rightleftarrows}$ $CH_3-C\overset{^{18}OH}{\underset{OH}{\diagup}}\cdots OR$

79a

$\longrightarrow$ $CH_3-C\overset{OH}{\underset{OH}{\diagup}}{:}^+$ $\longrightarrow$ $CH_3-C\overset{O}{\underset{OH}{\diagup\diagdown}}$

$+ ROH$

principle of microscopic reversibility, designated $A_{Ac}2$ (acid-catalyzed, acyl–oxygen, fission, bimolecular).

LFER

Unlike basic hydrolysis, acidic ester hydrolysis is not sensitive to electronic perturbations. Values of ρ, ρ^* are usually in the range 0 to $+0\cdot5$[207]. The reason for this is the composite nature of the observed rate constant,

$$k_{obs} = K_1 \cdot k_2;$$

Hence $\rho_{obs} = \rho_1 + \rho_2$.

Electronic demands on the pre-equilibrium step (ρ_1 negative) and the rate-determining step (ρ_2 positive) tend to cancel; that is, structural change which increases basicity of the ester makes it less electrophilic.

10.6.3 Stereoelectronic factors in the decomposition of the tetrahedral intermediate

The departure of a nucleophilic leaving group such as —OR from a hemiorthocarboxylate requires the molecule to be in the conformation **80**

80

with a *trans*-periplanar arrangement of the bond to be broken and an unshared pair on each of the other oxygen atoms[208].

The smooth transformation to products with retention of the overall symmetry of the molecular orbitals can then proceed and conjugation between the forming carbonyl and alkoxy groups is maximized. Exactly the same situation is required for the reverse reaction, the coordination of the nucleophile to the carbonyl group. Acyclic esters can invariably achieve a favorable conformation but certain pathways in cyclic systems may sometimes be inhibited. For example, lactones (cyclic esters) such as **81** do not exchange isotopic oxygen with the solvent since the ether oxygen always has an O–H bond instead of an unshared pair periplanar to the carbonyl oxygen.

10.6.4   Other mechanisms for ester hydrolysis

All eight possible combinations of the three mechanistic dichotomies in rate-determining steps (Fig. 10.18) may be considered, namely:

Base (B) or acid (A) promoted,
Acyl (Ac) or alkyl (Al)–oxygen fission,
Unimolecular (1) or bimolecular (2).

The $A_{Ac}2$ and $B_{Ac}2$ mechanisms are the favored pathways for hydrolyses of simple esters in aqueous or mixed aqueous–organic media, but other mechanisms may be identified in more extreme situations.

Hydrolyses in strongly acidic solution[209–211]

Rates of hydrolysis of many types of ester have been studied in sulfuric acid varying in composition from dilute aqueous to 100%. Their behavior is not simple in any case and highly irregular in many (Fig. 10.19). A plot of rate against activity of water in the medium (a Yates–McClelland plot) often shows a break (Fig. 10.20a) and Hammett plots for aromatic esters in highly acidic media may do so also (Fig. 10.20b). These points indicate regions in which the mechanism changes from the $A_{Ac}2$ to another which may be shown to be a unimolecular process. The protonated ester, **82**, may dissociate in either of two ways to give an acyl or an alkyl cation (**83, 84**), these being the rate-determining steps in $A_{Ac}1$ and $A_{Al}1$ mechanisms[212].

The $A_{Ac}1$ mechanism[212]

The hydrolysis of mesitoate esters, **85**, is a well-established example. Aromatic esters with *ortho* substituents are very resistant to hydrolysis by alkali and by dilute acid owing to severe steric hindrance. The carboxylate

Basic (B) Acidic (A): $E + H^+ \rightleftharpoons EH^+ \equiv R-\overset{+}{C}\cdots\overset{OH}{\underset{OR}{}}$

		Basic (B)		Acidic (A)	
Acyl–oxygen fission (Ac)	Bimolecular (2)	$B_{Ac}2$	common	$A_{Ac}2$	common
	Unimolecular (1)	$B_{Ac}1$	(not observed)	$A_{Ac}1$	aryl carboxylates in strong acid
Alkyl–oxygen fission (Al)	Bimolecular (2)	$B_{Al}2(S_N2)$	methyl esters of strong acids	$A_{Al}2$	(not observed)
	Unimolecular (1)	$B_{Al}1$ (S_N1) tertiary esters of strong acids		$A_{Al}1$	tertiary esters in ionizing media

Fig. 10.18 Eight possible mechanisms for ester hydrolysis, combinations of three dichotomies: Acid or Base catalysed Acyl or Alkyl-oxygen fission, unimolecular (1) or bimolecular (2) rate-determining step.

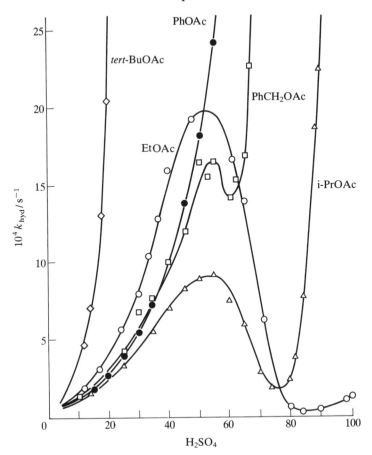

Fig. 10.19   Rates of hydrolysis of some esters in H_2SO_4 of varying concentration. The $A_{Ac}2$ mechanism in dilute aqueous acid undergoes a change to $A_{Al}1$ or $A_{Ac}1$ (PhOAc). Reductions in rate in the 50–60% region are due to reduction in the activity of water, kinetically implicated in the $A_{Ac}2$ process.

cannot lie in the plane of the ring and approach of the nucleophile is blocked. Hydrolysis may be accomplished with facility if the ester is first dissolved in concentrated sulfuric acid and the solution quenched with water. Esterification of the acid is similarly carried out by quenching its solution in sulfuric acid by alcohol. It may be shown by freezing-point depression (three-fold) and by NMR spectra that both the acid and its esters are transformed in sulfuric acid solution to the acyl cation:

$$ArCO \cdot OR + H_2SO_4 \rightleftarrows ArC{=\!=}O^+ + HSO_4^- + ROH.$$

The equilibrium may be displaced to the left by an excess of ROH (R = alkyl,H). This constitutes an $A_{Ac}1$ process, the slow step being the unimolecular ionization of protonated ester, **82→83**. Acyl cations are, of course, well-known species and reactive aromatic compounds introduced into the solutions will undergo Friedel–Crafts acylation. Even in more dilute acidic solution, the hydrolysis of mesitoate esters occurs by the same pathway: this is shown by the following characteristics, quite distinct from those of the $A_{Ac}2$ mechanism:

(a) Rates are proportional to H_0 rather than to $[H^+]$, which is taken to indicate a unimolecular slow step not involving solvent, (the Zucker–Hammett hypothesis, Section 9.1.3).

(b) $\Delta S^{\ddagger}$ is positive, $+70\,J\,K^{-1}\,mol^{-1}$ ($17\,cal\,K^{-1}\,mol^{-1}$) in 10M sulfuric acid, typical of a dissociative process.

(c) the Bunnett w-parameter (Section 9.1.3) is -1.1, a value characteristic of a reaction not involving solvent in the transition state[213].

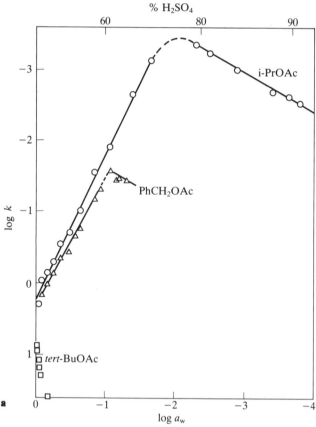

Fig. 10.20 Mechanistic change in ester hydrolysis. **a**, Mechanistic change in the hydrolyses of benzyl and isopropyl acetates (Yates–McClelland plot) shown by a discontinuity in the relationship between the rate coefficient and the activity of water. **b**, Mechanistic change indicated by a discontinuity in the Hammett plot for hydrolyses of aryl-substituted benzoate esters in 99% H_2SO_4.

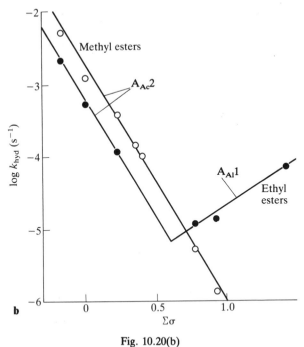

Fig. 10.20(b)

(d) There is no ^{18}O exchange of the carbonyl oxygen[214].
(e) The Hammett reaction constant for hydrolysis of ArCOOEt is negative, -3.2, and rates correlate with σ^+. This value is a composite one, $\rho_{obs} = \rho_1 + \rho_2$: since ρ_1 for protonation of ArCOOH is known to be less than -1, ρ_2 for dissociation of the acyl cation is about -2.5 and reflects the electronic demands for delocalization of charge on the carbonyl carbon.

The $A_{Ac}1$ mechanism occurs in the hydrolyses of aryl carboxylic esters in general if the acidity and ionizing capability of the medium permit this to compete successfully with the $A_{Ac}2$ pathway. Even acetates will react by this route; the changeover for phenyl acetate takes place in sulfuric acid of around 60%, the acetylium ion, CH_3CO^+, being the reactive intermediate (Fig. 10.19).

The $A_{Al}1$ mechanism

This requires unimolecular ionization of protonated ester to the carbocation derived from the alcohol, an acid-catalyzed solvolysis (S_N1 reaction) in which the leaving nucleophile is carboxylic acid (**86**). This will clearly be most favorable when the carbocation is tertiary or benzylic and the medium has good ionizing power[215–217]. *tert*-Butyl acetate is a typical case, a compound

86

$A_{Al}1(S_N1)$

known to ionize to the *tert*-butyl cation in superacid solution. Its rate of hydrolysis increases very rapidly with sulfuric acid concentration (Fig. 10.19), more so than could be accounted for by acidity of the medium alone. The initial $A_{Ac}2$ reaction changes over to $A_{Al}1$ at about 10% H_2SO_4 with the following characteristics.

(a) ^{18}O in the medium is found only in the alcohol product, confirming alkyl–oxygen fission[212]. In 60% aqueous dioxane there is clearly a mixture of the two mechanisms since the incorporation of solvent oxygen is less than 100% and is temperature-dependent (85% at 25°C and 95% at 70°C); the proportion of the unimolecular mechanism with higher $\Delta H^{\ddagger}$ is greater at the higher temperature).

(b) The activation parameters for the two competing processes are[216]:

Mechanism	$\Delta H/kJ\,mol^{-1}\,(kcal\,mol^{-1})$	$\Delta S/J\,K^{-1}\,mol^{-1}\,(cal\,K^{-1}\,mol^{-1})$
$A_{Ac}2$	67 (16)	−100 (−24)
$A_{Al}1$	108 (26)	+54 (13)

The positive entropy of activation and the Bunnett *w*-parameter of −1·2 both point to a unimolecular rate-determining step. The $A_{Al}1$ mechanism operates in the hydrolyses of esters of tertiary alcohols in moderately acidic solution and with secondary and primary esters in sulfuric acid of moderate concentration. The region of change is 60–70% H_2SO_4 for benzyl and allyl, 70–80% for secondary alkyl and 80–90% for primary alkyl, the order of carbocation stability. An $A_{Al}1$ mechanism is possible in esterification; the alcohol must ionize to a carbocation which is scavenged by solvent. An example is **87**, for which ρ (−3·69) is in accordance with this mechanism[218].

87

Other mechanisms

$B_{Al}2$[219]

This implies a bimolecular attack of OH⁻ on the alkyl group — simply an S_N2 reaction, accompanied by inversion. Carboxylate is not a very facile leaving group but electron-withdrawing substituents will increase the nucleofugacity. The highest reactivity will be for the methyl ester (Table 10.1), so that esters such as **88** and **89** are ideally constituted for hydrolysis by this route. Some proportion of reaction by this mechanism occurs even in methyl benzoate, shown by the isolation of dimethyl ether from the reaction with methoxide ion, **90**. In this case, the predominant $B_{Ac}2$ alcohol-exchange reaction will result in no net change so the S_N2 side reaction becomes the only route to a product[218]. Another example apparently is the lactone, **91**, which hydrolyzes with inversion of configuration[220], the S_N2 reaction being aided by strain release.

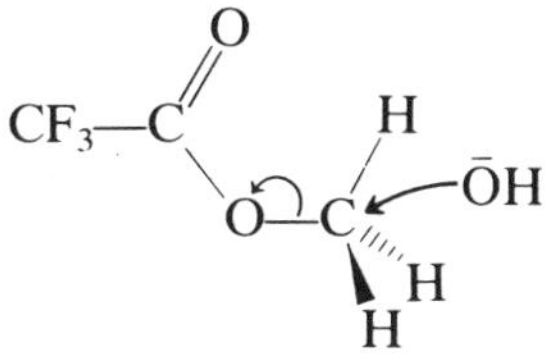

88

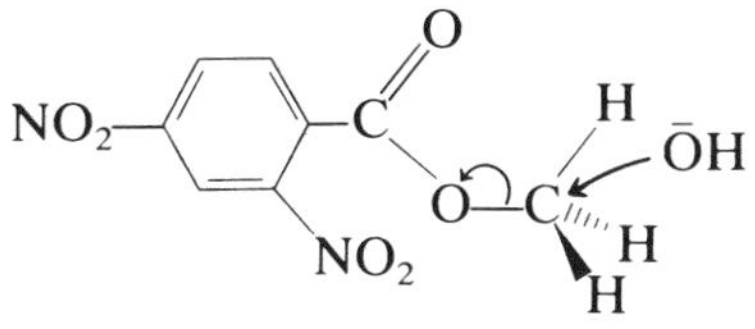

89

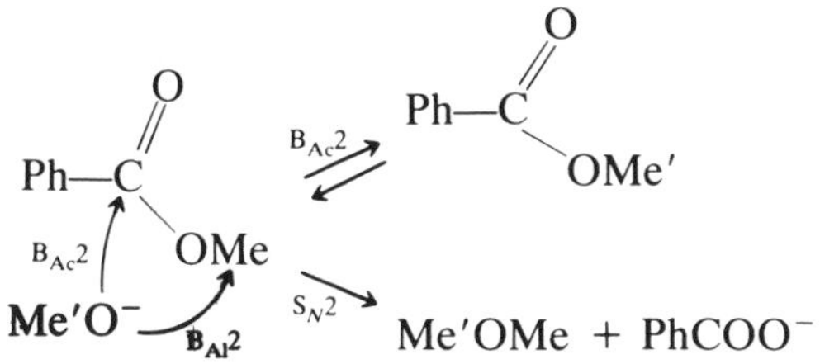

90

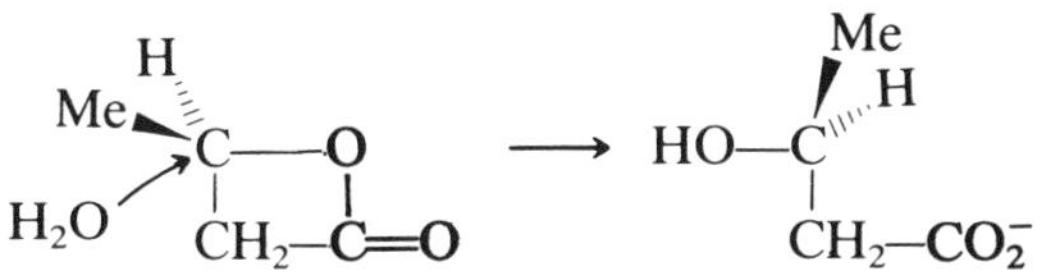

91

$A_{Al}2$

Nucleophilic attack by OH⁻ on the protonated ester is required by this mechanism; these measurements are clearly incompatible in any medium so this route is not observed. If the most likely conditions for such a reaction

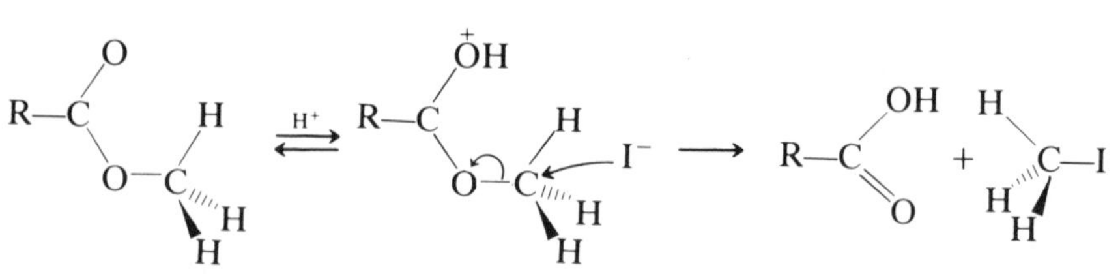

92

were to be devised, a reagent such as aqueous HI might be chosen since this combines high acidity with nucleophilicity of I^-. A methyl ester would be most susceptible to attack, for example 92.

$B_{Ac}1$

This requires unimolecular heterolysis to acyl cation and alkoxide ion, a very unusual type of ionization, and one not observed as a route to ester hydrolysis. An ideal system would require there to be an extremely stable acyl cation and a stabilized alkoxide, possibly along the lines of 93.

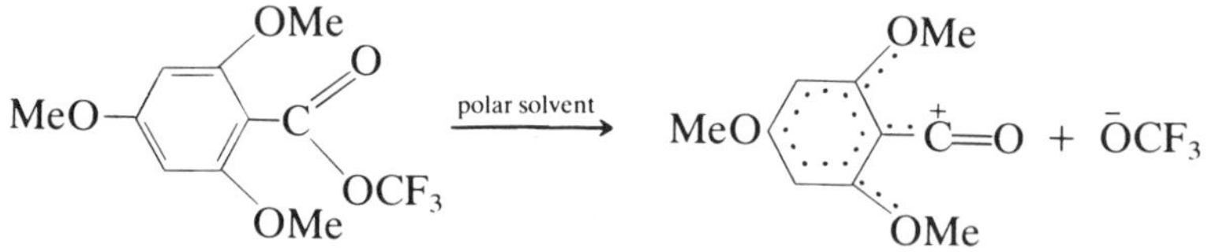

93

E1cb

This route is considered in Section 11.1.9.

10.6.5 Hydrolysis of amides, acyl halides and anhydrides

Amides

Similarly to esters, carboxamides may be hydrolyzed under either basic or acidic conditions to the carboxylic acid and amine, rates being first order in amide and in either OH^- or H_3O^+:

$$R-\overset{\displaystyle ||}{\underset{\displaystyle NR_2'}{C}}=O + (OH^- \text{ or } H^+) \rightarrow R-\overset{\displaystyle ||}{\underset{\displaystyle OH}{C}}=O + NHR_2'.$$

Amides differ from esters only in the nature of the leaving group, —NR$_2'$, which is more difficult to displace than is —OR. Also, the amino group has a smaller $-I$ effect so the coordination of a nucleophile to the carbonyl carbon is less facilitated. Under basic conditions, therefore, amides hydrolyze considerably more slowly than do esters. On the other hand, amides are much stronger bases, $pK_A \sim -1$ to -3, so they are more extensively protonated in acidic solution. Rates of acidic hydrolysis, therefore, are far greater than those of the corresponding esters[221-224]:

$$\frac{k_2(CH_3CONH_2)}{k_2(CH_3COOEt)} \quad \begin{aligned} &= 0.04(OH^-, H_2O, 25°C); \\ &= 1280\ (H_3O^+, H_2O, 25°C). \end{aligned}$$

From a practical point of view it is advantageous to hydrolyze amides under acidic catalysis, conditions normally employed for protein degradation, for example.

Basic hydrolysis of amides can be shown to occur by the $B_{Ac}2$ route accompanied by extensive ^{18}O exchange with solvent, since expulsion of OH^- is easier than expulsion of NH_2^-. The reaction constants, $\rho^* = +2.7$ for $RCONH_2$, are similar to those for ester hydrolysis. There is, however, a large solvent isotope effect, $k_{H_2O}/k_{D_2O} = 3$, interpreted as indicating a proton transfer from solvent in the rate-determining step. Therefore the mechanism may be expressed as in **94**.

When subjected to acidic hydrolysis, amides could in principle protonate either on oxygen or on nitrogen: either could lead to the orthoamide intermediate, **95**. Detailed investigations extending into the highly acidic range confirms that *O*-protonation is normal. This leads to a stabilized carbocation rather than to an acylammonium ion which would result from *N*-protonation.

(To add further confusion to the designation of mechanism, the $A_{Ac}2$ process is sometimes referred to as A_O^T2: Acid-catalyzed, Oxygen-protonated, Tetrahedral intermediate, bimolecular). The transition state, furthermore, can be shown to contain, besides a molecule of amide, one proton and three water molecules. A large, negative entropy of activation is the consequence of this[223].

β-Lactams hydrolyze by an A1 (A_N^D1) (Acid-catalyzed, Nitrogen proton-ated, Digonal intermediate, unimolecular) mechanism. Protonation occurs on nitrogen followed by unimolecular ring opening driven by relief of ring strain.

$$CH_2-C{=}O \text{ (ring } CH_2-NH) \rightleftharpoons CH_2-C{=}O \text{ (} CH_2-\overset{+}{NH_2}) \xrightarrow{slow} CH_2-\overset{+}{C}{=}O \text{ (} CH_2-NH_2) \longrightarrow CH_2-COOH \text{ (} CH_2-NH_2)$$

Rates of amide hydrolysis tend to a maximum as pH falls, corresponding to complete protonation, but no simple relationship between rate and pH or H_0 seems to exist. Evidently amides do not behave like Hammett indicators (aromatic amines). A possibility is the incursion of a reaction pathway from N-protonated amide $(A_N + A_O)^{221}$, leading to two parallel reactions with complex kinetics. Unlike ester hydrolysis, solvent isotope effects are very small and no isotopic exchange with the carbonyl oxygen occurs. This does not invalidate the $A_{Ac}2$ mechanism but probably indicates a product-determining intermediate such as **95** which loses amine much more rapidly than water.

Acyl Halides[223-227]

Halide ion is a far better leaving group than alkoxide or amide and reactions of acyl halides with all manner of nucleophiles are facile. Solvolysis takes place rapidly in neutral water or alcohol. In solvents of moderate polarity the rate-determining step appears to depend on the halide since typical rate ratios are $Cl:Br:I = 1\cdot0:200:1000$. There is considerable assistance from electron-withdrawing substituents such that the relative reactivities, $CCl_3COCl:CH_3COCl$, are around $10^4:1$ with

$$\rho^* \text{ for } RCOCl = 1\cdot25 \quad \text{and} \quad \rho \text{ for } ArCOCl = 2\cdot1 \quad \text{(in ethanol–ether).}$$

^{18}O-Exchange occurs in some cases and entropies of activation are mostly negative, ca $-80 \text{ J K}^{-1} \text{ mol}^{-1}$ $(-19 \text{ cal K}^{-1} \text{ mol}^{-1})$. A $B_{Ac}2$ mechanism best fits this pattern of reactivity. However, in more polar solvents such as acetone–water, $\Delta S^{\ddagger}$ is negative $(-0\cdot4)$, indicating a change of mechanism. Another indication of this is the abrupt change in selectivity between water and amine for reaction with benzoyl chloride as the water content of the medium is increased, while rates of ethanolysis in mixed ethanolic solvents seem to follow a polarity order. Entropies of activation can become positive, as in the case of isoPrO—COCl[228]. It seems likely that rate-

$$
R-C
\begin{array}{c}
\nwarrow O \\
\diagdown O{\cdot}COR'
\end{array}
\underset{(H^+)}{\overset{}{\rightleftharpoons}}
\quad
R-C
\begin{array}{c}
OH \\
\ddot{} + \\
H_2O \quad O{\cdot}COR'
\end{array}
$$

96

$$
\longrightarrow
\left[
\begin{array}{c}
OH \\
R-C \\
H_2O \quad OCOR \\
{\scriptstyle \delta+} \qquad {\scriptstyle \delta-}
\end{array}
\right]^{+\ddagger}
\longrightarrow RCOOH + R'COOH.
$$

determining ionization is occurring under these conditions and hydrolysis is by an S_N1 ($B_{Ac}1$) mechanism. In general, these routes seem to be fairly competitive and reaction by a combination of $B_{Ac}2$ and $B_{Ac}1$ may be common. Gold has estimated almost equal contributions from each route in the hydrolysis of benzoyl chloride in 50% aqueous acetone, going over to S_N2 in the 4:1 mixed solvent[226]. Similarities[229] in rate effects by mixed solvents to solvolyses of alkyl halides has been interpreted as typical of S_N2 processes. As a further illustration, the effect of added OH^- on the rate of hydrolysis of benzoyl chloride in 25% aqueous acetone is a 10^4-fold increase, typical of an S_N2 reaction. In 50% aqueous acetone the effect is only 600-fold, much closer to that expected for an S_N1 process, in which ionization is independent of added nucleophiles, but sensitive to ionic strength.

Carboxylic anhydrides[228,230,231]

Carboxylate is an easy leaving group and, again, hydrolysis occurs in neutral water though less rapidly than for the halides. Catalysis by both acids and

Table 10.32   Mechanisms of hydrolysis of some anhydrides

Anhydride	Medium	$\Delta H^{\ddagger}/kJ\,mol^{-1}$ $(kcal\,mol^{-1})$	$\Delta S^{\ddagger}/J\,K^{-1}\,mol^{-1}$ $(cal\,K^{-1}\,mol^{-1})$	Presumed mechanism
Acetic	Water	69 (16)	−145 (−35)	A2
Acetic	H_2O–dioxane, 60:40	64 (15)	−180 (−43)	A2
Benzoic	H_2O–dioxane, 60:40	76 (18)	−80 (−19)	A2
Mesitoic	H_2O–dioxane, 60:40	90 (21)	−30 (−7)	A1
Acetic–mesitoic	H_2O–dioxane, 60:40	75 (18)	−16 (−3·8)	A1?
Acetic–benzoic	H_2O–dioxane, 60:40	65 (15)	−94 (−22)	A2
Trimethylacetic	Water or H_2O–dioxane		−100 (−24)	A2

bases is observed with two distinct types of behavior; acetic–mesitoic anhydride hydrolyzes with acetate fission and $\Delta S^{\ddagger} = -16\,\text{J K}^{-1}\,\text{mol}^{-}$ ($3.8\,\text{cal K}^{-1}\,\text{mol}^{-1}$) while acetic–benzoic anhydride splits off benzoate and has $\Delta S^{\ddagger} = -94\,\text{J K}^{-1}\,\text{mol}^{-1}$ ($-22\,\text{cal K}^{-1}\,\text{mol}^{-1}$). Bunnett has interpreted these results as indicating initial protonation followed by unimolecular fission to the mesitoyl cation (A1 or S_N1 reaction) in the first case and synchronous displacement of benzoate (A2 or S_N2 reaction) in the second with no discrete tetrahedral intermediate **96** [232]. It seems likely that the major pathway is A2 going over to A1 only when an especially stable acyl cation can be formed and the polarity of the medium favors this route (Table 10.32).

Aquo complexes of many transition metals catalyze the hydrolyses of anhydrides: $Co(CN)_5OH^{3-}$ is almost as effective as OH^- and is presumed to transfer OH^- since rates depend upon the pK_A of the complex, though general base catalysis is not to be ruled out [229]:

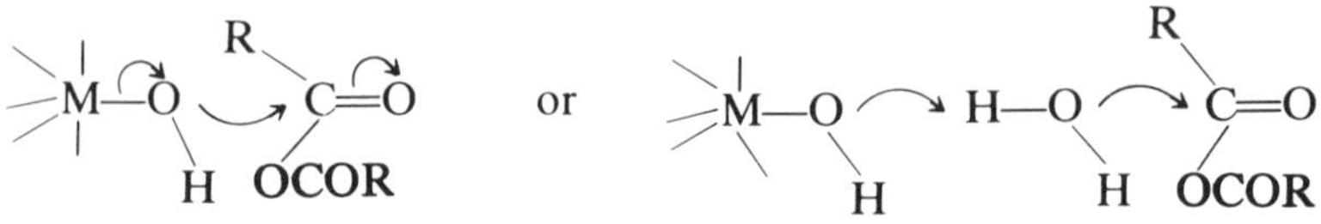

10.6.6 Properties of tetrahedral intermediates

While examples of especially stabilized hemiorthoesters (which are part of a polycyclic system or from a precursor which contains strong $-R$ groups, e.g. CF_3COOEt [233]) are known, there has been a search for direct observation of simple examples of these species, identical with the tetrahedral intermediates in ester hydrolysis. Despite their transient existence, several successful identifications have been made. The rates of decomposition of simple hemiorthoesters are much faster than their rates of formation in ester hydrolysis, which may be determined by thermochemical calculations [2] (Table 10.33), so there is no possibility of detecting them in this context. However, other reactions were devised leading to the same species at rates comparable to or faster than their rates of disappearance so that, by working at low temperatures or in flow systems, stationary concentrations were maintained which could be identified by NMR spectroscopy and their rates of disappearance could be followed by fast UV spectroscopy.

Dimethylhemiorthoformate, **97**, was obtained by hydrolysis of the acetate, **98**, in a flow system monitored by ^{13}C-NMR. Under the optimum conditions, resonances due to **97** were observed, rapidly transformed into the product, **99**. Addition of water to the ketene acetal, **100**, yields a cyclic hemiorthoacetal, **101** [234,235]. The single resonance for the equivalent methyl groups of **100** is transformed into two, corresponding to the two non-equivalent pairs in **101** before formation of the characteristic spectrum of the product, **102**. Other examples formed from orthoesters and of thoamides, **103**, have been observed. Table 10.33 sets out some reactivity

Table 10.33 Equilibrium constants for formation of hemiorthoesters and orthoamides from thermochemical data[197]

Equilibrium	K	$\Delta G°/kJ\,mol^{-1}\,(kcal\,mol^{-1})$
$HCOOH + H_2O \rightleftarrows HC(OH)_2$	$2\cdot8 \times 10^{-7}$	$38\,(9\cdot0)$
$CH_3COOH + H_2O \rightleftarrows CH_3C(OH)_3$	$4\cdot0 \times 10^{-9}$	$51\,(12\cdot2)$
$HCOOCH_3 + H_2O \rightleftarrows HC(OH)_2OCH_3$	$4\cdot7 \times 10^{-6}$	$30\,(7\cdot1)$
$CH_3COOCH_3 + H_2O \rightleftarrows CH_3C(OH)_2OCH_3$	$4\cdot0 \times 10^{-9}$	$51\,(12\cdot2)$
$HCOOCH_3 + OH^- \rightleftarrows HC(OH)(O^-)OCH_3$	$7\cdot2 \times 10^{-4}$	$18\,(4\cdot3)$
$CH_3COOCH_3 + OH^- \rightleftarrows CH_3C(OH)(O^-)OCH_3$	$1\cdot4 \times 10^{-7}$	$42\,(10\cdot0)$

Reaction	$R = H$		$R = CH_3$	
	K	$\Delta G°$	K	$\Delta G°$
$R-C(=O)-OH + HN(CH_3)_2 \rightleftarrows R-C(OH)_2-N(CH_3)_2$	$4\cdot7 \times 10^{-8}$	$42\,(10\cdot0)$	$7\cdot8 \times 10^{-11}$	$58\,(13\cdot9)$
	$1\cdot1 \times 10^{-10}$	$57\,(13\cdot6)$	$3\cdot5 \times 10^{-13}$	$71\,(16\cdot9)$
$R-C(=O)-OCH_3 + HN(CH_3)_2 \rightleftarrows R-C(OH)(OCH_3)-N(CH_3)_2$	$5\cdot0 \times 10^{-7}$	$36\,(8\cdot6)$	$9\cdot3 \times 10^{-11}$	$57\,(13\cdot6)$
	$7\cdot8 \times 10^{-11}$	$58\,(13\cdot8)$	$2\cdot8 \times 10^{-14}$	$77\,(18\cdot4)$
$R-C(=O)-N(CH_3)_2 + H_2O \rightleftarrows R-C(OH)_2-N(CH_3)_2$	$7\cdot3 \times 10^{-11}$	$81\,(19\cdot3)$	$4\cdot4 \times 10^{-15}$	$82\,(19\cdot6)$
	$1\cdot7 \times 10^{-17}$	$96\,(22\cdot9)$	$2\cdot0 \times 10^{-17}$	$95\,(22\cdot7)$
$R-C(=O)-N(CH_3)_2 + CH_3OH \rightleftarrows R-C(OH)(OCH_3)-N(CH_3)_2$	$2\cdot7 \times 10^{-15}$	$83\,(19\cdot8)$	$5\cdot8 \times 10^{-16}$	$87\,(20\cdot8)$
	$4\cdot1 \times 10^{-19}$	$105\,(25)$	$1\cdot8 \times 10^{-19}$	$107\,(25\cdot5)$
$R-C(OH)-N(CH_3)_2^+ + H_2O \rightleftarrows R-C(OH)_2-NH(CH_3)_2^+$	$1\cdot2 \times 10^{-9}$	$51\,(12\cdot2)$	$8\cdot8 \times 10^{-19}$	$46\,(11\cdot0)$
$R-C(=O)-N(CH_3)_2 + HO^- \rightleftarrows R-C(O^-)(OH)-N(CH_3)_2$	$1\cdot1 \times 10^{-12}$	$74\,(17\cdot7)$	$1\cdot7 \times 10^{-14}$	$78\,(18\cdot6)$

Rates for formation and reaction of hemiorthoesters in aqueous solution

Ortho acid derivative			Formation	Decomposition
Structure	R	L	$k_1/M^{-1}s^{-1}$	$k_2/M^{-1}s^{-1}$
$R-C(-OMe)(-L)-OMe$	H	OMe	$2\cdot6 \times 10^2$	$\left.\begin{array}{c}\\\end{array}\right\}2\cdot8 \times 10^4$
	H	OAc		
	Me	OMe	1×10^4	$>3 \times 10^4$
	Ph	OMe	56	$\left.\begin{array}{c}\\\end{array}\right\}1\cdot9 \times 10^4$
	Ph	NMePh	8×10^6	
	CycloPr	OMe	$8\cdot1 \times 10^4$	$5\cdot3 \times 10^3$

Table 10.33 (*continued*)

Rates for formation and reaction of hemiorthoesters in aqueous solution

	Ortho acid derivative			Formation	Decomposition
Structure	R		L	$k_1/\text{M}^{-1}\text{s}^{-1}$	$k_2/\text{M}^{-1}\text{s}^{-1}$
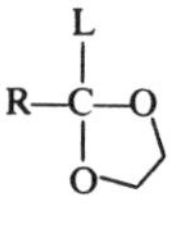	H		OMe	1.8×10^2	4.6×10^2
	H		OAc		
	CH_3		OMe	2.0×10^4	1.4×10^3
	:	$-CH_2-$	:	$>10^6$	
	Ph		OMe	5.4×10^3	3.0×10^2
	Ph		NMe_2	4×10^7	

data for both formation and disappearance of some examples[236]. Rates of formation are much greater when the leaving group is —NR_2 rather than —OR.

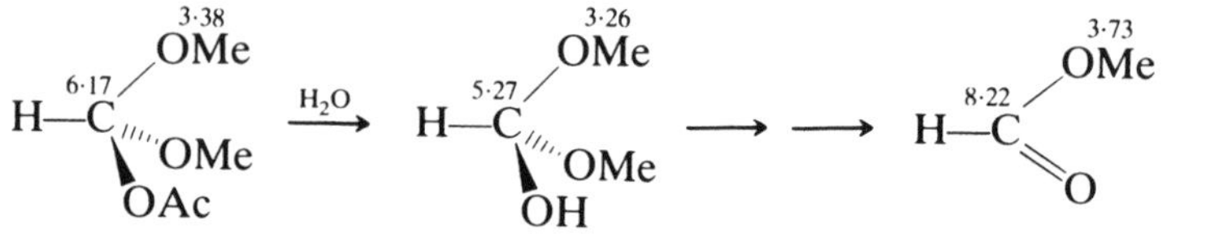

98   **97**   **99**

[Figures are ^{13}C chemical shifts/ppm]

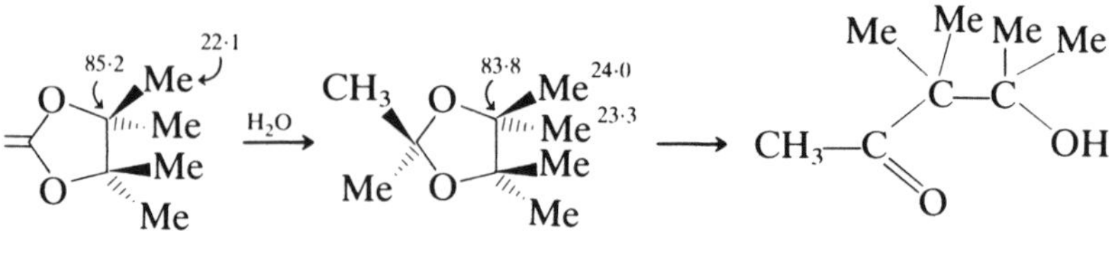

100   **101**   **102**

$$C(OR)_4, \quad R_2N \cdot C(OR)_3$$

103

10.6.7 Nucleophilic catalysis in carbonyl substitutions[185,187,188]

Nucleophiles other than OH^- in aqueous solution may accelerate rates of hydrolysis of esters and other acyl transfer reactions, a form of catalysis particularly significant in enzymic reactions. The catalysis may occur by one of several mechanisms.

490

Nucleophilic catalysis

The nucleophile undergoes a $B_{Ac}2$ substitution of the ester (**104**) to give an intermediate which itself undergoes rapid hydrolysis by a similar route. For catalysis to occur, the nucleophile must have a reactivity considerably greater than that of the ultimate acyl acceptor (i.e. water) and the catalytic intermediate must be more reactive than the substrate and thermodynamically less stable than the product, otherwise it will simply accumulate.

General base catalysis

The nucleophile accepts a proton from a water molecule as it attacks the carbonyl center and so increases the nucleophilicity of the water. It may also catalyze the decomposition to products of the tetrahedral intermediate.

Neighboring-group participation

Either of these forms of catalysis may be inter- or intra-molecular (neighboring group participation, Chapter 13). Figure 10.15a illustrates the wide variety of nucleophiles, both anionic and neutral, which catalyze the hydrolysis of *p*-nitrophenyl acetate. This Brønsted plot clearly shows that nucleophilic activity correlates poorly with basicity. Hard bases tend to fall below the least-squares line (OH⁻, for example, is less reactive by a factor of 5000), while 'α-nucleophiles' are above it. Hydroperoxide ion is some 30 times more reactive than its pK would suggest. This behavior indicates that the nucleophiles are attacking a center other than a proton of the solvent, indeed a relatively soft electrophilic center such as the carbonyl group. The catalytic action of imidazole, **105**, has been studied with especial care because of its relationship with histidine, **108**, believed to be implicated in enzymic catalysis, although imidazole is not the most active of catalysts. The hydrolysis of p-nitrophenyl acetate in the presence of imidazole, **104**, shows a rate law[237]:

$$\text{Rate} = k_2[\text{ester}][\text{imidazole}],$$

and the rate reaches a maximum at a pH corresponding to neutral imidazole, which evidently is the catalytic form. The entropy of activation is large and negative, $\Delta S^{\ddagger} = -114 \text{ J K}^{-1} \text{mol}^{-1}$ ($-27 \text{ cal K}^{-1} \text{mol}^{-1}$), which is typical of a bimolecular reaction and also one with a dipolar transition state, highly solvated[237]. The intermediate acetylimidazole, **106**, can be observed by its characteristic absorption at 245 nm since its hydrolysis, although faster than that of the ester, is not extremely fast. The reason for the high reactivity of acetylimidazole compared with other amides lies in the resonance energy which it acquires on regaining the free heterocyclic structure, reduced in the amide by the electron-withdrawing carbonyl group. This is an example of reactivity enhanced by raising the energy of the initial state. The hydrolysis, **107**, of acetylimidazole catalyzed by general acids and bases (including imidazole) occurs by a similar $B_{Ac}2$ process on the protonated compound which has a better leaving group than the neutral amide. This is shown by the very effective catalysis by N-methylimidazole, **109**, via an intermediate which does not accumulate since it cannot be stabilized by methyl loss as can **104** by proton loss. Presumably many catalyzed reactions shown in Fig. 10.15 occur by the same general mechanism as **104** but hard bases such as HPO_4^{2-} and $HAsO_4^{2-}$ are likely to act by general base catalysis. Nucleophilic catalysis of the hydrolysis of other esters including phosphates, acyl halides and anhydrides also occurs[238].

PROBLEMS

1 Nucleophilic displacements of halide at saturated carbon are greatly facilitated by an adjacent carbonyl group, for example:

	RCl	PhCH$_2$Cl	EtO·CO·CH$_2$Cl	RCOCH$_2$Cl	PhCOCH$_2$Cl
k_{rel}:	1	200	2000	35 000	105 000

[*J. Amer. Chem. Soc.*, **47**, 488 (1925); *Can. J. Chem.*, **42**, 1897 (1964); *J. Chem. Soc.*, 848 (1938).]

Suggest reasons for the activating effect of the carbonyl group.

2 $$CH_3O \cdot ClO_3 + Me_2NAr \rightarrow CH_3 \overset{+}{-} N(Me_2)ArClO_4^-.$$
The reaction of methyl perchlorate with dimethylanilines has the following characteristics. The kinetics are of second order and $\rho = -3 \cdot 0$ (variation of substituents in Ar); $\Delta S^{\ddagger} = -146 \, J \, K^{-1} \, mol^{-1}$ ($-35 \, cal \, K^{-1} \, mol^{-1}$); $k_H/k_D = 0 \cdot 94$ (for $CH_3O \cdot ClO_3$ compared with $CD_3O \cdot ClO_3$); $k(CH_3O \cdot ClO_3)/k(CH_3I) = 1170$ (for variation of leaving group).

Interpret the data and characterize the reaction pathway. [*J. Amer. Chem. Soc.*, **103**, 4515 (1981).]

3 Reagents which are effective in the cleavage of alkyl ethers include aqueous HI, trimethyliodosilane (Me_3SiI) and haloboranes such as Me_2BBr and BBr_3. Write mechanisms for the reactions of diethyl ether with each of these.

4 Neopentyl halides cannot be prepared from neopentanol ($Me_3C \cdot CH_2OH$) by the usual routes involving acidic conditions since rearrangement tends to occur. This undesirable side-effect may be suppressed by the following method:

$$Me_3C \cdot CH_2OH \xrightarrow[\text{(ii) } SOCl_2]{\text{(i) } Et_3SiCl/\text{pyridine}} Me_3C \cdot CH_2Cl.$$

Write mechanisms for the reactions involved.

5 Displacements by nucleophiles at silicon are a great deal more facile than at carbon analogues. Leaving groups fall in the order: —Cl > —OAc > —OMe, but even —OMe groups are readily displaced. The stereochemistry of displacement may be either retention or inversion of unequal mixtures of the two but not racemization, and it depends upon the medium. For

$$\underset{\substack{| \\ Ph \quad Me}}{\alpha\text{-naph}} Si-OMe + {}^-O-Bu \longrightarrow \underset{\substack{/ \\ Ph \quad Me}}{\alpha\text{-naph}} Si-OBu + OMe^-$$

the result is 100% retention in benzene, falling to 33% in butanol. Suggest mechanisms which can accommodate these findings.

6 The progress of solvolysis of an optically active halide may be monitored either by a change in rotation or release of halide with time, giving rate constants k_α and k_t respectively, which may or may not be equal. Some relative values are:

$$(R)\text{-}cis\text{-}5\text{-methyl-3-chlorocyclohexene} \begin{cases} \text{in EtOH} & k_\alpha/k_t = 1 \cdot 11 \\ \text{in AcOH} & k_\alpha/k_t = 3 \cdot 03 \end{cases}$$

Give the reason why $k_\alpha \neq k_t$.

What is the mechanistic interpretation of the ratio k_α/k_t? [*J. Amer. Chem. Soc.*, **77**, 5026 (1955).]

7 Free energies and entropies of activation and of reaction for the Menshutkin reaction between substituted pyridines and methyl iodide are as follows:

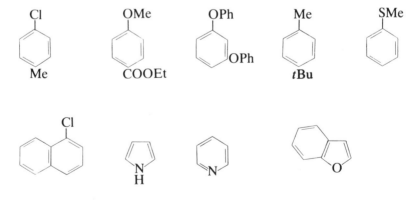

X	$\Delta G^{\ddagger}/kJ\,mol^{-1}$ $(kcal\,mol^{-1})$	$\Delta S^{\ddagger}/J\,K^{-1}\,mol^{-1}$ $(cal\,K^{-1}\,mol^{-1})$	$\Delta G°/kJ\,mol^{-1}$ $(kcal\,mol^{-1})$	$\Delta S°/J\,K^{-1}\,mol^{-1}$ $(cal\,K^{-1}\,mol^{-1})$
H	93 (22)	−130 (−31)	−50 (−12)	−160 (−38·5)
3-Br	98 (23)	−134 (−32)	−26 (−6)	−167 (−40)
3-Cl	98 (23)	−134 (−32)	−21 (−5)	−177 (−42)
4-CN	199 (24)	−135 (−32)	−21 (−5)	−187 (−45)
3,5-diCl	103 (24·6)	−139 (−33)	−11 (−2·7)	−183 (−44)
2-Cl	105 (25)	−130 (−31)	−11 (−2·7)	−140 (−33·5)

How can these figures be used to infer transition state properties? [*J. Amer. Chem. Soc.*, **102**, 5892 (1980).]

8 Both substituent constants and partial rate factors give information concerning reactivity in an aromatic ring. What relationship exists between σ and f?

Calculate the partial rate factors and product distributions for:

(a) molecular chlorination (Cl_2, AcOH) of toluene;
(b) nitration (HNO_3, Ac_2O) of chlorobenzene.

Use values from Tables 4.1 and 10.24, assuming *ortho* substituent constants, σ_o^+, to have the same value as those for the *para* position, σ_p^+.

9 Predict the major product(s) from nitration of the following.

10 1,2,3-Trimethylbenzene under strongly acidic conditions undergoes rearrangement to 1,3,5-trimethylbenzene. Write out a detailed pathway which plausibly leads to this transformation and account for the preference for the observed product.

11 *p*-Bromophenol undergoes the following rearrangements:

(+ small amount PhOH
and dibromophenol)

Suggest mechanisms for these reactions. [*J. Org. Chem.* **30**, 2301, (1965); **35**, 16 (1970).]

12 Methoxide is able to displace fluoride at an aromatic carbon. The following relative rates and positions of attack at polyfluorinated benzenes are observed.

$$ArF + MeO^-/MeOH \rightarrow ArOMe + F^-$$

C_6F_6				
k_{rel}: 1	0·23	$2·2 \times 10^{-3}$	$2·2 \times 10^{-2}$	2×10^{-9}

Attempt to rationalize this pattern of reactivity.

13 Suggest a mechanism for the reaction below.

[*Acc. Chem. Res.*, **5**, 139 (1972).]

14 Nitration of an aromatic substrate by pure nitric acid takes place according to the scheme:

$$2HNO_3 \underset{k_{-1}}{\overset{k_1}{\rightleftharpoons}} H_2\overset{+}{O}-NO_2 + NO_3^-;$$

$$H_2\overset{+}{O}-NO_2 \underset{k_{-2}}{\overset{k_2}{\rightleftharpoons}} H_2O + NO_2^+;$$

$$ArH + NO_2^+ \overset{k_3}{\longrightarrow} \overset{+}{Ar}\overset{H}{\underset{NO_2}{\diagdown}} \overset{fast}{\underset{B:}{\longrightarrow}} ArNO_2 + BH^+.$$

From this the following rate expression may be predicted to hold:

$$\text{Rate} = \frac{k_2 \cdot k_1/k_{-1}[HNO_3]^2}{[NO_3^-]\left\{1 + \dfrac{k_{-2}[H_2O]}{k_3[ArH]}\right\}} \cdot$$

Derive the expression and suggest experiments by which it might be tested. Under what circumstances might the rate be constant? Explain how the addition of nitrite can inhibit the nitration of relatively unreactive substrates but (in dilute nitric acid) accelerate the nitration of highly reactive ones.

15 1-Chloronaphthalene reacts with potassium amide in piperidine to give a mixture of two isomeric products:

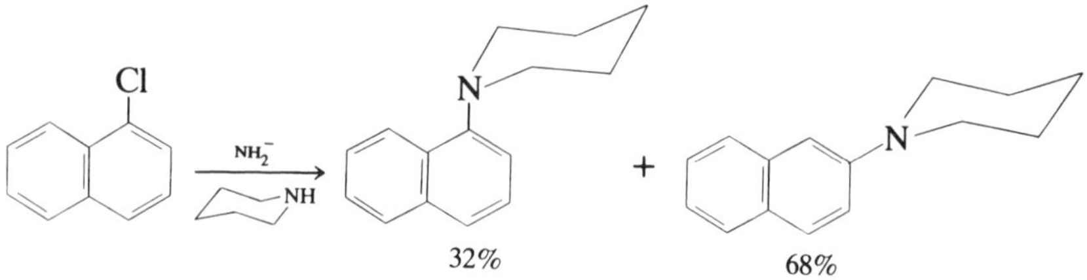

32%   68%

Account for the formation of these products and predict the product ratio from 1-bromonaphthalene.

16 Suggest pathways which might be available for the hydrolyses of

(a) phosphate esters, $(RO)_3P{=}O$;
(b) sulfonate esters such as $MeO{\cdot}SO_2Ph$
(c) nitrate esters such as $EtO{-}NO_2$.

The hydrolysis of methyl dihydrogen phosphate, $MeO{\cdot}P(OH)_2$, shows a pH–rate profile with a maximum at pH = 4; suggest an explanation.

17 The Mannich reaction is a condensation between an aldehyde, a primary or secondary amine and a nucleophile by the sequence:

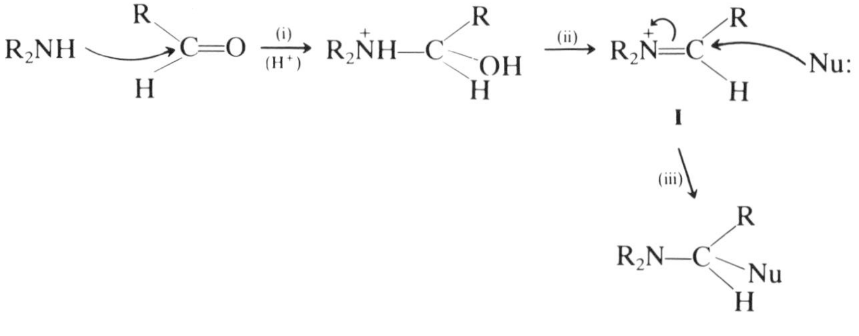

Comment upon the analogy between nucleophilic addition to the intermediate imidium ion, **I** (reaction iii) and that to carbonyl analogues. Which would be the more reactive?

Write mechanisms for the following reactions:

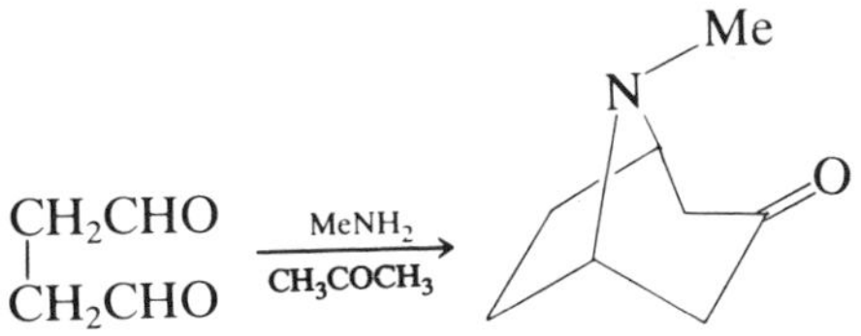

$$CH_2CHO \atop CH_2CHO \quad \xrightarrow[CH_3COCH_3]{MeNH_2}$$

Tropinone
(Robinson synthesis)

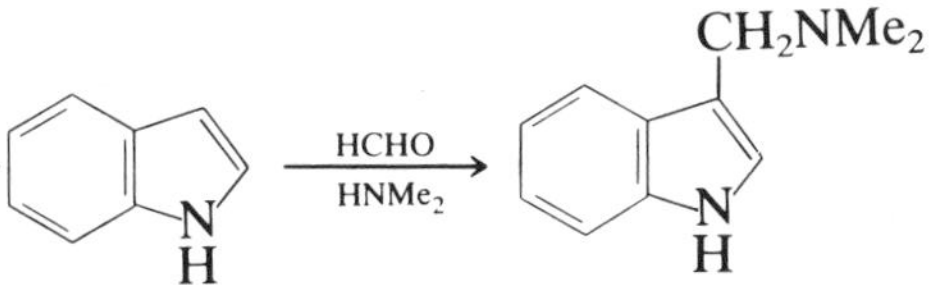

$$\xrightarrow[HNMe_2]{HCHO}$$

Gramine

$$\xrightarrow[\text{NH}]{HCHO} \quad ?$$

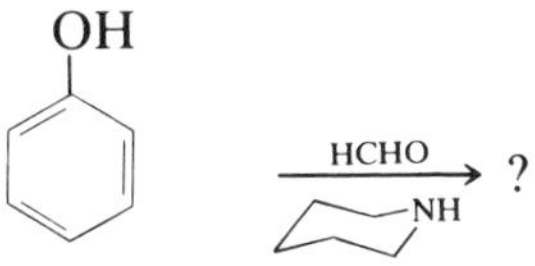

18 For the mechanisms set out in Fig. 10.18, suggest the most favorable structures for R,R′ and media which would be expected to promote the $B_{Ac}1$ and $B_{Al}1$ mechanisms.

19 4-Dimethylaminopyridine is a very efficient catalyst for trans-acylation:

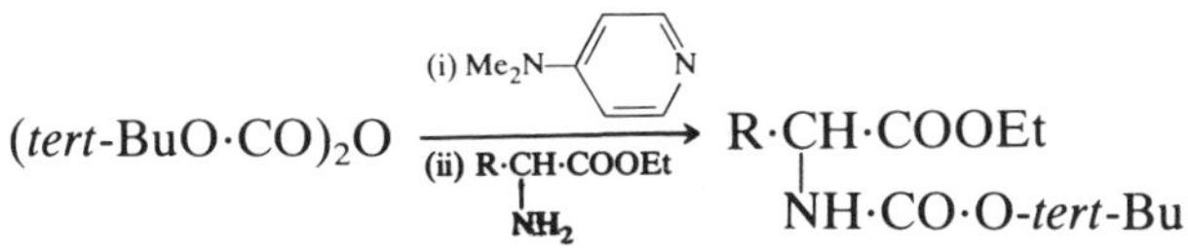

$$(tert\text{-BuO·CO})_2O \xrightarrow[\text{(ii) R·CH·COOEt} \atop \text{NH}_2]{\text{(i) Me}_2N-\text{pyridine}} R\cdot CH\cdot COOEt \atop NH\cdot CO\cdot O\text{-}tert\text{-Bu}$$

(protected aminoacid)

Suggest a mechanism for the involvement of this reagent. [*Synthesis*, 619 (1972).]

20 Thiocarbonyl chloride is a useful reagent for heterocyclic synthesis. What products would $S{=}CCl_2$ be expected to give with the following?

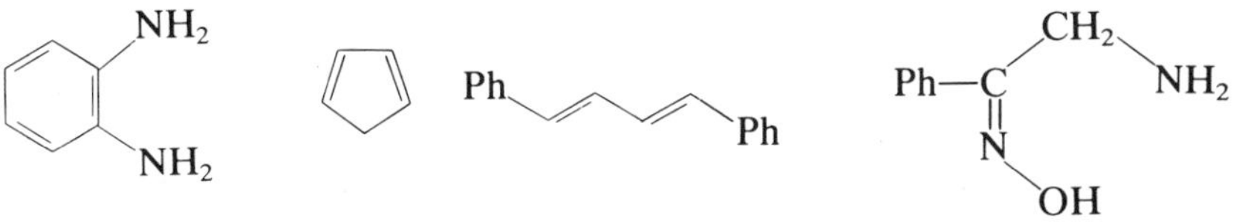

21 Treatment of (*R*)-butan-2-ol with phosgene, $O{=}CCl_2$, and thermal decomposition of the product resulted in the formation of (*R*)-2-chloro-butane. By which mechanism could this have occurred? [*J. Amer. Chem. Soc.*, **83**, 1955, 1959 (1961).]

REFERENCES

1 J. March, *Advanced Organic Chemistry*, McGraw-Hill, New York, 1977.
2 S. Winstein, *Quart. Rev.*, **23**, 141 (1969).
3 C. A. Bunton, *Nucleophilic Substitution at a Saturated Carbon Atom*, Elsevier, Amsterdam, 1963.
4 C. K. Ingold, *Structure and Mechanism in Organic Chemistry*, Bell, London, 1969.
5 D. N. Kevillc, *Chem. Comm.*, 421 (1981).
6 J. T. Burns and K. T. Leffek, *Can. J. Chem.*, **47**, 3725 (1969).
7 N. S. Isaacs, *Liquid Phase High Pressure Chemistry*, Wiley, Chichester, 1981.
8 S. Krishnamurthy and H. C. Brown, *J. Org. Chem.*, **47**, 276 (1982).
9 R. E. Parker and N. S. Isaacs, *Chem. Rev.*, 737 (1958).
10 J. Hine, *J. Amer. Chem. Soc.*, **72**, 2438 (1950).
11 L. A. Paquette, *J. Amer. Chem. Soc.*, **86**, 4096 (1964); R. F. Barsch, *J. Org. Chem.*, **34**, 627 (1969).
12 A. Streitweiser, *Solvolytic Displacement Reactions*, McGraw-Hill, New York, 1962.
13 T. W. Bentley and P. v. R. Schleyer, *Adv. Phys. Org. Chem.*, **14**, 1 (1977).
14 A. V. Willi and C. M. Won, *Can. J. Chem.*, **48**, 1452 (1970).
15 J. M. Harris, M. S. Paley and T. W. Presthofer, *J. Amer. Chem. Soc.*, **103**, 5915 (1981).
16 J. B. Stothers and A. N. Bourns, *Can. J. Chem.*, **40**, 2007 (1962).
17 H. Yamataka and T. Ando, *J. Phys. Chem.*, **85**, 2281 (1981).
18 E. D. Kaplan and E. R. Thornton, *J. Amer. Chem. Soc.*, **89**, 6644 (1967).
19 W. J. LeNoble and A. R. Miller, *J. Org. Chem.*, **44**, 889 (1979).
20 D. Graczyk, J. W. Taylor and C. R. Turnquist, *J. Amer. Chem. Soc.*, **100**, 7333 (1978).
21 H. Yamataka and T. Ando, *J. Amer. Chem. Soc.*, **101**, 266 (1979).
22 M. P. Friedberger, *Diss. Abs.*, 2222 (1975).
23 R. T. Hargreaves, *Diss. Abs.*, **36B**, 1709 (1975).
24 Brewster, *Topics in Stereochemistry* (N. L. Allinger, Ed.), Vol. 2, Interscience, New York, 1967.
25 D. P. Evans, V. G. Morgan and H. B. Watson, *J. Chem. Soc.*, 1173 (1935); E. D. Hughes, F. Juliusberger, S. Masterman, B. Topley and J. Weiss, *J. Chem. Soc.*, 1525 (1935); W. A. Cowdrey, E. D. Hughes, T. P. Nevell and C. L. Wilson, *J. Chem. Soc.*, 209 (1938); P. Brewster, F. Hiron, E. D. Hughes, C. K. Ingold and D. A. D. Rao, *Nature*, **166**, 178 (1950).
26 A. R. Stein, *J. Org. Chem.*, 519 (1976).
27 S. S. Shaik, *J. Amer. Chem. Soc.*, **103**, 3692 (1981); S. S. Shaik and A. Pross, *J. Amer. Chem. Soc.*, 3702.
28 N. L. Allinger, J. C. Tai and F. T. Wu, *J. Amer. Chem. Soc.*, **92**, 579, (1970).
29 J. Chandrasekar, S. F. Smith and W. L. Jorgensen, *J. Amer. Chem. Soc.*, **107**, 154, (1985).
30 W. Forster and R. M. Laird, *J. Chem. Soc., Perkin II*, 135 (1982).
31 H. H. Jaffe, *Chem. Rev.*, **53**, 191 (1953).
32 B. Bariou and M. Kerfanto, *Comptes. rend.*, **264C**, 1134 (1967).
33 D. Cook and A. J. Parker, *J. Chem. Soc. (B)*, 142 (1968).
34 P. Haberfield, L. Clayman and J. S. Cooper, *J. Amer. Chem. Soc.*, **91**, 787 (1969).
35 H. S. Golinkin, D. M. Parbloo and R. E. Robertson, *Can. J. Chem.*, **48**, 1296 (1970).
36 A. Pross and S. S. Shaik, *J. Amer. Chem. Soc.*, **103**, 3702 (1981).
37 T. M. Su, W. F. Sliwinsky and P. v. R. Schleyer, *J. Amer. Chem. Soc.*, **91**, 5386 (1969).

38 S. Hartman and R. E. Robertson, *Can. J. Chem.*, **38**, 2035 (1960).

39 D. N. Kevill, G. H. Johnson and W. Likhite, *Chem. Ind.*, 1555 (1969).

40 S. Winstein, B. Appel, R. Baker and A. Diaz, *Chem. Soc. Sp. Pub.* **19**, 109 (1965).

41 J. M. Harris, *Prog. Phys. Org. Chem.*, **11**, 89 (1974).

42 E. R. Thornton, *Solvolysis Mechanisms*, Ronald Press, New York, 1964.

43 D. J. Raber, J. M. Harris, and P. v. R. Schleyer, *Ions and Ion Pairs in Organic Chemistry* (M. Szwarc, Ed.), Wiley, New York, 1972.

44a T. W. Bentley and C. T. Bowen, *J. Chem. Soc., Perkin I*, 557 (1978).

44b F. P. Lossing and J. L. Holmes, *J. Amer. Chem. Soc.*, **106**, 6917 (1984).

45 T. W. Bentley, C. T. Bowen, W. Parker and C. I. F. Watt, *J. Amer. Chem. Soc.*, **101**, 2478 (1979).

46 I. Dostrovsky, E. D. Hughes and C. K. Ingold, *J. Chem. Soc.*, 173 and succeeding papers (1946).

47 T. W. Bentley and P. v. R. Schleyer, *J. Amer. Chem. Soc.*, **98**, 7658, 7667 (1976); **100**, 228 (1978); **101**, 2486 (1979).

48 K. T. Leffek, J. A. Llewellyn and R. E. Robertson, *Can. J. Chem.*, **38**, 2171 (1960).

49 K. T. Leffek, J. A. Llewellyn and R. E. Robertson, *Can. J. Chem.*, **38**, 1505 (1960).

50 J. A. Llewellyn, R. E. Robertson and J. M. W. Scott, *Can. J. Chem.*, **38**, 222 (1960).

51 J. O. Stoffer and J. D. Christen, *J. Amer. Chem. Soc.*, **92**, 3190 (1970).

52 S. L. Loukas, F. S. Varveri, M. R. Velkou and G. A. Gregoriou, *Tetrahedron Lett.*, 1803 (1971).

53 J. M. Harris, R. E. Hall and P. v. R. Schleyer, *J. Amer. Chem. Soc.*, **93**, 2551, 2553 (1971).

54 A. Streitweiser, R. H. Jagow, R. C. Fahey and S. Susuki, *J. Amer. Chem. Soc.*, **80**, 2326 (1958).

55 T. W. Bentley and P. v. R. Schleyer, *Tetrahedron Lett.*, 2335 (1974).

56 V. P. Vitullo, J. Grabowski and S. Sridharan, *Chem. Comm.*, 737 (1981);

57 T. W. Bentley and C. T. Brown, *J. Chem. Soc., Perkin II*, 557 (1978).

58 M. A. Abraham, D. H. Buisson and R. A. Schutz, *Chem. Comm.*, 693, (1975).

59 W. G. Young, S. Winstein and H. L. Goering, *J. Amer. Chem. Soc.*, **73**, 1958 (1951).

60 H. L. Goering, E. G. Briody and J. F. Levy, *J. Amer. Chem. Soc.*, **85**, 1257 (1963).

61 H. L. Goering and S. Chang, *Tetrahedron Lett.*, 3607 (1965).

62 S. P. McManus, K. K. Safavy and F. E. Roberts, *J. Org. Chem.*, **47**, 4388 (1982).

63 S. Winstein, E. Clippinger, A. H. Fainberg and G. C. Robinson, *J. Amer. Chem. Soc.*, **76**, 2597 (1954); S. Winstein and G. C. Robinson, *J. Amer. Chem. Soc.*, **80**, 169 (1958).

64 V. P. Vitullo and F. P. Wilgis, *J. Amer. Chem. Soc.*, **97**, 458 (1975); **103**, 1982 (1981).

65 H. Maskill, J. T. Thompson and A. A. Wilson, *Chem. Comm.*, 1239 (1981).

66 R. Fujiyama, S. Kiyooka, K. Funatsu, M. Fujio and Y. Tsuno, *Mem. Fac. Sci. Kyushu Univ.*, **13C**, 117 (1981).

67 C. G. Swain, C. B. Scottand, K. M. Lohmann, *J. Amer. Chem. Soc.*, **75**, 136 (1953).

68 J. L. Fry, C. J. Lancelot, L. K. M. Lom, J. M. Harris, R. C. Bingham, D. J. Raber, R. E. Hall and P. v. R. Schleyer, *J. Amer. Chem. Soc.*, **92**, 2539 (1970).

69 R. D. Howells and J. D. McCown, *Chem. Rev.*, **77**, 69 (1977).

70 P. v. R. Schleyer, J. L. Fry, L. K. M. Lom and C. J. Lancelot, *J. Amer. Chem. Soc.*, **92**, 2543 (1970).

71 P. v. R. Schleyer, *Adv. Alicyc. Chem.*, **1**, 283 (1966).

72 R. C. Bingham and P. v. R. Schleyer, *J. Amer. Chem. Soc.*, **93**, 3189 (1971).

73 P. v. R. Schleyer and R. D. Nicholas, *J. Amer. Chem. Soc.*, **83**, 2700 (1961).

74 G. A. Olah, *J. Amer. Chem. Soc.*, **107**, 2764 (1985).

75 T. T. Tidwell, *Ang. Chem., Int. Ed.*, **23**, 20 (1984).

76 C. J. Lancelot, J. J. Harper and P. v. R. Schleyer, *J. Amer. Chem. Soc.*, **91**, 4294 (1969).

77 Y. Yukawa, H. Morisaki, K. Tsuji, S. Kim and T. Ando, *Tetrahedron Lett.*, **22**, 5187 (1981).

78 Y. Seki, M. Fujio, M. Mishima and Y. Tsuno, *Mem. Fac. Sci, Kyushu Univ.*, **12C**, 197 (1980).
79 M. Fujio, Y. Seki, R. Fujiyama, M. Mishima and Y. Tsuno, *Mem. Fac. Sci, Kyushu Univ.*, **13C**, 71 (1981).
80 G. D. Sargent, *Quart. Rev.*, 301 (1966).
81 I. L. Reich, A. Diaz and S. Winstein, *J. Amer. Chem. Soc.*, **91**, 5636 (1969).
82 S. Winstein, B. K. Morse, E. Grunwald, K. C. Scheiber and J. Corse, *J. Amer. Chem. Soc.*, **74**, 1114 (1952).
83 R. L. Heppolette and R. E. Robertson, *Proc. Roy. Soc.*, **252A**, 273 (1959); R. D. Moelwyn-Hughes, R. E. Robertson and S. Sugamori, *J. Chem. Soc.*, 1965 (1965); *J. Amer. Chem. Soc.*, **91**, 5637, 5639 (1969); G. R. Cowie, H. J. M. Fitches and G. Kohnstam, *J. Chem. Soc.*, 1585 (1963); 3187 (1955).
84 R. Heck and S. Winstein, *J. Amer. Chem. Soc.*, **79**, 3106 (1957); S. W. Lundegren, H. Marshall and L. L. Ingraham, *J. Amer. Chem. Soc.*, **75**, 147 (1953).
85 T. Asano and W. J. LeNoble, *Chem. Rev.*, **78**, 407 (1978).
86 R. K. Mohanty and R. E. Robertson, *Can. J. Chem.*, **55**, 1319 (1977).
87 P. M. Laughton and R. E. Robertson, *Can. J. Chem.*, **37**, 1491 (1959).
88 P. W. C. Barnard and R. E. Robertson, *Can. J. Chem.*, **39**, 881 (1961).
89 G. A. Hamilton and R. E. Robertson, *Can. J. Chem.*, **37**, 967 (1959).
90 J. Albery and B. N. Robinson, *Trans. Farad. Soc.*, **65**, 980 (1969).
91 M. J. Blandamer, J. Burgess, P. P. Duce, R. E. Robertson and J. M. W. Scott, *J. Chem. Soc., Faraday I*, **77**, 1999 (1981)
92 M. J. Blandamer, J. Burgess, P. P. Duce, R. E. Robertson and J. M. W. Scott, *Chem. Comm.*, 12 (1981).
93 M. J. Blandamer, R. E. Robertson, P. D. Golding, J. M. McNeil, M. Joseph and J. M. W. Scott, *J. Amer. Chem. Soc.*, **103**, 2415 (1981).
94 T. W. Bentley, C. T. Bowen, D. M. Morton and P. v. R. Schleyer, *J. Amer. Chem. Soc.*, **103**, 5466 (1981).
95 E. S. Lewis and C. E. Boozer, *J. Amer. Chem. Soc.*, **74**, 308 (1952); **83**, 1955, 1959 (1961).
96 W. M. Olmstead and J. I. Braumann, *J. Amer. Chem. Soc.*, **99**, 4219 (1977).
97 M. J. Pellerite and J. I. Brauman, *J. Amer. Chem. Soc.*, **105**, 2672 (1983).
98 G. Caldwell, T. F. Magnera and P. Kebarle, *J. Amer. Chem. Soc.*, **106**, 959 (1984).
99 M. J. Pellerite and J. I. Brauman, *J. Amer. Chem. Soc.*, **102**, 5993 (1980).
100 S. S. Shaik, *J. Amer. Chem. Soc.*, **106**, 1227 (1984).
101 D. G. Hall, C. Gupta and T. H. Morton, *J. Amer. Chem. Soc.*, **103**, 2416 (1981).
102 D. K. Bohme and A. B. Raksit, *J. Amer. Chem. Soc.*, **106**, 3447 (1984).
103 M. H. Abraham, in *Comprehensive Chemical Kinetics* (C. H. Bamford and C. F. M. Tipper, Eds), Vol. 12, Elsevier, Amsterdam, 1972.
104 E. D. Hughes, C. K. Ingold, F. G. Thorpe and H. C. Volge, *J. Chem. Soc.*, 1133, 1142 (1961).
105 R. M. G. Roberts, *J. Chem. Soc.*, 3900 (1964).
106 M. Gielen and J. Nasielski, *Bull. Soc. Chim. Belg.* **71**, 32 (1962).
107 G. Modena and U. Tonellato, *Adv. Phys. Org. Chem.*, **9**, 185 (1971).
108 Z. Rappoport, *Acc. Chem. Res.*, **14**, 7 (1981); L. R. Subramanian and M. Hanack, *J. Chem. Ed.*, **52**, 520 (1975).
109 J. Hine, *J. Amer. Chem. Soc.*, **88**, 5525 (1966).
110 G. A. Olah, *J. Amer. Chem. Soc.*, **86**, 1039, 1044 (1964).
111 G. A. Olah, *J. Amer. Chem. Soc.*, **86**, 1046, 1060 (1964); P. Stang, *Acc. Chem. Res.*, **11**, 107 (1978).
112 Z. Rappoport, T. Bassler and M. Hanack, *J. Amer. Chem. Soc.*, **92**, 4985 (1970); G. A. Olah, *J. Amer. Chem. Soc.*, **86**, 1055, (1964).
113 G. A. Olah, *J. Amer. Chem. Soc.*, **86**, 1065 (1964).
114 R. O. C. Norman and R. Taylor, *Electrophilic Substitution in Benzenoid Compounds*, Elsevier, Amsterdam, 1965.

115 C. K. Ingold *Structure and Mechanism in Organic Chemistry*, Bell, London, 1969.

116 E. Berliner, *Prog. Phys. Org. Chem.*, **2**, 253 (1964).

117 L. M. Stock and H. C. Brown, *Adv. Phys. Org. Chem.*, **1**, 35 (1963).

118 E. Baciocci and G. Illuminati, *Prog. Phys. Org. Chem.*, **5**, 1 (1967).

119 K. Schofield, *Aromatic Nitration*, Cambridge U.P., 1980.

120 P. B. D. De la Mare, *Electrophilic Halogenation*, Cambridge U.P., 1976.

121 J. M. Brittain, P. B. D. De la Mare, N. S. Isaacs and P. D. McIntyre, *Tetrahedron Letts*, 4835 (1976).

122 L. M. Stock, *Prog. Phys. Org. Chem.*, **12**, 2 (1976).

123 P. B. D. De la Mare and J. H. Ridd, *Aromatic substitution—Nitration and Halogenation*, Butterworth, London, 1959.

124 C. L. Perrin, *J. Org. Chem.*, 420 (1971).

125 L. Melander, *Ark. Kim.*, **2**, 211 (1950).

126 T. G. Bonner, F. Bowyer and G. Williams, *J. Chem. Soc.*, 2650 (1953).

127 G. W. Wheland, *J. Amer. Chem. Soc.*, **64**, 900 (1942).

128 P. Pfeiffer and R. Wizinger, *Ann.*, **461**, 132 (1928).

129 F. G. Bordwell, *J. Amer. Chem. Soc.*, **82**, 3588 (1960); A. Fisher, A. J. Read and J. Vaughan, *J. Chem. Soc.*, 3692 (1964).

130 S. Sivakamasundarai and R. Ganesan, *Int. J. Chem. Kinet.*, **12**, 837 (1980).

131 D. M. Brouwer, E. L. Mackor and C. MacLean, *Carbonium Ions* (G. A. Olah and P. v. R. Schleyer, Eds), Vol. 2, Chap. 20, Wiley, New York, 1970.

132 H. Zollinger, *Adv. Phys. Org. Chem.*, **2**, 163 (1963).

133 J. K. Bosscher and H. Cerfontain, *J. Chem. Soc. (B)*, 1524 (1968).

134 M. Kobayashi, K. Honda and A. Yamaguchi, *Tetrahedron Lett.*, 487 (1968).

135 R. Taylor, 'Kinetics of electrophilic substitution', in *Comprehensive Chemical Kinetics* (C. H. Bamford and C. F. A. Tipper, Eds), Vol. 13, Chap. 1, Elsevier, Amsterdam, 1972.

136 H. W. Gibbs, L. Main, R. B. Moodie and K. Schofield, *J. Chem. Soc., Perkin II*, 848 (1981).

137 P. F. Christy, J. H. Ridd and N. D. Stears, *J. Chem. Soc. B*, 797 (1970).

138 L. Main, R. B. Moodie and K. Schofield, *Chem. Comm.*, 48 (1982).

139 W. J. Hehre, R. W. Taft and R. D. Topsom, *Prog. Phys. Org. Chem.*, **12**, 159 (1976).

140 N. S. Isaacs and D. Cvitas, *Tetrahedron*, **27**, 439 (1971).

141 S. Clementi, F. Genel and G. Marino, *Chem. Comm.*, 498 (1967).

142 P. Linda and G. Marino, *Chem. Comm.*, 499 (1967). R. O. C. Norman and G. K. Radda, *J. Chem. Soc.*, 3030 (1061).

143 D. A. Simpson, S. G. Smith and P. Beale, *J. Amer. Chem. Soc.*, **92**, 1072 (1970).

144 N. L. Allinger, J. Tai and F. T. Wu, *J. Amer. Chem. Soc.*, **92**, 579 (1970).

145 W. v. E. Doering, H. G. Sanders, H. W. Boyton, E. F. Earhart and W. R. Edwards, *Tetrahedron*, **4**, 178 (1958).

146 G. A. Olah, *Carbocations and Electrophilic Reactions*, Verlag Chemie, Weinheim, 1973.

147 D. M. Brouwer, E. L. Mackor and C. MacLean, *Carbonium Ions*, **2**, 837, 1970.

148 D. M. Brouwer, *Rec. Trav. Chim.*, **87**, 611 (1968).

149 J. M. Brittain, P. B. D. De la Mare, N. S. Isaacs and P. D. McIntyre, *J. Chem. Soc., Perkin II*, 933 (1973).

150 C. F. Bernasconi, *MTP Int. Rev. Sci; Org. Chem. Ser 1*, **3**, 33 (1973); C. Perrin and G. A. Skinner, *J. Amer. Chem. Soc.*, **93**, 3389 (1971).

151 J. Miller, *Aromatic Nucleophilic Substitution*, Elsevier, London, 1968.

152 S. D. Ross, *Prog. Phys. Org. Chem.*, **1**, 31 (1963).

153 J. F. Bunnett and R. E. Zahler, *Chem. Abs.*, **49**, 273 (1951).

154 F. Pietra, *Quart. Rev.*, **23**, 504 (1969).

155 A. L. Beckwith, G. D. Lehay and J. Miller, *J. Chem. Soc.*, 3552 (1952).

156 J. F. Bunnett and N. S. Nudelman, *J. Org. Chem.*, **34**, 2038 (1969).

157 J. F. Bunnett, *Quart. Rev.*, **12**, 1 (1958).

158 D. L. Miller, J. O. Lay and M. L. Gross, *Chem. Comm.*, 970 (1982).
159 J. D. Roberts, C. W. Vaughan, L. A. Carlsmith and L. Semenov, *J. Amer. Chem. Soc.*, **78**, 611 (1956); **79**, 601 (1956).
160 J. F. Bunnett, E. W. Garbisch and K. M. Pruitt, *J. Amer. Chem. Soc.*, **79**, 385 (1957).
161 B. Wakefield, *J. Chem. Soc. C*, 72 (1967).
162 G. Bartoli and P. G. Todesco, *Acc. Chem. Res.*, **11**, 125 (1977).
163 C. F. Bernasconi, *Acc. Chem. Res.*, **11**, 147 (1978).
164 D. M. Bevis, N. B. Chapman, J. S. Paine, J. Shorter and D. J. Wright, *J. Chem. Soc., Perkin II*, 1787 (1974).
165 J. F. Bunnett and J. J. Randall, *J. Amer. Chem. Soc.*, **80**, 6020 (1958).
166 J. F. Bunnett and R. H. Garst, *J. Amer. Chem. Soc*, **87**, 3875 (1965).
167 O. Banjoko and K. U. Rahman, *J. Chem. Soc., Perkin II*, 1105 (1981).
168 F. Pietra, *Tetrahedron Lett.*, 2405 (1965).
169 I. R. Bellobono, P. Beltrame, M. G. Cattania and M. Simonetta, *Tetrahedron Lett.*, 2673 (1968).
170 A. Meisenheimer, *Ann.*, **323**, 205 (1902).
171 M. J. Strauss and R. G. Johanson, *Chem. Ind.*, 242 (1969).
172 G. G. Messmer and G. J. Palenik, *Chem. Comm.*, 470 (1969).
173 S. Sekiguchi, T. Aizawa and M. Aoki, *J. Org. Chem.*, **46**, 3657 (1981).
174 M. L. Crossley, R. H. Kienle and C. H. Benbrook, *J. Amer. Chem. Soc.*, **62**, 1400 (1940).
175 T. L. Gilchrist and C. W. Rees, *Carbenes, Nitrenes and Arynes*, Nelson, London, 1969.
176 P. Caubere, *Acc. Chem. Res.*, **7**, 301 (1974).
177 J. D. Roberts, H. E. Simmons, L. A. Carlsmith and C. W. Vaughan, *J. Amer. Chem. Soc.*, **75**, 3290 (1953).
178 G. Wittig and L. Pohmer, *Ber.*, **89**, 1334 (1956).
179 H. C. VanderPlas, *Acc. Chem. Res.*, **11**, 462 (1978).
180 E. Berliner and L. C. Monack, *J. Amer. Chem. Soc.*, **74**, 1574 (1952); J. F. Bunnett, H. Moe and D. Knutson, *J. Amer. Chem. Soc.*, **76**, 3936 (1954); J. F. Bunnett and G. T. Davis, *J. Amer. Chem. Soc.*, **76**, 3011 (1954).
181 T. Okamoto and H. C. Brown, *J. Org. Chem.*, **22**, 485 (1957).
182 D. P. N. Satchell and R. S. Satchell, Eds, *The Chemistry of Carboxylic Acids and Esters*, Interscience, New York, 1969.
183 R. J. E. Talbot, Carbonyl Substitution Reactions, in *Comprehensive Chemical Kinetics* (C. H. Bamford and C. J. E. Tipper, Eds), Vol. 10, Elsevier, Amsterdam, 1972.
184 A. J. Kirby, in *Comprehensive Chemical Kinetics* (C. H. Bamford and C. J. E. Tipper, Eds), Vol. 10, Elsevier, Amsterdam, 1972.
185 W. Jencks, *Catalysis in Chemistry and Enzymology*, McGraw-Hill, New York, 1969.
186 S. L. Johnson, *Adv. Phys. Org. Chem.*, **5**, 237 (1967).
187 W. Jencks, *Prog. Phys. Org. Chem.*, **2**, 129 (1965).
188 M. L. Bender, *Chem. Rev.*, **60**, 53 (1960).
189 C. A. Bunton and T. Hadwick, *J. Chem. Soc.*, 3248 (1958); 943 (1961).
190 S. C. Datta, J. N. E. Day and C. K. Ingold, *J. Chem. Soc.*, 838 (1939).
191 M. L. Bender, *J. Amer. Chem. Soc.*, **73**, 1626 (1951); M. L. Bender and R. J. Thomas, *J. Amer. Chem. Soc.*, **83**, 4189 (1961).
192 C. A. Bunton and N. D. Spatcher, *J. Chem. Soc.*, 1079 (1956).
193 B. Holmberg, *Chem. Ber.*, **45**, 2997 (1912).
194 O. R. Quayle and H. M. Norton, *J. Amer. Chem. Soc.*, **62**, 1170 (1940).
195 H. H. Jaffe, *Chem. Rev.*, **53**, 191 (1953).
196 B. Zerner and M. L. Bender, *J. Amer. Chem. Soc.*, **83**, 2267 (1961).
197 R. A. McClelland and G. Patel, *J. Amer. Chem. Soc.*, **103**, 6912 (1981).
198 W. P. Jencks and J. Carriolo, *J. Amer. Chem. Soc.*, **83**, 1743 (1968).
199 S. L. Johnson, *Adv. Phys. Org. Chem.*, **5**, 237 (1967).
200 B. Capon, A. K. Ghosh and D. M. A. Grieve, *Acc. Chem. Res.*, **14**, 306 (1981); B. Capon, M. I. Dosunmu and M. Sanzhez, *Adv. Phys. Org. Chem.*, **21**, 37 (1985).

201 W. P. Jencks and M. Gilchrist, *J. Amer. Chem. Soc.*, **86,** 5616 (1964).
202 D. R. Robinson, *J. Amer. Chem. Soc.*, **92,** 3138 (1970).
203 W. P. Jencks, *Catalysis in Chemistry and Enzymology*, McGraw-Hill, New York, 1969.
204 H. Zimmerman and J. Rudolph, *Angew. Chem., Int. Ed.*, **4,** 40 (1965).
205 C. A. Bunton and T. Hadwick, *J. Chem. Soc.*, 3043 (1957); 943 (1961); 3248 (1958).
206 G. A. Olah, *Carbocations and Electrophilic Reactions*, Verlag Chemie, Weinheim, 1973.
207 P. W. Taft, *Steric Effects in Organic Chemistry* (M. Newman, Ed.) Chap. 13, Wiley, New York, 1956.
208 P. Deslongchamps, *Tetrahedron*, **31,** 2463 (1976).
209 K. Yates, *Acc. Chem. Res.*, **4,** 136 (1971).
210 K. Yates and R. A. McClelland, *J. Amer. Chem. Soc.*, **89,** 2686 (1967).
211 D. N. Kershaw and J. A. Leisten, *Proc. Chem. Soc.*, 84 (1960).
212 M. L. Bender, H. Ladenheim and M. C. Chen, *J. Amer. Chem. Soc.*, **83,** 123 (1961).
213 J. F. Bunnett, *J. Amer. Chem. Soc.*, **83,** 4956, 4968 (1961).
214 C. E. Meyer and J. F. Bunnett, *J. Amer. Chem. Soc.*, **85,** 1891 (1963).
215 A. G. Davies and J. Kenyon, *Quart. Rev.*, **9,** 203 (1955).
216 K. R. Adam, I. Lander and V. R. Stimson, *Austr. J. Chem.*, **15,** 467 (1962).
217 C. J. Gillan, A. C. Knipe and W. E. Watts, *Tetrahedron Lett.*, **22,** 597 (1981).
218 J. F. Bunnett, M. M. Robinson and F. C. Pennington, *J. Amer. Chem. Soc.*, **72,** 2378 (1950).
219 C. A. Bunton, E. D. Hughes, C. K. Ingold and D. F. Leigh, *Nature*, **166,** 680 (1950).
220 A. R. Olson and J. R. Miller, *J. Amer. Chem. Soc.*, **60,** 2687 (1938).
221 C. A. Bunton, C. O'Connor and T. A. Turney, *Chem. Ind.*, 1835 (1967).
222 A. Bruylants and F. Kezdy, *Rec. Chem. Progr.*, **21,** 213 (1960).
223 A. Queen, *Can. J. Chem.*, **45,** 1619 (1967).
224 E. W. Crundon and R. F. Hudson, *J. Chem. Soc.*, 501 (1956).
225 D. P. N. Satchell, *J. Chem. Soc.*, 3724 (1964).
226 V. Gold, J. Hilton and E. G. Jefferson, *J. Chem. Soc.*, 2756 (1954).
227 I. Ugi and F. Beck, *Chem. Ber.*, **94,** 1839 (1961).
228 C. A. Bunton and S. G. Perry, *J. Chem. Soc.*, 3070 (1960).
229 A. D. Buckingham, *J. Amer. Chem. Soc.*, **97,** 5915 (1975).
230 V. Gold, F. Hilton and E. G. Jefferson, *J. Chem. Soc.*, 2756 (1954).
231 V. Gold and J. Hilton, *J. Chem. Soc.*, 843 (1955).
232 R. E. Barnett, *Acc. Chem. Res.*, **6,** 41 (1973).
233 J. P. Guthrie, *J. Amer. Chem. Soc.*, **95,** 6999 (1973); **96,** 3608 (1974); *Acc. Chem. Res.*, **16,** 122 (1983).
234 B. Capon, A. K. Ghosh and D. M. A. Grieve, *Acc. Chem. Res.*, **14,** 306 (1981).
235 B. Capon and D. M. A. Grieve, *J. Chem. Soc., Perkin II*, 300 (1980); B. Capon and A. K. Ghosh, *J. Amer. Chem. Soc.*, **103,** 1765 (1981).
236 R. A. McClelland and L. J. Santry, *Acc. Chem. Res.*, **16,** 394 (1983).
237 T. C. Bruice and G. L. Schmir, *J. Amer. Chem. Soc.*, **79,** 1663 (1957); **80,** 148 (1958).
238 C. A. Bunton and J. H. Fendler, *J. Org. Chem.*, **32,** 1547 (1967).
239 A. Rahm and M. Pereyre, *J. Amer. Chem. Soc.*, **99,** 1672 (1977).
240 J. M. Fukui and F. R. Jensen, *Acc. Chem. Res.*, **16,** 177 (1983).

11 Elimination reactions

Olefin-forming eliminations encompass a wide variety of mechanistic types occurring under conditions ranging from strongly basic solution to gas-phase high-temperature reaction[1-8].

11.1 BASE-PROMOTED ELIMINATIONS IN SOLUTION

Polar eliminations include reactions in which an electrophilic and a nucleophilic leaving group depart from the same molecule.

$$E \overset{\frown}{\underset{|\beta}{C}} \overset{\frown}{\underset{|\alpha}{C}} Nu \longrightarrow E^+ \quad \overset{\backslash}{\underset{/}{C}} = \overset{/}{\underset{\backslash}{C}} \quad :Nu^-$$

If they are on adjacent atoms, usually but not necessarily carbon, an olefinic double bond results (β-elimination): if the leaving groups are further removed, a cyclic product is formed (γ- and δ-eliminations; see Chapter 13). Alkene-forming eliminations are frequently assisted by base since the electrophilic leaving group is usually a proton. The nucleophilic leaving groups include those which have been shown to take part in S_N reactions: —Hal, —OSO$_2$Ar, —NR$_3^+$, —SR$_2^+$, —OR$_2$ and, less frequently, —OCO·R, —OAr, —OR, —CN. Although eliminations usually occur under basic conditions, in which case the function of the base is to remove the β-proton, acid catalysis also occurs, by protonation of the nucleophilic leaving group (e.g. —OH to —OH$_2^+$). Therefore the compounds which undergo eliminations, for example alkyl halides and sulfonates, and the conditions also are those which are conducive to nucleophilic substitutions. Consequently elimination and substitution pathways frequently occur together, leading to mixtures of products. The base has two sites open to attack, the α-carbon (S_N2) or the β-hydrogen, leading to elimination (1). Solvents which are commonly used are normally polar (alcohols, water or DMSO), but the reactions will occur in non-polar media often with significant changes in the products. Strong hard bases are usually required — alkoxide or hydroxide — but in some circumstances weak bases which are carbon nucleophiles such as Cl$^-$ will bring about eliminations[9,10].

504

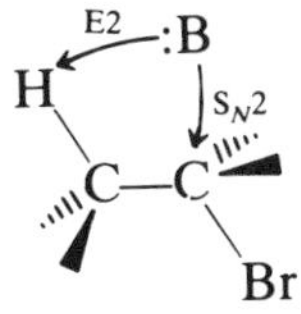

E2   :B

H

S_N2

C—C

Br

1

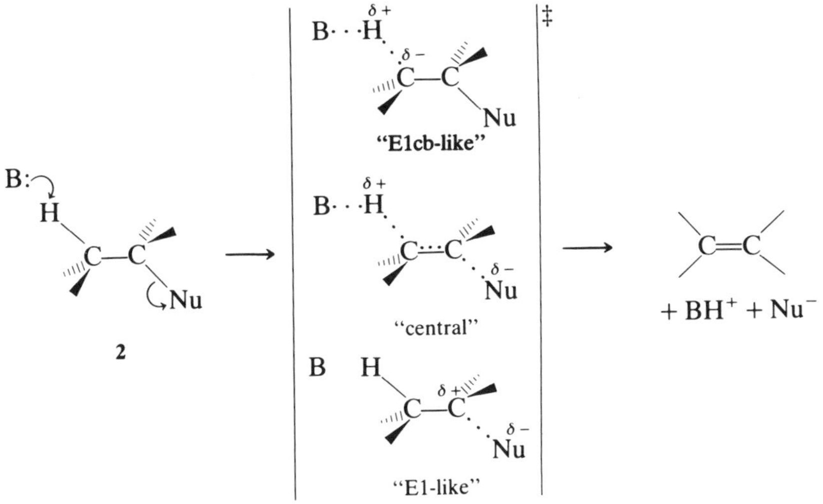

B···H $\delta+$

$\delta-$

C—C

Nu

"E1cb-like"

B:

H

C—C

Nu

2

B···H $\delta+$

C—C $\delta-$

Nu

"central"

B   H

C—C $\delta+$

$\delta-$

Nu

"E1-like"

‡

C=C

+ BH$^+$ + Nu$^-$

With so many bonds to be made and broken it is hardly surprising that these reactions are mechanistically diverse. The main division of mechanistic types is made according to the precedence of bond fission of C–Nu and of C–E bonds. Three extreme situations may be recognized (Scheme 2 p. 505).

(a) E1 reactions (unimolecular elimination): The rate-determining step is C–Nu bond fission, i.e. ionization of a nucleophilic group from carbon. This is identical to that of S_N1 solvolytic reactions (Section 10.1). Products of elimination may arise from each of the intermediates, the ion-pairs or free carbocation, in competition with substitution products. As far as transition state properties are concerned, therefore, this reaction has been discussed in Section 10.1 and only the product-determining stages remain to be considered.

(b) E2 reactions (bimolecular elimination): This group of mechanistic types is characterized by synchronous fission or C–Nu and C–E bonds brought about by attack of the base at the β-H and forming the alkene in a single step. A further distinction may be made according to whether the base coordinates solely to the proton (E2H) or initially to the α-carbon (E2C).

(c) E1cb reactions[11–13] (unimolecular elimination from the conjugate base): The initial step is β-C–H bond fission, which may be

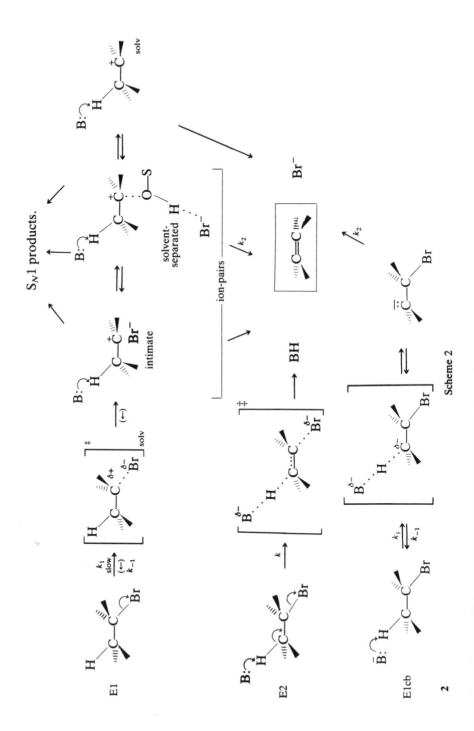

Scheme 2

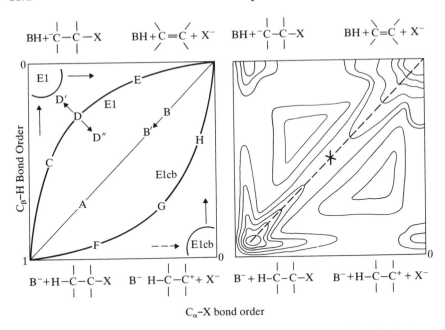

Fig. 11.1 **a**, Energy contours for E2 eliminations. The diagonal route indicated is the 'central' pathway with equal degrees of fission of C_β—H and C_α—X at the transition state. **b**, Jencks–More O'Ferrall plot for the E2 reaction showing divergence from 'central' pathway towards either 'E1-like' or 'E1cb-like' transition states.

rate-determining or may be reversible. The expulsion of Nu: occurs subsequently. These categories of mechanism are not clearly distinct[14]: the evidence points to a merging of mechanism so that an E2 reaction may be designated 'E1cb-like' (or '*paene*-carbanion') or 'E1-like' (*paene*-carbocation') according to whether C–E bond fission has progressed ahead of or lags behind C–Nu bond fission at the transition state[15] (**2**, Fig. 11.1).

11.1.1 Kinetic criteria of mechanism

Table 11.1 sets out some experimental criteria by which these reaction types might be differentiated.

E2 reactions may occur whether the nucleophilic leaving group is at a primary, secondary or even a tertiary center, and are of first order each in substrate and base; for instance,

$$PhCH_2CH_2Br + OEt^- \rightarrow PhCH{=}CH_2 + EtOH + Br^-;$$

$$Rate = k_2[PhCH_2CH_2Br][OEt^-].$$

The empirical rate law for E1 reactions is the same as that for S_N1 (Eq. 10.2) and does not include [base].

Table 11.1 Criteria for distinguishing elimination mechanisms

Mechanism	Kinetic order in:		β-proton exchange?	Base catalysis[a]	Isotope effect		ρ	
	Substrate	Base			β-H	Nu	C_α-subst.	C_β-subst.
E1	1	0	no	no	no	yes	large −	—
E2	1	1	no	gen.	2–8	yes	small + or −	small
(E1cb)$_R$	1	1	yes	sp.	no	yes	small −	small −
(E1cb)$_I$	1	1	no	gen. or sp.	2–8	no	nil	+
(E1cb)$_{anion}$	1	0	yes	gen.	no	yes	—	+

E1cb reactions offer some variety in kinetic equations[16]; three situations may be recognized:

$$B: + H-\overset{|}{\underset{|}{C}}-\overset{|}{\underset{|}{C}}-X \underset{k_{-1}}{\overset{k_1}{\rightleftharpoons}} BH^+ + \overset{|}{\underset{|}{\overset{..}{C}}}-\overset{|}{\underset{|}{C}}-X \overset{k_2}{\longrightarrow} \overset{|}{\underset{|}{C}}=\overset{|}{\underset{|}{C}} + X^-;$$

$$\text{Rate} = k_2[C^-] = \frac{k_1 k_2 [RX][B:]}{k_{-1}[BH^+] + k_2}.$$

(i) $k_2 \gg k_{-1}[BH^+]$—(E1cb)$_I$ ('irreversible'); ionization is not reversible and consequently $k_{-1} \ll k_1$:

$$\text{Rate} = k_1[RX][B:],$$

as for an E2 process.

(ii) $k_1[BH^+] \gg k_2$—(E1cb)$_R$ ('reversible'); the intermediate carbanion mostly returns to the initial state; $k_{-1} \gg k_2$:

$$\text{Rate} = k_1 k_2 [RX][B:]/k_{-1}[BH^+].$$

The rate varies with $[BH^+]$ and proton exchange with the solvent is rapid.

(iii) $k_1 \gg k_{-1}$, k_2—(E1)$_{anion}$; the carbanion is particularly stable and accumulates, decomposing unimolecularly in the rate-determining step:

$$\text{Rate} = k[RX],$$

and becomes independent of $[B:]$ at a value at which all substrate is converted to its conjugate base.

11.1.2 Structural effects on rates of elimination[17]

Rates of E2 elimination in simple alkyl halides are not greatly affected by alkyl substitution (Table 11.2). One needs to partition overall rates into those for each isomer. When the structure of the substrate affords two or more isomeric products there is a preference for elimination towards a more substituted center, but overall there is only a mild accelerating effect by alkyl groups adjacent to the developing double bond. There is a stronger driving force towards elimination adjacent to an aryl group, evidently a reflexion of product stability through conjugation.

Hammett relationships have been obtained for many 2-arylethyl systems reacting with a variety of bases and leaving groups (Table 11.3). Correlations of $\log k$ with σ°- and ρ-values are positive, moderate to large, lying in the range $+1\cdot5$ to $+4$. All are E2 reactions but the differences in sensitivity to electronic perturbation are quite marked. This can be attributed to varying amounts of carbanion character at the β-carbon in the transition state. The higher ρ, the more 'E1cb-like' is the reaction; $-R$ groups assist the reaction by stabilizing negative charge on the β-carbon atom. Leaving groups affect electronic demands on the reaction; the less labile the leaving group, the more negative the charge on the β-carbon which is needed to expel it. Consequently values of ρ for halide leaving groups are in the order $F > Cl > Br > I$ and the largest value of ρ in the Table is for —SOMe, a notably difficult group to replace and one which indicates a very 'E1cb-like' reaction. Leaving group ability (*nucleofugacity*) can be assessed in this way (Section 6.12) and would place toluenesulfonate between Cl and Br.

Electronic demands at the α-carbon are smaller than at the β-. Indeed, an 'E1-like' transition state might be expected to show a negative value of ρ (or, usually, ρ^+). There is evidently cancelation of effects similar to that deduced in S_N2 reactions, resulting in a small positive value for E2 reactions; negative values of ρ^* for fully E1 reactions, correlating with σ^+, are those for the S_N1 reaction (Tables 4.4 and 10.7). Reaction constants for leaving groups (—OSO_2Ar) are positive but quite small and may give curved Hammett plots on account of changes in the position of the transition state within the E2 spectrum. Multicorrelation analysis of rates of elimination with the 'electronic' parameters σ_I, σ_R and the steric parameter,

Table 11.2   Reactivity in E2 eliminations[2]

(a) Relative rates of elimination from alkyl bromides by ethoxide in ethanol
Each figure represents rate relative to 2-bromopropane for elimination towards the hydrogen indicated. Regiospecificity shown by underline.

Alkyl bromide	k_{rel}	Alkyl bromide	k_{rel}
CH_3CH_2Br	0·21	$CH_3CH_2CHBrCH_2CH_3$	1·70
$CH_3CHBrCH_3$	1·00	$CH_3CBr(CH_3)_2$	8·5
$CH_3CH_2CHBrCH_3$	2·38	$CH_3CH_2CBr(CH_3)_2$	35
$CH_3CH_2CHBrCH_3$	0·55	$CH_3CH_2CBr(CH_3)_2$	7·2
$CH_3CH_2CH_2CHBrCH_3$	1·66	$PhCH_2CH_2Br$	21
$CH_3CH_2CH_2CHBrCH_3$	0·68	$PhCHBrCH_3$	90

Table 11.2 (*Continued*)

(b) Effect of leaving group and base on E2 product ratios

2-Alkyl-X	Medium	Temp./°C	$X = I$		$X = Br$		$X = Cl$		$X = F$		$X = Otos$	
			1-Alkene/%	$\frac{trans}{cis}$	1-Alkene/%	$\frac{trans}{cis}$	1-Alkene/%	$\frac{trans}{cis}$	1-Alkene/%	$\frac{trans}{cis}$	1-Alkene/%	$\frac{trans}{cis}$
2-Hexyl	MeONa–MeOH	100	19	3·6	28	3·0	33	2·9	70	2·3	39	1·7
2-Pentyl	EtONa–EtOH	reflux	20	4·1	25	3·8	35	3·5	82	2·6		
2-Butyl	t-BuOK-t-BuOH	50	33	2·2	54	1·4	67	1·3			62	0·6
2-Hexyl	t-BuOK-t-BuOH	100	69	1·8	80	1·4	88	1·1	97	1·2	80	0·4
2-Decyl	t-BuOK-t-BuOH	100	68	1·8	79	1·3	85	1·1	92	0·8	77	0·4
2-Decyl	t-BuOK-18-crown-6-t-BuOH	100	42	5·4	53	5·1	65	3·6	91	2·9	75	2·2
2-Butyl	Et₃COK-Et₃COH	50	49	1·5	71	1·3	80	1·1				
2-Butyl	t-BuOK–DMSO	25	20	3·5	32	3·8	41	4·1				
2-Hexyl	t-BuOK–DMSO	25	35	5·2	47	4·9	59	4·9			73	2·9
2-Decyl	t-BuOK–DMSO	20	32	5·7	48	5·3	59	5·1	97	3·0	75	3·2
2-Decyl	t-BuOK-18-crown-6-DMSO	20	31	5·5	47	5·7	59	5·3	97	3·0	79	6·0
2-Butyl	Et₃COK–DMSO	25	21	3·8	33	3·9	44	4·2				
2-Decyl	t-BuOK–benzene	130	66	1·5	80	0·8	86	0·7	94	0·5	88	0·7
2-Decyl	t-BuOK-18-crown-6-benzene	20	47	3·2	62	2·7	71	2·8	95	1·4	82	1·7
2-Butyl	t-BuOK–toluene	50	36	1·7	52	1·4	67	1·0				
2-Butyl	t-BuOK–THF	25	20	3·6	34	3·3	49	2·9				
2-Decyl	t-BuOK–THF	50	40	5·3	56	4·0	70	3·2				
2-Butyl	Et₃COK–THF	25	25	4·1	34	3·1	49	2·7				

Table 11.2   *(Continued)*

(c) Effect of base concentration and ion-pairing

2-Alkyl group	Leaving group	0·10M tert-BuOK		0·25M tert-BuOK		0·50M tert-BuOK		1·00M tert-BuOK	
		1-Alkene/%	trans/cis	1-Alkene/%	trans/cis	1-Alkene/%	trans/cis	1-Alkene/%	trans/cis
2-Butyl, 50°C	I	24·4	2·36	28·6	2·14	29·9	2·09	33·3	1·95
2-Butyl, 50°C	Br	37·7	1·86	41·6	1·78	44·1	1·66	50·6	1·47
2-Butyl, 50°C	Otos	52·9	0·70	60·4	0·37	62·9	0·35	63·5	0·39
2-Decyl, 120°C	I	66·7	1·89	68·8	1·81	69·7	1·78	74·5	1·68
2-Decyl, 120°C	Br	77·0	1·44	76·5	1·37	79·7	1·25	84·1	1·12
2-Decyl, 120°C	Cl	83·1	1·17	84·0	1·11	84·5	1·10	87·2	1·00
2-Decyl, 120°C	Otos	76·7	0·36	78·7	0·30	79·9	0·28	81·8	0·24

(d) Effect of solvent: 2-halodecanes, $CH_3CHXC_9H_{17}$, with $tert\text{-}BuO^- K^+$

X	In Me$_2$SO at 20°C		In t-BuOH at 100°C		In benzene at 120°C	
	1-Decene/%	trans/cis	1-Decene/%	trans/cis	1-Decene/%	trans/cis
I	32·1	5·7	68·2	1·8	66·1	1·5
Br	48·0	5·3	79·4	1·3	79·9	0·8
Cl	59·4	5·1	84·5	1·1	86·0	0·7
F	96·8	3·0	92·4	0·8	94·0	0·5

Table 11.2 (*Continued*)

(e) Effect of structure: $RCH_2CH(Otos)$—C_5H_{11}, 110°C

Base	E/Z ratio, $RCH_2CH{=}CHC_4H_9$	
	tert-BuO⁻K⁺–DMF	*EtO⁻K⁺–EtOH*
Me	3·9	2·0
Et	3·4	1·6
n-Pr	3·7	1·7
isoPr	2·0	1·1
tert-Bu	0·6	0·6

(f) Effect of crown ether/2-butyl systems, $CH_3CHXC_2H_5$, 50°C

X	Crown ether present?	1- Butene/%	trans/cis- 2-Butenes
Br	no	44·1	1·66
Br	yes	32·5	2·92
Otos	no	63·5	0·40
Otos	yes	53·6	1·88

(g) Competition between *syn* and *anti* elimination pathways (*syn* + *anti* = 100%).

$$RCH{-}CHR' \xrightarrow{:B} RCH{=}CHR$$
$$^+NMe_3 \quad H$$

trans-Alkene formation

cis-Alkene formation

R	R'	Syn elimination/%		R	R'	Syn elimination/%	
		tert-BuOK/ tert-BuOH	*MeOK/ MeOH*			*tert-BuOH tert-BuOK*	*MeOK/ MeOH*
D	*n*-Oct	7	—	Me	*n*-Pr	≤5	≤5
Me	*n*-Pr	15	0	Et	Et	≤5	≤5
Et	Et	80	20	*n*-Bu	*n*-Bu	4·7	6·3
n-Bu	*n*-Bu	89	24	neo-Pe	*n*-Bu	≤5	≤5
tert-Bu	*n*-Pe	>97	83 •	*tert*-Bu	*n*-Pe	≤5	≤5
neo-Pe	*n*-Bu	>99	89				
n-Pe	*tert*-Bu	>90	>90				

(h) Effect of medium on geometrical isomer ratio of product [18]

$$CH_3CHCl{\cdot}CCl_2CH_3 + :B \longrightarrow$$

(E)   (Z)

	Medium	[Base]/M	[E]/[Z]	Comment
1	MeO⁻Na⁺/MeOH	0·5	6·7	non-associated
2	MeO⁻Na⁻/MeOH	1·0	6·7	non-associated
3	*tert*-BuO⁻K⁺/*tert*-BuOH	0·25	1·5	associated
4	*tert*-BuO⁻K⁺/*tert*-BuOH	0·5	1·1	more associated than 3
5	*tert*-BuO⁻K⁺/*tert*-BuOH	1·0	0·9	more associated than 4
6	*tert*-BuO⁻K⁺/*tert*-BuOH + 18-crown-6	0·5	3·8	less associated than 4

512

Table 11.3   Linear free-energy parameters
Hammett reaction constants for eliminations

Reaction	ρ		
	R = Et	R = tert-Bu	R = H
ArCH$_2$CH$_2$X + OR$^-$, 30°C			
X = Otos	2·27	3·39	
X = F	3·10		
X = Cl	2·60		
X = Br	2·15	2·08	
X = I	2·07	1·88	
X = $\overset{+}{N}$Me$_3$	3·77	3·04	
X = $\overset{+}{S}$Me$_2$	2·70		2·2
X = SOMe		4·4	
2-Ar-cyclopentyl-Otos (*cis*)		1·48	
2-Ar-cyclopentyl-Otos (*trans*)	2·4	0·99	
MeCH$_2$CH—OSO$_2$Ar			
ArCH$_2$CHPhCl	1·98		
ArCH$_2$CHArCl	0·77		
Ar$_2$CHCCl$_3$ (E1cb)	2·43		
ArSO$_2$CH$_2$Cl (E1cb)	1·08		

Brønsted coefficients, β[4]

Substrate	Base/solvent	T/°C	β
PhCH$_2$CH$_2$Br	ArO$^-$/EtOH	60	0·56
p-NO$_2$PhCH$_2$CH$_2$Br	ArO$^-$/EtOH	60	0·72
(*p*-ClPh)$_2$CHCCl$_3$	ArO$^-$/EtOH	45	0·88
(*p*-ClPh)$_2$CHCCl$_2$	ArS$^-$/EtOH	75	0·77
Cyclo-C$_6$H$_{11}$OTs	ArS$^-$/EtOH	35	0·27
Cyclo-C$_6$H$_{11}$Br	ArS$^-$/EtOH	55	0·36
Cyclo-C$_6$H$_{11}$Cl	ArS$^-$/EtOH	55	0·39
1,1-Cyclo-C$_6$H$_{10}$Br$_2$	ArS$^-$/EtOH	55	0·51
1,1-Cyclo-C$_6$H$_{10}$Cl$_2$	ArS$^-$/EtOH	55	0·58
tert-BuCl	ArS$^-$/EtOH	45	0·17
tert-BuSMe$_2^+$	ArS$^-$/EtOH	25	0·46
EtMe$_2$CCl	ArS$^-$/EtOH	55	0·19
EtMe$_2$CCl	ArS$^-$/i-PrOH	55	0·16
EtMe$_2$CCl	ArS$^-$/t-BuOH	55	0·13

v, seem to result in inconsistencies but with v alone significant rate correlations were obtained[19]:

$$\log k_{rel} = \psi v.$$

$$X—CH_2CH_2—\overset{+}{N}Me_3 + :B \rightarrow X—CH{=}CH_2.$$

$$:B = EtO^-; \qquad \psi = -2·36.$$

$$:B = tert\text{-Bu}; \qquad \psi = -3·75.$$

These values point to the strong influence of steric hindrance on E2 reactivity[19-27].

11.1.3 Kinetic isotope effects

There is considerable scope for the application of isotope effects in the investigation of elimination mechanisms. The E2 process should be accompanied by primary effects both for the β-hydrogen and for the nucleophilic leaving group (a heavy-atom effect) while secondary ^{2}H and ^{14}C effects should be observed at both α- and β-positions[23,24] since both centers are undergoing hybridization change. These have all been observed (Table 11.4).

Heavy-atom effects

Rate-determining loss of the nucleophilic groups SR_2 and NR_3 are confirmed by PKIEs (respectively, k_{32}/k_{34} and $k_{14}/k_{15} > 1$) for substrates

Table 11.4   Kinetic isotope effects in elimination reactions

(a) Primary, leaving group (heavy-atom effects)

Substrate	:B	$k_L/k_H{}^a$
$CH_3CH_2\overset{+}{N}Me_3$	O-*tert*-Bu$^-$	1·0141
$CH_3CH_2\overset{+}{N}Me_3$	OEt$^-$	1·0186
$PhCH_2CH_2\overset{+}{N}Me_3$	OEt$^-$	1·0142
$Me_3C\overset{+}{-}SMe_2$	OEt$^-$	1·0072
$Me_3C\overset{+}{-}SMe_2$	OEt$^-$	(*anti*) 1·0123
$Me_3C\overset{+}{-}SMe_2$	OEt$^-$	(*syn*) 1·0039

^{a} Light/heavy.

(b) Primary, β-hydrogen

$PhCL_2CH_2Nu + :B$

	k_H/k_D	
Nu	:B = OEt$^-$	:B = O-tert-Bu$^-$
Br	7·11	7·89
Otos	5·66	8·01
$\overset{+}{S}Me_2$	5·07	
NMe_3	2·98	6·96
SOMe		2·7

$PhCL_2CH_2\overset{+}{N}Me_3 + OH^-$ in aq. DMSO

Water in solvent/%	k_H/k_D
17	3·80
23	3·86
30	4·77
34	5·22
41	4·94
57	3·83

Table 11.4   (*Continued*)

Other reactions

Substrate	Base	k_H/k_D
PhCHLCH$_2$Otos	tert-BuO$^-$	(*syn*) 2·3
PhCHLCH$_2$Otos	tert-BuO$^-$	(*anti*) 3·9
	tert-BuO$^-$	(*syn*) 1·8
	tert-BuO$^-$	(*anti*) 5·4
(p-ClC$_6$H$_4$)$_2$CL·CCl$_3$	EtO$^-$	3·8
	EtO$^-$	4·47
	tert-BuO$^-$	7·53
	Cl$^-$ in acetone	2·7

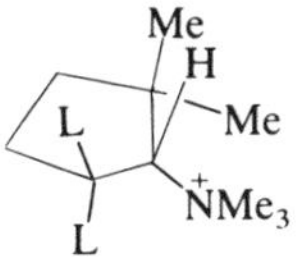

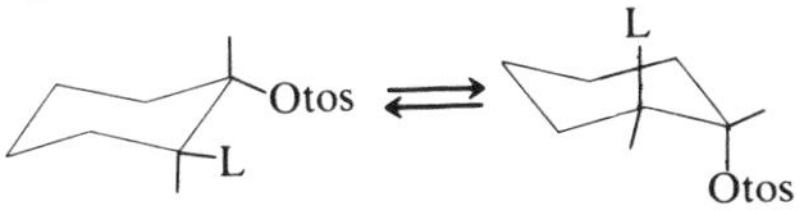

(c) Secondary deuterium isotope effects

Substrate	Base	k_H/k_D
Deuterium on α-carbon		
PhCH$_2$CL$_2$Br	EtO$^-$	1·09
PhCH$_2$CL$_2$Otos	tert-BuO$^-$	1·047
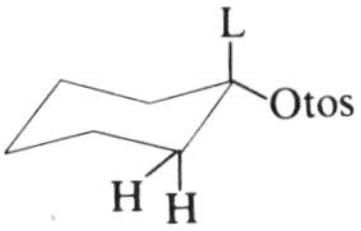	tert-BuO$^-$	1·15
Deuterium on β-carbon		
	EtO$^-$	1·36
	tert-BuO$^-$	1·51

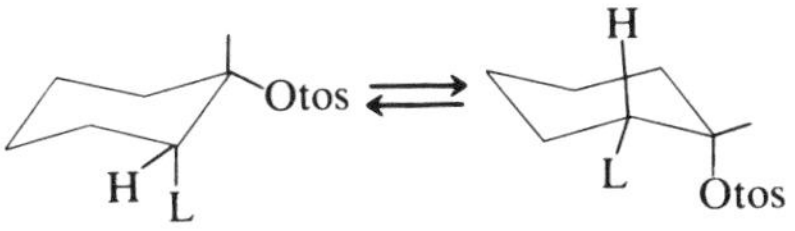

(d) Secondary carbon isotope effects

Substrate	Base	$100(k_{12}/k_{14} - 1)$
CH$_3$C*H$_2$N$^+$Me$_3$	EtO$^-$	3·3
Me$_3$C*N$^+$Me$_3$	EtO$^-$	5·2

(e) Solvent effects

Substrate	Base	k_{H_2O}/k_{D_2O}
PhCH$_2$CH$_2$N$^+$Me$_3$	OL$^-$	0·56
PhCH$_2$CH(OH)Ph	L$_2$SO$_4$	1·62

reacting both by E1 and E2 mechanisms. The E1cb route will be accompanied by a leaving-group isotope effect only when k_2 is rate-determining.

Primary β-H effects[2,3]

The transference of the proton from carbon to an external base in the E2 reaction should show a maximum PKIE(~8) if it is equally coordinated to both centers in the transition state (Fig. 7.2b). A smaller value would be expected for an unsymmetrical transition state in which the proton is either more or less than 50% transferred. Values of k_H/k_D are observed to range between 4 and 8 and may be interpreted to give a measure of the degree of proton transfer, determined by the interplay of several structural factors.

Variation of the leaving group results in an order of isotope effects: $-Br > -Otos > -\overset{+}{S}Me_2 > -\overset{+}{N}Me_3$. There is an ambiguity in the interpretation of this order since it may correspond to a decreasing order of the extent of C–H bond fission at the transition state, the early part of a curve such as Fig. 7.2, or it could accord with an increasing order of bond fission, all points being in the latter part of the curve after the maximum. Now reaction constants, ρ, for elimination of HX in the series $ArCH_2CH_2X$ increase in the order $X = -Br < -Otos < -\overset{+}{S}Me_2 < -\overset{+}{N}Me_3$, an order of increasing electronic demand and higher carbanion character in the transition state in order to expel leaving groups of increasing difficulty. This is consistent with a late transition state and a large degree of proton transfer to the base; consequently the second explanation of the isotope effect order appears to fit best. Isotope effects for formation of *cis*-alkenes are greater than those for *trans*-alkenes indicating significant differences in their respective transition states[25].

The strength of the base would obviously be expected to affect transition-state structure[20]. An example of this effect is the variation of the β-H isotope effect as dimethyl sulfoxide is added to aqueous sodium hydroxide. This increases the basic strength of the medium and the isotope effect rises and passes through a maximum value corresponding to 50% proton transfer between base and carbon at the transition state[26,27].

Brønsted coefficients and secondary kinetic isotope effects

Brønsted coefficients, β, the slopes of the plots of ln k against base strength, are a measure of transition-state structure and a few values have been measured for eliminations by a series of phenoxide ions (Table 11.3). Values of β span a wide range, between 0·13 and 0·88, which presumably reflects the extent of proton transfer, from a small degree to almost complete; a value of 1 would be expected for E1cb reactions (Section 11.1.9) proceeding by initial complete proton transfer[4]. In some cases values of β

for the parallel S_N2 reactions are available and are smaller than for E2. That is, elimination is more sensitive to base strength than is substitution. Acidifying groups such as *p*-nitro tend to increase β, which is consistent with an increase in the extent of proton transfer. Secondary isotope effects at both α and β carbons should reveal the extent of hybridization change towards sp^2 at the transition state[28]. That this is quite variable is indicated by SKIE values at the α-position which vary between 1·02 and 1·15. At the high end this would be consistent with considerable loosening of the leaving group. β-SKIE values can only be obtained for hydrogens in a *gauche* orientation to the leaving group since the *trans*-periplanar hydrogen is lost and its isotope effect is a combined primary and secondary one. The *gauche* orientation is not one to lead to a maximum value since it is much less favorable for hyperconjugation with the developing unsaturated center, yet isotope effects are larger than those at the α-carbon. This presumably indicates that β-C—H bond fission is more advanced than that of α-C—X. Cyclohexyl tosylate, for example, has $k_H/k_D = 1·36$ (B: = EtO⁻) and 1·51 (*tert*-BuO⁻) for a *cis*-β-hydrogen. Presumably it reacts as the diaxial conformer to achieve a periplanar geometry.

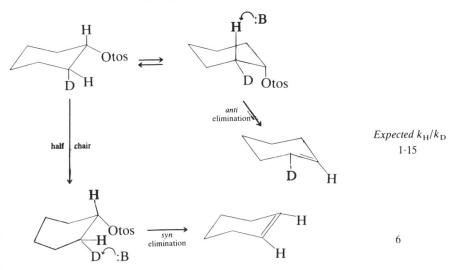

A possible reason for these values, which are exceptionally high for secondary effects, is the incursion of some *syn* elimination, for then the deuterium isotope effect would be primary for the *syn* component.

A difference is observed between secondary isotope effects for terminal and non-terminal olefin formation (the 'Hofmann' and 'Saytzev' pathways, see below); it can be discerned from the products of elimination of $Me_2CL \cdot CHOtos \cdot CL_3$ (L = H, D). Other evidence suggests the 'Hofmann' transition state is later than the 'Saytzev' one, in agreement with its larger SKIE.

11.1.4 Variation of the base–solvent system[29,30]

Rates of transference of a proton from a carbon acid usually correlate with the acidity function, H_-. Rates of E2 reactions also vary in accordance with this function, though seldom linearly, in media including alcohols, DMSO and their mixtures with water.

Ion-pairing adds further complication; rates may differ markedly between reactions promoted by K^+OtBu^- and by Na^+OtBu^-, the former being two- to three-fold faster. Potassium butoxide solutions have a higher H_- than equimolar solutions of the sodium base due to less ion-pairing to the larger cation[31,32]. A further test of the importance of ion-pairing is the change in product ratio upon addition of crown ether which complexes the cation[33,34]. Crown ethers are cyclic polyethers which are capable of complexing alkali-metal ions and thus reduce ion-pairing with the associated anions, e.g.

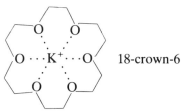

18-crown-6

In non-polar solvents higher aggregates of the base are known to exist, each of which may be expected to react at a unique rate.

For these reasons and others, Brønsted plots of log(rate of elimination) against pK_A of the base show some scatter. Perhaps a good correlation would not be expected in any case, since one is comparing behavior of carbon acids with that of oxygen acids and reactions in which rates are determined not by proton transfer alone but also by nucleofugacity. Interpretation of rate constants in terms of the degree of proton transfer is dubious in such cases, unless they are closely related. Thus, the greater value of β for cyclohexyl chloride, 0·39, compared with the tosylate, 0·27, is consistent with a greater degree of proton transfer and of carbanionic charge at the β-C, required for the expulsion of the less easy leaving group (Cl^-) than for the tosylate. The same principle seems to hold in the series of E2 reactions:

$$ArS^- + \quad \overset{\displaystyle H}{\underset{\displaystyle Br}{-C-C-}} \quad \longrightarrow \quad ArSH + \quad C{=}C \quad + \quad Br^-$$

Bromide:	Primary $PhCH_2CH_2Br$	Secondary	Tertiary *tert*-BuBr
β:	0·56	0·36	0·17

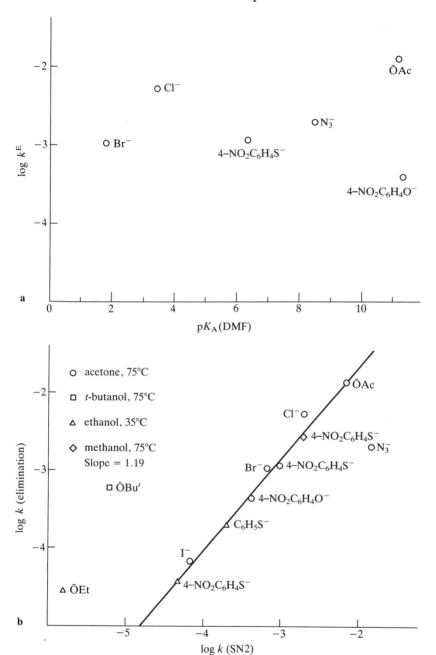

Fig. 11.2 **a**, Brønsted plot of rates of elimination from cyclohexyl tosylate by several anions versus their pK_A values. The complete lack of correlation indicates that interactions between substrate and nucleophile in the transition state differ from acid–base interactions. **b**, Plot of rates of elimination from cyclohexyl tosylate against rates of substitution (S_N2) by a variety of nucleophiles. The good correlation indicates a similarity of interactions between the nucleophiles and the substrates at the transition states for both processes.

Again, the proton transfer is more complete in the transition state for the most 'difficult' reaction, that of the primary compound, for which loss of Br⁻ receives least assistance from the structure. The ratio of geometrical isomers in the product olefins is also dependent upon the medium (Table 11.2h). In general, association of the metal counterion with the alkoxide base brings about an increase in the $Z(cis)$ product[18]. Elimination is promoted efficiently by carbon nucleophiles which are weak bases so that rates do not correlate with pK_A (Fig. 11.2a). At the same time, rates of elimination by these soft anionic bases correlate well with rates in S_N2 reactions which, by their nature, must occur by attack at carbon (Fig. 11.2b). This observation prompted the proposal that coordination both to α-C and to β-H (the E2C process) occurred in the initial stages of the reaction in such cases. The nature of the B:–α-C interaction was considered to be orbital overlap, as befits a soft nucleophile–electrophile attraction, and it may be important that the bases which are active in E2C reactions have more than one unshared pair of electrons. Other workers have suggested electrostatic stabilization between anionic base and cationic carbon (i.e. 'E1-like') as the basis for this unexpected form of elimination[35].

11.1.5 Competition between elimination and substitution

Elimination and substitution usually occur concurrently, the E1 and S_N1 pathways being coupled and the E2 with S_N2. In each case, product determination depends upon relative rates of attack of nucleophile at α-C or β-H, (Table 11.5). In the progression from primary → secondary → tertiary substrates the amount of olefin generally increases since the S_N2 reaction becomes subject to severe steric hindrance, while the E2 reaction is much less so since the peripheral hydrogens are exposed to attack. Alkyl substituents which stabilize the alkene product will also facilitate reaction. The same combination of steric and electronic factors guides the selectivity of attack in the carbocation intermediate of E1/S_N1 reactions in tertiary substrates.

A β-aryl group greatly promotes elimination over substitution due to the conjugation developed. Branching at the α-carbon has a similar effect though less pronounced.

Higher temperatures may favor elimination, which usually has a higher activation energy than substitution. For example,

$$CH_3CHCH_3 + OEt^- \longrightarrow CH_3CHCH_3 + CH_3CH{=\!=}CH_2.$$
$$\quad | \qquad\qquad\qquad\qquad\quad |$$
$$\quad Cl \qquad\qquad\qquad\qquad\quad OEt$$

Reaction	E_A	log A
S_N2	96(23)	10·0
E2	103(25)	11·3

Table 11.5 Competition between elimination and substitution in product determination[1,4,6]

(a) Effect of structure and medium in ethanolysis

Substrate, R—X	Medium		$T/°C$	E, %	S_N, %
	Ethanol, %	Water, %			
Et—Br	100	0	55	1	99
2-Pr—Br	100	0	55	3	97
2-Pr—Br (+EtO⁻)	100	0	55	80	20
tert-Bu—Cl	80	20	65	36	64
tert-Bu—Cl	80	20	25	16	84
tert-Bu—Cl	60	40	25	17	83
tert-Bu—Br	100	0	55	27	73
tert-Bu—Br (+EtO⁻)	100	0	55	92	8
EtCMe₂—Cl	80	20	25	34	66
Et₃C—Cl	80	20	25	40	60
(2-Pr)₂MeC—Cl	80	20	25	78	12
(*tert*-Bu)Et₂C—Cl	80	20	25	90	10

(b) Effects on rates

Reaction	Medium		$T/°C$	$k_2/\text{M}^{-1}s^{-1}$	$k_2/\text{M}^{-1}s^{-1}$
	Ethanol, %	Water, %			
Et—Br + EtO⁻	100	0	55	1·2	118
2-Pr—Br + EtO⁻	100	0	55	2·1	7·6
tert-Bu—Br + EtO⁻	100	0	55	50	10^{-5}

(c) Steric effects; solvolyses in *N*-butylcellosolve

$$RCH_2CMe_2—Br \rightarrow RCH=CMe_2 + RCH_2C(Me)=CH_2 + S_N \text{ products}$$

R	$RCH=CMe_2$, %	$RCH_2C(Me)=CH_2$, %	S_N, %
Me	21·4	5·6	73
Et	22·7	9·3	68
2-Pr	27·1	18·8	54·1
tert-Bu	10·8	46·1	43·1

(d) Alkene proportions; solvolysis in 80% ethanol, 20% water

Substrate ($X = —Otos$)	Alkene proportions, %		
	46·5	43·2	10·3
	49	35	16
	20	80	

Table 11.5 (*Continued*)

Substrate (X = —Otos)	Alkene proportions, %		

18   82

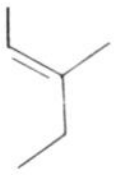

38·5   58·0   3·5

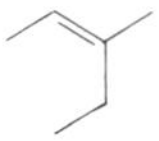

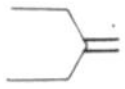

19   81

(e) Effect of base and of mechanism

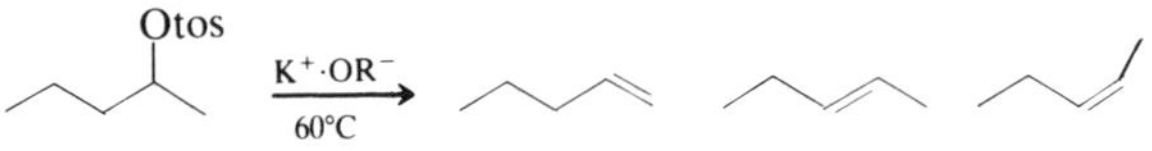

I   II   III

Conditions	R	I, %	II, %	III, %	S_N, %
E2	1-Bu	12·6	8·4	4·8	74·2
E2	*tert*-Bu	43·6	0·44	0·34	55·0
E1	1-Bu	0·65	3·6	1·6	94·2
E1	*tert*-Bu	0·4	2·4	1·1	96·1

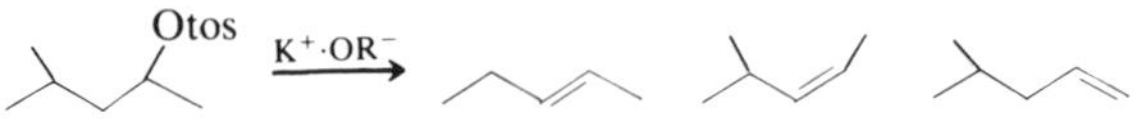

IV   V   VI

Conditions	R	IV, %	V, %	VI, %	S_N, %
E2	1-Bu	19	8·8	32	42·3
E2	*tert*-Bu	1·1	2·2	56	40·6
E1	1-Bu	8·5	1·9	1·3	88·2
E1	*tert*-Bu	5·3	1·5	1·1	92·0

E2 reactions lead to more alkene product than E1 reactions of the same substrates. It is often possible to manipulate the preferred route in tertiary systems by means of solvent variation: in methanol, the E1 route is favored but in the less polar *tert*-butanol, the E2 route predominates. The nature of the leaving group is important; treatment of octadecyl bromide with butoxide yields 85% elimination; with the corresponding tosylate, 99% substitution is obtained[36].

The factors which determine the E/S_N ratio are evidently complex. Conjugation and hyperconjugation can be discerned as favoring elimination, implying that the transition state resembles products. Steric interactions will have a similar result by disfavoring substitution.

11.1.6 Orientation in product formation

A substrate may possess up to three alkyl chains at the α-carbon, each of which may lead to a regioisomeric alkene (which in addition might contain E and Z stereoisomers). It is usual to find that one regioisomer is favored as indicated in the two empirical rules formulated in the early days of organic chemistry[37,38].

Saytzev's Rule (1875)[39].

The preferred alkene is the isomer bearing the greatest number of alkyl groups attached to the double bond[40,41], e.g.

$$CH_3CHBrCH_2CH_3 + OEt^- \rightarrow CH_2{=}CHCH_2CH_3 + CH_3CH{=}CHCH_3$$
$$\text{19\%} \phantom{CH_2{=}CHCH_2CH_3 + } \text{81\%}$$

Hofmann's Rule (1851)[42]

When an alkene is formed by heating a quaternary ammonium hydroxide, the preferred isomer is that bearing the smallest number of alkyl groups at the double bond, e.g.

$$CH_3CH_2\overset{+}{-}NMe_2{-}CH_2CH_2CH_3 \longrightarrow CH_2{=}CH_2 + CH_3CH{=}CH_2$$
$$\underset{\overline{O}H}{\big\lfloor \big\rfloor} \text{98\%} \text{2\%}$$

Table 11.2 gives other examples of the orientational behavior of E2 and E1 reactions. The Saytzev product is the most stable alkene and in this respect seems the natural one to be formed most rapidly. This would be the case if the transition state was a late one and partakes of the same type of stabilization as the product, conjugation and hyperconjugation of substituents with the double bond. The more powerful and bulkier the base, however, the higher is the proportion of the terminal alkene formed due to a combination of steric factors — the terminal proton is more accessible —

and an earlier transition state, which therefore derives less stabilization from conjugation. The same reasoning may account for the observation that the ratio of $Z:E$ (*cis*:*trans*) alkene product from Saytzev elimination also depends upon the base[41,43,44]. The stronger the base and the more 1-alkene produced, the greater is the ratio of $Z:E$ in the 2-alkene. This is consistent with an earlier transition state; with a stronger base, the unfavorable eclipsing interactions of the *cis*-alkene are less apparent than for a product-like transition state. Ion-pairing of the base also affects the *cis*/*trans* ratio, which as a consequence changes with the concentration of the base[45]. The situation is complex, however, and very small energy differences can affect product ratios (Table 11.2).

Trans- and *cis*-alkene products must arise, assuming *anti* elimination, from two different conformers of the substrate and consequently the *trans*/*cis* product ratio will depend upon both the conformational equilibrium constant and the relative rates of each conformer according to the Curtin–Hammett principle (Section 8.6):

$$trans/cis = K_e k_t/k_c; \qquad K_c = [\textbf{3b}]/[\textbf{3a}].$$

If we include also an incursion of *syn* elimination, the kinetic treatment becomes very complicated.

The Hofmann Rule applies to eliminations of ammonium leaving groups promoted by the strongest bases. *Syn* elimination tends also to favor these conditions and to lead to terminal alkenes. It seems likely that Hofmann elimination is largely a result of the *syn* route becoming the more favorable. At the same time, the *syn* transition state, **4**, is an eclipsed one and so subject to steric repulsions. These would be less severe from a terminal than from a non-terminal position. Brown has put forward evidence in support of the importance of steric effects in determining the ratio of Hofmann- to Saytzev-type products. Increasing the size of a β-alkyl substituent, of base or of leaving group is accompanied by an increase in the proportion of 1-alkene relative to the 2-isomer[20–22] (Table 11.6).

The most consistent picture which emerges is that both steric and electronic influences affect rates and product ratios, interacting in a complex manner difficult to disentangle[17,46,47]. Eliminations promoted by carbon nucleophiles such as Cl^-, OAc^- — E2C reactions — have a very strong tendency to undergo *anti* elimination and to give the Saytzev product, the first of these two requirements being the dominant one (see Refs *9, 48, 49*).

11.1.7 Stereochemistry of E2 reactions

The relative dispositions of the proton and leaving group with respect to the two carbon centers is not a matter of critical importance in either the E1 or E1cb reactions. The E2 mechanism, however, has very definite stereochemical requirements, such that all four centers must assume a coplanar (periplanar) geometry for maximum facilitation of the concerted movement

Table 11.6   Steric effects on elimination products

$$R—CH_2CMe_2Nu+B:→R—CH_2CMe=CH_2+R—CH=CMe_2$$

(a) Effect of β-alkyl group

B:=$\bar{O}$Et;   Nu:=Br

R	Percentage 1-alkene
Me	24
Et	48
n-Pr	54
tert-Bu	87

(b) Effects of base size

R=H, Nu:=Br

B:	Percentage 1-alkene
$\bar{O}$Et	30
$\bar{O}$t-Bu	72
$\bar{O}$t-Am	78
$\bar{O}$CEt$_3$	89

(c) Effect of size of leaving group

R=H; B:=OEt$^-$

Nu:	Percentage 1-alkene
—Br	31
—I	30
—Otos	48
—$\overset{+}{S}$Me$_2$	87
—$\overset{+}{N}$Me$_3$	98

of three electron pairs. Two conformations satisfy this prerequisite, the *anti*, **3**, and *syn*, **4**, geometries. In either case it may be seen that σ-orbitals to the leaving atoms can grow into the π-orbital of the alkene product with no change in symmetry. The *anti* geometry is apparently preferred in most cases but it is now known that *syn* elimination is not uncommon and may be the major pathway in some cases. The following examples will illustrate the principle.

Menthyl compounds, **5**, eliminate to give 2-menthene, **5a**, only whereas the neomenthyl isomers, **6**, under the same conditions give mixtures of 2- and 3-menthenes, **6a**. In cyclohexane rings, an *anti*-periplanar geometry is only achieved between 1,2-diaxial substituents. This is only possible in the unfavorable conformer of menthyl compounds, **7**, in which case there is

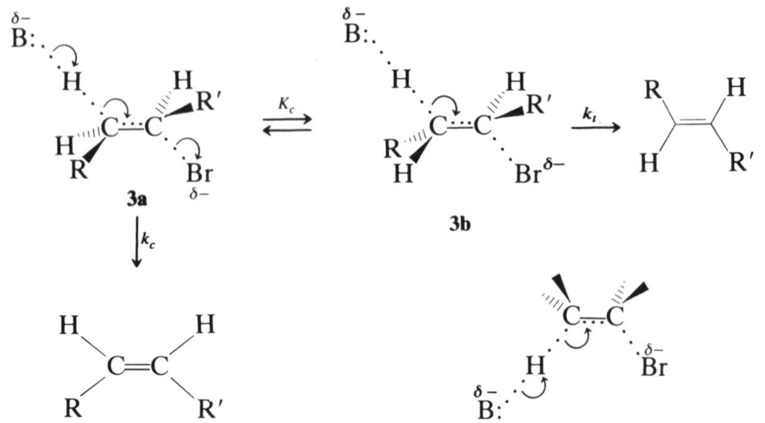

3a **3b**

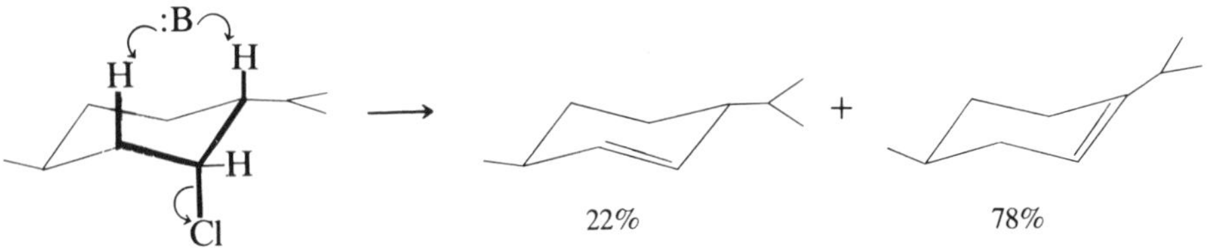

4

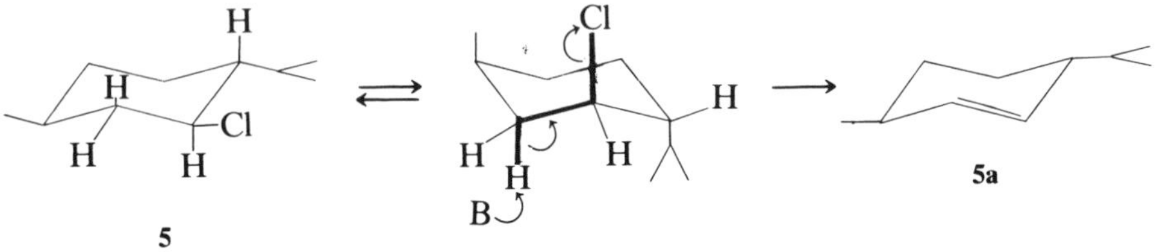

5 **7** **5a**

6 22% 78%

6a

only one axial β-hydrogen, that at C2. Neomenthyl compounds in their favored conformation (leaving group 3-axial) have axial hydrogens on both C2 and C4 so that elimination can take two routes. The preference for 3-elimination is in accordance with the Saytzev rule that the most substituted double bond is favored.

The stereoisomers **8** and **9**, (X = Br, Cl, $\overset{+}{N}Me_3$) have each only one conformer with *anti*-periplanar geometry and this results in specific formation of the geometrical isomers **10** and **11**, respectively, under the action of alkoxide bases in benzene. The exception is the quaternary ammonium compound (X = $\overset{+}{N}Me_3$) reacting with *tert*-butoxide in *tert*-butanol; a mixture of products is obtained from each, suggesting both *syn* and *anti* elimination to be occurring[50–51].

Cyclic substrates, stereospecifically deuterium-labeled on the β-carbon, e.g. **12**, may be used to determine reaction stereochemistry. In four- to

	Percentage of product	
n	*anti*	*syn*
2	10	90
3	54	46
4	96	4
5	63	37

seven-membered rings only *cis*-cycloalkenes can form, the pathway being characterized by loss of H or D (**13, 14**). Much *syn* elimination occurs from cyclobutyl compounds and in the rigid norbornyl system[52] in which reaction centers are naturally *syn*-periplanar. The *anti*-periplanar conformation would require ring deformation[53,54]. The same applies to cyclopentyl, though the *anti* conformation is more accessible.

Cyclohexyl systems have a naturally *anti*-periplanar geometry available when leaving group and β-proton are 1,2-diaxial (*trans*-diaxial) and this remains a more favorable geometry for E2 elimination than the conformational isomer, *trans*-diequatorial or a *cis* arrangement (equatorial–axial). No conformation is naturally *syn*-periplanar (a geometry which would force the molecule into a boat conformation). Hence *syn* elimination from cyclohexyl systems is normally very slight and limited to the strongest bases.

The most interesting study of *syn/anti* elimination was made by Sicher and colleagues, who studied quaternary ammonium derivatives of medium-sized rings (8- to 14-membered). These all yield mixtures of both *cis*- and *trans*-cycloalkenes originating from different conformers, e.g. **15→16**, **17→18**[59–61]. Specific deuteration was again used so that *syn* and *anti* pathways could be differentiated and an unexpected result was obtained,

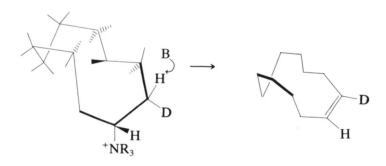

15   **16**

Syn elimination → *trans*-cycloalkene

anti elimination → *cis*-cycloalkene

17   **18**

known as the *syn–anti* dichotomy:

cis-Cycloalkene originated from *anti* elimination;

trans-Cycloalkene originated from *syn* elimination.

That is, the two products arise from separate mechanistic routes. This was substantiated by comparing 'rate profiles' of the series, in which rates are presented as a function of ring size (Fig. 11.3). The *syn* and *anti* elimination rates showed quite characteristic and different profiles confirming their different requirements. Table 11.2g sets out some *syn/anti* rate ratios which show that the *syn* route to elimination is not at all uncommon and is especially to be expected when the leaving group is an ammonium ion and the base is very strong. One explanation of the dichotomy, but probably not the whole reason, is that the leaving group must lie on the *exo* side of the ring on account of its large size ($-N\overset{+}{R}_3$). The conformer leading to *trans*-alkene then has the *anti* β-hydrogen sterically inaccessible. The *syn* β-hydrogen can become coplanar and is more vulnerable to attack, especially by a very strong base. The opposite is true for formation of *cis*-alkene. The *syn* $\rightarrow$ *cis* path may be further disfavored by the eclipsing of the large leaving group. Similar effects may be discerned in acyclic reactants such as **19**, though such examples are less stereochemically specific. The size of the leaving group appears to have an effect on the preferred route; the larger groups may make *syn* elimination more favorable by steric strain relief[62].

$$RCH_2\!-\!\underset{\underset{\displaystyle \text{Otos}}{|}}{CH}\!-\!(CH_2)_4Me \quad \overset{anti}{\underset{syn}{\curvearrowright}} \quad \overset{R\qquad H}{\underset{H\qquad (CH_2)_4Me}{\diagup\!\!\!=\!\!\!\diagdown}}$$

19

11.1.8   Frontier orbital considerations

The nucleophile which brings about E2 elimination is obviously contributing the HOMO to the interaction and, in consequence, the alkyl halide undergoing elimination contributes the LUMO. Reactivity in elimination has been examined theoretically using LUMO properties of various substrates. It is found that rates parallel the coefficient of the LUMO at the β-hydrogen atom (c^2, in fact) as the following example shows[55–58]:

	$Me_2CH\cdot CH_2Br$	$MeCH_2CH_2Br$	$EtCH_2CH_2Br$	$n\text{-}PrCH_2CH_2Br$
c^2_{LUMO} at β—H	0·1289	0·1020	0·0663	0·0551
k_{rel} (E2 elimination)	1·0	0·62	0·50	0·41

It is further found that the coefficient in the LUMO of a β-hydrogen which is *trans*-periplanar to halogen is far greater than that associated with a *gauche* hydrogen, accounting for the observed stereochemical preference for attack by the base.

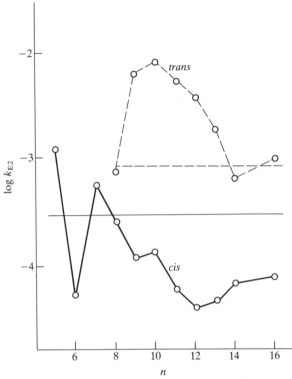

Fig. 11.3   Effect of ring size, *n*, upon rates of *trans*- and *cis*-cycloalkene formation in reactions of cycloalkyl ($-\overset{+}{\text{N}}\text{Me}_3$) with *tert*-BuO⁻. The differences in relative rates for each indicate that *cis*- and *trans*-cycloalkenes originate in different mechanistic pathways.

11.1.9   E1cb reactions [12,63,64]

For the stepwise mechanism to be preferred rather than E2 concerted processes there must be a considerable driving force towards β-proton loss and against leaving-group departure. This indicates that an acid-strengthening group on the β-carbon (a $-R$ type such as —COR, —SO$_2$R, —NO$_2$) and a rather difficult leaving group, e.g. —OR or —CN, will favor this route. Which of the three mechanistic variants on the E1cb reaction to be adopted will depend upon both substrate and the base/solvent system. The following are examples of each case.

(E1cb)$_R$ [65,66]

The compound **20** exchanges β-protons with solvent more rapidly than elimination occurs; the slow step is the loss of $\overline{O}$Me to give a nitroalkene. As with many such products, rapid addition of the solvent follows so that an equilibrium results.

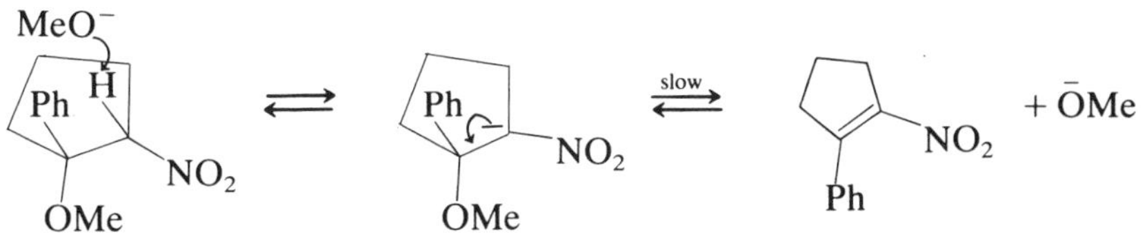

20

(E1cb)$_I$[67-69]

Rates of eliminations forming acenaphthene, **21**, are independent of the nature of the leaving group but show second-order kinetics and no appreciable deuterium exchange with solvent. Compound **22** reacts by a similar route, the evidence in this case being a large inverse solvent isotope effect, $k_{H_2O}/k_{D_2O} = 0.13$, which can be attributed to the PKIE on k_{-1}, i.e. D_2O as solvent shifts the initial equilibrium to the right and accelerates the overall rate. This is a sensitive test for E1cb reactions in which proton transfer is not entirely rate-controlling[70].

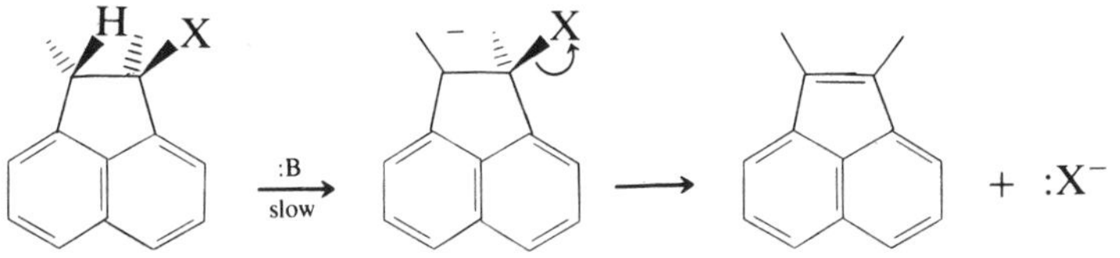

X = Cl, Otos

21

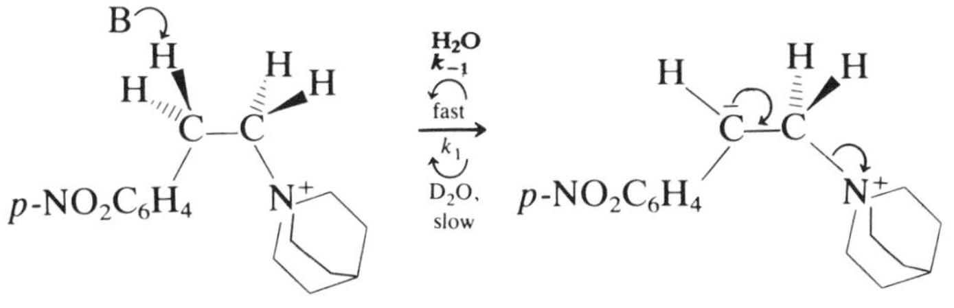

22

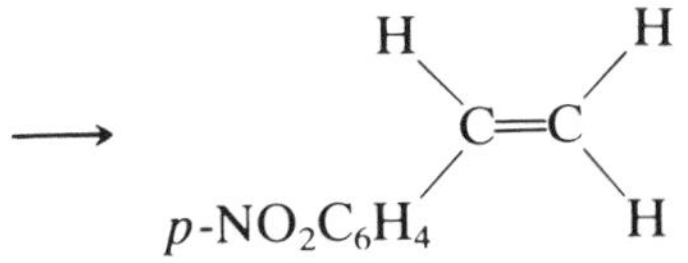

(E1cb)$_{anion}$[71]

The very acidic β-proton of **23** permits this compound to ionize almost completely under the action of amine bases. Slow expulsion of CN⁻ is then

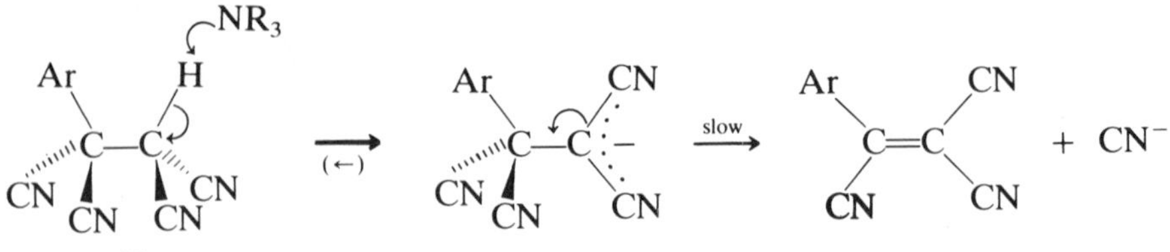

23

rate-determining. There is no β-isotope effect, as is required for this mechanism.

E1cb reactions in general tend to lack specificity towards *syn* or *anti* geometry of the substrate[72] and, in some cases, carbanions may be isolated and shown to undergo decomposition to alkenes[73], a reaction parallel to the expulsion of the leaving group in the second step of the $B_{Ac}2$ ester hydrolysis.

11.1.10 Ester hydrolysis by the E1cb mechanism[74-76]

Esters which have an acidic β-proton and an easy leaving group may undergo hydrolysis by a prior elimination, giving as intermediate a ketene, which rapidly adds water. Examples are the *p*-nitrophenyl esters of *p*-hydroxybenzoic and malonic acids (**24, 25**). These hydrolyze at rates faster than predicted from the Brønsted plot of pK_A (leaving group) against rate, giving a value of $\alpha = -1.3$, consistent with a unimolecular rate

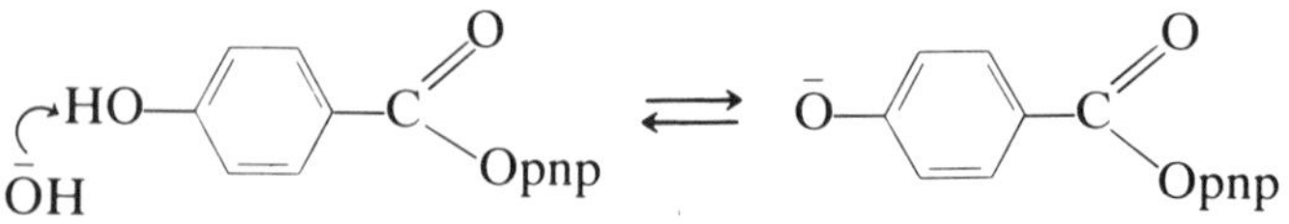

24

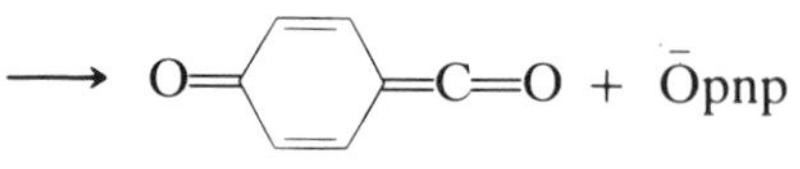

24a

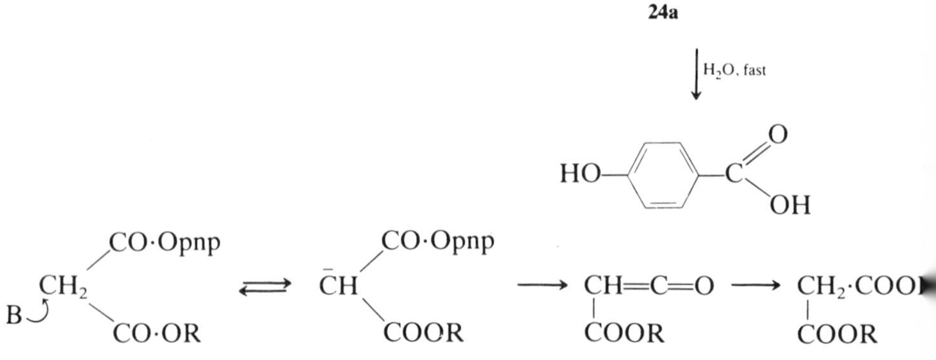

25

determining step. The entropy of activation is large and positive, ca +170 $J K^{-1} mol^{-1}$ (ca 40 cal $K^{-1} mol^{-1}$) which is also characteristic of a dissociative slow step and distinct from values for the $B_{Ac}2$ mechanism. The volume of activation, ΔV, is also positive, ca +20 $cm^3 mol^{-1}$, typical of a dissociative mechanism; whereas that for hydrolysis by the $B_{Ac}2$ route is negative (ca −20 $cm^3 mol^{-1}$), as is found for the hydrolysis of p-methoxybenzoate, which cannot undergo elimination though it is otherwise expected to be similar to the p-hydroxy compound.

11.2 INTRAMOLECULAR PYROLYTIC ELIMINATIONS (THE E_i REACTION)

Preparatively useful routes to alkenes include a family of unimolecular thermal decompositions which are summarized in Table 11.7[77,78]. These

Table 11.7 Summary of pyrolytic alkene-forming eliminations

$$\underset{\substack{|\quad|\\ H\quad Z}}{\diagdown C - C \diagup} \xrightarrow{E_i} \diagdown C = C \diagup + H - Z$$

Compound	Temp. range/°C	−Z
Esters (esp. acetates, R = Me)	400–500	O=C(O)(R)
Xanthates	150–250	S=C(O)(SR)
Amine oxides	70–150	$^+NR_2$, $\bar{O}$
Sulfoxides	80–150	^+S, $\bar{O}$, R
Selenoxides	0–50	^+Se, $\bar{O}$, Ph
Alkyl halides	400–500	Cl, Br
Wittig reaction	<0	C—C, R_3P^+ O^-

reactions are classed as pericyclic processes (Chapter 14) but it is convenient to treat them in this section.

11.2.1 Ester pyrolysis

Esters fragment to alkenes and carboxylic acid, often cleanly and in good yield, when passed through a heated tube at 400–450°C with a contact time of a few seconds. Reactions are of first order and uncontaminated with radical processes. The stereochemistry is that of a strictly *syn* elimination (**26, 27**)[79,80]. Electronic demands at the α-carbon are fairly small but rates correlate well with σ^+, suggesting carbocation character developing at that site. Indeed, this reaction has been extensively used for the determination of σ^+ values (Table 11.8)[81–83]. Rates increase rapidly with other carbocation-stabilizing structural features such as α-alkylation.

26   **26a**

Table 11.8 Structural effects on the pyrolyses of esters Brown–Okamoto reaction constants, ρ^+

Ester	$T/°C$	ρ^+
R—OAc		
R = CH_3CHAr	337	−0·66
R = $PhCH_2CHAr$	337	−0·62
R = $ArCH_2CH_2$	>350	+0·30
R = $ArCH_2CHPh$	>350	+0·08
EtOCOAr	>350	+0·20

Alkyl substitution

Ester	$k_{rel}(400°C)$
CH_3CH_2OAc	1
$CH_3CHMeOAc$	26
CH_3CMe_2OAc	1660
$CH_3CH_2CH_2OAc$	0·6

Values of ρ for substitution at the β-carbon are smaller and positive, in keeping with some assistance to the departure of a proton by an electron-withdrawing group. At the same time, a full primary isotope effect for the β-H is observed, $k_H/k_D = 2 \cdot 0 - 2 \cdot 5$ (at the high temperatures of the reaction these values are near maxima[80,84]):

$$CL_3CH_2OAc \longrightarrow CL_2{=}CH_2; \quad k_H/k_D(500°C) = 2\cdot1$$

$; \quad k_H/k_D(500°C) = 1\cdot7$

The mechanism is therefore deduced to occur by a concerted intramolecular fission involving a six-centered transition state and the synchronous reorganization of three electron pairs, as in **26a**. It is not unreasonable to ascribe 'quasi-aromatic' character to this species, accounting for its favorability. The Arrhenius log $A = 12\cdot5$ corresponds to a value of $\Delta S^{\ddagger} = -20 \, J \, K^{-1} \, mol$ ($-4\cdot8 \, cal \, K^{-1} \, mol^{-1}$) at $450°C$[85]. A rigid transition state in which three rotational modes had been immobilized would require about $-60 \, J \, K^{-1} \, mol$ ($-14 \, cal \, K^{-1} \, mol^{-1}$) in the limit. The observed value therefore indicates a compensating entropy gain in the partially broken C–H and C–O bonds. Further evidence for the concerted nature of the reaction is shown by the relative inertness of neophyl acetate (**28**), which, lacking a β-hydrogen, is stable up to $600°C$, above which the characteristic cationic rearrangement occurs[86].

28

$$\underset{\underset{\displaystyle 29}{\overset{\displaystyle |}{OAc}}}{PhCH\cdot CH\cdot CH_3} \longrightarrow \underset{75\%}{PhCH{=}CHCH_3} + \underset{25\%}{PhCH_2CH{=}CH_2}$$

30

Regiospecificity in acyclic esters, e.g., **29**, is not high and is determined statistically in the main by the numbers of β-hydrogens. Medium-ring esters give high yields of *trans*-cycloalkenes, presumably a result of the steric availability of the *syn–trans* conformer, **30**.

11.2.2 The Chugaev reaction

Xanthates (dithiocarbonates), **31**, undergo a pyrolytic decomposition analogous to that of acetates, but at a much lower temperature in the range 100–250°C[87,88]. Since the compounds may be readily prepared from alcohols, this reaction provides a facile route to dehydration avoiding acidic conditions, but it is limited in usefulness by the noxious by-products, COS and MeSH. Again, *syn* specificity is high and product ratios are close to statistical. There is, however, a preference for endocyclic rather than exocyclic alkene formation from cyclic xanthates, **33**[89,90]. The mechanism of xanthate pyrolysis is clearly a concerted fragmentation but products could conceivably arise from β-hydrogen abstraction by either the thiol or thione sulfur atoms (**32, 33**). That the latter alternative is correct is shown by a considerable sulfur isotope effect (k_{32}/k_{34}) for the C=S group but only a small one for the —SMe, consistent with a change in bonding at the C=S but not at the —SMe site; also, there is no carbon isotope effect for the xanthate carbon ($k_{12}/k_{13} = 1\cdot000$), which can only be the case if the C–S bond is intact at the transition state[91]. The mechanism is therefore the sulfur counterpart of carboxylate ester pyrolysis and the initial product must be an unstable dithiocarbonate, **34**, not isolable.

31

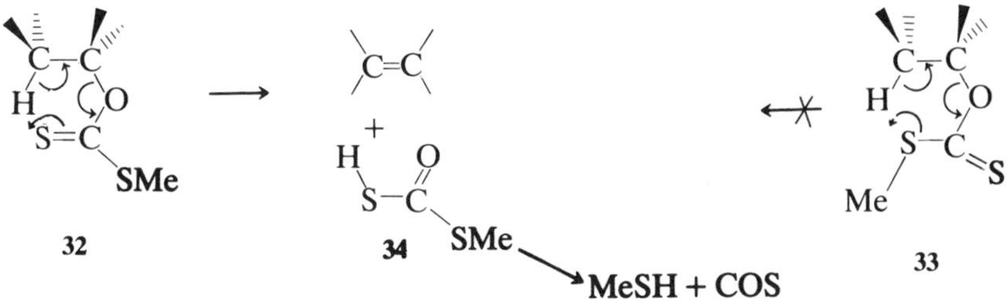

32   **34** SMe

MeSH + COS

33

11.2.3 Amine oxide, sulfoxide and selenoxide pyrolyses

Syn elimination via a five-centered transition state isoelectronic with that of
ester pyrolysis occurs on heating amine oxides **35** (the Cope reaction)[92] or
their sulfur or selenium analogs[93,94]. Because of the tighter steric stringency
of the five- than of the six-membered ring, *syn* specificity is even higher[95].
Exocyclic alkenes form in preference to endocyclic in the cyclohexane series
since the 1,2-diequatorial leaving groups in an endocyclic process are not
quite periplanar. Presumably something approaching a boat conformation is
needed in the transition state leading to a cyclohexene, **36**. Medium-ring
amine oxides give mainly *trans*-cycloalkenes and may be far more reactive
for conformational reasons; $k(C_{10})/k(C_6) \sim 10^{5}$ [96].

Sulfoxides with a β-hydrogen, **37**, decompose above 80°C to alkene and
presumably to a sulfenic acid which is not isolable and normally
polymerizes[97]. Strongly negative entropies of activation (-50 to -60
$J\,K^{-1}\,mol^{-1}$ (-12 to $-15\,cal\,K^{-1}\,mol^{-1}$)) are characteristic of a tight
transition state and the use of chiral sulfoxides enables the stereospecificity
to be determined. For **37** it is below 100% and drops rapidly with
temperature; this has been interpreted as indicating the presence of a
competing radical chain pathway.

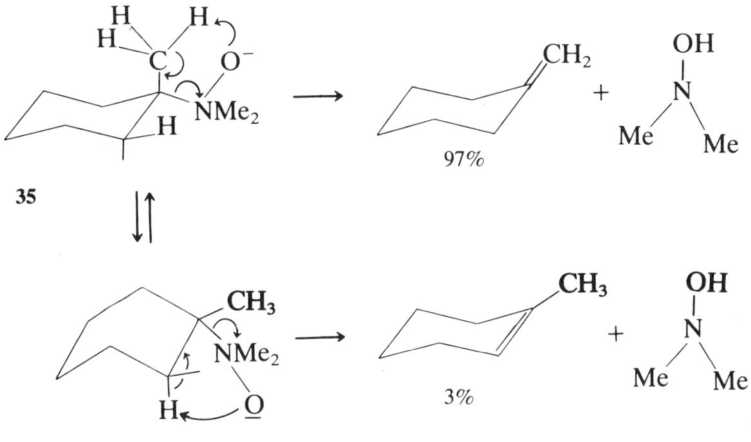

97%

35

36

3%

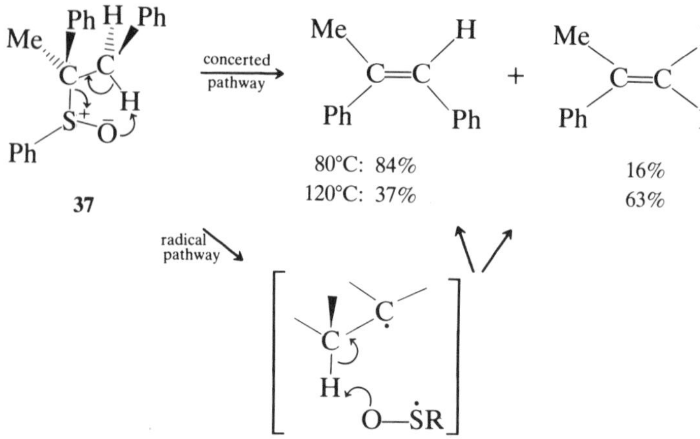
Ph H Ph
Me
C
C
H
S
O
Ph
37
concerted pathway
radical pathway
Me
H
C=C
Ph
Ph
+
Me
Ph
C=C
Ph
H
80°C: 84%
120°C: 37%
16%
63%
C
C
H
O—ṠR

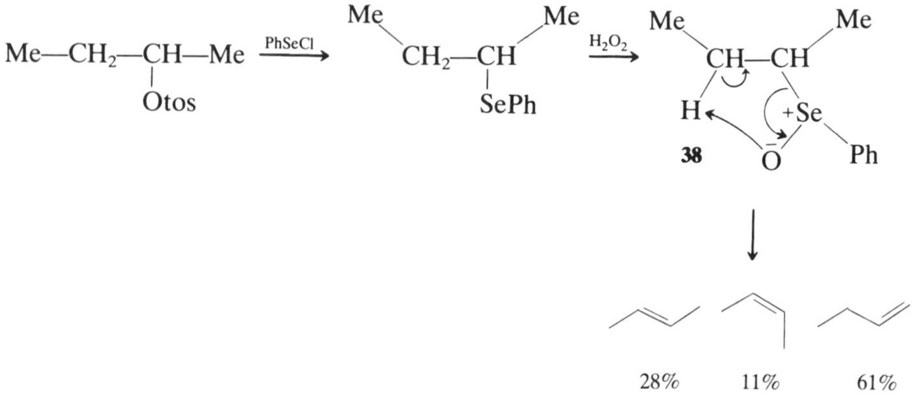
Me—CH2—CH—Me
Otos
PhSeCl
Me
Me
CH2—CH
SePh
H2O2
Me
Me
CH—CH
H
+Se
38
O
Ph
28%
11%
61%

Selenoxides, **38**, decompose rapidly at temperatures as low as 0°C, providing a preparative method of great value for sensitive alkenes under neutral conditions. The selenoxides are prepared *in situ* from the corresponding selenides and cleave to alkene with a preference for the Hofmann product, perhaps on account of the large size of the selenium group[98].

11.2.4 Pyrolysis of alkyl halides

Thermal cleavage to alkene and H—Hal occurs at temperatures in the range 400–600°C by a unimolecular gas-phase process (in the absence of polar surfaces). Activation energies are related to heterolytic but not to homolytic energies of dissociation of the C–Hal bond. Furthermore, electron-releasing substituents on the α-carbon (but not β-) increase rates strongly, more so than is the case for ester pyrolysis; electron-withdrawing groups behave conversely[98–102]

Alkyl halide: CH$_3$CH$_2$Br	CH$_3$CHMeBr	CH$_3$CMe$_2$Br	CH$_3$CH(OMe)Br	CH$_3$CH$_2$CH$_2$Br	CH$_3$CH(COCH$_3$)Br
k_{rel}(400°C): 1	170	32 000	10^9	3	0·064

1-Arylethyl chlorides pyrolyze with rates which correlate with σ^+ values of the substituents; $\rho^+ = -1\cdot36$, similar to but larger than that for acetate pyrolysis. There is a large β-hydrogen PKIE; for CHD$_2$CD$_2$Cl, $k_H/k_D = 2\cdot1$, a maximum at 400°C but this includes a secondary effect which may account for half the value. These observations lead to the conclusion that the transition state has a good deal of positive charge on the α-carbon and the C–Hal bond breaking is far advanced while C–H bond breaking is moderately advanced (40)[103]. A four-center, 4e system is anti-aromatic; consequently this reaction would not proceed readily by a non-polar pericyclic transformation but it is highly dipolar.

39   40

11.3 α-ELIMINATIONS [104,105]

Loss of nucleophilic and electrophilic groups from the same carbon atom result in the formation of a carbene, a compound containing two-coordinate carbon with only six valence electrons, :CR$_2$, clearly a highly reactive species. Intermediates of this type have been shown to form during the hydrolyses of trihaloalkanes, 41,[106–110] by the following evidence. The reactivity order for S$_N$2 reactions of the halomethanes is CH$_3$Hal $\gg$ CH$_2$Hal$_2$ > CHHal$_3$, due to increasing inductive effects. However, reactivity towards strong bases such as tert-BuO$^-$ is CH$_3$Hal $\gg$ CH$_2$Hal$_2$ $\ll$ CHHal$_3$. Evidently a different route is available for chloroform. Moreover, in D$_2$O, hydrogen exchange takes place more rapidly than hydrolysis:

	k_{exch}/k_{hydr}
CHCl$_3$	14 000
CHBr$_3$	420 000
CHI$_3$	1·05 × 10^9

and in the presence of an alkene, a dihalocyclopropane, 43, is obtained, the only plausible route to which is the pericyclic addition of dihalocarbene. Insertion is another characteristic reaction of carbenes and :CCl$_2$ also reacts in this way with acetals, 44 ($\rho^+ = -0\cdot63$). Such reactions are shown by carbenes generated by other routes. The requirements for carbene forma-

44

tion by α-elimination are an acidic α-hydrogen, the absence of a β-hydrogen and an easy leaving group. Metals, especially copper, can bring about 'carbenoid' reactions, though the intermediate is probably not a free carbene but a metal complex: The Simmons–Smith reagent (CH_2I_2–Cu) is particularly useful in generating ':CH_2' and behaves as I—CH_2—Cu, a powerful electrophilic reagent which reacts with substituted styrenes, Ar—CH=CH_2, with $\rho^+ = -2 \cdot 4$[110].

α-Eliminations from alkenes are also possible when a sufficiently good nucleophilic leaving group is available, such as trifluoromethanesulfonate (triflate, Tf, —OSO_2CF_3). For example[111],

11.4 OXIDATIVE ELIMINATIONS

An elimination across a carbon–oxygen bond comprising the loss of a proton from carbon and a nucleophilic leaving group from oxygen leads to the formation of a carbonyl group and is formally an oxidation.

Several important methods for the oxidation of alcohols to aldehydes and ketones employ this principle, the first step being to replace the alcoholic hydrogen with a suitable leaving group X.

11.4.1 Oxidations of alcohols by chromium(VI)

Chromic acid and its derivatives readily oxidize primary and secondary alcohols in acidic solution, the media commonly used being water or acetic acid. The rate law for the oxidation of isopropanol by chromic acid in water was found to be dependent upon pH in a way pointing to the involvement of one- and two-protonated species[112-114].

$$\text{rate} = k[\text{isoPrOH}][\text{HCrO}_4^-][\text{H}^+] + k''[\text{isoPrOH}][\text{HCrO}_4^-][\text{H}^+]^2.$$

There is a maximum primary kinetic isotope effect, $k_H/k_D = 7$ and evidently chromate esters are involved as reactive intermediates supported by the fact that such species as $R_2CH\text{-}O\text{-}CrO_2\text{-}O\text{-}CHR_2$ may be isolated and shown to decompose to the ketone, $R_2C{=}O$ upon addition of a weak base such as pyridine. The scheme **45** fits the evidence cited and the slow step is

45

seen to be an E2 elimination leading to a carbonyl function. The products originate from both protonated forms of the chromate ester. Substituent effects are rather small, the Hammett reaction constant, ρ, for oxidation of α-phenylethanols, ArCHMeOH being, -0.4 to -1.0, consistent with a compensating effect, electron-donation by the substituent favoring ester formation and electron-withdrawal assisting in the elimination as in the examples of Table 11.3[115]. The rate-determining step is evidently the final elimination as shown by the PKIEs; values of k_H/k_D as high as 12 have been observed.

11.4.2 The Moffatt oxidation[116,117]

A group of reactions which provide mild and specific conversions of alkyl halides and sulfonates to carbonyl compounds involve dimethyl sulfoxide, **46**, as the oxidant. The sulfoxide oxygen is sufficiently nucleophilic to displace halide in an initial S_N2 reaction with the formation of an alkoxysulfonium ion, **47**. This undergoes a ready E2 elimination to form the carbonyl compound. Alcohols cannot be oxidized directly since —OH is not a good nucleophilic leaving group. However, if the dimethyl sulfoxide is activated by dicyclohexylcarbodiimide, **48**, transfer of dimethyl sulfide to the alcohol can be carried out to give the same alkoxysulfonium ion, **47a**. The difference now is that the oxygen of the carbonyl product is derived from the original alcohol whereas, in the oxidations of halides, it is derived from the dimethyl sulfoxide. A similar type of activation is brought about by acetic anhydride in place of the carbodiimide, **49**.

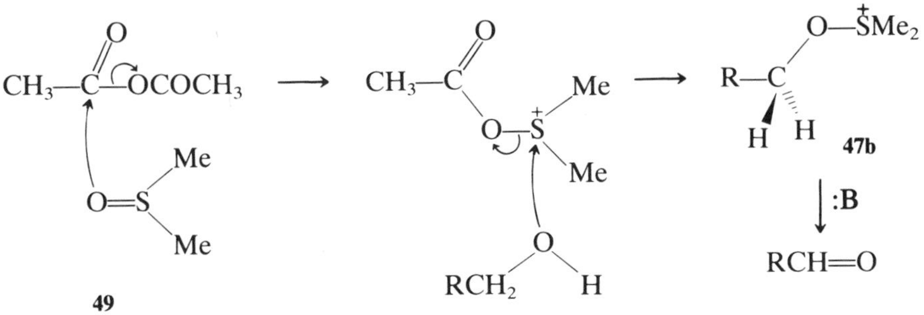

49

PROBLEMS

1 Predict the major products from the following eliminations:

(a)

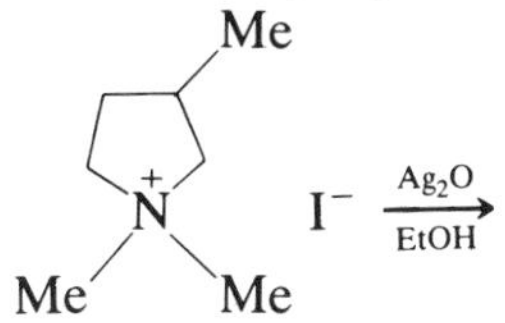

(b)

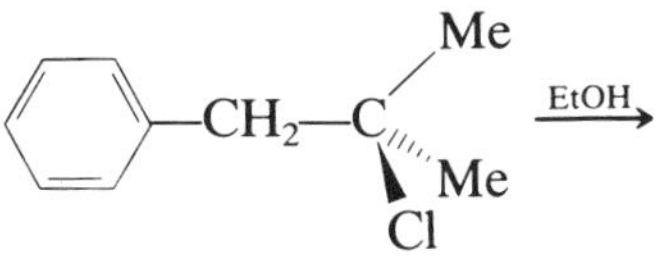

(c)

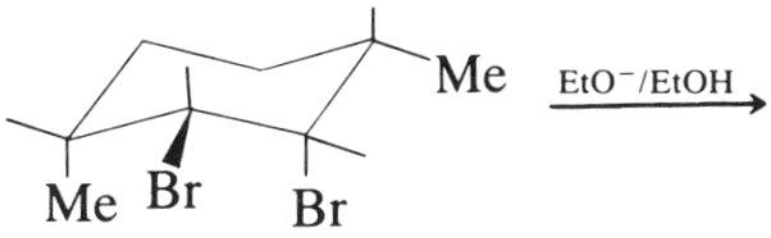

(d)

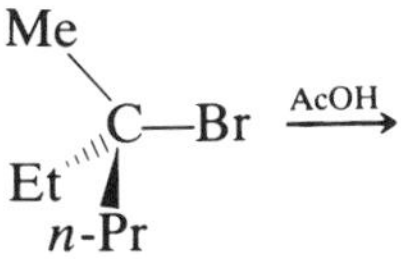

(e)

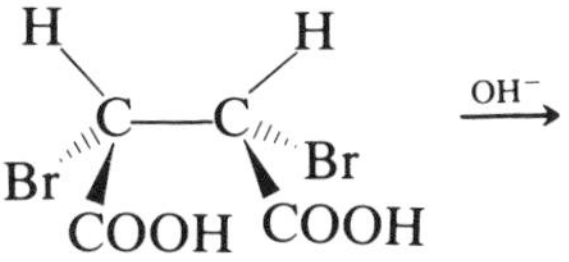

(f) 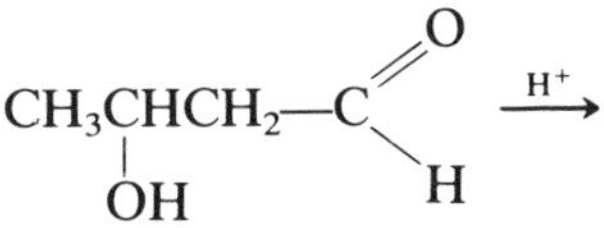 CH_3CHCH_2—C $\xrightarrow{H^+}$
          $|$ $\parallel$
          OH   H

(g) 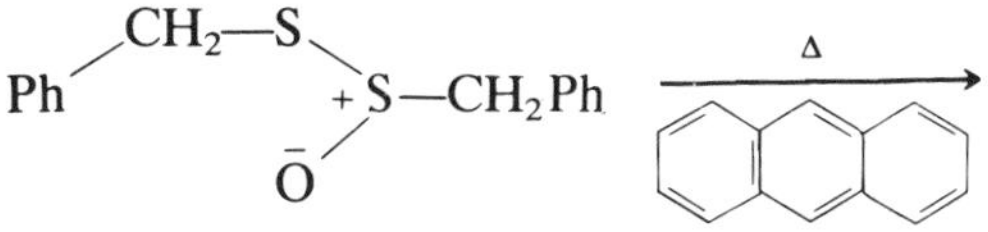 Ph

Elimination reactions

2 Dehalogenation of vicinal dihalides is a further example of a β-elimination. Suggest mechanisms for the following reactions and explain how the medium effect can result in a change in reaction pathway:

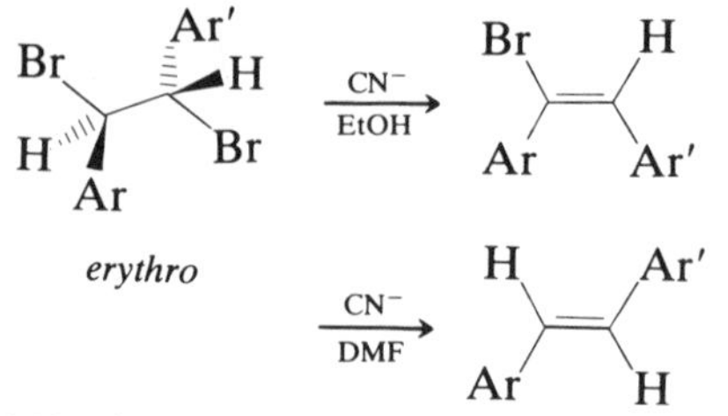

erythro

[*Tetrahedron Letts*, 4709 (1975).]

3 The following examples of the introduction of a double bond may be accomplished using selenium derivatives ('Sharpless reagents').

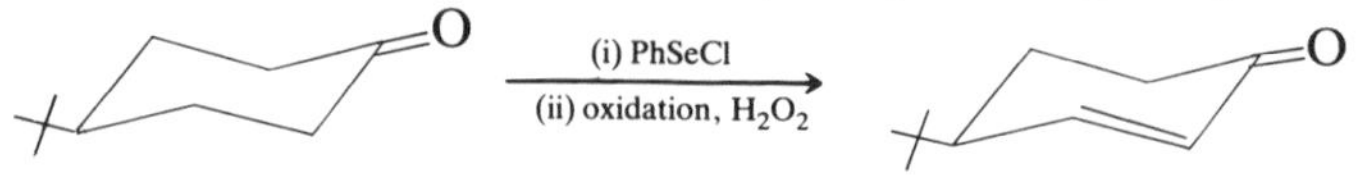

$PhCH_2CH_2COOEt$ $\xrightarrow[\text{(ii) PhSeSePh}]{\text{(i) } K^+OEt^-}$ $PhCH{=}CHCOOEt$
(iii) oxidation, H_2O_2

$ClCH_2COOEt$ $\xrightarrow[\text{(ii) EtBr}]{\text{(i) PhSe}^-}$ $MeCH{=}CHCOOEt$
(iii) oxidation

Suggest pathways for these reactions. [*J. Amer. Chem. Soc.*, **94**, 7154 (1972); **95**, 2697 (1973); **95**, 6137 (1973). *J. Org. Chem.*, **37**, 3973 (1972). *Tetrahedron Letts* 1979 (1973).]

4 Elimination from 9-fluorenylmethanol,

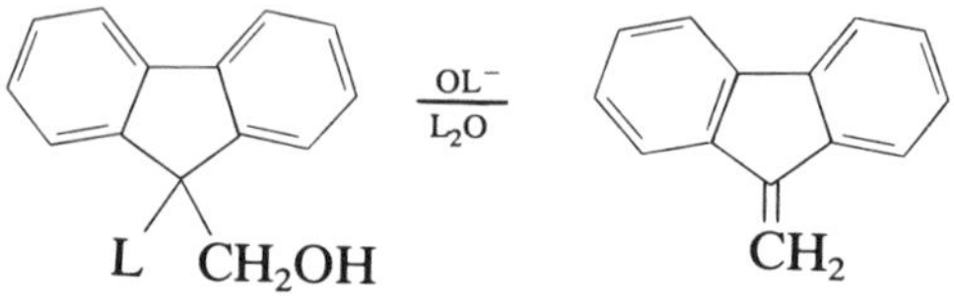

I

in aqueous sodium hydroxide has been carried out with L = H or D both in the compound and the medium. Rates of elimination, k_E, and for isotopic exchange of the 9-proton, k_x, have been measured, thus:

Solvent	$10^4 k_E(\mathbf{I}, L = H)$	$10^4 k_E(\mathbf{I}, L = D)$	$10^4 k_x$
D_2O	14·3	1·97	59·0
H_2O	1·81	0·37	6·0

Calculate isotope effects for elimination and exchange and show that the rates tabulated are consistent with an E1cb but not with an E2 elimination mechanism. [*Chem. Comm.* 486 (1969).]

5 Design an experiment to test the hypothesis that Hofmann elimination to give a terminal alkene occurs by a *syn* pathway while Saytzev elimination to the non-terminal isomer occurs by an *anti* pathway.
6 Eight stereoisomers of 1,2,3,4,5,6-hexachlorocyclohexane can exist. They may be partially distinguished by their behavior towards alkali and the number of HCl units each readily loses on elimination, thus:

No. of moles HCl readily lost:	3	2	1	0
No. of isomers:	2	4	1	1

Of the four isomers which lose 2HCl, three lose the second molecule more rapidly than the first. Identify as far as possible the isomers.

REFERENCES

1 E. S. Gould, *Mechanism and Structure in Organic Chemistry,* Holt–Rinehart–Winston, New York, 1959.
2 R. A. Bartsch and J. Závada, *Chem. Rev.,* **80,** 453 (1980).
3 W. H. Saunders, *Acc. Chem. Res.,* **8,** 19 (1976).
4 W. H. Saunders and A. F. Cockerill, *Mechanisms of Elimination Reactions,* Wiley–Interscience, New York, 1973.
5 F. G. Bordwell, *Acc. Chem. Res.,* **5,** 374 (1972).
6 C. K. Ingold, *Structure and Mechanism in Organic Chemistry,* 2nd edn, Bell, London, 1969.
7 W. T. Ford, *Acc. Chem. Res.,* **6,** 410 (1973).
8 F. Badea, *Reaction Mechanisms in Organic Chemistry,* Abacus Press, Tunbridge Wells, 1977.
9 S. Winstein, D. Darwish and N. J. Holness, *J. Amer. Chem. Soc.,* **78,** 2915 (1956).
10 J. Hayami, N. Omo and A. Kaji, *Tetrahedron Letts.,* 2727 (1970).
11 J. Hine, *J. Org. Chem.,* **27,** 4149 (1962); *J. Amer. Chem. Soc.,* **85,** 3894 (1963).
12 D. McLennon, *Quart. Rev.,* **21,** 490 (1967).
13 F. G. Bordwell, R. L. Arnold and J. B. Biranowski, *J. Org. Chem.,* **28,** 2496 (1963).
14 G. H. Fraser and H. M. R. Hoffmann, *J. Chem. Soc. (B),* 425 (1967).
15 J. F. Bunnett, *Angew. Chem. Int. Ed.,* **1,** 225 (1962); *Survey Prog. Chem.,* **5,** 53 (1969).
16 T. I. Crowell, A. A. Wall, R. T. Kemp and R. E. Lutz, *J. Amer. Chem. Soc.,* **85,** 2521 (1963).
17 K. A. Cooper, E. D. Hughes, C. K. Ingold, E. J. McNulty and M. L. Dhar, *J. Chem. Soc.,* 2038, 2049, 2058–2119 (1948).
18 R. A. Bartsch and A. P. Croft, *J. Org. Chem.,* **47,** 1364 (1982).
19 M. Charton, *J. Amer. Chem. Soc.,* **97,** 6159 (1975).
20 H. C. Brown and O. H. Wheeler, *J. Amer. Chem. Soc.,* **78,** 2199 (1956).
21 H. C. Brown and I. Moritani, *J. Amer. Chem. Soc.,* **78,** 2203 (1956).
22 H. C. Brown and I. Moritani, *J. Amer. Chem. Soc.,* **77,** 3607 (1955).
23 J. Banger, A. Jaffe, A. C.-Lin and W. H. Saunders, *J. Amer. Chem. Soc.,* **97,** 7177 (1975).
24 D. J. Miller, R. Subramanian and W. H. Saunders, *J. Amer. Chem. Soc.,* **103,** 3519 (1981).
25 R. A. Bartsch, *Acc. Chem. Res.,* **8,** 239 (1975).
26 D. J. Miller and W. H. Saunders, *J. Org. Chem.,* **46,** 4247 (1981).
27 K. C. Brown, F. J. Romano and W. H. Saunders, *J. Org. Chem.,* **46,** 4242 (1981).

28 D. J. McLennan and R. J. Wang, *J. Chem. Soc., Perkin, II,* **526,** 1373 (1974).
29 L. Steffa and E. R. Thornton, *J. Amer. Chem. Soc.,* **89,** 6149 (1967).
30 B. Feit and W. H. Saunders, *Chem. Comm.,* 610 (1967).
31 J. Závada, M. Pánková and A. Vitek, *Coll. Czech Chem. Comm.,* **46,** 3247 (1981).
32 V. Pechanek, O. Kocian, V. Halaska, M. Pánková and J. Závada, *Coll. Czech Chem. Comm.,* **46,** 2166 (1981).
33 M. Pánková and J. Závada, *Coll. Czech Chem. Comm.,* **45,** 3150 (1980).
34 M. Svoboda, J. Hapala and J. Závada, *Tetrahedron Letts,* 265, 1972.
35 D. J. McLennon, *Tetrahedron,* **31,** 2999 (1975).
36 P. Veeravagu, R. T. Arnold and E. W. Eigenmann, *J. Amer. Chem. Soc.,* **86,** 3072 (1964).
37 R. A. Bartsch and J. F. Bunnett, *J. Amer. Chem. Soc.,* **91,** 1382 (1969).
38 R. A. Bartsch, C. F. Kelly and G. M. Pruss, *J. Org. Chem,* **36,** 662 (1971).
39 A. Saytzev, *Ann. Chem.,* **179,** 296 (1975).
40 R. A. Bartsch, J. A. Allaway, D. D. Ingram and J.-G. Lee, *J. Amer. Chem. Soc.,* **97,** 6873 (1975).
41 D. M. Froemsdorf and M. D. Robbins, *J. Amer. Chem. Soc.,* **89,** 1737 (1967).
42 A. W. Hoffmann, *Ann. Chem.,* **78,** 253 (1851); **79,** 11 (1851).
43 R. A. Bartsch and K. E. Wiegers, *Tetrahedron Letts,* 3819 (1972).
44 P. D. Buckley, B. D. England and D. J. McLennan, *J. Chem. Soc. (B),* 98 (1967).
45 J. K. Barchardt and W. H. Saunders, *Tetrahedron Letts,* 3439 (1972).
46 J. L. Coke, G. D. Smith and G. H. Britton, *J. Amer. Chem. Soc.,* **97,** 4323 (1975).
47 A. K. Colter and R. E. Miller, *J. Org. Chem.,* **36,** 1898 (1971).
48 D. J. Lloyd and A. J. Parker, *Tetrahedron Letts,* 637 (1971).
49 P. Beltrame, G. Biale, D. J. Lloyd, A. J. Parker, M. Ruane and S. Winstein, *J. Amer. Chem. Soc.,* **94,** 2240 (1972).
50 I. N. Feit, F. Schmidt, J. Lubinkowski and W. H. Saunders, *J. Amer. Chem. Soc.,* **93,** 6606 (1971).
51 W. F. Bayne, *Tetrahedron Letts,* 571 (1971).
52 D. F. Froemsdorf, W. Dowd and K. E. Leimer, *J. Amer. Chem. Soc.,* **88,** 2345 (1966).
53 J. L. Coke and M. P. Cooke, *J. Amer. Chem. Soc.,* **89,** 6701 (1967).
54 J. L. Coke and M. P. Cooke, *J. Amer. Chem. Soc.,* **89,** 2779 (1967).
55 K. Fukui, *Tetrahedron Letts,* 2427 (1965).
56 K. Fukui and H. Fujimoto, *Tetrahedron Letts,* 4303 (1965).
57 K. Fukui and H. Fujimoto, *Bull. Chem. Soc. Japan* **40,** 2018 (1967).
58 K. Fukui, H. Hao and H. Fujimoto, *Bull. Chem. ʋoc. Japan,* **42,** 348 (1969).
59 J. Závada, J. Krupicka and J. Sicher, *Chem. Comm.,* 66 (1967).
60 J. Sicher, J. Závada and M. Pánková, *Coll. Czech Chem. Comm.,* **33,** 1278, 1393, 1415 (1968); **36,** 3140 (1971).
61 M. Pánková, M. Svoboda and J. Závada, *Tetrahedron Letts,* 2465 (1972).
62 J. Závada and M. Pánková, *Coll. Czech Chem. Comm.,* **45,** 2171 (1980).
63 R. P. Radman and C. J. P. Stirling, *Chem. Comm.,* 633 (1970).
64 Z. Rappoport, *J. Chem. Soc. (B),* 171 (1971).
65 M. Albortz and K. T. Douglas, *J. Chem. Soc., Perkin Trans. II,* 331 (1982).
66 F. G. Bordwell, A. C. Knipe and K. C. Yee, *J. Amer. Chem. Soc.,* **92,** 5945 (1970).
67 E. Baciocci, R. Ruzziconi and G. V. Sabastiani, *Chem. Comm.,* 80 (1980).
68 D. H. Hunter, Y. Liu, A. L. McIntyre, D. J. Shearing and M. Zvagulis, *J. Amer. Chem. Soc.,* **95,** 8327, 8333 (1973).
69 D. H. Hunter and D. J. Shearing, *J. Amer. Chem.Soc.,* **93,** 2348 (1971).
70 J. R. Keefe and W. P. Jencks, *J. Amer. Chem. Soc.,* **103,** 2457 (1981).
71 Z. Rappoport and E. Shohamy, *J. Chem. Soc. (B),* 2060 (1971).
72 J. Hine and O. B. Ramsey, *J. Amer. Chem. Soc.,* **84,** 973 (1962).
73 A. Berndt, *Angew. Chem.,* **81,** 567 (1969).
74 S. Thea, G. Guanti, G. Petrillo, A. Hopkins and A. Williams, *Chem. Comm.,* 577 (1982).

75 G. Cevasci, G. Guanti, S. Thea and A. Williams, *Chem. Comm.*, 783 (1984).
76 N. S. Isaacs and T. Najem, *Chem. Comm.*, 1361 (1984).
77 C. H. DePuy and R. W. King, *Chem. Rev.*, **60,** 431 (1960).
78 D. V. Banthorpe, *Elimination Reactions*, Elsevier, Amsterdam, 1963.
79 E. R. Alexander and A. Mudrak, *J. Amer. Chem. Soc.*, **72,** 3194 (1950); **73,** 59 (1951).
80 D. Y. Curtin and D. B. Kellam, *J. Amer. Chem. Soc.*, **75,** 6011 (1953).
81 G. C. Smith, F. D. Bagley and R. Taylor, *J. Amer. Chem. Soc.*, **83,** 3647 (1961).
82 R. Taylor, G. C. Smith and W. H. Wetzel, *J. Amer. Chem. Soc.*, **84,** 4817 (1962).
83 R. Taylor, *J. Chem. Soc. (B)*, 1559 (1968).
84 C. H. DePuty, R. W. King and D. H. Froemsdorf, *Tetrahedron*, **7,** 123 (1959).
85 D. L. Stull, F. Westheimer and G. F. Sinke, *Chemical Thermodynamics of Organic Compounds*, John Wiley, New York, 1969.
86 H. Kwart and D. P. Hoster, *Chem. Comm.*, 1155 (1967).
87 L. Chugaev, *Ber.*, **32,** 3332 (1899).
88 H. R. Nace, *Organic Reactions*, Vol. 12, Wiley–Interscience, New York, 1962.
89 D. J. Cram, *J. Amer. Chem. Soc.*, **71,** 3883 (1949).
90 R. A. Benkeser and J. J. Hazdra, *J. Amer. Chem. Soc.*, **81,** 228 (1959).
91 R. F. W. Bader and A. N. Bourns, *Can. J. Chem.*, **39,** 348 (1961).
92 A. C. Cope and E. R. Trumbell, *Organic Reactions*, Vol. 11, Wiley–Interscience, New York, 1960.
93 D. W. Emerson and T. J. Korniski, *J. Org. Chem.*, **34,** 4115 (1969).
94 C. A. Kingsbury and D. J. Cram, *J. Amer. Chem. Soc.*, **82,** 1810 (1960).
95 A. C. Cope, C. L. Bumgardner and E. E. Schweizer, *J. Amer. Chem. Soc.*, **79,** 4729 (1957).
96 J. Závada, J. Krupicka and J. Sicher, *Coll. Czech Chem. Comm.*, **31,** 4273 (1966).
97 D. W. Emerson, A. P. Craig and I. W. Potts, *J. Org. Chem.*, **32,** 102 (1967).
98 K. B. Sharpless, M. W. Young and R. F. Laver, *Tetrahedron Letts*, 1979 (1973).
99 A. T. Blades, *Can. J. Chem.*, **36,** 1043 (1958).
100 A. T. Blades and G. W. Murphy, *J. Amer. Chem. Soc.*, **74,** 6219 (1952).
101 A. Maccoll and S. C. Wong, *J. Chem. Soc.*, 1492 (1968).
102 S. W. Benson and A. N. Bose, *J. Chem. Phys.*, **39,** 3463 (1963).
103 A. Maccoll and P. J. Thomas, *Nature*, **176,** 392 (1955).
104 R. A. Moss and M. Jones, Eds., *Reactive Intermediates*, Vols 1, 2, Wiley, New York, 1981.
105 T. Gilchrist and C. W. Rees, *Arenes, Nitrenes and Carbenes* Nelson, London, 1969; B. Capon, J. Perkins and C. W. Rees, Eds, *Organic Reaction Mechanisms*, Wiley–Interscience, London, 1970–1980.
106 J. Hine, *Physical Organic Chemistry*, McGraw-Hill, New York, 1962.
107 J. Hine, *et al.*, *J. Amer. Chem. Soc.*, **79,** 1406, 5497 (1957); **80,** 824, 4282 (1958); *J. Org. Chem.*, **25,** 606 (1960).
108 W. P. Weber and G. W. Gokel, *Phase Transfer Catalysis in Organic Synthesis,* Springer, New York, 1978.
109 C. M. Starks and C. Liotta, *Phase Transfer Catalysis*, Academic Press, New York, 1978.
110 N. Kawabuta, I. Kamenura and M. Naka, *J. Amer. Chem. Soc.*, **101,** 2139 (1979); *Tetrahedron*, **27,** 1799 (1971).
111 P. J. Stang, *Acc. Chem. Res.*, **11,** 107 (1978).
112 K. B. Wiberg, Ed., *Oxidation in Organic Chemistry*, Academic Press, New York, 1965.
113 F. H. Westheimer and A. Novick, *J. Phys. Chem.*, **11,** 506 (1943).
114 J. Schreiber and A. Eschenmoser, *Helv. Chem. Acta*, **38,** 1529 (1955).
115 H. Kwart and P. S. Francis, *J. Amer. Chem. Soc.*, **77,** 4907 (1955).
116 N. Kornblum, W. J. Jones and G. J. Anderson, *J. Amer. Chem. Soc.*, **81,** 4113 (1959).
117 K. E. Pfitzner and J. G. Moffatt, *J. Amer. Chem. Soc.*, **87,** 5661, 5670 (1965); A. H. Feneslaw and J. G. Moffatt, *J. Amer. Chem. Soc.*, **88,** 1972 (1966).

12 Polar addition reactions

The reverse of an elimination is the addition of a nucleophile and an electrophile, in either order, across a double or triple bond[1]. Thus, three entities need to come together to form the product; but additions normally take place in two steps since termolecular processes are of low probability. However, the termolecular mechanism is by no means unknown. The reactions are denoted electrophilic additions, Ad_E, if the electrophilic part of the reagent adds first, and nucleophilic additions, Ad_N, if the nucleophilic moiety coordinates first to the double bond. The simplest two-step processes (**1, 2**) will give rise to second-order kinetics (Ad_E2, Ad_N2). More complex kinetic schemes, however, are frequently encountered; they include third-order reactions, which may arise if the initial interaction permits the

Table 12.1   Polar additions: scope of the reaction

(a) Electrophilic additions of E—Nu

$$Nu\!-\!E \quad \longrightarrow \quad \left[\overset{+}{\underset{N\bar{u}}{}} \;\; E \right] \quad \longrightarrow \quad \left[\underset{Nu}{} \;\; E \right]$$

H—Nu	Hal—Nu	Other E—Nu
H—OH	Hal-Hal	epoxidation by
H—OCOR	Hal-Hal′ (e.g. Br—Cl)	RCO·OOH
H—OSO$_3$R	Hal-NR·COR	RS—Hal
H—OR (H$^+$)	I—OCOR	RSe—Hal
H—OAr	I—N$_3$	HO—Hal
H—SR	I—ONO$_2$	O=N—Hal
		O$_2$N—OCOR
		R$^+$ (polymerization)
		R$_2$B—H
		R$_2$Al—H
		AcOHg—Hal

(b) Nucleophilic additions

$$Nu\!:\!\overset{\frown}{}\!\!\diagup\!\!\underset{Z}{} \quad \longrightarrow \quad \underset{Nu}{}\!\!\diagup\!\!\underset{Z}{\ddot{}} \quad \overset{H^+}{\longrightarrow} \quad \underset{Nu}{}\!\!\diagup\!\!\underset{H}{\overset{Z}{}}$$

Nu:	—Z
CN$^-$	—CN
R$_3$C$^-$	—NO$_2$
RO$^-$	>C=O
	>SO$_2$

approach of a nucleophile while the π-bond is still intact (the Ad$_E$3 reaction, **3**). The scope of polar additions is illustrated in Table 12.1.

12.1 ELECTROPHILIC ADDITIONS TO ALKENES[2,3]

12.1.1 Kinetics

The acidic hydration of alkenes has frequently been studied:

$$R\!\cdot\!CH{=}CH_2 + H_2O \xrightarrow{\;H^+\;} RCH(OH)\!\cdot\!CH_3.$$

Second-order empirical rate expressions are often observed,

$$\text{Rate} = k_2[\text{alkene}][H^+],$$

for which the initial step is irreversible protonation[4,5]. Rates of these reactions correlate with H_0 and show specific acid catalysis[6,7], which has been found to be the case also for hydrations of alkynes[8] and allenes[9]. The Bunnett ϕ-parameter is negative, indicating solvation of the transition state to be less than that of the reactants, which is understandable considering the high degree of solvation of both the proton and the water molecule. Additions of molecular chlorine across an olefinic double bond are usually of second order but the corresponding brominations are of third order, their rates being proportional to $[Br_2]^2$. Evidently, electrophilic catalysis by a second molecule of bromine is occurring, facilitating removal of Br^- (4). Additions of HBr in acetic acid may show both second- and third-order terms; the latter is unlikely to be due to electrophilic catalysis (i.e. removal of Br^- from H—Br by a proton) since this would imply a hard–soft interaction;

$$\text{Rate} = k[\text{alkene}][\text{HBr}] + k'[\text{alkene}][\text{HBr}]^2.$$

The second term is due to the Ad_E3 component and the first may similarly be a consequence of this type of mechanism but with solvent acetic acid as the nucleophile, 5[10]. Entropies of activation for halogen addition are large and negative; for the bromination of styrenes, $ArCH{=}CH_2$, $\Delta S^{\ddagger} = -170\,\text{J}\,\text{K}^{-1}\,\text{mol}^{-1}$ ($-41\,\text{cal}\,\text{K}^{-1}\,\text{mol}^{-1}$) due both to association between the reagents in forming the transition state and to increased solvation around the developing charges. This makes rates of reaction very sensitive to the solvent polarity, (the Grunwald–Winstein parameter, $m_s = 1{\cdot}16$ for bromination, is even greater than for solvolysis)[11] and volumes of activation are highly negative (for hydration of propene, $\Delta V^{\ddagger} = -30\,\text{cm}^3\,\text{mol}^{-1}$)[12]. Each of these measurements is in accordance with a very polar transition state.

550

12.1.2 Effect of structure

The frontier orbital supplied by the alkene is its HOMO: consequently all reactions show a strong acceleration by electron-donating groups adjacent to the double bond which raise the HOMO energy. Hammett reaction constants are large and negative. Rates of additions to arylethenes usually correlate with σ^+ and values of ρ^+ range between -2 and -5[4,13]. Ethenes with substituents directly on the double bond show even larger structural effects on rates; Taft reaction constants ρ^* are as high as -11 (Table 12.2). The simplest LFER which correlates additions to many types of alkene is given by Eq. (12.1)[4]:

$$\log k/k_0 = \rho^+ \sum \sigma^+, \qquad (12.1)$$

while a more complex but more precise expression, Eq. (12.2), separates electronic effects at each carbon[14]:

$$\log k = 1.64 - 7.7 \sum_\alpha \sigma^+ - 13.7 \sum_\beta \sigma^+ \quad (+\text{ minor terms}). \qquad (12.2)$$

These values clearly point to the need to stabilize positive charge on carbon

Table 12.2 Linear free-energy parameters for polar additions

(a) Electrophilic additions

Substrate	Reaction	ρ^+	Ref.
$ArCH{=}CH_2$	hydration	-3.6	85
$ArCH{=}CH_2$	bromination (2nd order)	$-4.05, -4.71$	87
		$(\rho_I)\ -5.1$	
$ArCH{=}CH_2$	bromination (3rd order)	-4.6	86
		$(\rho_I)\ -7.1$	
$ArCH{=}CH_2$	epoxidation	-1.3	88
$ArCH{=}CH_2$	epoxidation (variation of $ArCO \cdot OOH$)	$+1.4$	88
$ArCH{=}CH_2$	cationic polymerization	-2.3	89
$ArCH{=}CH_2$	oxidation by CrO_3	-2.9	89
$ArCH{=}CH_2$	addition of $CrO_2Cl{-}Cl$	-2.0	90
$ArCH{=}CH_2$	addition of $NO{-}Cl$	-2.1	91
$ArCH{=}CH_2$	ozonization	-0.6 to -0.9	92
$ArCMe{=}CH_2$	$+ N$-bromosuccinimide	-5.69	22
$ArCMe{=}CH_2$	$+ N$-chlorosuccinimide	-3.0	22
$Ar\underset{\beta}{C}H{=}\underset{\alpha}{C}HPh$	bromination at C_α	-1.5	93
$ArCH{=}CHPh$	bromination at C_β	-5.0	93
$ArCH{=}CHPh$	$+ I{-}SCN$	-2.6	93
$ArCH{=}CHCOOH$	chlorination	-3.9	94
$R_1R_2C{=}CR_3R_4$	hydration	$(\rho^*)\ -10.5$	4
$R_1R_2C{=}CR_3R_4$	chlorination	$(\rho^*)\ -6.7$	95
$RCH_2C{=}CH_2$	bromination	$(\rho^*)\ -8.2$	95
$ArC{\equiv}CH$	hydration	-4.3	96
$ArC{\equiv}CCOOH$	hydration	-4.79	96
$ArCH{=}C{=}CH_2$	hydration	-4.2	9

Table 12.2 *(Continued)*

(b) Two-parameter correlations [97]:

$$\log k_X = \rho_I \sigma_I + \rho_R \sigma_R + \ln k_H$$

	Reaction	ρ_I	ρ_R	$\ln k_H$
1	(R$_4$ alkene) + Cl$_2$/AcOH	−9·72	−5·60	2·28
2	(R$_4$ alkene) + Cl$_2$/aq.AcOH	−13·5	−3·38	0·912
3	(Ar, R alkene) + Cl$_2$/AcOH	−7·14	−3·02	3.41
4	(R$_4$ alkene) + Br$_2$/aq.H$^+$	−22·4	2·29	6·10
5	(R, R' alkene) + Br$_2$/aq.H$^+$	−10·7	−5·82	4·7
6	(R, Me alkene) + H$_3$O$^+$	−4·32	−32·9	−0·158
7	(R, Me alkene) + CF$_3$COOH	−8·48	−13·7	0·672
8	(R$_4$ alkene) + Hg^{2+}	−18·3	−0·554	0·835
9	(R, Z alkene) + MeO$^-$	2·13	−18·5	−3·09
10	(R, Z alkene) + HN(morpholine)	1·28	+32·7	−2·02

adjacent to the center to which the electrophile becomes attached (C_β) but also to a lesser extent at the reaction center itself (C_α). While conjugation will stabilize the reagent alkene, it will stabilize the cationic character of the Ad$_E$ transition state even more [15]. The large difference between ρ_α^+ and ρ_β^+ emphasizes the asymmetry of charge borne by the two alkene carbons in the transition state, characteristic of an open carbocation.

Two-parameter correlations of relative rates with σ_R and σ_I show that the relative sensitivities (ρ_R, ρ_I) of different addition reactions to these two components of the substituent effect can be very variable (Table 12.2). Acid-catalyzed hydration must proceed via an open carbocation, **1a**, for which resonance stabilization of the charge is most important, leading to $\rho_R \gg \rho_I$. On the other hand electrophiles which lead to bridged intermediates (in halogenation or mercuration for example, **1b**) are mainly influenced by inductive stabilization since in these intermediates the charge center is not in direct conjugation with substituents and $\rho_I \gg \rho_R$. This analysis can be used as additional evidence of bridging.

The data in Table 12.3 summarize the effect of alkene structure on rates of additions. Series (a) suggests that relief of bond angle strain on hydration (as estimated from equilibrium data) is especially favorable for the four-membered rings, **6**, but this is tempered at the same time with an increase in eclipsing interactions as the addition proceeds. This series shows how the very large effect of $-R$ substituents is attenuated when transmitted through an aromatic ring. Series (c) illustrates the effect of an $-I$ substituent transmitted through an alkyl chain; after four bonds it is still able to halve the rate of addition of bromine which, arguably, could be attributed to an electric field effect, transmitted through space.

Steric effects are undoubtedly of importance, as shown by the following study. Relative rates of addition to a series of alkenes of bromine and of mercuric acetate are not parallel (Fig. 12.1a). They become linearly related when corrected for differences in steric interactions by the use of Eq. (12.3),

Table 12.3   Reactivities of alkenes towards electrophilic additions

(a) Acid-catalyzed hydration

Alkene:	$\diagup\!\!=$	⬠	⬜	⬜
k_{rel}	1	0·007	30	200

$XC_6H_4CH{=}CH_2$; X: =	MeO	Me	H	Cl	NO_2
k_{rel}[4]: =	602	20	1	0·4	0·0013

$XCH{=}CH_2$; X: =	MeO	Ph	Me	H
k_{rel}: =	5×10^{14}	$1·6 \times 10^8$	$1·6 \times 10^7$	1

(b) Addition of CF_3COOH to $CH_2CH{-}CH_2CH_2X$

X:	H	$OCOCF_3$	OAc	Cl	OMe
k_{rel}[85]:	1000	0·2	2·2	2·3	3·8

Table 12.3 *(Continued)*

(c) Bromination

$XCH_2CH{=}CH_2$; X: =	Me	H	OH	Ph	OAc	Cl	CN
k_{rel}[14]: =	1230	1000	160	130	21	0·4	0·07

$CH_2{=}CH(CH_2)_nBr$; n: =	1	2	3	4	5
k_{rel}[85]: =	0·06	3·5	24	56	100

(d) Epoxidation by $CH_3CO{\cdot}OOH$ in AcOH

Alkene: =	 $\diagdown\!{=}$	$\diagdown\!{=}$	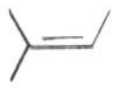 $\diagup\!{=}\diagup$	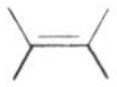 $\diagup\!{=}\diagdown$
k_{rel}:	1	22	484	6526	v. fast

(e) Addition of PhSCl to p-$XC_6H_4CH{=}CH_2$

X:	MeO	Me	H	Cl	NO_2
k_{rel}[53]:	340	3·1	1·0	0·29	0·019

(f) Ad–E and E–Ad reactions

Reaction	$10^4k_2/$ $M^{-1}s^{-1}$	$\Delta H^{\ddagger}/$ $kJ\,mol^{-1}$ $(kcal\,mol^{-1})$	$\Delta S^{\ddagger}/$ $J\,K^{-1}\,mol^{-1}$ $(cal\,K^{-1}\,mol^{-1})$	Mechanism
cis-p-$NO_2C_6H_4CH{=}CHBr$ + PhS⁻, MeOH	0·34	76 (18·2)	−48 (−11·5)	Ad_N–E
trans-p-$NO_2C_6H_4CH{=}CHBr$ + PhS⁻, MeOH	1·26	74 (17·7)	−48 (−11·5)	Ad_N–E
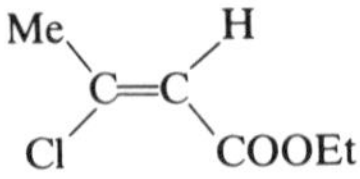 Me, H / C=C / Cl, COOEt + PhS⁻, MeOH	2·23	64 (15·3)	−69 (−16·5)	Ad_N–E
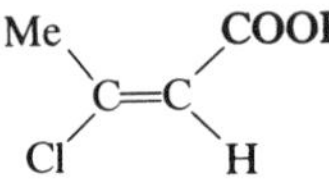 Me, COOEt / C=C / Cl, H + PhS⁻, MeOH	0·55	61 (14·6)	−98 (−23·4)	Ad_N–E
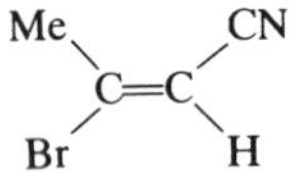 Me, CN / C=C / Br, H + PhS⁻, MeOH	1 (k_{rel})	75 (17·9)	+4 (−1·0)	Ad_N–E
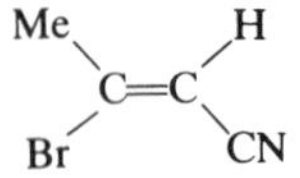 Me, H / C=C / Br, CN + PhS⁻, MeOH	83 (k_{rel})	83 (19·6)	+63 (15·1)	E–Ad_N
$PhSO_2$, H / C=C / H, Cl + MeO⁻, MeOH	440 / 51	89 (21·3) / 100 (23·9)	+33 (7·9) / +110 (26·3)	addition / elimination

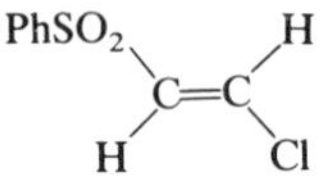

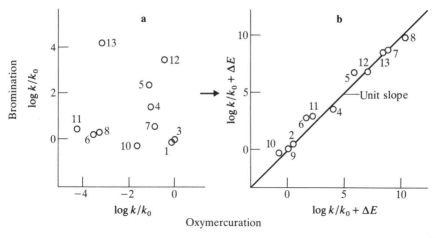

Fig. 12.1 Comparison of the reactivity of alkenes towards mercuration and bromination, (a) before and (b) after inclusion of a steric parameter according to Eq. (12.3)[99].

Fig. 12.1b[99].

$$\ln k_{rel}(Br_2) + \Delta E_{Br}/RT = \ln k_{rel}(HgX_2) + \Delta E_{HgX}/RT \qquad (12.3)$$

in which ΔE for each electrophile, a relative steric effect, is defined as:

$$\Delta E = -\Delta I_D + \Delta h\nu_{CT}, \qquad (12.4)$$

where ΔI_D is the difference in ionization potential between two alkenes with different steric requirements and $\Delta h\nu_{CT}$ the difference in charge-transfer excitation energy for the electrophile interacting with each alkene; this depends upon closeness of approach and hence upon the size of both reactants. The less bulky the electrophile, the smaller is ΔE; Br_2 is evidently less bulky than $Hg(OAc)_2$ the difference being accentuated the more sterically hindered is the approach to the alkene.

12.1.3 Isotope effects

The transfer of a proton from a strong acid to the alkene should be accompanied by a primary isotope effect. In aqueous solution this is complicated by the possibility that proton transfer may be rate-determining or a pre-equilibrium event. In the latter case, inverse isotope effects, $k_{H_2O}/k_{D_2O} < 1$, would be observed for reasons discussed in Section 7.3. Furthermore, small or even inverse primary isotope effects could be obtained if the proton were transferred from an acid for which the X–H bond was of low frequency to form a C–H bond of high frequency with a net gain of zero-point energy. This would be the case for proton transfers from HCl, HBr and especially HI[16]. These factors may explain the values in Table 12.4 which, it is supposed, all reflect initial protonation of an alkene

Table 12.4 Isotope effects for polar addition reactions[4,5,7,15]

Reaction[a]	k_H/k_D
(a) Primary + solvent	
$RCH{=}CH_2 + L_3O^+$	0·3–0·5
$RCH{=}CH_2 + LF$	3·3
$CH_2{=}CH{-}CH{=}CH_2 + L_3O^+$	1·3–2·0
$CH_2{=}CHOR + L_3O^+$	2·95
$CH_2{=}CHOR + HCOOL$	6·8
(b) Primary	
$RCH{=}CHR + LCl$	1·8
$RCH{=}CHR + LBr$	0·8
(c) Secondary	
$PhCL{=}CH_2 + Br_2$	0·98
$PhCL{=}CH_2$ epoxidation	0·90
(d) Elimination–addition $(E{-}Ad_N)$[58]	

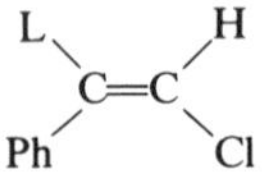

	2·2

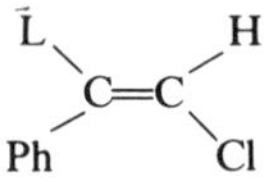	3·2

[a] L = H or D.

in the various additions depicted. PKIE values decrease in the order $RCOOH > H_3O^+ > HCl > HBr$ together with H—X stretching frequencies. Secondary isotope effects of hydrogen attached to the double bond should reveal the degree of hybridization change at the transition state. The greater the conversion from sp^2 to sp^3, the lower is the value of k_H/k_D. For the β-protons, the SKIE is very small (0·98) but for the α-protons it is much larger (0·90). This is consistent with a change in hybridization ($sp^2 \rightarrow sp^3$) at C_α but not at C_β, the incipient carbocationic center.

12.1.4 Orientation and stereochemistry

The Markownikoff Rule (1870)[17] states that the addition of HX to an unsymmetrically substituted ethene results in the adduct in which H has bonded to the carbon bearing the greater number of hydrogens. This, of course, is simply an expression of the mechanistic requirement that H$^+$ adds initially to the alkene generating a carbocation, the most stable one possible (7, 8). There are some exceptions to the rule as originally formulated, but they disappear if it is expressed in a more modern version:

'The major product originates from the more stable intermediate formed by the addition of an electrophile to the double bond'.

556

$$\overset{+}{EtCH_2-CH_2} \xleftarrow{\;H^+I^-\;}{\not\!\!\!} EtCH=CH_2 \xrightarrow{\;H^+I^-\;} Et\overset{+}{CH}-CH_3 \longrightarrow EtCHI-CH_3$$
$$\qquad\qquad\qquad\qquad\qquad\qquad\qquad\qquad\qquad\qquad\qquad\qquad 100\%$$

7

$$Me_2CH-\overset{+}{CH_2} \xleftarrow{}{\not\!\!\!} Me_2C=CH_2 \xrightarrow{\;H^+Br^-\;} Me_2\overset{+}{C}-CH_3 \longrightarrow Me_2CBr-CH_3$$

8

Such is the difference in energy between primary, secondary and tertiary carbocations that electrophilic additions, at least those initiated by protonation, adhere strictly to the Markownikoff Rule. Apparent exceptions to the original version are due to the more stable carbocation having positive charge on the less substituted carbon; for example, in reactions **9, 10**[18,19] the 'contra-Markownikoff' cations are thus more stable by virtue of being more remote from electron-withdrawing groups (compare orientation in aromatic electrophilic substitution, Section 10.4.6).

$$Me_3\overset{+}{N}-CH_2-\overset{+}{CH_2} \xleftarrow{\;H^+\;} Me_3\overset{+}{N}-CH=CH_2 \xrightarrow{\;H^+\;}{\not\!\!\!} Me_3\overset{+}{N}-\overset{+}{CH}-CH_3$$
$$\downarrow{\scriptstyle I^-} \qquad\qquad\qquad\qquad\quad \textbf{9}$$
$$Me_3\overset{+}{N}-CH_2-CH_2-I$$

$$F_3C-CH_2-\overset{+}{CH_2} \xleftarrow{\;H^+\;} F_3\overset{\delta+}{C}-CH=CH_2 \xrightarrow{}{\not\!\!\!} F_3\overset{\delta+}{C}\rightarrow\overset{+}{CH}-CH_3$$
$$\downarrow{\scriptstyle I^-} \qquad\qquad\qquad\qquad\quad \textbf{10}$$
$$F_3C + CH_2-CH_2-I$$

Less extreme regioselectivity is found in additions of bromine under conditions in which the intermediate reacts with some other nucleophile than Br⁻ (**11**). This behavior indicates that products have arisen not from a carbocation but from a bridged species, a bromonium ion (**12, 4a**), on which subsequent S$_N$2 attack leads to the two products with relatively low selectivity. A bridged intermediate is not possible for proton addition but it is possible and, indeed, more stable than the corresponding secondary alkyl carbocation when an adjacent Br, I or (probably) oxygen is available. This results in the proportions of 'Markownikoff' products, **13a**, in the additions of hypohalous acids being quite low and almost independent of the halide. Both products most likely arise through the intermediacy of **13**.

$$ClCH_2CH{=}CH_2 \longrightarrow \left[ClCH_2\underset{\diagdown\;\;O\;\;\diagup}{CH{-}CH_2} \right] X^-$$

$$\underset{HO{-}X}{\Big\downarrow} \qquad\qquad \underset{\overset{|}{H}}{\overset{+}{O}} \qquad \mathbf{13}$$

$$ClCH_2\underset{\underset{\mathbf{13a}}{X \quad OH}}{CH{-}CH_2} \;+\; ClCH_2\underset{\underset{\mathbf{13b}}{OH \quad X}}{CH{-}CH_2}$$

	13a	13b
X = Cl:	31%	69%
X = Br:	27%	73%
X = I:	29%	71%

Further stereoisomerism may arise depending on whether E and Nu add to the same (*syn*) or opposite (*anti*) sides of the double bond. This will depend on the type of intermediate formed.

12.1.5 The nature of the intermediates in Ad_E reactions

Addition of an electrophile to an alkene could lead initially to a π-complex of the type formed by Ag^+ and other metal ions, **14**. Though alkene–proton π-complexes do not appear to be intermediates, there is spectroscopic evidence for the existence of charge-transfer complexes between bromine and alkenes, and kinetic evidence that these lie on the reaction path to dibromoalkane[101]. The disappearance of these complexes is second-order, requiring attack by another molecule of bromine. This means that the Ad_E3 process is a three-step mechanism. The choice of structure for the second intermediate is either an open carbocation or a cyclic 'bridged' cation. Bridging of the two alkene carbons is a possibility for all Group V–VII electrophiles since they contain an unshared electron pair in addition to an accessible vacant orbital. Bridged intermediates are revealed by the *anti* stereochemistry of addition which must accompany the opening of the intermediate three-membered ring, an S_N2 reaction, whereas open carbocations add with little stereochemical preference. The addition of DCl or DBr

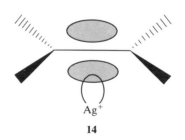

14

to cyclic alkenes yields *syn* and *anti* products in equal ratio[20], indicative of an open carbocation, but addition of halogen usually shows a strong preference for *anti,* the more so when structures disfavor the open carbocation, for example by the presence of electron-withdrawing groups in reaction **15**[21]; see Table 12.5

Table 12.5 Stereochemistry of addition of halogen to ethyl *trans*-cinnamates

$$p\text{-}X\text{---}C_6H_4CH\text{==}CHCOOEt + Hal_2$$
$$(15)$$

X	Anti/syn ratio in product	
	Br_2	Cl_2
Me	1·4	3·4
H	6·8	5·9
Cl	5·9	8·4
NO$_2$	13·3	>20

Additions of ArS—Cl to alkenes[22], and of Br—OAc[23] and I—N$_3$[24], are mainly of *anti* geometry, especially in solvents of low ionizing power; again this is probably due to bridging.

The Ad$_E$3 mechanism, **3**, would also tend to give *anti* products and this seems to be the case in the addition of ClBr to certain alkenes in the presence of LiCl.

NOCl gives mixtures of *syn* and *anti* products, plausibly since the cyclic product is not a stabilized one, **16**[25,26]. Norbornene, **18**, tends to add

16

X	18a : 18b
Cl	1 : 1
OAc	3 : 1

19 $Me_3C \cdot CH = CH_2 \xrightarrow[\text{AcOH}]{\text{HCl,}} Me_3C \cdot CH - CH_3 + Me_2C - CH - CH_3$

65%   35%

exo–syn, **18a**, whatever the reagent; this is due partly to steric effects blocking the *endo* side and partly to especially favorable stereoelectronic factors[27–29].

The intermediate formation of carbocations is often revealed by 1,2-shifts (Wagner–Meerwein rearrangements) which they characteristically undergo. However, the acid-catalyzed addition of water usually occurs without rearrangement. Quenching of the initial carbocation by water must in this case be extremely rapid, in the limit tending towards the S_E3 mechanism. Wagner–Meerwein rearrangements are common in solvolytic reactions in which it has been suggested that they provide some of the driving force towards ionization. This is evidently not beneficial for protonation of an alkene and it is likely to occur only when the carbocation is sufficiently long-lived. Some examples of additions accompanied by carbocation rearrangement are shown in **17–19**. Even in those cases most prone to 1,2-shifts, such as neopentyl cations, substantial proportions of product are formed before rearrangement can occur[30,31]. Nucleophilic groups within the alkene molecule can readily quench the intermediate cation if a 5- or 6-membered ring results[32,33].

12.2 MISCELLANEOUS ADDITIONS

12.2.1 Hydroboration

Boron hydrides, either as diborane $(BH_3)_2$ or its organic derivatives, $(RBH_2)_3$ and $(R_2BH)_2$, add B—H across an olefinic double bond, the boron being the electropositive atom so that contra-Markownikoff addition occurs. Successive addition occurs for each B–H bond present in the borane (**20**). The trialkylborane products can undergo a host of synthetically useful transformations resulting from the subsequent displacement of boron, **21**[34]. For example, oxidation to the alcohol results in the effective contra-Markownikoff hydration of the alkene. Mechanistic studies of hydroboration are hampered by several factors. Rates of addition of $(BH_3)_2$ are mostly too fast for convenient measurement and the three available hydrogens would give rise to complex sequential reaction kinetics. Secondly, the boranes exist as hydrogen-bridged dimers, though often in equilibrium with monomers. The two species may react at their individual rates but several extreme conditions may be defined (Table 12.6)

Table 12.6   Molecularity of hydroboration

	Rate-determining step	Molecularity	Rate-law rate
a	k_1	Unimolecular	$k_1[(R_2BH)_2]$
b	$k_{2'}$	$\frac{3}{2}$	$k_{\frac{3}{2}}[\text{alkene}][(R_2BH)_2]^{\frac{1}{2}}$
c	k_2	Bimolecular	$k_2'[\text{alkene}][(R_2BH)_2]$

Only in cases b and c can one learn anything of the structural effects of the alkene on rates of reaction. Mechanistic studies have been carried out using the secondary boranes disiamylborane, **22**, and 9-borabicyclo[3,3,1]nonane, **23**, for which no sequential reactions are possible and rates are reasonably slow[35–38]. Many alkenes react with first-order kinetics and at the same rates, clearly showing that dissociation of the dimeric borane, k_1, is the slow step: relative rates with various alkenes can be obtained by a competition method by analysis of the products obtained from a mixture of alkenes in large excess of the borane. Terminal alkenes appear to be favored, presumably by steric effects (Table 12.7)[100].

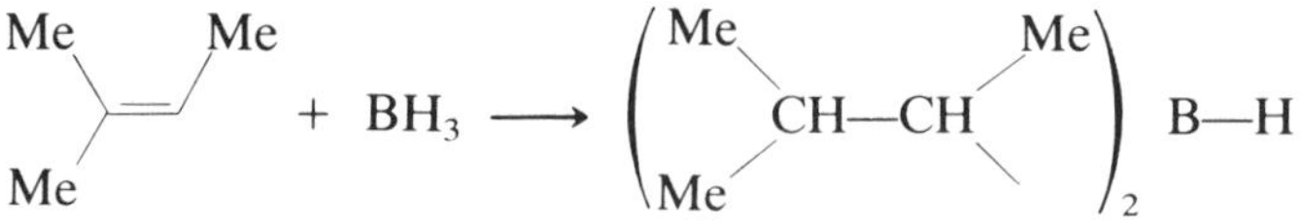

22

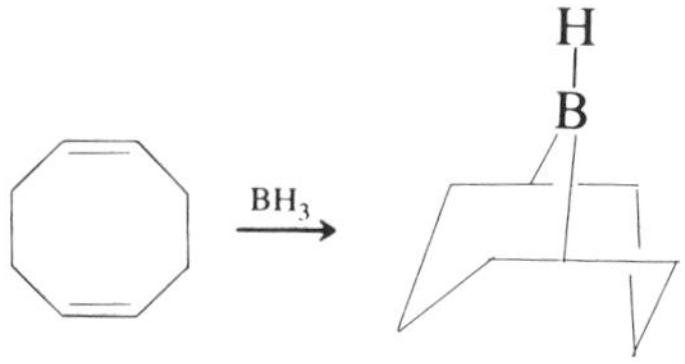

23

Table 12.7   Some alkene reactivities in hydroboration

Alkene	k_{rel}	$k(CCl_4)$	$k(THF)$	$k(Et_2O)$
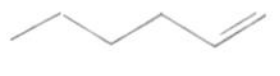	1·00	1·54	12·9	2·83
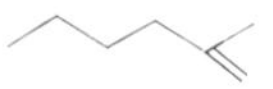	1·96	1·46	13·7	2·81
	0·233	1·45	14·0	2·80
	0·072	1·54	11·8	2·77

24   Cl—CH₂⁔ + **23** $\xrightarrow{\text{Ad2}}$ Cl⏜CH₂⌇C⟨B⟩
                                                                          CH₃

$\xrightarrow{\text{E2}}$ CH₂=⟨CH₃⟩ $\xrightarrow[\text{Ad2}]{\text{23}}$ CH₃⟨B⟩
                                                                                      CH₃

Allylic halides often undergo addition followed by elimination (**24**). Less reactive alkenes, such as cyclohexene, and haloalkynes tend to react with **23** exhibiting $\frac{3}{2}$-order kinetics of the type b reaction as the Ad2 process becomes rate-controlling.

Epoxidation is carried out by reaction of an alkene with a peroxy carboxylic acid and has all the characteristics of an electrophilic addition, necessarily with *syn* stereochemistry. Rates are of second order, are enhanced by electron-donating groups and correlate with σ^+ though reaction constants are smaller than those for other additions; $\rho^+ = 1\cdot2$ for ArCH=CH₂ with peroxyacetic acid, and -4 for substituents directly on the double bond (Table 12.8).

At the same time, steric effects can be seen to reduce the rate of epoxidation of triphenylethene. No prior dissociation of the peracid to OH⁺ occurs since there is no exchange of oxygen between peracetic acid and acetate ion. The transition state is supposed to be similar to **25a**, forming epoxide in a single concerted step.

The **Prévost** and **Woodward reactions** are essentially additions of iodine carboxylate formed *in situ*. The Prévost procedure[39] uses iodine and silver carboxylate. The initially formed *trans*-iodocarboxylate then undergoes internal displacement and formation of the diester, **26**, with retention of configuration, an example of neighboring-group participation. In the Woodward procedure[40,41] a little water is present which intercepts the intermediate, leading to *cis*-hydroxyester, **27**. Cambie's modification uses thallium carboxylates and addition to a vinyl ester leads to formation of the β-iodoketone, **28**[42].

Table 12.8   Rates of epoxidation of some alkenes

Alkene	k	Alkene	k	Alkene	k
=	1	⚊⟋	2·2	=⟨	484
⟩⟨	6526	⟩=⟨	v. fast		
⟍⟍	59	Ph⟩=⟨Ph	252	Ph⟩=⟨Ph	30

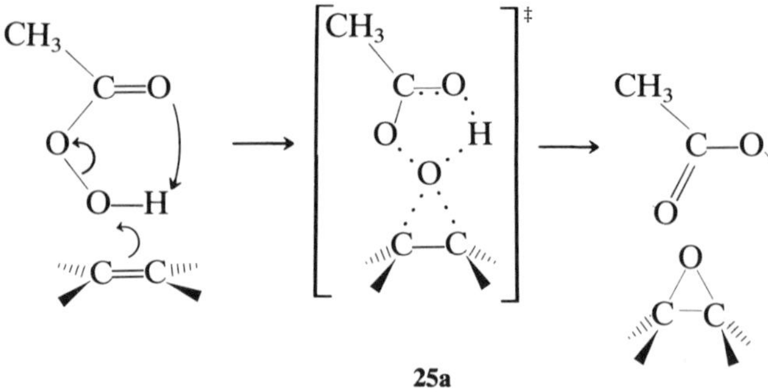

25 → 25a

26

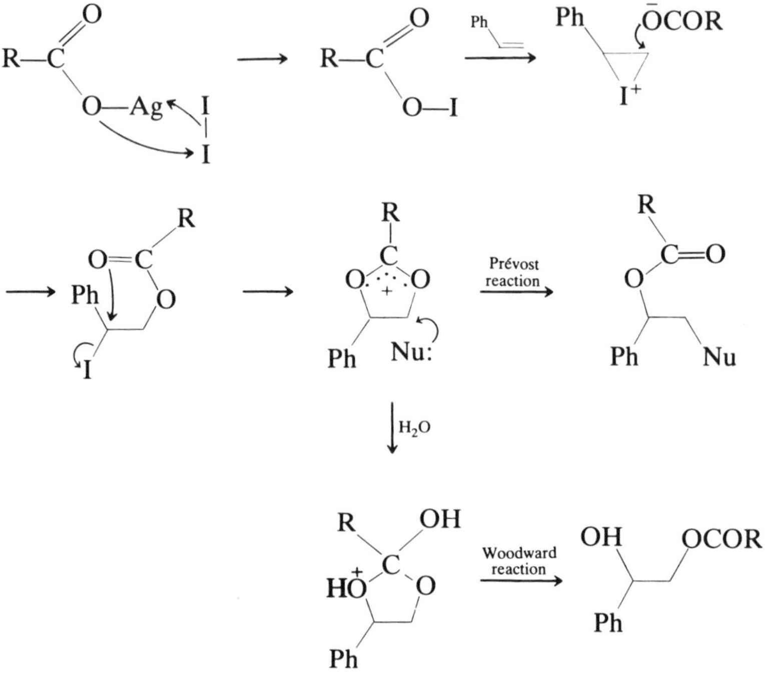

Prévost reaction

$\downarrow$ H₂O

Woodward reaction

27

28

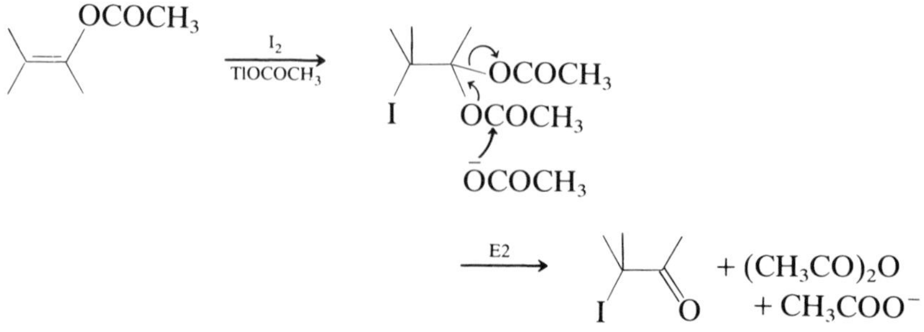

$\xrightarrow[\text{TlOCOCH}_3]{\text{I}_2}$

$\xrightarrow{\text{E2}}$ + (CH₃CO)₂O + CH₃COO⁻

12.2.2 Addition with ring closure; halolactonization[43]

Addition of iodine or bromine to a 3-, 4- or 5-alkenoic acid commences as usual but the intermediate carbocation may then be captured by the carboxylic acid group and a lactone formed. Reactions of this type can be very useful for diagnosis of the stereochemistry of a molecule which contains both functions. Of the stereoisomers **29** and **30**, only the former can iodolactonize, for obvious reasons[44].

12.2.3 Addition of carbocations

Carbocations, as powerful electrophiles, add to alkenes. Protonation of the alkene may be followed by addition to another molecule, setting up a cationic polymerization chain reaction, **31**. Addition of alkyl and acyl cations (produced from the corresponding halides by Lewis acid) can also occur. The addition of protonated formaldehyde, **32**, constitutes the Prins reaction[45].

12.2.4 Additions to dienes, alkynes and allenes

The initial adduct of an electrophile with a 1,3-diene is an allylic cation, **33**, which can attach a nucleophile at either end giving products of 1,2- or 1,4-addition which may themselves interchange. Butadiene and isoprene give 75% of the 1,2-adduct with HCl at low temperatures although prolonged reaction leads to the equilibrium proportion 25% 1,2- to 75% 1,4-isomers. Cyclohexadiene, however, adds DBr to give 80% of the 1,4-adduct. The additions of the halogens give mixtures which depend much on the solvent, for reasons which are not well understood.

75%   25% initial ratio

25%   75% equilibrium ratio

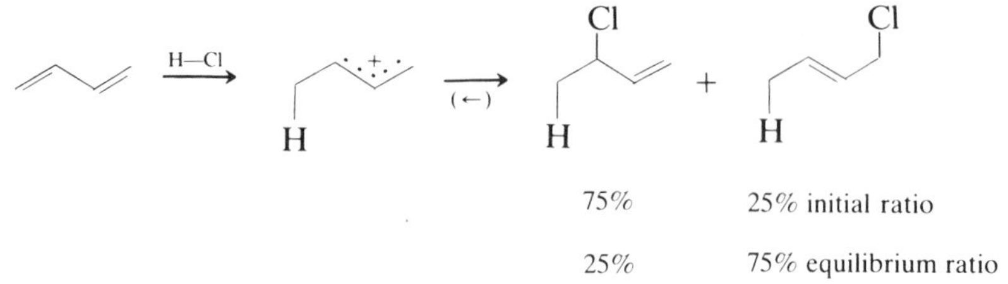

Occasionally aromatic systems undergo addition rather than substitution, especially when benzenium ions are highly stabilized by +*R* groups such as —OH and after *ipso* attack. Subsequent eliminations may produce complex and unexpected products, e.g. **34**.

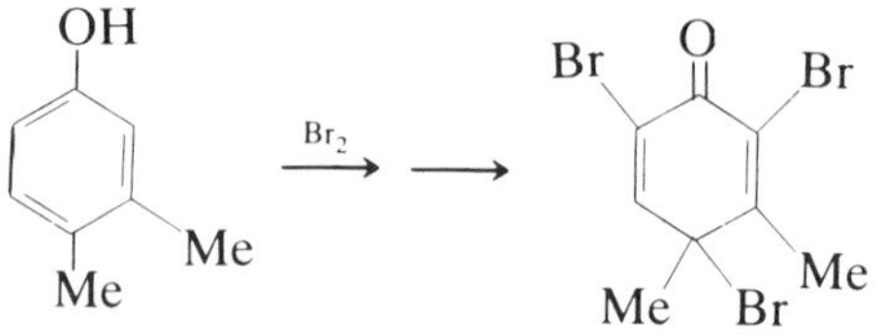

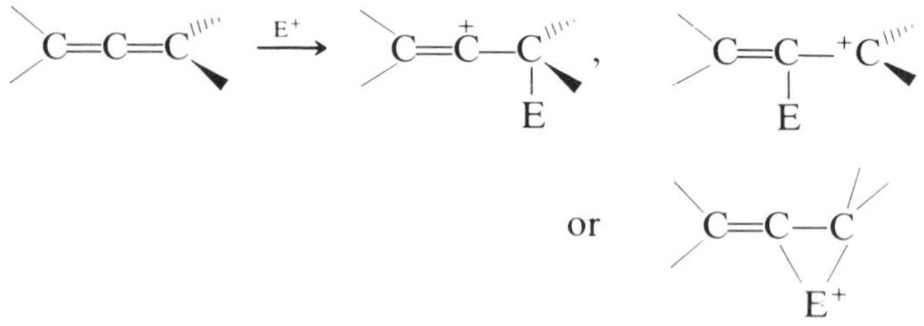

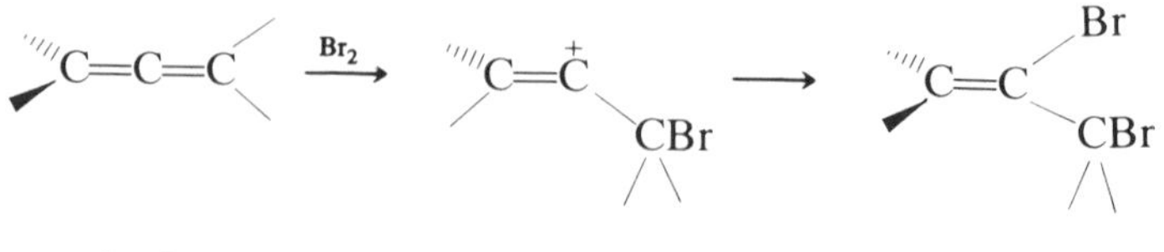

Allenes and alkynes may undergo addition via vinyl cations, which are not resonance-stabilized species, **35**, **36** or via cyclic intermediates, **35a**[46]: 2:1 adducts may result from addition of HOCl but mono-addition can be achieved with bromine. *Trans*-addition is usual[99] though diphenylethyne adds mercuric acetate to give the *cis* (Z) stereoisomer exclusively[98]. Despite the much greater stabilities of alkyl cations than of vinyl cations relative to their saturated precursors (as inferred from solvolytic rate studies), the hydrations of alkynes and the analogous alkenes take place at similar rates. Because of the high positive heats of formation of alkynes, formation of a vinyl cation is more favorable than is ionization to the same species from a saturated precursor. Again, rate dependence upon H_0 is observed. The addition of acetic acid may result in the formation of a detectable enol acetate which undergoes disproportionation[6]:

$$PhC\equiv CH + AcOH \rightarrow PhC(OAc)=CH_2 \xrightarrow{AcOH} PhCOCH_3 + Ac_2O$$

The reaction is related to the mercury-catalyzed hydration of alkynes. Alkynes are more prone to nucleophile addition, however, and the reactivity sequence for hydration of HCl,

$$PhC\equiv CH \ll MeC\equiv C \cdot CO_2Me < MeO_2C\equiv C \cdot CO_2Me$$

is consistent with initial addition of Cl^-, the Ad_N2 reaction. Mechanisms of addition to alkynes and to alkenes then may differ in detail (**37**).

37 $EtO_2C-C\equiv C-CO_2Et \longrightarrow EtO_2\overset{..\,\bar{}}{C}-C=C\underset{Cl}{\overset{CO_2Et}{<}}$
            H^+Cl^-

$$\xrightarrow{H^+} \underset{EtO_2C}{\overset{H}{>}}C=C\underset{Cl}{\overset{CO_2Et}{<}}$$

12.3   NUCLEOPHILIC ADDITIONS TO MULTIPLE BONDS[47]

The preceding sections illustrated the tendency of the carbon–carbon double bond to display donor activity and to coordinate electrophiles. Simple alkenes do not normally coordinate nucleophiles since the resulting localized carbanions would be high-energy species. Alternatively it may be stated that the LUMO of an alkene is too high-lying for it to function easily as an acceptor molecule. It follows that nucleophilic additions would be more likely to occur if the energy of the LUMO were decreased and if negative charge could become delocalized. This would be the case if a substituent with a $-R$ electronic effect were introduced at the double bond, **38**. There are, indeed, three classes of unsaturated systems which are well adapted to nucleophilic additions:

(a) α,β-unsaturated carbonyl, cyano, nitro, sulfonyl and related species, including those with more extensive conjugated systems

$$\left(\begin{array}{l} \text{C}{=}\text{C}{-}\text{Z}; \quad \text{Z} = \text{C}{=}\text{O}, \quad -\text{C}{\equiv}\text{N}, \quad -\text{NO}_2, \quad \text{S}{=}\text{O}, \\[2ex] \overset{\text{O}}{\underset{\text{O}}{\overset{\|}{\underset{\|}{-\text{S}-}}}} \quad, \text{ also } \quad -\text{CF}_3 \end{array} \right)$$

and polyhalogenated alkenes;

(b) heterenes, $\text{C}{=}\text{X}$ where X = O, N, S (electronegative atoms able to bear negative charge) as in

$$\text{C}{=}\text{O} \qquad \text{C}{=}\text{N}{-} \qquad \text{C}{=}\overset{+}{\text{N}} \qquad -\text{C}{\equiv}\text{N} \qquad \text{C}{=}\text{S};$$

(c) heterocumulenes, X=Y=Z in which one at least terminal atom is not carbon, e.g.

$$-\text{N}{=}\text{C}{=}\text{O}, \qquad \text{C}{=}\text{C}{=}\text{O} \qquad -\overset{+}{\text{S}}{=}\text{C}{=}\overset{-}{\text{N}} \qquad -\text{N}{=}\text{C}{=}\text{S}$$

$$\text{O}{=}\text{C}{=}\text{O} \qquad \text{S}{=}\text{C}{=}\text{S}$$

Alkynes are somewhat more prone to undergo nucleophilic addition than alkenes since vinyl anions are more stable than alkyl anions on account of the greater s-character at carbon; $-R$ groups activate these also. Nucleophiles which will add include ROH, RO$^-$, NHR$_2$, RSH, and a variety of carbanions from structurally stabilized species such as N$\equiv$C—CH$^-$— COOEt and organometallic compounds, R—M. Basic catalysts may be necessary to provide the more reactive conjugate base of the nucleophile: CH$_3$CN will add to a propenoate ester in the presence of methoxide ion, by which the active nucleophile $^-$CH$_2$CN is generated. The addition is normally completed by the addition of a proton as the electrophilic entity.

12.3.1 Michael addition

This is the name given to nucleophilic addition to the β-carbon of α,β-unsaturated ketones, esters and similar compounds, **39**, vinylogues of the carbonyl group. There is competition between attack at β-C or C=O and final protonation might occur at α-C or O, the latter yielding an enol which would rapidly ketonize. Acrylonitrile, itself made by nucleophilic addition of HCN to acetylene, readily adds a variety of nucleophiles

39 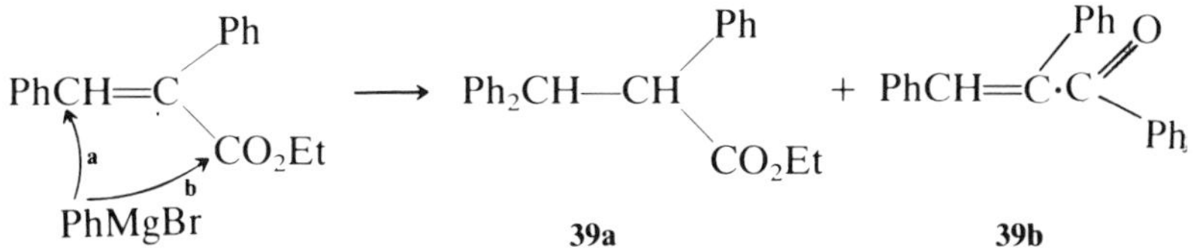

39a **39b**

40 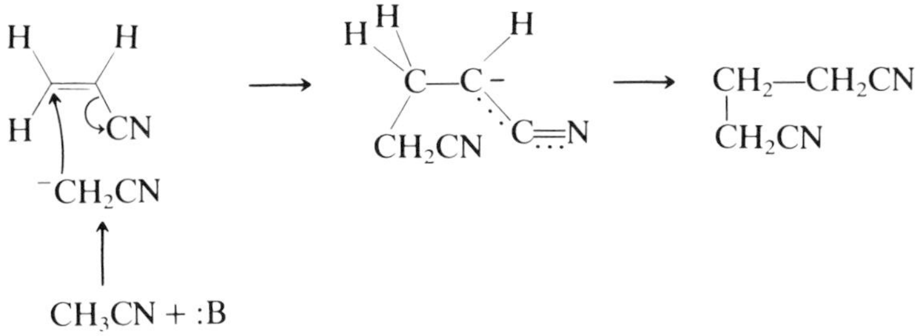

('cyanoethylation'), a synthetic route of importance for lengthening a carbon chain (**40**).

Regioselectivity in addition is strictly that determined by the more stable carbanion. Stereoselectivity appears to be similarly controlled, the most stable conformation of this carbanion determining the orientation of protonation. For example, addition of thiophenol to the cyclohexene sulfone, **41**, is *anti* but of the opposite stereochemistry for the five-

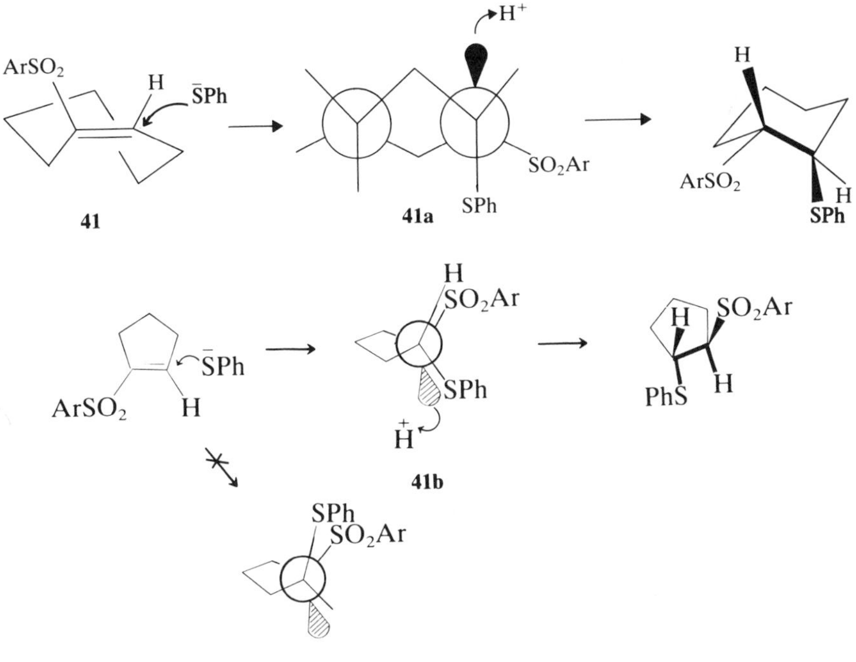

41 **41a** **41b** **41c**

membered analog. The most stable carbanions appear to be the all-staggered *axial–equatorial* structure, **41a**, in the cyclohexene case, but in the five-membered ring eclipsing interactions force the sulfonyl and thiophenyl groups *anti* (**41b** in preference to **41c**).

12.3.2 Carbonyl additions

Nucleophilic coordination to the carbonyl group is the initial step of the $B_{Ac}2$ ester hydrolysis (Section 10.6.4). If no nucleophilic leaving group is available, protonation at oxygen will occur, completing the addition reaction. This is typical behavior of aldehydes or ketones, which will add water or alcohols, amines, HCN, hydride, or bisulfite (for example, **42**). Hydration and acetal formation are acid-catalyzed (as are the reverse reactions, the hydrolyses of these compounds, Section 9.2.3). Hydration equilibria usually lie in favor of the carbonyl compound except for lower aldehydes and those with strongly electron-withdrawing groups. Chloral, **43**, for example, forms a stable hydrate, a *gem*-diol, **43a**. Addition of OH⁻ to an aromatic aldehyde may cause expulsion, even of hydride, provided a receptor is available such as another molecule of aldehyde, as in **44**, the Cannizzaro reaction.

The Aldol reaction

Aldol reactions are a family of nucleophilic additions of enolate ions to a carbonyl group[48]. Both reagents may arise from the same aldehyde or ketone resulting in dimerization (the 'Aldol Reaction' proper, **45**). Dehydration of the β-hydroxyketone is facile and gives the α,β-unsaturated aldehyde or ketone but at higher temperatures the attack of a base on a β-hydroxyketone can result in elimination of the enolate ion — a reversal of aldol formation known as a retro-aldol process.

The following variants may all be classified as rate-determining additions of carbanions to carbonyl groups.

Claisen reaction

The enolate of an ester adds to an ester carbonyl to form a β-ketoester, **46**, which will be cyclic if a dibasic ester is used (Dieckmann reaction)[49], **47**. A ketone may sometimes provide the acceptor carbonyl group. The mechanism is basically of the $B_{Ac}2$ type with a carbanion as the attacking nucleophile.

571

Thorpe Reaction [49]

This is a Claisen-type addition between two enolizable cyanoalkanes which, after hydrolysis, yields an α-cyanoketone, **48**. This in turn may be hydrolyzed and decarboxylated to give a ketone.

Knoevenagel reaction [50]

Addition of carbanion from an 'active methylene' compound to a non-enolizable aldehyde or ketone, usually with subsequent elimination of water, can be used to obtain a variety of alkene or alcohol products, **49**.

Stobbe condensation [51]

Addition of the enolate from a succinate ester to a ketone may be considered a type of Knoevenagel reaction (**50**).

Perkin reaction

The enolate from an anhydride undergoes addition to a non-enolizable aldehyde. The base used is the salt of the corresponding acid and, since this is weak and the anhydride is only very weakly acidic, high temperatures are needed. This is the standard route to cinnamic acids, **51**.

50 $\underset{\text{CH}_2\cdot\text{COOEt}}{\overset{\text{CH}_2\cdot\text{COOEt}}{|}} \xrightarrow[]{\overset{-OtBu}{\rightleftharpoons}} \underset{\text{CH}_2\text{COOEt}}{\overset{\bar{\text{C}}\text{HCOOEt}}{|}}$

51 $\text{CH}_3-\overset{\overset{\text{O}}{\|}}{\text{C}}-\text{O}^-$ $\text{CH}_3\overset{\text{CO}}{\diagdown}\text{O}\overset{\text{CO}}{\diagup}\text{CH}_3 \rightleftharpoons \bar{\text{C}}\text{H}_2\overset{\text{CO}}{\diagdown}\text{O}\overset{\text{CO}}{\diagup}\text{CH}_3$

The Wittig reaction[52-54]

The carbanionic center of a phosphorus ylide, **52**, readily adds to a wide variety of aldehydes and ketones in one of the most versatile alkene syntheses. The resultant betaine, **52a**, undergoes thermal elimination even at low temperatures. The stability and consequently the ease of formation of ylides from the corresponding phosphonium salts varies with the acidity of the α-proton but the most stable ylides are the least reactive. Acyl ylides, **53**, which can exist in water will react with aldehydes but not with ketones unless they are converted to their dianions with greatly enhanced nucleophilicity. Products may be under kinetic or thermodynamic control. A stabilized ylide adds reversibly to the carbonyl compound to form the more stable intermediate, which then determines the stereochemistry of the alkene, also the more stable. Highly reactive ylides will tend to add irreversibly and may then form *cis* and *trans* mixtures of alkenes. Phosphine

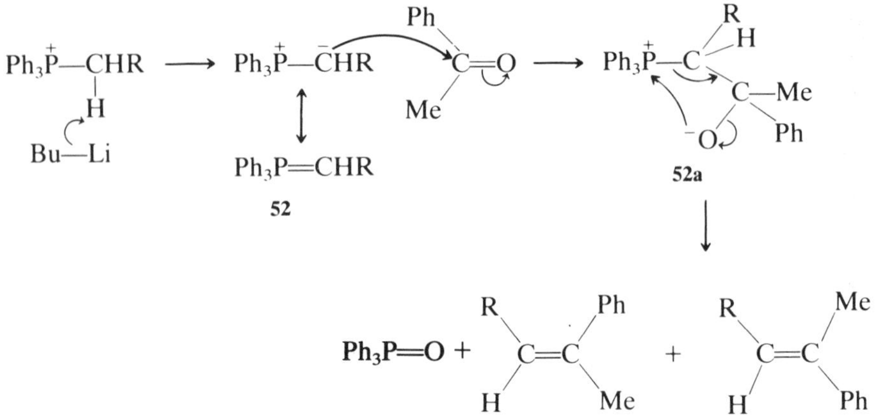

$$Ph_3\overset{+}{P}-CHR \longrightarrow Ph_3\overset{+}{P}-\overset{-}{C}HR \quad \overset{Ph}{\underset{Me}{>}}C=O \longrightarrow$$

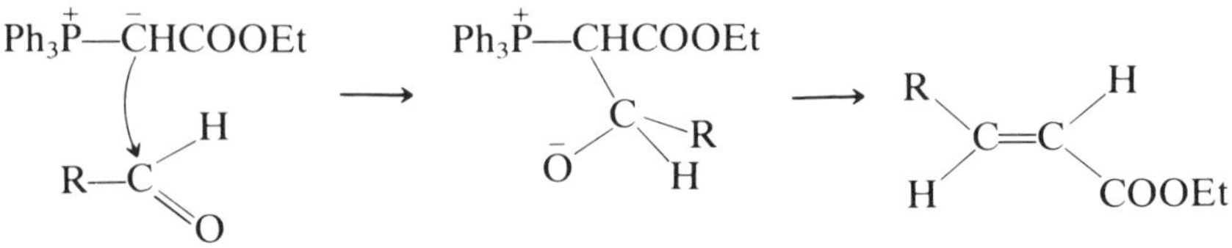

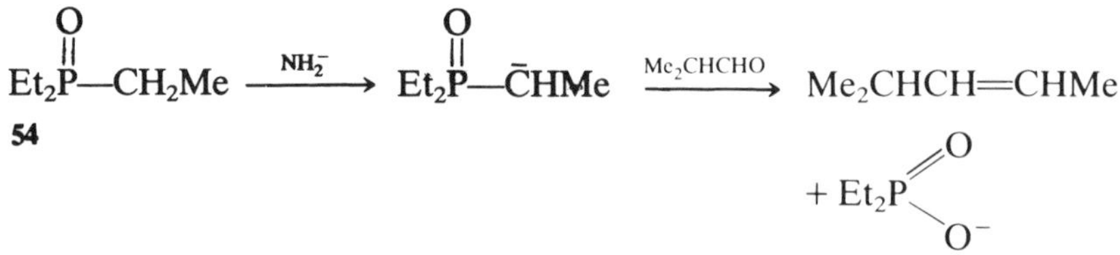

Ph₃P=CHR

52

52a

$$Ph_3P{=}O \; + \; \underset{H}{\overset{R}{>}}C{=}C\underset{Me}{\overset{Ph}{<}} \; + \; \underset{H}{\overset{R}{>}}C{=}C\underset{Ph}{\overset{Me}{<}}$$

53 $Ph_3\overset{+}{P}{-}\overset{-}{C}HCOOEt$ $\qquad Ph_3\overset{+}{P}{-}CHCOOEt \longrightarrow \underset{H}{\overset{R}{>}}C{=}C\underset{COOEt}{\overset{H}{<}}$

$$\underset{\textbf{54}}{\overset{O}{\underset{\|}{Et_2P}}}{-}CH_2Me \overset{NH_2^-}{\longrightarrow} \overset{O}{\underset{\|}{Et_2P}}{-}\overset{-}{C}HMe \overset{Me_2CHCHO}{\longrightarrow} Me_2CHCH{=}CHMe$$

$$+ \; Et_2P\overset{\overset{O}{\|}}{\underset{O^-}{}}$$

oxides may be used in place of phosphonium ions, **54**, the dipolar P=O bond having a strongly acid-strengthening effect on α-protons.

Sulfur ylides, **55**, undergo similar reactions initially but the intermediate betaines have less tendency to thermal elimination of sulfoxide since the S=O bond is less strong than P=O. Instead, neighboring-group participation of the oxide group occurs with the displacement of sulfur, forming an epoxide, **55a**[55].

55

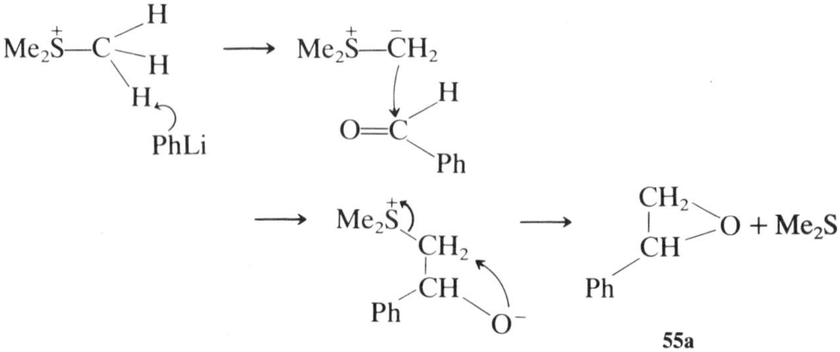

55a

The Mannich Reaction[56]

This is the addition of a carbonyl compound, an amine and an 'active hydrogen' species, **56**. The initial step is addition of amine to the carbonyl group with the formation of an imine or iminium ion, **56a**, which is even more susceptible to nucleophilic addition than is the carbonyl group and adds the available carbanion. Addition of cyanide and hydrolysis leads to an α-aminoacid (the Strecker synthesis, **57**).

$$\textbf{56} \quad Me_2NH \quad C{=}O \longrightarrow Me_2N{-}CH_2OH \xrightarrow{-H_2O} Me_2\overset{+}{N}{=}CH_2$$
$$\textbf{56a}$$

gramine

$$\textbf{57} \quad NH_3 \quad \underset{H}{\overset{R}{\diagdown}}C{=}O \xrightarrow{-H_2O} \underset{H}{\overset{R}{\diagdown}}C{=}\overset{+}{N}H_2 \qquad \overset{-}{C}N$$

$$\longrightarrow \underset{H\ CN}{\overset{R}{\diagdown}}C{-}NH_2 \xrightarrow{\text{hydrolysis}} \underset{H\ COOH}{\overset{R}{\diagdown}}C{-}NH_2$$

Carbonyl condensation reactions

Identification of aldehydes and ketones is accomplished by conversion to crystalline imino derivatives by reagents of the type X—NH₂; those listed in Table 12.9 are in common use. All reactions occur by nucleophilic addition of the amine group, **58**, followed by elimination of water (E2) from the β-hydroxyamine, **59**, initially formed, a facile reaction analogous to the dehydration of ketone hydrates (1,1-diols).

Table 12.9 Carbonyl condensing reagents

$$R_2C{=}O + H_2N{-}X \rightarrow R_2C{=}N{-}X + H_2O$$

X	$H_2{-}X$	Product
—OH	hydroxylamine	oxime
—NH$_2$	hydrazine	hydrazone
—NHPh	phenylhydrazine	phenylhydrazone
—NH—⟨benzene⟩NO$_2$ (NO$_2$)	2,4-dinitrophenylhydrazine (Brady's reagent)	
—NHCOCH$_2\overset{+}{N}$Me	Girard's reagent T	
—NHCOCH$_2\overset{+}{N}$⟨pyridine⟩	Girard's reagent P	
	Girard's reagent P	
—NHCONH$_2$	semicarbazide	semicarbazone
—R	primary amine	imine

12.3.3 Additions to heterocumulenes

Ketenes, $R_2C{=}C{=}O$, and isocyanates, $R{-}N{=}C{=}O$, are very reactive towards nucleophilic addition across the C=C and N=C bonds. Most reagents containing —OH, —NH or —SH groups will add under neutral conditions. The electrophilic reactivity stems from a combination of dipolar character, with positive charge at the central atom:

$$\overset{\diagdown}{\diagup}C{=}C{=}O \longleftrightarrow \overset{\diagdown}{\diagup}C{=}\overset{+}{C}{-}O^- \left[\longleftrightarrow \overset{\diagdown}{\diagup}\overset{+}{C}{-}\overset{-}{C}{=}O \right]$$

and a low-lying unoccupied molecular orbital (HOMO) with a large coefficient on the central atom (Fig. 12.2).

Ketene and its derivatives (most of which are highly reactive towards dimerization, Section 14.3, and need to be generated and used *in situ*) are therefore acylating agents, **60**: the sulfur analogues, sulfenes, $R_2C{=}SO_2$, are also transient species similarly reactive towards nucleophilic addition.

60 structure: $H_2C{=}C{=}O$ with HOEt adding to give $CH_3{-}C({=}O){-}OEt$

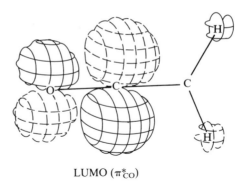

LUMO (π^*_{CO})

HOMO (π_2)

Fig. 12.2 Frontier orbitals of ketene.

Among commercially important reactions of isocyanates are additions of alcohols and phenols to give carbamate esters (urethanes), **61**.

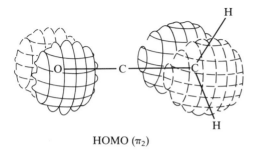

Isonitriles, R—$\overset{+}{N}$≡$\overset{-}{C}$, add hydrolytic reagents under acid catalysis initially in an 'α-addition', both nucleophile and proton bonding to carbon. The imide formed may then hydrolyze (**62**).

Carbonation of strong bases such as carbanions (in the form of Grignard reagents or alkyllithiums) is another example of this type of reaction:

$$Me\text{—}MgI + O\text{=}C\text{=}O \longrightarrow Me\text{—}C\overset{\displaystyle O}{\underset{\displaystyle O^-\overset{+}{M}gI}{}} \xrightarrow{\ H^+\ } Me\cdot COOH.$$

$$R-\overset{+}{N}\equiv\overset{-}{C} \xrightarrow{H^+} R-\overset{+}{N}\equiv C-H \longrightarrow R-\underset{H}{\overset{H}{N}}-\underset{OH}{\overset{\cdot\cdot}{C}}\overset{+}{\cdot} \longrightarrow R-\underset{H}{\overset{H}{N}}-\overset{H}{C}\overset{}{\underset{}{}}\!\!=\!O$$

62

$$\xrightarrow{H^+} R-\underset{\overset{+}{|}\;H}{\overset{H}{N}}-C\overset{O}{\underset{}{}} \longrightarrow R-NH_2 + H-C\overset{O}{\underset{OH}{}}$$

12.4 FRONTIER ORBITAL CONSIDERATIONS

Simple alkenes are non-polar and therefore frontier orbital control is likely to be important in determining rates and regiospecificity. An alkene contributes the HOMO when undergoing an electrophilic addition and its energy and distribution will depend upon the pattern of substitution. To take two extreme cases, the allyl cation and anion may be regarded as ethene substituted with the most powerful $+R$ and $-R$ substituents (X- and Z-types). The MO pattern is:

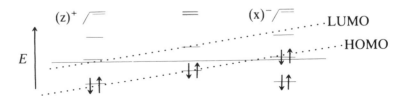

The HOMO energy is raised by a $+R$ group and lowered by a $-R$ group, so that reactivity would be expected to increase by the substitution of alkyl groups or of alkoxyl (vinyl ethers) and amino (enamines). This order is consistent with a large and negative ρ-value for substitution at the alkene center, as is observed. Polyenes are ambident nucleophiles and consequently the coefficients of the HOMO at each carbon should determine regiospecificity.

$H^+ \bullet\!=\!\bullet$ $\qquad c^2(\psi_1)$: C1, 0·5

$H^+ \bullet\;\bullet\;\bigcirc$ $\qquad c^2(\psi_2)$: C1, 0·362; $\qquad$ C2, 0·138.

$H^+ \bullet\;\bullet\;\bullet$ $\qquad c^2(\psi_3)$: C1, 0·2715; $\qquad$ C2, 0·053; $\qquad$ C3 = 0·17

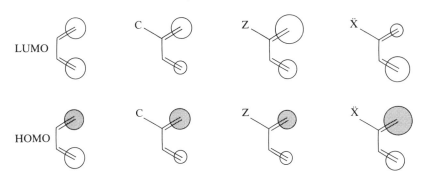

Fig. 12.3 Frontier orbital properties of substituted butadienes. The open or shaded circles denote relative phases of the lobes and their sizes indicate magnitudes of the MO coefficients.

In this series, the highest coefficient of the HOMO is on the terminal carbon which will be the site at which the electrophile bonds leading to Markownikoff addition. An ambient electrophile remains as an intermediate with two or more potential sites to which the nucleophile is attracted. Butadiene, for example, will form a perturbed allyl cation with C2 and C4 as potential sites for attack. The coefficient of the LUMO is somewhat larger at C4, which means from purely frontier orbital control that 1,4-addition would be most likely but not exclusive.

In practice, both 1,2- and 1,4-addition of HCl is usually observed but the additional factor of thermodynamic control may distort the picture.

Nucleophilic additions to alkenes usually occur only when at least one Z-substituent is present. This can be seen to lower the LUMO energy and increase reactivity at the α-position. Z-substituted dienes will have a LUMO even lower in energy with coefficients of greatest magnitude at the terminal atom, at which nucleophilic attack will often occur (conjugate addition):

$$Nu: \diagup\!\!\!\!\diagdown\!\!\!\diagup\!\!\!\diagdown\!\!\!\diagup\!\!\!\diagdown C \overset{O}{\diagdown} \longrightarrow Nu-\diagup\!\!\!\diagdown\!\!\!\diagup\!\!\!\diagdown \overset{\bar{O}}{|}$$

Figure 12.3 summarizes the effects of —X and —Z substituents at both 1- and 2-positions of butadiene, upon the coefficient of the frontier orbitals.

12.5 VINYL SUBSTITUTION VIA ADDITION/ELIMINATION[57,58]

A sequential addition to a double bond followed by elimination, or elimination followed by addition, will result in overall substitution which may be either nucleophilic or electrophilic, designated Ad_N–E, E–Ad_N and Ad_E–E, E–Ad_E, respectively (**63**), as determined by the initial interaction. The second stage in each case is normally fast so that the intermediates are

63

rarely isolable. The Ad_N–E reaction begins as a Michael addition and so requires $-R$ substituents to be present. The following characteristics of Ad_N–E reactions may be observed.

(a) Rates will not correlate with the nucleofugacity of the leaving group since it departs in a fast step. In fact, there should be an inverse leaving-group effect, $F \gg Cl > Br$, because the rate-determining coordination of the nucleophile is assisted by electron-withdrawing properties (i.e. electronegativity) of —Hal. An analogous sequence is observed in the S_NAr process, Section 10.5, for the same reasons. A normal leaving-group effect ($F \ll Cl < Br$) would indicate a rate-determining C—Hal bond fission which could signal a mechanism proceeding by an initial E2, S_N2 or S_N1 reaction.
(b) There should be an inverse secondary α-deuterium isotope effect since hybridization at the carbon center changes from sp^2 to sp^3. The E–Ad reactions should all show normal primary isotope effects for the rate-determining loss of the β-hydrogen.
(c) Negative reaction constants (ρ^-) for E–Ad reactions of β-aryl systems, ArCH=CHX, should be observed, corresponding to the need to stabilize negative charge at C_β. The magnitude of ρ will depend on the demands made upon the aryl group (the amount of negative charge developed at β-carbon) as a result of other electron-withdrawing groups which might be present: it will be rather small if the elimination step is E2 but larger negative if that step is E1cb.
(d) The intermediates are stable species though reactive, and should be capable of detection or isolation under the right conditions.
(e) E–Ad reactions can occur only when a β-hydrogen is present and the elimination step is more facile when it is *trans*-periplanar to the leaving group, X, than when it is *cis*. Reactions occurring by this mechanism should therefore show a large rate ratio k_{trans}/k_{cis}, whereas Ad–E reactions should be much less sensitive to the geometry of the reagent. This can be seen from the examples in Table 12.3. Entropies of activation for E–Ad processes are more positive than for Ad–E.

12.5.1   Examples

Halide displacement from 2,2,-diarylethenyl halides, **64**, is of the Ad_N–E type[77,78]. Rates are controlled by electronegativity, not leaving-group ability (F:Cl:Br = 300:1:1). Substitutions of the vinyl sulfones, **65**, show similar trends[64,65] but the analogous compounds, **66**, which possess a β-hydogen react with a normal leaving-group effect, $k_{Br}/k_{Cl} > 100$, indicating the E–Ad_N route to be preferred when both are possible. This only applies to the Z-isomer, which has a *trans* β-hydrogen; the E-isomer, **67**, reacts by the Ad–E route, with $k_{Br}/k_{Cl} < 1$. Ad_N–E mechanisms are considered to occur in displacements of X from many compounds: *trans*-ZCH=CHX, in which Z is a strong $-R$ group and X a poor leaving group such as —OR, —CN, —NH$_2$. These reactions are formally analogous to the $B_{Ac}2$ mechanism of carbonyl compounds (Section 10.6.1; **68a,b**), a carbanion rather than an oxide ion providing a powerful electron 'push' to expel the leaving group. Nucleophilic aromatic displacement (Section 10.5) is also of this type, the carbanion being a pentadienide anion and the expulsion of the leaving group being driven by the recovery of aromatic character in the product, **68c**.

The Ad$_N$–E mechanism operates in substitutions of halocyclobutenes[66–76]. In the two-step example, **69**, the more stable carbanion is formed at each of the three steps at which alkoxide is coordinated to the ring. Reactions of this type are very facile when multiple anion-stabilizing groups are present. For example, in **70** even an activated aromatic ring may act as the nucleophile in a substitution analogous to a Friedel–Crafts alkylation by an alkene but without the necessity of supplying a Lewis acid catalyst[59,60]. In this respect the cyano groups may be said to act as internal Lewis acids. Rates of these reactions correlate with σ^0 [61–63]. Reaction constants, ρ (Table 12.10), are positive but quite small for additions to the sulfones, **65**, since the —SO$_2$— group fulfills the electronic demands of the reaction almost entirely, leaving little demand for electron withdrawal from the more distant aromatic ring.

Table 12.10   LFER for some nucleophilic additions

	Reaction	ρ
64	Ar$_2$C=CHCl + OEt$^-$ → Ar$_2$C=CHOEt	+3·38
65	ArSO$_2$CH=CHCl + R$_2$NH → ArSO$_2$CH=CHNR$_2$	+1·15
66	ArSO$_2$CH=CHCl(*cis*) + PhS$^-$ → ArSO$_2$CH=CHSPh	+1·6
67	ArSO$_2$CH=CHCl(*trans*) + PhS$^-$ → ArSO$_2$CH=CHSPh	+1·84
67	ArSO$_2$CH=CHCl(*trans*) + N$_3^-$ → ArSO$_2$CH=CHN$_3$	+1·8

The order of nucleophilicity in reactions of this type is similar to that found in nucleophilic aromatic substitutions (S_NAr), namely:

$$\text{Hal}^- \ll N_3^- \sim RNH_2 < R_2NH < OR^- < ArS^- < RS^-,$$

and does not correspond well with the Swain–Scott order of nucleophilicity. This emphasizes differences between reactivity towards saturated and unsaturated carbon, relatively hard and soft electrophilic centers, respectively.

12.5.2 Stereochemistry

Either a retention or an inversion of geometrical isomerism is possible in Ad–E or E–Ad sequences. In many cases reported the preference seems to be towards retention[57]. Since there is a plane of symmetry through the alkene reagent, there is no preferred side from which the nucleophile will approach and stereochemical selection must arise at the elimination step, for which a *trans*-periplanar arrangement of leaving-group and unshared-pair orbital is required (**65**). A rotation of 60° (or 120°) is required for retention of geometry, while for inversion a rotation of 180° is required. It may be that the preference for retention is a consequence of the principle of least motion. However, the picture is complicated by steric effects which will differ in the two possible elimination geometries.

The E–Ad$_N$ mechanism is favored by substrates which lack a carbanion-stabilizing substituent but possess a β-hydrogen and, preferably, the nucleophilic leaving group *trans* to it[77,78]. Nucleophilic additions to alkynes substituted by $-R$ groups are usually facile and complete the overall reaction in a fast step. Strong bases are necessary to bring about an unactivated E2 elimination but *cis*-1,2-dichloroethene reacts with thiolate by this route, **71**. The intermediate chloroethyne may be independently synthesized and shown to yield identical products under the reaction conditions. β-Halostyrenes, ArCl=CHCl, also undergo E–Ad substitution with alkoxide, as shown by the primary isotope effect, $k_H/k_D = 2$[79].

Surprisingly, the reactions of 1-chlorocyclohexene with strong bases such as butyllithium occur by elimination–addition despite the highly strained nature of the intermediate cyclohexyne, **72**. The incorporation of a ^{14}C label at C2 and determination of the product labeling pattern shows conclusively that C1,C2 have become equivalent at an intermediate stage in the reaction[80,81]. A similar result has been obtained from cyclopentyl chloride, which must react via an even more strained cyclopentyne. With a better

71

72

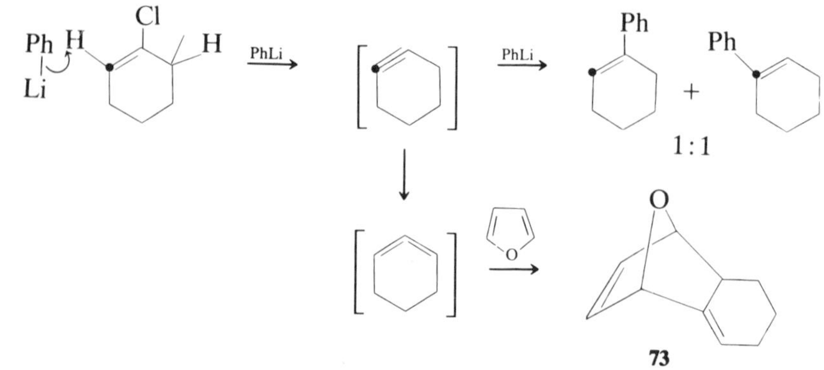

73

74

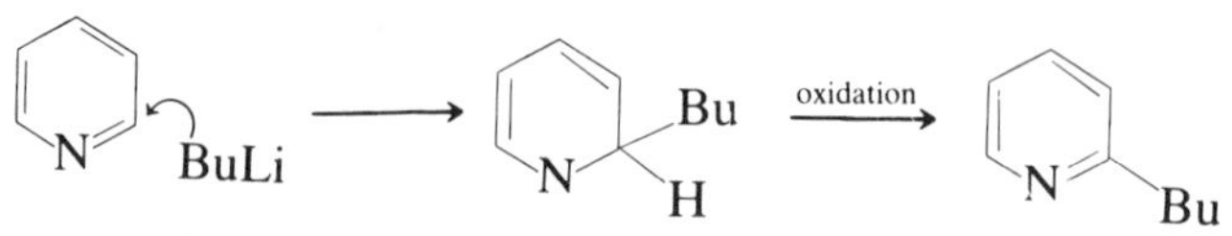

75

leaving group, elimination can be achieved with less strong bases and the product trapped as a Diels–Alder adduct with furan[82,83]. The structure **73** suggests that it arises not from cycloalkyne but from the isomeric cyclic allene. Taken together with the evidence from the labeling experiment, it seems likely that this might be a secondary product formed by isomerization of the cycloalkyne under (relatively) long lifetime conditions. The alkylation of pyridine by alkyllithium reagents, **74**, is a further example of the Ad_N–E reaction; the intermediate dihydropyridine (**75**) may be isolated[84].

PROBLEMS

1 The addition of bromine to 1,2-diphenylethene (DPE) is of second order, rate $= k[DPE][Br_2]$, but rates are depressed by the addition of Br^- due to the formation of the less reactive Br_3^-. Derive a rate expression which takes into account $[Br^-]$. If the reaction is carried out in methanol what products in addition to dibromide would you expect to obtain?

2 Suggest mechanisms by which the following transformations are effected by thallium(III) nitrate or acetate:

$$PhCOCH_3 \xrightarrow[MeOH]{Tl(NO_3)_3} PhCH_2CO\cdot OMe;$$

$$PhC\equiv CPh \xrightarrow[MeOH]{Tl(NO_3)_3} PhCO\cdot CH(OH)Ph \rightarrow PhCOCOPh;$$

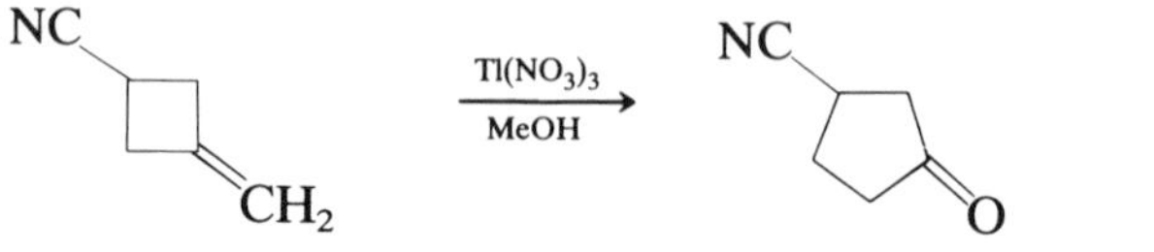

584

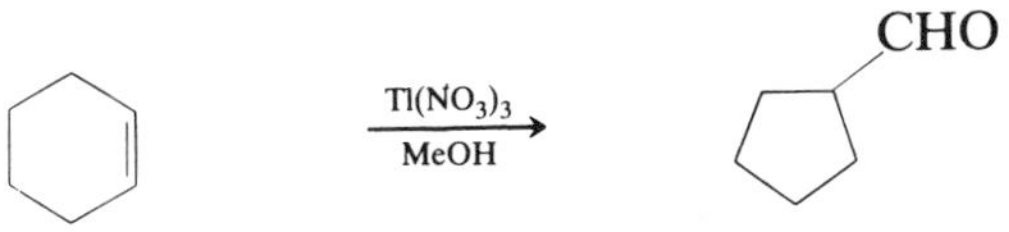

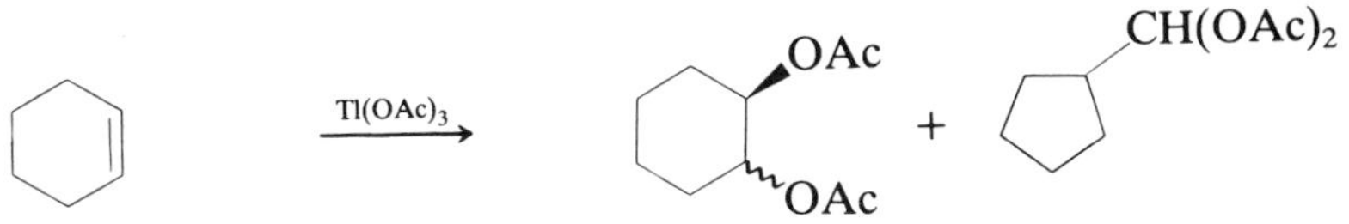

3 While fluorine is too reactive to add smoothly to alkenes, the 'inert gas' compound XeF$_2$ is capable of addition to give 1,2-difluoroalkanes; suggest a mechanism accounting for the products,

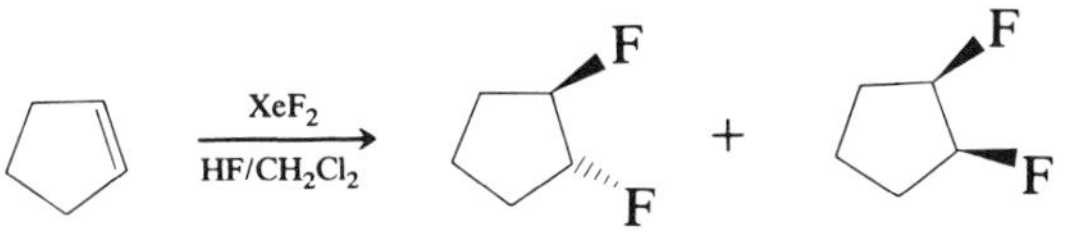

80%   20%

4 Deduce mechanisms of the following additions from the effects of substitution upon rates.

(a) XCH=CH$_2$ + Cl$_2$, aq. AcOH.

X:	Me	H	CH$_2$F	CH$_2$Cl	CH$_2$CN	CHCl$_2$	Br	CCl$_3$
k_{rel}:	2	1	0·034	0·019	0·0027	$2·6 \times 10^{-5}$	$1·3 \times 10^{-5}$	$2·9 \times 10^{-7}$

(b)   XCH=CH$_2$ + morpholine, O⟨ ⟩NH, MeOH, 45°C.

Z:	PO(OEt)$_3$	CONH$_2$	CN	CO$_2$Me	SO$_2$Ph	CO$_2$Ph	Ac	CHO	PhCO
k_{rel}:	1	2	3·2	4·9	36	68	700	515	1500

[*J. Org. Chem.*, **38**, 1631 (1973).]

5 Account for the products in the following additions:

(a)   Me$_2$C=CH$_2$ $\xrightarrow{\text{AcO—NO}_2}$ Me$_2$C=CH·NO$_2$

(b)   Me$_2$C=CH$_2$ $\xrightarrow{\text{Cl}_2}$ $\underset{\text{CH}_2}{\overset{\text{Me}}{>}}$C—CH$_2$Cl.

(c)

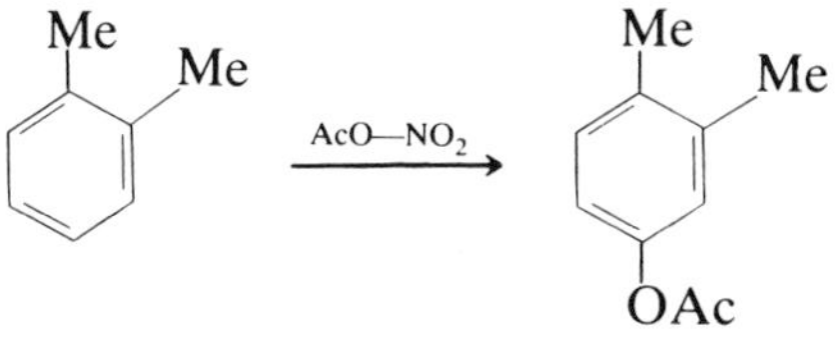

(d)   CH$_2$=CH·CH$_2$Br $\xrightarrow{\text{aq . HBr}}$ CH$_3$·CHBr·CH$_2$OH.

(e) 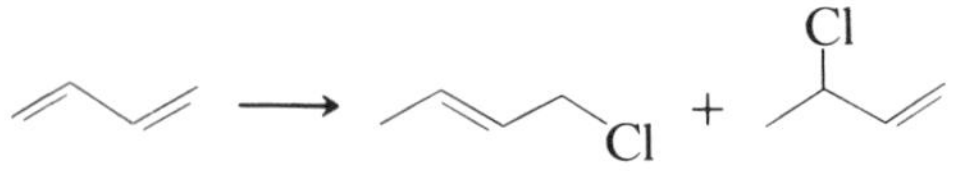

(f) $$PhCH{=}CH\cdot COOEt \xrightarrow[I_2,\ catalyst]{Cl_2} PhCHCl\cdot CHCl\cdot COOEt.$$

(g) $$HC{\equiv}C\cdot COOH \xrightarrow{aq.\ I_2} ICH_2CO\cdot COOH.$$

(h) $$Me_2C{=}C{=}CH_2 \xrightarrow{Cl_2} Me_2C{=}CCl\cdot CH_2Cl + Me_2CCl\cdot CCl{=}CH_2 + \underset{Me}{\overset{CH_2}{\backslash\!\!/}}\!CCl{=}CH_2$$

REFERENCES

1 E. Gould, *Mechanism and Structure in Organic Chemistry,* Holt–Rinehart–Winston, New York, 1959.
2 P. B. D. De la Mare and R. Bolton, *Electrophilic Additions to Unsaturated Systems,* 2nd edn, Elsevier, Amsterdam, 1982.
3 M. J. S. Dewar and R. C. Fahey, *Angew. Chem. Int. Ed.,* **3,** 245 (1964).
4 T. Tidwell, *Acc. Chem. Res.,* **10,** 252 (1977).
5 W. M. Schubert and J. R. Keefe, *J. Amer. Chem. Soc.,* **94,** 559 (1972).
6 J. P. Montheard, M. Camps and A. Benzaid, *Bull. Soc. Chim. France,* 33 (1981).
7 E. L. Purlee and R. W. Taft, *J. Amer. Chem. Soc.,* **78,** 5807, 5811 (1956).
8 P. Cramer and T. T. Tidwell, *J. Org. Chem.,* **46,** 2683 (1981).
9 A. D. Allen, Y. Chiang, A. J. Kresge and T. T. Tidwell, *J. Org. Chem.,* **47,** 775 (1982).
10 R. C. Fahey, *Topics in Stereochemistry,* **3,** 237 (1968).
11 F. Garnier, R. H. Donnay and J.-E. Dubois, *Chem. Comm.,* 829 (1971).
12 H. Takaya, N. Todo, T. Hosoya and T. Minegishi, *Bull. Chem. Soc. Japan,* **44,** 1175 (1971); S. K. Bhattacharya and C. K. Das, *J. Amer. Chem. Soc.,* **91,** 6715 (1969).
13 J. H. Rolston and K. Yates, *J. Amer. Chem. Soc.,* **91,** 1469, 1477 (1969).
14 E. Bienvenue-Goetz and J.-E. Dubois, *J. Amer. Chem. Soc.,* **103,** 5388 (1981).
15 K. Yates, G. H. Schmidt, W. Regulski, D. G. Garrett, H. W. Leung and R. McDonald, *J. Amer. Chem. Soc.,* **95,** 160 (1973).
16 D. J. Pasto, G. R. Meyer and B. Lepeska, *J. Amer. Chem. Soc.,* **96,** 1858 (1974).
17 W. Markownikoff, *Ann.* **153,** 256 (1870).
18 Z. Schmidt, *Ann.* 267, 300 (1891).
19 A. L. Henne and S. Kaye, *J. Amer. Chem. Soc.,* **72,** 3369 (1950).
20 M. J. S. Dewar and R. C. Fahey, *J. Amer. Chem. Soc.,* **85,** 3645 (1963).
21 M. A. Wilson and P. D. Woodgate, *J. Chem. Soc., Perkin II,* 141 (1976).
22 I. V. Badrikov, N. G. Bronnikova and I. S. Okrokova, *J. Org. Chem.* (*USSR*), **9,** 978 (1973).
23 T. G. Traylor, *Acc. Chem. Res.,* **2,** 152 (1969).
24 R. C. Cambie, J. L. Jurlina, P. S. Rutledge and P. D. Woodgate, *J. Chem. Soc., Perkin I,* 315, 413, 961 (1982).
25 J. Meinwald, Y. C. Meinwald and T. N. Baker, *J. Amer. Chem. Soc.,* **86,** 4074 (1964).
26 J. Meinwald, Y. C. Meinwald and T. N. Baker, *J. Amer. Chem. Soc.,* **85,** 2513 (1963).
27 H. C. Brown and J. H. Kawkami, *J. Amer. Chem. Soc.,* **92,** 201 (1970).
28 J. Spanget-Larsen and R. Gleiter, *Tetrahedron Letts.,* **23,** 2435 (1982).

29 N. G. Rondan, M. N. Paddon-Row, P. Caramella, J. Mareda, P. H. Mueller and K. N. Houk, *J. Amer. Chem. Soc.*, **104**, 4974 (1982).

30 J. M. Brown and M. C. McIvor, *Chem. Comm.*, 238, 1969.

31 R. O. C. Norman and C. B. Thomas, *J. Chem. Soc. (B)*, 598, (1967).

32 E. Bienvenue-Goetz and J.-E. Dubois, *J. Org. Chem.*, **40**, 221 (1975); D. L. M. Williams, E. Bienvenue-Goetz and J.-E. Dubois, *J. Chem. Soc. (B)*, *517* (1969); E. Bienvenue-Goetz, J.-E. Dubois, F. B. Pearson and D. L. H. Williams, *J. Chem. Soc. (B)*, 1275 (1970).

33 W. L. Orr and N. Kharasch, *J. Amer. Chem. Soc.*, **75**, 6030 (1953); **78**, 1201 (1956).

34 H. C. Brown, *Boranes in Organic Chemistry*, Cornell UP, 1972; *Organic Synthesis via Boranes*, Wiley, New York, 1975; *Adv. Organometall. Chem.*, **11**, 1 (1973); *Pure Appl. Chem.*, **47**, 49 (1976).

35 L. C. Vishwakarma and A. Fry, *J. Org. Chem.*, **45**, 5306 (1980).

36 K. K. Wang, C. G. Scouten and H. C. Brown, *J. Amer. Chem. Soc.*, **104**, 531 (1982).

37 K. K. Wang and H. C. Brown, *J. Org. Chem.*, **45**, 5303 (1980).

38 H. C. Brown, J. Chandrasekharan and K. K. Wang, *Pure. Appl. Chem.*, **55**, 1387 (1983).

39 C. Prevost and J. Wiemann, *Compt. Rend.*, **204**, 989 (1937).

40 F. V. Brutcher and G. Evans, *J. Org.-Chem.*, **23**, 618 (1958).

41 R. B. Woodward and F. V. Brutcher, *J. Amer. Chem. Soc.*, **80**, 209 (1958).

42 R. C. Cambie, *J. Chem. Soc., Perkin I*, 1864 (1974); 840 (1976).

43 T. A. Degurko, Y. I. Gevasa and V. I. Staninets, *Ukran. Khim. Zh.*, **47**, 649 (1981).

44 O. Diels and K. Alder, *Ber.*, **62**, 557 (1929).

45 E. Arundale and L. A. Mikeska, *Chem. Rev.*, **51**, 505 (1952).

46 A. Z. Shikhmamedbekova, *Chem. Abstr.*, **82**, 155626 (1975).

47 S. Patai and Z. Rappoport, *The Chemistry of Alkenes*, S. Patai, Ed., Vol. 1, Interscience, New York, p. 469.

48 A. T. Nielsen and W. J. Houlihan, *Org. React.*, **16**, 1 (1968).

49 J. P. Schaefer and J. J. Bloomfield, *Org. React.*, **15**, 1 (1967).

50 G. Jones, *Org. React.*, **15**, 204 (1967).

51 W. S. Johnson and G. H. Daub, *Org. React.*, **6**, 1 (1951).

52 W. S. Johnson, *Ylide Chemistry*, Academic Press, New York, 1966.

53 A. Maerker, *Org. React.*, **14**, 270 (1965).

54 G. Wittig, *Pure Appl. Chem.*, **9**, 245 (1964).

55 E. J. Corey and M. Chaykovsky, *J. Amer. Chem. Soc.*, **87**, 1353 (1965).

56 F. F. Blicke, *Org. React.*, **1**, 303 (1942).

57 Z. Rappoport, *Adv. Phys. Org. Chem.*, **7**, 1 (1969).

58 G. Modena, *Acc. Chem. Res.*, **4**, 73 (1971).

59 N. S. Isaacs, *J. Chem. Soc. (B)*, 1053 (1966).

60 Z. Rappoport, *J. Chem. Soc.*, 4498 (1963); 1348, 1360 (1964).

61 G. Modena, P. E. Todesco and S. Tanti, *Gazz. Chim. Ital.*, **89**, 878 (1959).

62 L. Maioli and G. Modena, *Gazz. Chim. Ital.*, **89**, 854 (1959). G. Modena and P. E. Todesco, *Gazz. Chim. Ital.*, **89**, 866 (1959).

63 P. Beltrame, P. L. Beltrame and L. Bellotto, *J. Chem. Soc. (B)*, 932 (1969).

64 A. Compagni, G. Modena and P. E. Todesco, *Gazz. Chim. Ital.*, **90**, 694 (1960).

65 G. Modena, F. Taddei and P. E. Todesco, *Ric. Sci.*, **30**, 894 (1960).

66 J. D. Park and E. W. Cook, *Tetrahedron Letts.*, 4853 (1965).

67 J. D. Park and W. C. Frank, *J. Org. Chem.*, 1333 (1967).

68 J. D. Park and R. J. McMurtry, *Tetrahedron Letts.*, 1301 (1967).

69 J. D. Park, M. L. Sharrah and J. R. Lacher, *J. Amer. Chem. Soc.*, **71**, 2337 (1949).

70 J. D. Park and C. M. Shaw, *J. Amer. Chem. Soc.*, **73**, 2342 (1951).

71 J. D. Park and J. R. Dick, *J. Org. Chem.*, **28**, 1154 (1963).

72 J. D. Park, J. H. Wilson and J. R. Lacher, *J. Org. Chem.*, **28**, 1008 (1963).

73 J. D. Park, J. R. Dick and J. H. Adams, *J. Org. Chem.*, **30**, 400 (1965).

74 J. D. Park, J. R. Lacher and J. R. Dick, *J. Org. Chem.*, **31**, 1116 (1966).

Polar addition reactions

75 J. R. Park, G. Grapelli and J. H. Adams, *Tetrahedron Letts.*, 103 (1967).
76 J. R. Park, R. Sullivan and R. J. McMurtry, *Tetrahedron Letts.*, 173 (1967); *J. Org. Chem.*, **33**, 33 (1968).
77 P. Beltrame, G. Favini, M. G. Cattania and F. Guella, *Gazz. Chim. Ital.*, **98**, 380 (1968); P. Beltrame, I. R. Bellobono and A. Fere, *J. Chem. Soc. (B)*, 1165 (1968).
78 G. Modena, *Ric. Sci.*, **28**, 341 (1958).
79 W. E. Truce, H. G. Klein and R. B. Kruse, *J. Amer. Chem. Soc.*, **83**, 4636 (1961).
80 W. E. Truce, W. Banuster, B. Groten, H. G. Klein, R. B. Kruse, A. Levy and E. Roberts, *J. Amer. Chem. Soc.*, **82**, 3799 (1960).
81 W. E. Truce, *Organo-Sulfur Compounds*, Vol. 1, N. Kharasch, Ed., Pergamon, London, 1961.
82 F. Scardiglio and J. D. Roberts, *Tetrahedron*, **1**, 343 (1957).
83 L. K. Montgomery, F. Scardiglio and J. D. Roberts, *J. Amer. Chem. Soc.*, **87**, 1917 (1965).
84 G. Fraenkel and P. Cooper, *Tetrahedron Lett.*, 1825 (1968).
85 P. E. Peterson, *J. Amer. Chem. Soc.*, **82**, 5834 (1960); **85**, 3608 (1963); *Tetrahedron Letts.*, 1569 (1963); *J. Org. Chem.*, **27**, 1505, 2290 (1962); **37**, 1770 (1972); **38**, 493 (1973).
86 J. A. Pincock and K. Yates, *Can. J. Chem.*, **48**, 2944 (1970).
87 P. B. D. De la Mare and R. D. Wilson, *J. Chem. Soc., Perkin II*, 2048 (1977).
88 R. P. Hanslik and G. O. Shearer, *J. Amer. Chem. Soc.*, **97**, 5231 (1975).
89 Y. Okamoto and H. C. Brown, *J. Org. Chem.*, **22**, 485 (1957).
90 F. Freeman, *Chem. Rev.*, **75**, 439 (1975).
91 I. Hahnemann and W. Pritzkew, *J. Prakt. Chem.*, 315, 345 (1973).
92 J. Klutsch and S. Fliszar, *Can. J. Chem.*, **50**, 2841 (1972).
93 J. E. Dubois and M. H. F. Ruasse, *J. Org. Chem.*, **37**, 1770 (1972); **38**, 493 (1973).
94 P. B. D. De la Mare, *J. Chem. Soc.*, 3823 (1960).
95 M. L. Poutsma, *J. Amer. Chem. Soc.*, **87**, 2162, 2172, 4285, 4293 (1965); *J. Org. Chem.*, **31**, 4167 (1966); **33**, 4080 (1968).
96 R. W. Bott, C. Eaborn and D. R. M. Walton, *J. Chem. Soc.*, 384 (1965).
97 M. Charton and B. I. Charton, *J. Org. Chem.*, **38**, 1631 (1973).
98 R. D. Bach, R. A. Woodward, T. J. Anderson and M. D. Glick, *J. Org. Chem.*, **47**, 3707 (1982).
99 S. Fukuzumi and J. K. Kochi, *J. Amer. Chem. Soc.*, **103**, 2783 (1981).
100 G. D. Graham and S. C. Friedrich and W. N. Lipscomb, *J. Amer. Chem. Soc.*, **103**, 2546 (1981).
101 G. Bellucci, R. Bianchini and R. Ambrosetti, *J. Amer. Chem. Soc.*, **107**, 2464 (1985).

13 Intramolecular reactions

13.1 NEIGHBORING-GROUP PARTICIPATION [1-5]

The solvolysis of a primary sulfonate, RCH_2OSO_2Ar, in a protic solvent of moderate polarity, such as ethanol or acetic acid, would be expected to occur by an S_N2 route and to be rather insensitive towards electronic substituent effects since bond-making and bond-breaking are concurrent (**1**). This expectation is borne out for reactions of compounds **2–4** (Table 13.1), part of a series of ω-methoxyalkyl sulfonates with increasing chain length [6,7]. Compared with the reference compound (**2**), **3** and **4** are somewhat slower; this is due, no doubt, to an inductive effect of the methoxyl oxygen, an effect which is damped with distance from the reaction center. So far this is unexceptional. The surprise comes with the next members of the series, for on removing the methoxyl group even further (and thus rendering the small inductive effects quite negligible), large rate increases are suddenly observed for compounds **5** and **6** but are no longer apparent in the solvolysis of **7**. Evidently the reactivities of **5** and **6** are abnormal. Neither inductive nor resonance effects can explain this phenomenon yet in some way the methoxyl group must be involved. A somewhat similar rate profile is found in ω-aminoalkyl halides, **8–12**. These compounds react spontaneously in protic solvents to form the cyclic amines, **13**, at rates varying with ring size in the order $5 \gg 6 > 3 \gg 4$. This suggests a convincing explanation of the abnormal solvolytic rates of **5** and **6**: they must be due to a mechanistic change in which initial cyclization occurs, despite the fact that the internal nucleophile is an ether oxygen which would normally be expected to show only feeble nucleophilicity, **14**. The product, a cyclic oxonium ion, **15**, is not observed since it undergoes immediate S_N2 attack by solvent leading to the normal acyclic product. This is the usual behavior of oxonium ions since positive oxygen is one of the most nucleofugic groups known. This function

1

Table 13.1 Relative solvolytic rates for ω-methoxyalkyl sulfonates, **1**[1]

Compound	k_{rel}			Estimated k_Δ/k_s		
	EtOH	*AcOH*	*HCOOH*	*EtOH*	*AcOH*	*HCOOH*
2 Me(CH$_2$)$_3$OBros	1·00	1·00	1·00	0·93	0·81	1·07
3 MeO(CH$_2$)$_2$OBros	0·25	0·28	0·10	1·33	1·51	0·85
4 MeO(CH$_2$)$_3$OBros	0·67	0·63	0·33	1·14	1·24	1·84
5 MeO(CH$_2$)$_4$OBros	20·4	657·0	461·0	22·0	425	610·0
6 MeO(CH$_2$)$_5$OBros	2·8	123·0	32·6	2·42	47·2	30·6
7 MeO(CH$_2$)$_6$OBros	1·2	1·2	1·1	0·87	0·71	0·85

Relative rates of ring closure of ω-aminoalkyl bromides, H$_2$O, 25°C [28,29]

H$_2$N—(CH$_2$)$_n$CH$_2$Br	k_{rel}
8 ($n = 1$)	70
9 ($n = 1$)	1
10 ($n = 3$)	6×10^4
11 ($n = 4$)	1×10^3
12 ($n = 5$)	17

of the methoxy group in promoting reaction is an example of *neighboring-group participation* [1,2] (NGP), and is defined[8] as 'the direct interaction of the reaction center (usually, but not necessarily an incipient carbocation) with the lone pairs of an atom or with π- or σ-electrons contained within the parent molecule but not conjugated with the reaction center'. The associated rate increase is known as *anchimeric assistance*[9]. Equation (13.1), applicable to a solvolysis:

$$k_{obs} = k_s + Fk_\Delta + k_c, \qquad (13.1)$$

expresses partitioning of the observed rate into three components; k_s is the rate constant associated with nucleophilic attack by solvent (S$_N$2), k_Δ is that

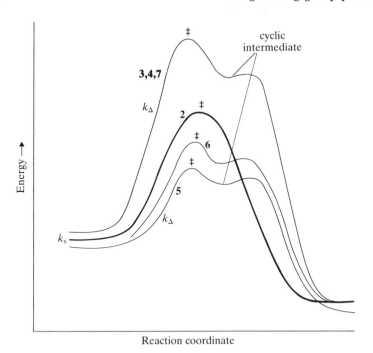

Fig. 13.1   Schematic energy profiles for solvolyses of compounds **2–7**: k_s is the rate of solvent attack. The neighboring-group assisted route, k_Δ, is of lower activation energy for compounds **5**, **6** so will be preferred; k_Δ is of higher activation energy than k_s for compounds **3**, **4**, **7** and will not contribute.

for the neighboring-group promoted component and k_c is that for ionization without assistance (often ignored from lack of information). This analysis does not take into account internal return, which can contribute an important proportion of the ionization rate. Therefore, the factor F $(0 < F < 1)$, is included, representing the fraction of the intermediate resulting from neighboring group participation which proceeds to products rather than returning to reagent.

The energy profiles of these components for compounds **2**, **3**, **4** and **7**, and for **5** and **6**, are shown in Fig. 13.1; the pathway adopted is, of course, in each case that of lower activation energy. The energy barrier for k_s is not much affected by structure; that of k_Δ is very sensitive to the size of the ring formed. Five- and six-membered rings are usually the most rapidly formed; the reasons for this order will be discussed below. Neighboring-group participation is designated by specifying (participating group, ring size). The examples mentioned would be (MeO,5) and (MeO,6) participation. Intramolecular reactivity evidently has quite special structural requirements which, when satisfied, can result in astonishingly high rates by opening up an additional channel to products.

Table 13.2 The scope of neighboring-group participation

Type	Participating groups		Reaction centers

Type of participation	Equation

13.1.1 The scope of neighboring-group effects

In addition to assistance in solvolysis, neighboring-group nucleophilic activity may be observed in carbonyl reactions and in additions and eliminations (Table 13.2). Electrophilic participation (i.e. intramolecular proton transfer) is common also, and participation through a solvent molecule (general acid–base catalysis) can be intramolecular. Participating groups may include any with unshared pairs (O, N, S, P and halogen), also the π-electrons of alkenes and arene groups and, occasionally σ-electrons in certain strained molecules.

13.1.2 Methods for recognizing neighboring-group participation

The formation of the cyclic intermediate in the rate-determining step is the crucial feature of this type of reaction. The observation of cyclic products, as in the reactions of **8–12**, is highly suggestive that this is the case but one may need to consider the possibility that the ring is formed subsequently to the slow step (i.e. after ionization process k_c). The intermediates themselves are usually too fugitive for direct observation though they may be observable under different conditions.

13.1.3 The kinetic criterion

All compounds of the series **2–7** have direct solvolytic routes available and, apart perhaps from the first member which feels the inductive effect of the oxygen, all will have similar values of k_s. The availability of an assisted pathway which opens an alternative route to products can only increase the observed rate although the increment, k_A, can be very small. Anchimeric assistance, A, in a test compound is then measured by the rate ratio:

$$A = \frac{k - k_0}{k_0} = \frac{[(k_s + k_A) - k_s]}{k_s} = k_A/k_s, \tag{13.2}$$

(k, k_0 are observed rates of test and model compounds, respectively). The model is chosen to be as similar structurally to the test compound as possible, though incapable of reacting by the assisted route. It may lack the participating atom as in **2** or **20** or it may be a stereoisomer unable to permit rearface attack at the reaction center as are **23**, **25**[10] (Table 13.3). Some examples of anchimerically assisted reactions are given (**16–19**).

The choice of a suitable model is always fraught with difficulty since any structural change will lead to subtle if small changes in reactivity from numerous causes, including changes in solvation. Compound **2** is a reasonable model for the series **3–7** but since it cannot be assumed that values of k_s are exactly equal for all compounds, confidence that A is a measure of anchimeric assistance depends on the magnitude of A. This becomes a somewhat subjective matter but a value of $1 < A < 10$ might

16 'mustard gas' $\log A = 2^{15}$ **16a**

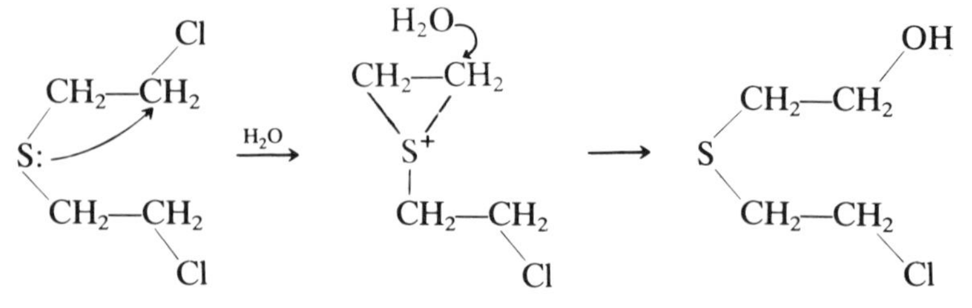

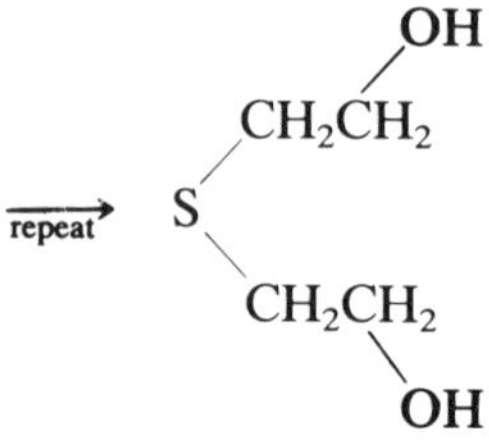

17

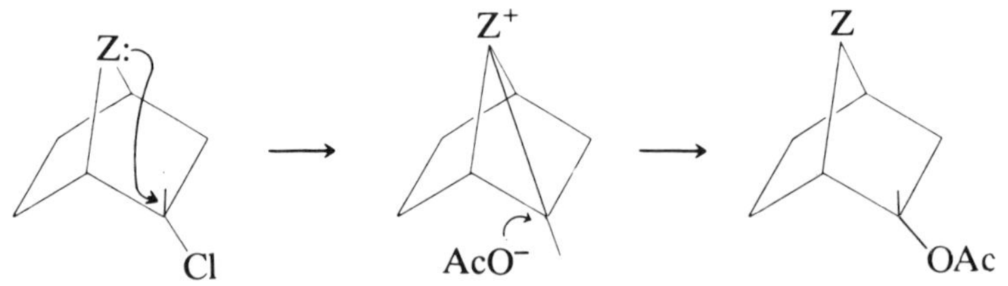

$Z = S, \log A = 9$, rate $\propto [OAc^-]^{46,\,47}$
$Z = O, \log A = 0$.

18 $\log A = 7 \cdot 5^6$ (Refs   *48, 49*)

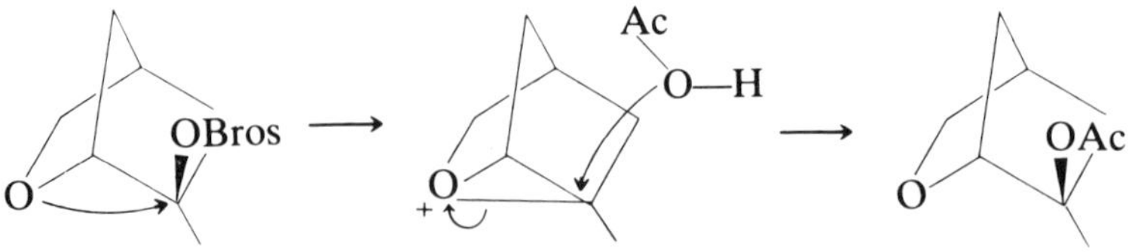

19

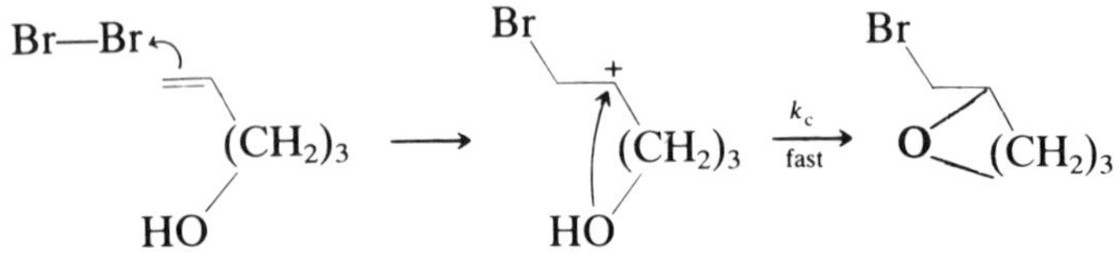

19a

Table 13.3 Brown–Okamoto constants ρ^+ for some assisted reactions

Reaction	ρ^+	References
20 (Ar, Opnb → **20a** Ar)	−5·27	11
21 (Ar, Opnb → **21a** Ar)	−2·30	11
22, *exo* (MeO, Opnb, Ar)	−3·72	12, 13
23, *endo* (Ar, Opnb)	−4·5	12, 13
24, *exo* (Me, Opnb, Ar)	−3·27	14, 15
25, *endo* (Me, Ar, Opnb)	−4·19	14, 15

originate in causes other than nucleophilic participation. On the other hand, extreme values of $A(\sim10^{11})$ are known for which little doubt can be entertained as to their origin. In the example above, some (MeO, 3) participation might be expected by analogy with the cyclization of **8**, yet this is incapable of being discerned by rate measurements above the 'background noise' of uncertainty in k_s.

The kinetics of an internally assisted ionization should be the same as for an S_N1 reaction, first order in substrate and independent of added nucleophiles. Occasionally one finds examples of anchimeric assistance so powerful (or rather an intermediate so long-lived) that the slow step becomes the reaction of the cyclic intermediate with a nucleophile, in which case second-order kinetics may be observed, e.g. **40**.

13.1.4 Linear free-energy relationships

The Hammett and related equations may be used inferentially in two ways. For two related series of compounds which respond well to substituent effects, e.g. in solvolysis by the ionization route, and if one only is capable of neighboring-group participation, then that one will have a reaction constant smaller in magnitude than the other (Table 13.3). This is because the neighboring group will partly fulfill the electronic requirements of the reaction center which then exerts a smaller demand upon substituents or, in other words, the charge developed on carbon will be less in the assisted series. Of the series **20** and **21**, π-participation is available only for the latter, which accordingly shows a greatly reduced susceptibility to electron donation from the aryl group, indicating much less positive charge on the reaction center at the transition state, **21a** compared with **20a**[11]. Of the pairs **22**, **23** and **24**, **25**, π-participation is available only to the *exo* compounds **22** and **24** since attack by the participating group, as in any S_N2 reaction, must come from the rearface[11–15]. Consistently with this, ρ^+ for the *exo* isomers is lower than that for the *endo* isomers. This approach, the 'tool of increasing electron demand'[16], provides qualitative evidence for the existence of participation. Moreover, the competition between the two modes of stabilization of the carbocation, electron release from the $+R$ substituent and intramolecular nucleophilic participation, can be seen by the sudden change in slope of the plot, Fig. 13.2, at the point where the former becomes more favorable. The relative rates of compounds **20**, **21** indicate that the resonance effect of a *p*-anisyl group can overcome the 10^{11}-fold rate increase of π-participation and 'level' rates of ionization of these two systems by a factor $10^{11}/3$[9]. A more quantitative approach can sometimes be achieved when a linear free-energy relationship defining k_s can be drawn up and deviations attributed to k_Δ (Fig. 13.3). The rates of acetolysis of a series of secondary tosylates, RCHMe—Otos, were measured. The rate-determining step is solvent-assisted ionization, k_s, and rates correlate with the Taft σ^* values of R. Large positive deviations are observed for R = —CH$_2$Ar, particularly when the aryl group contains $+R$ substituents[16,17]. This is attributable to aryl participation, **26**, ionization assisted by π-electrons from the aromatic ring with the formation of a bridged intermediate, a species of benzenium ion known as a phenonium ion, **26a**. This undergoes rapid solvolysis to **27a,b**. Differences in rate

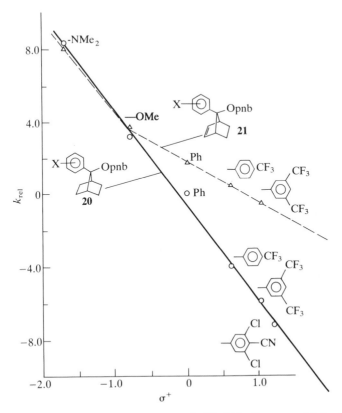

Fig. 13.2 LFER plots for solvolyses of compounds **20** which show normal behavior and no neighboring-group assistance for a range of substituents, X, and compounds **21** in which assistance from the neighboring double bond becomes operative when the aryl group is not strongly electron-donating. The sudden change in ρ is indicative of the incursion of a second route to products[9].

597

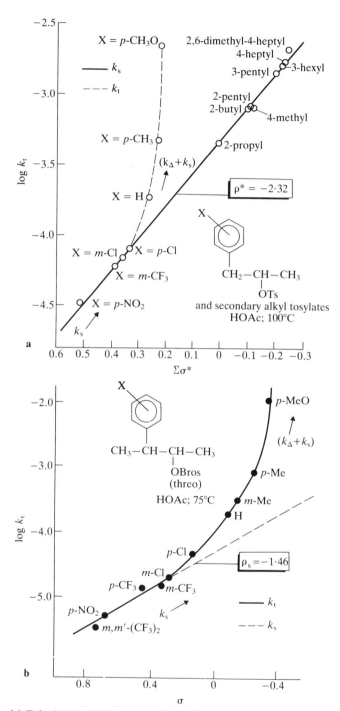

Fig. 13.3 (a) Taft plot for the acetolyses of secondary toluenesulfonates showing deviation of 1-aryl-2-propyl compounds which receive neighboring-group assistance from the aryl group (unless it contains an electron-withdrawing substituent[16,17]. (b) Hammett plot for acetolyses of 3-aryl-2-butyl bromobenzenesulfonates showing the incursion of aryl-group assistance as the substituents become progressively more electron-donating.

between observed values and those predicted from the LFER line are equated to k_Δ:

26; Ar $= p$-X—C_6H_4, R $=$ Me

X	k_Δ/k_s
H	1·5
Me	3
OMe	16

A similar break is observed in the Hammett plot for solvolysis of 3-aryl-2-butyl compounds, **26** (R═Me), Fig. 13.3b, for which deviations of $+R$ substituents are greater than can be accounted for by σ^+ values: moreover, the aryl group is not even conjugated with the reaction center[7,18].

The Schleyer–Foote relationship, Eq. (10.5), was discussed earlier as an aid to calculating solvolytic rates. Large deviations of rates of acetolysis from the calculated values (k_s) enable k_Δ to be estimated (Table 10.16) and may be seen to reach values as high as $10^{19,20}$. Deviations are observed not only for structures which, by other criteria, are judged to receive internal assistance, such as cholesteryl, **43**[21,22], and *anti*-norbornenyl, **66**, but also for strained systems such as cyclopropyl and cyclobutyl. A component at least of the calculated value of A ($= \log k_{\text{obs}}/k_{\text{calc}}$) in all such compounds and especially such structures as quadricyclyl*, may be due to relief of strain outside that given by Eq. (10.5) rather than actual neighboring-group participation. This is also suggested to be the case for *syn*-norbornenyl, **68**, unable to receive internal π-assistance to ionization but with $\log A = 5·6$ and a solvolytic rate 10^4 times that of norbornyl.

13.1.5 Kinetic isotope effects

The distinction between processes k_s and k_Δ is basically one of the extent of nucleophilic participation, which is necessarily high for neighboring-group participation to occur. Any of the criteria which reveal departures from limiting behavior can be used as evidence for internal participation. Secondary α- and β-deuterium isotope effects have been shown to be of the order 1·15–1·25 for limiting solvolyses and close to 1·0 for S_N2 reactions. The latter value should also be appropriate to reactions undergoing neighboring-group attack, as can be seen to be the case for the solvolyses of

$$\text{CH}_2\text{—CL}_2$$
$$\beta$$
$$\text{Me—CH} \qquad \text{CL}_2\text{—OBros}$$
$$\text{O} \qquad \alpha$$
$$\text{Me}$$

k_H/k_D: α, 1·00
β, 1·00

28a

$$\text{CH}_2\text{—CH}_2$$
$$\text{Me—CH} \qquad \text{CL—OBros}$$
$$\text{O} \qquad \alpha$$
$$\text{Me} \qquad {}_\beta\text{CL}_3$$

α, 1·08
β, 1·03

28b

compounds **28a,b**, which are already known to undergo (MeO, 5) participation[7]. More revealing of the involvement of the methoxyl group would be secondary isotope effects α- to the participating atom, evidence of which has not at present been reported.

13.1.6 Solvent effects

Neighboring-group participation in solvolysis is in competition with solvent nucleophilic participation in stabilizing a carbocation. The k_Δ/k_s ratio should therefore fall with increasing nucleophilicity of the solvent. On the other hand, k_Δ should not be very sensitive to solvent polarity and should have a Winstein–Grunwald m_s of 0·1–0·2, similar to that of other S_N2 reactions. Figure 13.4 shows rate effects on hydrolyses of *o*-carboxybenzal chloride in dioxane–water mixtures[23]. The slope at the dioxane-rich end is indeed 0·13 (CO-participation) but rapidly increases at the water-rich end when H_2O participation takes over and polarity becomes much more important. The rate profiles for compounds **2–7** in three solvents are a further illustration (Table 13.1). The value of k_Δ/k_s is largest in formic acid and becomes quite small in ethanol. This is because the alcohol is far more nucleophilic than the carboxylic acids and can compete more effectively with nucleophilic participation by the methoxyl group. It follows that we should find the largest values of k_Δ/k_s, the kinetic criterion of neighboring-group participation, in solvents which induce limiting behavior. It is for this reason that solvolytic rates of primary compounds of the neopentyl type are so high in FSO_3H (Table 10.9). Ionization is facilitated by methyl migration in such cases (**29**).

29

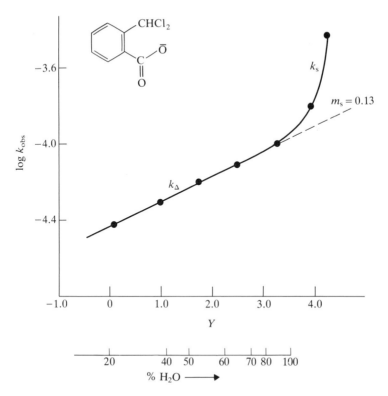

Fig. 13.4 Solvent effect on rate of solvolysis of *o*-carboxybenzal chloride in dioxane–water. Neighboring-group assistance occurs at the dioxane-rich end but the reaction goes over to S$_N$1 ionization when the medium becomes highly aqueous[23].

13.1.7 Participation in carbonyl reactions[24]

Neighboring-group reactions are as important in carbonyl displacements as in solvolytic reactions[25,26].

Rates of hydrolysis of a series of phenyl esters, **30–33**, at pH = 7 in water, reveal exceptionally high reactivity of the monosuccinate (Table 13.4). This compound is favorably disposed for (—COO$^-$, 5) participation (**34**), which leads to the anhydride **35** as intermediate. Since these reactions occur by general nucleophilic catalysis, one can compute (notionally!) the concentration of an external carboxylate catalyst, such as acetate, which would give the same rate as does the intramolecular one. This quantity, the Effective Concentration, EC, is a graphic expression of anchimeric assistance. Values of EC are often in excess of 10^3 and may be as high as 10^5 for the phthalate, **36**, in which the rigid molecular is always in a favorable conformation. The concept of effective molarity in intramolecular catalysis can, of course, be extended to reactions other than those at a carbonyl group; see Table 13.4.

Table 13.4   Anchimeric assistance in ester hydrolysis[25-27]

	Reaction	Reference reaction	$T/°C$	Effective concn, EC/M
1	Enolization of $MeCOCH_2CH_2CO_2^-$	$AcO^- + MeCOCH_2CH_2CO_2Et$	25	0·5
2	Aminolysis of AcIm by $H_2NCH_2CH_2NH_2$	$RNH_3^+ + RNH_2 + AcIm$	25	0·6
3	Hydrolysis of $PhO_2CCH_2CH_2CO_2^-$	$AcO^- + MeCO_2Ph$	25	$5·1 \times 10^4$
4	Lactonization of $PhO_2CCH_2CH_2CH_2O^-$	$MeO^- + MeCO_3Ph$	25	$5·0 \times 10^3$
5	Hydrolysis of $PhO_2CCH_2CH_2CH_2CH_2NMe_2$	$NMe_3 + MeCO_2Ph$	20	$1·3 \times 10^3$
6	Lactonization of $H_2\overset{+}{O}_2CCH_2CH_2CH_2OH$	$EtOH + MeCO_2H_2$	25	80
7	Cyclization of $^-O(CH_2)_4Cl$	$EtO^- + EtCl$	30	$8·1 \times 10^4$
8	Hydrolysis of $HO(CH_2)_4Cl$	$H_2O + C_2H_5Cl$	55	$2·4 \times 10^4$
9	Hydrolysis of $^-O_2C(CH_2)_3Cl$	$AcO^- + C_2H_5Cl$	37·5	$8·5 \times 10^2$
10	Cyclization of $H_2N(CH_2)_4Br$	$NEt_3 + C_2H_5Br$	25	$1·4 \times 10^4$
11	Cyclization of $(EtO_2C)_2C(CH_2)_4Br$	$HC(CO_2Et)_2 + C_2H_5Br$	25	$4·6 \times 10^3$
12	Lactonization of [ortho-substituted benzene: CH_2CO_2Ph, O^-]	$PhO^- + CH_3CO_2Ph$	25	$2·5 \times 10^3$
13	Hydrolysis of [ortho-disubstituted benzene: CO_2Ph, CO_2^-]	$AcO^- + CH_2CO_2Ph$ $RCO_2^- + PhCO_2Ph$	30 30	$1·8 \times 10^5$ $n\sim1·0 \times 10^9$
14	Hydrolysis of [alkene: $CO_2\text{-}p\text{-}C_6H_4OMe$, CO_2^-]	$AcO^- + MeCO_2\text{-}p\text{-}C_6H_4OMe$ $RCO_2^- + MeCH{=}CHCO_2\text{-}p\text{-}C_6H_4OMe$	25 25	$1·8 \times 10^4$ $\sim2·8 \times 10^9$

15 Hydrolysis of [structure with CO_2-p-C_6H_4OMe and CO_2^-]

$$AcO^- + MeCO_2\text{-}p\text{-}C_6H_4OMe \qquad 25 \qquad 9.1 \times 10^7$$
$$RCO_2^- + MeCO_2\text{-}p\text{-}C_6H_4OMe \qquad 25 \qquad {\sim}1.1 \times 10^9$$

16 Equilibrium anhydride formation of [structure: Me, CO_2H, Me, CO_2H]

$$MeCO_2H + MeCO_2H \qquad 25 \qquad 2 \times 10^{12}$$

17 Lactonization of [structure]

$$PhOH + Me\overset{+}{C}O_2H_2 \qquad 30 \qquad 3 \times 10^{15}$$

18 [structure] $\longrightarrow$ [structure]

$$6 \times 10^6$$

19 *Hydrolysis of R—$CH_2CO\cdot OPh$*

	k_{rel}
R = H (**30**)[a]	30 1
R = COO⁻ (**31**)[a]	50
R = CH₂COO⁻ (**32**)	23 000
R = CH₂CH₂COO⁻ (**33**)	150

[a] General base catalyzed.

34

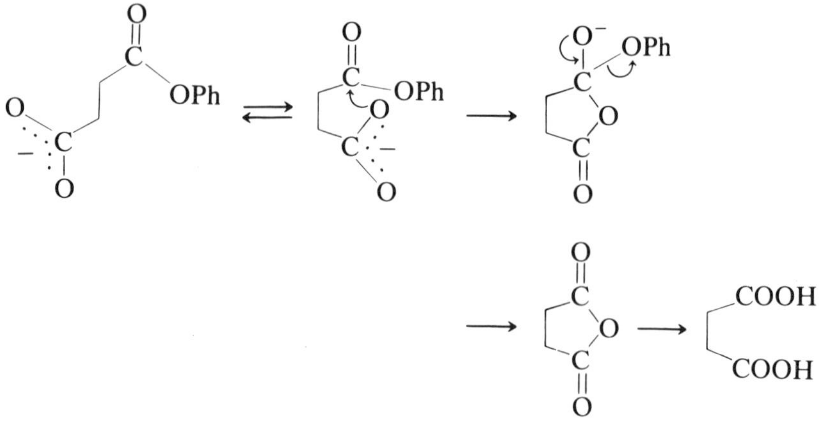

35

36

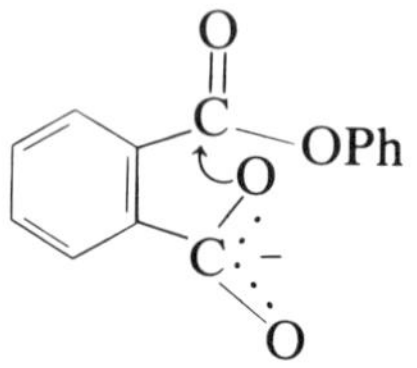

$(-CO\bar{O}, 5): EC = 10^5$

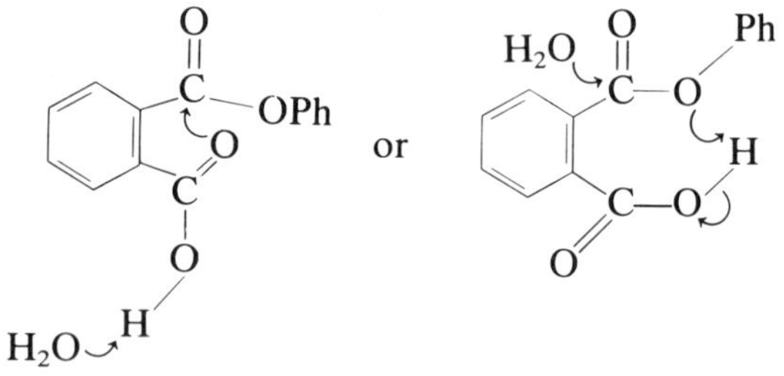

$(-COOH, 5): EC = 10^5$

37

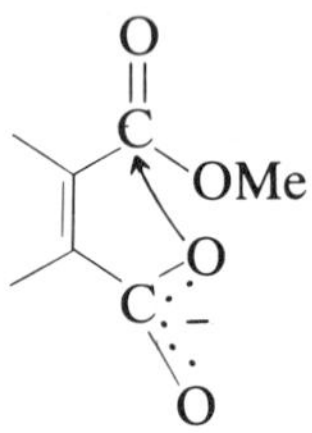

$(-COO^-, 5)$ or $(-COOH, 5)$
$EC = 10^{10}$

38a

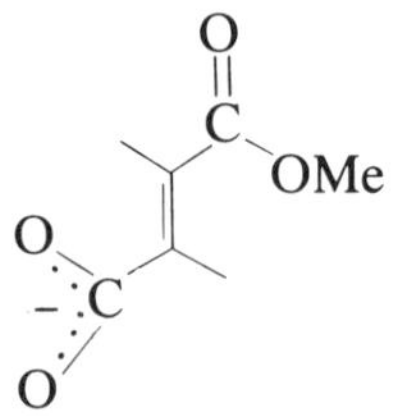

38b

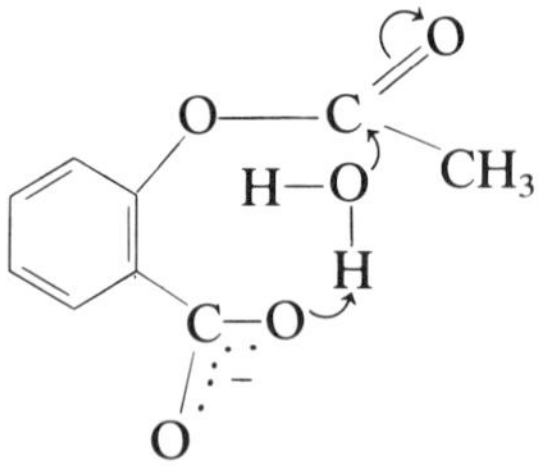

39

Participation by the unionized carboxyl group, not normally observed in intermolecular catalysis, is very effective in hydrolyses of phthalates, **37**, and maleates, **38a**, (but not fumarates, **38b**); hydrolysis of aspirin (acetyl salicylate) occurs by intramolecular general base catalysis[30], the neighboring carboxylate ion causing OH$^-$ from water to attack the carbonyl center, **39** — a favorite strategy in enzymic reactions.

13.1.8   The stereochemical criterion

Neighboring-group participation in solvolytic reactions such as those of **5** and **6** requires two sequential S_N2 displacements, each of which must be accompanied by inversion of configuration at the reaction center. The net stereochemical result therefore should be 100% retention. This is indeed the case, as the following examples, **40–46**, testify. In the cyclohexane series, which has been well examined,[31], NGP requires a 1,2-diaxial (*trans*) geometry. When reactions proceed by a mixture of solvolytic attack, k_s, and internal participation, k_Δ, the proportions of the two may be inferred as the ratio of inverted to retained product. This criterion requires the reaction center to be chiral or in a non-symmetrically substituted ring.

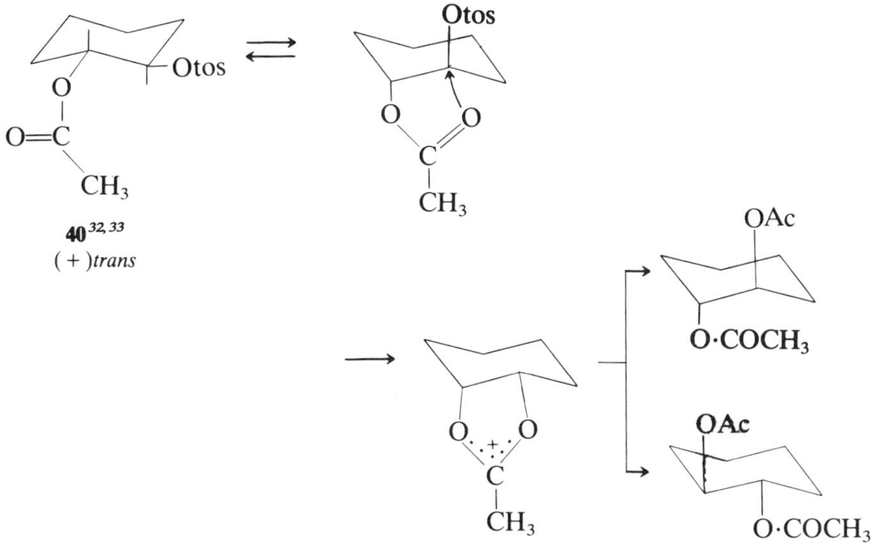

40[32,33]
(+)*trans*

40a

(±)*trans*

41[11]

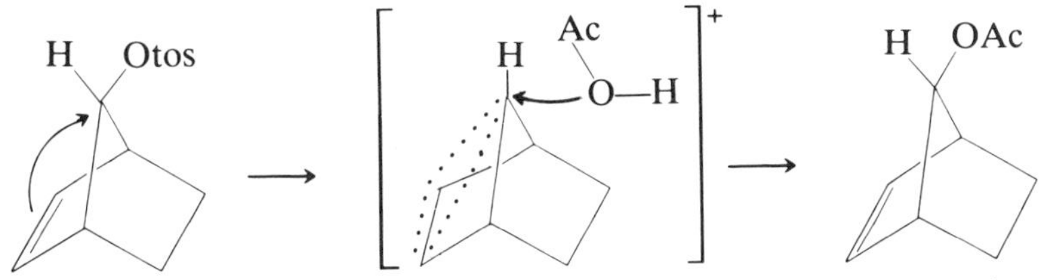

100% retention

13.1.9   The rearrangement criterion

The cyclic intermediates which have been proposed in previous examples frequently possess symmetry of bonding which permits them to open in two directions e.g. **16a**, **26a**, **40a**, **44**, **47**. One of these will lead to an overall

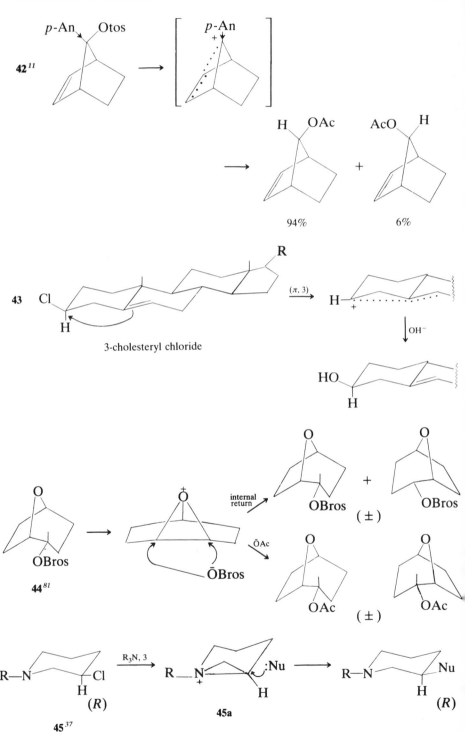

94%   6%

43   3-cholesteryl chloride

44^{81}

45^{37} 45a

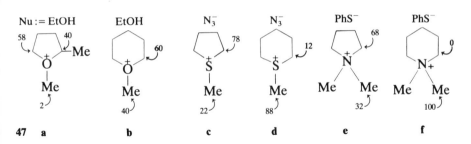

(S)-Bromopropanoate

46 [82,83]

(S)-lactate

47

Nu := EtOH

EtOH

N_3^-

N_3^-

PhS^-

PhS^-

47   a   **b**   **c**   **d**   **e**   **f**

transfer of the participating group to the reaction center, frequently across four or five atoms, a result which strongly argues for the cyclic intermediate. In order to recognize the occurrence of rearrangements, the symmetry of the cyclic intermediate must be perturbed by the pattern of substitution such as by the incorporation of a substituent in order that the rearranged and unrearranged products may be distinguished, e.g. **47**. In this case one could expect unequal amounts of the isomeric products. It is interesting to note that the order of preference of attack by solvent on the presumed intermediate depends upon several factors; the size of the ring, the nature of the heteroatom in the ring and the nucleophile/solvent system in the environment. In general, five-membered rings in protic solvents tend to open at the methylene (or methine) position and exhibit S_N1-like reactivity in their subsequent transformations. This pathway would be preferred for relief of strain in the ring. Analogous six-membered cyclic intermediates which are strain-free by contrast tend to exhibit S_N2-like behaviour favoring attack at methyl, **47a–f**[31]. Attack at methyl relative to methylene also increases when the heteroatom changes from oxygen to sulfur to nitrogen and when non-polar solvents are used (e.g. lithium aluminum hydride in ether).

In several cases (**26, 40, 44**), initial chirality in the reagents is lost in the cyclic intermediate and therefore also in the products, which can include those of internal return[32]. The 100% rearrangement accompanying solvolysis of $PhCH_2CMe_2S$—$(CH_2)_2$—Otos reveals an unusual case of four-membered ring (RS-4) participation[33].

13.1.10 Factors influencing neighboring-group participation[27,34]

Two questions central to the understanding of this topic are: why is an intramolecular reagent move effective compared with intermolecular attack and, secondly, how are ring size and rates of ring closure related? The figures of Table 13.1 reveal that in certain instances intramolecular nucleophilic attack by an ether oxygen can be faster than that by solvent acetic acid despite being a far less powerful nucleophile (based on their relative strengths as hydrogen-bond donors) and at a lower actual concentration. An intramolecular reaction evidently possesses some considerable advantage. Rate effects originate from differences in $\Delta H^\ddagger$ and $\Delta S^\ddagger$ which together constitute $\Delta G^\ddagger$. We may compare intra- and inter-molecular processes in which the same types of bonds are made and broken in which case the $\Delta H^\ddagger$ values, determined by bond strengths and strain energies, will be similar.

A bimolecular reaction in general is accompanied by the loss of three translational and three rotational degrees of freedom, which means a negative contribution to the entropy of the process of perhaps $100 \, J \, K^{-1} \, mol^{-1}$ ($25 \, cal \, K^{-1} \, mol^{-1}$). In contrast, an intramolecular cyclization is accompanied by no change in the translational freedom, which must mean a considerable entropic advantage. At the same time there is an entropy cost in converting a mobile chain of atoms into a rigid ring due to the conversion of some low-frequency internal rotations into high frequency vibrations. This amounts to approximately $15\,J\,K^{-1}\,mol^{-1}$ $(3\cdot6\,cal\,K^{-1}\,mol^{-1})$ per CH_2 rotor immobilized in ring formation and becomes less as rings become larger and more mobile. Enthalpy changes, on the other hand, are very unfavorable for small ring formation on account of strain energy, which is minimized at the six-membered ring. These quantities determine the free-energy changes. Table 13.5 sets out entropies and enthalpies for some hypothetical gas-phase reactions forming cycloalkenes and for the formation of some oxygen and sulfur heterocycles, which illustrate these principles[35]. In all cases the most favorable free-energy change comes with the five- or six-membered ring formation, but there are considerable differences in the energy values when heteroatoms are present. In particular, heats of formation of oxirane and especially of thiirane rings are much lower than that for cyclopropanes. Differences in activation parameters between anchimerically-assisted and unassisted reactions would on this basis be expected to reside in a less negative entropy of activation. This does seem to be the case, but the effect may be accompanied in addition by a lower enthalpy of activation as the examples in Table 13.6 show.

Theories of the origins of the intramolecular rate advantage have been reviewed by Menger[36] who concludes that 'the rate of reaction between functionalities A and B is proportional to the time that A and B reside within a critical distance'. This simple principle is also an expression of an entropy advantage. The 'critical distance' at which attractive interactions are likely to lead to reaction is of the order 250 pm or less than 1·4 times the sum of the atomic radii of the interacting atoms but the angular dependence is not very critical.

Table 13.5 Thermodynamic functions for some hypothetical cyclizations[39]
(a) *n*-Alkane, $C_nH_{2n+2}\rightarrow$ cycloalkane, $C_nH_{2n}+H_2$

Ring size, *n*	$\Delta G/kJ\,mol^{-1}$ $(kcal\,mol^{-1})$	$\Delta H/kJ\,mol^{-1}$ $(kcal\,mol^{-1})$	$\Delta S/J\,K^{-1}\,mol^{-1}$ $(cal\,K^{-1}\,mol^{-1})$
3	128	157	98
	(30·5)	(37·5)	(23)
4	127	153	86
	(30·3)	(36·5)	(20)
5	47	69	74
	(11·2)	(16·5)	(18)
6	32	44	40
	(7·6)	(10·5)	(9)
7	55	68	45
	(13)	(16·2)	(11)
8	73	82	30
	(17)	(19·6)	(7)
	48	42	−20
9	(11·5)	(10·0)	(−5)

Table 13.5   (*Continued*)
(a) *n*-Alkane, $C_nH_{2n+2} \rightarrow$ cycloalkane, $C_nH_{2n} + H_2$
gem-Dimethyl-substituted

Compound	$\Delta G/kJ\,mol^{-1}$ $(kcal\,mol^{-1})$	$\Delta H/kJ\,mol^{-1}$ $(kcal\,mol^{-1})$	$\Delta S/J\,K^{-1}\,mol^{-1}$ $(cal\,K^{-1}\,mol^{-1})$
2,2-Dimethylpentane	39	68	97
	(9·3)	(16·2)	(23)
3,3-Dimethylpentane	36	63	90
	(8·6)	(15·0)	(21)
2,2-Dimethylhexane	24	44	64
	(5·7)	(10·5)	(15)
3,3-Dimethylhexane	22	39	58
	(5·2)	(9·3)	(14)

(b) 1-Alkene, $C_nH_{2n} \rightarrow$ cycloalkane, C_nH_{2n}

Ring size, *n*	$\Delta G/kJ\,mol^{-1}$ $(kcal\,mol^{-1})$	$\Delta H/kJ\,mol^{-1}$ $(kcal\,mol^{-1})$	$\Delta S/J\,K^{-1}\,mol^{-1}$ $(cal\,K^{-1}\,mol^{-1})$
3	41	32.8	−29
	(9·8)	(7·8)	(−7)
4	39	26·7	−43
	(9·3)	(6·4)	(−10)
5	−40	−56·2	−54
	(−9·5)	(−13·4)	(−13)
6	−55	−81·4	−87
	(−13·1)	(−19·4)	(−21)
7	−32	−57·0	−81
	(−7·6)	(−13·6)	(−19)
8	−19	−42·8	−78
	(−4·5)	(−10·2)	(−19)

(c) Heterocycle formation, $HO-(CH_2)_n-ZH \rightarrow \overset{Z}{\overset{\diagup \diagdown}{(CH_2)_n}} + H_2O$

Z	*n*	$\Delta G/kJ\,mol^{-1}$ $(kcal\,mol^{-1})$	$\Delta H/kJ\,mol^{-1}$ $(kcal\,mol^{-1})$	$\Delta S/J\,K^{-1}\,mol^{-1}$ $(cal\,K^{-1}\,mol^{-1})$
O	3	63	95	107
		(15)	(23)	(25)
O	4	57	86	99
		(14)	(20)	(24)
O	5	−22	3	88
		(−5·2)	(0·7)	(21)
O	6	−32	−15	57
		(−7·6)	(−4)	(14)
S	3	9	39	101
		(2·1)	(9)	(24)
S	4	11	38	92
		(2·6)	(9)	(22)
S	5	−59	−36	77
		(−14)	(−9)	(18)
S	6	−60	−44	51
		(−14)	(−10)	(12)

Table 13.6 Activation parameters for solvolyses (80% ethanol) involving neighboring-group participation

Substrate	$\Delta H^{\ddagger}/kJ\,mol^{-1}$ $(kcal\,mol^{-1})$	$\Delta S^{\ddagger}/J\,K^{-1}\,mol^{-1}$ $(cal\,K^{-1}\,mol^{-1})$	NGP?
ArSCH$_2$CHPhCl	64 (15)	−54 (−13)	Yes[28]
ArSCH$_2$CHPhCl	71 (17)	−79 (−19)	No[28]
MeC$_6$H$_4$CHMe·CHMe·OBros	106 (25)	1(0·2)⎫ −8 ⎬	More[18]
O$_2$NC$_6$H$_4$CHMe·CHMe·OBros	121 (29)	(−2)⎭ +30	Less[18]
exo-2-norbornyl-OBros	105 (25)	(7·2) −6	Yes[28]
endo-2-Norbornyl-OBros	109 (26)	(−1·4)	No[28]

One other important effect which influences cyclization is steric in origin. It appears that alkyl substitution favors the cyclic isomer over the acyclic, both in rates of formation and equilibria (often known as the '*gem*-dimethyl effect' but not limited to this structural feature). Table 13.5a shows that both enthalpy and entropy changes become more favorable to cyclization. The effect can be very large when many methyl groups are present, as in the rates and equilibria of cyclization of succinic acids to the anhydrides (Fig. 13.5). The larger the substituents the greater is the rate of cyclization (**48**):

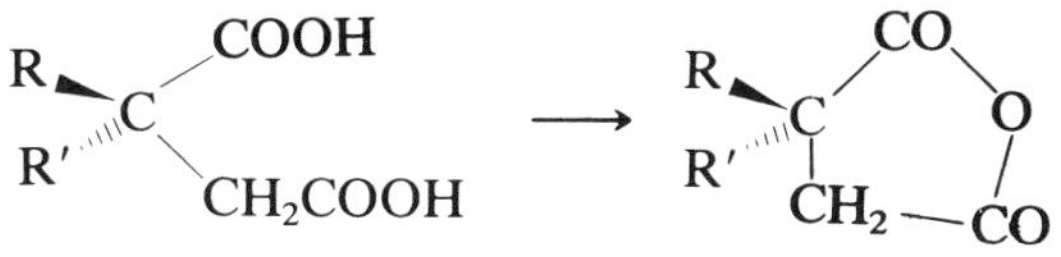

48

R,R':	H,H	H,Me	H,Et	H,isoPr	Et,Et
k_{rel}.	1	4·8	13	33	178

The position of alkylation is important: the cyclization of **49**, for example, is actually retarded by a *gem*-dimethyl group α to the leaving group. However, this creates a neopentyl structure, which is notoriously inert

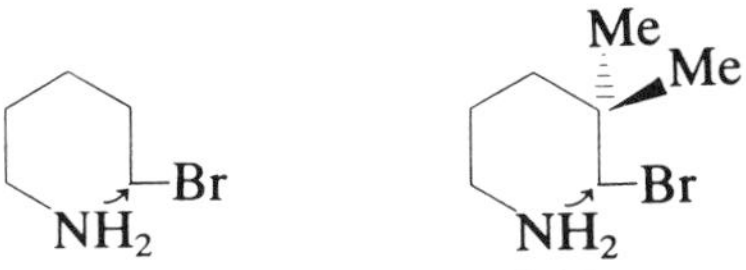

k_{rel}: 1   0.16

49

611

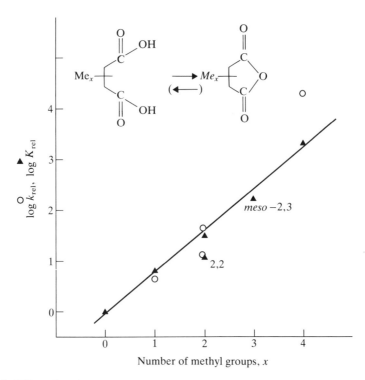

Fig. 13.5 Effect of methyl substitution on rates of anhydride formation of succinic acids: on the ordinate, ▲, $\log k_{rel}$; ○, $\log K_{rel}$.

towards S_N2 reactions for steric reasons. The origin of this steric acceleration on cyclization is probably complex and may be due in part to changes in the populations of rotational conformers resulting in an increase in those favorable toward cyclization. There is likely also to be relief of strain in the form of *gauche* interactions present in the acyclic form.

Turning now to some results, ease of cyclization, although frequently in the order of ring size $5>6>3>4, 7$, does show exceptions. It is found, for instance, that three-membered ring closure is fastest with participation by sulfur and by aryl groups (Tables 13.7, 13.8). Ring strain is less in such cases but solvation may play a part, an aspect which has not been considered and concerning which little is known at present.

Three-membered ring formation is also fastest in the cyclization of stabilized carbanions (Table 13.8c):

$$(PhSO_2)_2\bar{C} - (CH_2)_n - Hal \rightarrow (PhSO_2)_2C\,\,(CH_2)_n + Hal^-.$$

The driving force for cyclopropane formation compared with cyclobutane formation is both a favorable entropy of activation ($+42\,J\,K^{-1}\,mol^{-1}$ ($10\,cal\,K^{-1}\,mol^{-1}$), corresponding to a $\Delta G_s^{\ddagger}$ of $-12\,kJ\,mol^{-1}$ ($-29\,kcal\,mol^{-1}$) at 25°C), and a more favorable enthalpy. Strain

Table 13.7   Ease of ring closure of some anchimerically assisted reactions

$$\text{Nu} \overset{(CH_2)_n}{\diagdown} X \xrightarrow{k} \left(\overset{(CH_2)_n}{\underset{Nu^+}{}} \right) X^-$$

Participating group, Nu—	Leaving group, —X	k_{rel}		
		$n = 3$	$n = 5$	$n = 6$
MeO⁻	OBros, AcOH	0·043	100	18·7
HO—	—Cl, H₂O	0·10	100	41
⁻O—	—Cl, aq. OH⁻	20	100	0·1
H₂N—	—Br, H₂O	0·12	100	1·7
Cl—	—OSO₂Ar, TFA	0·2	100	2·8
ArS—	—Cl, MeOH	115	100	3·3
⁻O₂C—	—Cl, aq. OH⁻	0·008	100	90
O=C<	—OBros, TFA		100	360
⁻O—C₆H₄—	—Br, MeOH	110, 000	100	
MeO—C₆H₄—	—OBros, AcOH	26, 200	100	1·4
HO—	—COOH lactonization	0	100	260
⁻O—	—CONH₂ lactonization	0	100	139

energy evidently does not greatly enter into the composition of the transition state, which argues for a structure in which little C–C bond formation but much C–Hal bond fission has taken place. Ease of ring formation is evidently going to depend on the extent of ring formation at the transition state, and this may vary along a homologous series and in quite unrelated ways between different series. This and changes in solvation must account for the wide range of values of $\Delta S^{\ddagger}$ observed (Table 13.8).

13.1.11   Observation and isolation of cyclic intermediates

Evidence for reactive intermediates can sometimes be obtained from trapping experiments in which either the reactive species is diverted to give some characteristic product or it is generated under conditions in which it is stable. In the latter category are phenonium ions, norbornenyl, norbornyl and many other cations which can be generated as long-lived entities in superacid solutions at low temperatures[3,43–45].

Under non-nucleophilic conditions cyclic bromonium and iodonium ions can exist[37] as stable species. The aziridinium ion **45a** is also isolable[37]. Cyclic intermediates from —OH participation can frequently be trapped as stable products since they may very rapidly lose a proton to form epoxides such as **19a** or tetrahydrofurans[38]. Phenolate participation may lead to an isolable cyclohexadienone **51**, by loss of a proton; other phenonium ions may be captured by addition of a nucleophile **52** → **53**[39,40]. Several 'tetrahedral intermediates' analogous to those formed during carbonyl participation have been isolated[40]. A few α-lactones and α-lactams have now been synthesized and their properties shown to be consistent with transient existence in

Table 13.8 Thermodynamic parameters for some cyclizations [85]

(a) Lactone formation: $Br-(CH_2)_n-COO^- \rightarrow$ cyclic $(CH_2)_n\,C{=}O$

Ring size, n	k_{intra}/s^{-1}					k_{intra}/s^{-1} at 50·0°C	$\Delta H^{\ddagger}$ kJ mol^{-1} (kcal mol^{-1})	$\Delta S^{\ddagger}$ J K^{-1} mol^{-1} (cal K^{-1} mol^{-1})
	18·0°C	26·0°C	34·0°C	42·0°C	50·0°C			
4		0·0582	0·136	0·280	0·587	~2·6	74 (17·7)	−4·9 (−1·17)
5	10·1	21·8	45·1	87·9	170	~310	66 (15·9)	−5·6 (−1·3)
6	0·0879	0·210	0·447	0·922	1·88	~2·9	72 (17·2)	−4·2 (−1·0)

(b) Cycloalkane formation [86]: $(PhSO_2)_2\bar{C}-(CH_2)_n\text{-Hal} \xrightarrow[\text{NaOEt/EtOH}]{25°C} (PhSO_2)_2C\,(CH_2)_n + \text{Hal}^-$

n	Hal	k_r/s^{-1}	E_A kJ mol^{-1} (kcal mol^{-1})	$\Delta H^{\ddagger}$ kJ mol^{-1} (kcal mol^{-1})	$\Delta S^{\ddagger}$ J K^{-1} mol^{-1} (cal K^{-1} mol^{-1})	$\Delta G^{\ddagger}$ kJ mol^{-1} (kcal mol^{-1})	% Carbocycle isolated	Ring strain kJ mol^{-1} (kcal mol^{-1})	ΔS J K^{-1} mol^{-1} (cal K^{-1} mol^{-1})
2	Cl	$9·05 \times 10^{-1}$	88·2 (21)	85·7 (20)	−42 (−10)	73 (17)	99	115 (27·5)	−29 (−7)
	I	359							
3	Cl	$6·05 \times 10^{-6}$	93·7 (22)	91·2 (22)	−39 (−9)	103 (25)	97	107 (25·5)	−43 (−10)
3	I	$4·05 \times 10^{-3}$							
4	Cl	$1·49 \times 10^{-2}$	70·8 (17)	68·3 (16)	−51 (−12)	83 (20)	95	23 (5·5)	−55 (−13)
4	I	3·58							

(c) Entropies of activation for cyclization [85]

Substrate	Ring size, n	Solvent	$\Delta S^{\ddagger}/J\ K^{-1}\ mol^{-1}\ (cal\ K^{-1}\ mol^{-1})$
$H_2N(CH_2)_{n-1}Cl$	3	50% dioxane	−63 (−15)
	5		−54 (−13)
$C_6H_5NH(CH_2)_{n-1}Br$	3	60% EtOH	−46 (−11)
	4		−46 (−11)
	6		−71 (−17)
$C_6H_5NH(CH_2)_{n-1}Cl$	3	50% dioxane	−71 (−17)
	5		−62 (−15)
$H_2NC(C_6H_5)H(CH_2)_{n-2}Cl$	3	50% dioxane	−30 (−7)
	5		−42 (−10)
$p\text{-}CH_3C_6H_4S(CH_2)_{n-1}Cl$	3	80% EtOH (w/w)	−100 (−24)
	5		−83 (−20)
$CH_3O(CH_2)_{n-1}OBros$	5	EtOH	−43 (−10.4)
	6		−63 (−15.1)
$C_6H_5C(=O)(CH_2)_{n-2}Cl$	4	80% EtOH, Ag$^+$	−67 (−16.0)
	5		−66 (−15.8)
	6		−85 (−20.4)
	7		−97 (−23.1)
$(CH_3)_2CHC(=O)(CH_2)_{n-2}Cl$	5	80% EtOH, Ag$^+$	−76 (−18.2)
	6		−87 (−20.8)
	7		−94 (−22.6)
$o\text{-}{}^-OC_6H_4(CH_2)_{n-4}Br$	6	75% EtOH	+26 (+6.2)
	7		+5 (+1.2)
$o\text{-}{}^-OC_6H_4(CH_2)_{n-3}Br$	5	75% EtOH	+17 (+4.1)
	6		+22 (+5.3)
	7		−13 (−3.2)
$HO(CH_2)_{n-3}OCH_3$	5	CCl$_4$	−12 (−3.0)
	6		−29 (−6.9)
	7		−45 (−10.8)
$H_2N(CH_2)_{n-3}NH_3^+$	5	Gas phase	−53 (−12.7)
	6		−86 (−20.6)

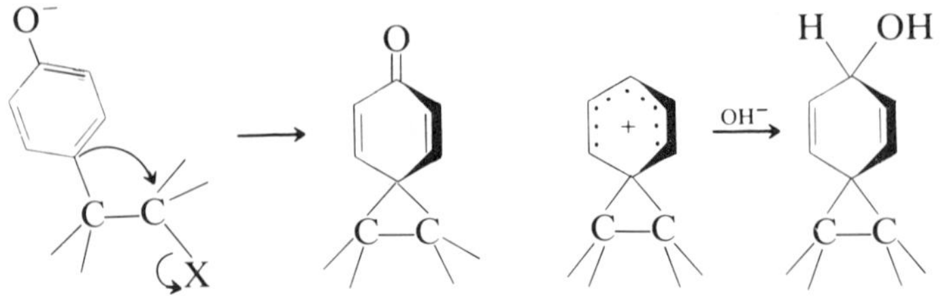

50　　　　　　　51　　　　　52　　　　　53

reaction **46**[41,42]. These experiments add circumstantial evidence to the proposed mechanisms.

13.1.12 π- and σ-participation: the question of non-classical ions [35,50-54]

There is little choice in the assignment of structures for the cyclic intermediates from *n*-participation since they are 'onium ions which, if reactive, have stable analogs. π-Participation leads to structures which may be less easy to express adequately. The stabilization of a carbocation by an adjacent alkene double bond is by π-overlap and the formation of a more extended π-system, an allyl cation. Many examples of nucleophilic participation require π-electrons to interact when they are located on non-contiguous atoms. If the formalism is extended to this situation, then π-overlap must bridge a larger distance (and be correspondingly poorer) and the result must be written as **54** or **55**. The question of structure hinges on whether these are resonance-contributing structures or real species in equilibrium, designated respectively *non-classical* and *classical* carbocations. This type of structure is denoted a *homoallyl cation*, a type which recurs in many examples of neighboring-group participation, three such systems will be mentioned, cholesteryl (**43**), norbornenyl (**66**) and cyclopropylmethyl. Many such cations may be directly observed by NMR in superacid media[55] though the spectra do not unambiguously resolve the structural problem.

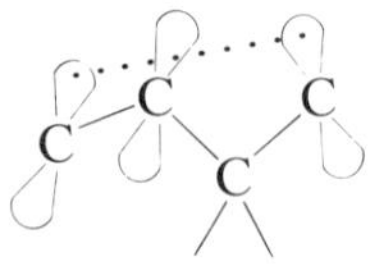

54 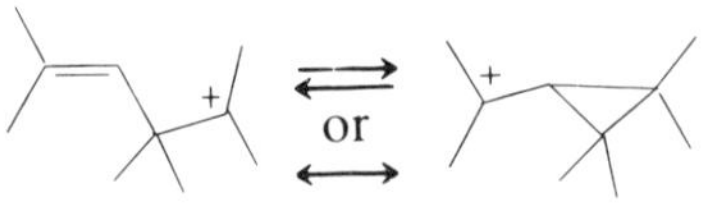　　　　　　　　　　　　55

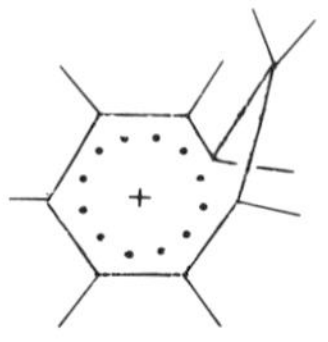

56 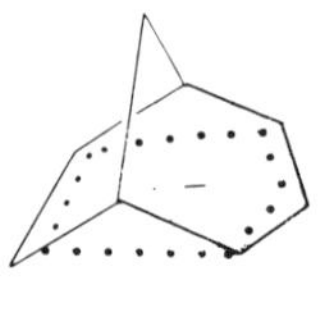　　　　　　　　　　　57

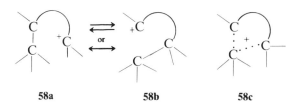

58a   58b   58c

The non-classical aspect of these species is of π-overlap between atoms which are not σ-bonded such as undoubtedly occurs in 'homoaromatic' ions such as **56** ('homotropylium') and, possibly, **57**, whose spectra can be observed in solution[55,56]. σ-Participation requires carbocation stabilization via delocalized σ-bond electrons (**58**). The classical view is of ions which undergo a Wagner–Meerwin type of rearrangement, **58a,b**; the non-classical representation requires the writing of three-center, two-electron bonds, **58c**, a situation closely related to hyperconjugation. The question as to whether the non-classical representation of certain species is more correct than classical alternatives has been the subject of fierce controversy in recent years. This topic provides a fascinating study of the development and testing of a chemical theory: but, in brief, the non-classical ideas propounded by Winstein[57,58], Schleyer, Olah and others led to the representation of almost all carbocations as partaking of extensive π- and σ-delocalization with the result that structures such as **59** became commonplace (the 'rococo period' of non-classical structure!). These interpretations were opposed principally by Brown, who stimulated such a volume of experimental work that today the necessity of invoking remote σ-participation, with a possible few

59

60   AcOH   63

64

        AcOH

OAc
                (⟶)   AcO
62   61

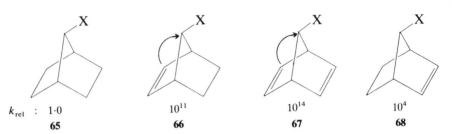

k_{rel} : 1·0 10^{11} 10^{14} 10^4

65   **66**   **67**   **68**

exceptions, has largely disappeared[35]. The controversy has become muted but not entirely resolved. At the same time, high-level MO calculations stress that orbitals occupied even by single-bonding σ-electrons are not completely localized between the atoms which they bond. There is good evidence that cholesteryl compounds, **43**, solvolyze with assistance due to the double bond, rates being faster than expected values by 10^3 or more and retention of stereochemistry in substitution occurring[53,55,56,59–63]. Also, it is possible to isolate 3,5-bonded derivatives, *i*-cholesteryl products, **62**, which are some $10^{7·4}$ times more reactive than the cholesteryl compounds to which

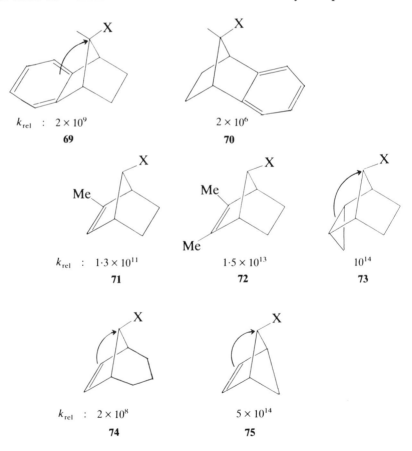

k_{rel} : 2×10^9 2×10^6

69   **70**

k_{rel} : $1·3 \times 10^{11}$ $1·5 \times 10^{13}$ 10^{14}

71   **72**   **73**

k_{rel} : 2×10^8 5×10^{14}

74   **75**

they revert. The mechanisms may be expressed in terms of classical **63**, or non-classical **64**, structures.

Anchimeric assistance in solvolyses of norbornenyl compounds is an experimental fact attested by a value of $A \simeq 10^{11}$ based on the saturated analog as a model[64-67]. π-Assistance can only apply to the 7-*anti* isomer, **66**, but the *syn* compounds, **68**, are highly reactive compared with the parent norbornyl. Also, the additional double bond of norbornadienyl, **67**, contributes further to reactivity though it cannot be involved in direct participation[68]. It seems likely therefore that strain relief in ionization makes a contribution to these rates and also to that of the cyclopropyl analog, **73**, which appears to be a more efficient electron donor than a double bond. The cyclopropane ring behaves as a π-system with σ_p-value, -0.21, more negative than that of vinyl, -0.08, or phenyl, $+0.02$, and will certainly introduce additional strain. Further evidence pointing to this aspect of reactivity comes from rates of the corresponding bicyclo[3,2,1]nonenyl compounds, **74**, which are less strained and far less reactive in solvolysis though the double bond is positioned similarly to that of *anti*-7-norbornenyl. The more strained **75** is correspondingly more reactive[69]. There is a high *anti–syn* reactivity ratio of $k(\mathbf{66}):k(\mathbf{68}) \simeq 10^7$ and electronic demands at C7 on ionization are at least partly satisfied by the π-electrons according to LFER criteria (Table 13.3). Again, products analogous to the *i*-cholesteryl compounds can be obtained by trapping the ion with hydride, e.g. to form **76**, **77**, which, surprisingly can be the major products[70]. Under solvolytic conditions species such as **78** are minor

$$(\mathbf{67};\ X = \text{Opnb}) \xrightarrow[\text{BH}_4^-]{\text{MeOH}_1} \quad \mathbf{76} \quad + (\mathbf{67};\ X = \text{H})$$

83%   12%

$$(\mathbf{66};\ X = \text{Opnb}) \xrightarrow[\text{BH}_4]{\text{MeOH}} \quad \mathbf{77} \quad + (\mathbf{66};\ X = \text{H})$$

15%   70%

MeO   **78** $\longrightarrow$ OMe   **78a**

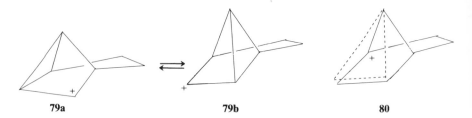

79a   79b   80

products of high solvolytic reactivity. It is possible that these tricyclic compounds are the kinetic products which (except for **76**, **77**) may revert to the more stable form, **78a**. This homoallylic system differs from the cholesteryl one in that both ends of the double bond are symmetrically disposed to the carbocationic center. Classical and non-classical representations of the cation are therefore **79a,b** and **80**. Experimental evidence may be interpreted in terms of either. The ion is directly observable in superacid solution[55] and NMR chemical shifts confirm that considerable positive charge resides on C2, C3 which could fit either representation providing the equilibration of **79a** and **79b** is fast. Similar π-participation in solvolyses of *exo*-5-norbornenyl compounds (**81**; R = H) has also been discussed. Both *exo* and *endo* isomers can yield the same tricyclic product, indicating double-bond participation either during or after ionization[70,71]. There is only a moderate acceleration, $A = 22$, compared with the saturated analog but there is high *exo/endo* rate ratio ($>10^3$) increasing to $>10^8$ in the methyl derivative (**81**; R = Me), which presumably has better donor properties. There may be other driving forces favoring *exo* reactivity; the *exo* side of the molecule is generally considered to be less crowded than the *endo*, and solvation of the departing anion might be sterically more facile and able to lend electrophilic assistance. A *p*-anisyl group for R swamps this *exo/endo* effect completely down to a value *ca* 300 which could be due to different steric environments to ionization, but an aryl group at C2 (i.e. at the reaction center) removes any further electronic demand by π-participation[14,15]. σ-Participation in the ionization of *exo*-2-norbornyl compounds has been the subject of more controversy than has that of any other compound. It is in this area, if anywhere, that 'true' non-classical or σ-bridged ions having three-center, two-electron bonds are to be contemplated. This type of bonding is well known in the chemistry of boranes, with which carbocations are isoelectronic; the point is made by Schleyer that

R   OBros   AcO   R

fast → R ← slow

81 (exo)   (endo)   OBros

Table 13.9   Relative solvolytic rates of cyclopropyl and related compounds

	RCH$_2$X	CH$_2$=CHCH$_2$X	PhCH$_2$X	$\triangleright$CH$_2$X
k_{rel}:	1	36	2×10^3	1.4×10^5

	$\overset{\mid}{\underset{\mid}{Y}}$C—X	C—X	C—X	C—X
k_{rel}:	1	250	2000	1.4×10^5

non-classical structures must lie close in energy to classical bonding ones. The question is: which form has the lower energy?

The capacity of the cyclopropyl group for assisting and stabilizing ionization at an adjacent carbon is very large, considerably greater than that of vinyl or phenyl, as shown in Table 13.9. During solvolysis of **82**, rearrangements occur, scrambling the four carbons, though incompletely, and products contain cyclobutyl and butenyl compounds as well as cyclopropylmethyl[71]. These observations can be explained in terms of both classical and non-classical representations of the intermediate cation, **82**. It has been pointed out that, whereas the most favorable geometry for cyclopropyl participation is **83**, large rate enhancements are found when this cannot be attained and when no rearface assistance is possible, as in **84**, while in other cases, no acceleration is observed when the geometry should be favorable as in **85**[72–74]. The three-membered ring does increase strain

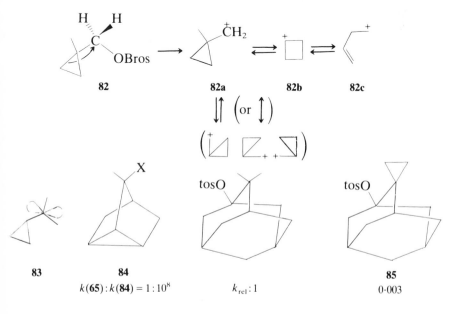

82	82a	82b	82c

$$\Updownarrow \left(\text{or } \Uparrow \right)$$

83	84	85
	$k(\mathbf{65}) : k(\mathbf{84}) = 1 : 10^8$	
	$k_{rel} : 1$	0·003

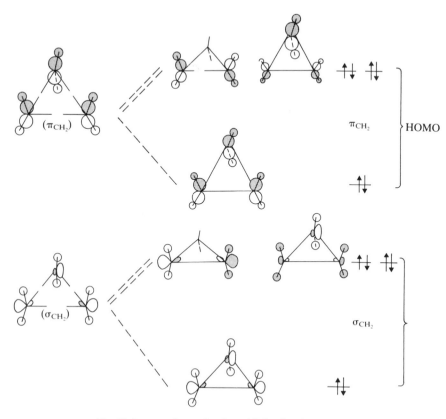

Fig. 13.6 π- and σ-molecular orbitals of cyclopropane.

energy in the molecule and, it is argued, is the cause of the rate effect. In acyclic derivatives, however, this cannot be the case and the cyclopropane ring **86** must be inferred to exhibit conjugative properties as expected from the MO representation of this ring (Fig. 13.6), the HOMO having π-character[75]. Perhaps the species which has drawn most attention and which has remained as the most likely candidate for non-classical formulation is the 2-norbornyl cation[53,76], expressed either as structures **88** or **89**. The cation is characterized by unusually rapid formation from its esters and

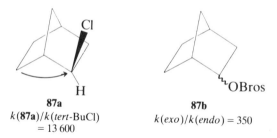

87a	**87b**
$k(\mathbf{87a})/k(tert\text{-BuCl})$	$k(exo)/k(endo) = 350$
$= 13\,600$	

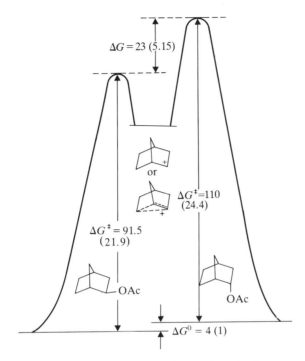

Fig. 13.7   Energy profile of the 2-norbornyl cation.

halides in protic solvents, **87**, together with high *exo/endo* rate ratios (10^2-10^3) [17], a result of the energy profile (Fig. 13.7) and difficult to explain in terms of the classical ion. There is complete racemization of *endo* compounds during solvolysis, consistent with a symmetrical intermediate, **90**, or with rapid degenerate rearrangement of **88**. There are in addition variable amounts of 6:2 and 3:2 hydride migration which also result in racemization. The interpretation in non-classical terms is of rates accelerated by σ-participation with **90** as intermediate. In classical terms these structures represent a transition state ($\rightarrow$**89**) and rates are attributed to relief of steric strain. The problem is not an easy one to resolve satisfactorily and most of the data can be interpreted plausibly by either model [77]. The 6-oxa analogue appears to be a relatively stabilized species, arguing some degree of overlap across the ring [78]. Transannular participation of π-electrons in monocyclic systems, notably cyclo-octenes, has been well explored by Cope [79,80].

88   **89**   **90**

13.2 ENZYMIC REACTIONS[87-92]

All living organisms are capable of the most complex chemistry, though individual reactions are those familiar to organic chemistry. The special features of biological reactions include extremely high reaction velocities at low temperatures and high stereospecificities. Virtually all such natural reactions are mediated by enzymes, catalysts which specifically promote each reaction necessary for life. Enzymes have evolved by the processes of natural selection over the course of the geological time-scale, the most sophisticated examples of neighboring-group participation known, whereby complex reactions can be broken down into a sequence of simple steps, each requiring so little activation energy that rates can approach the diffusion limit. It is only recently that some insight into the mechanisms of such fascinating processes has appeared, revealing sequences which have their counterparts in familiar chemistry and which can be understood in terms of the principles of physical organic chemistry as well as can laboratory reactions.

13.2.1 The structures of enzymes

Enzymes are protein molecules of definite molecular structure, linear poly(α-aminoacids), bearing sidechains of some 20 different compositions (Table 13.10) in a particular sequence. Each chiral center is of the same configuration (usually S). While the sequence of the aminoacid residues constitutes the primary structure, enzymic activity arises from intramolecular bonding (secondary structure) which results from local regions of the polymer associating by weak interactions (mainly hydrogen bonding) into helical or other conformations. Higher-order organization (tertiary structure) results from further coiling of the structure which is secured in a complex three-dimensional and completely essential conformation by a combination of many weak interactions together with some covalent linking via —S—S— bridges. These are brought about by the oxidation of two cysteine thiol groups, which may be on remote parts of the polymer chain but have been brought into close proximity by the coiling. Some enzyme molecules are now known to associate in groups of two, four or more units.

Lysozyme is one of the smallest and simplest of all enzymes. Its molecular weight is only 14 500 and there are 129 aminoacid residues in the chain (Fig. 13.11). The complexity of enzyme structures can be enormous, however, molecular weights being more commonly in the range above 10^5 and even above 10^6. Their structural determination has posed great problems. By careful hydrolysis (in part using enzymic methods), a peptide chain may be cleaved into smaller fragments which can be identified by chromatographic techniques. The information can then be combined so that a sequence of aminoacids can be deduced. The three-dimensional structure can only be obtained from X-ray crystallography, which requires the enzyme be isolated

Table 13.10 Structures of some natural aminoacids, $R-\overset{\displaystyle COOH}{\underset{\displaystyle NH_2}{C}}$

Name	Symbol	R
Neutral		
Glycine	Gly	H—
Alanine	Ala	Me—
Valine	Val	isoPr—
Leucine	Leu	Me_2CHCH_2—
Isoleucine	Ile	EtCHMe—
Phenylalanine	Phe	$PhCH_2$—
Polar		
Serine	Ser	$HOCH_2$—
Threonine	Thr	MeCHOH—
Cysteine	CysH	$HSCH_2$—
Methionine	Met	$MeSCH_2CH_2$—
Tyrosine	Tyr	$HOC_6H_4CH_2$—
Proline	Pro	
Hydroxyproline	Hyp	
Basic		
Lysine	Lys	$NH_2(CH_2)_4$—
Arginine	Arg	$HN{=}C{-}NH(CH_2)_3$—
Asparagine	Asn	NH_2COCH_2—
Glutamine	Gln	$NH_2COCH_2CH_2$—
Histidine	His	
Tryptophan	Trp	
Acidic		
Aspartic acid	Asp	$-CH_2COOH$
Glutamic acid	Glu	$-CH_2CH_2COOH$

and obtained in pure crystalline form. Structures of more than 100 enzymes have now been obtained but many more are too unstable for isolation. Furthermore, the molecular structure alone does not necessarily reveal the mode of action; occasionally by good fortune a crystal structure of the enzyme bound to its substrate or a closely related molecule can be obtained, though only a few such examples have so far been achieved. As a result only a few enzymic mechanisms have been established with any confidence at the present.

13.2.2 A model for enzyme action

The complex structure of enzymes is essential for their action and evidently serves the purpose of maintaining a very few aminoacid residues in a specific and very precise orientation. Almost the whole of the enzyme molecule then is 'scaffolding' with which to generate an *active site,* a region which may bind to the substrate by weak forces and bring various neighboring groups into the correct orientation in order for the intramolecular catalysis to be most effective. The simplest model, basically due to Michaelis and Menten[93], assumes an equilibrium association between enzyme (E) and substrate (S) to form a complex (ES) which consists of the substrate weakly bound to the active site. Reaction then occurs to form an enzyme–product complex (EP) which then dissociates to free enzyme and product for the sequence to continue. The enzyme is regenerated after each cycle so it acts as a true catalyst:

$$\text{E} + \text{S} \underset{k_{-1}}{\overset{k_1}{\rightleftharpoons}} \overset{K_s}{} \text{ES} \underset{}{\overset{k_2}{\rightleftharpoons}} \overset{k_{cat}}{} \text{EP} \longrightarrow \text{E} + \text{P}.$$

For this simplest of schemes,

$$\text{Rate} = d[\text{P}]/dt = V = k_{cat}[\text{ES}]. \tag{13.3}$$

Now $[\text{ES}] = [\text{E}_0] - [\text{E}]$ where E_0 is the total enzyme concentration added and E is the concentration of free enzyme; also, $K_s = [\text{E}][\text{S}]/[\text{ES}]$, the dissociation constant of the enzyme–substrate complex. Eliminating [E] and [ES] terms gives:

$$V = \frac{k_{cat}[\text{E}_0][\text{S}]}{[\text{S}] + K_s}, \tag{13.4}$$

the basic equation for an enzymic reaction in which the ES complex is in thermodynamic equilibrium with free enzyme and substrate, $k_2 \ll k_{-1}$. This is not always the case and if, for instance $k_2 \sim k_{-1}$, K_s is replaced by K_M where

$$K_M = (k_2 + k_{-1})/k_1. \tag{13.5}$$

Here K_M is the 'operational' Michaelis constant, the concentration of substrate producing a reaction velocity $V_{max}/2$. Strictly $K_M = K_s$ when $[\text{S}] = K_s$ or $[\text{E}] = [\text{ES}]$.

Many enzymic reactions obey this modified equation even when their mechanism is relatively complex and the step ES→EP is replaced by a succession of steps. The equilibrium constant K_M is somewhat inexact in meaning but the term K_s will be retained to have the meaning defined above.

The characteristics of this type of reaction differ from those of bimolecu-

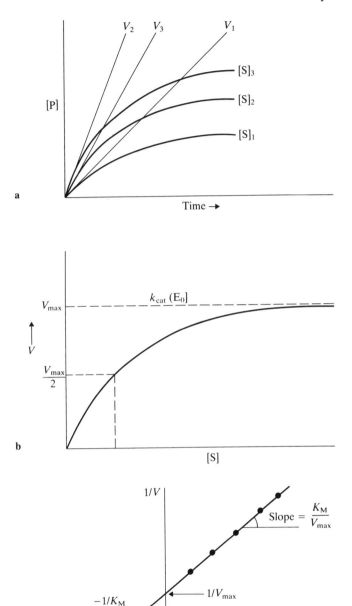

Fig. 13.8 Michaelis–Menten kinetics: (**a**) effect of initial substrate concentrations on initial reaction velocities; (**b**) effect of initial substrate concentration on initial reaction velocities showing the maximum reached at saturation of the enzyme, V_{max}; (**c**) Lineweaver–Burke plot from which K_M and V_{max} can be evaluated.

lar processes in that the rate will not increase with [S] indefinitely[95,96]. In fact, as [S] becomes large, V reaches a constant, maximum value at $V \rightarrow V_{max} = k_{cat}[E_0]$, corresponding to saturation of all active sites by substrate molecules, Fig. 13.8. The evaluation of the various constants may be accomplished by linearization of Eq. (13.4), for example by the Lineweaver–Burke plot[94]) as exemplified in Fig. 13.8c; values of V at $t = 0$ for a series of substrate concentrations are first determined (Fig. 13.8a,b). These are then plotted in the reciprocal form of Eq. (13.6),

$$1/V = 1/V_{max} + K_M/V_{max} \cdot 1/[S]. \qquad (13.6)$$

The form of the Michaelis–Menten equation also allows predictions to be made as to the best strategy for accomplishing the fastest rates of reaction, a target which may be assumed to be important, in many cases at least, under biological conditions. The highest value of k_{cat} and the lowest K_M, corresponding to a high proportion of bound substrate, will clearly increase the rate. That is, the highest k_{cat}/K_M ratio will be most favorable. However, rates and maximum rates will depend not only on this ratio but on the absolute values (Fig. 13.9). It may be seen that the rate rises to a maximum at fixed k_{cat}/K_M as K_M becomes greater than [S]. The larger is K_M, the larger can be [S] and k_{cat} and the observed rate of reaction. So there is the somewhat paradoxical situation that weak binding of the substrate is most conducive to reaction.

This can be understood from another point of view. The free-energy profile of the simplest reaction will consist of two steps, those of binding and

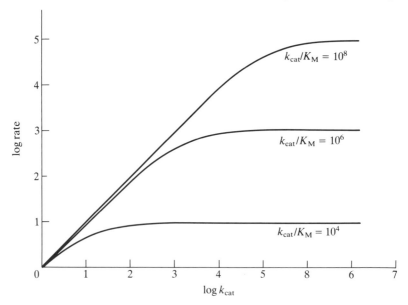

Fig. 13.9 Variation of the rate of an enzymic reaction with the value of k_{cat} as a function of the ratio, k_{cat}/K_M.

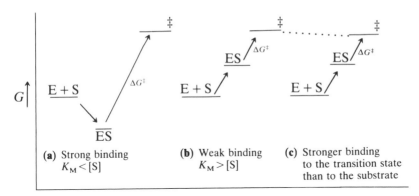

Fig. 13.10 Effect of enzyme–substrate binding energy on the reaction rate, and the importance of strong binding of the enzyme to the transition state rather than to the substrate.

of reaction (Fig. 13.10). If the substrate binds tightly to the enzyme and $K_M < [S]$, the free energy of the complex ES is lower than that of E and S in the initial state. The activation process therefore has to surmount the potential-energy well created by the binding. When the substrate is bound weakly $(K_M > [S])$, the free energy of ES is actually higher than the initial state so that the binding energy has been used as a step towards the transition state. Table 13.11 gives values of k_{cat} and K_M for several enzymes and their substrates; most values of K_M are in the range $10^2–10^3$, corresponding to only 2–0·2% of the substrate bound (assuming $[E] = [S]$) but all of the bound substrate pre-activated towards reaction. It should be mentioned that not every enzyme is required to work at a very high intrinsic rate and the cell can exercise a considerable degree of control over processes by suitable manipulation of the two constants under evolutionary pressure. An example of a very low K_M, 0·1 mM, is that of hexokinase relative to its substrate, glucose, which it phosphorylates in the first step of glycolysis. At normal cell glucose concentration of about 5 mM, this enzyme

Table 13.11   Constants for some enzyme-catalyzed reactions

Enzyme	*Reaction*	K_M/M	k_s/s^{-1}
Acetylcholinesterase	Hydrolysis of $Me_2NCH_2CH_2OAc$	9×10^{-5}	1.4×10^4
Carbonic anhydrase	$CO_2 \rightleftarrows HCO_3^-$	0·0012	10^6
Catalase	$H_2O_2 \rightleftarrows O_2$	1·1	4×10^7
Fumarase	Fumarate hydration	5×10^{-6}	800
Triosephosphate isomerase	Glyceraldehyde-3-phosphate isomerization	5×10^{-4}	4300
Chymotrypsin	Ac—Trp Et ester hydrolysis	0·1	27
	Ac—Trp amide hydrolysis	7·3	0·026
	Ac—Phe Et ester hydrolysis	37	0·039
	PhCO—Tyr Et ester hydrolysis	0·022	86
	PhCO—Val Et ester hydrolysis	4·2	0·064

is near the constant rate region and rates hardly change with a ten-fold change in glucose concentration. This buffering of the consequences of substrate fluctuation has a regulatory effect on the metabolic rate.

A further aspect of enzyme strategy which is now clear is that the enzyme is evolved to bind with better complementarity to the transition state of the reaction than to the substrate itself. Binding produces a lowering of free energy and this consequently creates an additional driving force towards the transition state (Fig. 13.10c). There is experimental evidence for this general principle; a synthetic substrate which has a geometry or other features which make it resemble the supposed transition state often has stronger binding (lower K_M) than the true substrate. The lactone, **91,** is a transition-state analog of the polyglucosamine substrate of lysozyme, **92,** which is presumably hydrolyzed via a carbocation, **93,** whose preferred conformation is the half-chair form rather than the chair conformer of the cyclohexane ring precursor. The binding constant of **91** is 100 times greater than that of the substrate[97], **92.** Proline, **94,** racemizes under the action of proline racemase and presumably via a planar intermediate. The binding of planar analog **96** to the enzyme is 160 times greater than that of proline[98], and it acts as a potent inhibitor.

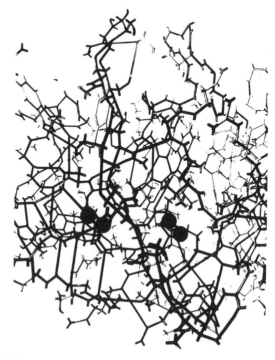

Fig. 13.11 (**a**) Model of the lysozyme molecule showing binding to a polysaccharide. (**b**) Mechanism of action of lysozyme.

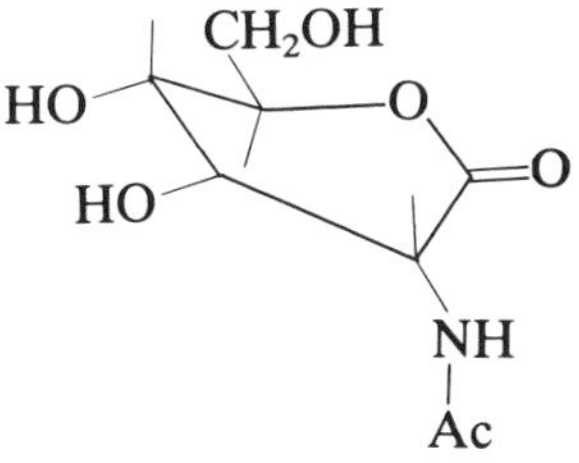

91

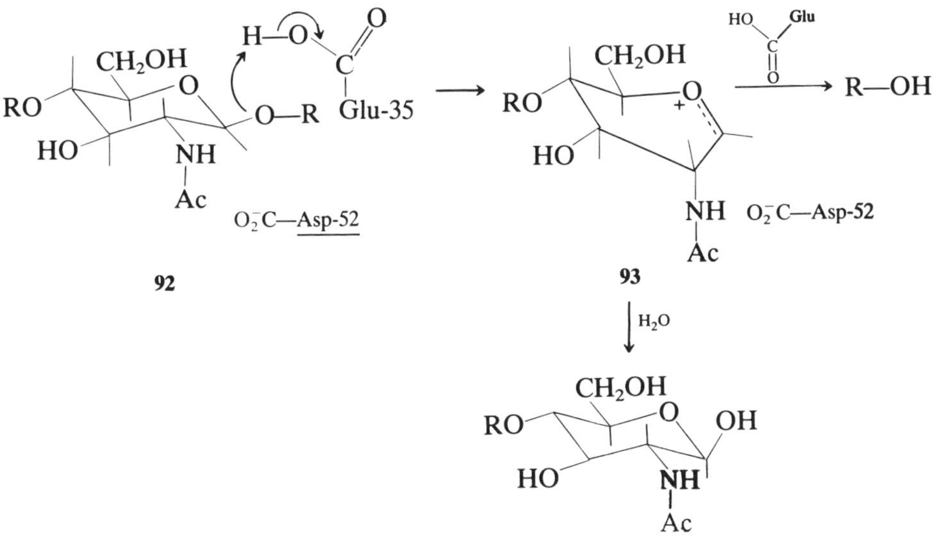

92   **93**

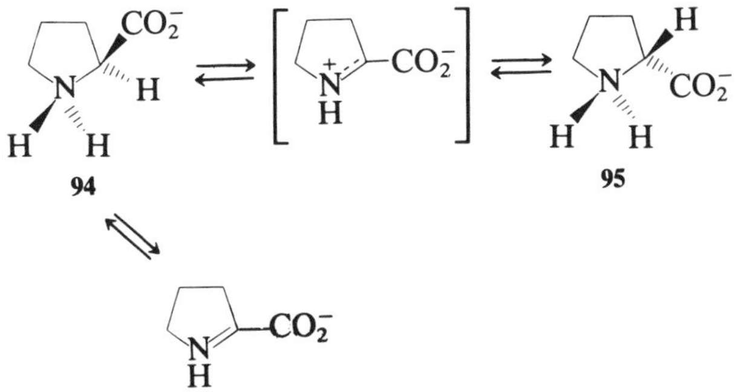

94   **95**

b   **96**

Fig. 13.11   (*Continued*)

13.2.3 Mechanisms of some enzyme-catalyzed reactions

Lysozyme

This small enzyme hydrolyzes polysaccharide **92** and possesses six binding sites into which may be fitted six units of the substrate (Fig. 13.11). The glycoside bond between the fourth and fifth is hydrolyzed, cleaving off a disaccharide. The reaction is the acid-catalyzed hydrolysis of an acetal which is brought about by the neighboring glutamic acid (Glu-35). The carbocation intermediate, **93**, binds more tightly to the enzyme than does the substrate and its charge is stabilized also by a neighboring asparate (Asp-52), either ionically or by covalency formation. Water then completes the reaction, loosening the binding, and the product separates from the enzyme[99,100].

Serine proteinases

Proteinases which hydrolyze protein to its constituent aminoacids are widespread digestive enzymes. One of the best understood is chymotrypsin which is readily obtained in a crystalline state. The molecular structure is known from crystallographic studies and so also is the disposition of the substrate on the active site since some enzyme–substrate complexes can indeed be crystallized. The enzyme has a preference for aromatic amino acids such as phenylalanine and tyrosine but is rather unspecific in its structural requirements for substrates. The aromatic ring fits into a hydrophobic cleft while several hydrogen bonds orient the peptide bond which is to be cleaved (Fig. 13.12)[102,103]. This must be of the correct chirality (L-aminoacids) to fit[101]. The cleavage reaction is then accomplished by an intramolecular $B_{Ac}2$ hydrolysis in which the nucleophile is a serine hydroxyl group (Ser-195). This is brought into play by rotating towards the carbonyl of the amide bond. In doing so it becomes activated by a remarkable 'charge-relay' system by which the proton is abstracted by the basic nitrogen of histidine-57 which in turn passes on its acidic proton to a deeply hidden aspartate-102. The histidine acts as a proton relay while the more remote aspartate is a general base catalyst[107]. The result is the familiar tetrahedral intermediate **97**, to which the enzyme has evolved to bind tightly and to reduce its chemical potential. The reaction proceeds by release of the free amino group of the product, repelled by the newly formed serine ester and leaving behind an enzyme molecule acylated on serine-195. Direct evidence for this type of intermediate comes from isolation at low pH or during hydrolysis of certain unnatural substrates such as aminoacid esters. Deacylation, which can be the slowest step in the sequence, then occurs by attack of water or an added alcohol, presumably assisted by the proton transfers now occurring in reverse[104] (Fig. 13.13).

Metalloenzymes

Many enzymes contain metal ions which are part of the active site and may be used in binding of the substrate by ligation or as electrophilic catalysts in the reaction. Carboxypeptidase contains zinc which acts in both capacities[105,106]. The enzyme is another type of proteinase which cleaves

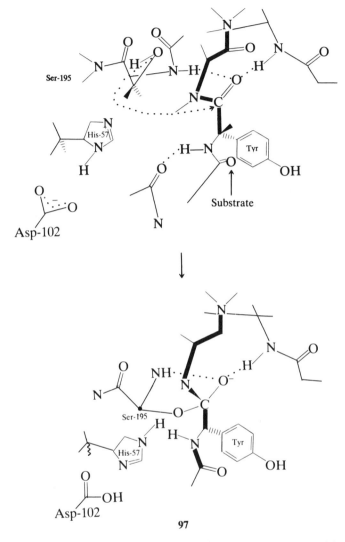

97

Fig. 13.12 Binding of a protein chain to chymotrypsin and of the tetrahedral intermediate formed during hydrolysis.

carboxyl-terminal aminoacids with large, hydrophobic sidechains such as phenylalanine (Fig. 13.14). The active site binds this residue by hydrophobic interactions. It fits into a pocket lined with aliphatic groups expelling water as it enters. Hydrogen bonding to several polar groups in the vicinity of the reaction center completes the substrate binding, the zinc attaching to the amide carbonyl. This is activated to nucleophilic attack by the carboxyl group of glutamate-270, either directly or via a water molecule as a general

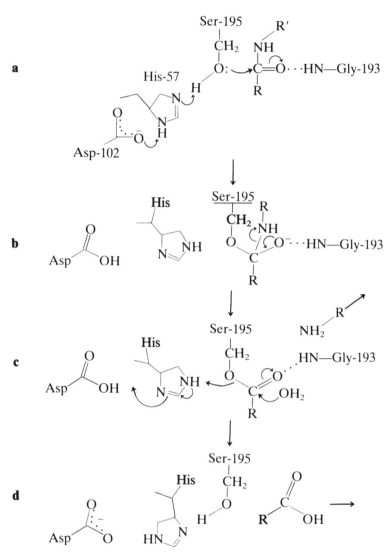

Fig. 13.13 Mechanism of peptide-bond hydrolysis by chymotrypsin. Histidine-57 acts as a general base catalyst to serine-195.

base as the rate-determining step. The tetrahedral intermediate then breaks down and the products dissociate.

Ribonuclease is another very small enzyme of only 124 aminoacid units which cleaves RNA (ribonucleic acid) at an alcohol–phosphate bond. This is brought about by concerted nucleophilic and electrophilic catalysis by two histidines in the active site. The adjacent 2′-hydroxyl group on the ribose residue is brought into play as a neighbouring group and the first product is the cyclic phosphate, **98**, (Fig. 13.15). This is isolable and hydrolyzes

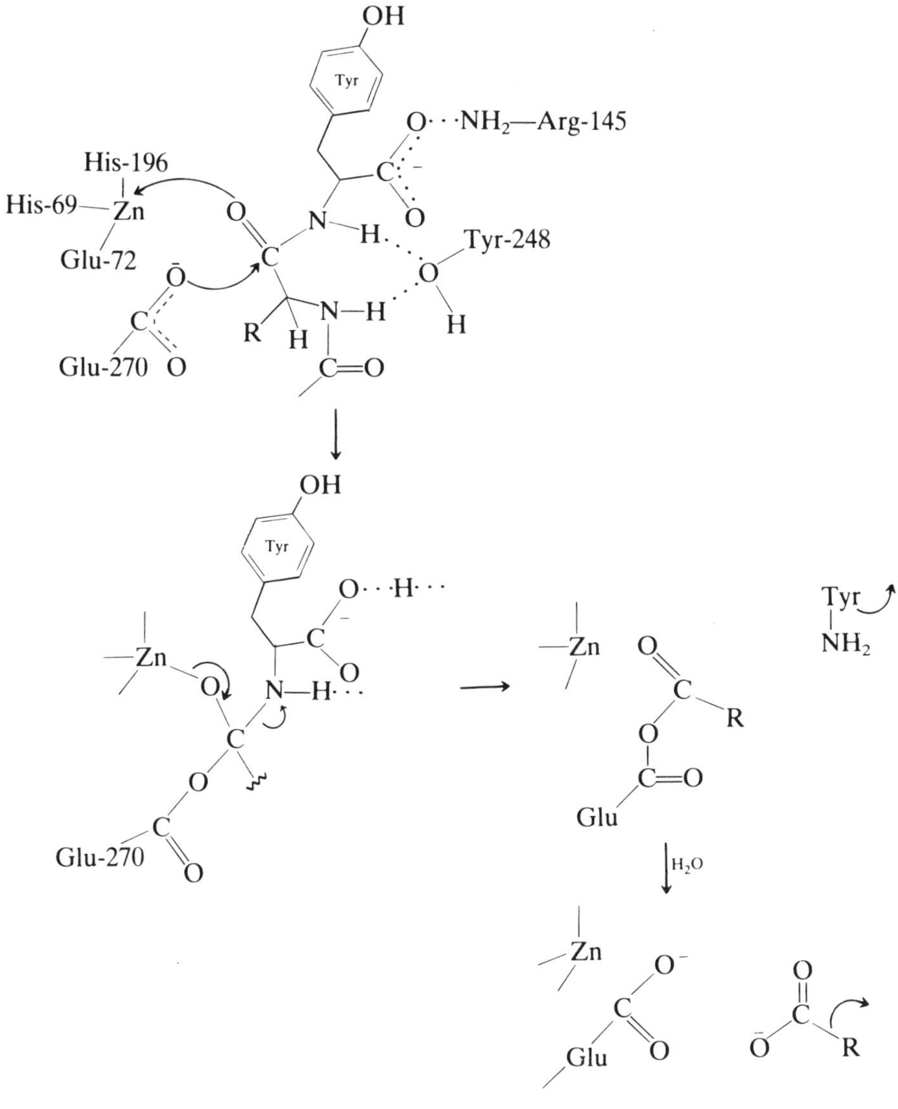

Fig. 13.14 Mechanism of action of carboxypeptidase. Hydrolysis of the peptide bond occurs with zinc acting as an electrophilic catalyst.

relatively slowly by the reverse process after replacement of the departing group by water[108].

13.2.4 Enzymes which use cofactors

Many types of hydrolytic reaction can be carried out utilizing only the nucleophilic and electrophilic properties of functional groups present in

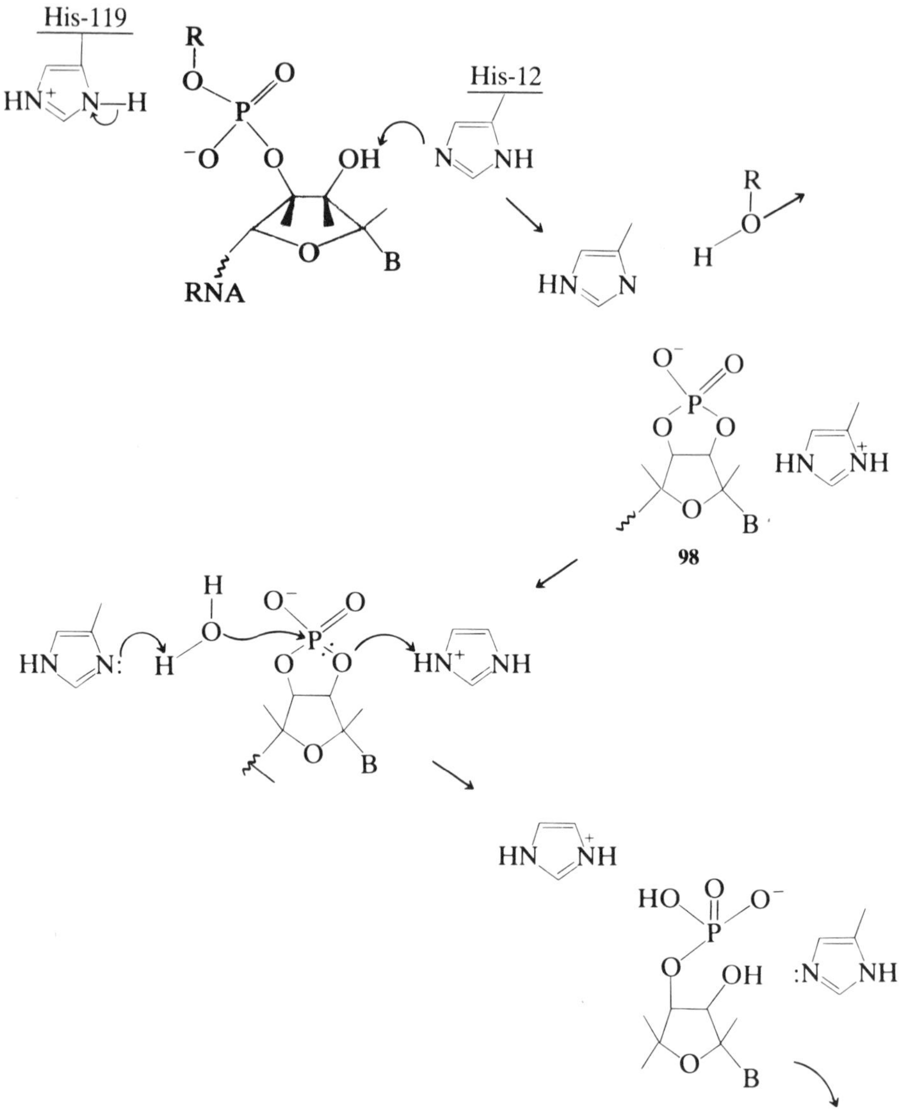

Fig. 13.15 Mechanism of hydrolysis of ribose phosphate in RNA by ribonuclease.

protein and the advantages of neighboring-group effects. There are limits, however, to reactions which can be mediated in this way; many redox reactions, carbon–carbon bond formation and energetically 'uphill' reactions often require reagents which must be brought into the reactive site. These are called cofactors and bind to the active site near the substrate so that a bimolecular reaction between the two, with possible assistance from the protein, may readily occur.

Nicotinamide adenine dinucleotide

NAD^+, **99**, is biological oxidizing agent acting as an acceptor of H^- which becomes attached to the pyridine moiety. It is the cofactor for alcohol oxidases, zinc-containing enzymes of broad specificity which convert many alcohols to aldehydes and ketones *in vivo*. The general mechanism is shown in Fig. 13.16, in which it can be seen that the alcohol binds to the zinc and transfers H^- to NAD^+ (converting it to the reduced form, NADH) and a proton to a nearby serine which relays it to histidine-51. After reaction, both products and NADH depart, the latter to be oxidized for re-use[109].

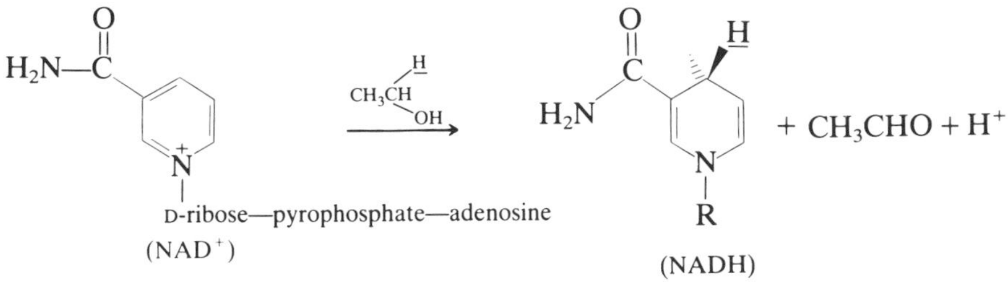

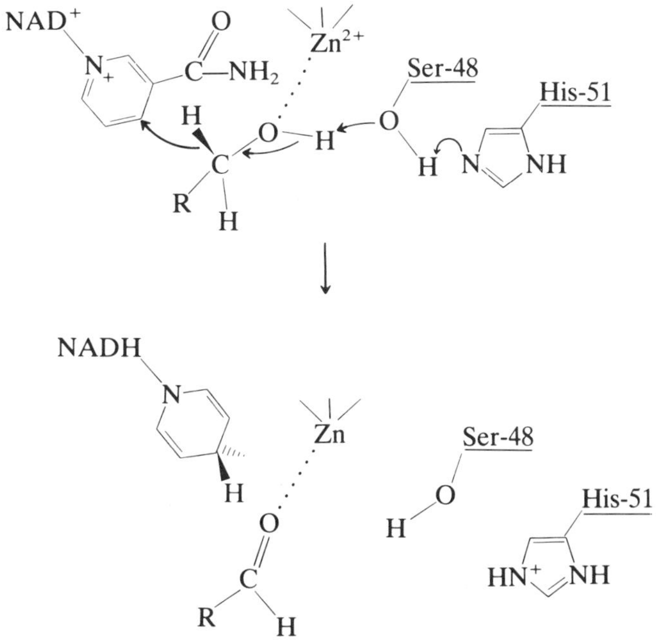

Fig. 13.16 Mechanism of oxidation of ethanol by alcohol dehydrogenase, mediated by the cofactor NAD^+.

Reversible redox reactions of such important molecules as glyceraldehyde occur by a basically similar pathway. In this case the redox change is between aldehyde and carboxylic acid levels; glyceraldehyde-3-phosphate, **100**, reacts with its dehydrogenase at the thiol group of a cysteine residue to form a thiohemiacetal, which is readier to transfer H^- than the aldehyde group (Fig. 13.17). The hydride transfer takes place leaving the substrate in the form of a thiol ester of the enzyme. The deacylation is then accomplished by phosphate, but only after the NADH has been replaced by a further molecule of NAD^+. The result is the mixed anhydride, 1,3-diphosphoglycerate [110].

These few examples merely illustrate some of the fascinating chemistry at

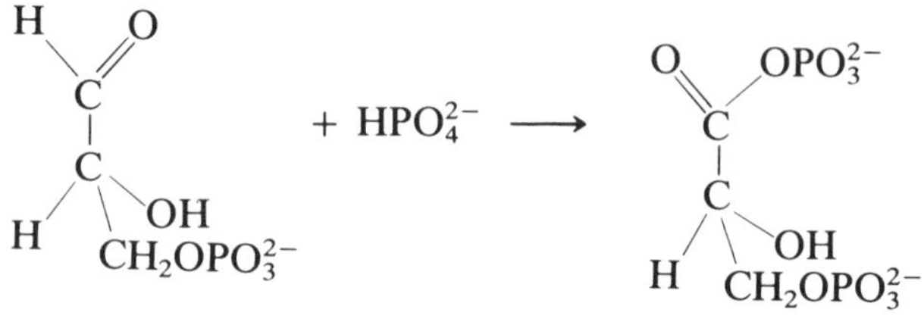

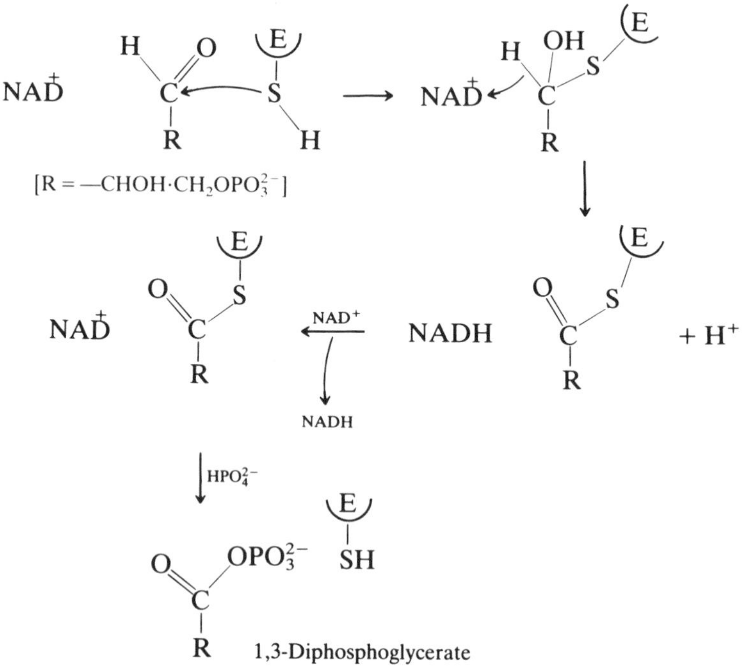

Fig. 13.17 Oxidation of glyceraldehyde to glyceric acid (phosphate) by glyceraldehyde hydrogenase, also mediated by NAD^+.

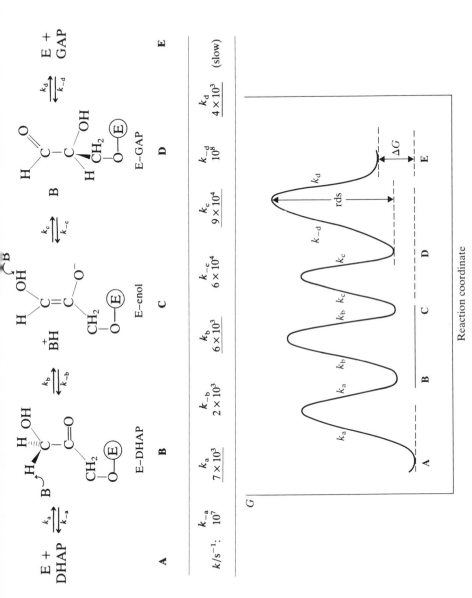

Fig. 13.18   Mechanism of action of triosephosphate isomerase and its energetics (rds, rate-determining step).

the disposal of the living cell in which familiar chemical principles are used in systems of great complexity, while by suitable positioning of reagents a concerted movement of electrons cascades down an energy gradient minimizing the activation energies far below those required for normal laboratory processes.

Triosephosphate isomerase (TIM)

This is a ubiquitous enzyme which converts dihydroxyacetone phosphate (DHAP) to glyceraldehyde-3-phosphate (GAP): see Fig. 13.18. The biological function of this enzyme is in the glycolytic process during which glucose in the form of glycogen is broken down to lactate in muscle. TIM enables DHAP from one half of the glucose molecule to be fed into the energy-producing pathway, doubling the number of ATP molecules generated in the process which are then available for energy conversion.

The mechanism of the isomerization is known in general outline and the total structure of the crystalline enzyme has been elucidated[112]. More interestingly, the energy profile of the system has been well characterized[113]. The isomerization proceeds by way of an enol intermediate and requires three proton transfers overall. Four small energy barriers exist between reagent and product state, two for the actual reaction and one each for substrate binding and for release of product. Rates of each step are indicated and are all very fast. A condition for optimum catalysis in such a situation must be the superposition of no greater barriers to reaction than exist by the physical nature of the process. These are the slightly endothermic nature of the isomerization $(K < 1)$ and the necessity of coordinating the substrate to the enzyme and removing the product. The former is a diffusion process and hence fast, so the latter must be rate-limiting if the enzyme is to operate with the maximum throughput. This has been achieved in the design of the enzyme TIM, which by its nature cannot be further improved by any structural change. The efficiency of the enzymic catalysis is shown in step b, the enolization of DHAP, which occurs on the enzyme 10^9 times faster than in neutral aqueous solution while the proton shift to give product is almost diffusion-controlled.

The unfavorable equilibrium for this endothermic process is no bar to a fast rate process since the uphill work is accomplished in several small steps for which there is ample thermal energy available, and the product is rapidly consumed, so forcing the reaction to proceed efficiently to products. This strategy is most probably very common in the evolution of enzymes.

13.2.5 Enzyme model systems

Certain molecules are known to exert a catalytic effect upon reactions by binding the reagents in such a way that their reactivity is altered, thus mimicking the function of an enzyme, though far less efficiently[111]. One such group of systems is the cyclodextrins, cyclopolyglucose molecules which contain a central cavity in which molecules of a specific size and shape may enter.

PROBLEMS

1 Unsaturated acids react with iodine to form iodolactones:

From the following rate data, suggest a mechanism for this reaction:

Acid	k_{rel}
$CH_2\!\!=\!\!CH(CH_2)_n COO^-$	
$n = 0$	0
$n = 1$	0·01
$n = 2$	84
$n = 3$	11
$n = 4$	0·17
$n = 5$	0
$CH\!\!=\!\!CH\cdot CH_2CH(COO^-)_2$	169
$RCH\!\!=\!\!CH(CH_2)_2COO^-$	
$R = Me$	142
$R = Ph$	61
$R = Cl$	0·025

[*Russ. Chem. Rev.* **40**, 272 (1971).]

2 The Diels–Alder reaction between furan and ethyl *trans*-crotonate leads to two isomeric products. They can be distinguished since one undergoes iodolactonization and one does not. Explain this.

3 Suggest reasons why 'mustard gas', **1**, and 'nitrogen mustards' such as Chlorambucil, **2**, are highly toxic, the latter being useful in cancer therapy and the former as a war gas.

What precautions would you take if you were synthesizing **3**?

4 Suggest explanations for the following observations.
(**a**) Acetolysis of **4** is faster than that of **5** but the ratio $k(\mathbf{4})/k(\mathbf{5}) = 58$ for $X = m$-Br though for $X = p$-OMe it is only 7.

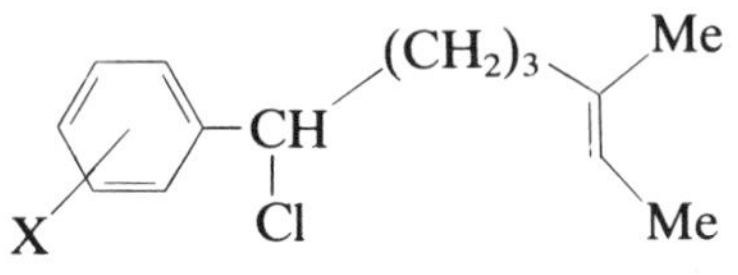

4

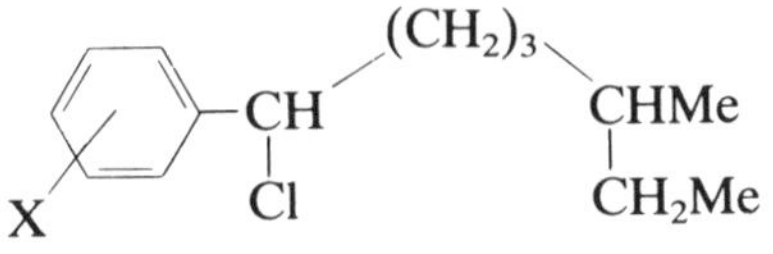

5

[*J. Org. Chem.*, **44**, 4086 (1980).
(**b**) Hydrolysis of the epoxide **6** gives the lactone, **7**, with stereochemistry and labeling pattern as shown.

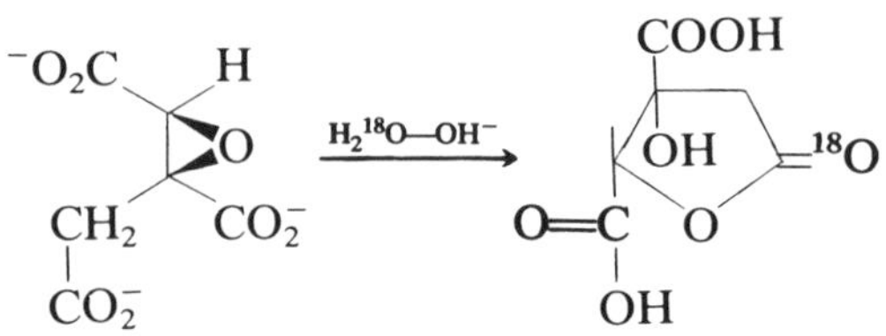

6 **7**

[*Chem. Comm.*, 555 (1980).]
(**c**) The rate of acetolysis of **8** is 2×10^6 times faster than that of 2-chloroadamantane.

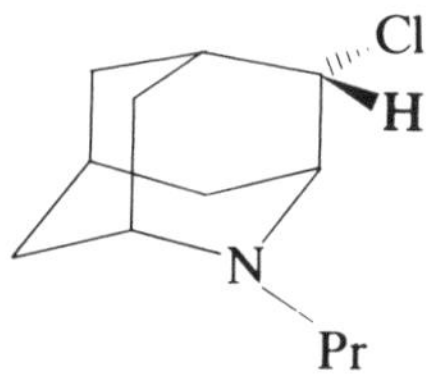

8

[*J. Org. Chem.*, **46**, 4953 (1981).]
(**d**) The oxidation of sulfides to sulfoxides is catalyzed by dibasic acids:

$$R_2S \xrightarrow[(CH_2)_2(COO^-)_2]{\text{aq. } I_2} R_2\overset{+}{S}-\bar{O}.$$

[*J. Org. Chem.*, **47,** 1416 (1982).]
(**e**) Acetolyses of the tosylates **9–11** are very different in rate.

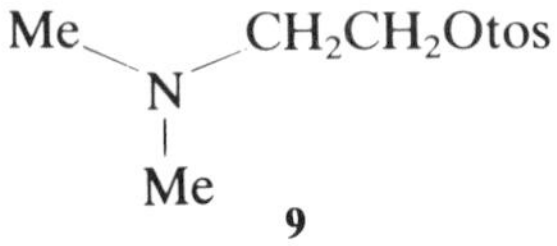

9
No reaction

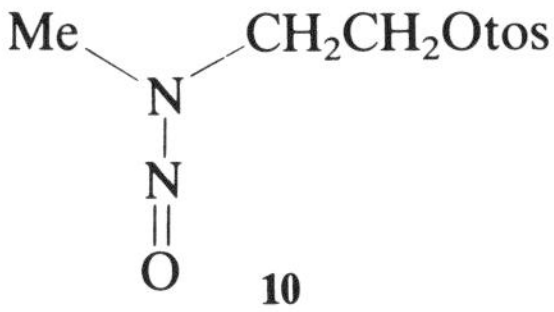

Me＼   ＼CH₂CH₂Otos
  ＼N
     |
     N
     ‖
     O   **10**

$t_{\frac{1}{2}} \sim 20\ \text{min}$

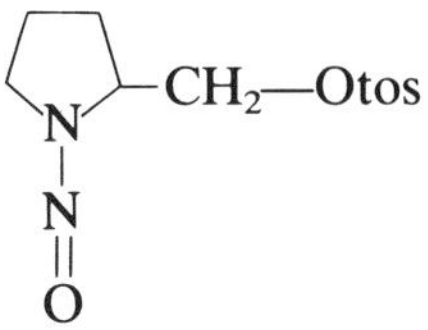

⎡N⎤–CH₂—Otos
  |
  N
  ‖
  O   **11**

$t_{\frac{1}{2}} \ll 30\ \text{s}$

[*J. Amer. Chem. Soc.*, **100**, 1959 (1978).]
(f) Hydrolyses of *ortho*- and *para*-cyanobenzoic esters occur at similar rates above pH 2, but in more strongly acidic solution the *ortho* isomer hydrolyzes 10^5 times faster than the *para*. [*Finn. Chem. Lett.*, 9 (1978).]
(g) The selenide, **12**, undergoes solvolysis 200 times faster than does the corresponding sulfide, **13**. [*J. Org. Chem.*, **43**, 650 (1978).]

$$\text{PhSeCH}_2\text{CH}_2\text{Cl} \qquad\qquad \text{PhSCH}_2\text{CH}_2\text{Cl}$$
$$\textbf{12} \qquad\qquad\qquad \textbf{13}$$

(h) Acetolysis of **14** is accompanied by 82% inversion and 18% retention of configuration

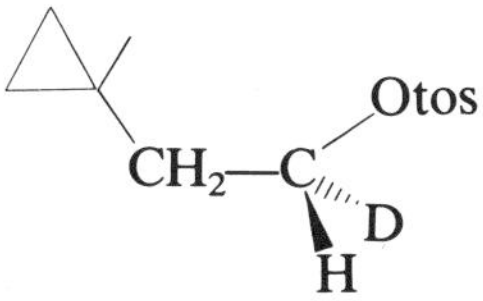

$$\text{CH}_2-\overset{\displaystyle \text{Otos}}{\underset{\underset{\text{H}}{\displaystyle \text{D}}}{\text{C}}}$$

14

[*Tetrahedron Letts* **24**, 4959 (1983).]
(i) Disalicyl phosphate, **15**, hydrolyzes 10^{10} times faster than does diphenyl phosphate, $(\text{PhO})_2\text{P}\overset{\displaystyle \text{OH}}{\underset{\displaystyle \text{OH}}{\diagdown}}$

$$\text{OH}\ \ \text{OH}$$
$$\text{O—P—O}$$
$$\text{COOH}\ \ \text{COOH}$$

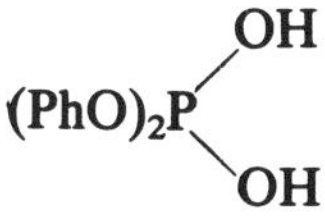

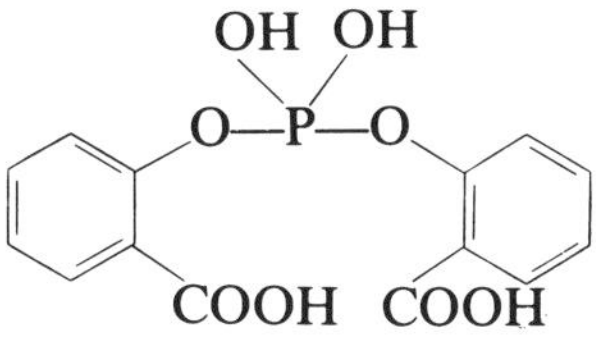

15

[*J. Chem. Soc., Perkin II*, 1171 (1983).]
(j) The thiironium ion, **16**, undergoes cyclization to the *cis*-decalin, **17**, upon solvolysis whereas its isomer **18** undergoes elimination.

Intramolecular reactions

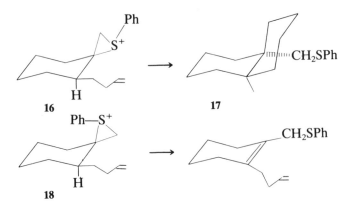

16   17

18

[*Rec. Trav. Chim.* **99**, 53 (1980).]

REFERENCES

1 B. Capon and S. P. McManus, *Neighboring Group Participation*, Plenum, New York, 1976.

2 B. Capon, *Quart. Rev.*, **18**, 45 (1964).

3 G. A. Olah and P. v. R. Schleyer, Eds, *Carbonium Ions*, Vol. 3, John Wiley, New York, 1972.

4 A. J. Kirby, *Organic Reaction Mechanisms*, Vol. 12, B. Capon, J. Perkins and C. W. Rees, Eds, Academic Press, New York, 1976, p. 97.

5 W. J. LeNoble, *Highlights of Organic Chemistry*, Marcel Dekker, New York, 1974.

6 S. Winstein, E. Allred, R. Heck and R. Glick, *Tetrahedron*, **3**, 1 (1968).

7 E. L. Allred and S. Winstein, *J. Amer. Chem. Soc.*, **89**, 3991, 3998 (1967).

8 V. Gold, Ed., *Pure. Appl. Chem.*, 1285 (1983).

9 C. K. Ingold, *Structure and Mechanism in Organic Chemistry*, 2nd edn, Bell, London, 1969.

10 S. Winstein and D. Trifan, *J. Amer. Chem. Soc.*, **74**, 1147, 1154 (1952).

11 P. G. Gassmann and A. F. Fentiman, *J. Amer. Chem. Soc.*, **92**, 2581 (1970).

12 H. C. Brown and K.-T. Lim, *J. Amer. Chem. Soc.*, **91**, 5909 (1969).

13 H. C. Brown, S. Ikegami, K.-T. Lim and G. L. Tritle, *J. Amer. Chem. Soc.*, **98**, 2511 (1976).

14 H. C. Brown, M. Ravindranathan and E. N. Peters, *J. Amer. Chem. Soc.*, **96**, 7351 (1974).

15 E. N. Peters and H. C. Brown, *J. Amer. Chem. Soc.*, **95**, 2398 (1973).

16 M. Ravindranathan, C. G. Rao and E. N. Peters, *Proc. Ind. Acad. Sci.*, **90**, 353 (1981); C. J. Lancelot and P. v. R. Schleyer, *J. Amer. Chem. Soc.*, **91**, 4291, 4296, 4297 (1969).

17 C. J. Lancelot, J. J. Hooper and P. v. R. Schleyer, *J. Amer. Chem. Soc.*, **91**, 4294 (1969).

18 H. C. Brown and C. J. Kim, *J. Amer. Chem. Soc.*, **93**, 5765 (1971); S. L. Loukas, M. L. Velkon and G. A., Gregoriou, *Chem. Comm.*, 1199 (1969); 251 (1970).

19 P. v. R. Schleyer, *J. Amer. Chem. Soc.*, **86**, 1854, 1856 (1964).

20 G. D. Sargent, *Quart. Rev.*, **20**, 1966 (301).

21 C. W. Shoppee, *J. Chem. Soc.*, 1147 (1964).

22 S. Winstein and R. Adams, *J. Amer. Chem. Soc.*, **70**, 838 (1948).

23 V. P. Vitullo and F. P. Wilgis, *J. Amer. Chem. Soc.*, **97**, 5616 (1975).

24 W. P. Jencks, *Catalysis in Chemistry and Enzymology*, McGraw–Hill, New York, (1969).

25 A. R. Fersht and A. J. Kirby, *Chem. Brit.*, **16**, 136 (1980).

26 A. J. Kirby, in *Comprehensive Organic Chemistry*, D. H. R. Barton and W. D. Ollis, Eds, Vol. 5, p. 389, Pergamon, Oxford, 1979.

27 M. I. Page, *Chem. Soc. Rev.*, **2**, 295 (1973); A. M. Davis, M. I. Page, S. C. Mason and I. Watt, *Chem. Comm.*, 1671 (1984).

28 R. Bird and C. J. M. Stirling, *J. Chem. Soc. (B)*, 1218 (1968); H. C. Brown, *The Nonclassical Ion Problem*, Plenum, New York, 1977.

29 R. Bird, A. C. Knipe, and C. J. M. Stirling, *J. Chem. Soc., Perkin II*, 1215 (1973).

30 A. R. Fersht and A. J. Kirby, *J. Amer. Chem. Soc.*, **89**, 4853, 4857, 5960, 5961 (1967).

31 E. L. Eliel, R. O. Hutchins, R. McBane and R. L. Willer, *J. Org. Chem.*, **41**, 1052, 1976.

32 H. C. Lucas, F. W. Marshall and H. K. Garbner, *J. Amer. Chem. Soc.*, **72**, 2138 (1950); S. Winstein and R. E. Buckles, *J. Amer. Chem. Soc.*, **64**, 812, 2796 (1942); S. Winstein, R. E. Buckles, E. Grunwald and C. Hansen, *J. Amer. Chem. Soc.*, **70**, 816, 828 (1950).

33 E. L. Eliel and D. E. Knox, *J. Amer. Soc.*, **107**, 2946, 1985; T. C. Bruice and U. K. Pandit, *J. Amer. Chem. Soc.*, **82**, 5858 (1960).

34 M. A. Winnick, *Chem. Rev.*, **81**, 491 (1981).

35 D. L. Stull, E. F. Westheimer and G. F. Sinke, *Chemical Thermodynamics of Organic Compounds*, John Wiley, New York, 1969.

36 F. M. Menger, *Acc. Chem. Res.*, **18**, 128 (1985).

37 C. F. Hammer and S. R. Heller, *Chem. Comm.*, 919 (1966); J. G. Traynham, *J. Chem. Ed.*, 392 (1963).

38 R. E. Parker and N. S. Isaacs, *Chem. Rev.*, **53**, 737 (1953).

39 L. Eberson, J. P. Petrovich, R. Bairo, D. Dyckes and S. Winstein, *J. Amer. Chem. Soc.*, **87**, 3504 (1965).

40 G. A. Rogers and T. C. Bruice, *J. Amer. Chem. Soc.*, **95**, 4452 (1973).

41 H. E. Baumgarten, *J. Amer. Chem. Soc.*, **85**, 3524 (1963).

42 O. L. Chapman, P. W. Wojtkowski and W. Adam, *J. Amer. Chem. Soc.*, **94**, 1365 (1972).

43 G. A. Olah, *Carbocations and Electrophilic Reactions*, Verlag Chemie/Wiley, 1973.

44 D. J. Cram, *J. Amer. Chem. Soc.*, **86**, 3767 (1964).

45 G. A. Olah, C. L. Jewett, D. P. Kelly and R. D. Porter, *J. Amer. Chem. Soc.*, **94**, 146 (1972).

46 I. Tabushi, Y. Tamaru, Z. Yoshida and T. Sugimoto, *J. Amer. Chem. Soc.*, **97**, 2886 (1975).

47 J. C. Martin and P. D. Bartlett, *J. Amer. Chem. Soc.*, **79**, 2533 (1957).

48 D. L. H. Williams, *Tetrahedron Letts*, 2001 (1967).

49 D. L. H. Williams, E. Bienvenue-Goetz and J.-E. Dubois, *J. Chem. Soc. (B)*, 517 (1969); S. R. Hooley and D. L. H. Williams, *J. Chem. Soc., Perkin II*, 503 (1975).

50 H. C. Brown, *Pure Appl. Chem.*, **54**, 1783 (1982); G. D. Sargent, in Ref. 3, Chapter 24.

51 P. D. Bartlett, *Non-Classical Ions*, Benjamin, New York, 1965.

52 G. A. Olah, C. L. Jewett, D. P. Kelly and R. D. Porter, *J. Amer. Chem. Soc.*, **94**, 147 (1972).

53 J. A. Berson, *Molecular Rearrangements*, P. DeMayo, Ed., Wiley, New York, 1963.

54 H. C. Brown, *The Nonclassical Ion Problem*, Plenum, New York, 1977.

55 G. A. Olah, *Carbocations and Electrophilic Reactions*, Wiley, New York, 1974.

56 J. M. Brown and L. V. Occlolowitz, *J. Chem. Soc. (B)*, 411 (1968).

57 S. Winstein, E. Clippinger, R. Howe and E. Vogelfanger, *J. Amer. Chem. Soc.*, **87**, 376 (1965); A. Colter, E. C. Friedlich, N. J. Holness and S. Winstein, *ibid.*, **87**, 378 (1965).

58 S. Winstein, *J. Amer. Chem. Soc.* **87**, 376, 379, 381 (1965).

59 W. Stoll, *Z. Physiol.*, **207**, 147 (1932).

60 C. W. Shoppee, *J. Chem. Soc.*, 1147 (1946).

61 S. Winstein and A. H. Schlessinger, *J. Chem. Soc.*, 3528 (1948).

62 S. Winstein and R. Adams, *J. Amer. Chem. Soc.*, **70**, 838 (1948).

63 S. Winstein and E. M. Kosower, *J. Amer. Chem. Soc.*, **81**, 4399 (1959).

64 S. Winstein and D. Trifan, *J. Amer. Chem. Soc.*, **74**, 1147, 1152 (1952).

65 S. Winstein, M. Shatavsky, C. Norton and R. B. Woodward, *J. Amer. Chem. Soc.*, **77**, 4183 (1955).

66 S. Winstein and C. Ordronneau, *J. Amer. Chem. Soc.*, **82**, 2084 (1950).

67 S. Winstein and E. T. Stafford, *J. Amer. Chem. Soc.*, **79**, 505 (1957).

68 B. Hess, *J. Amer. Chem. Soc.*, **91**, 5657 (1969).

69 H. C. Brown and H. M. Bell, *J. Amer. Chem. Soc.*, **85**, 2324 (1963).

70 J. J. Tufariello, T. F. Mich and R. J. Lorence, *Chem. Comm.*, 1202 (1967).

71 J. D. Roberts, W. Bennett and R. Armstrong, *J. Amer. Chem. Soc.*, **72**, 3329 (1950); K. Wiberg, B. A. Andes and A. J. Ashe, in *Ref. 3*, Chapter 26.

72 H. G. Richey and N. C. Buckley, *J. Amer. Chem. Soc.*, **65**, 3057 (1963).

73 B. R. Ree and J. C. Martin, *J. Amer. Chem. Soc.*, **92**, 1660 (1970).

74 V. Russ, R. Gleiter and P. v. R. Schleyer, *J. Amer. Chem. Soc.*, **93**, 3927 (1971).

75 W. L. Jorgensen and L. Salem, *The Organic Chemist's Book of Orbitals*, Academic Press, New York, 1973.

76 G. D. Sargent, in *Ref. 3*, Chapter 24.

77 H. C. Brown, *Acc. Chem. Res.*, **6**, 377 (1973).

78 L. A. Spurlock and R. G. Fayter, *J. Amer. Chem. Soc.*, **94**, 2707 (1972).

79 A. C. Cope, *J. Amer. Chem. Soc.*, **87**, 3107, 3111, 3122, 3125, 3130, 3644 (1965).

80 A. C. Cope and R. B. Kinnel, *J. Amer. Chem. Soc.*, **88**, 752 (1966).

81 L. A. Paquette, I. R. Dunkin, J. P. Freeman and P. C. Storm, *J. Amer. Chem. Soc.*, **94**, 8124 (1972).

82 E. D. Hughes and C. K. Ingold, *J. Chem. Soc.*, 1196 (1937).

83 C. B. Anderson, E. C. Friedrich and S. Winstein, *Tetrahedron Letts*, 2037 (1963).

84 E. R. Novak and D. S. Tarbell, *J. Amer. Chem. Soc.*, **89**, 73, 3086 (1967); A. W. Friederang and D. S. Tarbell, *J. Org. Chem.*, **33**, 3797 (1968).

85 L. Mandolini, *J. Amer. Chem. Soc.*, **100**, 550 (1978).

86 F. Benedetti and C. J. M. Stirling, *Chem. Comm.*, 1374 (1983).

87 J. Westley, *Enzymic Catalysis*, Harper and Row, New York, 1969.

88 P. D. Boyer, Ed., *The Enzymes*, 3rd edn, Academic Press, New York, 1970.

89 M. Dixon and E. C. Webb, *Enzymes*, 3rd edn, Longmans, 1964.

90 A. Fersht, *Enzyme Mechanism and Structure*, Freeman, San Francisco, 1977.

91 J. A. Hartsuck and W. N. Lipscombe, *The Enzymes*, **3**, 1 (1971).

92 M. L. Bender *Mechanisms of Homogeneous Catalysis*, Wiley, New York, 1971.

93 L. Michaelis and M. L. Menten, *Biochem. Z.*, **2**, 333 (1913).

94 H. Lineweaver and D. Burke, *J. Amer. Chem. Soc.*, **56**, 658 (1934).

95 W. W. Cleland, *Adv. Enzym.*, **45**, 273 (1977).

96 W. P. Jencks, *Adv. Enzym.*, **43**, 219 (1975).

97 I. I. Secemski and G. E. Lienhard, *J. Biol. Chem.*, **249**, (1974).

98 M. V. Keenan and W. L. Alwath, *Biochem. Biophys. Res. Commun.*, **57**, 500 (1974).

99 D. C. Phillips, *Scientific American*, 78 (Nov. 1986).

100 C. C. F. Blake, L. N. Johnson, G. A. Mair, A. C. T. Orth, D. E. Phillips and V. R. Sorma, *Proc. Roy. Soc. B*, **167**, 378 (1967); C. A. Vernon, *ibid*, **167**, 389 (1967).

101 C. Niemann, *Science*, **143**, 1287 (1967).

102 T. A. Steitz, D. Blow and R. Henderson, *J. Molec. Biol*, **46**, 337 (1969).

103 R. Henderson, *J. Molec. Biol.*, **54**, 341 (1970).

104 N. S. Isaacs and C. Niemann, *Biochem. Biophys. Acta*, 196 (1969).

105 W. N. Lipscomb, *Tetrahedron*, **30**, 1725 (1974).

106 W. N. Lipscomb, *Proc. Nat. Acad. Sci.*, **70**, 3797 (1973).

107 M. Eigen and G. G. Hammes, *Adv. Enzym.* **25**, 1 (1963).

108 A. Deavin, A. P. Mathias and B. R. Rabin, *Biochem. J.*, **101**, 146 (1966).

109 J. W. Cornforth, R. H. Cornforth, C. Donninger, G. Popjak, G. Ryback and G. J. Schroepfer, *Proc. Roy. Soc. B*, **163**, 436 (1966); C. S. Hanes, P. M. Bronshill, P. A. Gurr and J. T. F. Wong, *Can. J. Biochem.*, **50**, 1385 (1972).

110 C. J. Gray, *Enzyme-Catalysed Reactions,* Van Nostrand–Reinhold, London, 1971, p. 121.

111 M. L. Bender and M. Komiyama, *Cyclodextrin Chemistry,* Springer–Verlag, Berlin, 1978; R. Breslow, *Chem. Brit.,* **19,** 126 (1983).

112 D. C. Phillips, *et al., Nature,* **255,** 609 (1975); *J. Mol. Biol.,* **119,** 329 (1978).

113 J. R. Knowles and W. J. Albery, *Acc. Chem. Res.,* **10,** 105 (1977).

14 Pericyclic reactions

14.1 CLASSIFICATION OF PERICYCLIC REACTIONS

The importance of stereo-electronic factors, i.e. the availability of molecular orbitals with the correct phase relationships, has been mentioned in connection with concerted reactions such as S_N2 and E2 processes. These considerations become of overwhelming importance in the area of *pericyclic reactions* [1-3], a term which includes several families of reactions having in common the following two properties: (i) covalency change occurs in a single synchronous step over many (usually between three and ten) centers and (ii) reagents pass through a cyclic conjugated transition state. In all examples to be discussed, the electrons which undergo reorganization remain spin-paired throughout, as far as can be told. Reactions of this type cannot be described in terms of electrophile–nucleophile interactions although there may be an element of charge-separation involved in certain cases.

Three principal classes of pericyclic reaction will be discussed:

(a) Cycloadditions, the linking of two conjugated systems at either end to form a cyclic product. Convincing reasons may now be put forward to explain why the Diels–Alder reaction, **1**, takes place with ease while alkene dimerizations, **2**, are not observed as thermal reactions. Strain energy in the four-membered ring is not the whole answer since the regressions occur in the same order of ease; cyclohexenes fission more readily than do cyclobutanes.

(b) Electrocyclic reactions, valence tautomerism between extended conjugated molecules and their cyclic isomers exemplified by the interconversion of buta-1,3-diene and cyclobutene, **3**.

(c) Sigmatropic rearrangements, rearrangements in which a molecular fragment migrates from one end of a conjugated system to the other. The 1,5-hydrogen shift of a penta-1,4-diene, **4**, is an example.

In each case, the conversion of reagents to products is indicated by a cyclic train of curved arrows. The direction of these arrows is purely notional, the important factor throughout being the smooth transition between MOs of reagents and products. The central tenet of the theory of pericyclic reactions is the Principle of Orbital Symmetry Conservation[1], which implies that reactions in general will be of lower activation energy and therefore more facile if the two sets of bonding molecular orbitals have matching symmetry properties than if this is not the case. It is in many-center reactions that this principle can be most readily tested. If orbital symmetry conservation holds, a reaction may take advantage of synchronicity in which the energy required for bond fission is released by that of bond making; this can keep the activation energy low.

14.2 THE THEORY OF PERICYCLIC REACTIONS

Before 1969, the paradox of the occurrence of the widespread and facile reaction **1** and the non-occurrence of **2** was an intriguing curiosity, but it is now considerably better understood largely due to the systematization introduced by R. B. Woodward and R. Hoffmann[1,4-8], whose theory of pericyclic reactions underlies this chapter. There are, in fact, three ways in which these observations may be rationalized, each having its particular advantages in a given context[3], and each depending upon the properties of π-orbitals and the favorable or unfavorable interactions between frontier orbitals.

14.2.1 Conservation of orbital symmetry: correlation diagrams[1,9]

The individual molecular orbitals of a conjugated system may be designated as either symmetric (S) or antisymmetric (A) with respect to any symmetry elements which the molecule possesses (Section 1.2.2). If bonding orbitals of the reagents and bonding orbitals of the product have matching symmetry characteristics, they may transform one into the other with facility and the reaction is said to be *orbital symmetry allowed*. If, however, for symmetry reasons bonding orbitals have to be matched with antibonding, then a high activation energy is predicted and the reaction is *orbital symmetry forbidden*.

The Diels–Alder reaction is the cycloaddition of a 1,3-diene to an alkene ('dienophile') to give a cyclohexene. The orbital symmetry relationships are given in Fig. 14.1 with respect to the mirror plane *m* of the whole reacting system whose geometry is specified in the Figure. The Hückel MO approximation is adequate for this purpose and tends to emphasize orbital nodal and symmetry properties. Furthermore, substituents are held not to affect orbital symmetries. Penta-1,3-diene is assumed to have π-orbitals with the same mirror-plane symmetry as does buta-1,3-diene although the coefficients of these orbitals at each atom, expressed in the diagrams by the relative sizes of the lobes, will be changed somewhat. The other point which needs explanation is the method whereby σ-orbitals may be incorporated into this scheme. In fact, the two σ-bonds of the product are considered as a

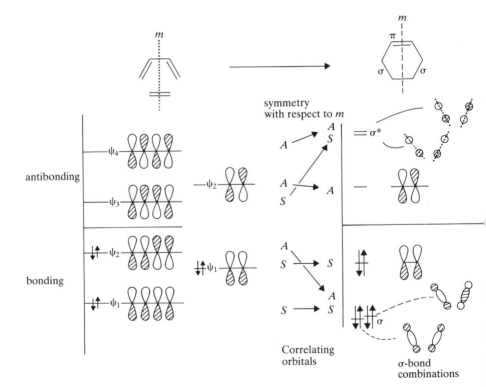

Fig. 14.1 Correlation diagram for the Diels–Alder Reaction, a $[_\pi 4_s + _\pi 2_s]$ cycloaddition, orbital symmetry allowed.

The theory of pericyclic reactions

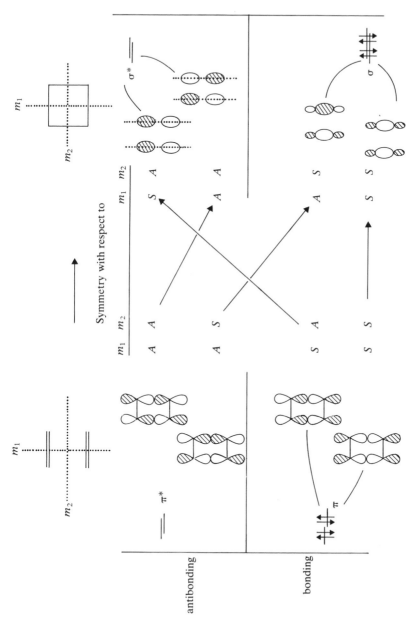

Fig. 14.2 Correlation diagram for alkene dimerization, a $[_\pi 2_s + _\pi 2_s]$ cycloaddition. This is orbital-symmetry forbidden.

symmetric and an antisymmetric combination ($\psi_a + \psi_b$ and $\psi_a - \psi_b$). This will be the case whenever a degenerate pair of either σ- or π-orbitals is involved. It may be seen from the diagram that orbital symmetry conservation is maintained within the bonding orbitals and also within the antibonding set and no crossover between the two sets occurs. This is therefore an allowed reaction. By contrast, the dimerization of ethene, Fig. 14.2, requires that one of the reagent π-orbital combinations must pass towards an antibonding σ-orbital combination. This reaction is therefore symmetry forbidden. This analysis assumes the approach geometry implied in Figs. 14.1 and 14.2. Other possibilities will be considered below.

14.2.2 The frontier orbital concept[10-12]

Perturbation MO theory gives a measure of the interaction energy, E, between two molecules (Eq. (14.1)); the second term contains the attractive forces due to overlap of filled orbitals of one molecule with vacant orbitals of the other (of matching symmetry).

$$E_i = \sum \frac{q_i q_j}{\varepsilon R} + \sum \frac{(2c_i c_j)\beta}{\Delta E_\psi} \tag{14.1}$$

in which q_i, q_j are charges on species i, j at a separation distance R (this first (electrostatic) term will be relatively unimportant for reactions between neutral molecules, so that frontier-orbital control assumes prominence in most pericyclic reactions); c_i, c_j are coefficients of molecular orbitals at atoms in contact, permutating all filled/vacant orbital combinations. ΔE_ψ is the energy difference between each pair in energy units and β, the overlap integral. While all permutations of reagent orbitals i, j must be considered, the largest term will come from interactions between the highest occupied molecular orbital (HOMO) of one reagent and the lowest unoccupied molecular orbital (LUMO) of the other which are those closest in energy (smallest ΔE_ψ). These MOs were designated the *frontier orbitals* by Fukui[10,11]. Qualitatively one may describe a pericyclic reaction as allowed or forbidden by identifying the frontier orbitals and examining their nodal properties at the interacting centers in order to determine whether or not bonding overlap occurs at both. This means that the interacting p-orbital lobes at each reacting center must have their wave functions of the same sign (Fig. 14.3).

The frontier orbitals of reactants in the Diels–Alder reaction are ψ_2 of the diene with ψ_2 of the dienophile, and also ψ_3 of the diene with ψ_1 of the dienophile. In each case, bonding overlap occurs at both bonding centers and the reaction is allowed (Fig. 14.4). This situation contrasts with alkene dimerization, Fig. 14.5a. If positive overlap between one bonding pair of centers is arranged, there is antibonding character at the other ($c_i c_j$ is negative) and hence this is a symmetry-forbidden reaction. Quantitatively

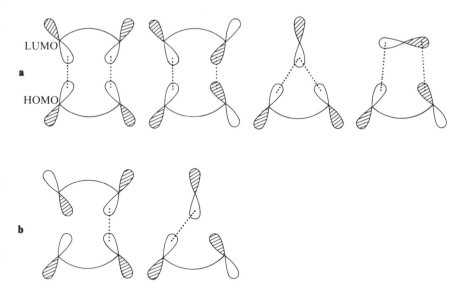

Fig. 14.3 HOMO–LUMO interactions: (**a**) leading to cycloaddition; (**b**) not leading to cycloaddition.

one may calculate the orbital overlap energy term knowing the magnitudes of the coefficients c_i, c_j in the frontier orbital terms. This will be a useful way of rationalizing regiospecificity.

Twisting of the p-orbitals along a sufficiently long conjugated chain might, in principle, result in a change of nodal symmetries at the termini, **5**. This is referred to as a Möbius system[12] (in contrast to the normal Hückel geometry) but it is not clear whether this has any practical significance in permitting relaxation of orbital symmetry restrictions and will not be further discussed.

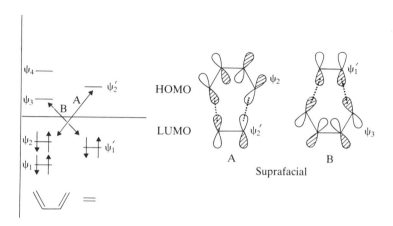

Fig. 14.4   Frontier orbital interactions in the $[_\pi 4_s + _\pi 2_s]$ cycloaddition.

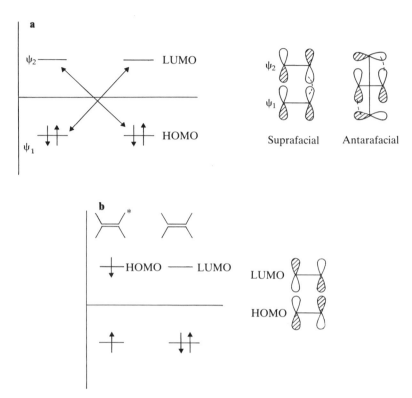

Fig. 14.5 Frontier orbital interactions in the $[_\pi 2_s + _\pi 2_s]$ cycloaddition: (**a**) thermal reaction; (**b**) photochemical reaction.

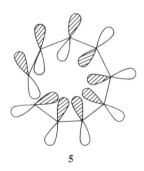

5

14.2.3 The aromaticity concept[13,14]

The division of cyclic conjugated systems into the stable, aromatic types with $(4n + 2)$ π-electrons and the unstabilized, anti-aromatic members with $(4n)$ π-electrons has been previously described (Section 1.2.2). Transition states of pericyclic reactions may be classified similarly. Those reactions with aromatic transition states $(2, 6, 10, 14 \ldots (4n + 2)$ delocalized electrons) are permitted while the anti-aromatic systems (4, 8, 12 . . . (4*n*) delocalized electrons) are the forbidden ones. It is clear that the Diels–Alder reaction has a 6π aromatic transition state isoelectronic with benzene while the forbidden cyclobutane formation has the unfavorable 4π transition state, isoelectronic with cyclobutadiene.

The three views outlined in Sections 14.2.1, 14.2.2 and 14.2.3 are, of course, complementary and all yield the same conclusions since they are all aspects of the same molecular orbital theory.

14.2.4 Suprafacial and antarafacial geometries

It was noted above that orbital symmetry constraints might differ according to the particular reaction geometry adopted. This may best be seen from the frontier orbital properties. With respect to a single π-system, bonding is denoted as *suprafacial* if overlap occurs at two lobes located on the same side of the nodal plane, and *antarafacial* if on opposite sides (Fig. 14.6). The terms are also applied to σ-bonds and to single *p*-orbitals. A single *s*-AO, however, can only take part in suprafacial overlap (as in hydrogen-transfer reactions) since it has a spherical geometry. The bonding stereochemistry of

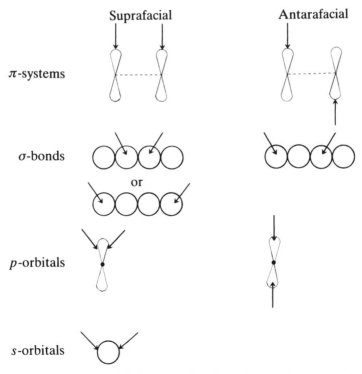

Fig. 14.6 Suprafacial and antarafacial geometries of attack on various types of molecular orbital.

each component taking part in a cycloaddition must be specified: the convention is [$_{type}$number of electrons$_{faciality}$]. The Diels–Alder reaction as shown in Fig. 14.1 is then described as a [$_\pi 4_s + _\pi 2_s$] cycloaddition and the forbidden ethene dimerization as [$_\pi 2_s + _\pi 2_s$]. Frontier orbital overlap in either case is reversed if one component approaches antarafacially, as shown in Fig. 14.5a. As far as orbital symmetry restrictions are concerned, the [$_\pi 4_s + _\pi 2_a$] or [$_\pi 4_a + _\pi 2_s$] geometries become forbidden, while the [$_\pi 2_s + _\pi 2_a$] process becomes allowed. That the latter is still not observed is due to severe steric compression in forming this transition state. Orbital symmetry correlations will not guarantee that a reaction will occur but it may be fairly stated that a truly concerted forbidden reaction will not be observed.

Photochemical reactions [15]

The excitation of an electron from the HOMO of one component to the next higher orbital makes this in turn the HOMO, and one having the opposite symmetry. This change will permit a previously forbidden reaction to become allowed and conversely (Fig. 14.5b). Photochemical formation of cyclobutanes and oxetanes from 2π components, e.g. **6**, is well known[13] but will not be further discussed here as the chemistry of excited states is a vast subject which requires a book to itself (Chapter 16).

A correlation table for cycloadditions may now be drawn up (Table 14.1). The Woodward–Hoffmann rule summarizes these cycloadditions and all other pericyclic reactions in a succinct, if somewhat abstract, statement;

'A ground-state pericyclic reaction is symmetry-allowed when the total number of $(4q + 2)$ suprafacial and $(4r)$ antarafacial components is odd"

Table 14.1 Correlation table for cycloadditions

$(p + q)^a$	Mode^b	Thermal^c	Photochemical^c
$4n + 2$	S,S	✓	×
$4n + 2$	A,S	×	✓
$4n$	S,S	×	✓
$4n$	A,S	✓	×

^{a} n = integer, 0, 1, 2, ...
^{b} A, antarafacial; S, suprafacial.
^{c} ✓, Allowed; ×, forbidden.

Table 14.2   Analysis of cycloadditions by the Woodward–Hoffmann rule

Bonding stereochemistry	Reaction example	Suprafacial		Antarafacial		Sum of components
		$(4q+2)_s$	$(4r)_s$	$(4q+2)_a$	$(4r)_a$	
			(ignored)			
$[_\pi 4_s + {}_\pi 2_s]$ $a(4r)_s$ $a(4q+2)_s$ component component $r=1$ $q=0$		1	(1	0)	0	1 (odd, allowed)
$[_\pi 4_s + {}_\pi 4_s]$		0	(2	0)	0	0 (even, forbidden)
$[_\pi 4_s + {}_\pi 4_a]$		0	(1	0)	1	1 (odd, allowed)
$[_\pi 2_s + {}_\pi 2_s]$		2	(0	0)	0	2 (even, forbidden)
$[_\pi 2_s + {}_\pi 2_a]$		1	(0	1)	0	1 (odd, allowed)
$[_\pi 2_s + {}_\sigma 2_s + {}_\sigma 2_s]$ (retro-Diels–Alder reaction)		3	(0	0)	0	3 (odd, allowed)

in which q, r are integers. The use of this rule is best illustrated by examples (Table 14.2). One counts the number of electrons participating in covalency change in each reactant, of which there will be two in a cycloaddition. Each will be either a $(4q+2)$ number (i.e. 2, 6, 10...) or a $(4r)$ number (0, 4, 8...). The faciality of attack at each reactant is determined and thus each is placed in one of the four categories; $(4q+2)S$, $(4r)A$, $(4q+2)A$ and $(4r)S$. The number of entries in the first two categories are summed and any in the latter two ignored. The sum is either odd (allowed) or even (forbidden).

14.3 THERMAL CYCLOADDITIONS; THEIR SCOPE AND CHARACTERISTICS

The most favorable stereochemistry for a cycloaddition is that which involves suprafacial bonding on both components, limited to systems with overall 2, 6, 10... delocalized electrons. In practice $(4\pi + 2\pi)$cycloadditions, the Diels–Alder reaction and its extensions, are by far the most thoroughly studied[1,16–20].

Concerted cycloadditions as expressed in **1** should have the following characteristics.

(a) Reactions must be bimolecular and consequently will show second-order kinetics.

(b) There should be a relatively low enthalpy of activation since there will inevitably be a very negative entropy of activation due to the loss of translational freedom in the activation step.

(c) There will be large negative volumes of activation and of reaction, as a consequence of bond formation.

(d) The rates should be rather insensitive to solvent change since both reagents and transition states are non-polar.

(e) The stereochemical integrity of each component must be maintained throughout the reaction.

(f) Since a change of hybridization from sp^2 to sp^3 is occurring at all four termini, small inverse secondary kinetic isotope effects $(k_H < k_D < 1)$ should be observed at each (Table 14.14).

It will be seen how well these criteria are met in some important types of cycloaddition.

14.3.1 The Diels–Alder Reaction[8,21]

The facile reaction between a 1,3-diene and a dienophilic alkene which usually bears electron-withdrawing groups, forming a cyclohexene, was discovered in 1928[16]. Various heteroatom analogs of both diene and dienophile can also take part; the scope of the reaction is shown in Table 14.3, and some typical examples in the reactions **7–11**. Diels–Alder reactions normally exhibit precise second-order kinetics and reactivities of the diene which are quite sensitive to structure. Rates of cycloadditions to

7

Table 14.3 Components taking part in Diels–Alder type reactions

4π-Components	2π-Components
	$C=C$ $-C\equiv C-$
	$C=N$ $C=O$
	$N=N$ $C=S$
	$N=O$ $O=O$
	singlet oxygen $^1\Delta_g$ State

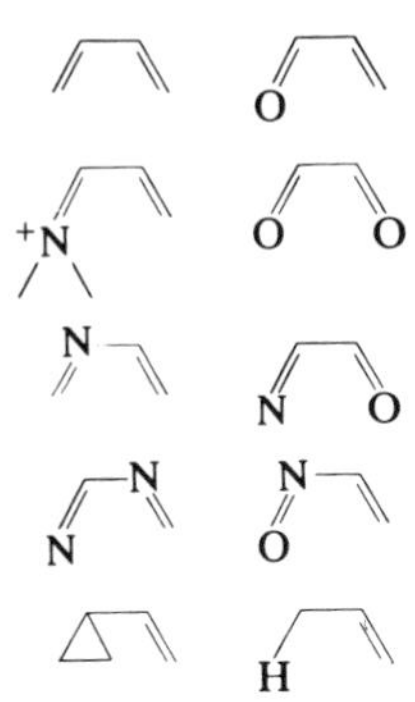

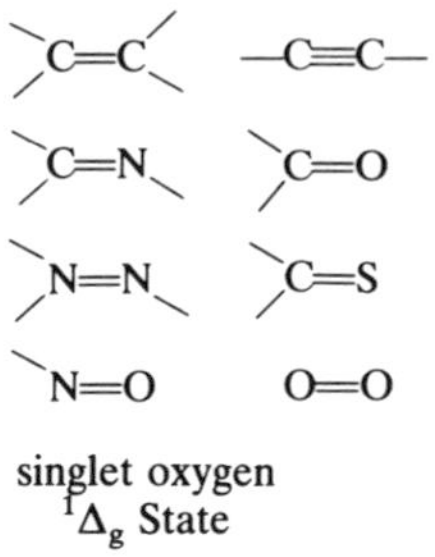

Some common dienophiles

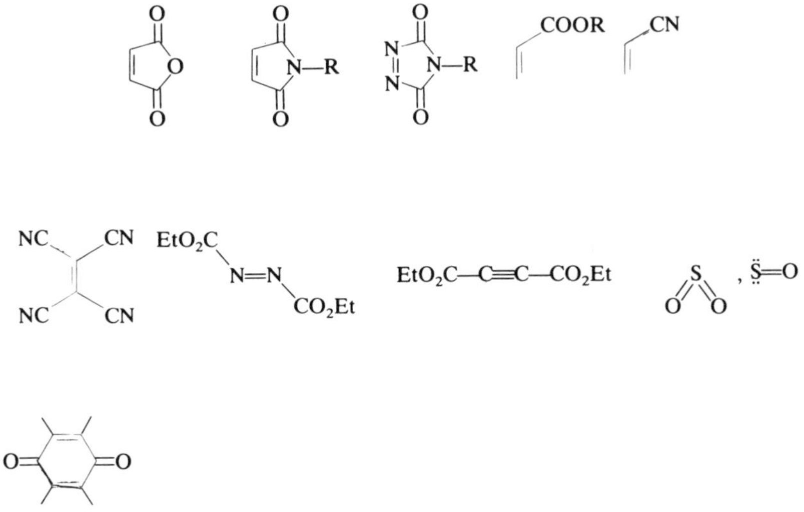

COOR   CN

NC, CN   NC, CN   EtO₂C N=N CO₂Et

$EtO_2C-C\equiv C-CO_2Et$

$S=O$

simple dienes, for example 2-methyl-*trans*-pentadiene and 2-methyl-*cis*-pentadiene, can vary by a factor of 10^7 [22] (Table 14.4). *Cis*-substituents at the diene termini inhibit reactions strongly by steric hindrance to the approach of the dienophile. Substituents at both C2 and C3 often bring about a reduction of rates by increasing the conformational energy of the *cis*-planar form of the diene, the only reactive conformer. The importance of this conformation is manifest in the very high reactivities of exocyclic

8

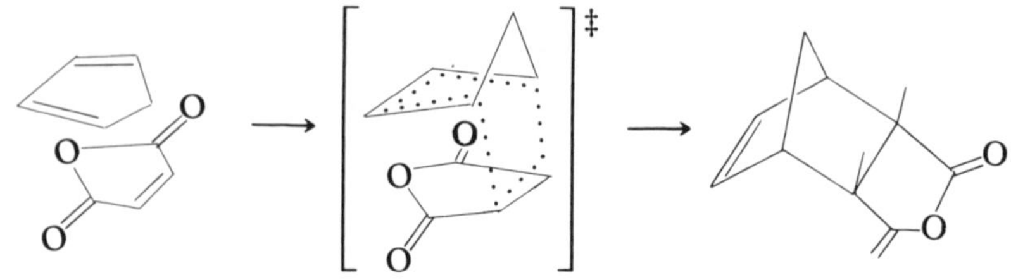

8a

endo-**8b**

9

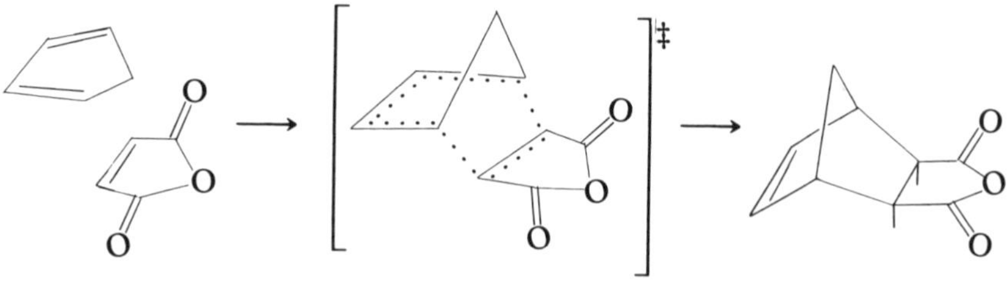

9a

exo-**9b**

10

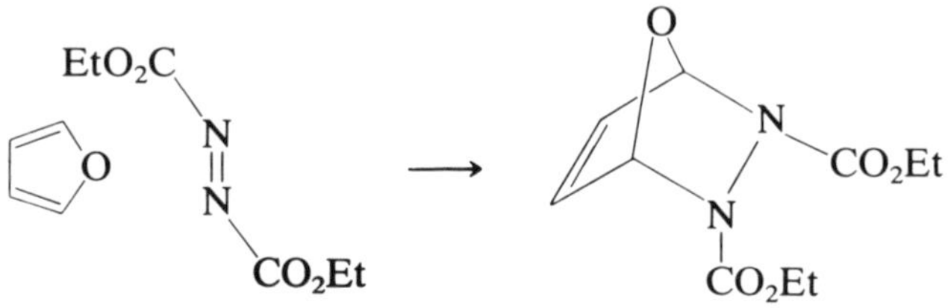

11

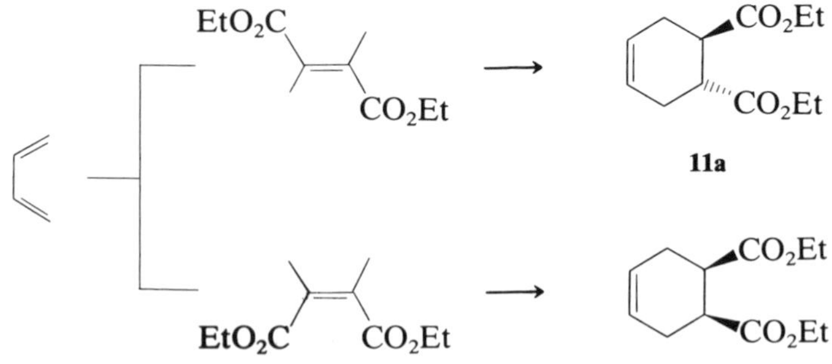

11a

11b

dienes, **12**, and of cyclic *cis*-dienes such as cyclopentadiene. Electronic influences are important, rates being increased by electron-donating groups in the diene and electron-withdrawing groups on the dienophile. The most powerful dienophiles (Table 14.3) are those which have highly electron-deficient double bonds bearing $-R$ substituents, e.g. tetracyanoethene, maleic anhydride and *N*-phenyltriazenedione, and also especially reactive molecules such as benzyne.

Table 14.4   Reactivity in Diels–Alder reactions

Compound	k_{rel} *(TCNE)*	Compound	k_{rel} *(cyclopentadiene)*
	1000		V. fast
	570		
	240		430×10^6
	200		
	27		5×10^6
	20		4.8×10^6
	15		
	0.025		455×10^3
	0.018		115×10^3
	0.009		70×10^3
	0.002		
	10^{-3}		55×10^3
	10^{-4}		40×10^3
			13.6×10^3
			9×10^3
			1.9×10^3
			910

661

Table 14.4 (*Continued*)

Compound	k_{rel} (*TCNE*)	*Compound*	k_{rel} (*cyclopentadiene*)
		CN / NC	806
		PhCO·C≡C·COPh	690
		EtO$_2$C·C≡C·CO$_2$Et	313
			180
		PhCO COPh	67
		CO$_2$Me	12
		CN	10
		MeO$_2$C CO$_2$Me	6
			0·9
12			

Electronic factors in reactivity [11]

Concerted cycloadditions proceed by interaction between two π-systems for which the perturbation equation (14.1) predicts that the rate will be greater, the closer in energy are the frontier MOs. In order to ascertain how this bonding energy term changes with structure, the energy levels of both diene and dienophile, bearing appropriate substituents, must be computed and, as a secondary matter, changes in the coefficients of each frontier orbital at the bonding termini must also be determined. Table 14.5 shows HOMO and LUMO energy trends for both 1- and 2-substituted butadienes, substituents being both $+R$ and $-R$ types. The energy differences between these and the frontier orbitals of a typical dienophile are also shown for both pairs of interactions:

(A) HOMO(diene)–LUMO(dienophile) and
(B) LUMO(diene)–HOMO(dienophile).

Table 14.5 Frontier orbital energies, E/kJ mol^{-1} (kcal mol^{-1}), for substituted butadienes and values of the energy gap ΔE for the two possible HOMO–LUMO combinations with ethene

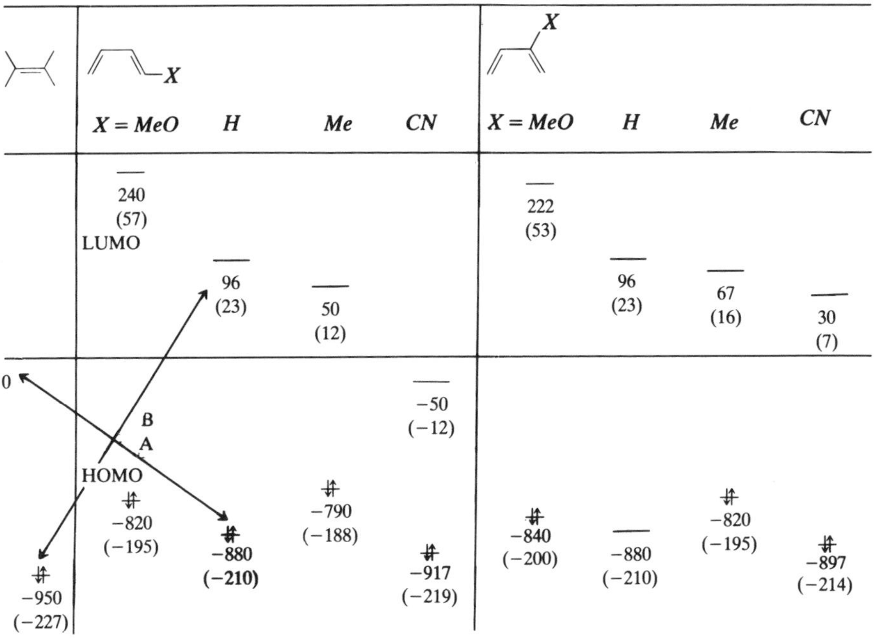

	X = MeO	H	Me	CN	X = MeO	H	Me	CN
ΔE(B):	1190 (308)	1046 (250)	1000 (239)	900 (215)	1172 (280)	1046 (250)	1017 (243)	980 (234)
ΔE(A):	820 (195)	880 (210)	790 (188)	917 (219)	840 (200)	880 (210)	820 (195)	897 (214)

The lower this difference, the greater is the orbital interaction. In all the usual cases interaction A is dominant, but it becomes less so as the electron-withdrawing character of the diene increases, while B at the same time becomes increasingly important. The result is a complicated relationship between rate and structure which may be curved (Fig. 14.7a) and may even have a minimum (Fig. 14.7b). In such a case, the increasing rate towards the left is due to the increasing effect of term A while the increase on the right hand side is due to increasing importance of term B, this type of situation being known as *reverse electronic demand.* When, in addition to the orbital energy differences, changes in the coefficients, c, with substitution, electrostatic attractions due to dipoles present and steric factors are added in, it is easy to understand that no very precise rate correlations are to be found except over a very narrow reactivity range with closely related dienes or dienophiles, in which case ρ values are small. Even a series of reactions with substituted styrenes, $ArCH{=}CH_2$, as the dienophiles give non-linear Hammett plots.

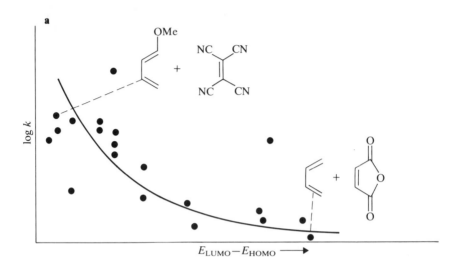

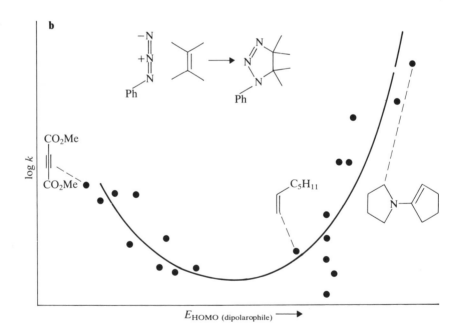

Fig. 14.7 Rates of cycloaddition as a function of the energy difference between HOMO and LUMO in the frontier orbitals: (**a**) Diels–Alder Reactions; those at the left are with 'inverse electron demand'; (**b**) dipolar cycloadditions of phenyl azide with various dipolarophiles (after Fleming[11]).

14.3.2 Stereo- and regio-specificity in Diels–Alder reactions

The stereochemistry of the original diene and dienophile is retained during the cycloaddition with complete integrity (**11, 14**). It must be inferred that the reaction is either concerted or, if it is a two-step process, factors are operating which completely prevent any rotations during its course. This is an unlikely situation in the case, for instance, of a diradical intermediate, **13**, which would surely lead to a mixture of **11a** and **11b**.

Cycloaddition of an unsymmetrically substituted diene and dienophile can, in principle, lead to two regioisomers (**16** and **17**) [17]. It is found that there is usually a strong preference for one only, as indicated, and the reasons for this may be sought in the changes in the coefficients of the frontier orbitals brought about by substitution and calculated by the Hückel procedure.

Figure 14.8 shows how a $+R$ and a $-R$ substituent can effect the magnitudes of the coefficients of the HOMO of butadiene (ψ_2) and the LUMO of ethene (ψ_2). Both molecular orbitals become polarized so that C1 and C4 of the diene and the two carbons of the dienophile are no longer

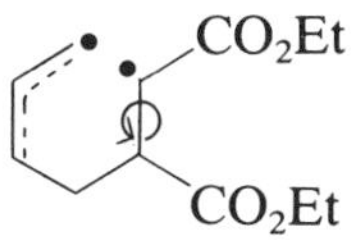

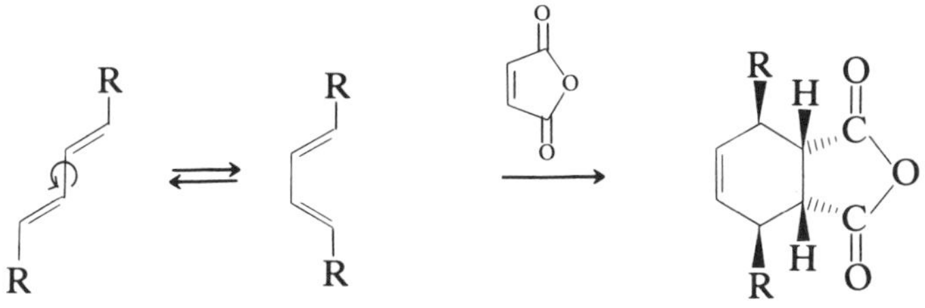

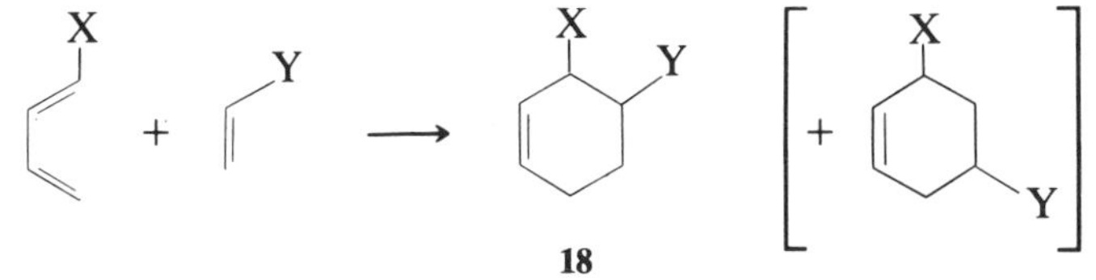

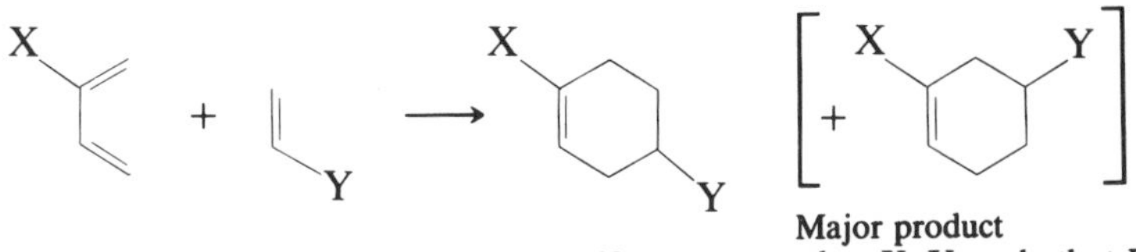

Major product
when X, Y are both $+R$ groups

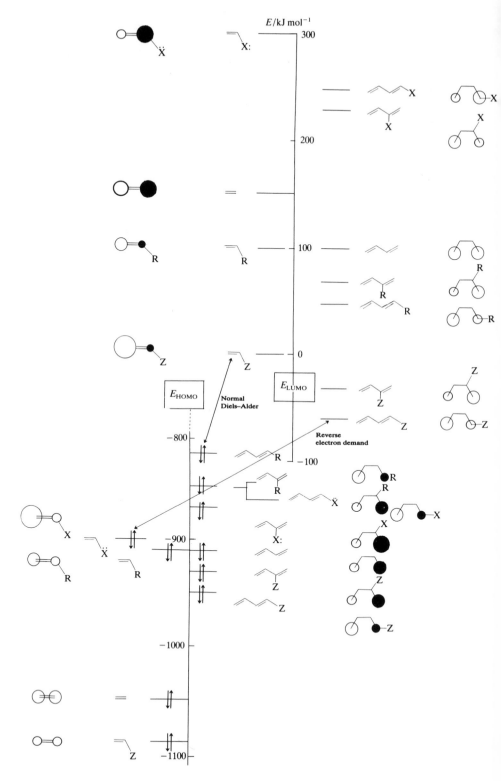

identical. The polarization of a diene is in the same sense for both $+R$ and $-R$ substituents, though it is opposite in a dienophile. Now both diene and dienophile have reaction termini with a large and a small coefficient in the frontier orbitals. The greatest bonding interaction is achieved by juxtaposition of the two centers with the large coefficients and of those with the small, rather than by pairing large with small (this is simply a result of the inequality, $LL + SS > 2LS$ where L is a large and S a smaller quantity).

Figure 14.8 and Table 14.6 summarize the orientation preferences which result from the principle of achieving maximum overlap. For cases of reverse demand, the analysis would need to consider the other pair of frontier orbitals whose coefficients would be affected by substitution in the manner shown in Fig. 14.8.

A cyclic diene may add to a dienophile which is unsymmetrical about the C=C axis in two sterically distinct modes forming *exo* and *endo* isomers which have the dienophile substituents respectively on the same side as the diene bridge and on the opposite side (**20a,b**). The *endo* isomer is normally the major or even the exclusively formed product, a feature of this reaction recognized by Diels and Alder who observed preference for the product formed 'by the maximum juxtaposition of double bonds'. Steric considerations make the *exo* isomer the more stable, however, so that when the reaction is under thermodynamic control this product (**21a**) may be extensively formed. The preference for *endo* product under kinetic control was explained by Woodward as due to 'secondary overlap' between the orbitals of the substituent on the dienophile moiety with those at C2,C3 in the diene. The LUMO of a dienophile such as maleic anhydride does indeed have lobes at the carbonyl carbons of the appropriate polarity to add further bonding interactions in this way and to reduce the activation energy, **22a**[23]. No such interactions are possible in the *exo* transition state, **22b**. Another factor which may increase discrimination between *exo* and *endo* modes of

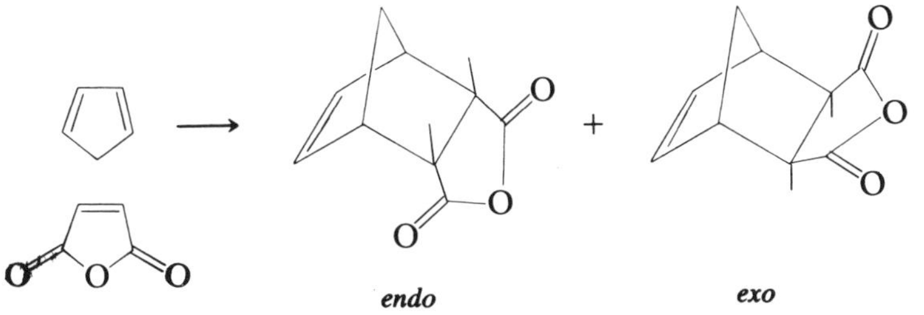

	endo	*exo*
20	**20b** (99%)	**20a** (1%)

Table 14.6 Products of Diels–Alder reactions expected

Dienophile	Diene	Product expected	Dienophile	Diene
C (mildly +R)	C	C / C	Z (strongly −R)	C
	Z	Z / C		Z
	Ẍ	Ẍ / C		Ẍ
	C	C / C		C
	Z	Z / C		Z
	Ẍ	Ẍ / C		Ẍ

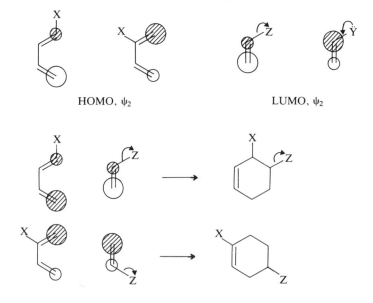

HOMO, ψ_2 LUMO, ψ_2

on the basis of maximum frontier orbital overlap.

Product expected	Dienophile	Diene	Product expected
	(strongly $+R$)		

addition is the mixing-in of σ-orbitals of the correct symmetry to the π-system, which becomes perturbed, the terminal lobes being twisted from their normal alignment and the symmetries of π-orbitals with respect to the

21b (*endo*)
Kinetic product

21a (*exo*)
Thermodynamic product

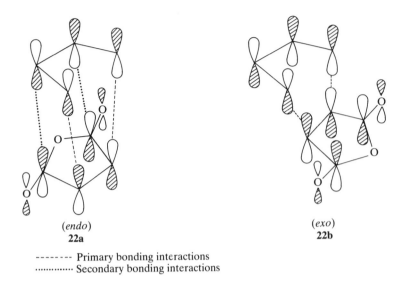

(*endo*) (*exo*)
22a **22b**

‐‐‐‐‐‐‐ Primary bonding interactions
·············· Secondary bonding interactions

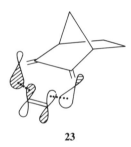

23

plane of the four carbon atoms being disturbed. The preferred *endo* addition to **23** is attributed to minimization of repulsive interactions between filled orbitals, the HOMOs of diene and dienophile[23]. It is not possible to be general in assessing the importance of these effects. They must be calculated for each specific case using high-level MO theory.

14.3.3 Retro Diels–Alder reactions

Thermal reversion of an allowed cycloaddition must also be allowed, since both reactions partake of the same reaction path and have the same transition state (principle of microscopic reversibility). The dissociation of cyclohexene into buta-1,3-diene and ethene is denoted $[_\pi 2_s + _\sigma 2_s + _\sigma 2_s]$ which, having an odd number (three) of two-electron suprafacial components, is allowed. Many such reactions occur cleanly and with retention of stereochemistry at moderate temperatures. The splitting out of such stable molecules as nitrogen (**24**)[24] or SO_2[25], or recovery of aromatic character of a pyrrole[26], **25**, facilitates the reaction. Entropies of activation are small in

24

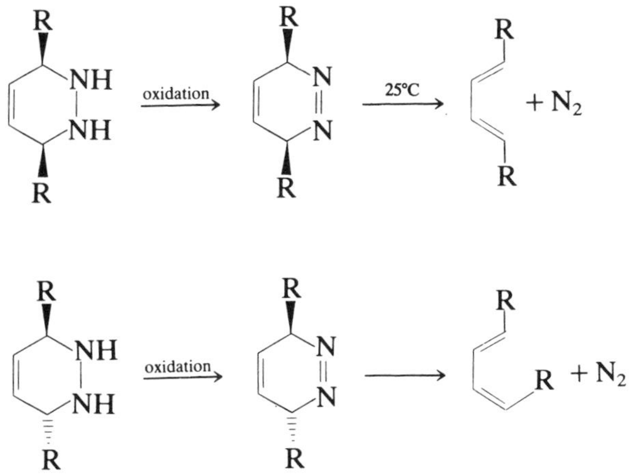

25

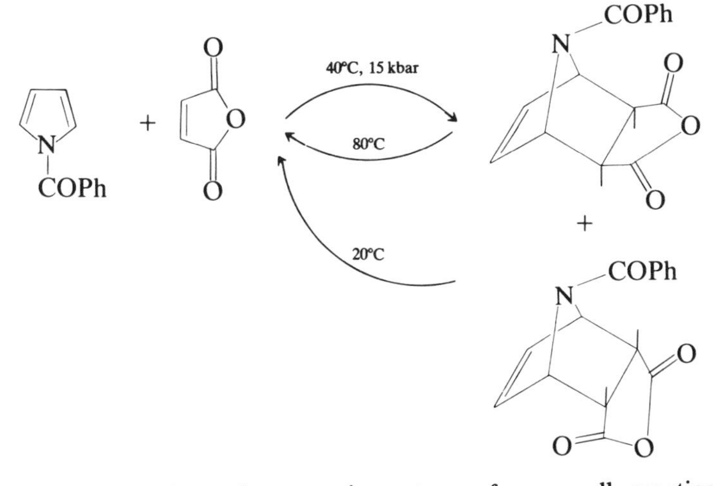

comparison with the large increase in entropy for overall reaction[27-29].
Retention of stereochemistry must occur and *endo* adducts will decompose
faster than *exo*, since the secondary interactions accessible to the former
isomers will come into play just as in the forward reaction. Cyclohexene
pyrolysis is frequently used in synthesis and can be an elegant route to
otherwise inaccessible compounds, e.g. in reactions **26–28**[30,31].

26

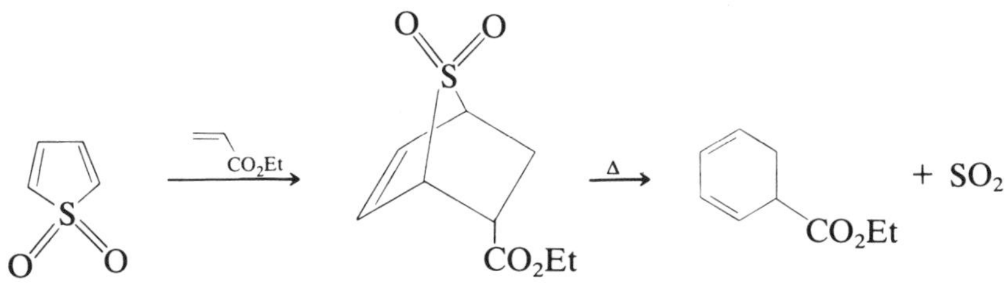

27 $\xrightarrow{EtO_2C\cdot C\equiv C\cdot CO_2Et}$

EtO$_2$C   CO$_2$Et

$\xrightarrow{\Delta}$ +

CO$_2$Et

CO$_2$Et

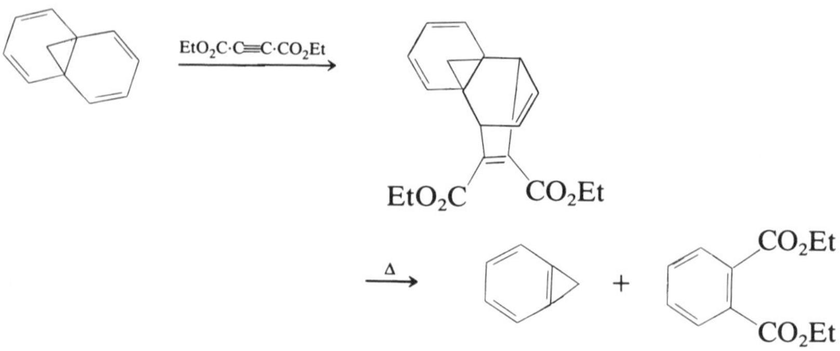

28 $\longrightarrow$

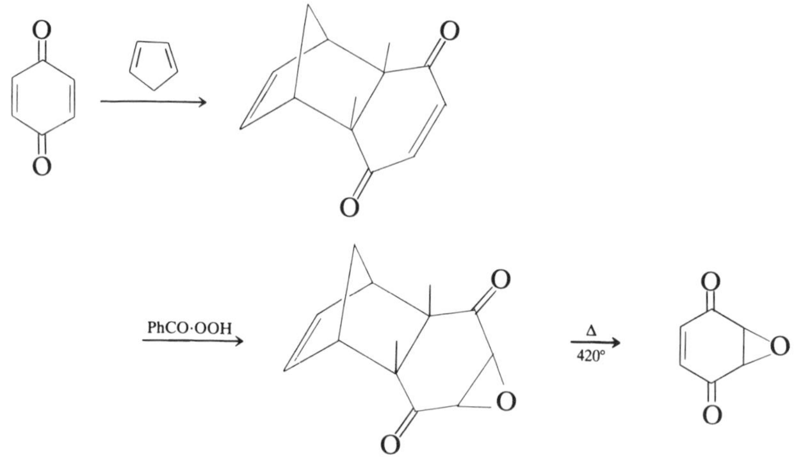

$\xrightarrow{PhCO\cdot OOH}$ $\xrightarrow[420°]{\Delta}$

14.3.4 The nature of the Diels–Alder transition state[32-34]

The activation parameters for a wide range of $(4\pi + 2\pi)$ cycloadditions, Table 14.7, are similar in showing a highly unfavorable entropy of activation, usually -150 to $-200\,\mathrm{J\,K^{-1}\,mol^{-1}}$ (-35 to $-48\,\mathrm{cal\,K^{-1}\,mol^{-1}}$) but this is often accompanied by a fairly low activation enthalpy by way of compensation. They are rather insensitive to the solvent. The entropy of activation of the retro-reaction by contrast is very small, either positive or negative. This indicates a transition state which is very cyclohexene-like as regards entropy, the diene and dienophile being tightly bound with additional rigidity being imparted to the structure by the secondary interactions. This picture is reinforced by the volume profile. Volumes of activation are large and negative, -30 to $-45\,\mathrm{cm^3\,mol^{-1}}$, while volumes of reaction are often slightly less negative, pointing to a slight relaxation of rigidity as the product bonding develops. This property makes high-pressure conditions attractive for driving difficult cycloadditions[35]. Enthalpies of reaction are often large and negative; the conversion of two π-bonds to two σ-bonds is exothermic to approximately $-180\,\mathrm{kJ\,mol^{-1}}$ ($-43\,\mathrm{kcal\,mol^{-1}}$)

Table 14.7 Relative reactivities and activation energies for reactions of dimethylketene with alkenes at 100°C

Alkene	$k_{rel.}$	E_A/kJ mol^{-1} (kcal mol^{-1})
cis-But-2-ene	1·00	
cis-Pent-2-ene	1·00	
But-1-ene	0·44	
Isobutane	0·124	
trans-But-2-ene	8×10^{-4}	
trans-Pent-2-ene	10^{-3}	
Butadiene	2·64	
2-Methylbut-2-ene	0·016	
2,3-Dimethylbut-2-ene	V. small	
Styrene	0·01	
Cyclobutene	0·009	
Cyclopentene	0·190	
Cyclohexene	0·068	
Cycloheptene	0·06	
Cyclo-octene	0·92	
Methylenecyclopropane	1·81	
Ethyl vinyl ether	8·9	67 (16)
Butyl vinyl ether	9·0	60 (14)
Ethoxyacetylene	23·7	83 (20)
p-Methoxystyrene	2·6	
p-Methylstyrene	1·8	
p-Chlorostyrene	0·36	

(Section 1.3.6). This, according to the Hammond postulate, would suggest an early transition state with respect to free-energy change. The apparent contradiction implies that the changes in entropy and in free energy along the reaction pathway are not necessarily parallel, though changes in volume and in entropy frequently are. Small secondary deuterium isotope effects, **29**[36-38], confirm that bonding is taking place at all four centers simultaneously. Intramolecular Diels–Alder reactions such as **30** occur readily in

$$\frac{k_H}{k_D} = 1 \cdot 14$$

suitable systems since the entropy change is less negative than is the case for intermolecular reactions **30**. This is because no translational freedom needs to be lost. Correspondingly, the volume of activation is less negative than is typical for bimolecular reactions[39,40].

14.3.5 Related six-electron cycloadditions

The ene reaction[41]

This addition with hydrogen transfer of an allylic compound to an alkene ('enophile'), **31**, occurs mainly at high temperatures, e.g. **32**, **33**, since activation energies are, on the whole, higher than are those of the Diels–Alder reaction. It is designated $[_\pi2_s + _\pi2_s + _\sigma2_s]$[42–44]. Reactivities are higher at strained double bonds, e.g. for **34**, and may compete with Diels–Alder addition[45–47]. Maleic anhydride, acetylenedicarboxylic ester and especially azodicarboxylic ester are efficient enophiles, though tetra-cyanoethene usually prefers to react by a $(2\pi + 2\pi)$ route. The concerted

nature of the reaction is shown by the large negative entropy of activation
and the absence of solvent effects on the rate, for example in **35**, for which
$k(\text{MeCN})/k(\text{cyclohexane}) \sim 2$[48]. Furthermore, the product of addition of
limonene **36**, to singlet oxygen contains no **37**, which would have been the
case were an allylic radical an intermediate[49]. Intermolecular examples
occur at rather high temperatures in hydrocarbon systems (**38**)[50,51].

Retro-ene reactions are rather common in hydrocarbon pyrolyses, especially of strained allylic systems such as **39–41**[52,53]. The pyrolytic eliminations of esters, amine oxides and related compounds discussed in Section
11.2 are processes of this type.

Concerted hydrogen transfers

Preparatively useful reductions, consisting of the transfer of two hydrogens from a donor to an alkene acceptor, are known which have the characteristics of six-center concerted reactions. In all cases the donor acquires some special stability which drives the reaction. A typical example is diimide, HN=NH, a highly reactive, though isolable, compound normally generated *in situ*, which will reduce alkenes to alkanes with complete *syn*-stereospecificity (**42**)[54-57]. Dihydroaromatic compounds similarly have a predilection to transfer hydrogen and regain aromaticity (**43**)[58,59].

14.4 THERMAL (2 + 2) CYCLOADDITIONS

While $[_\pi 2_s + _\pi 2_a]$ concerted cycloadditions of ground-state reagents are orbital-symmetry forbidden and are, indeed, inaccessible to simple alkenes, many reactions do occur in which formal double bonds cyclize to four-membered rings. These appear to fall into two classes:

(a) Concerted processes involving cumulenes, especially ketenes $R_2C=C=O$, but also allenes, $\overset{\diagdown}{}C=C=C\overset{\diagup}{}$. The reactions probably take place by an antarafacial approach of the alkene.

(b) Non-concerted cyclizations via diradicals or dipolar intermediates.

14.4.1 Cycloadditions of cumulenes

Stable ketenes such as diphenylketene have long been known to react with electron-rich alkenes to form cyclobutanones, **44**[60-62]. Ketene itself and its alkyl derivatives dimerize to four-membered cyclic products in one of two modes, **45** and **46**, usually the latter. Reactive and inisolable ketenes formed by E2 elimination are the presumed intermediates in reactions of acyl halides with base. Even when a cisoid diene is available, $(2\pi + 2\pi)$ addition (**47**) is preferred to $(2\pi + 4\pi)$ by these species[63,64]. The reaction is facilitated by electron-donating groups on the alkene (vinyl ethers are particularly reactive) and electron-withdrawing groups on the ketene. Evidently the LUMO of the ketene is the important frontier orbital. It is not adequate to consider the π-electrons of ketene either as occupying two orthogonal ethene-like orbitals or as implied in the vinylium oxide contributing structure, **48**. Mixing of the various atomic orbitals can yield three-center molecular orbitals, **49**, of which the LUMO has a large coefficient at the carbonyl carbon[65]. It is evidently this LUMO which can interact with the HOMO of the alkene at the terminal carbon of a monosubstituted ethene, **50a**. The nitrogen analogs, keteniminium ions, **51**, are even more reactive[66].

The evidence is consistently in favor of the concerted formation of the two bonds[67-70]. Retention of stereochemistry of the alkene moiety is usual, there is the expected negative entropy of activation (Table 14.6[71]) and large negative volumes of activation have been reported ($-50\,\text{cm}^3\,\text{mol}^{-1}$ [72]). Cycloaddition of styrene and its α- and β-deuterated analogs to diphenylketene is accompanied by secondary isotope effects[73,74]; $k_H/k_D = 1\cdot23(\alpha)$, $0\cdot91(\beta)$. The inverse effect is expected for this hybridization change: the normal effect observed at the α-carbon is unexpected but at least it seems to indicate that some bonding change is in progress at both centers in the transition state, but to different extents. To this may be added a modest solvent effect; $k(\text{MeCN})/k(\text{cyclohexane}) \sim 50$ for the diphenylketene–butyl vinyl ether cycloaddition. It may be concluded that while these reactions are concerted, bonding at the carbonyl center is considerably in advance of that at the other (**50b**).

The reactivity pattern of simple alkenes seems to be controlled more by steric effects than by electronic (Table 14.6)[75,76]. *Trans*-1,2 and 1,1-disubstituted ethenes are far less reactive than *cis*-1,2-disubstituted or monosubstituted ethenes. This is consistent with an antarafacial approach of the alkene, sterically hindered if an alkene substituent is forced to approach a ketene substituent (**52**). *Endo* adducts are preferred over *exo*, as was the case with Diels–Alder reactions[77].

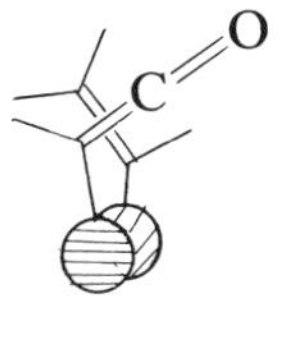

52

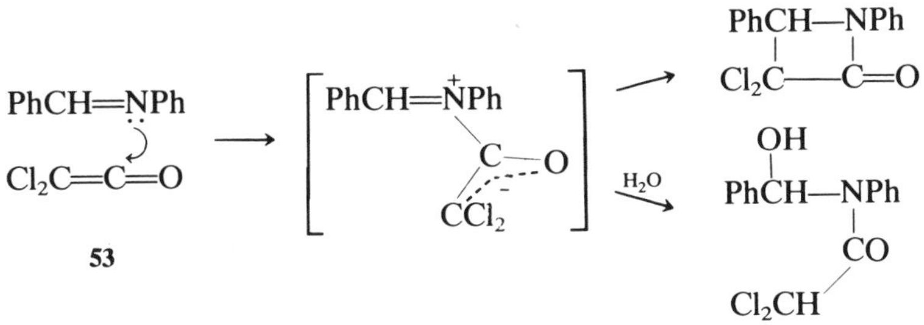

53

Cycloaddition of a ketene to an imine, **53**, appears to be a two-step process, plausible in any case since the charges in a dipolar intermediate are highly stabilized. Acyclic products may be trapped by addition of water[78,79]. The reaction of an isocyanate with a vinyl ether or of chlorosulfonyl isocyanate, with an alkene, **54**, is probably analogous[80,81]. Additions of ketenes to 1,3-dipoles, e.g. **55**, will occur, though now across the carbonyl double bond. β-Lactones may be formed analogously from ketones[82]. Allenes will dimerize and add to alkenes at elevated temperatures, **56**, in a head-to-head fashion[83–89]. This is the expected product for a two-step process via a diradical intermediate, **57**. However, some additions of chiral allenes such as **58** are highly stereospecific[86] and the products are those expected from a $[_{\pi}2_a + _{\pi}2_s]$ process, while others are not[87,88]. Though there are grounds for assigning the concerted mechanism, the diradical alternative may not at present be ruled out: the highly crowded transition state could be of such a short lifetime that it does not permit the expected rotations of

54

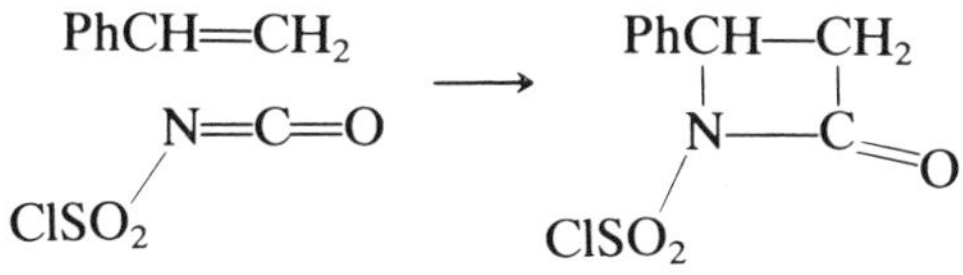

55

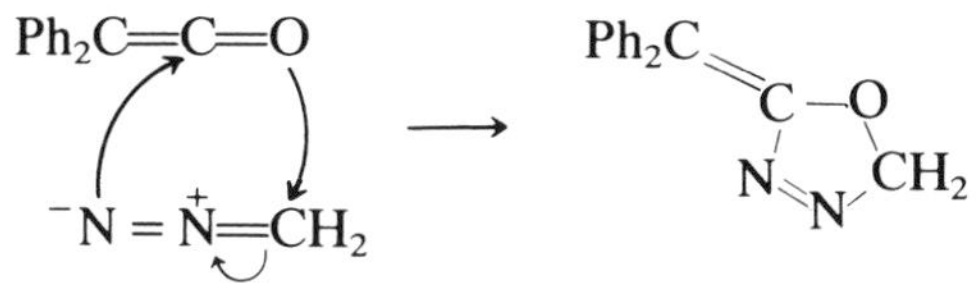

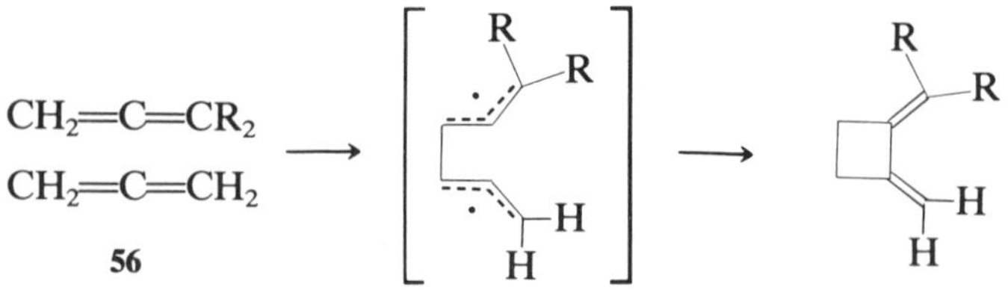

$$CH_2=C=CR_2$$
$$CH_2=C=CH_2$$

56 → **57**

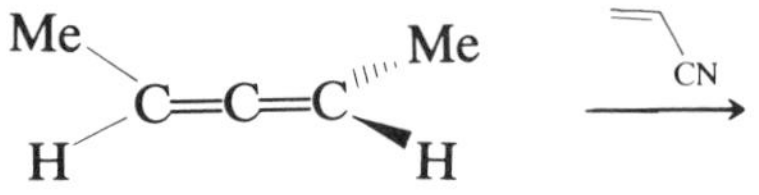

58

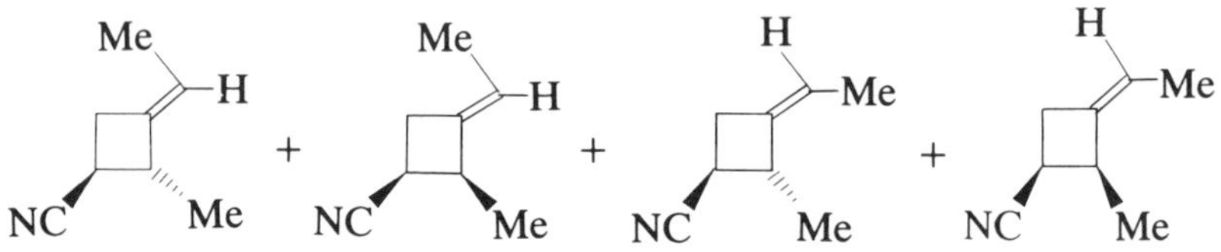

all optically active

an intermediate allylic radical which would confirm its existence[89], though indifference to solvent precludes a dipolar intermediate. In fact, the demonstration (Table 14.13c) of a difference between the intra- and inter-molecular secondary isotope effect requires the product-forming step to be different from the rate-determining step and hence is indicative of a two-step mechanism. This appears to be true both of the dimerization and of the addition of acrylonitrile to allene[90].

14.4.2 Two-step cycloadditions[91–93]

The restrictions of orbital symmetry conservation can be avoided if formation of the two bonds in a cycloaddition take place sequentially. Many such examples are known. Lack of stereochemical integrity is good evidence for such a route while the existence of dipolar intermediates is indicated by high sensitivity of the rates to solvent change and to electronic factors[94].

Halogenated alkenes dimerize or add to other alkenes (**59–61**) by diradical pathways as indicated by the stereochemistry and the mixed $(2 + 2)$ and $(2 + 4)$ products derived from dienes, **61a,b**. Strained alkenes such as *trans*-cycloheptene will also dimerize (**62**)[95]. These compounds are far more reactive than normal alkenes since their π-bonds cannot achieve a planar

59 $2\,F_2C{=}CF_2 \xrightarrow{>200°C}$ ☐—F_8

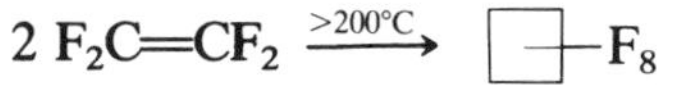

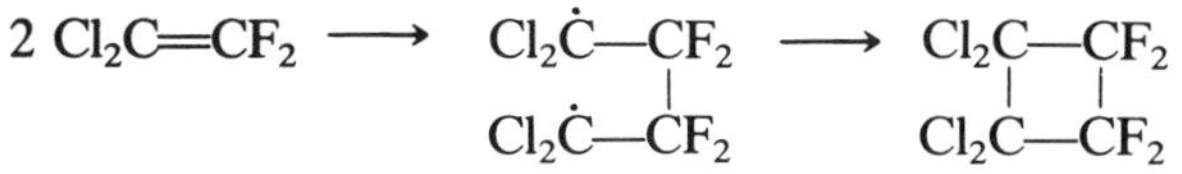

60 $2\,Cl_2C{=}CF_2 \longrightarrow$ $\begin{array}{c}Cl_2\dot{C}{-}CF_2\\ Cl_2\dot{C}{-}CF_2\end{array} \longrightarrow \begin{array}{c}Cl_2C{-}CF_2\\ |\quad\ \ |\\ Cl_2C{-}CF_2\end{array}$

61a $\begin{array}{c}PhCH{=}CH_2\\ Cl_2C{=}CF_2\end{array} \longrightarrow$

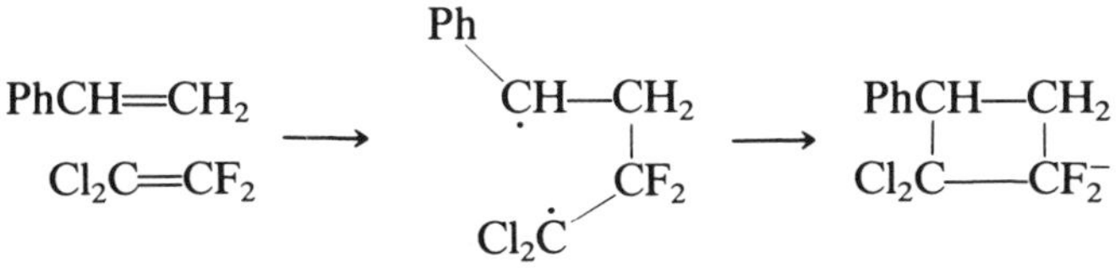

61b

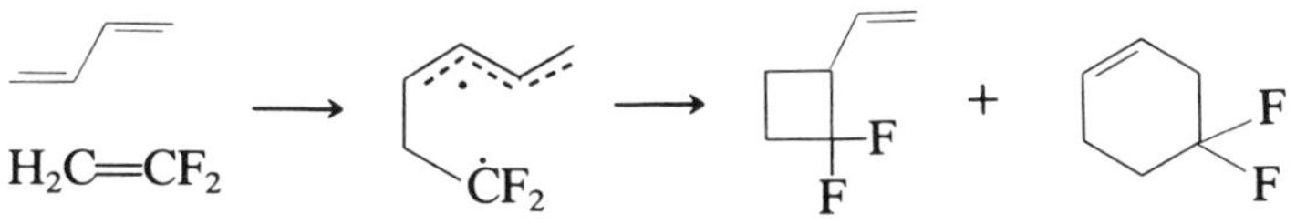

62

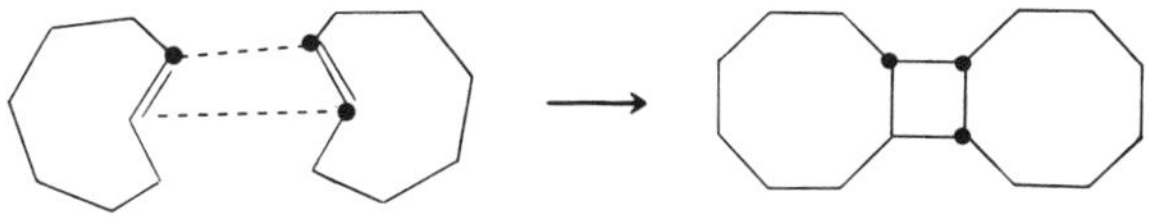

geometry. The dimers, however, are not formed with complete stereospecificity, and diradicals are again likely intermediates.

Polar cycloadditions occur with highly electron-poor alkenes such as
tetracyanoethene, which readily adds to vinyl ethers and, less readily, to
unactivated alkenes, **63**[96-98]. These cycloadditions are highly sensitive to
the solvent, rates correlating with E_T and allowing a transition state
dipole moment, 15D, to be estimated; typically $k(MeCN)/k(cyclohexane) \sim$
10^4 [99-101]. Large, negative Hammett reaction constants, ρ, for additions to
$ArCH{=}CH_2$ point to the need to stabilize positive charge. This is also the
explanation of the higher reactivity of 1,1-disubstituted ethenes compared
with *cis*-1,2-disubstituted ones in contrast to that of the analogous ketenes;

$$k_{rel}\!:1 \qquad\qquad\qquad\qquad 10^5$$

Large negative and solvent-sensitive entropies of activation are observed
due to increased solvation of the dipolar transition state[102] and corresponding volumes of activation, $-60\,cm^3\,mol^{-1}$ [72,103]. There seems no doubt that
intermediates of the type **64**, well equipped to stabilize the charges, are

63

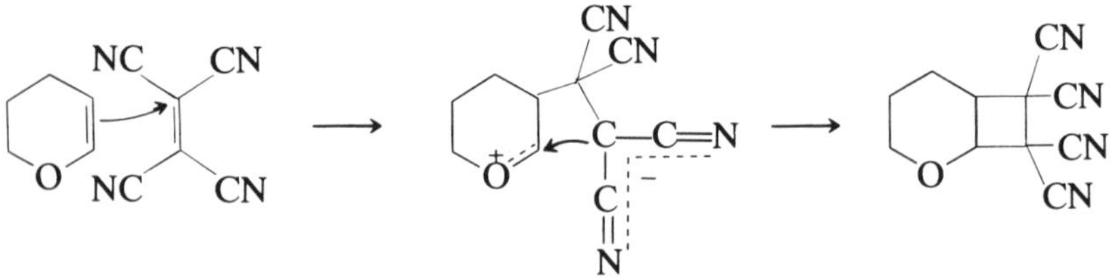

64

65

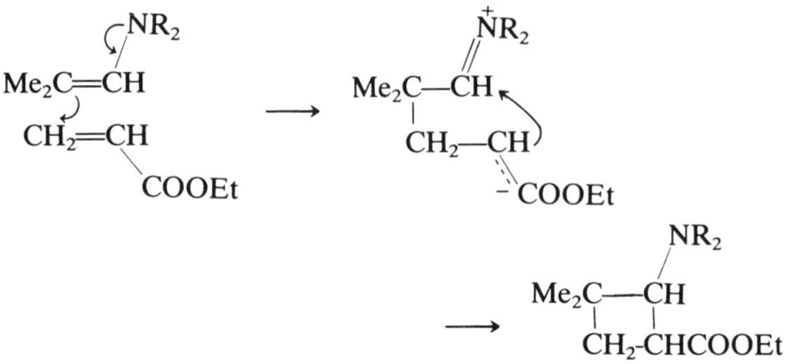

66

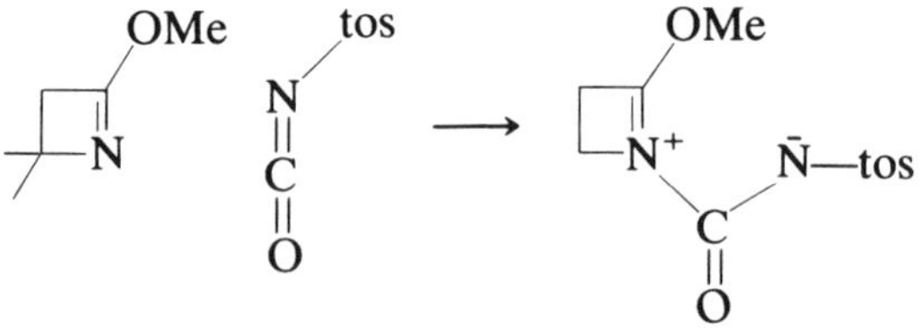

67

involved. Other polar cycloadditions are shown (65, 66). The dipolar intermediate, 67, may even be isolated. Despite the apparent open structure of the transition state, high stereospecificity is often observed. This could be due to dipolar forces preventing rotation before cyclization, in which case concerted and two-step mechanisms become merged. The distinction is often difficult to make and, moreover, it has been suggested that the orbital symmetry restrictions applicable to simple alkenes break down with such perturbed derivatives as tetracyanoethene[104].

14.4.3 (2 + 2)Cycloreversions[105]

Calculations predict that the pyrolysis of cyclobutane to two molecules of ethene will occur via a diradical intermediate with an activation energy of $260 \, kJ \, mol^{-1}$ ($62 \, kcal \, mol^{-1}$) despite the reaction being exothermic by $80 \, kJ \, mol^{-1}$ ($19 \, kcal \, mol^{-1}$). The strained analogs 68–70 have lower

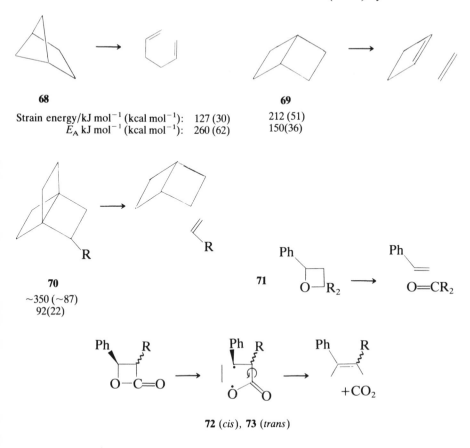

68

Strain energy/kJ mol^{-1} (kcal mol^{-1}):　127 (30)
E_A kJ mol^{-1} (kcal mol^{-1}):　260 (62)

69
212 (51)
150(36)

70
~350 (~87)
92(22)

71

72 (*cis*), **73** (*trans*)

activation energies as a result of strain release[106]. Similarly, oxetanes will undergo thermolysis to alkene and ketone (**71**), steric factors tending to retard reaction of the *cis* series, **72**, particularly. Mixtures of stereoisomeric products are obtained though there is a distinct trend towards excess retention of configuration suggesting the second bond breaks before sufficient time has elapsed for extensive rotation.

Compound	k_{ref}		
	R = Me	R = isoPr	R = tert-Bu
cis (**72**)	1	0·31	0·028
trans (**73**)	11	4	2

The β-lactone **74** is structurally well adapted for decomposition via a dipolar intermediate. This mechanism is indeed favored, as confirmed by the high dependence of the rates on solvent polarity and the negative volume of activation due to solvent electrostriction despite this being a bond-breaking reaction[107]. Orbital symmetry conservation appears to be maintained in decompositions of dioxetanes, **75→76**. These have no inclination towards a $[_\pi 2_s + _\pi 2_a]$ geometry and instead use the $[_\pi 2_s + _\pi 2_s]$ route which leads to an excited state driven by the formation of a highly exothermic product. The products then fluoresce or transfer their energy to a second molecule such as anthracene which in turn fluoresces. This is the phenomenon of chemiluminescence, which appears to be based on reactions which are highly exothermic (such as the decomposition of the CO_2 dimer, **76**) yet orbital symmetry forbidden, there being a symmetry-allowed path which leads to an energetically accessible excited state. The oxidation of luminol, **77**, is another such reaction[108].

14.5   1,3-DIPOLAR CYCLOADDITIONS[109–111]

There are many examples of compounds with the basic structure $(\overset{+}{x}=y-\overset{=}{z})$, x, y, z being C, N, O or S, containing four π-electrons and therefore

Table 14.8   Classification of 1,3-dipoles

(a) With octet stabilization

$-\overset{+}{C}=\overset{..}{N}-\overset{..}{C}\big\langle \;\leftrightarrow\; -C\equiv\overset{+}{N}-\overset{..}{C}\big\langle$	Nitrile ylides
$-\overset{+}{C}=N-\overset{..}{N}\big\langle \;\leftrightarrow\; -C\equiv\overset{+}{N}-\overset{-}{N}-$	Nitrileimines
$-C=N-\overset{..}{O}{}^{-} \;\leftrightarrow\; -C\equiv\overset{+}{N}-\overset{-}{O}$	Nitrile oxides (cyanates)
$\overset{+}{N}=N-\overset{..}{C}\big\langle \;\leftrightarrow\; N\equiv\overset{+}{N}-\overset{..}{C}\big\langle$	Diazoalkanes
$\overset{+}{N}=N-\overset{..}{N}- \;\leftrightarrow\; N\equiv\overset{+}{N}-\overset{-}{N}-$	Azides
$\overset{+}{N}=N-\overset{..}{O} \;\leftrightarrow\; N\equiv\overset{+}{N}-\overset{-}{O}$	Nitrous oxide

(b) With no orthogonal π-orbital

$\big\rangle\overset{+}{C}-\overset{\|}{N}-\overset{..}{C}\big\langle \;\leftrightarrow\; \big\rangle C=\overset{+}{N}-\overset{-}{C}\big\langle$	Azomethine ylides
$\big\rangle\overset{+}{C}-\overset{\|}{N}-\overset{..}{N}\big\langle \;\leftrightarrow\; \big\rangle C=\overset{\|+}{N}-\overset{-}{N}\big\langle$	Azomethineimines
$\big\rangle\overset{+}{C}-\overset{\|}{N}-\overset{-}{O} \;\leftrightarrow\; \big\rangle C=\overset{\|+}{N}-\overset{-}{O}$	Nitrones
$-\overset{+}{N}-\overset{\|}{N}-\overset{-}{O} \;\leftrightarrow\; -N=\overset{\|+}{N}-\overset{-}{O}$	Azoxy compounds
$\big\rangle\overset{+}{C}-O-\overset{..}{C}\big\langle \;\leftrightarrow\; \big\rangle C=\overset{+}{O}-\overset{-}{C}\big\langle$	Carbonyl ylides
$\overset{+}{O}-O-\overset{-}{O} \;\leftrightarrow\; O=\overset{+}{O}-\overset{-}{O}$	Ozone

(c) Without octet stabilization

$-\overset{+}{C}=\overset{\|}{C}-\overset{-}{C}\big\langle \;\leftrightarrow\; -\overset{..}{C}-\overset{\|}{C}=C\big\langle$	Vinyl carbenes
$-\overset{+}{C}=\overset{\|}{C}-\overset{-}{O} \;\leftrightarrow\; -\overset{..}{C}-\overset{\|}{C}=O$	Keto carbenes
$\overset{+}{N}=\overset{\|}{C}-\overset{-}{C}\big\langle \;\leftrightarrow\; \overset{..}{N}-\overset{\|}{C}=C\big\langle$	Vinyl nitrenes
$\overset{+}{N}=\overset{\|}{C}-\overset{-}{O} \;\leftrightarrow\; \overset{..}{N}-\overset{\|}{C}=O$	Keto nitrenes

isoelectronic with butadiene. Most of these species are highly reactive in cycloadditions and act as the $[_\pi 4_s]$ component. The name 1,3-dipole is appropriate from the structure above, but in most cases more than one valence-bond contributing structure needs to be written in order to express bonding adequately and the compounds may possess little or no permanent dipole moment. Examples are given in Table 14.8. They include types with and without full octet stabilization. Some, such as the sydnones, **78**, and nitrones, are stable species. Others, such as nitrile oxides and azomethine-imines, are highly reactive and need to be generated *in situ* in the presence of a suitable 'dipolarophile'. Some examples of the many cycloadditions possible are shown in reactions **79–82**. All result in the formation of a five-membered heterocyclic system. Despite the great variety of types, similar characteristics such as large negative entropies, small solvent effects and stereospecificity of addition of the dipolarophile all point to concerted cycloadditions. Volumes of activation appear to be even more negative than

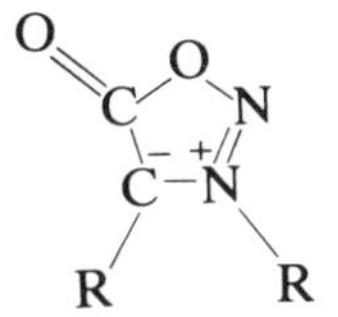

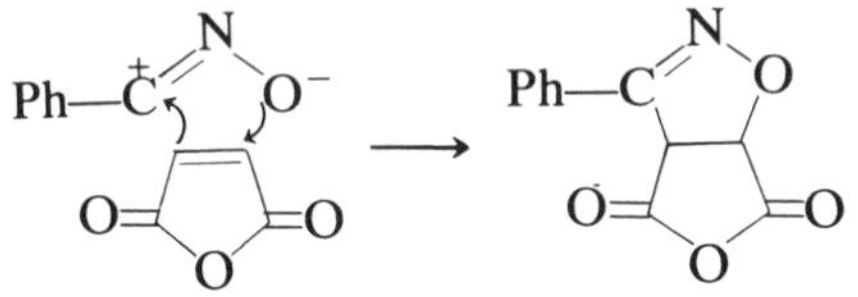

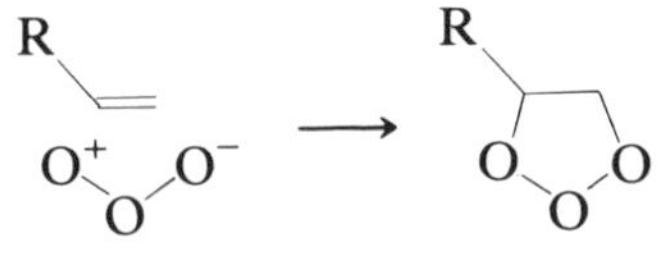

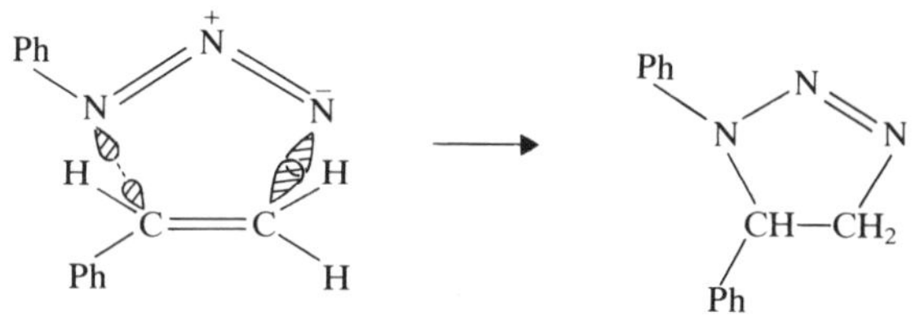

Table 14.9   Linear free-energy relationships for 1,3-dipolar cycloadditions

Reaction	ρ
PhCH=$\overset{+}{\text{N}}$Me—$\bar{\text{O}}$ (aryl nitrone) + ArCH=CH$_2$	+0·83
Ar—N=$\overset{+}{\text{N}}$=$\bar{\text{N}}$ (aryl azide) + maleic anhydride	−1·2
Ar—N=$\overset{+}{\text{N}}$=$\bar{\text{N}}$ + N-phenylmaleimide	+0·8
Ar—N=$\overset{+}{\text{N}}$=N$^-$ + enamine	+2·6

those of Diels–Alder reactions and are among the most negative values known (Table 14.7). Some reasonably linear Hammett plots have been obtained; the reaction constants, ρ, are small and of variable sign (Table 14.9), as would be expected for a concerted reaction, since each component is both a donor and an acceptor of electrons. The relatively large value of the last entry in the Table suggests some dipolar character of the transition state, for this case at least. Reactivity and regiospecificity[3,11] appear to be controlled by frontier orbital energies and coefficients. It is difficult to be general here since each case needs to be considered separately but Table 14.10 compiled by Houk[112,113] sets out these values for the more important

Table 14.10 Molecular orbital energies of the HOMO and LUMO of 1,3-dipoles and coefficients at the terminal atoms [112,113]

Dipole	HOMO		LUMO[a]	
	Energy/eV[a]	$(c\beta)^2/15$[b]	Energy/eV[a]	$(c\beta)^2/15$[b]
Nitrile ylides	−7.7	$HC\equiv\overset{+}{N}-\overset{-}{CH_2}$ 1·07 1·50	0·9	$HC\equiv\overset{+}{N}-\overset{-}{CH_2}$ 0·69 0·64
$PhC\equiv\overset{+}{N}-CH_2^-$	−6·4		0·6	
Nitrile imines	−9·2	$HC\equiv\overset{+}{N}-\overset{-}{NH}$ 0·90 1·45	0·1	$HC\equiv\overset{+}{N}-\overset{-}{NH}$ 0·92 0·36
$PhC\equiv\overset{+}{N}-\overset{-}{NPh}$	−7·5		−0·5	
Nitrile oxides	−11·0	$HC\equiv\overset{+}{N}-\overset{-}{O}$ 0·81 1·24	−0·5	$HC\equiv\overset{+}{N}-\overset{-}{O}$ 1·18 0·17
$PhC\equiv\overset{+}{N}-\overset{-}{O}$	−10·0		−1·0	
Diazoalkanes	−9·0	$H_2C=\overset{+}{N}=\overset{-}{N}$ 1·37 0·85	1·8	$H_2C=\overset{+}{N}=\overset{-}{N}$ 0·66 0·56
Azides	−11·5	$HN=\overset{+}{N}=\overset{-}{N}$ 1·55 0·72	0·1	$HN=\overset{+}{N}=\overset{-}{N}$ 0·37 0·76
$PhN=\overset{+}{N}=\overset{-}{N}$	−9·5		−0·2	
Nitrous oxide	−12·9	$\overset{-}{O}-\overset{+}{N}\equiv N$ 1·33 0·67	−1·1	$\overset{-}{O}-\overset{+}{N}\equiv N$ 0·19 0·96
Azomethine ylides	−6·9	$H_2C\diagup\overset{+}{N}(H)\diagdown CH_2^-$ 1·28 1·28	1·4	$H_2C\diagup\overset{+}{N}(H)\diagdown CH_2$ 0·73 0·73
$ROOCCH\diagup\overset{+}{N}(Ar)\diagdown \overset{-}{CHCOOR}$	−7·7		−0·6	
Azomethine imines	−8·6	$H_2C\diagup\overset{+}{N}(H)\diagdown \overset{-}{NH}$ 1·15 1·24	−0·3	$H_2C\diagup\overset{+}{N}(H)\diagdown NH$ 0·87 0·49
$PhCH\diagup\overset{+}{N}\diagdown \overset{-}{NPh}$	−5·6		−1·4	

[a] 1 eV = 96·5 kJ mol^{-1} = 23·1 kcal mol^{-1}.
[b] $(c\beta)^2/15$ is given rather than c as a convenient measure of electronic distribution.

Table 14.10   (*Continued*)

Dipole	HOMO		LUMO[a]	
	Energy/eV[a]	$(c\beta)^2/15$[b]	*Energy/eV*[a]	$(c\beta)^2/15$[b]
H₂C=N⁺(−)−N̄COR	−9·0		−0·4	
Nitrones H₂C=N⁺(H)−Ō	−9·7	H₂C=N⁺(H)−Ō 1·11 1·06	−0·5	H₂C=N⁺(H)−Ō 0·98 0·32
H₂C=N⁺(R)−Ō	−8·7		0·3	
PhCH=N⁺(H)−Ō	−8·0		−0·4	
Carbonyl ylides H₂C=O⁺−C̄H₂	−7·1	H₂C=O⁺−C̄H₂ 1·29 1·29	0·4	H₂C=O⁺−C̄H₂ 0·82 0·82
Ar(CN)C=O⁺−C̄(CN)Ar	−6·5		−0·6	
(NC)₂C=O⁺−C̄(CN)₂	−9·0		−1·1	
Carbonyl imines	−8·6	H₂C=O⁺−N̄H 1·04 1·34	−0·2	H₂C=O⁺−N̄H 1·06 0·49
Carbonyl oxides	−10·3	H₂C=O⁺−Ō 0·82 1·25	−0·9	H₂C=O⁺−Ō 1·30 0·24
Ozone	−13·5		−2·2	

systems. As for Diels–Alder reactions, reactivities depend, at least in part, upon the frontier-orbital energy differences (Figs. 14.5, 14.7b) and the regiospecificity may be understood to stem from the preference for the centers bearing the largest coefficients of HOMO(1,3-dipole) and LUMO (dipolarophile), respectively, to bond together (reaction **81**). Reactions may be further driven in this favorable direction by the greater heats of reaction available, e.g. in reaction **83**, and by the ability of the structure to stabilize some degree of dipolar character (**84**).

$$\Delta H = -170 \, \text{kJ mol}^{-1} \, (-41 \, \text{kcal mol}^{-1})$$

$$\Delta H = -118 \, \text{kJ mol}^{-1} \, (-28 \, \text{kcal mol}^{-1})$$

14.6 ELECTROCYCLIC REACTIONS[114-119]

These are molecular rearrangements which involve the formation of a σ-bond between the termini of a fully conjugated linear π-system, resulting in the formation of a ring with one fewer π-bond. The reverse of this process is included and examples **85–99** in Table 14.11 are typical.

The orbital-symmetry requirement for this process is the necessity for achieving bonding overlap between the lobes at the terminal atoms in the HOMO of the π-system. This is always possible and requires a rotation of 90° at each center. However, the direction of rotation is prescribed by the orbital properties and gives rise to definite stereochemical relationships between reagents and products.

Buta-1,3-diene → cyclobutene:

HOMO = Ψ_2, **85**

Bonding overlap at the termini is achieved by rotation in the same sense. This is described as *conrotation*; the reaction is allowed when conrotatory.

Hexa-1,3,5-triene → cyclohexa-1,3-diene:

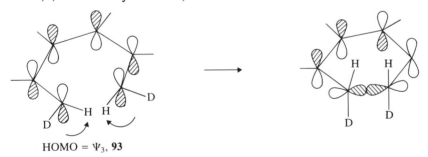

HOMO = Ψ_3, **93**

Bonding overlap is now achieved by rotation of the termini in opposite directions, a *disrotatory* mode.

The selection rule becomes

$4n$ π-electrons	$(4n + 2)$ π-electrons
conrotation	disrotation

Examples which illustrate the strict stereochemical integrities of these processes are shown in Table 14.11. Intramolecular analogs such as **96** are usually referred to as *valence tautomers* and can result in unexpected products even if the reactive tautomer cannot be observed.

Table 14.11 Examples of electrocyclic reactions

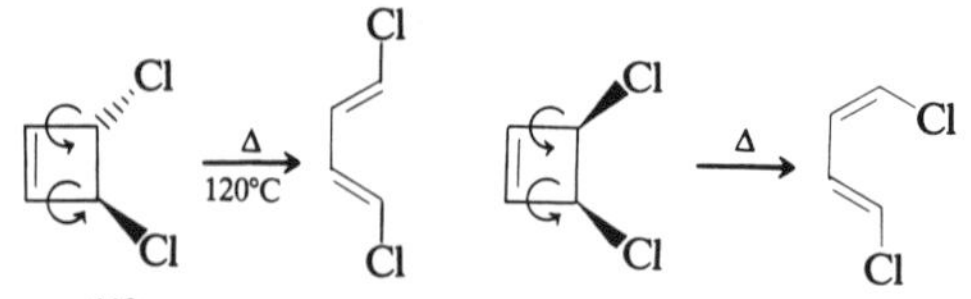

85[112]   **86**

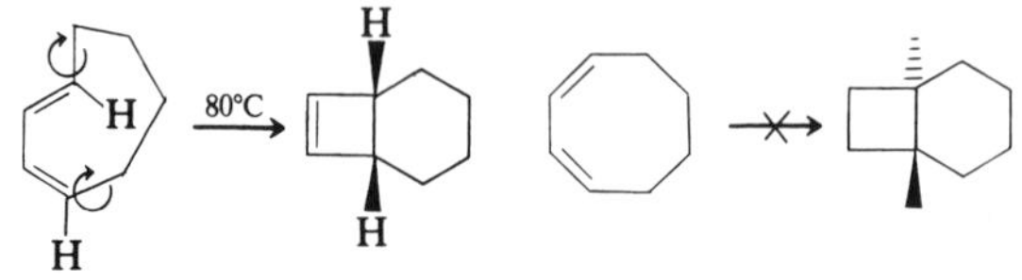

87   **88**   **89**   (highly strained)

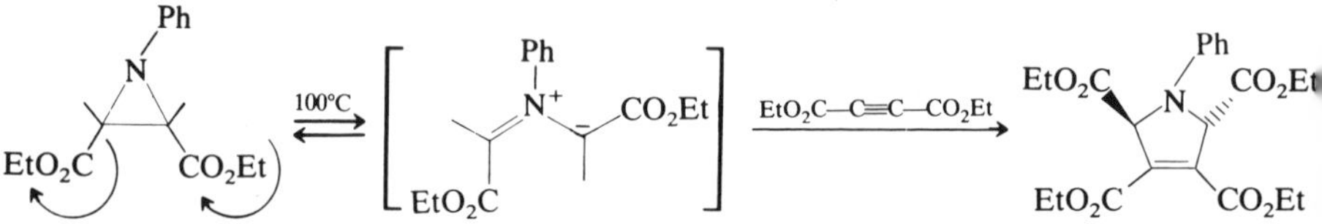

90[113]   **91**

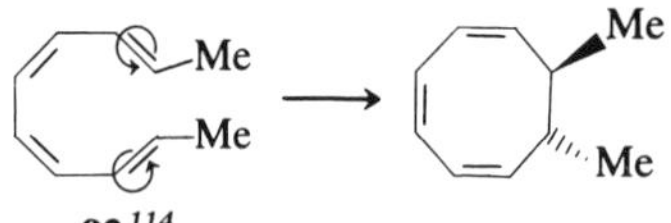

92[114]

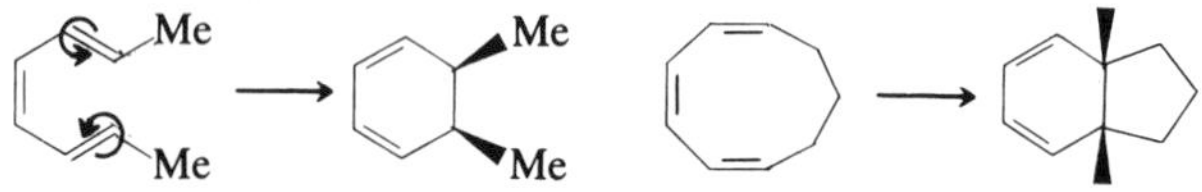

93[115]   >99·9%   **94**[116]

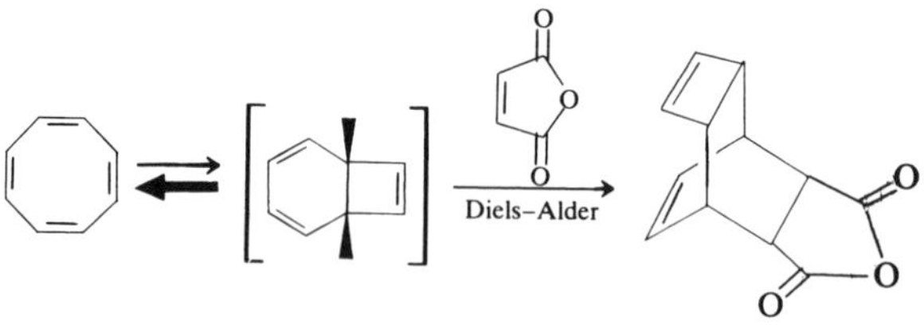

95[117]   **96**

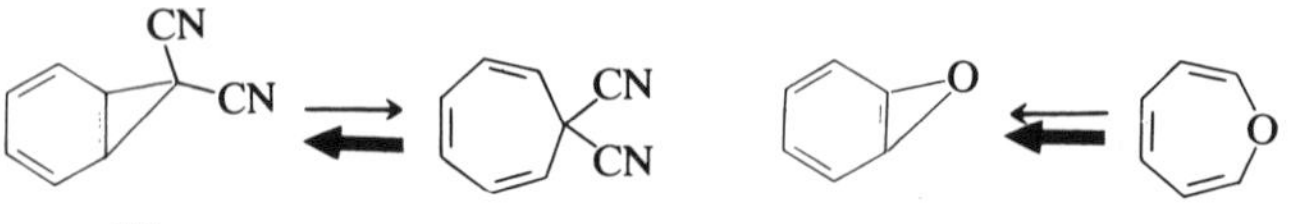

97[118]   **98**[119]   **99**

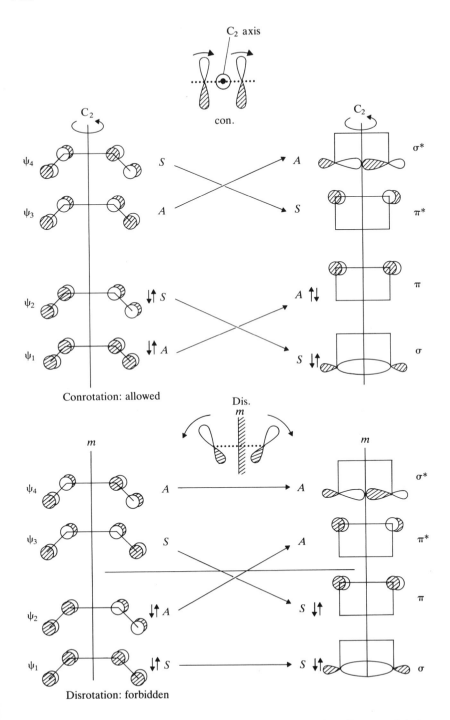

Fig. 14.9   Correlation diagram for electrocyclic reactions of butadiene ⇌ cyclobutene.

The correlation diagram which can be drawn for the butadiene cyclization is shown in Fig. 14.9. The symmetry element now present when a conrotation is carried out is the two-fold axis, C_2, and symmetries of reagent and product orbitals are expressed with respect to 180° rotation. Symmetry correlation between the bonding orbitals of reagent and product occurs and the process is allowed. Disrotation of the termini occurs with the preservation of a mirror plane but no rotation axis. Symmetries of the molecular orbitals with respect to this element no longer show ground-state correlation and the process is forbidden. A similar analysis may be carried out on analogous electrocyclic reactions. As always, the same considerations apply to either direction of reaction.

14.7 CHELOTROPIC REACTIONS

This term refers to cycloadditions in which both termini of a π-system are linked by the same atom. Addition of SO_2 to dienes, **100**[120–123], and its regression (thermolysis of sulfolenes) and of carbenes to alkenes, **101**, are examples. Among examples of retro-chelotropic reactions are the facile expulsions of small stable molecules such as CO, N_2O or SO_2 from three-membered rings, **102, 103**, or from norbornadienes, **104**, a process driven by the aromatic stabilization of the product. These reactions are stereospecific, they are associated with large entropies of activation, their rates are insensitive to solvent and they have the characteristics of concerted

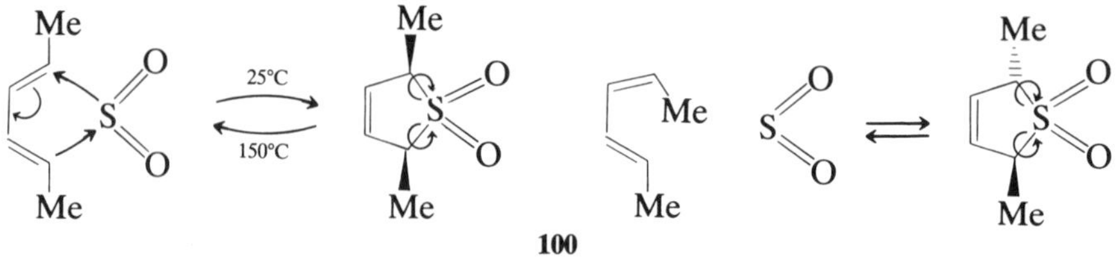

100

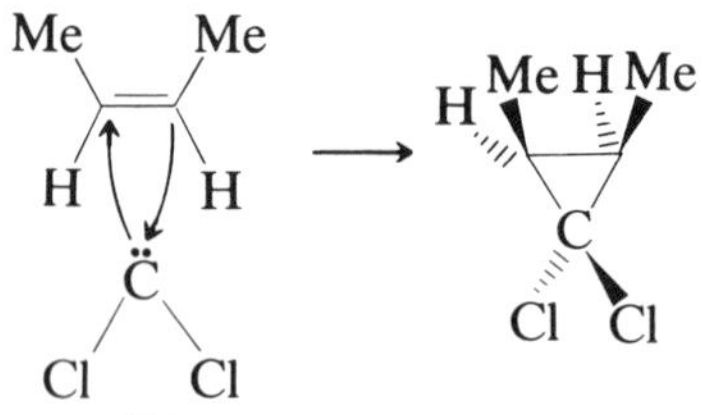

101

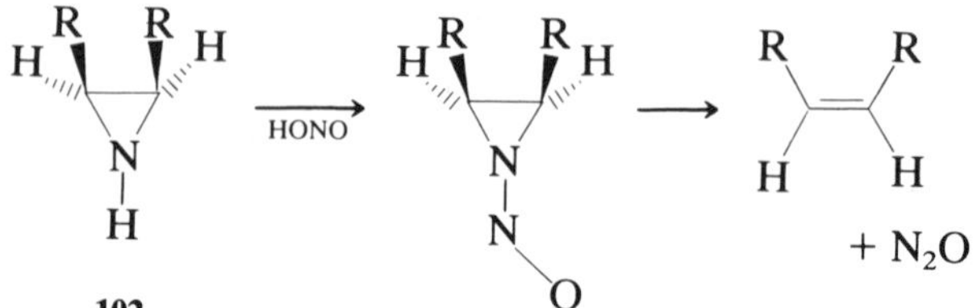

102   **103**

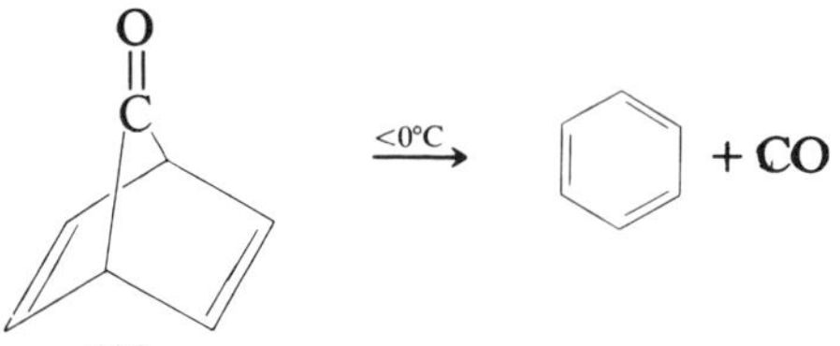

104

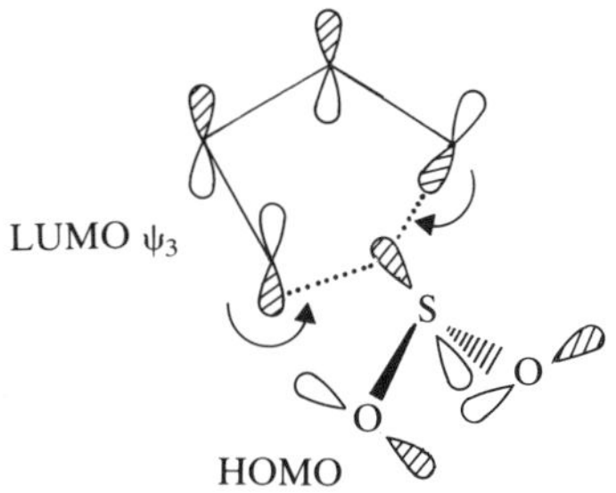

LUMO ψ_3

HOMO

105
Linear approach: disrotatory

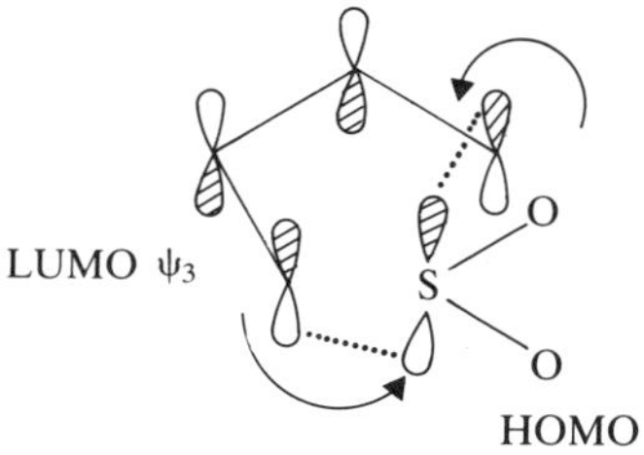

LUMO ψ_3

HOMO

106
Non-linear approach: conrotatory

cycloadditions[124]. The same orbital symmetry considerations apply as for other cycloadditions, though for the examples cited it is less easy to specify the frontier orbitals of the one-atom component, which must have available both a filled and a vacant orbital of the correct symmetry. The termini of the π-system must rotate in forming the cyclic product and they will do so in the sense, conrotatory or disrotatory, which maintains overlap and which depends upon the mode of approach of the reagents. This may be either 'linear', giving a coplanar geometry (105), or 'non-linear', an orthogonal approach followed by rotation of one reacting species relative to the other as bonding occurs (106). Selection rules deduced by Woodward are as follows:

Total 'moving' electrons	Linear approach	Non-linear approach
$4n + 2$	disrotatory	conrotatory
$4n$	conrotatory	disrotatory

The decomposition of a sulfolene, 100, to SO_2 and diene, a six electron (6e) process, is disrotatory and hence is presumed to have a linear transition state, as is also the decomposition of 107 (8e, conrotatory). However, decomposition of the episulfone 108, and other three-membered rings (4e), is also disrotatory so that a non-linear mode of reaction is likely[125].

Structural effects on reactivities are given in Table 14.12. There is only a mild response of rates to electronic influences but steric effects are marked. In particular, large substituents at C2 which favor the *cis* conformer of the diene cause rate accelerations, while the most reactive dienes are the

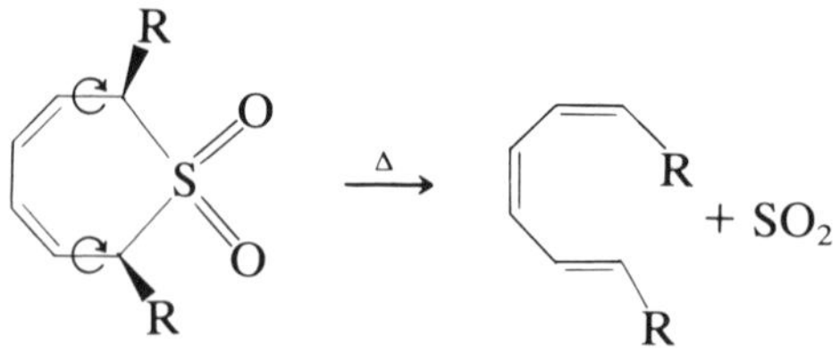

107

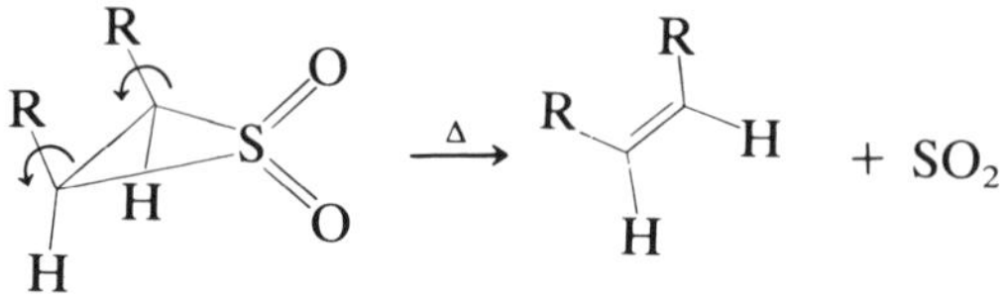

108

exocyclic compounds, **109a** and **109b**, particularly the former in which the diene system is coplanar.

Carbenes are highly reactive species of divalent carbon, $R_2C:$, having an unfilled octet. Various examples may be generated in solution by α-elimination (Section 11.3), **110**, or by photolysis of a diazoalkane, **111**. Two

Table 14.12 Relative rates of addition of SO_2 to butadienes and of thermal decomposition of the corresponding sulfolenes [121,122]

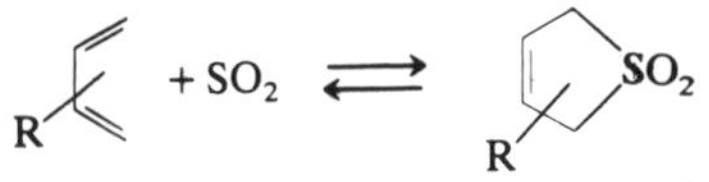

Butadiene	k_{rel}, addition (303 K)	k_{rel}, sulfolene decomposition (388 K)
Butadiene	—	1·7
2-Me	1·00	1·00
2-Et	2·6	0·91
2-isoPr	7·4	0·34
2-*tert*-Bu	20·7	0·54
2-neoPe	9·4	1·03
2-Cl	0·13	1·81
2-CH$_2$Br	0·39	1·22
2-Ph	9·4	0·91
2-*p*-tol	13·5	—
2-*p*-BrC$_6$H$_4$	4·9	—
2,3-diMe	1·9	0·30
cis-1-Me	0·1	2·0
trans-1-Me	0·38	
2,3-(CH$_2$)$_4$	135	1·1
2,3-(CH$_2$)$_3$	175	54

electrons and two orbitals give rise to two possible spin states. For many carbenes singlet (spin-paired; **112a**) and triplet (diradical; **112b**) electronic states are fairly close in energy, the singlet being lower for dichlorocarbene and the triplet for carbene (methylene), :CH$_2$. Cycloaddition may be concerted for the singlet but not for the triplet state, which must undergo spin inversion at some stage. This explains how it is that some carbene additions are stereospecific and some are not (**113, 114**). Similar reactions are available to nitrenes, RN:, though the greater reactivity of these species

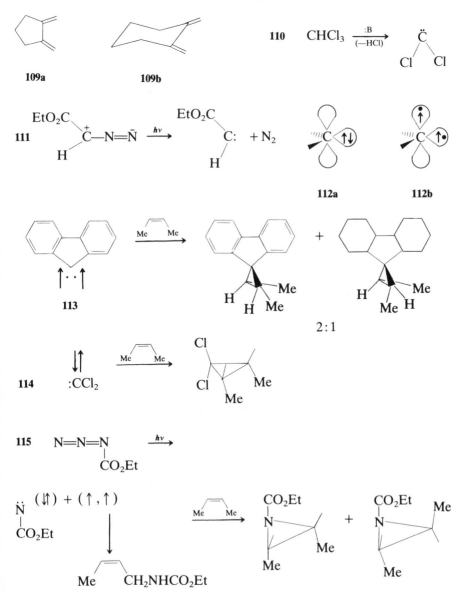

116

usually leads to bond-insertion reactions with σ-orbitals rather than chelotropic reactions (**115**). It is assumed that nitrenes approach the alkene in the non-linear mode, **116**.

14.8 SIGMATROPIC REACTIONS

This is the term describing concerted reactions in which a σ-bonded entity (atom or π-system) migrates from one terminus of a π-system to the other. Examples are the 1,5-hydrogen shift, **117**, and the Cope rearrangement, **118**. The order of a sigmatropic rearrangement, [i, j] expresses the number of atoms, i and j, across which the σ-bond migrates; according to this formalism the examples mentioned are [1, 5] and [3, 3] sigmatropic shifts. If

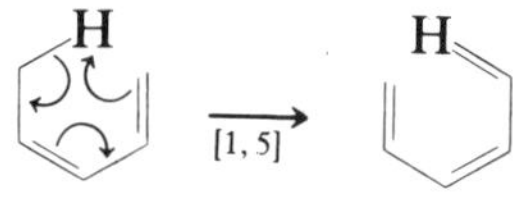

117

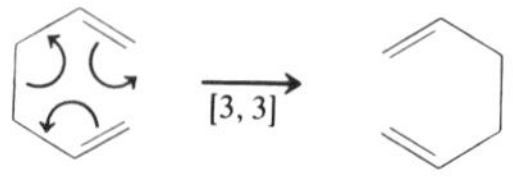

118

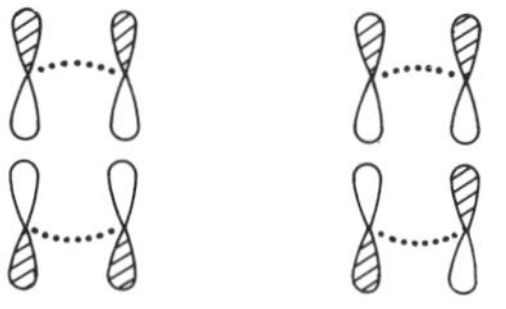

allowed   forbidden

119

one notionally considers the molecule to split and then recombine, i and j are the numbers of atoms in the two conjugated fragments. Some examples of sigmatropic reactions are listed in Table 14.13.

The theory of sigmatropic reactions is based on the assumption that a bonding type of overlap must exist at both the bond-breaking and bond-making termini simultaneously in the cyclic transition state, **119**. The

Table 14.13   Some sigmatropic rearrangements

Reaction		Order
Cope rearrangement	**130**	[3, 3]
Ylide rearrangement	**133**	[2, 3]
Claisen rearrangement	**132**	[3, 3]
Wagner–Meerwein rearrangement		[1, 2]
Pyrolytic *cis*-eliminations		[3, 3]
Wittig rearrangement	**134**	[2, 3]
1,5-Hydrogen shift		[1, 5]
Oxy-Cope rearrangement	**131**	[3, 3]

possibilities may first be examined for hydrogen migrations across π-systems of varying length, i.e. $[1, j]$ sigmatropic rearrangements. The rearrangement of propene requires a hydrogen atom $1s$ orbital to interact simultaneously with C1 and C3 of an allyl system (it makes no difference whether we consider H^+ with the allyl anion or H^- with the allyl cation, the interacting orbitals are $H(1s)$ and allyl (ψ_2)). This can only be achieved by overlap with lobes on opposite sides of the nodal plane, **120**. The allowed rearrangement would then require the hydrogen to migrate to the opposite side of the allyl system, an antarafacial migration. This happens to be sterically difficult and so is not normally observed. Compare this with the situation in penta-1,3-diene, **121**. Bonding overlap is achieved on the same side of the pentadienyl moiety and the migration can proceed by the sterically facile suprafacial mode. This kind of reaction is very frequently observed.

It will be appreciated that the 1,5-hydrogen shift is a six-electron process, a $4n + 2$ number, while the 1,3-shift involves four moving electrons $(4n)$.[126] In general, $(4n + 2)$ sigmatropic rearrangements will be able to partake of the sterically favored suprafacial geometry and will be the observed reactions. Alkyl migrations are also $[1, j]$ sigmatropic shifts and the same restrictions apply. In addition, the migration geometry of the alkyl group must be considered. The [1,3] shift can now be suprafacial if the migrating group does so antarafacially, i.e. with inversion of configuration, **122**. The allowed suprafacial [1, 5] shift will require the alkyl group to migrate suprafacially — with retention of configuration, **123**. These predictions are borne out experimentally. The following selection rule covers all sigmatropic reactions of order [i,j] in the ground state[1]:

	$i+j$ Charge state:		
+	neutral	−	*Permitted geometry*
$4n + 1$	$4n$	$4n - 1$	s, a or a, s
$4n + 1$	$4n + 2$	$4n + 1$	s, s or a, a

where s, a indicates suprafacial and antarafacial geometry on the i, j components. The following examples will illustrate the power of the orbital symmetry requirement.

The Wagner–Meerwein rearrangement, order [1, 2], occurs in carbocations because of the allowed s, s pathway, **124**, but not in carbanions which would require s, a. The migrating group retains its chirality[127]. By contrast, the [1, 3] shift, **125**, is accompanied by inversion[128].

The hydrogens in the five-membered ring of indene become scrambled on heating by a series of [1, 5] migrations, **126**[129].

124

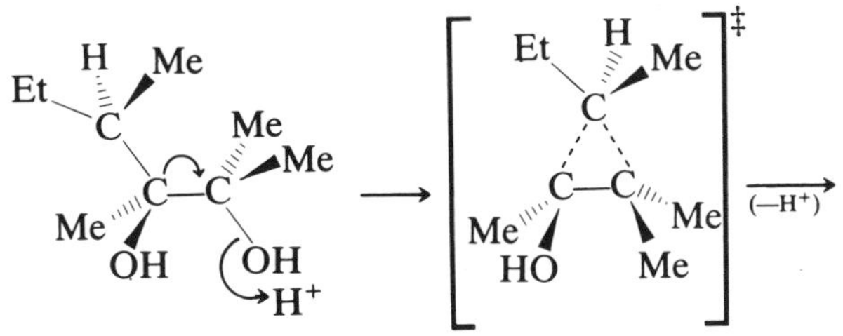

125

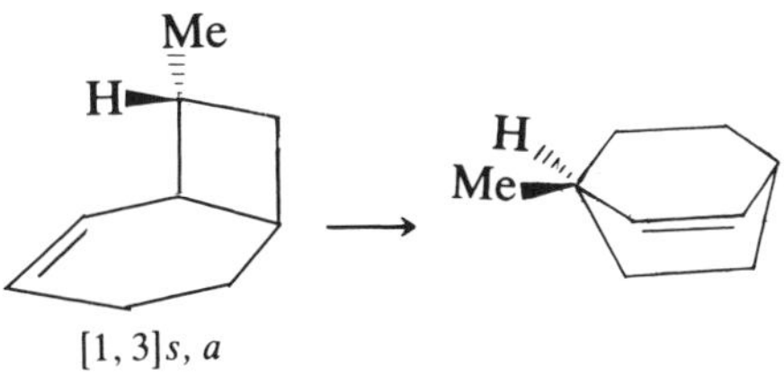

$[1, 3]s, a$

126

$200°C$

—D   etc.

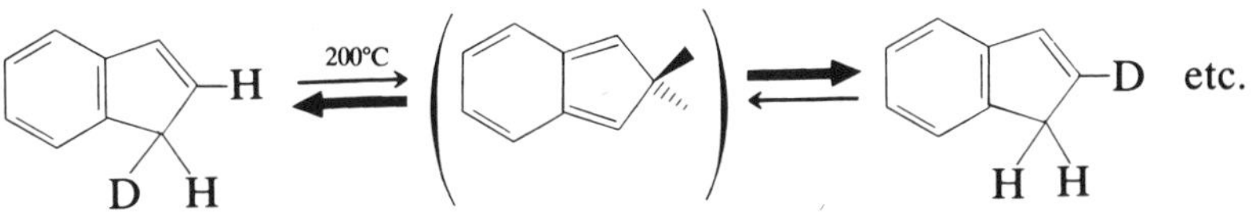

Calciferol (vitamin D), **127**, thermally isomerizes by a [1, 7] antarafacial shift to precalciferol, **128**, which in turn undergoes electrocyclic ring closure to pyrocalciferol, **129**[130].

[3, 3] Sigmatropic rearrangements of neutral substrates are of common occurrence as they may proceed by the *s, s* pathway. The Cope, **130**,[131,132] oxy-Cope, **131**, and Claisen, **132**, rearrangements are all of this type. No examples seem to occur of the [2, 4] mode for these reactions which should

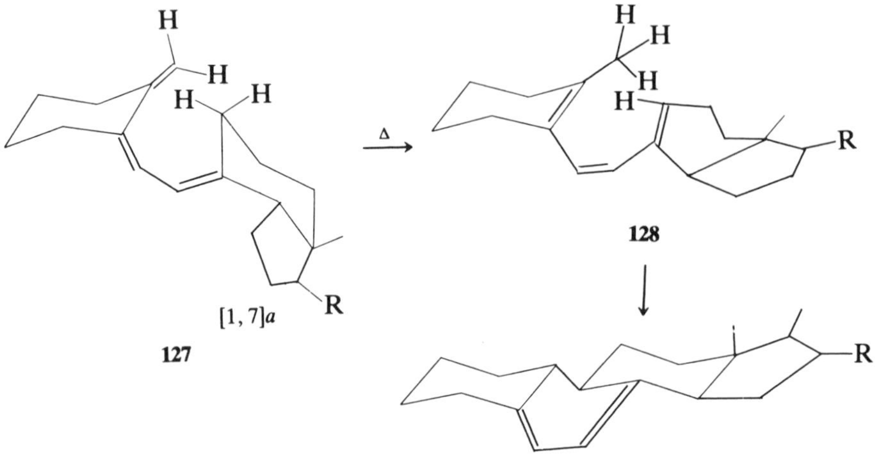

[1, 7]*a*

127

128

129

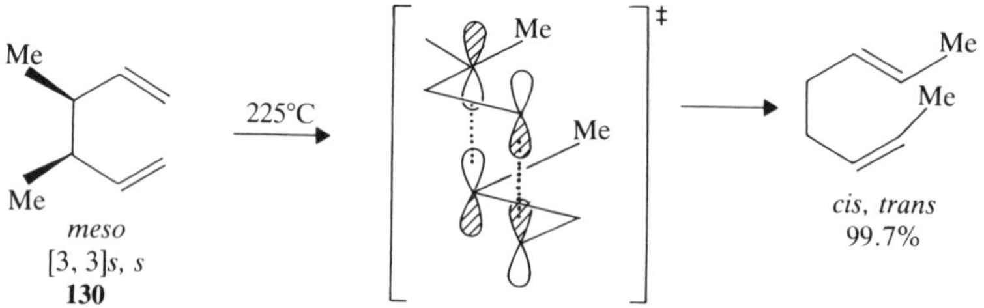

meso
[3, 3]*s, s*
130

225°C

‡

Me
Me

cis, trans
99.7%

also be permitted. [2, 3] Rearrangements of anions (ylide rearrangements) likewise are favorable [(4*n* + 1), −], **133**, and include the Wittig reaction, **134**[133, 134].

14.8.1 Concertedness in sigmatropic rearrangements

The precise timing of the two bond changes, bond formation at C1–C6 and bond breaking at C2–C3, gives rise to a situation which may be expressed diagrammatically in terms of a More O'Ferrall–Jencks energy surface, Fig. 14.10. Two extreme pathways which proceed from reagents to products go via the corners B and D representing two types of diradical intermediate; the diagonal would represent the truly concerted process. The case of hexa-1,5-diene may be examined from the viewpoint of its energetics. The sigmatropic rearrangement proceeds with an activation energy of 170 kJ mol^{-1} (41 kcal mol^{-1}) while the calculated energy of cleavage to two

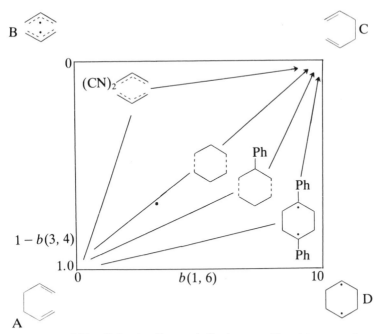

Fig. 14.10 More O'Ferrall–Jencks diagram indicating transition-state properties of some sigmatropic rearrangements; b is defined by Eq. (14.2).

allyl radicals (B) is 238 kJ mol^{-1} (57 kcal mol^{-1}) and of 1,6-bonding to the diradical (D) is 200 kJ mol^{-1} (53 kcal mol^{-1}). It follows that the reaction must proceed with some degree of concertedness. The position of the transition state with respect to the diagonal AC may be probed by secondary deuterium isotope effects[90]. The SKIE at C1, C6 is inverse (i.e. <1) due to the conversion of these centers from sp^2 to sp^3 hybridization, while that at C2, C3 is normal (>1) for the converse reason. Assuming that the degree of bond change at center (n, m) is measured by the ratio $b(n, m)$, where

$$b(n, m) = \frac{\ln k_H/k_D - 1}{\ln K_H/K_D - 1},$$ (14.2)

i.e. the ratio of the SKIE to the equilibrium isotope effect, then bonding changes at centers (n, m) can be ascertained and a plot of $b(1, 6)$ against $1 - b(2, 3)$ $(1 - b$ because the SKIE is inverse) gives a quantitative energy-surface diagram. It is found that values of $b(1, 6)/b(2, 3)$ are around 1·8 for the parent compound and alkyl derivatives, pointing to a fairly concerted transition state. This ratio rises to around 8 for 2,5-diphenyl substitution, indicating a transition state much more resembling D, with bond making far ahead of bond breaking. On the other hand, the value 0·3 for 3,3-dicyano substitution points to closer resemblance with B, in which bond breaking is ahead of bond making (Table 14.14, Fig. 14.10).

Table 14.14 Secondary kinetic isotope effects in pericyclic reactions

(a) Diels–Alder reactions

Reaction	k_H/k_D per D
Maleic anhydride-d$_2$ + butadiene	0·99
Maleic anhydride-d$_2$ + cyclopentadiene	0·97
Maleic anhydride-d$_2$ + anthracene	0·95
Maleic anhydride + anthracene-9,10-d$_2$	0·94
Maleic anhydride + butadiene-1,1,4,4-d$_4$	0·93
Cyanoethene + anthracene-9-d	1·07 (4,10:1,9 product)

(b) Retro-diene reactions

Reaction	k_H/k_D per D
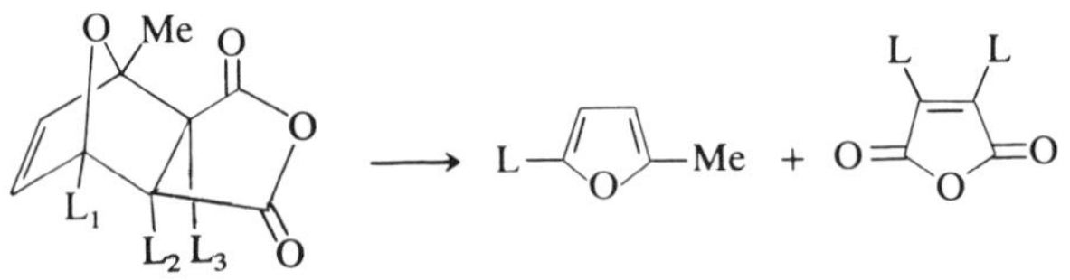 (structure)	(2,3-d$_2$) 1·16 (1-d) 1·08 (Me-d$_3$) 1·03

(c) 2 + 2 Cycloadditions

Reaction	k_H/k_D per D
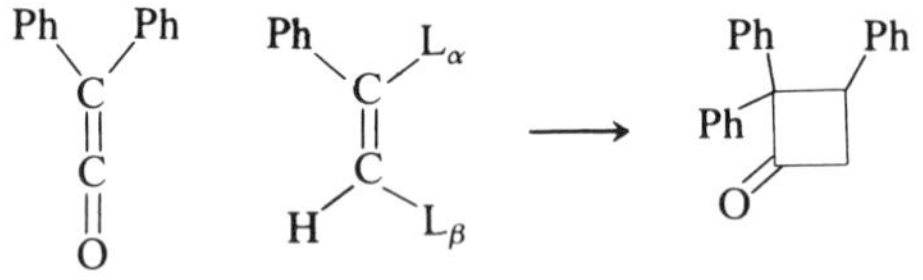 (structure)	(α-d) 1·23 (β-d) 0·91

	k_H/k_D	
	Intra-molecular[a]	Inter-molecular[b]
CH$_2$=C=CH$_2$ 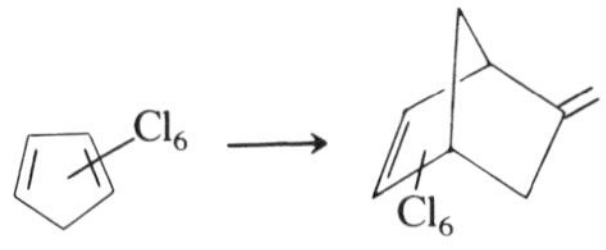 + (structure) allene → (structure)	0·90	0·90
CH$_2$=C=CH$_2$ 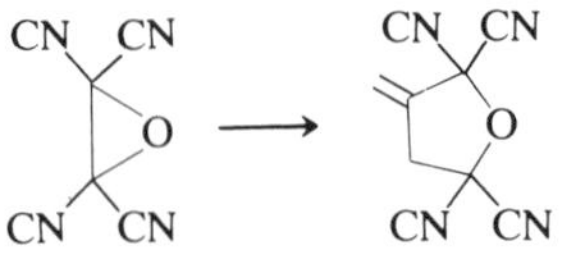 + (structure) → (structure)	0·93	

[a] CD_2=C=CH$_2$ [b] CH_2=C=CH$_2$; CD_2=C=CD$_2$

Table 14.14 (*Continued*)

(c) 2 + 2 Cycloadditions

	k_H/k_D	
	Intra-molecular[a]	Inter-molecular[b]
$CH_2=C=CH_2$ 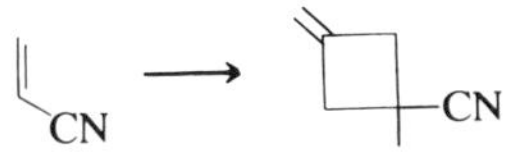 + (acrylonitrile, CN) → (cyclobutane-CN)	1·13	1·04
$CH_2=C=CH_2$  + (allene) → (dimethylenecyclobutane)	1·14	1·02

(d) Sigmatropic reactions

Reaction	Definition of SKIE	k_H/k_D Forward	k_H/k_D Reverse	b (Eq. (14.2))
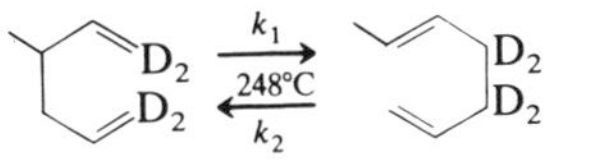 (1,5-H shift, D_2/D_2, k_1/k_2, 248°C)	$k_1^{D_4}/k_1^{H}$ $k_2^{H}/k_2^{D_4}$	1·129	1·07	1·85
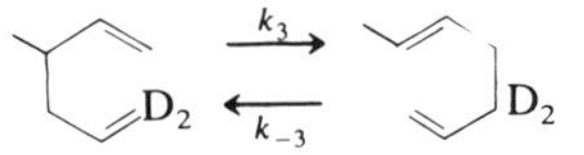 (k_3/k_{-3}, D_2)	$k_3^{D_2}/k_3^{H}$ $k_{-3}^{H}/k_{-3}^{D_2}$	1·052	1·09	1·71
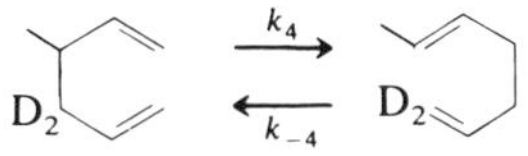 (k_4/k_{-4}, D_2)	$k_4^{H}/k_4^{D_2}$ $k_{-4}^{D_2}/k_{-4}^{H}$	1·047	1·03	
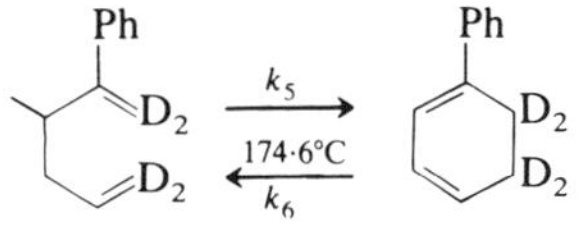 (Ph, D_2/D_2, k_5/k_6, 174·6°C)	$k_5^{D_4}/k_5^{H}$ $k_6^{H}/k_6^{D_4}$	1·295	1·09	3·3
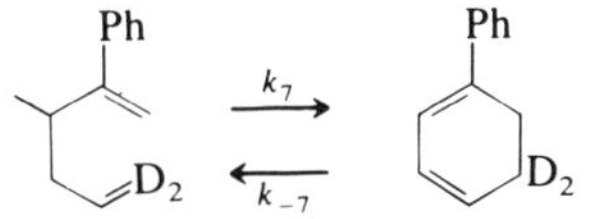 (Ph, D_2, k_7/k_{-7})	$k_7^{D_2}/k_7^{H}$ $k_{-7}^{H}/k_{-7}^{D_2}$	1·105	0·992	2·7
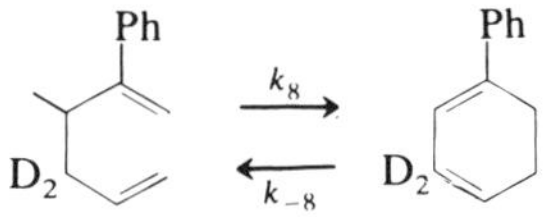 (Ph, D_2, k_8/k_{-8})	$k_8^{H}/k_8^{D_2}$ $k_{-8}^{D_2}/k_{-8}^{H}$	1·11	1·039	

703

Table 14.14 *(Continued)*

(d) Sigmatropic reactions

Reaction	Definition of SKIE	k_H/k_D Forward	k_H/k_D Reverse	b (Eq. (14.2))
(structure) $\xrightarrow[k_{10}]{k_9}$ (structure) at 55.5°C	$k_9^{D_4}/k_9^H$	1·57	1·07	
	$k_{10}^H/k_{10}^{D_4}$			8·1
(CN)$_2$ (structure) $\longrightarrow$ (CN)$_2$ (structure)	$k_{11}^{4\text{-}D_2}/k_{11}^H$	1·06	1·19	
	$k_{11}^H/k_{11}^{6\text{-}D_2}$			0·31

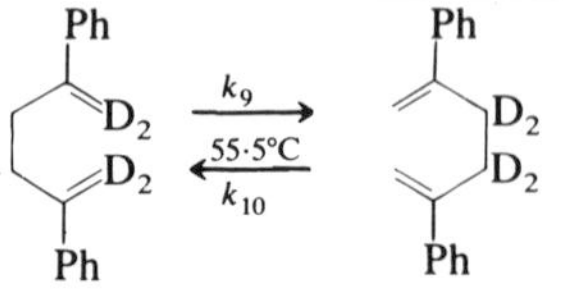

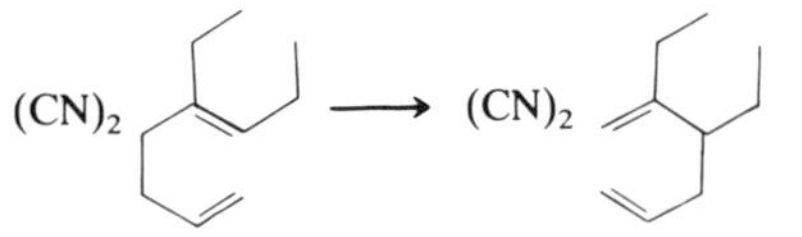

(e) Hydrogen shifts

Reaction	PKIE	SKIE
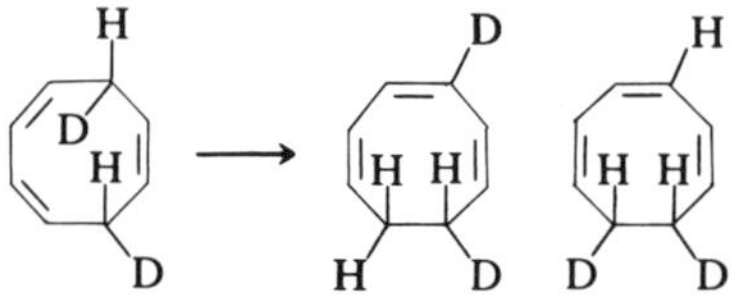 (structure) $\longrightarrow$ (structures)	5·0	1·02
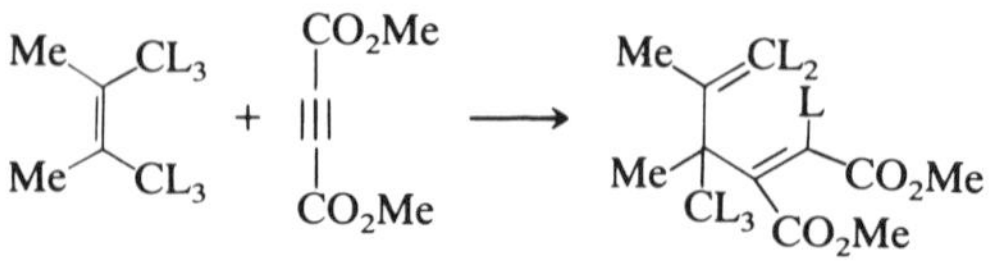 Me CL$_3$ / Me CL$_3$ + CO$_2$Me / CO$_2$Me $\longrightarrow$ (structure)	2·33	
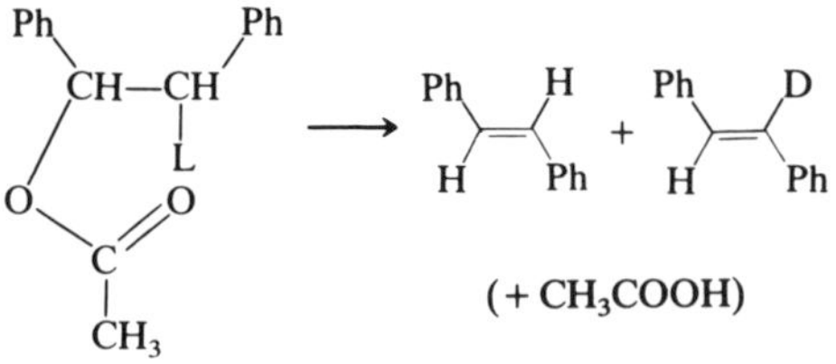 (structure, Ph CH—CH Ph) $\longrightarrow$ (structures) (+ CH$_3$COOH)	2·8	

14.9 ACID CATALYSIS OF THE DIELS–ALDER REACTION

Since concerted cycloaddition is a non-polar reaction little influenced by solvent, it is perhaps surprising that catalysis by Lewis acids takes place. Species such as aluminum chloride, diethylaluminum chloride, zinc and copper salts can markedly affect rates, yields and regiospecificity of

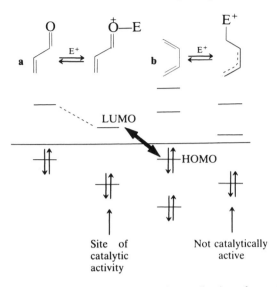

Fig. 14.11 Changes in frontier orbital energies with coordination of an acid catalyst of (**a**) dienophile and (**b**) diene.

addition. The explanation of this effect lies in changes brought about in the electronic properties of the reagents when coordinated to the electrophile[135]. As models for this, the diene and dienophile may be compared with their protonated counterparts, **135** (Fig. 14.11). Protonated diene is an allyl cation but the dienophile is coordinated to an electrophile at the carbonyl substituent. In either case, energies of both HOMO and LUMO are reduced. This would result in a favorable change in interaction

Catalyst	Product	
	exo	endo
None	12%	88%
AlCl$_3$	1%	99%

energy between the HOMO (diene) and LUMO (dienophile/Lewis acid) pair of orbitals; this is the dominant interaction for the uncatalyzed reaction. Furthermore the LUMO becomes more polarized which increases the regioselectivity of the cycloaddition. The unfavorable coordinated diene with lowered HOMO would not take part in reaction but would only constitute a small fraction of the total diene at equilibrium. Examples of the effectiveness of electrophilic catalysis are shown in **135, 136**[136-138].

PROBLEMS

1 Ketenes such as $Me_2C=C=O$ will add to alkenes stereospecifically:

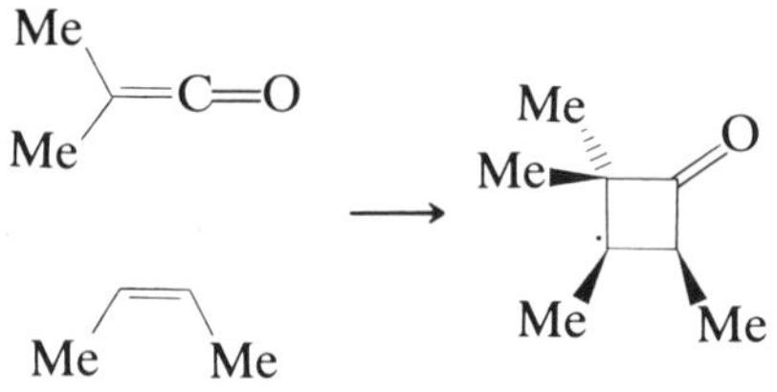

The HOMO and LUMO have geometries as shown below.

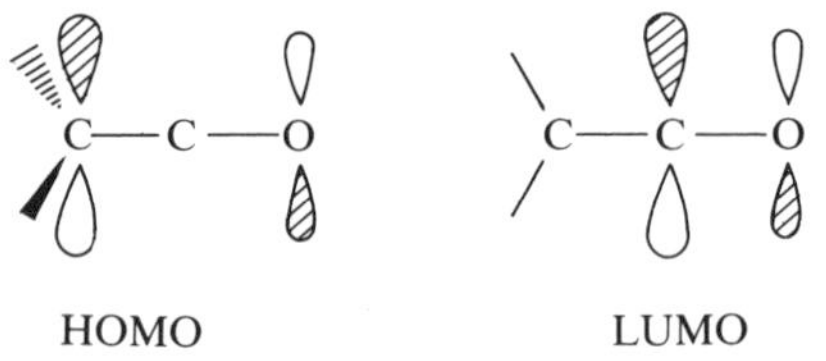

HOMO   LUMO

Is a concerted cycloaddition feasible for this system and, if so, what approach geometry would be favored?

2 Predict the regiospecificity for cycloadditions of the following reactants:

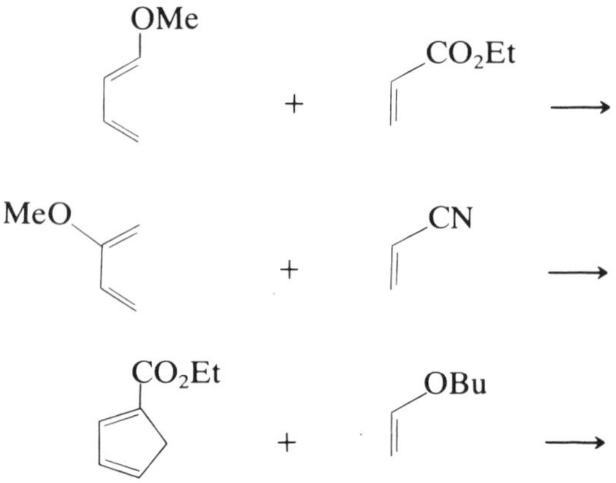

3 Draw up correlation diagrams for cycloadditions between

(a) allyl cation and anion with ethene;
(b) allyl cation and anion with butadiene.

Interpret these reactions in terms of the generalized Woodward–Hoffmann Rule.

4 Suggest mechanisms for the following reactions:

(a)

[*J. Amer. Chem. Soc.*, **91**, 5668 (1969).]

(b)

[*J. Amer. Chem. Soc.*, **91**, 777 (1969).]

(c)

[*Org. Reactions*, **12**, 1 (1962).]

(d)

(Paterno–Buchi Reaction)

[*Chem. Rev.* **66**, 373 (1966).]

(e) $PhCH{=}CH_2 + Cl_2C{=}CF_2 \xrightarrow{\Delta}$

[*Science*, **159**, 833 (1968).]

Pericyclic reactions

(f)

$\xrightarrow{\text{H}^+}$

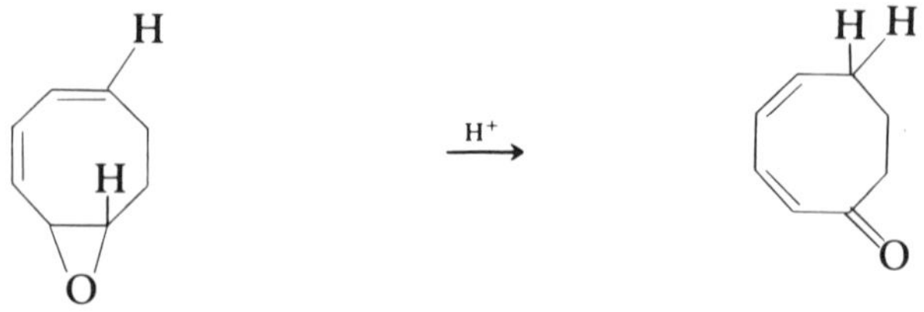

(g)

$\longrightarrow$

[*J. Amer. Chem. Soc.*, **90**, 7146 (1968).]

(h)

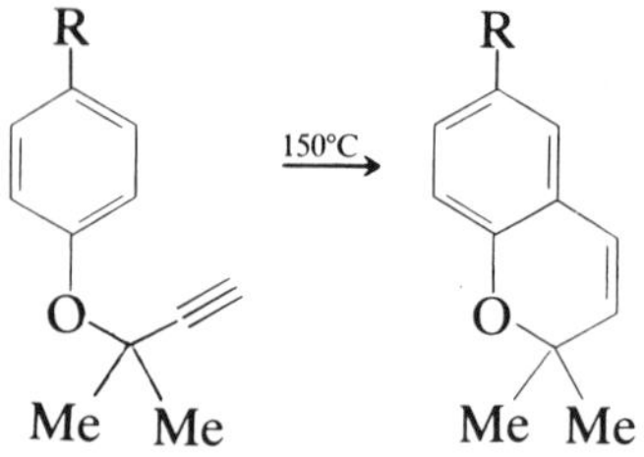

$\xrightarrow{150°C}$

[*Gazzetta*, **101**, 833 (1971); *Chem. Comm.*, 896 (1972); *J. Org. Chem.*, **37**, 2858 (1972)].

(i)

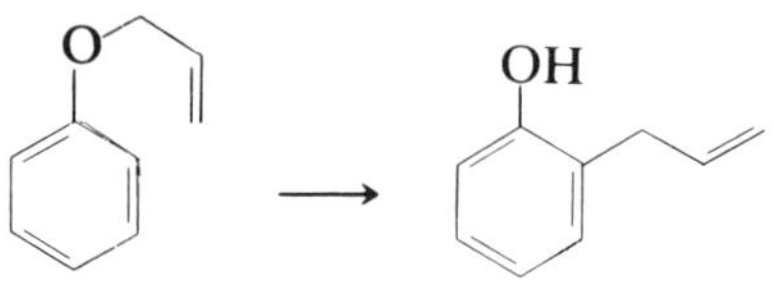

$\longrightarrow$

The reaction proceeds 10^5 times faster in CF_3COOH than in an inert solvent. [*Chem. Comm.*, 645 (1972).]

(j)

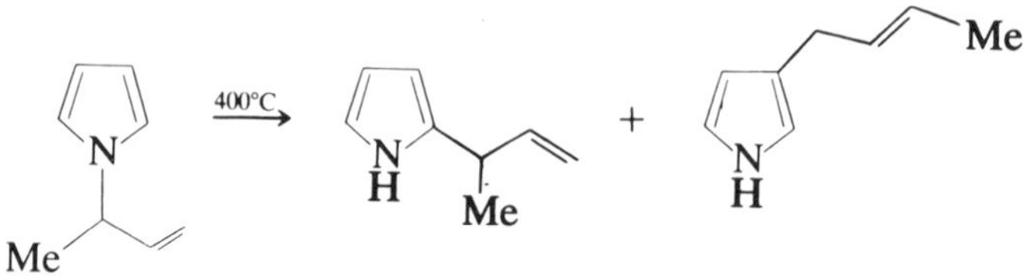

$\xrightarrow{400°C}$

$+$

[*J. Amer. Chem. Soc.*, **94**, 2487 (1972).]

(k)

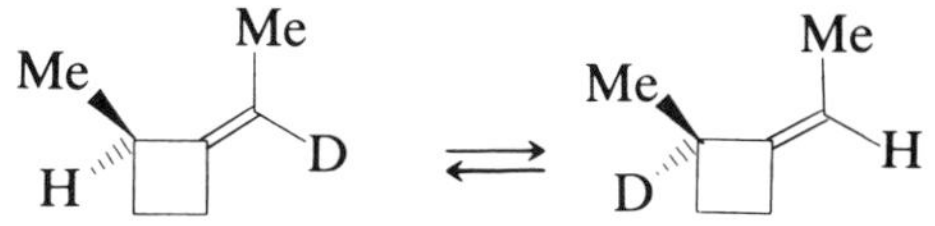

[*J. Amer. Chem. Soc.*, **94,** 2140 (1972).]

(l)

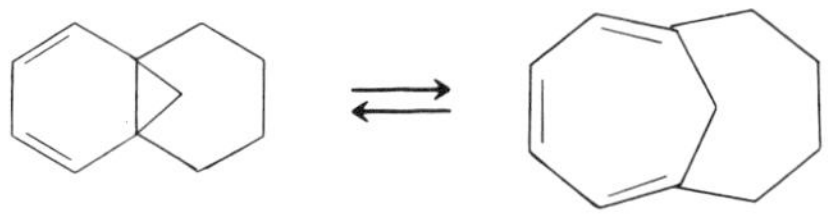

[*Annalen,* **759,** 1 (1972).]

(m)

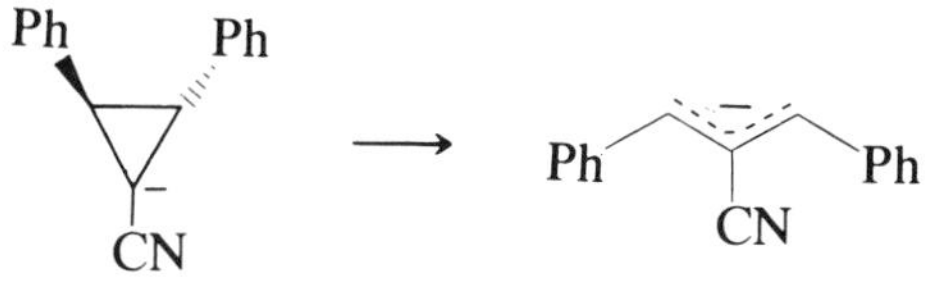

[*Angew. Chem. Int. Edn.*, **11,** 724 (1972).]

(n)

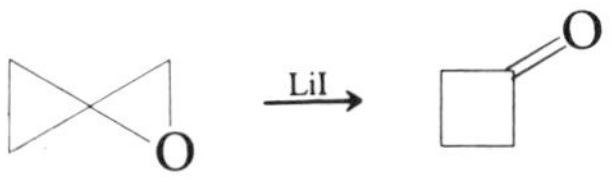

[*Tetrahedron,* **30,** 1413 (1974).]

(o)

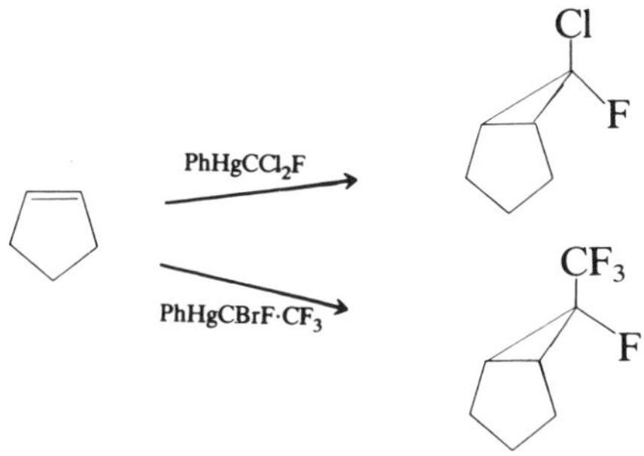

[*J. Organometall. Chem.*, **49,** 117 (1973); **52,** C1 (1973).]

(p)

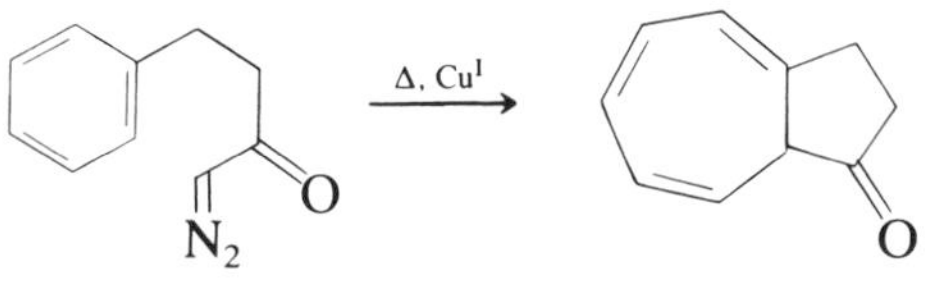

[*Chem. Comm.*, 882 (1973).]

Pericyclic reactions

(q)

[*J. Amer. Chem. Soc.*, **98**, 1875 (1976).]

(r)

(label randomized)

[*Tetrahedron*, **20**, 2697 (1964).]

REFERENCES

1 R. B. Woodward and R. Hoffmann, *The Conservation of Orbital Symmetry*, Verlag Chemie/Academic Press, 1970; *Angew. Chem. Int. Edn.*, **8**, 781 (1969).
2 A. P. Marchand and R. E. Lehr, *Pericyclic Reactions*, Academic Press, New York, 1977.
3 T. Gilchrist and R. C. Storr, *Organic Reactions and Orbital Symmetry*, Cambridge U.P. 1979.
4 R. B. Woodward and R. Hoffmann, *J. Amer. Chem. Soc.*, **87**, 395 (1965).
5 R. B. Woodward and R. Hoffmann, *J. Amer. Chem. Soc.*, **87**, 2046 (1965).
6 R. B. Woodward and R. Hoffmann, *J. Amer. Chem. Soc.*, **87**, 2511 (1965).
7 R. Hoffmann and R. B. Woodward, *Acc. Chem. Res.*, **1**, 17 (1968).
8 A. Wasserman, *Diels–Alder Reactions*, Elsevier, Amsterdam, 1965.
9 H. C. Longuet-Higgins and E. W. Abrahamson, *J. Amer. Chem. Soc.*, **87**, 2045 (1965).
10 K. Fukui, *Theory of Orientation and Stereoselection*, Springer Verlag, Berlin, 1975.
11 I. Fleming, *Frontier Orbitals and Organic Chemical Reactions*, John Wiley, London, 1976.
12 H. E. Zimmerman, *Acc. Chem. Res.*, **4**, 272 (1971).
13 C. L. Perrin, *Chem. Brit.*, **8**, 163, 1972.
14 M. J. S. Dewar and R. C. Dougherty, *The PMO Theory of Organic Chemistry*, Plenum, New York, 1975.
15 J. G. Calvin and J. N. Pitts, *Photochemistry*, Wiley, New York, 1966; D. C. Neckers, *Mechanisms of Organic Photochemistry*, Reinhold, New York, 1967.
16 O. Diels and K. Alder, *Ann.*, **460**, 98 (1928).
17 J. G. Martin and R. K. Hill, *Chem. Rev.*, **61**, 537 (1961).
18 J. Sauer, *Angew. Chem. Int. Edn.*, **5**, 211 (1966); **6**, 16 (1967).
19 J. Sauer and R. Sustmann, *Angew. Chem. Int. Edn.*, **19**, 779 (1980).

20 H. M. R. Hoffmann, *Angew. Chem. Int. Edn.*, **8**, 556 (1969).

21 R. Huisgen, R. Greshey and J. Sauer, *Chemistry of Alkenes*, Interscience, London, 1964.

22 C. A. Stewart, *J. Org. Chem.*, **28**, 3320 (1963).

23 R. Gleiter and M. C. Bohm. *Pure Appl. Chem.*, **55**, 237 (1983); R. Gleiter and L. A. Paquette, *Acc. Chem. Res.*, **16**, 328 (1983).

24 M. J. Goldstein and G. L. Thayer, *J. Amer. Chem. Soc.*, **87**, 1925, 1933 (1965).

25 P. Chao and D. M. Lemal, *J. Amer. Chem. Soc.*, **95**, 920 (1973).

26 N. S. Isaacs and A. V. George, *J. Chem. Soc.*, *Perkin I*, 1985, 1277.

27 W. C. Herndon and L. H. Hall, *Tetrahedron Letts*, **32**, 3095 (1967).

28 N. Rieber, J. Alberts, J. A. Lipsky and D. M. Lemal, *J. Amer. Chem. Soc.*, **91**, 5668 (1969).

29 H.-D. Martin, T. Urbanek, R. Brown and R. Walsh, *Int. J. Chem. Kinet.*, **16**, 117 (1984).

30 K. Alder, F. H. Block and H. Beumling, *Ber.* **93**, 1896 (1960).

31 E. Vogel, W. Grimme and S. Korte, *Tetrahedron Letts*, **30**, 3625. (1965).

32 H. Kwart and K. King, *Chem. Rev.*, **68**, 415 (1968).

33 J. L. Ripoll, A. Rouessac and F. Rouessac, *Tetrahedron*, **34**, 19 (1978).

34 R. B. Woodward and T. J. Katz, *Tetrahedron*, **5**, 70 (1959).

35 N. S. Isaacs, *Liquid Phase High Pressure Chemistry*, Wiley, Chichester, 1981.

36 P. Brown and R. C. Cookson, *Tetrahedron*, **21**, 1977, 1993 (1965).

37 W. R. Dolbier and S.-H. Dai, *J. Amer. Chem. Soc.*, **90**, 5028 (1968).

38 D. E. Van Sickel and J. O. Rodin, *J. Amer. Chem. Soc.*, **86**, 3091 (1964).

39 G. Brieger and J. Bennett, *Chem. Rev.*, **80**, 63 (1980).

40 N. S. Isaacs and P. G. Van der Beeke, *J. Chem. Soc.*, *Perkin II*, 1205 (1982).

41 H. M. R. Hoffmann, *Angew. Chem. Int. Edn.*, **8**, 556 (1969).

42 R. T. Arnold, R. W. Amidon and R. M. Dodson, *J. Amer. Chem. Soc.*, **72**, 2871 (1950).

43 R. T. Arnold and J. S. Showell, *J. Amer. Chem. Soc.*, **79**, 419 (1957).

44 J. P. Bain, *J. Amer. Chem. Soc.*, **60**, 638 (1946).

45 A. Nickon, *J. Amer. Chem. Soc.*, **77**, 1190 (1955).

46 R. Breslow and P. Dowd, *J. Amer. Chem. Soc.*, **85**, 2729 (1963).

47 P. Dowd and A. Gold, *Tetrahedron Letts.*, 85 (1969).

48 B. Franzus, *J. Org. Chem.*, **28**, 2954 (1963).

49 G. O. Schlenck, K. Gollnick, G. Buchwolds, S. Schroeter and G. Ohloff, *Ann.*, **674**, 93 (1964).

50 W. D. Huntsman, V. C. Salomon and D. Eros, *J. Amer. Chem. Soc.*, **80**, 5455 (1958).

51 A. T. Blomquist and P. R. Towssiq, *J. Amer. Chem. Soc.*, **79**, 3505 (1951).

52 W. R. Roth, *Chimia*, **20**, 229 (1966).

53 A. T. Blades and G. W. Murphy, *J. Amer. Chem. Soc.*, **74**, 1039 (1952); *Can. J. Chem.*, **31**, 418 (1953).

54 K. Wiberg, *Ber.*, **107**, 1456 (1974).

55 S. Hunig, H. R. Miller and W. Thier, *Angew. Chim. Int. Edn*, **4**, 271 (1965).

56 C. E. Miller, *J. Chem. Ed.*, **42**, 254 (1965).

57 H. O. House, *Modern Synthetic Reactions*, Academic Press, New York, 1973, p. 248.

58 E. E. Van Tamelen and R. J. Timmons, *J. Amer. Chem. Soc.*, **84**, 1067 (1962).

59 E. J. Corey, D. J. Pasto and W. Mock, *J. Amer. Chem. Soc.*, **83**, 2957 (1961).

60 H. Staudinger, *Helv. Chem. Acta*, **8**, 306 (1925).

61 R. Huisgen and L. A. Feiler, *Ber.*, **102**, 3391 (1969).

62 J. S. Grossert, *Chem. Comm.*, 305 (1970).

63 H. C. Stevens, D. A. Reich, D. R. Brandt, K. R. Fountain and E. J. Gaughan, *J. Amer. Chem. Soc.*, **87**, 5257 (1965).

64 W. T. Brady and J. P. Hible, *Tetrahedron Letts*, 3205 (1970).

65 H-D. Scharf and J. Fleischauer, *Tetrahedron Letts*, 5867 (1968).

66 I. Marko, B. Ronsmans, A.-M. Hesbain-Frisque, S. Dumas and L. Ghosez, *J. Amer. Chem. Soc.*, **107**, 2192 (1985).

67 R. Huisgen, L. A. Feiler and P. Otto, *Ber.*, **102**, 3405 3444 (1969).

68 R. Huisgen, L. A. Feiler and G. Busch, *Ber.*, **102**, 3460 (1969).

69 R. Huisgen and L. A. Feiler, *Ber.* **102**, 3391, 3428 (1969).

70 R. Huisgen and P. Otto, *Ber.*, **102**, 3475 (1969).

71 R. Huisgen and P. Otto, *J. Amer. Chem. Soc.*, **90**, 5342 (1968).

72 N. S. Isaacs and E. Rannala, *J. Chem. Soc., Perkin II*, 1555 (1975).

73 T. Katz and R. Desson, *J. Amer. Chem. Soc.*, **85**, 2172 (1963).

74 J. E. Baldwin and J. A. Kapecki, *J. Amer. Chem. Soc.*, **92**, 4868, 4874 (1970).

75 N. S. Isaacs and P. Stanbury, *J. Chem. Soc., Perkin II*, 168 (1973).

76 R. Huisgen, L. A. Feiler and P. Otto, *Tetrahedron Letts*, 4485 (1968).

77 W. T. Brady, F. H. Parry, R. Roe and E. F. Hoff., *Tetrahedron Letts*, 819 (1970).

78 R. Huisgen, B. A. Davies and M. Morikawa, *Angew. Chem. Int. Edn.*, **7**, 862 (1968).

79 H. B. Kagan and J. C. Luche, *Tetrahedron Letts*, 3093 (1968).

80 R. Graf, *Ann.*, **661**, 111 (1963); H. Bestian, H. Biener, K. Clauss and H. Heya, *Ann.*, **718**, 94 (1969).

81 H. Suschitzky, R. E. Waldrond and R. Hull, *J. Chem. Soc., Perkin I*, 47 (1977).

82 W. T. Brady and L. Smith, *Tetrahedron Letts*, 2963 (1970).

83 W. R. Dolbier and S.-H. Dai, *J. Amer. Chem. Soc.*, **90**, 5028 (1968).

84 H. N. Cripps, J. K. Williams and W. H. Sharkey, *J. Amer. Chem. Soc.*, **81**, 2723 (1959).

85 T. L. Jacobs, J. R. McClenon and O. J. Muscio, *J. Amer. Chem. Soc.*, **91**, 6038 (1969).

86 J. E. Baldwin and U. V. Roy, *Chem. Comm.*, 1225 (1969).

87 J. J. Gajewski and W. A. Blackman, *Tetrahedron Letts*, 899 (1970).

88 O. J. Muscio and T. L. Jacobs, *Tetrahedron Letts*, 2867 (1969).

89 W. R. Dolbier and S.-H. Dai, *J. Amer. Chem. Soc.*, **92**, 1776 (1970).

90 J. J. Gajewski and N. D. Conrad, *J. Amer. Chem. Soc.*, **101**, 6693 (1979); J. J. Gajewski, *Acc. Chem. Res.*, **13**, 142 (1980).

91 J. D. Roberts and C. M. Sharts, *Org. React.*, **12**, 1 (1962).

92 P. D. Bartlett, *Science*, 833 (1968).

93 P. D. Bartlett, *Quart. Rev.*, **24**, 473 (1970).

94 P. D. Bartlett, G. M. Cohen, S. P. Elliott, K. Hummel, R. A. Minns, C. M. Sharts and J. Y. Fukunaga, *J. Amer. Chem. Soc.*, **94**, 2899 (1972).

95 W. R. Moore, R. D. Bach and T. M. Ozretich, *J. Amer. Chem. Soc.*, **91**, 5918 (1969).

96 J. K. Williams, D. W. Wiley and B. C. McKusick, *J. Amer. Chem. Soc.*, **84**, 2210, 2216 (1962).

97 R. Huisgen and G. Steiner, *Tetrahedron Letts*, 3763 (1973).

98 E. Koerner von Gustorf, D. V. White, J. Listich and D. Henneberg, *Tetrahedron Letts*, 3113 (1969).

99 J. R. Dombroski, M. L. Hallensteben and W. Regel, *Tetrahedron Letts*, 3881 (1971).

100 R. Huisgen and G. Steiner, *J. Amer. Chem. Soc.*, **93**, 5654 (1973).

101 G. Steiner and R. Huisgen, *J. Amer. Chem. Soc.*, **95**, 5056 (1973).

102 F. K. Fleischmann and H. Kelm, *Tetrahedron Letts*, **39**, 3773 (1973).

103 J. v. Jouanne, D. A. Palmer and H. Kelm, *Bull. Chem. Soc. Japan*, **51**, 463 (1978).

104 N. Epiotis, *Angew. Chem. Int. Edn*, **13**, 751 (1974); *J. Amer. Chem. Soc.*, **94**, 1824 (1972).

105 E. Schwann and R. Ketchum, *Angew. Chem. Int. Edn.*, **21**, 225 (1982).

106 H. M. Frey and R. Walsh, *Chem. Rev.*, **69**, 103 (1969); T. Imai and S. Nishida, *J. Org. Chem.*, **45**, 2354 (1980).

107 N. S. Isaacs and A. H. Laila, *Tetrahedron Letts*, **24**, 2897 (1983).

108 F. McCapra, *Prog. Org. Chem.*, **8**, 231 (1973); *Essays in Chem.*, **3**, 101 (1972); M. A. DeLuca, Ed., *Methods of Enzymology*, Vol. 57, Academic Press, New York, 1978; T. Carrington and J. C. Polanyi, *MTP Int. Rev. Sci. Phys. Chem.*, Vol. 9, Butterworth, London, 1972, p. 135.

109 R. Huisgen, *Bull. Soc. Chim. France*, 3431 (1965); R. Huisgen and W. Mack, *Tetrahedron Letts*, 583 (1961); *Acc. Chem. Res.*, **10**, 117 (1977).

110 R. Huisgen, *Angew. Chem. Int. Edn.*, **2,** 633 (1963).
111 R. Huisgen, *Proc. Chem. Soc.*, 357 (1961); R. Criegee, D. Siebach, R. E. Winter, B. Borretzen and H. A. Brune, *Ber.*, **98,** 2339 (1965); R. Huisgen, *Angew. Chem. Int. Edn.*, **16,** 572 (1977).
112 K. N. Houk, J. Sims, R. E. Duke, R. W. Strozier and J. K. George, *J. Amer. Chem. Soc.*, **95,** 7287 (1973).
113 K. N. Houk, J. Sims, C. R. Watts and L. J. Luskins, *J. Amer. Chem. Soc.*, **95,** 7301 (1973).
114 R. Huisgen, A. Donen, and H. Huber, *J. Amer. Chem. Soc.*, **89,** 7130 (1967).
115 E. N. Marvell, G. Capel and B. Schatz, *Tetrahedron Letts,* 385 (1965).
116 D. S. Glass, J. W. H. Watthey and S. Winstein, *Tetrahedron Letts,* 377 (1965).
117 K. Alder, M. Schumacher and O. Wolff, *Ann.*, **564,** 79 (1949).
118 E. Ciganek, *J. Amer. Chem. Soc.*, **89,** 1454 (1967).
119 G. Schroder, *Angew. Chem. Int. Edn,* **4,** 752 (1965).
120 S. D. Turk and R. L. Cobb, *1,4-Cycloadditions,* J. Homer, Ed., Academic Press, New York, 1967.
121 N. S. Isaacs and A. H. Laila, *J. Chem. Res.,* (S) 10, (M) 188 (1977).
122 N. S. Isaacs and A. H. Laila, *Tetrahedron Letts,* 715 (1976).
123 W. L. Mock, *J. Amer. Chem. Soc.,* **97,** 3666 (1975).
124 R. F. Heldeberg and H. Hogeveen, *J. Amer. Chem. Soc.,* **98,** 2341 (1976).
125 F. G. Bordwell, J. M. Williams, E. B. Hoyt and B. B. Jarvis, *J. Amer. Chem. Soc.,* **90,** 429 (1968).
126 C. W. Spangler, *Chem. Rev.,* **70,** 187 (1976).
127 J. J. Beggs and M. B. Meyers, *J. Chem. Soc (B),* 930 (1970).
128 J. A. Berson, *Acc. Chem. Res.,* **5,** 406 (1972).
129 J. A. Berson and G. B. Aspelin, *Tetrahedron,* **20,** 2697 (1964).
130 E. Havinga and J. L. M. A. Schlatmann, *Tetrahedron,* **16,** 146 (1961).
131 S. J. Rhoads and N. R. Raulins, *Org. React.,* **22,** 1 (1975).
132 G. B. Bennett, *Synthesis,* 589 (1977).
133 J. E. Baldwin, A. H. Andrist and R. K. Oinschmidt, *Acc. Chem. Res.,* **5,** 402 (1972).
134 H. E. Zimmerman, *Molecular Rearrangements,* Vol. 1, P. de Mayo, Ed., Interscience, New York, 1963, Vol. 1, p. 345.
135 F. D. Mango and J. H. Schachtschneider, *J. Amer. Chem. Soc.,* **89,** 2484 (1967).
136 E. Viera and P. Vogel, *Helv. Chem. Acta,* **65,** 1700 (1982).
137 J. A. Moore and E. M. Partain, *J. Org. Chem.,* **48,** 1105 (1983).
138 E. J. Corey, N. M. Wainshenker, T. K. Schaaf and W. Huber, *J. Amer. Chem. Soc.,* **91,** 5675 (1969).

15 Reactions via free radicals

The chemistry discussed in previous chapters concerns bond making and breaking by the rearrangement of spin-paired electrons ('polar' reactions). In contrast, a radical, e.g. **1–6** (Table 15.1), contains an odd unpaired electron which may be more or less localized on a carbon, oxygen, nitrogen or other center or extensively delocalized over a π-system. In most cases the odd electron is associated with extremely high reactivity towards spin-paired products, so frequently the radical occurs as a reactive intermediate formed as a result of homolysis or some other spin-uncoupling process. Radicals may be neutral or charged (ion-radicals) or possess two unpaired electrons (diradicals). The characteristics of their reactions are quite distinct from those of polar processes and afford an equally varied and important chemistry[1–6a].

15.1 THE GENERATION OF RADICALS

Reactions by which radicals are created from spin-paired molecules may be referred to as primary processes, those by which they are generated from preformed radicals as secondary.

15.1.1 Primary processes

Thermolysis

Homolysis of any covalent bond will occur at sufficiently elevated temperatures, the rate depending upon the bond dissociation energy (Section 1.4)[8]. The most easily cleaved bonds are homopolar bonds, –O–O– (**7**), F–F, I–I, and N–Hal bonds, C–N bonds of azo compounds[238] R—N=N—R from which dinitrogen is expelled (**8**), and carbon–metal bonds of, for instance, R₄Pb (**9**)[7]. This last reaction is of historic interest in that it was used in the first unequivocal demonstration of gaseous radicals in 1929[9]. The vapor of tetramethyllead, a volatile, covalent liquid, was passed through a tube heated to 450°C at one point (Fig. 15.1). Decomposition of the compound occurred depositing a mirror of the metal on the glass surface and producing

Table 15.1 Some representative free radicals

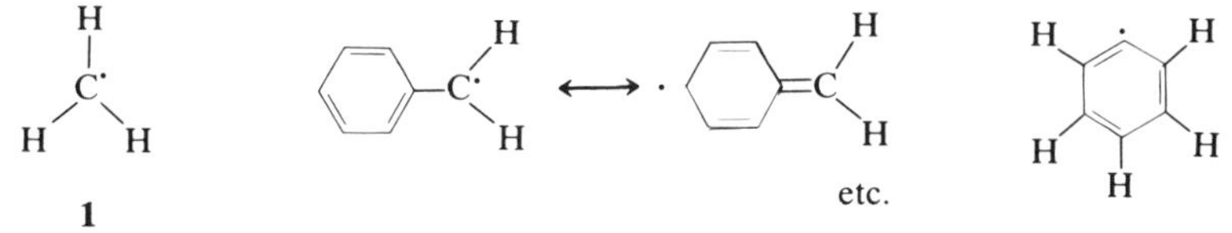

ethane. When the point of heating was moved upstream, however, while a new mirror was formed, the first disappeared and tetramethyllead appeared in the product. The inference was correctly made that pyrolysis of the organometal generated methyl radicals which either dimerized to ethane or reacted at a lower temperature with the metal, regenerating the starting material. Of the commonly used radical sources, acyl peroxides decompose more readily than alkyl peroxides, but among these two types there is little

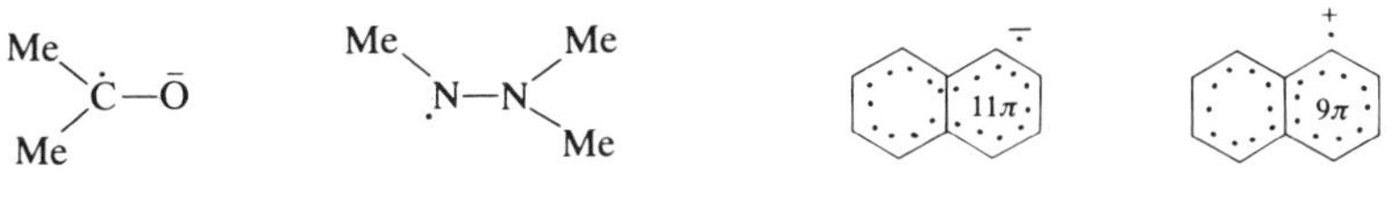

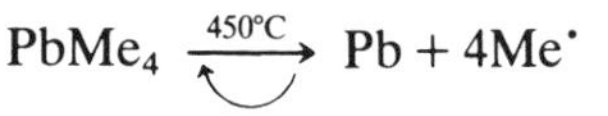

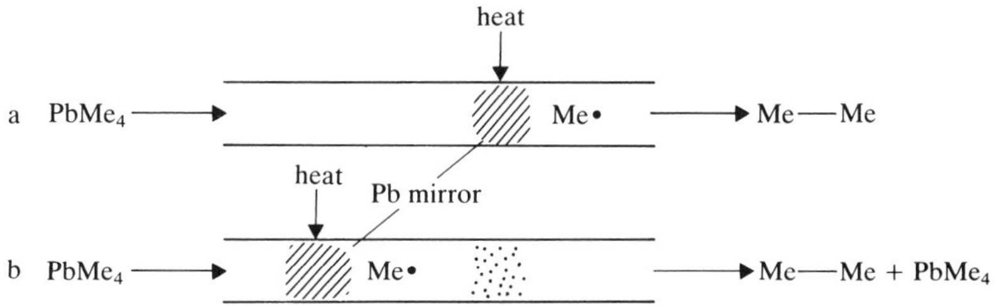

Fig. 15.1 Scheme of Paneth's experiment: (**a**), tetramethyllead is pyrolyzed in a flow apparatus leaving a metal mirror on the surface. (**b**) On applying heat upstream, a second mirror is deposited but the first then disappears with the regeneration of tetramethyllead.

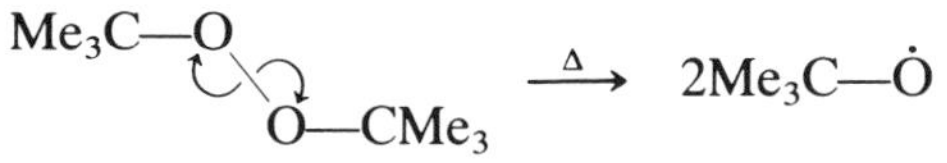

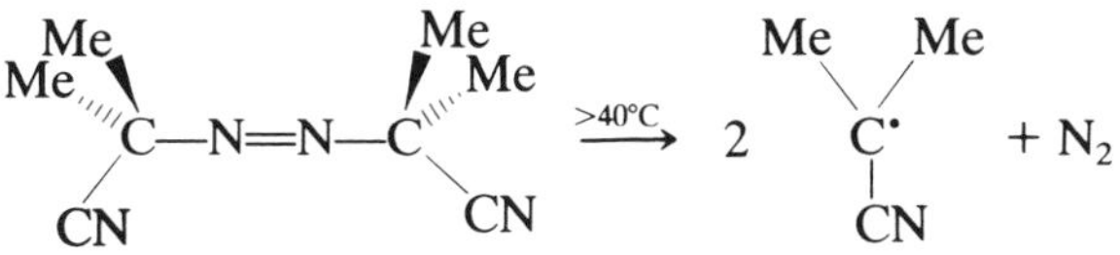

715

sensitivity to structure. Rates of decomposition of azo compounds, on the other hand, are much more sensitive to substituents (Table 15.2), which can stabilize the radicals. A notable example is azoisobutyronitrile (AIBN), **8**, which decomposes to 2-cyanopropyl radicals above 40°C and is industrially important as a source of initiator radicals in polymerizations. Entropies of

Table 15.2 Rates and activation parameters for some homolytic reactions (gas phase)[8,10]

Reaction	k_{rel}	$\Delta H^{\ddagger}/kJ\,mol^{-1}$ $(kcal\,mol^{-1})$	$\Delta S^{\ddagger}/J\,K^{-1}\,mol^{-1}$ $(cal\,K^{-1}\,mol^{-1})$
Peroxides, $RO{-}OR \rightarrow 2RO^{\bullet}$			
R = Me	1	160 (38)	+45 (11)
R = *tert*-Bu	0·3	165 (39)	+45 (11)
R = CH_3CO	1600	130 (31)	+48 (11·5)
R = PhCO	60	139 (33)	+12 (2·9)
R = *tert*-BuO—OCOPh		142 (34)	
Azo compounds[10], $R{-}N{=}N{-}R \rightarrow 2R^{\bullet} + N_2$			
R = Me	1	240 (57)	+78 (19)
R = CF_3	0·06	245 (58)	+59 (14)
R = isoPr	903	180 (43)	+11 (2·6)
R = *tert*-Bu	2500	190 (45)	+76 (18)
R = allyl	$7·7 \times 10^6$	156 (37)	+17 (4·1)
R = $PhCH_2$	10^6	155 (37)	+17 (4·1)
R = $Me_2C{\cdot}CN$	V. fast	125 (30)	
$CH_3{-}CH_3 \rightarrow 2CH_3^{\bullet}$		372 (89)	+133 (32)

	$E_A/kJ\,mol^{-1}$ $(kcal\,mol^{-1})$	
$CH_3CH_2CH_2^{\bullet} \longrightarrow CH_2{=}CH_2 + CH_3^{\bullet}$	15·3	(3·65)
$CH_3\dot{C}H_2CH_3 \longrightarrow CH_3CH{=}CH_2 + H^{\bullet}$	18·7	(4·5)
$CH_3CH_2\dot{C}HCH_3 \longrightarrow CH_3CH_2CH{=}CH_2 + H^{\bullet}$	18·3	(4·3)
$\longrightarrow CH_3CH{=}CH_2 + CH_3^{\bullet}$	14·7	(3·5)
$\longrightarrow \begin{array}{c} CH_2{-}CH_2 \\ \mid \qquad \mid \quad + H^{\bullet} \\ CH_2{-}CH_2 \end{array}$	17·5	(4·2)
$\longrightarrow CH_2{=}C(CH_3)_2 + H^{\bullet}$	17·1	(4·1)
$(CH_3)_2\dot{C}CH_3 \longrightarrow (CH_3)_2C{=}CH_2 + H^{\bullet}$	18·9	(4·5)
$(CH_3)_2\dot{C}CH_2CH_3 \longrightarrow (CH_3)_2C{=}CH_2 + CH_3^{\bullet}$	15·0	(3·6)

activation for homolyses of both peroxides and azo compounds are positive, as expected for a bond-breaking process.

Photolysis

A more specific method of endowing a molecule with sufficient energy for homolysis is by the absorption of a quantum of radiation if the structure contains a suitable chromophore. The fission of a bond requiring $400 \, kJ \, mol^{-1}$ ($100 \, kcal \, mol^{-1}$) will require light of wavelength $<300 \, nm$ which it must be able to channel into a decomposition mode. Ketones are good subjects for photolysis since they have a strong absorption band in this region of the ultraviolet spectrum associated with a $(n \rightarrow \pi^*)$ transition which leads to C–CO bond cleavage, providing a source of alkyl and acyl radicals, **10**. Radiolysis by X-rays, electron beams and γ-radiation falls also into this category. While this may seem a somewhat recondite method, it is an elegant tool for probing radical reactions in water due to OH˙, the primary radical produced in all cases and of relevance to the biochemical effects of high-energy radiation[11].

Redox methods

One-electron oxidizing and reducing agents remove and add, respectively, one electron to a substrate, thereby creating a radical **(11–15)**[12–22]. The feasibility of an electron transfer is determined by the requirement of a

14 $Ar\!-\!\overset{+}{N}\!\!\equiv\!\!N + Cu^{I} \longrightarrow Ar^{\cdot} + N_2 + Cu^{II}$

(Ref. 76)

15 $Br(CH_2)_4\!-\!C\!\!\overset{\displaystyle O}{\underset{\displaystyle O^-}{\diagup}} \xrightarrow[\text{anode, }-e]{\text{electrolysis,}} \left[Br(CH_2)_4\!-\!C\!\!\overset{\displaystyle O}{\underset{\displaystyle O^{\cdot}}{\diagup}} \right]$

$$\longrightarrow [Br(CH_2)_4^{\cdot}] \longrightarrow Br(CH_2)_8Br$$

favorable free-energy change, as for any other reaction. This is expressed as the difference in redox potentials between the reactants, $\Delta E_r{}^{23}$, the redox potential being the free-energy change associated with a one-electron transfer to or from a species. It is usually expressed in electron volts, where $1\,\text{eV} = 96\cdot45\,\text{kJ mol}^{-1}$ ($23\cdot05\,\text{kcal mol}^{-1}$); examples are listed in Table 15.3.

Table 15.3   Some redox potentials for one-electron reduction, E_r/electron volt[24,25]

Arenes (in MeCN)	E_r/eV^a	Quinones (in water)	E_r/eV^a
			0·099
Benzene	2·30	p-Benzoquinone	0·023
Toluene	1·98	2-Methylbenzoquinone	−0·067
m-Xylene	1·91	2,5-dimethylbenzoquinone	−0·235
o-Xylene	1·89	Duroquinone	−0·33
Mesitylene	1·80	O_2	−0·447
p-Xylene	1·77	N,N-Dimethyldipyridinium (paraquat)	
Phenanthrene	1·50		
Acenaphthene	1·21		
Pyrene	1·16		
Anthracene	1·09		
Tetracene	0·77		
Azulene	0·71		
9,10-Dimethylanthracene	0·87		

Miscellaneous	E_r/eV^a	Metals[12–22,26]	E_r/eV^a
		Li	−3·04
		K	−2·925
		Na	−2·714
Br_2	1·065	Zn	−0·763
Cl_2	1·36	Ag^+	0·78
MnO_4^-, H^+	1·51	Cr^{2+}	−0·41
PbO_2, H^+	1·68	Ti^{2+}	−0·37
$S_2O_8^{2-}$	2·01	Tl^{3+}	1·25
$OH^{\cdot}$	2·8	Mn^{3+}	1·51
Anisole	1·76	Phenol	1·04
Phenyl acetate	1·30	p-Cresol	0·543
Aniline	0·7	p-Nitrophenol	0·924
p-Toluidine	0·78	Acetamide	2·0
p-Nitroaniline	0·97	Chlorobenzene	2·07
p-Chloroaniline	0·60	Iodobenzene	1·77

^{a} $1\,\text{eV} = 96\cdot45\,\text{kJ mol}^{-1} = 23\cdot05\,\text{kcal mol}^{-1}$.

A one-electron transfer from some species to another with a more positive redox potential is energetically feasible but will only occur provided there is a suitably low-energy pathway available and is of use provided a second electron transfer does not occur to destroy the radical. Electron transfer from metals will take place within the coordination sphere which the substrate must enter. An applied voltage, i.e. electrolysis, may be used to provide any redox potential up to that of the medium[27,28]; the Kolbe alkane synthesis provides an example (also **15**):

$$n\text{-}C_{15}H_{31}COO^- \rightarrow n\text{-}C_{15}H_{31}COO^{\cdot} \rightarrow n\text{-}C_{15}H_{31} + CO_2$$
$$\downarrow{}_{2\times}$$
$$n\text{-}C_{30}H_{62}$$

15.1.2 Secondary routes

If a radical reacts with an all-spin-paired molecule the product (or at least one product) must still be a radical. Two main types of reaction occur; either the reactive primary radical abstracts an atom or group (very commonly a hydrogen or halogen atom) from a neutral molecule by a displacement reaction which may synchronous (denoted S_H2) (substitution, second order, homolytic), **16**, or it may add to an unsaturated system to form some more stable species (**17**, **18**). The hydroxyl radical is particularly useful in aqueous solution since it has a very large capacity for hydrogen

16

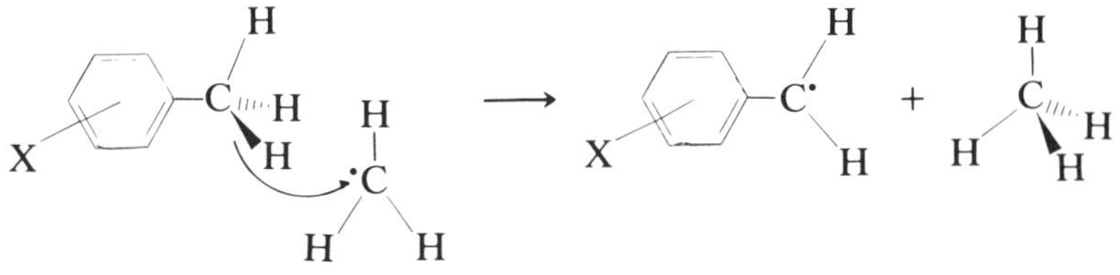

17

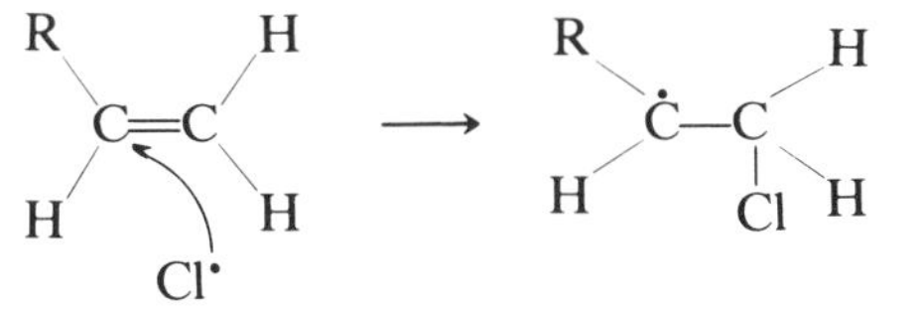

18   *tert*-Bu—N=O ⟶

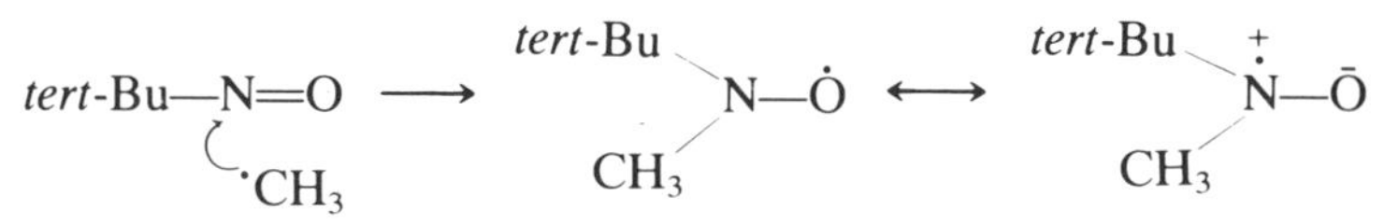

abstraction ($D(HO\text{—}H) = 500\,\text{kJ mol}^{-1}$ ($119\,\text{kcal mol}^{-1}$)) and can be generated not only by radiolysis but by the redox systems H_2O_2–Fe^{3+} (Fenton's reagent) and H_2O_2–Ti^{4+}, **11**[29].

Fragmentation

Some primary radicals are intrinsically unstable and will fragment to simpler radicals, ejecting a stable molecule. Carboxylate radicals typically lose CO_2 rapidly, providing a route to alkyl and aryl radicals, **19**.

15.2 THE DETECTION OF RADICALS

15.2.1 Direct observation

Electron spin resonance (ESR) spectroscopy is the most powerful single technique for the detection and characterization of radicals[30]. Almost all types can be studies in this way though the techniques for detection depend upon the lifetime of the radical (Table 15.4). Electron spin resonance is used specifically for radicals since it detects the spin inversion of unpaired electrons in a sample.

The orientation of the spin vector of an unpaired electron in a magnetic field is quantized in two states ('with' or 'against' the applied field, of magnetic flux density B_0), and the energies of these states are split. Transitions between spin states are induced by resonance radiation when the condition

$$h\nu = g\beta B_0 \tag{15.1}$$

applies (g = gyromagnetic ratio, a property of the radical and β = Bohr magneton, a proportionality factor).

Electrons in the lower energy state will be promoted to the upper on absorbing a quantum, $h\nu$, while those in the upper state will emit a quantum when they move to the lower state. Since the former are in slight excess as determined by the thermal Boltzmann distribution, there will be a net absorption of resonance radiation. This, typically, is observed in a magnetic field of 0·3 T and a frequency of 9 GHz (3 cm^{-1}). The observation of resonance under these conditions — electron spin resonance (ESR) or electron paramagnetic resonance (EPR) — is diagnostic of the presence of unpaired electrons, i.e. radicals. The ESR spectrometer requires a static concentration of radicals of only $>10^{-7}$ M for detection, so that the real possibility arises of being able to observe these species as transient intermediates. Additional structural information is obtained from an ESR spectrum, since the electron spin may interact and couple with nuclear spins, ^1H, in the neighboring structure. Multiplet spectra are thereby obtained, the interpretation of which is analogous to that of spin-coupled NMR spectra. Some examples are given in Fig. 15.2[79]. The separation

Table 15.4   Examples of radicals detectable by ESR

(a) Persistent radicals

Class	Example	Refs
20 Triarylmethyl, Ar$_3$C$^{\cdot}$		*31–34*
21 Diphenylpicrylhydrazyl (DPPH)		*35*
22 Semiquinone		*36*
23 Würster's salts and pyridyl radicals		*37–39*
24 Diaryl ketyl		*40*
25 Cation and anion radicals of polycyclic aromatics		*42*
26 Galvinoxyl		*41*
'Paraquat' radicals		*43*
ArN̊SO$_2$R		*44*
RO—N̊R		*45*

Table 15.4 (*Continued*)

(b) Reactive radicals

Source	Examples	Refs
Photolysis	$RO^{\cdot}$ $Me_2\dot{C}OH$ $R^{\cdot}$ $R^{\cdot}, R\dot{C}O$ $Ph\dot{C}O$	46–60
Flow systems	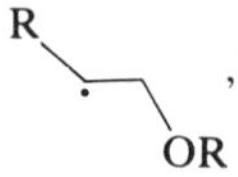 , $R\overset{+}{N}O_2$ $R\dot{C}HOH$ $RS^{\cdot}$ 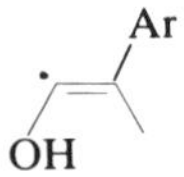 	61–66
Redox methods	$\cdot COOH$ $OH^{\cdot} \longrightarrow \dot{C}H_2COOH, \dot{C}H_3$  $R_3N^{\ddagger}$ $PhCOS^{\overline{}}$ $Ar\cdot$	67–70
Matrix isolation	*tert*-$BuO^{\cdot}$ 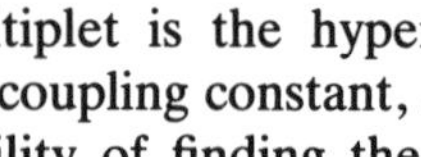 $-\equiv\cdot$ $==\cdot$ $R_2\dot{C}CH_2Hal$ $Et^{\cdot}$ $CS_2^{\overline{}}$	71–75

between resonances in a multiplet is the hyperfine coupling constant, a (analogous to the internuclear coupling constant, J, in NMR spectra), which is proportional to the probability of finding the unpaired electron on the adjacent atom (the *spin density*, ρ), expressed by McConnell's equation:

$$a = Q\rho. \quad (Q = \text{constant}) \quad (15.2)$$

Fig. 15.2 Detection of radicals by ESR. (**a**) Principle of the ESR experiment. (**b**) Spectrum of $Me_2\dot{C}$—OH produced from the flow systems $H_2O_2/Me_2CHOH + Ti^{3+}$. (**c**) Radicals formed from *in situ* photolysis of acetone at $-180°C$. The high-intensity triplet signal is due to $\dot{C}H_2COCH_3$, the low-intensity quartet to $\dot{C}H_3$. (**d**) The spectrum of $CH_3\dot{C}H_2$ from electron-beam irradiation of liquid ethane at $-180°C$. (**e**) The naphthalene anion radical; there is a quintet of quintets due to the coupling of the unpaired spin with four α- and four β-protons; $a_\alpha = 4\cdot95$, $a_\beta = 1\cdot865$ gauss, respectively.

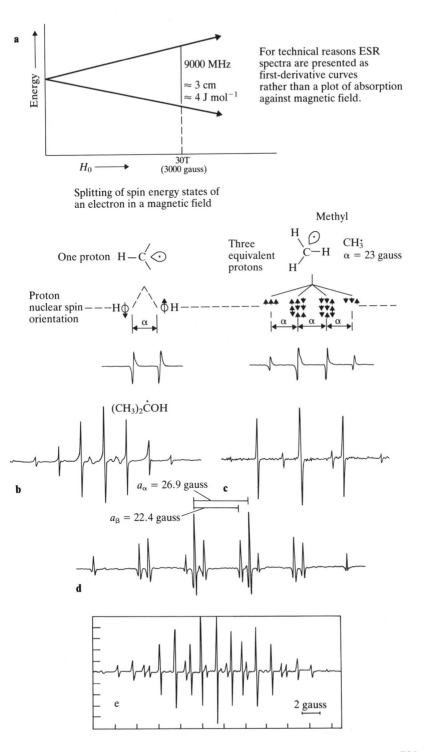

a

Energy

9000 MHz

≈ 3 cm

≈ 4 J mol^{-1}

For technical reasons ESR
spectra are presented as
first-derivative curves
rather than a plot of absorption
against magnetic field.

H_0 ⟶

30T
(3000 gauss)

Splitting of spin energy states of
an electron in a magnetic field

One proton H—C⟨⊙⟩

Three
equivalent
protons

Methyl

CH$_3$
α = 23 gauss

Proton
nuclear spin
orientation

——H⊕ /⟨^⟩\ ⊕H————

α

Three
equivalent
protons

α α α

$(CH_3)_2\dot{C}OH$

b

$a_\alpha = 26.9$ gauss

c

$a_\beta = 22.4$ gauss

d

e

2 gauss

Table 15.5 Spin densities and g-factors[78] of some radicals

Radical	ρ/e	Radical	g
MeĊHMe	0·844	Free electron	2·002322
MeĊHEt	0·837	Carbon-centered	2·002–2·009
MeĊHOOH	0·854	Nitrogen-centered	2·0035
MeĊHCN	0·786	Oxygen-centered	2·015
MeĊHOH	0·759	Sulfur-centered	2·006
ĊH(COOH)₂	0·74	CH₃	2·00255
FĊHCONH₂	0·907	CF₃	2·0031
		CCl₃	2·0091
		ĊH₂OH	2·00334
		CH₂=ĊH	2·00220
		Ph	2·00234
		MeĊ=O	2·0007

Allyl radical $CH_2 \cdots CH \cdots CH_2$: α 0·45, β 0·5

Phenyl/benzyl-type aromatic radical: α 0·57, o 0·14, p 0·14

Naphthalene-type radical: α 0·181, β 0·069

This enables estimates to be made of the distribution of the unpaired electron in a radical (Table 15.5). Spin densities in conjugated radicals such as allyl, benzyl and aromatic radical ions at any carbon, i, can be calculated from Eqs (15.3):

$$\rho = \text{constant} \times c_i^2(\psi^*) \quad \text{and} \quad \sum c_i^2(\psi^*) = 1, \tag{15.3}$$

in which c_i is the coefficient at carbon i in the molecular orbital in which the unpaired electron resides, ψ^*, while the spin densities at all positions must sum to unity. Equations (15.3) may be used as an experimental test for the validity of MO theory[80].

Persistent radicals

Many examples are now known of radical species which may be isolated and are completely stable or at least exist in equilibrium with their dimers. Their structures may be elucidated by the usual techniques of organic chemistry and radical character may be confirmed by their ESR spectra; Table 15.4a gives example of radicals which have no tendency to dimerize. Some, such as the ketyls, **24**, and aromatic radical ions, **25,** are nonetheless extremely reactive, for example to oxygen or water, and need to be handled with the exclusion of the atmosphere. Even a species such as diphenylpicrylhydrazyl,

21, which is stable enough towards dimerization to exist as purple crystals, combines readily with reactive radicals, which may be 'titrated' by the disappearance of the purple color.

Radicals with a lifetime of minutes to milliseconds may often be observed if they can be generated in or near the cavity of the ESR spectrometer. Fenton's reagent (generating OH˙) may be allowed to mix with a suitable substrate such as an alcohol in a continuous-flow apparatus which permits the formation of a sufficient static concentration of the secondary radicals produced by hydrogen abstraction within the detection apparatus. Photolysis within the cavity may achieve the same result. The primary radicals, e.g. Ph˙ from dibenzoyl peroxide, will abstract hydrogen and generate secondary radicals from a host of other solutes which may be present.

Short-lived radicals

Very short-lived radicals cannot be observed in this way if their concentrations cannot be raised to the level of detectability, so various methods have been devised to prolong their lifetime. The creation or trapping of the radicals in a solid, inert matrix at low temperature effectively prevents dimerization. Solid argon has been used, also adamantane[81,82] and zeolites[83], while radical sources may be photolyzed *in situ*. Alternatively, radicals generated at low pressure in the gas phase and carried on a stream of argon may be frozen out by impinging on a rotating cylinder cooled with liquid nitrogen (*rotating cryostat* technique[84]). Even infrared spectra of reactive radicals constrained in this may be observed. Sodium may be used as the matrix material and itself generates radical ions from embedded molecules by one-electron reduction, for example:

$$CS_2 + Na \rightarrow CS_2^- Na^+$$

If a reactive radical is generated in solution in the presence of a nitroso compound such as nitroso-*iso*-butane, **27**, it will add to it with the formation of a nitroxyl radical, **27a**. These are fairly stable types (half-life from minutes to hours) and their ESR spectra may be examined at leisure. Furthermore, their hyperfine structure is similar to that of the original radical since the unpaired electron is still spin-coupled to its protons. This technique is known as spin-trapping[85–97].

15.2.2 Indirect methods

The technique of Chemically Induced Dynamic Nuclear Polarization (CIDNP) applies to reactions in which products arise from *geminate* radical pairs, i.e. a pair of radicals which at some stage are in close proximity within

27   *tert*-Bu—N=O ⟶ *tert*-Bu
           $\overset{\cdot}{)}$ N—O
    CH$_3$—$\overset{\cdot}{C}$H$_2$ CH$_3$—CH$_2$

                                    27a

ESR spectrum of **27a**

the same solvent cavity or *cage*. In such cases, product formation will depend upon some of the radicals escaping from the cage[98–100], the alternative being recombination to starting materials within the cage. The recombination process is to some extent spin-selective, i.e. certain spin states of the radical pair are able to recombine faster than others. This comes about when at least one of the unpaired electrons is spin-coupled to adjacent magnetic nuclei, usually protons. The simplest system is illustrated in Fig. 15.3. Each of the unpaired electrons and an adjacent proton have two spin states giving $2^3 = 8$ possible combinations of which four are illustrated, the other four being the mirror image. Recombination requires the two electron spins to be opposite, so for combinations c and d this condition is met and the process is fast. Combinations a and b require a spin inversion of one unpaired electron initially; in a, this is facile since the spins of the electron and proton are not coupled as they are opposite. It is situation b which is responsible for the selection; spin inversion requires the spins of the electron and proton first to be uncoupled before recombination can occur. This is a slow process and consequently there is a higher chance of a radical pair in spin state b diffusing apart from the solvent cage than for the other spin states. This results in the proton which is observed in the product having a preferred spin state different from the normal Boltzmann distribution due to thermal motions. Now an NMR observation of the products by proton resonance during their formation will reveal this abnormal spin polarization as abnormal intensities of the resonance lines. The nuclear resonance experiment, irradiating the nuclei with resonance radiation, promotes both excitation of the ground state and de-excitation of the upper state (the former is only in excess by *ca* 1 in 10^6). What is normally observed is the difference spectrum between absorption and emission processes, absorption being only very slightly in excess. If the balance is shifted in favor of the upper state, then emission of radiation will be greater than absorption and the observed spectrum, or parts of it, will appear reversed (Fig. 15.3a,b). Therefore, in order to detect products arising from radical processes, a reaction is carried out within the probe of an NMR spectrometer and the spectrum is monitored at intervals. The observation of resonances which have abnormal intensities, including emission, is conclusive that the species observed have arisen from a

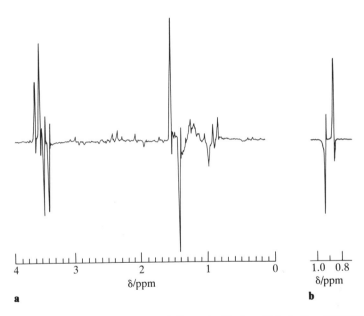

Fig. 15.3 Principle of Chemically Induced Dynamic Nuclear Polarization (CIDNP) for a three-spin system. (**a**) The spin-polarized NMR spectrum from the photolysis of dipropionyl peroxide in carbon tetrachloride.

(**b**) The spin-polarized spectrum of isobutane from the cage disproportionation of *tert*-butyl radicals (methyl resonance):

$$\text{tert-Bu—N=N—tert-Bu} \rightarrow [\text{tert-Bu}^{\cdot\,\cdot}\text{tert-Bu}] \rightarrow \begin{matrix} \text{Me} \\ \diagdown \\ \diagup \\ \text{Me} \end{matrix} \text{C=CH}_2 + \textit{iso}\text{-C}_4\text{H}_{10}$$

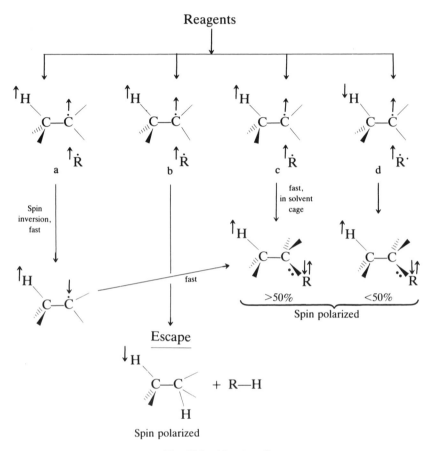

Fig. 15.3 (*Continued*)

geminate radical pair. Table 15.6 gives a list of reactions from which CIDNP signals have been observed. Any magnetic nuclei may be generated from radical-pair precursors in a polarized state so that CIDNP spectra have been observed from $^{19}F^{145}$, $^{13}C^{146,147}$, ^{15}N and ^{31}P ^{148} as well as from ^{1}H. Polarized ESR spectra (CIDEP) are also known in which spin-polarized radicals, including triplet carbenes and biradicals, are directly observed[149-152].

Whether spin polarization is such as to result in emission (E) or enhanced absorption (A) depends upon the origins of the species observed[153-157]. Predictions are summarized in Kaptein's rules, which are as follows.

The geminate radicals are designated a, b and have g-factors g_a, g_b, respectively; $\Delta g = (g_a - g_b)$. The radicals contain magnetic nuclei i (located in a) and j (located in b) which have hyperfine coupling constants a_i, a_j in the respective radicals. In the product these two nuclei become coupled with a nuclear spin coupling constant J_{ij}. The CIDNP spectrum is characterized

Table 15.6 Reactions which show CIDNP of the products

Reaction	Reference
(a) Thermal decompositions	

$$Ar—N{=}N—SO_2Ar \rightarrow Ar^\cdot + N_2 + \overset{\cdot}{S}O_2Ar \rightarrow ArSO_2Ar$$

101

$$\underset{\underset{N_3}{|}}{Ar—CH—OPh} \longrightarrow \underset{\underset{N:}{|}}{ArCHOPh} + N_2$$

102

103–105

$$PhCO{\cdot}O—O{\cdot}COPh + BrCCl_3 \rightarrow PhBr$$

106, 107

$$\overset{\text{┬┬}}{\underset{O—O}{\text{││}}} \longrightarrow \overset{\text{┬┬}}{\underset{O^\cdot\ O^\cdot}{\text{││}}}$$

108

$$tert\text{-}BuO—OCO—O^\cdot tert\text{-}Bu \rightarrow tert\text{-}BuO^\cdot + CO_2 + tert\text{-}BuO^\cdot$$

109

$$Ph_2CH—\underset{\underset{O}{\|}}{S}\text{-tol} \rightarrow Ph_2\overset{\cdot}{C}H + O\overset{\cdot}{S}tol$$

| **(b) Abstractions** | |

$$Ph_2C{=}\overset{*}{O} + PhOH \rightarrow Ph_2\overset{\cdot}{C}—OH$$

110

$$PhH + Ar^\cdot \rightarrow Ph^\cdot + ArH$$

111

$$Ph_2C{=}\overset{*}{O} + PhCH_3 \rightarrow Ph_2\overset{\cdot}{C}—OH$$

112

113

$$\diagup\!\!\!\diagdown I \longrightarrow I \diagdown\!\!\!\diagup$$

| **(c) Organometal reactions (Wurtz–Grignard and related reactions)** | |

$$EtLi + PhCHCl_2 \rightarrow PhCHCl{\cdot}Et$$

114–117

$$RMgBr + R'\text{-}Hal \rightarrow R—R'$$

118–121

$$R—I + Mg \text{ in} \diagup\!\!\!\diagdown OPh \rightarrow \text{``R—Mg—I''}$$

122, 123

$$R'MgBr + R_2C{=}O \rightarrow R_2R'C—OMgBr$$

124

$$RLi + R'I \rightarrow R—R'$$

125

| **(d) Carbene reactions** | |

$$Ph_2C{:} + PhCH_2F \rightarrow \text{various insertions}$$

126

$$Ph_2C{:} \rightarrow Ph_2C{=}CPh_2$$

127

$$Ph_2C{:} + PhCH_2COOMe \longrightarrow \underset{\underset{COOMe}{|}}{Ph_2CH—CHPh}$$

128

$$R_2C{:} + Cl—R' \longrightarrow \underset{\underset{R'}{|}}{R_2\overset{\cdot}{C}Cl}$$

129, 130

$$:CHCOOEt + CBr_2Cl_2 \rightarrow \text{various insertions}$$

131

| **(e) Diazonium reactions** | |

$$Ar—\overset{+}{N}{\equiv}N + O\overset{-}{H} \rightarrow ArN{=}N—OH \rightarrow ArOH$$

132

$$Ar—\overset{+}{N}{\equiv}N + OMe^- \rightarrow Ar—OMe$$

133

$$Ar—\overset{+}{N}{\equiv}N + ArNHNH_2 \rightarrow Ar—N{=}N—NHNHAr \rightarrow \text{products}$$

134

$$Ar—\overset{+}{N}{\equiv}N + HO\!\!\left\langle\bigcirc\right\rangle\!\!OH$$

135

| **(f) Rearrangements** | |

$$\underset{\underset{Ph}{|}}{Ph\overset{-}{C}H—\overset{+}{S}CH_2Ph} \text{ (Stevens rearrangement)} \longrightarrow \underset{\underset{CH_2Ph}{|}}{PhCH—SPh}$$

136, 137

$$Ph_3C—\overset{\cdot}{C}H_2 \rightarrow Ph_2\overset{\cdot}{C}—CH_2Ph$$

138

$$Me_3C—O—OH \xrightarrow{h\nu} Me_2C{=}C\!\!\underset{\diagdown OH}{\overset{\diagup Me}{}}$$

139

Table 15.6 *(Continued)*

	Reaction	*Reference*
(f) Rearrangements		

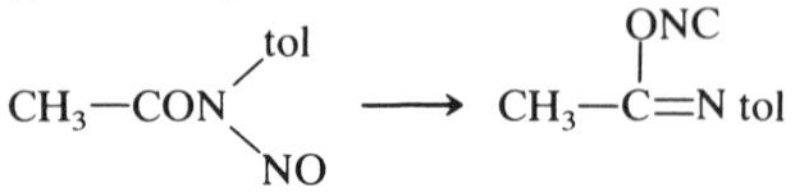

		140

(g) Miscellaneous
Nitration of PhNMe$_2$ (HNO$_3$ + HNO$_2$-catalyzed) — 141
ArH + Ph$^{\cdot}$ → Ar—Ph — 111, 142

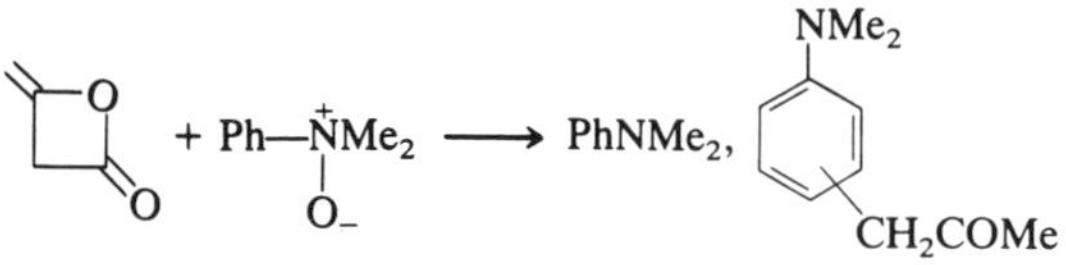

143

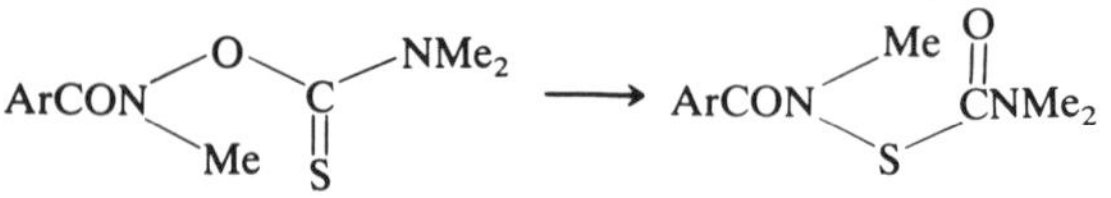

144

by two parameters Γ_{ne} (the net effect) and Γ_{me} (the multiplet effect):

$$\Gamma_{ne} = \mu\varepsilon\Delta g a_i,$$
$$\Gamma_{me} = \mu\varepsilon a_i a_j J_{ij}\sigma_{ij}, \qquad (15.4)$$

in which signs of the quantities, σ_{ij}, μ and ε are allocated as follows:

σ_{ij} is positive if i, j belong to the same radical,
negative if to different radicals;

μ is positive if a, b are free (spin-uncoupled) or triplet (coupled),
negative if singlet (coupled);

ε is positive if solvent cage products are observed,
negative if cage escape products are observed.

The net signs of Γ_{ne}, Γ_{me} can be determined if those of all the parameters can be assigned and have the significance:

	+	−
Γ_{ne}:	A	E
Γ_{me}:	E–A	A–E

(E–A means, in a multiplet, the low field half E, the high field half A; A–E is the converse of this).

Some typical results are shown in Fig. 15.3. Multiplet NMR signals of polarized nuclei are often half E and half A, and have a very characteristic appearance. As an example of the application of Kaptein's rules, the spectrum of methyl acetate and methyl chloride, products of the photolysis of diacetyl peroxide in CCl$_4$ (**28**), may be considered. Their origins are shown below Scheme **28**.

28

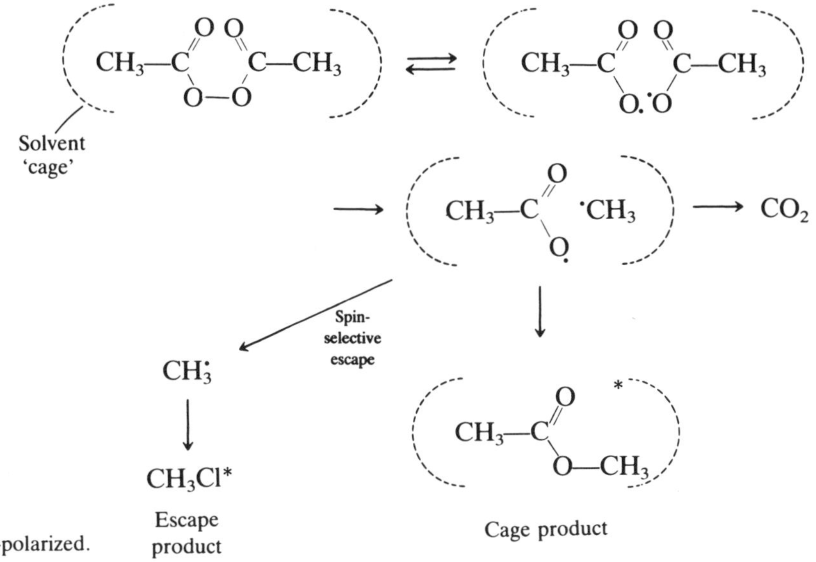

$$CH_3—C(=O)—O—O—C(=O)—CH_3 \rightleftharpoons CH_3—C(=O)—O\cdot\;\cdot O—C(=O)—CH_3$$

Solvent 'cage'

$$\longrightarrow \left(CH_3—C(=O)—O\cdot \quad \cdot CH_3 \right) \longrightarrow CO_2$$

Spin-selective escape

$CH_3^{\cdot}$

$\downarrow$

CH_3Cl^*

$\left(CH_3—C(=O)—O—CH_3 \right)^*$

Escape product   Cage product

*Spin-polarized.

CH₃Cl	CH₃COOMe
$\mu \times \varepsilon \times \Delta g \times a_i$	$\mu \times \varepsilon \times \Delta g \times a_i$
$- \times - \times - \times -$	$- \times + \times - \times -$
$= +$ (hence absorption).	$= -$ (hence emission).

The methyl signal of CH_3Cl, a cage escape product, shows enhanced absorption while the methyl acetate, a cage recombination product, shows emission (**28**; Fig. 15.3a,b)[100]. The amount of A or E observed depends upon the magnetic field strength of the NMR spectrometer and hence upon the frequency of the instrument used. More complex examples are analyzed analogously[98]. The interpretation of this information must be made with the *caveat* that certain reactions may proceed by parallel polar and radical pathways, the latter perhaps making only a small contribution, but they will still exhibit CIDNP spectra.

15.2.3 By chemical characteristics

Radical reactions differ qualitatively from polar reactions in many aspects which together may be compelling evidence for assigning a radical mechanism.

Chain reactions

Radicals are for the most part highly reactive intermediates. Reaction with a spin-paired molecule must of necessity yield a product which is still a radical

29   Initiation $\text{Cl—Cl} \overset{h\nu}{\rightleftharpoons} 2\text{Cl}^{\bullet}$

Propagation
$\times n$

$$\begin{cases} \text{PhCH}_3 + \text{Cl}^{\bullet} \xrightarrow{S_H2} \text{PhCH}_2^{\bullet} + \text{H—Cl} \\ \text{PhCH}_2^{\bullet} + \text{Cl—Cl} \xrightarrow{S_H2} \text{PhCH}_2\text{—Cl} + \text{Cl}^{\bullet} \end{cases}$$

Termination

$$\begin{cases} 2\text{PhCH}_2^{\bullet} \longrightarrow \text{PhCH}_2\text{CH}_2\text{Ph} \\ \text{Ph}\dot{\text{C}}\text{H}_2 + \dot{\text{C}}\text{l} \longrightarrow \text{PhCH}_2\text{—Cl} \end{cases}$$

and may be as reactive as the initial one, so a chain reaction propagates. An example is the chlorination of toluene, **29**. No reaction will occur in the cold and dark and with provision to exclude radicals, but if light is admitted or radical sources such as decomposing peroxides are added, chlorine atoms may be generated which initiate the reaction. A sequence of steps then follows, conserving the radical. The chlorine atoms abstract a benzylic hydrogen from toluene, to form the most stable radical possible which in turn abstracts chlorine from Cl_2, reforming chlorine atoms. These two steps are the chain propagation steps; the two radicals $\text{PhCH}_2^{\bullet}$ and $\text{Cl}^{\bullet}$ are the chain carriers. The reaction continues until a side reaction removes the chain carriers, for instance by dimerization or by the presence of inhibitors, species which can react with a radical to form a stable and unreactive radical. This is known as termination. The kinetic analysis of a chain reaction with a single propagation step (e.g. polymerization of a monomer, M) is as follows.

Initiator dissociation	$\text{In} \rightarrow 2\text{In}^{\bullet}$	k_d
Initiation	$\text{In} + \text{M} \rightarrow \text{M}^{\bullet}$	k_i
Propagation	$\text{M}^{\bullet} + \text{M} \rightarrow \text{M}^{\bullet}$	k_p
Termination	$2\text{M}^{\bullet} \rightarrow \text{products}$	k_t

$$\text{Rate} = -d[\text{M}]/dt = k_i[\text{In}^{\bullet}][\text{M}] + k_p[\text{M}][\text{M}^{\bullet}]$$

of which the first term is negligible.

Assuming $[\text{M}^{\bullet}]$ and $[\text{In}^{\bullet}]$ to be small and constant (steady-state assumption),

$$d[\text{M}^{\bullet}]/dt = k_i[\text{In}^{\bullet}][\text{M}] - 2k_t[\text{M}^{\bullet}]^2.$$

Hence

$$[\text{M}^{\bullet}] = [(k_i[\text{In}^{\bullet}][\text{M}])/2k_t]^{0.5},$$

and

$$[\text{In}^{\bullet}] = (2k_d[\text{In}])/k_i[\text{M}].$$

Hence

$$\text{Rate} = k_p[\text{M}](k_d[\text{In}]/k_t)^{0.5}.$$

In reactions for which $\text{In} = \text{M}$, i.e. the reaction is self-initiated by thermolysis of the substrate, or if there are two different propagation steps as in **29** and the second is rate-determining,

$$\text{Rate} \propto [M]^{1.5}.$$

If the first propagation step is slow, the rate dependence is

$$\text{Rate} \propto [R_1][R_2]^{0.5},$$

where R_1, R_2 are the two reagents in the first and second step, respectively.

The chain length is the average number of cycles promoted by each initiating radical, which may vary from 2–3 to above 10^6. Typically, rates of radical reactions are governed by the presence of initiators and inhibitors. Common examples of initiators are:

   (a) peroxides R—O—O—R′, e.g. dibenzoyl, diacetyl, di-*tert*-butyl, and hydroperoxides R—O—O—H, e.g. cumyl;
   (b) azo compounds, e.g. $Me_2C(CN)$—N=N—$C(CN)CMe_2$ (azo-isobutyronitrile; AIBN);
   (c) light (even laboratory lighting)[158,159].

Common inhibitors are
   (a) oxygen;
   (b) polyphenols such as hydroquinone;
   (c) thiols;
   (d) amines[160].

Induction is a phenomenon often encountered in chain reactions. A reaction mixture frequently contains inadvertent inhibitors, especially molecular oxygen. Initiator radicals will be rapidly scavenged by such species so that the reaction will not get under way until they are removed. There is frequently found to be an induction period, a delay after the mixing of reagents before the desired reaction begins, during which impurities are removed by initiator radicals. This behavior is characteristic of radical chain reactions.

The high reactivity typical of many radicals often gives rise to formation of polymers, tars and other undesirable products.

Cage effects

It has been mentioned above that special properties accompany products which are derived from geminate radical-pairs which need to escape from each other's company in order to show their individual reactivity. These include the following.

(a) CIDNP (see Section 15.2.2).
(b) If homolysis is partly or largely reversed, rates of observed reaction will be much smaller than the true values, just as is the case for ionization to an ion-pair. In the extreme, rates of homolysis can become purely rates at

which radicals can escape from the solvent cage. This effect seems to be responsible for the curiously low volumes of activation observed for the decompositions of peroxides, about $+5 \, \text{cm}^3 \, \text{mol}^{-1}$, a bond-breaking process being expected to be accompanied by a positive but somewhat larger $\Delta V^{\ddagger}$[161, 162].

$$k_{\text{obs}} = k_{\text{hom}}/k_{\text{esc}};$$

$$\Delta V^{\ddagger}_{\text{obs}} = \Delta V^{\ddagger}_{\text{hom}} - \Delta V^{\ddagger}_{\text{esc}};$$

$$15 - (+10) \simeq +5 \, \text{cm}^3 \, \text{mol}^{-1}.$$

Pressure tends to slow the escape rate because of an increase in viscosity of the medium; hence $\Delta V^{\ddagger}_{\text{esc}}$ is positive and will reduce $\Delta V^{\ddagger}_{\text{obs}}$. By contrast azo compounds, for which recombination is less likely since N_2 is lost, have $\Delta V^{\ddagger} \simeq +20 \, \text{cm}^3 \, \text{mol}^{-1}$.

(c) Rates of reaction tend to vary inversely with the viscosity of the medium. Peroxides can decompose in the gas phase faster than in solutions at the same temperature by factors as great as 10^6 since recombination in the absence of a medium is very improbable[163, 164].

(d) Quantum yields of radicals (the average number of radical products produced per photon of radiation absorbed) may be near unity in the gas phase, in which recombination is unlikely, but fall to very low values in solution on account of recombination; for example, I_2 has a quantum yield of 0·14 (14%) in CCl_4[165] for the production of iodine atoms.

(e) Radical recombination can show up in isomerization of the substrate which is reisolated before complete reaction. The chiral azo compound, **30**, when partially photolyzed is found also to be partially racemized[161]. ^{18}O-labeled peroxides, **31**, show some oxygen scrambling when allowed partially to decompose[162].

(f) 'Crossover' reactions tend not to occur in solution. The pyrolysis of a mixture of azomethane, $CH_3N{=}NCH_3$, and the d_6 analog gives, in the gas phase, the statistical mixture of ethanes ($CH_3CH_3 : CH_3CD_3 : CD_3CD_3 \sim 1:2:1$) but in solution none of the mixed product, CH_3CD_3, is obtained since recombination is much faster than escape[165, 166].

30

31

15.3 REACTIONS OF RADICALS

Six main types of reaction available to radicals are:
  (a) radical coupling, including dimerization;
  (b) abstraction, i.e. bimolecular substitution (S_H2), particularly on hydrogen and halogen;
  (c) addition to a π-system (alkene, alkyne, polyene, arene, or an unsaturated heteroatom system);
  (d) fragmentation, particularly the mode known as β-scission;
  (e) one-electron oxidation or reduction to a cation or anion;
  (f) rearrangement of the radical to a more stable isomer.

15.3.1 Radical coupling

For the majority of radicals, coupling of unpaired electrons (of opposite spin) between two similar or dissimilar species is very facile. There is no activation barrier since the process is the reverse of homolysis and the potential function is represented by the right-hand side of Fig. 1.2, only opposed by an unfavorable entropy change (-60 to $-70\,\mathrm{J\,K^{-1}\,mol^{-1}}$ (-14 to $-17\,\mathrm{kcal\,mol^{-1}}$)). Heats of reaction are essentially negatives of the bond dissociation energies. Rates of coupling, Table 15.7, vary widely from

Table 15.7   Rates of some radical coupling reactions (dimerization or disproportionation) [167]

$$2R^\bullet \rightarrow \text{products}$$

$R^\bullet$	$10^{-9}k_2/\mathrm{M^{-1}\,s^{-1}}$
H$^\bullet$, Br$^\bullet$, I$^\bullet$	~10
MeĊO	0·1–0·6
PhO$^\bullet$	0·6
PhṄH	0·7
Me$_3$Si$^\bullet$	2·5
Me$_3$Sn$^\bullet$	1·5
Ph$_3$Sn$^\bullet$	1·4
Cl$_3$C$^\bullet$	0·05
Me$_2$Ċ—CN	4
MeĊHOH	0·7
Ph$_2$Ċ—OH	0·11
p-R̲—C$_6$H$_4$O$^\bullet$	0·005
HOĊHCOOH	0·6
HOĊHCOO$^-$	0·04
Ō—ĊHCOO$^-$	0·007

R = H	5×10^{-2}
R = isoPr	$1\cdot1 \times 10^{-9}$
R = *tert*-Bu	stable

32   Ph₃C—CPh₃ $\xrightarrow[25°C]{\Delta}$ 2Ph₃C˙ $\rightleftharpoons$

33

34   2

diffusion-controlled rates for the simple alkyl radicals to a failure of the reaction altogether as with triarylmethyls, **20**, and diphenylpicrylhydrazyl, **21**, for which two factors can be held responsible. An increase in the instrinsic stability of the radical, for example, as a result of delocalization of radical character, will reduce its reactivity. Steric compression in the product will also tend to reduce the rates of dimerization and this is likely to be the major factor in maintaining **20** and **21** as stable radicals. Triphenyltin, which forms a Sn–Sn bond, weaker but longer than a C–C bond, by contrast dimerizes rapidly since the product is not as sterically congested. It is notable that triphenylmethyl dimerizes not to hexaphenylethane, **32**, from which it may be obtained, but to the quinoid isomer, **33**, which is less sterically strained[168,169]. The phenoxy radicals **34** dimerize in an analogous fashion. Charge repulsion effects are evident in the dimerization of anion and dianion radicals (Table 15.7; compare HOĊHCOOH, HOĊHCOO⁻ and ⁻OĊHCOO⁻). Coupling of dissimilar reactive radicals in solution is likely only when they are simultaneously created within a solvent cage since their lifetimes will usually be too short in solution for bimolecular collisions to occur. This is observed in the photolysis of acyl peroxides to a pair of acetoxy radicals, one of which may decarboxylate before recombination (**28**). The observation of dimeric species is diagnostic of a relatively long-lived radical precursor. Cumene, **35**, may be added to a reaction suspected of generating transient radicals. If these are capable of abstracting the tertiary hydrogen from cumene, its dimer (**36**; dicumyl) may be formed as an easily identifiable product, as for instance during the course of a Wurtz–Grignard reaction **37**[170].

35

37

15.3.2 Displacement (abstraction, transfer) reactions

If a reactive radical fails to dimerize it will most likely acquire stability by abstracting a hydrogen atom or halogen, as available, from the medium if in so doing a more stable radical results, i.e. if the reaction is exothermic. The feasibility of this reaction may be estimated from the difference in the dissociation energies of the bonds involved (Table 15.8). Relative rates of hydrogen abstraction decrease in the order tertiary > secondary > primary hydrogen, and the order of ability to stabilize a neighboring radical center by conjugation is $Ar > -CH=CH_2 > R$. However, the spread of rates of abstraction is remarkably small for radicals such as phenyl, which are highly reactive and hence very unselective in their attack. Less reactive radicals such as $\dot{C}Cl_3$ are much slower and more selective. From a practical point of view, radical bromination is likely to lead to fewer products than is chlorination (Table 15.8a,c). The effect of heteroatom substituents on radical formation is quite different than that on polar reactions. Both $+R$ and $-R$ types appear able to stabilize a radical to a greater or lesser degree. Abstraction of halogen follows a similar trend to hydrogen abstraction; rates decrease with increasing C–Hal bond strength, $-I > -Br > -Cl (\gg -F)$. Polyhalogeno compounds such as CCl_4 are very effective halogen-atom donors.

Hydrogen abstraction by one radical from another results in the formation of a saturated and an unsaturated product with the removal of both radicals. This is known as disproportionation and is an important terminating process in radical polymerization, **38**. Hydrogen abstraction by a growing polymer radical from a non-radical species will terminate the polymer but generate a radical which can initiate another, **39**. The effect is to limit the molecular weight of the polymer and the process is known as *chain transfer*. Some chain transfer rates are given in Table 15.8d.

Primary kinetic isotope effects should be obtained for hydrogen-atom abstraction and this is the case. Values of k_H/k_D (Table 15.9) vary widely. The maximum effect (7–9 at 25°C) presumably results from a central transition state with hydrogen equally bound to both radicals. Lesser values indicate an asymmetric transition state, ambiguous in that it may be either early or late. Very large values such as that for **40** have been attributed to

Table 15.8 Rates for some radical displacement reactions[167]

(a) Hydrogen abstraction, $R-H \ \cdot R' \xrightarrow{\ k\ } R\cdot \ H-R'$

R—H	k_{rel} ($\cdot R'$)					
	$Ph\cdot$	$Me\cdot$	$tert\text{-}BuO\cdot$	$Cl\cdot$	$\cdot CCl_3$	$Br\cdot$
RCH_3	0·12	0·4	0·1	0·8		0·0001
R_2CH_2	1·0	4·0	1·2	3	5	0·005
R_3CH	4·8	9·4	4·4	3·6	160	0·02
⟋CH₃	1·6	0·6	1·6			
⟋CH₂R	3·3	6·4	7		40	
⟋CHR₂	13	26	18			
		1·0	1·0	1·0	1·0	1·0
$PhCH_3$	1·0	4·1	3·2	2·5	50	17
$PhCH_2R$	4·6	13	6·9	5·5	260	37
$PhCHR_2$	9·7		4·7	2·0	50	10
Ph_2CH_2	7·7		9·6	7·2	160	18
Ph_3CH	39					

$R_2C\overset{H}{\underset{X}{\diagup\diagdown}}$

X:	H	NO₂	Cl	Br	Me	OAc	CN	MeOCO	RC=O
k_{rel}:	1	1	2	2	10	8	33	40	87

(b) Halogen abstraction, $Me\cdot + R-Hal \rightarrow Me\text{-}Hal + R\cdot$

R	k_{rel}		
	RI	RBr	RCl
Me	$1·3 \times 10^3$		
Et	$5·4 \times 10^4$		
isoPr	$2·5 \times 10^5$		
tert-Bu	5×10^6		
$PhCH_2$	$2·3 \times 10^7$	$2·0 \times 10^3$	~0·1

(c) Hydrogen abstraction in a carbon chain, $X-\overset{\overset{H}{|}}{\underset{|}{C}}\hspace{-2pt}\alpha-\overset{\overset{H}{|}}{\underset{|}{C}}\hspace{-2pt}\beta-\overset{\overset{H}{|}}{\underset{\diagdown}{C}}\gamma$ $\cdot R$

X	$R\cdot$	Percentage attack at			
		X	α	β	γ
Me—	Br· (gas)	1	49	49	1
Me—	Cl· (gas)	15	35	35	15
ClCH₂—	Br· (gas)	23	22	54	1
ClCH₂—	tert-BuO·	21	20	44	16
PhCH₂—	tert-BuO·	19	28	10	42
Cl₃C—	ṠO₂Cl		8	42	50
F₃C—	Cl·		0	45	55
HOOC—	Cl·		8	52	40
N≡C—	tert-BuO·		22	44	50
O₂N—	tert-BuO·		13	33	54

Table 15.8 (*Continued*)

(d) Rates of chain transfer: polystyrene radical + solvent,

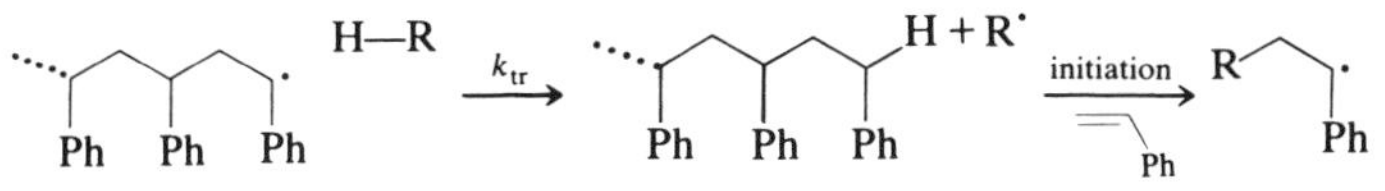

Solvent, RH	$10^4 k_{tr}/\text{M}^{-1} s^{-1}$ (60°C)
 ⬡ (cyclohexane)	0·024
PhH	0·018
PhCH₃	0·125
PhEt	0·67
PhisoPr	0·82
n-BuCl	0·04
CHCl₃	0·5
CCl₄	90

quantum-mechanical tunneling[178]. This appears to be especially well established for the intramolecular hydrogen transfer, **41**, for which a PKIE in excess of 100 at low temperatures has been reported[179].

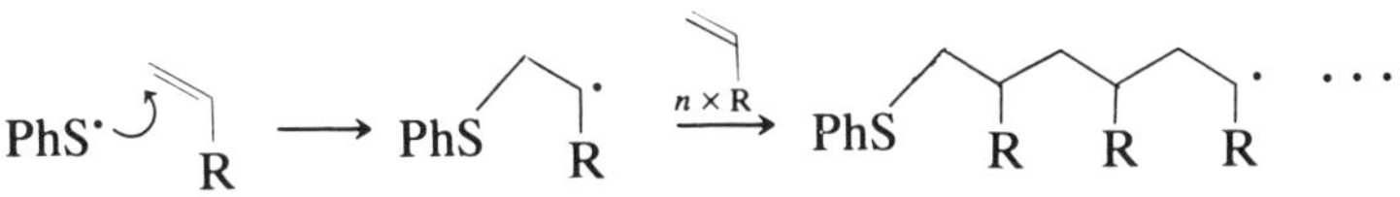

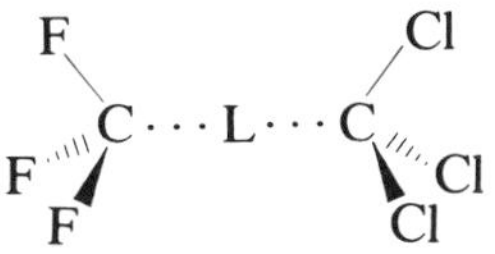

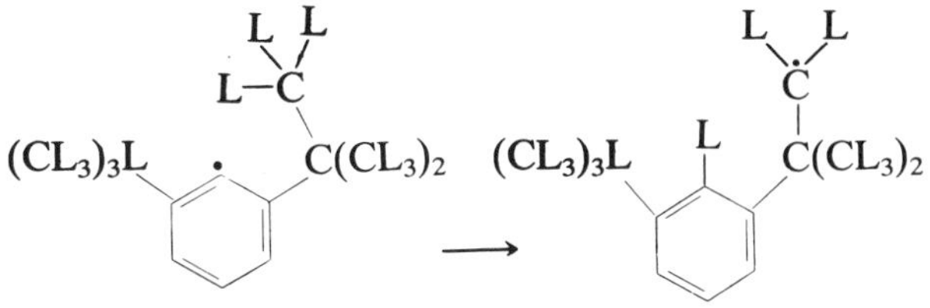

$k_H/k_D = 10^4$ at 23 K

40   **41**

Table 15.9   Primary isotope effects in radical reactions[171]

$$R\text{—}L + {}^{\bullet}R' \rightarrow R^{\bullet} + L\text{—}R'$$

$R\text{—}L$	${}^{\bullet}R'$	$T/°C$	k_H/k_D	Reference
PhCH₃	Ph•	60	4·5	172
PhCH₃	CH₃•	164	7·9	173
PhCH₃	tert-BuO•	40	5·5	174
PhCH₃	Cl•	77	1·3	175
PhCH₃	Br•	77	4·9	176
PhCH₃	Me₂N•	136	4·0	176
CH₄	Cl•	0	12·1	177
CHCl₃	Cl•	127	2	177
CHCl₃	CF₃•	164	32	178

739

15.3.3 Additions to π-systems

Reactive radicals will add to a wide variety of unsaturated compounds, both to carbon and to other elements. The addition is regiospecific and tends to give attack at the least substituted position [180] (Scheme 42). In this the reaction can be said to obey the Markownikoff Rule in its modern version (Section 12.1.4), controlled by the tendency to form the more stable radical intermediate. Among the more important examples of this behavior are vinyl polymerizations, **42**, and additions across alkene double bonds of the species X—Y shown in Table 15.10 by chain mechanisms analogous to **29**.

42 $\quad$ In—In $\xrightarrow[h\nu]{\Delta \text{ or}}$ 2In·

$$R = CN, Cl, OAc, Ph, \text{/\\}$$

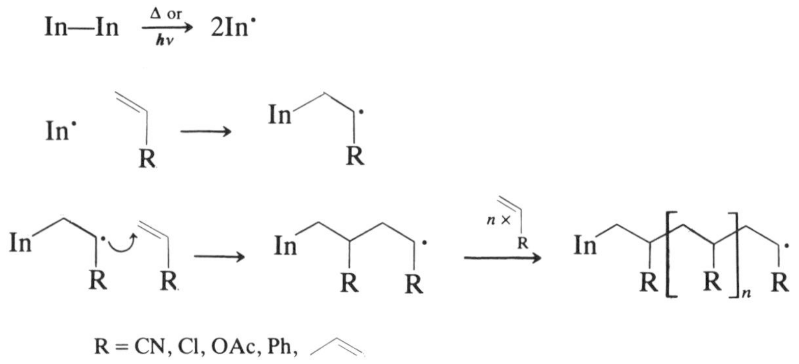

Table 15.10 Some radical chain additions to alkenes [1,181]

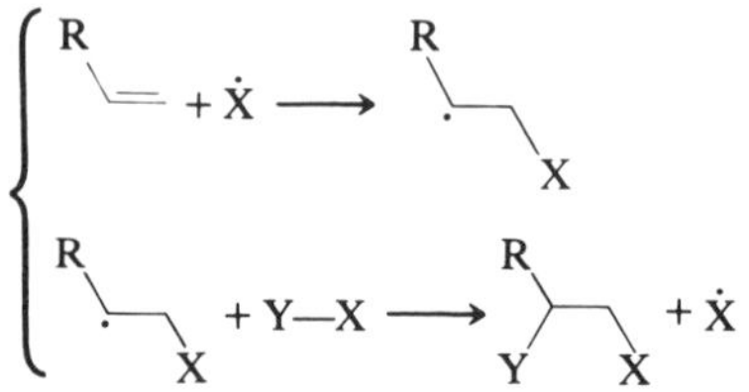

H—X	Y—X
H—Br	Cl—CCl$_3$
(H—Cl)	Br—CBr$_3$
H—SR	Br—CCl$_3$
H—O·COR	Cl—CCl$_2$
H—CHR	$\quad\quad$ |
$\quad$ |	$\quad\quad$ CN
$\quad$ OH	I—CF$_3$
H—CHR	Br—Br
$\quad$ |	
$\quad$ NH$_2$	Cl—Cl $\left(\text{from SO}_2\diagup\!\!\!\!\diagdown{}^{Cl}_{Cl}\right)$
H—CR	
$\quad$ ||	
$\quad$ O	
H—SO$_3^-$	
H—SiCl$_3$	

740

Table 15.11 Relative rates of addition of Me$^{\bullet}$ to alkenes at 65°C[182]

Alkene	k_{rel}	Alkene	k_{rel}
(ethylene)	1·0	Ph—CH=CH₂	31
(propene)	0·85		
(isobutylene)	1·4	Ph₂C=CH₂	58
(neopentyl-type)	0·23	Ph—CH=CH—Ph	1·1
(pentene)	0·12		
(pentene isomer)	0·27	PhCH=CH—CH₂Ph	4
CH₂=CH—OEt	0·23	Ph₂C=CHPh (Ph, Ph / Ph, H)	1·8
CH₂=CH—CH=CH₂	77		
(isoprene)	85	Ph₂C=CPh₂ (Ph, Ph / Ph, Ph)	0·3
Ph—CH=CH—CH=CH—Ph	15		

Rates of addition of methyl radicals (from decomposing diacetyl peroxide) may be compared by a competition method and found to be parallel with product-radical stabilities, indicating a product-like transition state (Table 15.11). The efficient progress of a radical chain addition reaction requires both chain-carrying steps to be exothermic. This principle provides an explanation of the familiar 'peroxide effect' in the addition of HBr to alkenes, whereby in the presence of peroxides (i.e. radical initiators), the normal polar addition route is supervened by the radical chain pathway[183]. Enthalpies of each step for the four hydrogen halides are as follows.

		H—F	H—Cl	H—Br	H—I	
R₂C=CR₂ + Ḣal ⟶ R₂Ċ—CR₂Hal		−167	−67	−12	+21	/kJ mol⁻¹
		(−40)	(−16)	(−2·9)	(+4.2)	/(kcal mol⁻¹)
R₂Ċ(Hal)—CR₂ + H-Hal ⟶ R₂C(H)—CR₂Hal		+155	+21	−46	−113	/kJ mol⁻¹
		(+37)	(+5)	(−11)	(−27)	/(kcal mol⁻¹)

HF will not take part in radical chain addition since the carbon-centered radical cannot abstract hydrogen from HF; HI will not take part either, but this is because the iodine atom is too unreactive to add to the alkene. Only HBr (and to some extent HCl at high temperatures which can overcome the unfavorable barrier at the second step) can readily participate in both steps since both are exothermic.

Additions to an alkene are completed by the product radical abstracting hydrogen or some other atom to give, finally, a saturated species. However, additions to arenes, **43**, lead mainly to substitution since the intermediate pentadienyl radical, **44** (analogously to the benzenium ion in electrophilic substitution), is more ready to donate a hydrogen atom and regain aromatic stability than to abstract H˙ and form a cyclohexadiene. A PKIE, $k_H/k_D = 3.0$, for the arylation of benzene shows the initial addition to be reversible and loss of H˙ to be rate-limiting, **45**, **46**.

Orientational effects of substituents in the ring are quite different than those experienced in S_EAr reactions (Table 15.12). If any generalizations may be made, *ortho, meta* and *para* substitutions all occur with a slight bias towards the order $o > m > p$, whatever the nature of the substituents[184]. This is presumably a reflection on the relative stabilities of the respective pentadienyl radical intermediates, but no obvious explanation can be put forward unless it is that initial attack occurs randomly at each carbon including the *ipso* position (carbon bearing the substituent), **44a**, the latter rearranging to *ortho* which becomes favored overall[185]. Other products such as biphenyls and benzoic acid are also formed as indicated in Scheme **43**.

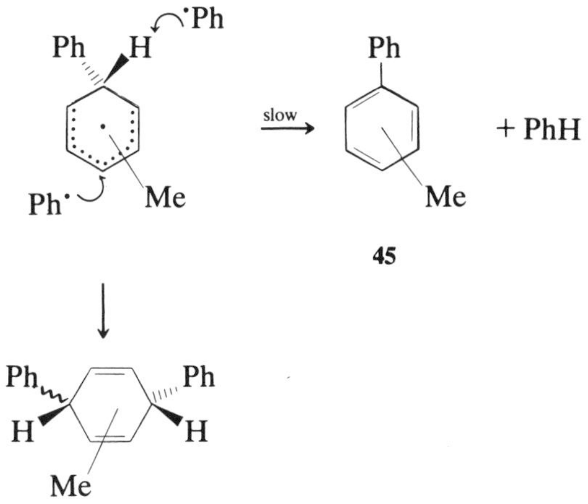

45

46

Table 15.12 Isomer ratios in aromatic homolytic arylation[184] Ph· +

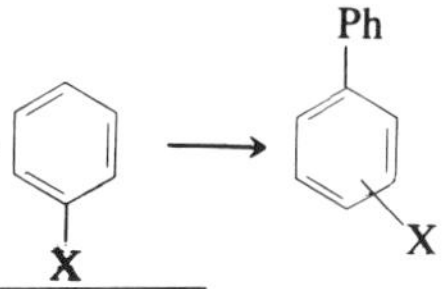

X	Percentage		
	o	*·m*	*p*
Me	66	20	14
Et	53	28	19
isoPr	31	42	27
tert-Bu	24	49	27
Ph	48	23	29
F	54	31	15
Cl	50	32	18
Br	50	33	17
I	52	31	17
CF$_3$	29	42	29
COOMe	58	17	25
CN	60	10	30
NO$_2$	62	10	28

15.3.4 Fragmentation of radicals[186,187]

The reverse of a radical addition to a double bond is the fission of a fragment radical located β- to the radical center. Tertiary alkoxy radicals are prone to decompose in this way to a ketone (**47**); carboxylate radicals rapidly lose CO$_2$ (**48**), and acyl radicals CO (**49**). Reactivity in this mode seems to follow the order of stability of the radical ejected: this is illustrated by the decomposition pathways available to **50** and to **51**, which shows the

47 $\underset{\underset{Me}{\overset{Me}{|}}}{\overset{R}{\underset{}{\diagdown}}}C\!-\!O^{\bullet} \longrightarrow \underset{Me}{\overset{Me}{\diagdown}}C\!=\!O \;+\; R^{\bullet}$

48 $R\!-\!\overset{\overset{\textstyle O}{\|}}{C}\diagdown_{O^{\bullet}} \longrightarrow R^{\bullet} + CO_2$ **49** $R\!-\!O\!-\!\overset{\bullet}{C}\!=\!O \longrightarrow RO^{\bullet} + CO$

50 $R'\!-\!O^{\bullet} + PR_3 \longrightarrow [R'\!-\!O\!-\!\overset{\bullet}{P}R_3] \quad \overset{\alpha}{\underset{\beta}{\Longrightarrow}} \begin{array}{l} R\!-\!OPR_2 + R^{\bullet} \\[4pt] R'_{\cdot} + R_3P = O \end{array}$

$$\underset{\underset{(96\%)}{\overset{\underset{\textstyle 2\text{-Pr}}{\diagup}}{Et\overset{\cdots\cdots}{\diagdown}C\!-\!\overset{\bullet}{O}}}}{\overset{(3\%)}{}}\overset{Me\;(<0\!\cdot\!5\%)}{\overset{\textstyle \diagdown}{}}$$

51

preference to eject alkyl radicals in the order: secondary ≫ primary ≫ Me[·][188]. The reactions are driven by an overall exothermicity, for example

$$MeCO_2^{\cdot} \rightarrow Me^{\bullet} + CO_2$$

$$E_A = 27\,kJ\,mol^{-1}\,(6\!\cdot\!4\,kcal\,mol^{-1}), \quad \Delta H = -42\,kJ\,mol^{-1}\,(-10\,kcal\,mol^{-1}).$$

Other examples are given in Table 15.2 and illustrate the low energies of activation, even for hydrocarbon radicals.

15.3.5 Radical rearrangements[189]

Radical rearrangements are, for the most part, intramolecular examples of the reactions discussed above which will be favored over their intermolecular counterparts by the usual neighboring-group advantages (Chapter 14)[190,191]. Hydrogen abstraction via a five- or six-membered cyclic transition state can be faster than the same reaction with solvent by a factor of 10^3, for example in the Barton reaction, **52**, a useful method of functionalizing an alkyl group when located near a hydroxyl, and in the transannular cyclization **53**. 1,2-Shifts of hydrogen, alkyl or aryl groups are frequently observed, the product necessarily being a more stable radical, **54**[192]. An intramolecular addition, **55**, is used diagnostically as evidence for radical intermediates, which are assumed to be involved if methylcyclopentane is obtained among the products.

Autoxidation is the process whereby a molecule, often a hydrocarbon, becomes converted to oxidation products in air. Molecular oxygen is a

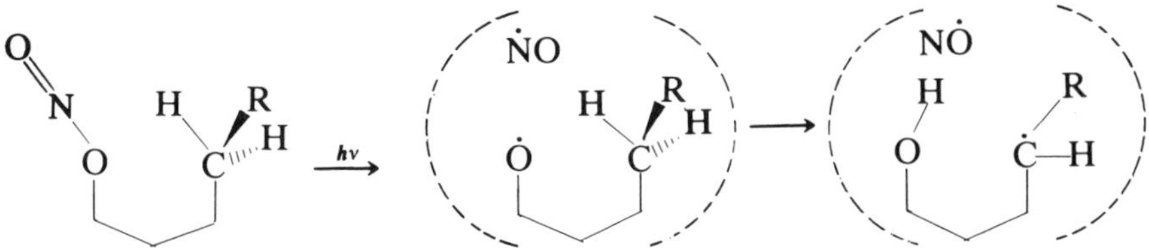

52

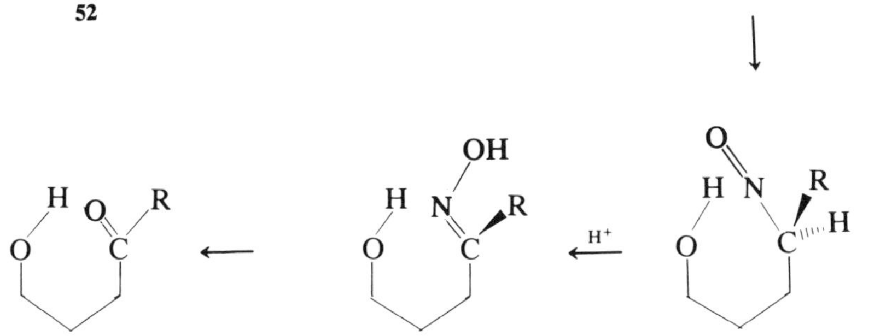

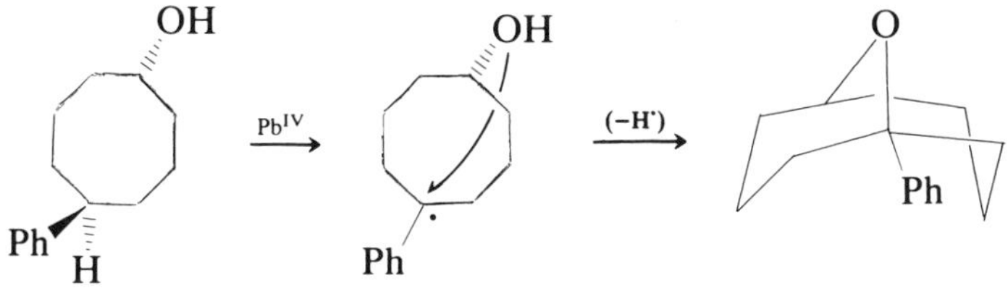

53

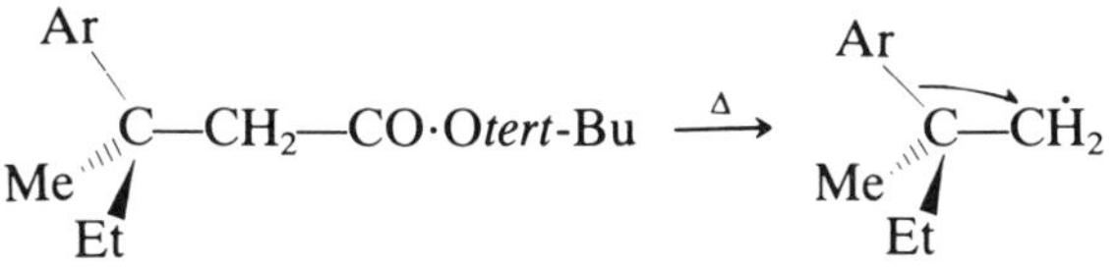

54

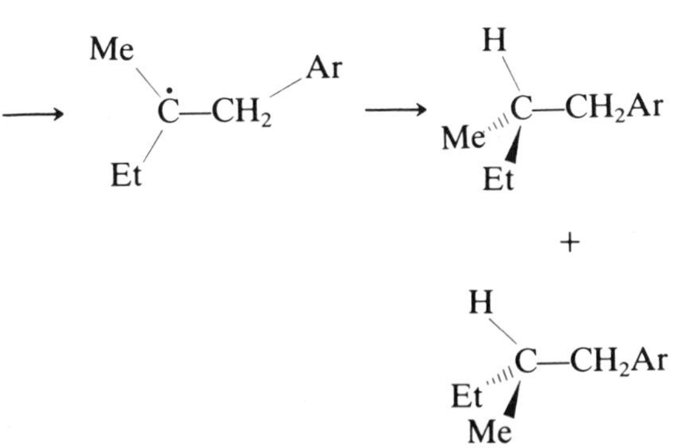

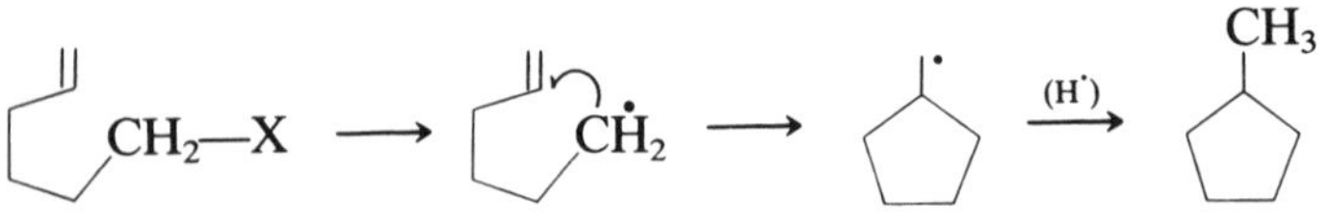

55

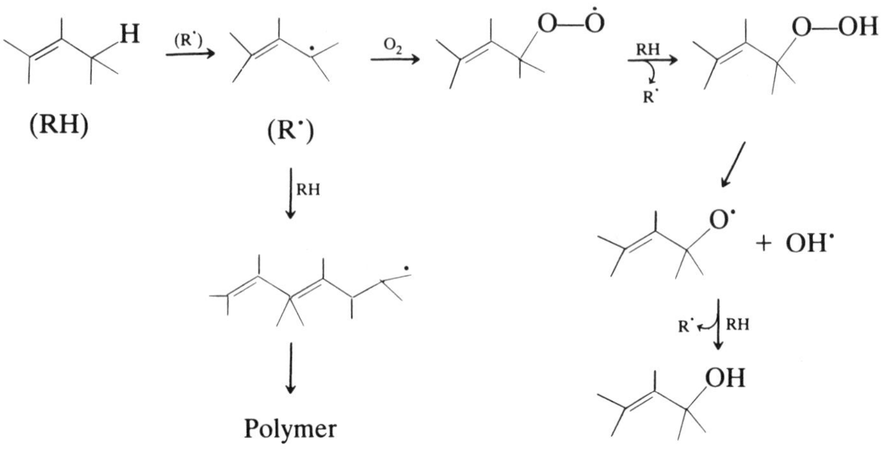

56

(RH)   (R')

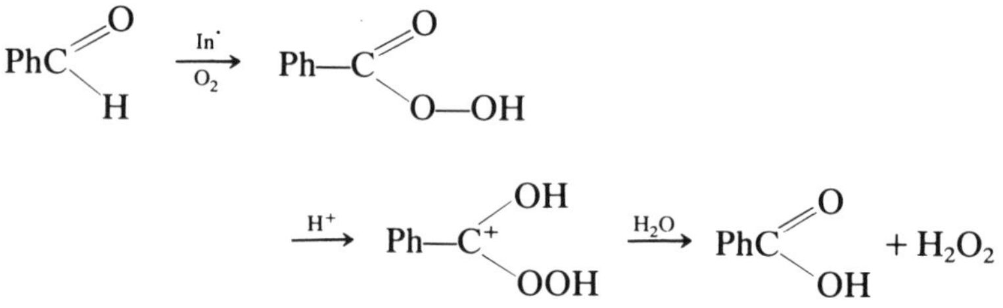

Polymer

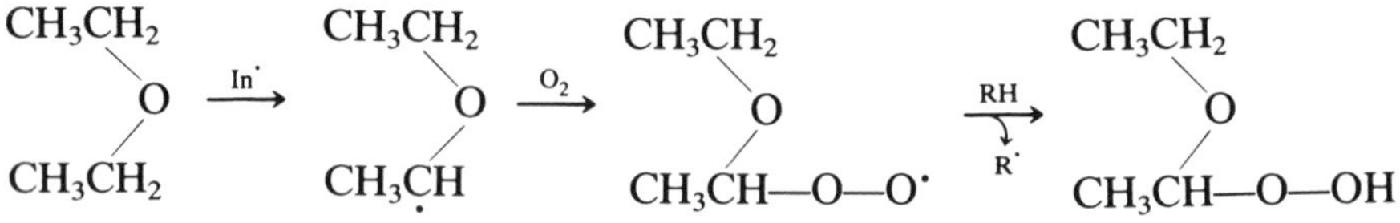

57

58

59

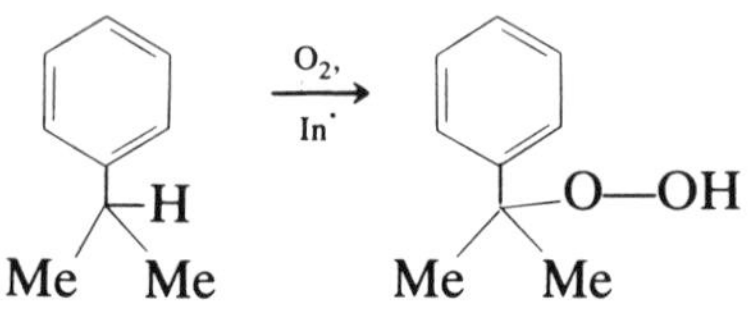

diradical and readily adds to radicals forming peroxides. Initiation may be brought about by light or adventitious radicals and a chain reaction may develop, sustained by the thermolysis of peroxide products (**56**). In this way, benzaldehyde is oxidized (**57**) to benzoic acid in air via perbenzoic acid (which undergoes acid-catalyzed polar hydrolysis), and ethers (**58**) and cumene (**59**) are converted to their hydroperoxides, those from **58** being explosive impurities. Autoxidation is responsible for the polymerization of 'drying oils' in paints.

15.3.6 Some preparatively useful radical chain reactions

The reduction of a secondary alcohol to alkane is not readily accomplished if reducible groups are present (which would preclude hydride reduction of the tosylate, for example). A technique introduced by Barton begins with the xanthate, **60** (Section 11.2.2) which reacts with $Bu_3Sn^{\bullet}$ radicals from tributyltin hydride. Addition to sulfur is followed by fragmentation to an alkyl radical which abstracts hydrogen from the tin hydride to continue the chain[193,194].

Another useful reaction is the radical chain reductive decarboxylation of acids commencing with esterification to 2-thiopyridone-*N*-oxide to form the thiohydroxamic ester, **61**. A tin radical will again bond to sulfur and fragmentation, releasing a pyridine with re-aromatization, is even more facile than with the xanthate. The carboxylate radical fragment loses CO_2 and is reduced by tin hydride or may be intercepted by solvent CCl_4 or $BrCCl_3$ yielding the halides. This reaction gives good yields of these halides when the Hunsdiecker reaction fails, e.g. with *p*-methoxybenzoic acid.

15.3.7 Linear free-energy relationships

The information discussed above suggests that concepts used in explaining polar reactions and electron supply and demand have little significance in radical chemistry. Indeed the rates of many radical reactions respond to polar substituent effects quite randomly[195]. This is not always the case, however; many others do give respectably linear Hammett plots but ρ-values are nearly always low and usually negative (Table 15.13). Interestingly, for reactions which show little response to electron donation, the best fit is usually with σ^+ rather than with σ^0 [198–201], but it seems possible that this reflects the ability of $+R$ groups to accept radical character, not necessarily positive charge. Nonetheless, radicals which respond to electron-donating groups in this way are referred to as 'electrophilic' and, similarly, the few which seem to correlate with a positive ρ-value are 'nucleophilic' and include *tert*-butyl and phenyl radicals.

A scale of substituent effects, $\sigma^{\bullet}$, is defined from relative rates of thermal decomposition of dibenzylmercury derivatives:

$$(ArCH_2)_2Hg \rightarrow 2ArCH_2^{\bullet} + Hg$$
$$\log k_X/k_H = \sigma^{\bullet} \ (\rho^{\bullet} = 1) \tag{15.5}$$

All values of $\sigma^{\bullet}$ are positive, indicating that any group is better than hydrogen in stabilizing a radical center. The reasons for this are to be sought in the frontier orbital interactions discussed in Section 15.4.5.

15.3.8 Electron-transfer catalysis[201–208]

Transference of an electron to or from a molecule (one-electron reduction or oxidation) will form an ion-radical which may have a very different

748

Table 15.13   LFER for some radical reactions[171]

Reaction	$T/°C$	ρ^+
$ArCH_3 + R^• → ArC\dot{H}_2 + RH$		
$R^• = Br^•$	80	−1·4
$R^• = ROO^•$	30	−0·6
$R^• = tert\text{-}BuO^•$	40	−0·4
$R^• = {}^•CCl_3$	55	−1·5
$R^• = Me_3C^•$	48	+0·50
$RH + Br^• → R^• + H\text{-}Br$		
$RH = Phr_2C\underline{H}_2{}^a$	75	−1·0
$RH = ArC\underline{H}_2CH_3$	70	−0·7
$RH = ArC\underline{H}(CH_3)_2$	70	−0·4
$RH = (ArC\underline{H}_2)_2O$		−0·1
$RH = Ar_2C\underline{H}OMe$		0
$ArI + Ph^• → Ar^• + PhI$		+0·57

^{a} Underlined $\underline{H}$ indicates position of hydrogen abstraction.

The $\sigma^•$ scale[196] and the $\sigma_H^•$ scale[197]

Substituents	$\sigma^•$
H, m-F, m-OMe	0
p-F	0·12
p-Cl	0·18
p-Me	0·39
p-Br	0·26
p-I	0·31
p-OMe	0·42
p-Ph	0·42
p-CN	0·71
p-NMe$_2$	0·61
p-NO$_2$	0·76

reactivity than the parent compound in substitutions, eliminations or other polar processes. Electron transfer, analogous to proton transfer, can bring about catalysis by lowering leaving-group bond strengths. Radical nucleophilic reactions of this sort may take place by chain transfer of the electron, **62**. The substitution step may be unimolecular or bimolecular, designated respectively $S_{RN}1$ or $S_{RN}2$. Initiation may be due to adventitious radicals or light. Bunnett found the displacement of iodide by enolate anions, **63**, occurred only in the presence of light or radical sources including the solvated electrons available in solutions of sodium in liquid ammonia or during electrolysis[209]. The ease of displacement of halogen in this reaction is in the order $I > Br > Cl > F$, inversely as the order of bond strengths but opposite to that experienced in S_N2Ar reactions (Section 10.5).

The reaction between benzyl chlorides and the nitropropane anion, **64**, normally gives *O*-alkylation and a linear Hammett plot except for the *p*-nitro compound, **65**, which deviates markedly and forms the *C*-alkyl product. Radical intermediates may be detected by ESR spectroscopy for this latter case and the presence of inhibitors reduces *C*-alkylation, pointing strongly to electron transfer.

A variety of leaving groups may be displaced from an aromatic ring via the radical anion, e.g. PhX where $X = I$, Br, Cl, F, SPh, NMe_3^+, groups not normally able to be displaced from unactivated aromatic rings. Moreover, when each of these compounds was allowed to react under irradiation with a mixture of diethyl phosphite and enolate, the ratio of the two products was found to be independent of the leaving group[210]. This is consistent with an

67

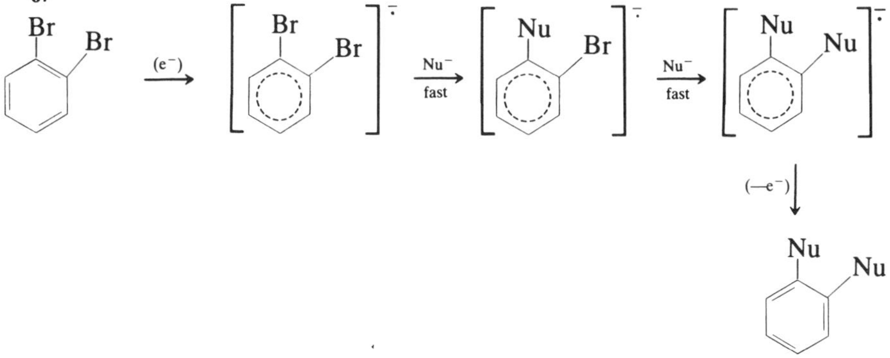

$S_{RN}1$, but not an $S_{RN}2$, mechanism. Alkyl transfer from the pyridinium compounds, **66**, is inferred to be $S_{RN}2$ from the complex kinetics observed[211]. Displacements of both halogens in dihalobenzenes may occur under radical-ion conditions without the monohalo product being observed. Presumably the formation of the radical-ion is rate-determining and displacement fast, continuing until termination occurs by electron loss, **67**[212].

Even nucleophilic displacements from phenacyl halides, $ArCOCH_2Hal$, may proceed by a radical chain mechanism. The following reaction is found to be completely inhibited by the addition of a stable radical (10% ($tert$-$Bu_2NO^{\bullet}$)[213]:

$$ArCO{\cdot}CH_2Br \xrightarrow{In^{\bullet}} ArCOCH_2Br^{\bar{\cdot}} \rightarrow ArCOCH_2^{\bullet} + Br^- \quad (S_{RN}1).$$

$$ArCOCH_2^{\bullet} + Nu^- \rightarrow ArCOCH_2Nu^{\bar{\cdot}} \xrightarrow{In} ArCOCH_2Nu + In^{\bullet}$$

$$(Ar = p\text{-}NO_2C_6H_5; \quad Nu = Me_2C^-NO_2.)$$

Analogous possibilities may be envisaged for reactions of cation-radicals, $S_{ON}1$ or $S_{ON}2$ processes and the electrophilic counterparts[214]. An example is **68**; the normally deactivating p-methoxyl group will allow the displacement of fluoride by the weakly nucleophilic acetate after oxidative initiation of a chain process[215,216]. It seems likely that investigation will reveal electron catalysis as being more common than has been hitherto suspected[217–219].

Radical intermediates have been implicated in the Cannizzaro reaction,

68

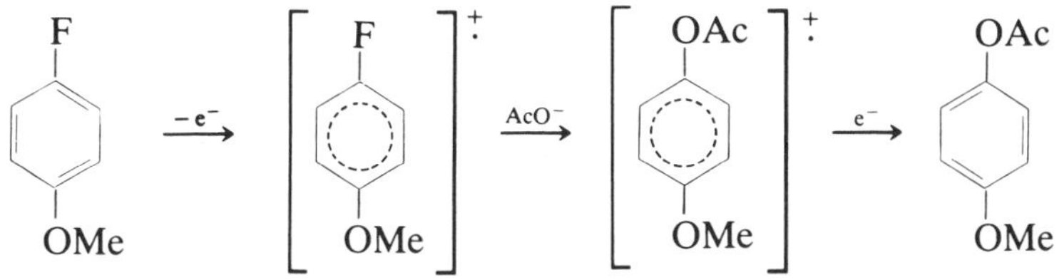

classically interpreted as occurring by a hydride transfer between two aldehyde molecules, **69** (path A). Significant incorporation of hydrogen from the solvent into the alcohol product suggested an alternative hydrogen atom transfer, path B, to be actually preferred.

15.4 FACTORS INFLUENCING THE REACTIVITIES OF RADICALS[220,221]

The formulation of a coherent qualitative model of reactivity appears simpler for ionic reactions than for radical processes. The reason is that, in the former case, reactivity is dominated by polar factors, the 'supply and demand' of electrons, while radical processes are controlled by a more equable and hence more complex interplay of polar, steric and bond-strength terms.

15.4.1 Radical stability

One measure of the stability of R· is the bond dissociation energy of R—H (Table 15.14). This quantity, however, contains terms due to differences in steric strain between RH and R· (usually relieved on forming a planar

752

Table 15.14 Steric effects upon radical reactions[222]

(a) Thermal decomposition of hydrocarbons $R^1R^2R^3C$—$CR^1R^2R^3$: temperature T for $t_{\frac{1}{2}} = 1\,h$, free energy of activation $\Delta G^{\ddagger}$ at 300°C

No.	R^1	R^2	R^3	$T/°C$ ($t_{\frac{1}{2}} = 1h$)	$\Delta G^{\ddagger}(300°C)/kJ\,mol^{-1}$ ($kcal\,mol^{-1}$)
1	CH_3	CH_3	CH_3	490	253 (60·5)
2	CH_3	CH_3	C_2H_5	420	231 (55·3)
3	CH_3	CH_3	$i\text{-}C_3H_7$	411	224 (53·6)
4	CH_3	CH_3	$i\text{-}C_4H_9$	412	225 (53·9)
5	CH_3	CH_3	$i\text{-}C_4H_9$	384	217 (51·9)
6	CH_3	CH_3	$2\text{-}C_3H_7$	329	194 (46·4)
7	CH_3	CH_3	$(CH_3)_3CCH_2$	321	194 (46·3)
8	CH_3	CH_3	$c\text{-}C_6H_{11}$	315	192 (45·8)
9	C_2H_5	C_2H_5	C_2H_5	285	180 (43·1)
10	CH_3	C_2H_5	$c\text{-}C_6H_{11}$	250	166 (39·6)
11	CH_3	CH_3	$tert\text{-}C_4H_9$	195	141 (33·7)
12	CH_3	CH_3	H	565	284 (68)
13	$c\text{-}C_6H_{11}$	$c\text{-}C_6H_{11}$	H	384	218 (52·1)
14	$c\text{-}C_6H_{11}$	$tert\text{-}C_4H_9$	H(D, L)	329	195 (46·7)
15	$c\text{-}C_6H_{11}$	$tert\text{-}C_4H_9$	H(meso)	285	178 (42.6)
16	$tert\text{-}C_4H_9$	$tert\text{-}C_4H_9$	H	141	124 (29·6)
17	CH_3	H	H	590	289 (69)
18	H	H	H	695	330 (79)
19	2,2,4,4-Tetramethylpentane			502	267 (63·9)
20	2,2,3,4,4-Pentamethylpentane			415	233 (55·8)
21	2,2,3,3,4,4-Hexamethylpentane			350	204 (48·8)
22	CH_3	C_6H_5	H	365	209 (50·0)
23	C_2H_5	C_6H_5	H	363	208 (49·7)
24	$i\text{-}C_3H_7$	C_6H_5	H	335	198 (47·4)
25	$tert\text{-}C_4H_9$	C_6H_5	H	289	176 (42·1)
26	$tert\text{-}C_4H_9$	C_6H_5	H	303	187 (44·6)
27	$tert\text{-}C_5H_4$	C_6H_5	H	259	168 (40·3)

Table 15.14 (*Continued*)

(b) Thermolyses of azo-compounds, $R_1R_2R_3C-N{=}N-CR_1R_2R_3 \xrightarrow{k} 2R_1R_2R_3C^{\cdot} + N_2$

			180°C	−28°C
			$\overset{\diagdown}{N}{=}N\diagdown$	$\overset{\diagdown}{N}{=}\overset{\diagup}{N}$
			trans	*cis*
R^1	R^2	R^3	k_{rel}	k_{rel}
CH_3	CH_3	CH_3	≡1·00	≡1·00^a
CH_3	CH_3	C_2H_5	1·19	4·4
CH_3	CH_3	i-C_3H_7	3·00	64
CH_3	CH_3	i-C_4H_9	7·51	153
CH_3	C_2H_5	C_2H_5	1·87	37
C_2H_5	C_2H_5	C_2H_5	3·65	1428
CH_3	CH_3	tert-C_4H_9	5·30	<1600

$^a k_1 = 0.615 \times 10^{-4}\,s^{-1}$.

(c) Thermolyses of peroxy esters, $R_1R_2R_3C\overset{\displaystyle O}{\underset{\displaystyle O-O tert\text{-}Bu}{\diagup\diagdown}}$, 60°C

R^1	R^2	R^3	k_{rel} (60°C)	$\Delta H^{\ddagger}$ $kJ\,mol^{-1}$ ($kcal\,mol^{-1}$)	$\Delta S^{\ddagger}$ $J\,K^{-1}\,mol^{-1}$ ($cal\,K^{-1}\,mol^{-1}$)
CH_3	CH_3	CH_3	≡1·00	118 (28·3)	22 (5·3)
CH_3	CH_3	C_2H_5	1·29		
CH_3	CH_3	1-C_8H_{16}	1·73		
CH_3	CH_3	$(CH_3)_2CHCH_2$	2·30		
C_2H_5	C_2H_5	C_2H_5	3·19		
C_2H_5	C_2H_5	i-C_3H_7	6·50		
CH_3	CH_3	tert-C_4H_9	3·4	114 (27·2)	19 (4·6)
CH_3	CH_3	$(CH_3)_3CCH_2$	2·6	111 (26·5)	8 (2·0)
i-C_3H_7	i-C_3H_7	i-C_3H_7	32	111 (26·6)	28 (6·7)

radical) together with any stabilization of the radical center *per se* by delocalization of the unpaired electron. Within a series of isomeric radicals, heats of formation may be used as an index of stability: as an example, values for the isomeric butyl radicals may be compared by Eq. (15.6):

$$\Delta H_f(R^{\cdot}) = \Delta H_f(RH) - \Delta H_f(H^{\cdot}) - \Delta H^0 \qquad (15.6)$$

using $D_{R-H} = 418$ (100), 405 (96·7) and 391 (93·5) kJ mol^{-1} (kcal mol^{-1}) for primary, secondary and tertiary alkyl positions, $\Delta H_f(H^{\cdot}) = 218$ kJ mol^{-1}, (52) (kcal mol^{-1}).

The greater the number of alkyl groups involved in the tetrahedral → trigonal stereochemical change, the more negative is the heat of formation in this series.

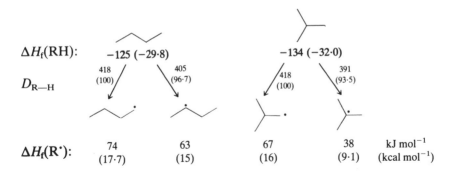

$\Delta H_f(\text{RH})$: $-125\ (-29\cdot8)$ $-134\ (-32\cdot0)$

$D_{\text{R—H}}$ 418   405   418   391
                    (100)   (96·7)   (100)   (93·5)

$\Delta H_f(\text{R}^{\boldsymbol{\cdot}})$:   74   63   67   38   kJ mol^{-1}
                    (17·7)   (15)   (16)   (9·1)   (kcal mol^{-1})

Reactivity scales are also frequently assigned significance in the context of stability (e.g. Tables 15.8 and 15.11), low reactivity and high selectivity being implicitly associated with stability, but in most cases these characteristics are also greatly dependent upon steric factors. For example, the lower rates and greater selectivity in hydrogen abstraction by $CCl_3^{\boldsymbol{\cdot}}$ compared with $CH_3^{\boldsymbol{\cdot}}$ could be interpreted as due to unpaired electron delocalization or greater bulk of the $CCl_3^{\boldsymbol{\cdot}}$ which creates B strain upon forming a tetrahedral product. It is less plausible to attribute rates of hydrogen abstraction from cycloalkanes to electron delocalization. These fall in the order cyclobutane > cyclopentane > cyclohexane, which parallels the order of strain release (relief of eclipsing interactions within the four- and five-membered rings on forming a planar radical), and the magnitude of this effect is at least as great as the differences in rates of abstraction of tertiary, secondary and primary hydrogens in acyclic systems in which case B-strain release accompanies radical formation. Strain is apparently the most important factor in the radical chemistry of saturated hydrocarbon systems. Despite the observation that the rates of formation of cations, **a**, and of radicals, **b**,

a $R\text{—Br} \rightarrow R^{+} + Br^{-}$,
b $R\text{—H} + X^{\boldsymbol{\cdot}} \rightarrow R^{\boldsymbol{\cdot}} + X\text{—H}$,

show similar structural dependence (tertiary > secondary > primary), the reasons are different. Ionization is assisted by hyperconjugation, while $H^{\boldsymbol{\cdot}}$ transfer is assisted by strain relief. Hyperconjugation does not seem to be important in saturated radicals such as *tert*-butyl but electron delocalization is a major stabilizing factor in allylic and benzylic radicals. Radical additions to styrenes are much faster than to aliphatic alkenes, yet the low reactivity of 1,2-diphenylethene points to steric hindrance by the α-phenyl group. This factor determines regioselectivity in radical additions to alkenes, and is reinforced by conjugation only when π-groups or strongly polar groups are present. The following figures illustrate this point.

$$R^• + CH_2\!\!=\!\!CHF \rightarrow R\!\!-\!\!CH_2\dot{C}HF + \dot{C}H_2\!\!-\!\!CHFR$$

$$\underset{\alpha}{} \quad \underset{\beta}{}$$

$R^•$:	$CF_3^•$	$CF_3CF_2^•$	$(CF_3)_2CF^•$	$(CF_3)_3C^•$
k_α/k_β	10	17	50	200

At the same time, the rate of attack at the α-position by all four radicals (relative to that at ethene) remains constant so that these figures represent increasing selectivity at the β-position, and are only compatible with steric control of the rates.

In many radical displacement reactions, there is a direct relationship between the energy of activation and the bond strength of the bond being broken (the Evans–Polanyi equation):

$$E_A = \alpha[D_{R-H}] + \text{constant}.$$

This indicates that the same blend of factors controls both atom transfer and bond fission but the susceptibility parameter (rather analogous to the Brønsted α) is a guide to transition-state structure. The larger is "α", the later the transition state should be (i.e. more product-like) which, according to the Hammond postulate, would mean a trend towards a more endothermic reaction. Figure 15.4 illustrates this parallel behavior.

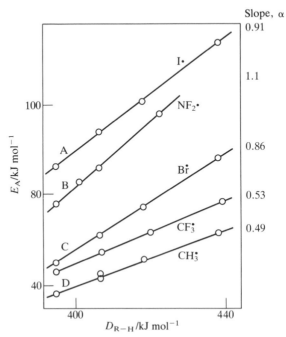

Fig. 15.4 Reactivity–selectivity relationship between energies of activation for hydrogen abstraction from X—H and D_{X-H}: the slopes, α, are a measure of the selectivity of the abstracting radical, $R^•$.

15.4.2   Polar influences

The correlations between radical abstraction rates and Hammett constants is evidence that some kind of polar influence can be important. This is also evident in departures from the Evans–Polanyi relationship, for instance in abstractions of hydrogen from H—Cl, for which polarity of the attacking radical would be expected to be important if anywhere:

$$R^{\cdot} + H\!-\!Cl \rightleftarrows R\!-\!H + Cl^{\cdot}$$

$R^{\cdot}$:	$CH_3^{\cdot}$	$H^{\cdot}$	$CF_3^{\cdot}$
E_A (forward)^a:	10·5	16·7	20·7
	(2·5)	(4·0)	(4·9)
E_A (back)^a:	14·6	20·9	33·0
	(2·5)	(5·0)	(7·0)
ΔH_a:	−4·1	−4·2	−13·7
	(−1·0)	(−1·0)	(3·3)

^{a} Values in kJ mol^{-1} (kcal mol^{-1}).

The most exothermic reaction, that of $CF_3^{\cdot}$, now has the highest activation energy, which can be explained in terms of a polar transition state, **70**, in which $CH_3^{\cdot}$, being electron-donating or 'nucleophilic', can disperse charge while $CF_3^{\cdot}$, an 'electrophilic' radical, resists charge separation.

Captodative stabilization is the term applied to radicals which have both electron-donating and electron-withdrawing substituents attached to the radical center and are particularly stable[121,223–230]. Examples of captodative radicals already described are **21–24, 44, 45**. The effectiveness of this type of structure in engendering stability is associated with the ability of the unfilled orbital to transmit electronic charge between the donor and acceptor groups, **71**, stabilizing structures such as **72**. This type of polar effect may be invoked to explain the effectiveness of substituents in promoting α-attack in 1-X-butanes, Table 15.8. For structures of the type X—CH$_2$—Pr, the reaction is most favored by $-R$ groups (CN, COOR, though not NO$_2$) and disfavored by $+R$ groups (OCOR, R) and $-I$ groups (CF$_3$), although the data are overlaid with steric effects which make this interpretation less clear.

Tedder has summarized stereoelectronic effects on radical reactions in five rules.

70

$$\overset{\rightarrow}{CH_3} \cdots \overset{\delta+}{H} \cdots \overset{\delta-}{Cl}$$

$$\overset{\leftarrow}{CF_3} \cdots \overset{\delta+}{H} \cdots \overset{\delta-}{Cl}$$

$^+X\!-\!C\!-\!Y$

71

$$\begin{matrix} Me \\ \\ Me \end{matrix} \!\!\! \dot{C}\!\rightarrow\!CN$$

$$\begin{matrix} Me \\ \\ Me \end{matrix} \!\!\! \dot{C}\!\rightarrow\!O^-$$

72

(1) In non-polar transition states, the relative rates of an S_H2 reaction will depend upon the bond strengths of bonds being broken (Evans–Polanyi principle). Bond strengths are the resultant of resonance stabilization of an incipient radical and relief of steric strain (i.e. the differences in electronic and steric effects between the radical and its precursor).

(2) Selectivity is determined mainly by bond strengths and consequently depends upon the heat of reaction. If ΔH is large and positive (endothermic) the reaction will be very selective, while if ΔH is large and negative (exothermic) the reaction be very undiscriminating.

(3) Rates of atom transfers which are thermoneutral ($\Delta H \approx 0$) are very dependent upon polarity of the transition state. For example,

$$CH_4 + H^{\cdot} \rightarrow CH_3^{\cdot} + H_2;$$

$$\Delta H^0 = -4(-1), \qquad E_A = 50 \ (12) \ \text{kJ mol}^{-1} \ (\text{kcal mol}^{-1}).$$

$$CH_3 + H\!\!-\!\!Cl \rightarrow CH_4 + Cl^{\cdot};$$

$$\Delta H^0 = -4 \ (-1), \qquad E_A = 10 \ (2{\cdot}4) \ \text{kJ mol}^{-1} \ (\text{kcal mol}^{-1}).$$

(4) It is possible to denote radicals as having electrophilic character ($Hal^{\cdot}$, $CHal_3^{\cdot}$, etc.) or nucleophilic (*tert*-Bu$^{\cdot}$, $Me_3Ge^{\cdot}$). Then electron-withdrawing substituents will facilitate atom transfer by nucleophilic radicals and conversely, electron-donating substituents will facilitate that of electrophilic radicals. This is the same principle that is applied in polar reactions (though effects are much smaller); compare the effects of electron-withdrawing X in reactions **a** and **b** below with those is **c**.

	$k_X/k_H{}^a$
a $X\!\leftarrow\!C\!\!-\!\!\overset{\frown}{H} \ (Nu^{\cdot} \rightarrow X\!\leftarrow\!\dot{C}$ $+$ $\quad$ H—Nu	>1
b $X\!\leftarrow\!C\!\!-\!\!\overset{\frown}{H} \ (E^{\cdot} \rightarrow X\!\leftarrow\!\dot{C}$ $+$ $\quad$ H—E	<1
c $X\!\leftarrow\!C\!\!-\!\!H\frown\!:Nu^- \rightarrow X\!\leftarrow\!\bar{C}$ $+$ $\quad$ H—Nu	≫1

^{a} k_X/k_H = rate ratio for X substituent : H substituent.

Fluorine as a substituent is found to facilitate hydrogen abstraction (relative to H) by nucleophilic radicals such as $Me_3Ge^{\cdot}$ but to retard abstraction by an electrophilic radical ($Br^{\cdot}$).

(5) Exothermic reactions, according to the Hammond postulate, have an early transition state; consequently neither steric strain release nor unpaired electron delocalization effects are as important as in endothermic reactions, which are associated with a late transition state. Of the two endothermic hydrogen transfers,

$$Br^{\cdot} + H\!\!-\!\!CCl_3 \rightarrow Br\!\!-\!\!H + {}^{\cdot}CCl_3; \qquad \Delta H = +37 \ \text{kJ mol}^{-1}, \ (8{\cdot}8 \ \text{kcal mol}^{-1}),$$

$$Br^{\cdot} + H\!\!-\!\!CH_3 \rightarrow Br\!\!-\!\!H + {}^{\cdot}CH_3, \qquad \Delta H = +72 \ \text{kJ mol}^{-1} \ (17 \ \text{kcal mol}^{-1}),$$

the first is faster by a factor of 10^3 despite unfavorable polarization, on account of the release of steric strain in chloroform being greater than that from methane. Steric considerations will override polarity when the reaction has a late transition state. The interpretation of radical reactivity along these lines will be more difficult when the dominant influences on rates are less marked.

15.4.3 Solvent effects on radical reactions

With highly reactive radical intermediates, it is difficult to vary the solvent widely: so many of the possible solvent media become involved in subsequent reactions, often by hydrogen abstraction, that product comparisons are meaningless. On the whole, thermolyses of peroxides and azo compounds are remarkably insensitive to the medium, rates of decomposition of dibenzoyl peroxide varying by a factor of only four between CCl_4 and acetic acid[1]. Chlorination of alkanes has been examined in solvents necessarily limited to benzene, carbon disulfide, carbon tetrachloride and the alkanes: significant changes in both rate and selectivity were brought about (Table 15.15). Solvents which solvate chlorine atoms (benzene, CS_2) diminish their reactivity and increase the selectivity of attack. Polarizable solvents (CCl_4) tend to solvate the polar transition state (in which the polar molecule H—Cl is forming) and level out selectivities[222].

15.4.4 Steric effects in radical reactions[231]

Since polar influences on radical processes are greatly reduced compared with those on ionic reactions, steric effects are often more clear cut.

Table 15.15 Solvent effects upon radical chlorinations[222]

Chlorination of *n*-heptane

Conditions	Relative rate of chlorination CH_3——CH_2——CH_2——C_4H_9		
Cl_2 neat, 20°C	1	3·0	2·9
Cl_2 in CS_2, 20°C	1	27	35

Chlorination at different sites in 4-chloropentanoyl chloride (20°C)

Solvent	Relative rate of chlorination[a] $Cl-CH_2$——CH_2——CH_2——CH_2——$COCl$			
(Gas phase)	0·7	1·4	1·4	0·12
CCl_4	0·4	0·6	0·2	0·07
C_6H_6	0·5	1·0	0·7	0·03
CS_2	0·7	1·5	0·9	0·06

[a] Rate of attack on primary position of *n*-butene is taken as unity.

Homolysis of ethane derivatives relieves front strain from which the steric parameter, $\mathscr{S}$, is derived (Section 8.2.7). This is the major factor which weakens the central bond of hexaphenylethane to the point that it homolyzes at room temperature, and similar effects are evident in homolyses of azo compounds and peroxides (Table 15.14). In fact, rates of thermolysis of azo compounds, $R_3C-N{=}N-CR_3$, correlate with rates of solvolysis of $R_3C{\cdot}pnb$, strongly suggesting the relief of strain to be important in each process. Steric parameters also correlate well with reactions in the opposite direction, e.g. radical stability measured as the

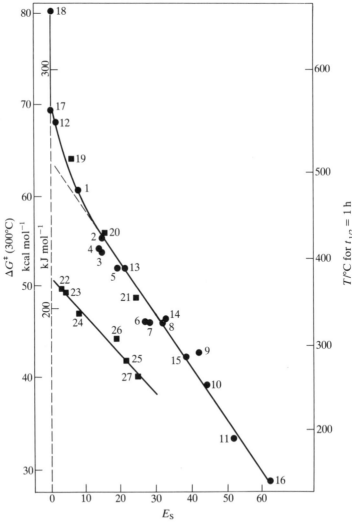

Fig. 15.5 Relation between thermal stability of substituted ethanes, and their strain energy[231]; the numbers correspond to those in Table 15.14a.

half-lifetime for dimerization (Fig. 15.5);

$$\Delta G^{\ddagger}(300°C)/\text{kJ mol}^{-1} = -2 \cdot 51 E_s + 272. \qquad (15.6)$$
$$(\text{kcal mol}^{-1}) \quad (-0 \cdot 6) \quad (65)$$

The same degree of F strain is created in additions to or abstractions by radicals which cause a reversion to the tetrahedral geometry. A plot of enthalpies of halogen abstraction against $\mathscr{S}$, Fig. 15.6, divides radicals into two classes according to whether they are rigid (aryl or vinyl) or flexible (alkyl). Rates of reaction of the former show a greater sensitivity to the amount of F strain created than do flexible radicals.

In relative rates of hydrogen abstraction by alkyl radicals and in the absence of steric factors, the order primary < secondary < tertiary is observed, i.e. the order of stability of the radical formed. When the attacking

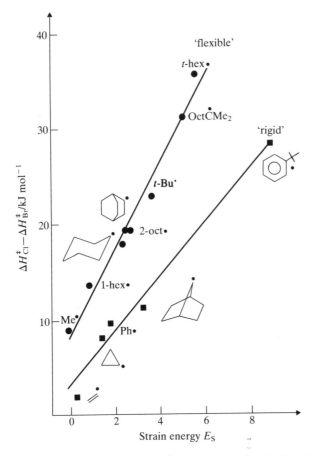

Fig. 15.6 Correlation between enthalpy difference for abstraction of Cl· and Br· from CCl_4 and CCl_3Br, respectively, and steric constants, $\mathscr{S}$. The enthalpy difference is used since it cancels bond-dissociation energy terms and focuses attention upon differences in steric energy.

Table 15.16 Relative rates of hydrogen abstraction from 2-methylbutane by various radicals

$$R^{\cdot} + Me_2CH-CH_2-CH_3 \rightarrow \overset{\cdot}{C}H_2CH-CH_2-CH_3 + Me-CH\overset{\cdot}{C}H-CH_3 + Me\overset{\cdot}{C}-CH_2CH_3$$

with Me substituents below respectively.

R	k_{ref}		
	Primary radical	*Secondary radical*	*Tertiary radical*
Me⁺	0·32	0·93	1·0
isoPr⁺	0·25	0·70	1·0
neoPe⁺	0·71	2·70	1·0
tert-Bu⁺	1·70	6·00	1·0

radical becomes large the steric order of rates – primary > secondary – is superimposed, with the results shown in Table 15.16. Selectivities can therefore be affected both by size and frontier orbitals of the attacking radical.

Nevertheless, radical dimerization is more sensitive to steric effects than is abstraction, since twice the F strain is created. This can be seen in the correlation between the ratio of disproportionation to dimerization[221] which, for alkyl radicals, obeys Eq. (15.7):

$$\log\left(\frac{k_{dis}}{k_{dim}} \cdot \frac{1}{n}\right) = -0·48E_s^c - 1·73 \qquad (15.7)$$

where n is the number of β-hydrogens.

Radical additions to alkenes are not very susceptible to polar effects but are severely affected by steric properties of substituents on the alkene which can be the major influence on regioselectivity; this is demonstrated by comparing k_{rel} values for attack by $CF_3^{\cdot}$:

Alkene:	$CH_2{=}CH_2$	$CH_2{=}CHMe$		$CH_2{=}CMe_2$		$MeCH{=}CMe_2$	
k_{rel}:	1·00	2·3	0·2	6·0	0·5	2·8	0

In aromatic homolytic substitution, *ortho* partial rate factors are reduced by bulky substituents in the aromatic ring (**73**).

Me   *tert*-Bu

(63, 21, 16)   (24, 49, 27)

Products (%) from attack of Ph⁺ are indicated

73

15.4.5 Frontier orbital considerations

Radical reactions occur usually between neutral species so that the frontier orbital overlap term of Eq. (6.36) is likely to be dominant in the early stages of the reaction. For a transfer reaction between a radical and an electron-paired species, the most common type of homolytic reaction, one must consider interactions between the singly-occupied orbital (SOMO) of the radical and the frontier orbitals of the other reagent. Both HOMO and LUMO will lead to bonding interactions through a lowering of energy but the relative importance of the two interactions depends upon the energies of the interacting orbitals. It is for this reason that an adjacent electron pair, n or π, or a vacant orbital is able to stabilize a radical center (**74**).

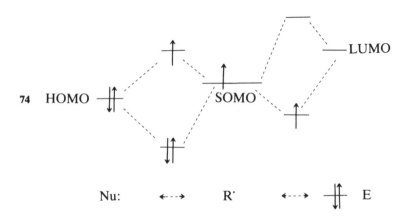

If the SOMO–LUMO interaction is the more important the radical is said to show nucleophilic character, while electrophilic radicals owe their properties to a major SOMO–HOMO term. The difference between nucleophilic and electrophilic radicals lies in the energy of the SOMO. There are interesting consequences of this generalization. A phenyl substituent adjacent to a carbon-centered radical makes it relatively nucleophilic, while a carbonyl group engenders electrophilic character. In the copolymerization of styrene and maleic anhydride these two types respectively are produced at the growing end of the polymer. A styrene-terminated polymer radical will interact most strongly with the LUMO of maleic anhydride while a maleic anhydride-terminated radical will preferentially interact with the HOMO of styrene leading to a strong tendency towards alternation. This follows without invoking polar forces (**75**).

Hydrogen-atom abstractions from substituted toluenes, such as **16**, lead to Hammett reaction constants ρ which, though small, are generally negative (Table 15.13). This implies that the radical is displaying electrophilic character, the exceptions apparently being *tert*-Bu· and $R_3Si·$. Reaction

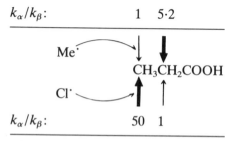

75

constants correlate roughly with SOMO energies and enable a scale of radical character to be drawn up as follows.

$E(SOMO)$/kJ mol^{-1}(kcal mol^{-1}):							*tert-*	*tert-*			
tert-Bu˙	Et$_3$Si˙	2-Hex˙	Cl$_3$C˙	Ph˙	Me˙	H˙	BuO˙	BuOO˙	CH$_2^-$	Br˙	Cl˙
−670	−675	−750	−850	−890	−950	−1310	−1160	−1110	−1050	−1140	−1250
(−160)	(−161)	(−179)	(−203)	(−212)	(−227)	(−313)	(−227)	(−265)	(−251)	(−272)	(−298)

Nucleophilic ⟶ Electrophilic

Stereospecificity of radical attack can often be explained on this basis. Hydrogen abstraction from propanoic acid can take place at C2 or C3. The carboxyl substituent, a Z-type, will lower both HOMO and LUMO at C2 and make it accessible to nucleophilic radicals such as Me˙: C3 is unaffected by the carboxyl group and its high-energy HOMO is favorable to attack by electrophilic radicals such as Cl˙ (**76**). The result is a change in selectivity.

$$k_\alpha/k_\beta: \qquad 1 \quad 5\cdot2$$

Me˙

CH_3CH_2COOH

Cl˙

$$k_\alpha/k_\beta: \qquad 50 \quad 1$$

Regioselectivity of attack at a vinyl group may be frontier-orbital determined. R- or Z-substituted ethenes have the larger coefficient of both HOMO and LUMO at C2 (Fig. 14.8). X-substituted ethenes such as vinyl ethers, $CH_2{=}CHOR$, have the HOMO principally at C2 but the LUMO mainly at C1. In most cases, therefore, a radical will attack a monosubstituted ethene at C2 (the less substituted end). A violation of this rule would require a substrate CH_2=CHX with an especially low LUMO. This condition is met in vinyl fluoride, the inductive effect of fluorine lowering the energy level of frontier orbitals while still retaining the reversal of polarity between them. It is found that radicals tend to add to this compound at C1 (76).

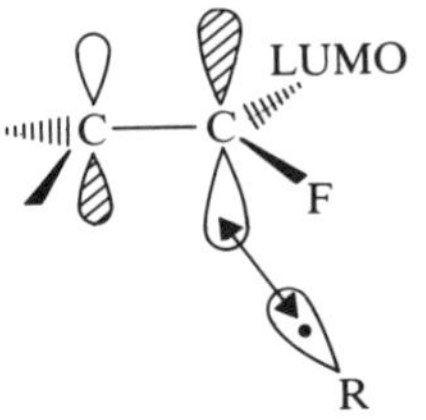

76

$$CH_2=CHF + R^{\cdot} \rightarrow {}^{\cdot}CH_2-\underset{\underset{R}{|}}{C}HF.$$

It is implicit that these considerations assert the transition state to be very early on the reaction coordinate and polar effects can be treated by frontier orbital theory. This is likely to be so for such a reactive, unselective reagent as a free radical but the question of regioselectivity is undoubtedly complex and the treatment given above oversimplified. Steric effects, for example, are very marked and tend to favor attack at the less substituted end of the double bond[221,231-234]. Molecular orbital methods, until recently, have not been as readily applied to radicals as to 'closed-shell' molecules. The former, having a half-filled orbital, require 'configuration-interaction' methods which consume a great deal more computer time than is the case for all-spin-paired molecules. This situation is now changing[235]: a recent computation of barrier heights for addition of a hydrogen atom to X- and Z-substituted ethenes gave the following results:

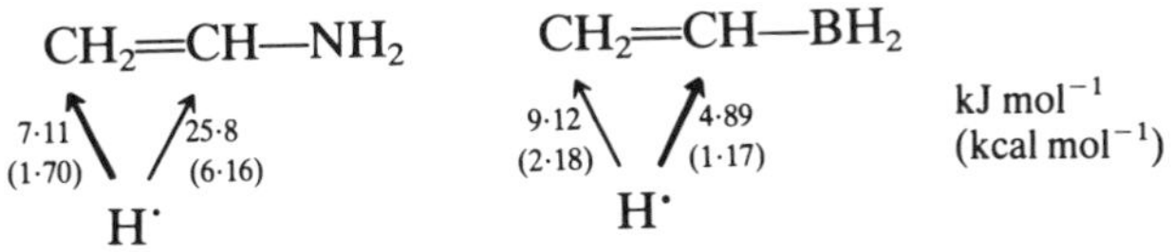

Some aspects of aromatic homolytic substitution are amenable to a frontier orbital explanation. Nucleophilic radicals tend to substitute anisole in a rather individualistic manner, mainly *ortho* and *meta*. As the attacking radical becomes more electrophilic the pattern changes to *ortho* + *para,* similar to an electrophilic polar substitution. Evidently then the low-energy SOMO is better able to interact with the HOMO of the substrate, for which the electron population is highest at *ortho* and *para* positions (Section 10.4.10).

15.5 THE STEREOCHEMISTRY OF RADICALS

Techniques such as matrix isolation and gas-phase measurements have allowed spectra — IR, UV and especially ESR — to be obtained for a great variety of radicals including the simplest, such as $CH_3^{\cdot}$. These data allow the geometry of the species to be investigated: there appears to be a consensus of opinion that carbon-centered radicals are planar or, if pyramidal, less than 10° from the planar geometry and freely inverting[234]. The evidence for this comes from coupling constants in the ESR spectra. Coupling between the unpaired electron and ^{13}C or ^{1}H in the methyl radical depends upon the amount of s-character in the SOMO. If this is a pure p-orbital (implying a planar geometry) there will be zero electron density at the nucleus and a very small coupling constant, a. At the other extreme, if the SOMO were a pure s-orbital, a would be 1200 gauss and, proportionately, 300 gauss for pyramidal sp^3. In fact the value of $a(^{13}C)$ in $CH_3^{\cdot}$ is 23 gauss, this originating from an alternative coupling mechanism, spin polarization of the σ-bonds. Consequently $CH_3^{\cdot}$ is planar but easily deformed into a pyramidal conformation[236]. Radicals formed from chiral substrates invariably lose their chirality since the electrostatic forces which aid configuration-holding in ion-pairs are not available for radical-pairs. The cyclohexyloxy radicals formed from either *cis*- or *trans*-di(4-*tert*-butylcyclohexyl) peroxide, **77**, appear to be identical in their reactions, both as to rates and selectivities[237]. Vinyl radicals do, however, retain their integrity and do not undergo geometrical isomerization, **78**.

PROBLEMS

1 Identify the radicals which give the following ESR spectra:

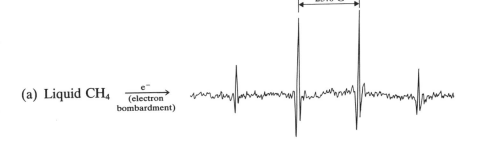

(a) Liquid CH₄ $\xrightarrow[\substack{\text{(electron} \\ \text{bombardment)}}]{e^-}$

23.0 G

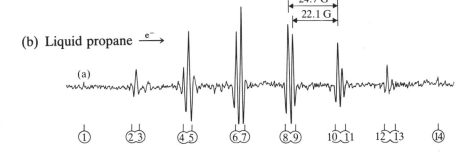

(b) Liquid propane $\xrightarrow{e^-}$

24.7 G
22.1 G

(a)

① ②③ ④⑤ ⑥⑦ ⑧⑨ ⑩⑪ ⑫⑬ ⑭

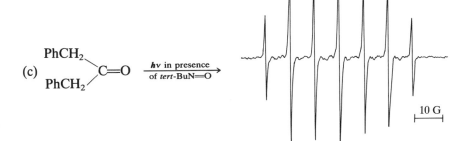

(c) $\begin{array}{c} \text{PhCH}_2 \\ \diagup \\ \text{PhCH}_2 \end{array}\hspace{-0.3em}\text{C}=\text{O}$ $\xrightarrow[\text{of }tert\text{-BuN}=\text{O}]{h\nu \text{ in presence}}$

10 G

(d) Me₃N $\xrightarrow{\text{H}_2\text{O}_2}$

+S

5 G

(e) CD$_3$—$\overset{\displaystyle O}{\overset{\|}{S}}CD_3$, OH$^-$
    +*tert*-BuNO →

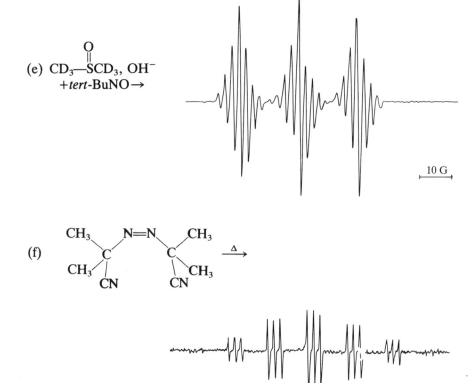

10 G

(f)

Δ →

(g) HC(NO$_3$)$_3$ $\xrightarrow{\text{electrolysis}}$

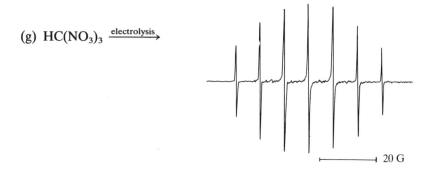

20 G

[*J. Chem. Phys.*, **39**, 2147 (1963); *Can. J. Chem.*, **60**, 1495, 1489 (1982).]

2 The vinyl radical shows the expected eight-line ESR spectrum when trapped in a solid matrix but in solution only four lines may be observed. Suggest an explanation.

3 Investigate the effect of β-methyl substitution on the ease of hydrogen abstraction.

4 Comment upon the following relative rates of hydrogen abstraction by halogen atoms.

Radical	Substrate			
	CH_4	$-CH_3$	$>CH_2$	$-\overset{\mid}{\underset{\mid}{C}}H$
F⋅	0·7	1	1	1·3
Cl⋅	0·03	1	2	3
Br⋅	0·003	1	640	$2{\cdot}5 \times 10^4$
I⋅	4×10^{-4}	1	16	6200

[*J. Amer. Chem. Soc.*, **91**, 1877, 1879 (1969).]

5 Write chain mechanisms for the peroxide-initiated additions of (a) $BrCCl_3$, (b) MeSH, to styrene.

6 Suggest mechanisms for the following transformations.

(a)

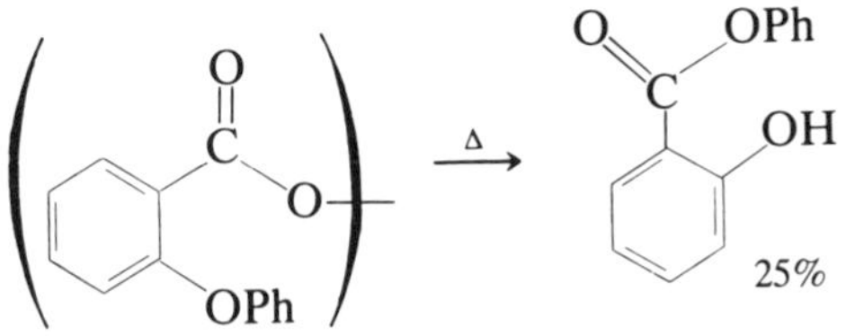

25%

[*J. Amer. Chem. Soc.*, **77**, 4411 (1955).]

(b)

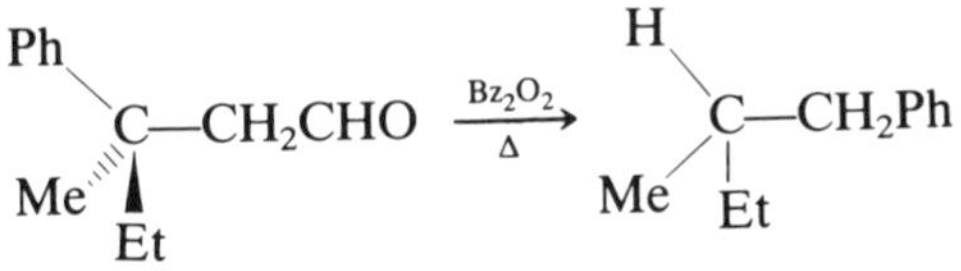

(optically active)   (racemic)

[*Ber.* **98**, 2478 (1965).]

(c)

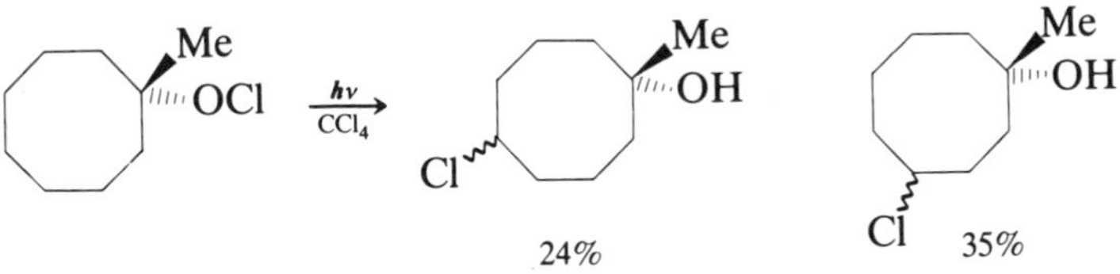

24%   35%

[*J. Amer. Chem. Soc.*, **87**, 3111 (1965).]

(d)

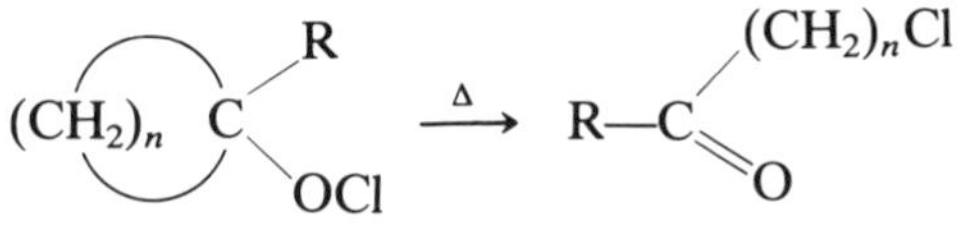

[*J. Org. Chem.*, **21**, 140 (1956).]

Reactions via free radicals

(e)

$RCOCH_3 + Me_2C{=}CH_2$
[Norrish II products)

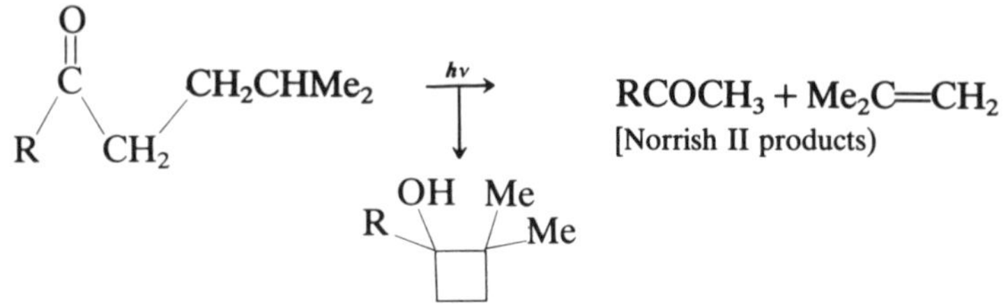

[*J. Amer. Chem. Soc.*, **80,** 2913 (1958).]

(f) $R{\cdot}COOH + Pb^{IV}(OAc)_4 \xrightarrow{\Delta} ROAc + CO_2 + Pb(OAc)_2 + AcOH$

(g)

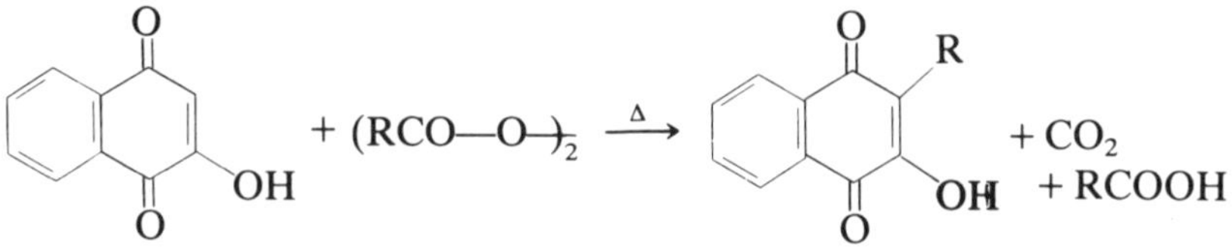

$+ (RCO{-}O)_2 \xrightarrow{\Delta}$

$+ CO_2$
$+ RCOOH$

[*J. Amer. Chem. Soc.*, **70,** 3174 (1948).]

(h)

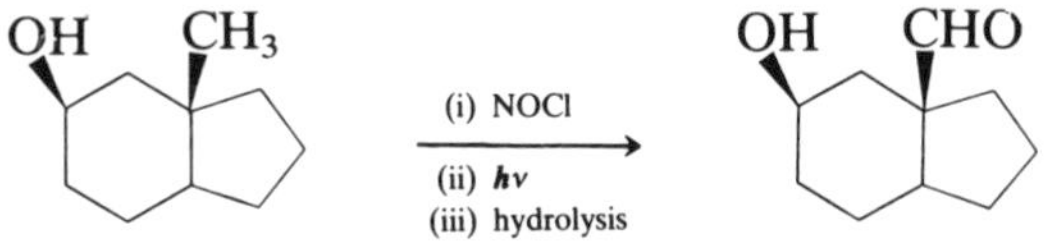

$\xrightarrow[\substack{(ii)\ \boldsymbol{h\nu} \\ (iii)\ \text{hydrolysis}}]{(i)\ NOCl}$

(i)

$R{-}Br \xrightarrow[\substack{\text{electrolysis,} \\ \text{cathode}}]{SO_2} R{-}\overset{\displaystyle O}{\underset{\displaystyle O}{S}}{-}R$

(j)

$\xrightarrow[\text{MeOH}]{\text{electrolysis}}$

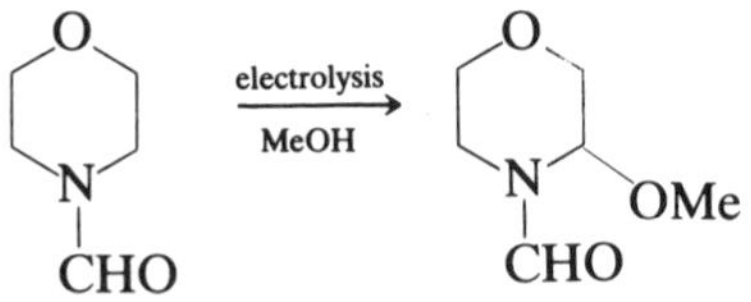

[*Phil. Trans. Roy. Soc.*, **302A,** 237 (1981).]

(k)

$Ar{\cdot}CMe_2Cl + Ar'S{-}O^- \rightarrow ArCMe_2{-}\overset{\displaystyle O}{\underset{\displaystyle O}{S}}{\cdot}Ar'$

and

$Ar{\cdot}CMe_2{-}O{-}\overset{\displaystyle O}{S}{\cdot}Ar' + Ar''S{-}O^- \rightarrow \ \ Ar{\cdot}CMe_2{\cdot}\overset{\displaystyle O}{\underset{\displaystyle O}{S}}{\cdot}Ar'' + Ar{\cdot}CMe_2{\cdot}\overset{\displaystyle O}{\underset{\displaystyle O}{S}}{\cdot}Ar'$

($Ar = p\text{-}NO_2C_6H_4$; $Ar' = Ph$; $Ar'' = p\text{-}MeC_6H_4$). [*J. Org. Chem.*, **45**, 5294 (1980).] N.B. The second reaction is found to be inhibited by nitroso compounds.

REFERENCES

1   C. Walling, *Free Radical Chemistry*, Wiley, New York, 1966.
2   W. A. Pryor, *Free Radicals*, McGraw-Hill, New York, 1966.
3   J. K. Kochi, Ed., *Free Radicals*, Wiley–Interscience, New York, 1973.
4   R. C. Sealy, *Electron Spin Res.*, **4**, 30 (1977).
5   W. A. Waters, *Chemistry of Free Radicals*, Oxford U.P., 1948.
6   L. Kaplan, *React. Intermed.*, **1**, 163 (1978).
6a   L. Kaplan, *Free Radicals as Reactive Intermediates*, Vol. 2, Wiley, New York, 1981, p. 251.
7   A. G. Davis, *J. Organometal. Chem.*, **200**, 87 (1980).
8   T. Koenig, Ref. *3*, Vol. I, Chapter 3.
9   F. Paneth and W. Hofeditz, *Chem. Ber.*, **62**, 1335 (1929).
10   W. Tsang, *J. Amer. Chem. Soc.*, **107**, 2872 (1985).
11   G. E. Adams, *Ann. Rep. Prog. Chem.*, **65B**, 223 (1968).
12   M. W. Rathke and A. Lindert, *J. Amer. Chem. Soc.*, **93**, 4605 (1971).
13   J. K. Kochi, A. Bemis and C. L. Jenkins, *J. Amer. Chem. Soc.*, **90**, 4616 (1968).
14   T. Cohen and T. Poeth, *J. Amer. Chem. Soc.*, **94**, 4363 (1972).
15   S. C. Dickermann and K. Weiss, *J. Org. Chem.*, **22**, 1070 (1957).
16   P. M. Nave and W. S. Trahanovsky, *J. Amer. Chem. Soc.*, **93**, 4536 (1971).
17   J. K. Kochi, *J. Amer. Chem. Soc.*, **87**, 1811 (1965).
18   L. F. Fieser, R. Clapp and W. Daudt, *J. Amer. Chem. Soc.*, **64**, 2060 (1942).
19   E. J. Behrman, *J. Amer. Chem. Soc.*, **85**, 3478 (1963).
20   C. Walling and S. Kato, *J. Amer. Chem. Soc.*, **93**, 4275 (1971).
21   J. K. Kochi and P. E. Mocadlo, *J. Amer. Chem. Soc.*, **88**, 4094 (1966).
22   T. Kawamura, M. Ushio, T. Fujimoto and T. Yonezawa, *J. Amer. Chem. Soc.*, **93**, 908 (1971).
23   D. Meisel and R. W. Fessenden, *J. Amer. Chem. Soc.*, **98**, 7505 (1976).
24   N. L. Weinberg and H. R. Weinberg, *Chem. Rev.*, **68**, 449 (1968).
25   P. Neta, *J. Chem. Ed.*, **58**, 110 (1981); E. S. Pish and N. C. Yang, *J. Amer. Chem. Soc.*, **85**, 2124 (1963).
26   E. S. Fawcett, *Chem. Rev.*, **47**, 219 (1950); C. S. G. Phillips and R. J. B. Williams, *Inorganic Chemistry*, Oxford U.P., 1966.
27   R. G. Woolford, *Can. J. Chem.*, **40**, 1846 (1962).
28   R. G. Woolford, W. Arbic and A. Rosser, *Can. J. Chem.*, **42**, 1788 (1964).
29   W. J. Dixon and R. O. C. Norman, *J. Chem. Soc.*, 3119, 4850, 4857 (1963).
30   W. Gordy, *Theory and Applications of Electron Spin Resonance*, Wiley, New York, 1980.
31   M. Gomberg and L. H. Cone, *Ber.*, **37**, 2037 (1904).
32   M. Gomberg and C. S. Schoepfle, *J. Amer. Chem. Soc.*, **39**, 1652 (1917).
33   E. M. Kosower, *J. Amer. Chem. Soc.*, **86**, 5515, 5524, 5528 (1964).
34   M. Ballester, J. Castaner, J. Riera, A. Ibanez and J. Pujadas, *J. Org. Chem.*, **47**, 259 (1982).
35   P. D. Bartlett and H. Kwart, *J. Amer. Chem. Soc.*, **72**, 1051 (1950).
36   P. S. Gill and T. E. Gough, *Can. J. Chem.*, **45**, 2112 (1967).
37   J. Hermolin, M. Levin and E. M. Kosower, *J. Amer. Chem. Soc.*, **103**, 4808, 4813 (1981).
38   M. Mohammad and E. M. Kosower, *J. Amer. Chem. Soc.*, **93**, 2709 (1971).
39   M. Ballester and J. Riera, *J. Amer. Chem. Soc.*, **86**, 4504 (1964).

40 T. Fujisawa, B. M. Monroe and G. S. Hammond, *J. Amer. Chem. Soc.*, **92,** 542 (1970).
41 B. Kirste, W. Harrer, H. Kurneck, K. Schubert, H. Bauer and W. Gienke, *J. Amer. Chem. Soc.*, **102,** 6280 (1981).
42 A. Carrington, *Quart. Rev.*, **17,** 67 (1963).
43 A. Rembaum, V. Hadek and S. P. S. Yen, *J. Amer. Chem. Soc.*, **93,** 2532 (1971).
44 Y. Miura, Y. Nakamura and M. Kinoshita, *Bull. Chem. Soc. Japan*, **54,** 3217 (1981).
45 H. Waynar and K. U. Ingold, *J. Amer. Chem. Soc.*, **102,** 3813 (1980).
46 L. H. Piette and W. C. Landgraf, *J. Chem. Phys.*, **32,** 1107 (1960).
47 H. Paul and H. Fischer, *Chem. Comm.*, 1038 (1971).
48 R. A. Sheldon and J. K. Kochi, *J. Amer. Chem. Soc.*, **92,** 4395 (1970).
49 P. J. Krusic and J. K. Kochi, *J. Amer. Chem. Soc.*, **90,** 7155 (1968).
50 H. Zeldes and R. Livingston, *J. Chem. Phys.*, **45,** 1946 (1966).
51 H. Zeldes and R. Livingston, *J. Amer. Chem. Soc.*, **93,** 1082 (1971).
52 J. K. Kochi and P. J. Krusic, *J. Amer. Chem. Soc.*, **92,** 4110 (1970).
53 J. K. Kochi and P. J. Krusic, *J. Amer. Chem. Soc.*, **90,** 7157 (1968).
54 P. J. Krusic and T. A. Rettig, *J. Amer. Chem. Soc.*, **92,** 722 (1970).
55 W. A. Noyes, G. B. Porter and J. E. Jolly, *Chem. Rev.*, **56,** 49 (1956).
56 G. S. Hammond and W. M. Moore, *J. Amer. Chem. Soc.*, **81,** 6334 (1959); **83,** 2789 (1960).
57 E. A. C. Lucken and B. Paucioni, *J. Chem. Soc., Perkin II,* 777 (1976).
58 H. Zeldes and R. Livingston, *J. Chem. Phys.*, **44,** 1245 (1966).
59 H. Zeldes and R. Livingston, *J. Chem. Phys.*, **45,** 1946 (1966).
60 (a) J. G. Calvert and J. N. Pitts, *Photochemistry,* Wiley, New York, 1966; (b) W. A. Waters (Ed.), *Int. Rev. Sci., Series 2,* Vol. 10, Butterworth, London, 1975.
61 T. Shiga, *J. Chem. Phys.*, **69,** 3807 (1965).
62 W. T. Dixon, J. Foxall, G. H. Williams, D. J. Edge, B. C. Gilbert, H. Kazarios-Moghaddon and R. O. C. Norman, *J. Chem. Soc., Perkin II,* 827 (1977).
63 W. T. Dixon and R. O. C. Norman, *J. Chem. Soc.,* 3119, 4850, 4857 (1963).
64 W. E. Griffiths, G. F. Longster, J. Myatt and P. F. Todd, *J. Chem. Soc. B,* 530, 533 (1967).
65 J. Q. Adams, *J. Amer. Chem. Soc.*, **92,** 4535 (1970).
66 D. J. Edge and R. O. C. Norman, *J. Chem. Soc. B,* 1083 (1970).
67 W. T. Dixon and R. O. C. Norman, *J. Chem. Soc.,* 3119 (1963); 3625 (1964).
68 T. Shona, T. Toda and R. Oda, *Tetrahedron Letts,* 369 (1970).
69 T. J. Kemp, P. Moore and G. R. Quick, *J. Chem. Res.,* S. 33 (1981).
70 T. J. Kemp, *Chem. Brit.*, **14,** 255 (1978).
71 J. E. Bennett, B. Mile, A. Thomas and B. Ward, *Adv. Phys. Org. Chem.*, **8,** 1 (1970).
72 R. V. Lloyd and D. E. Wood, *J. Amer. Chem. Soc.*, **97,** 5986 (1975).
73 C. U. Morgan and K. J. White, *J. Amer. Chem. Soc.*, **92,** 3309 (1970).
74 D. Griller and B. P. Roberts, *Chem. Comm.*, 1035 (1971).
75 J. E. Bennett, B. Mile and A. Thomas, *Trans. Farad. Soc.*, **63,** 262 (1967).
76 A. Citterio, M. Serravalle and E. Vismara, *Tetrahedron Letts,* 1831 (1982).
77 D. M. Chipman, *J. Chem. Phys.*, **71,** 761 (1979).
78 H. Fischer, Ref. 3, Vol II, p. 435.
79 R. O. C. Norman, *Chem. Soc. Rev,* **8,** 1, (1979).
80 K. Higasi, H. Baba and A. Rembaum, *Quantum Organic Chemistry,* Wiley–Interscience, New York, 1965.
81 M. E. Jacox, *J. Phys. Chem.*, **86,** 670 (1982).
82 M. V. Merritt and R. W. Fessenden, *J. Chem. Phys.*, **56,** 2353 (1972); D. E. Wood and R. V. Lloyd, *J. Chem. Phys.*, **53,** 3932 (1970); *J. Amer. Chem. Soc.*, **92,** 4115 (1970).
83 A. E. Lemire and H. D. Gesser, *Chem. Comm.*, 1175 (1983).
84 A. Thomas, *Adv. Phys. Org. Chem.*, **8,** 1 (1970).
85 K. J. Klabunde, *J. Amer. Chem. Soc.*, **92,** 2427 (1970).
86 M. J. Perkins, P. Ward and A. Horsfield, *J. Chem. Soc. B,* 395 (1970).

87 A. Ferruti, D. Gill, H. P. Klein, H. H. Wang, G. Entine and M. Calvin, *J. Amer. Chem. Soc.*, **92**, 3704 (1970).

88 M. J. Perkins, P. Ward and A. Horsfield, *J. Chem. Soc. B*, 395 (1970).

89 J. F. W. Keana, S. B. Keanna and D. Beetham, *J. Amer. Chem. Soc.*, **89**, 3055 (1967).

90 S. Terabe and R. Konaka, *J. Amer. Chem. Soc.*, **93**, 4306 (1971).

91 M. J. Perkins, *Adv. Phys. Org. Chem.*, **17**, 1 (1980).

92 E. G. Janzen and B. J. Blackburn, *J. Amer. Chem. Soc.*, **91**, 1181 (1969).

93 H. Kaur and M. J. Perkins, *Can. J. Chem.*, **60**, 1587 (1982).

94 W. Kermers, G. W. Koroll and A. Singh, *Can. J. Chem.*, **60**, 1597 (1982).

95 R. Konaka, S. Terabe, T. Mizuta and S. Sakata, *Can. J. Chem.*, **60**, 1532 (1982).

96 D. L. Haire and E. D. Janzen, *Can. J. Chem.*, **60**, 1514 (1982).

97 T. J. DeBoer, *Can. J. Chem.*, **60**, 1602 (1982).

98 H. R. Ward, Ref. *3*, Vol. I, Chapter 6.

99 R. G. Lawler, *Acc. Chem. Res.*, **5**, 25 (1972); R. Kaptein, *Acc. Chem. Res.* **5**, 18 (1972).

100 R. Kaptein and L. J. Oosterhoff, *Chem. Phys. Letts*, **4**, 214 (1969).

101 M. Yoshida, N. Furuta and M. Kobayashi, *Bull. Chem. Soc. Japan*, **54**, 2356 (1981).

102 B. B. Jarvis, P. E. Nicholas and J. O. Midiwo, *J. Amer. Chem. Soc.*, **103**, 3878 (1981).

103 R. Kaptein, F. W. Verheus and L. J. Oosterhoff, *Chem. Comm.* 877 (1971).

104 S. V. Rykov, A. L. Buchachenko and V. I. Baldin, *Chem. Abstr.*, 42564 (1970).

105 C. Walling and A. R. Lepley, *J. Amer. Chem. Soc.*, **94**, 2007 (1972).

106 P. D. Bartlett and N. Shimiju, *J. Amer. Chem. Soc.*, **97**, 6253 (1975).

107 G. F. Lehr and N. J. Turro, *Tetrahedron*, **37**, 2411 (1981).

108 E. M. Schulman, R. D. Bertrand, D. M. Grant, A. R. Lepley and C. Walling, *J. Amer. Chem. Soc.*, **94**, 5972 (1972).

109 H. Mizuno, M. Matsuda and M. Iino, *J. Org. Chem.*, **46**, 520 (1981).

110 M. L. M. Schilling, *J. Amer. Chem. Soc.*, **103**, 3077 (1981).

111 S. R. Fahrenholtz and A. M. Trozzolo, *J. Amer. Chem. Soc.*, **94**, 282 (1972).

112 G. L. Closs and L. E. Closs, *J. Amer. Chem. Soc.*, **91**, 4550 (1969).

113 R. G. Lawler, H. R. Ward, R. B. Allen and P. E. Ellenbogen, *J. Amer. Chem. Soc.*, **93**, 789 (1971).

114 H. R. Ward, R. G. Lawler, H. Y. Laken and R. A. Cooper, *J. Amer. Chem. Soc.*, **90**, 4928 (1969).

115 J. W. Rakshys, *Chem. Comm.*, 578 (1970).

116 B. J. Schaart, C. Blomberg, O. S. Akkerman and F. Bickelhaupt, *Can. J. Chem.*, **58**, 932 (1980).

117 A. R. Lepley, *J. Amer. Chem. Soc.*, **91**, 748, 749 (1969).

118 E. C. Ashley and A. B. Goch, *J. Amer. Chem. Soc.*, **103**, 4983 (1981).

119 E. C. Ashby and J. R. Bowers, *J. Amer. Chem. Soc.*, **103**, 2242 (1981).

120 H. R. Ward, R. G. Lawler and T. A. Marzelli, *Tetrahedron Letts*, 521 (1970).

121 H. G. Viehe, Z. Janousek, R. Merenyi and L. Stella, *Acc. Chem. Res.*, **18**, 148 (1985).

122 H. W. J. J. Bodewitz, B. J. Schart, J. D. Van der Niet, C. Blomberg, F. Bickenhaupt and J. A. den Hollander, *Tetrahedron*, **34**, 2523 (1978).

123 A. R. Lepley, *Chem. Comm.*, 64 (1969).

124 V. I. Savin, *Zh. Org. Khim.*, **14**, 2097 (1978).

125 A. R. Lepley, *Chem. Comm.*, 64 (1969).

126 D. Bethell, M. R. Brinkman and J. Hayes, *Chem. Comm.*, 475 (1972).

127 G. L. Closs and L. E. Closs, *J. Amer. Chem. Soc.*, **91**, 4549 (1969).

128 G. L. Closs and L. E. Closs, *J. Amer. Chem. Soc.*, **91**, 4550, (1969); G. L. Closs, *J. Amer. Chem. Soc.*, **91**, 4552 (1969); G. L. Closs and A. D. Trifunac, *J. Amer. Chem. Soc.*, **91**, 4554 (1969).

129 H. D. Roth, *Acc. Chem. Res.*, **10**, 85 (1977).

130 G. D. Sargent, *Carbonium Ions*, Vol. III, Ed. G. A. Olah and P. v. R. Schleyer, Wiley, New York, 1972.

131 H. Iwamura, Y. Imahashi, K. Kuskida, K. Aoki and S. Satoh, *Chem. Letts* 357 (1976).

132 E. L. Dreher, P. Nederer, R. Rieker and W. Schwartz, *Helv. Chim. Acta,* **64,** 488 (1981).

133 A. V. Dushkin, *Tetrahedron Letts,* 1309 (1977).

134 W. T. Evanochko and P. V. Shevlin, *J. Amer. Chem. Soc.,* **101,** 4668 (1979).

135 L. A. Rykova, L. A. Kiprianova and I. P. Gagarov, *Teor. Eksp. Khim.,* **17,** 542 (1981).

136 H. Iwamura, M. Iwamura, T. Nishida, M. Yoshida and J. Nakayama, *Tetrahedron Letts,* 63 (1971).

137 W. D. Ollis, M. Rey and I. O. Sutherland, *Chem. Comm.,* 543, 545 (1975).

138 P. B. Shevlin and H. J. Hansen, *J. Org. Chem.,* **42,** 3011 (1977).

139 J. W. Wilt, in Ref. *3,* Vol. 1, Chapter 8.

140 R. Scheel, *Diss. Abstr, B,* **40,** 1724 (1979).

141 J. H. Ridd and J. P. B. Sandall, *Chem. Comm.,* 402 (1981).

142 J. Borgon, *J. Amer. Chem. Soc.,* **93,** 4630 (1971).

143 G. A. Taylor, *J. Chem. Soc., Perkin I,* 376 (1979).

144 W. B. Ankersa, C. Brown, R. F. Hudson and A. J. Lawson, *J. Chem. Soc., Perkin II,* 127 (1978).

145 D. Bethell and M. R. Brinkman, *J. Chem. Soc., Perkin II,* 603 (1979).

146 W. B. Moniz, C. F. Poranski and S. A. Sojka, *Top. Carbon-13 NMR Spectr.,* **3,** 361 (1979).

147 I. P. Bleeker and J. B. F. N. Engberts, *J. Org. Chem.,* **46,** 1012 (1981).

148 J. A. Potenza, E. H. Poindexter, P. J. Caplan and R. A. Dwek, *J. Amer. Chem. Soc.,* **91,** 4356 (1969).

149 H.-D. Roth and M. L. B. Schilling, *J. Amer. Chem. Soc.,* **103,** 7210 (1981).

150 S. K. Wong, T.-M. Chiu and J. R. Bolton, *J. Phys. Chem.,* **85,** 12 (1981).

151 A. D. Trefunnac, *J. Amer. Chem. Soc.,* **98,** 5202 (1976).

152 F. J. J. DeKanter and R. Kaptein, *Chem. Phys. Letts,* **58,** 340 (1978).

153 R. Kaptein, *J. Amer. Chem. Soc.,* **94,** 6251, 6262, 6269, 6280 (1972).

154 G. L. Closs, *J. Amer. Chem. Soc.,* **91,** 4552 (1969).

155 K. Muller, *Chem. Comm.,* 45 (1972).

156 R. Kaptein, *Chem. Comm.,* 732 (1971).

157 G. L. Closs and M. S. Czeropski, *J. Amer. Chem. Soc.,* **99,** 6127 (1977).

158 C. Chatilialoglu and K. U. Ingold, *J. Amer. Chem. Soc.,* **103,** 4833 (1981).

159 N. S. Isaacs and O. H. Abed, *Tetrahedron Letts,* **23,** 2799 (1982).

160 O. L. Harle and J. R. Thomas, *J. Amer. Chem. Soc.,* **79,** 2973 (1957).

161 R. C. Neuman, *Acc. Chem. Res,* **5,** 138 (1972).

162 J. C. Martin and J. H. Hargis, *J. Amer. Chem. Soc.,* **91,** 5399, 6904 (1969).

163 A. Rembaum and M. Szwarc, *J. Chem. Phys.,* **23,** 907 (1955).

164 W. Pryor and K. Smith, *J. Amer. Chem. Soc.,* **92,** 5403 (1970).

165 J. Calvert and J. Pitts, *Photochemistry,* Wiley, New York, 1966; R. K. Lyon and D. H. Levy, *J. Amer. Chem. Soc.,* **83,** 4290 (1961).

166 D. Gegion, K. Muszkat and E. Fischer, *J. Amer. Chem. Soc.,* **90,** 12 (1968), R. E. Roberts and P. Ausloos, *J. Phys. Chem.,* **66,** 2253 (1962).

167 K. U. Ingold, in Ref. *3,* Volume I, Chapter 5.

168 R. D. Guthrie, *Chem. Comm.,* 1316 (1969).

169 K. Ziegler, A. Seib, K. Knoevenagel, P. Herte and F. Andreas, *Ann.,* **551,** 150 (1942).

170 D. Bryce-Smith, *J. Chem. Soc.,* 1603 (1956).

171 G. A. Russell, in Ref. *3,* Vol. I, Chapter 7.

172 R. F. Bridger and G. A. Russell, *J. Amer. Chem. Soc.,* **85,** 3754 (1964).

173 M. Cher, C. S. Hollingsworth and F. Sicilio, *J. Phys. Chem.,* **70,** 877 (1966).

174 C. Walling and J. A. McGuinness, *J. Amer. Chem. Soc.,* **91,** 2053 (1969).

175 G. A. Russell, *J. Amer. Chem. Soc.,* **80,** 5002 (1958).

176 K. B. Wiberg and L. H. Slough, *J. Amer. Chem. Soc.,* **80,** 3033 (1958).

177 K. B. Wiberg and E. L. Motell, *Tetrahedron,* **19,** 1053 (1963).

178 H. Carmichael and H. S. Johnson, *J. Phys. Chem.,* **41,** 1975 (1964).

179 V. Malatesta and K. U. Ingold, *J. Amer. Chem. Soc.*, **103**, 3094 (1981); G. Brunton, D. Griller, L. R. C. Barclay and K. U. Ingold, *J. Amer. Chem. Soc.*, **98**, 6803 (1976); **100**, 4197 (1978).

180 N. C. Deno, R. Fishbein and J. C. Wyckoff, *J. Amer. Chem. Soc.*, **93**, 2065 (1971).

181 A. Horowitz and G. Baruch, *Int. J. Chem. Kinet.*, **12**, 883 (1980).

182 L. S. Marcoux, R. N. Adams and S. W. Feldberg, *J. Phys. Chem.*, **73**, 2611, 2623 (1969).

183 C. Walling, *J. Amer. Chem. Soc.*, **61**, 2693 (1939).

184 G. H. Williams, *Homolytic Aromatic Substitution*, Pergamon, London, 1960.

185 M. Tiecco, *Pure Appl. Chem.*, **53**, 239 (1981).

186 J. W. Wilt, in Ref. *3*, Vol. I, Chapter 8.

187 W. A. Pryor, *Prepr. Div. Pet. Chem. Amer. Chem. Soc.*, **25**, 18 (1980).

188 F. D. Greene, M. L. Savitz and F. D. Osterhaltz, *J. Amer. Chem. Soc.*, **83**, 2196 (1961).

189 A. L. J. Beckwith and K. U. Ingold, *Rearrangements in Ground and Excited States*, Vol. I (P. DeMayo, Ed.), Academic Press, New York, 1980.

190 K. Hensler and J. Kalroda, *Angew. Chem., Int. Edn*, **3**, 525 (1964).

191 A. Clerici and O. Porta, *J. Chem. Soc., Perkin II*, 1234 (1980).

192 W. Rickatson and T. S. Stevens, *J. Chem. Soc.*, 3960 (1963).

193 D. H. R. Barton, J. M. Beaton, L. E. Geller and M. M. Pechet, *J. Amer. Chem. Soc.*, **82**, 2640 (1960).

194 R. H. Hesse, *Adv. Free Radical Chem.*, **3**, 83 (1969).

195 J. E. Packer, C. J. Heighway, H. M. Miller and B. C. Dobson, *Austr. J. Chem.*, **33**, 965 (1980).

196 S. Dincturk, R. A. Jackson, M. Townson, H. B. Agirbas, N. C. Billingham and G. Mach, *J. Chem. Soc., Perkin II*, 1121 (1981).

197 T. Funahashi, *Chem. Abstr.*, **95**, 6212 (1981); B. Maillard, J. J. Villnave and C. Filliatre, *Thermochem. Acta*, **39**, 205 (1980); C. Ruchardt, V. Golzke and G. Range, *Ber.*, **114**, 2769 (1981).

198 K. F. O'Driscoll, *Pure Appl. Chem.*, **53**, 617 (1981).

199 G. A. Russell and R. C. Williamson, *J. Amer. Chem. Soc.*, **86**, 2357 (1964); W. A. Pryor, D. F. Church, F. Y. Tang and R. H. Tang, *Front. Free Rad. Chem. 1979* (W. A. Pryor, Ed) Wiley, New York, 1980, p. 355; H. R. Duetsch and H. Fischer, *Int. J. Chem. Kinet.*, **14**, 195 (1982); W. C. Danen and D. G. Saunders, *J. Amer. Chem. Soc.*, **91**, 5924 (1969).

200 H. G. Kuivala and E. J. Walsh, *J. Amer. Chem. Soc.*, **88**, 571 (1966); W. A. Pryor, F. Y. Tang, R. H. Tang and D. F. Church, *J. Amer. Chem. Soc.*, **104**, 2885 (1982).

201 O. Ito and M. Matsuda, *J. Amer. Chem. Soc.*, **103**, 5871 (1981).

202 M. Chanon and M. L. Tobe, *Angew. Chem. Int., Edn.* **21**, 1 (1982); L. Eberson, *Adv. Phys. Org. Chem.*, **18**, 79 (1981).

203 K. J. Kim and J. F. Bunnett, *J. Amer. Chem. Soc.*, **92**, 7463, 7464 (1970).

204 E. C. Ashby, A. B. Goel and R. N. DePriest, *J. Amer. Chem. Soc.*, **102**, 7779 (1980).

205 R. A. Rossi, R. H. de Rossi and A. F. Lopez, *J. Amer. Chem. Soc.*, **98**, 1252 (1976).

206 G. A. Russell, B. Mudryk and F. Ros and M. Jawdosiuk, *Tetrahedron*, **38**, 1059 (1982).

207 M. Tilset and V. D. Parker, *Acta Chem. Scand.*, **B36**, 311 (1982).

208 C. Amatore, J. Pinson, J. M. Saveant and A. Thiebault, *J. Amer. Chem. Soc.*, **104**, 817 (1982).

209 R. G. Scamehorn and J. F. Bunnett, *J. Org. Chem.*, **42**, 1449 (1977).

210 C. Galli and J. F. Bunnett, *J. Amer. Chem. Soc.*, **103**, 7140 (1981).

211 A. R. Katritzky, G. Z. DeVille and R. C. Patel, *Tetrahedron*, **21**, 1723 (1980).

212 J. F. Bunnett and P. Singh, *J. Org. Chem.*, **46**, 5022 (1981).

213 G. A. Russell and F. Ros, *J. Amer. Chem. Soc.*, **107**, 2506 (1985).

214 R. W. Alder, *Chem. Comm.*, 1184 (1980).

215 L. Eberson, L. Joensson and L. G. Wistrand, *Tetrahedron*, **38**, 1087 (1982).

216 L. Eberson and L. Joensson, *Chem. Comm.*, 1187 (1980).

217 J. V. Hey and J. F. Wolfe, *J. Amer. Chem. Soc.*, **97**, 3702 (1975).

218 L. Eberson and F. Radner, *Acta Chem. Scand.*, **B34,** 739 (1980).
219 D. R. Carver, A. P. Komin and J. S. Hubbard, *J. Org. Chem.*, **46,** 294 (1981).
220 J. M. Tedder and J. C. Walton, *Tetrahedron*, **36,** 701 (1980).
221 J. M. Tedder, *Angew. Chem., Int. Edn*, **21,** 401 (1982).
222 A. Potter and J. M. Tedder, *J. Chem. Soc., Perkin II*, 1689 (1982).
223 H. G. Viehe, R. Merenyi, L. Stella and Z. Janousek, *Angew. Chem., Int. Edn.*, **18,** 917 (1979); *Acc. Chem. Res.*, **18,** 148 (1985).
224 R. W. Bennett, D. L. Wherry and T. H. Koch, *J. Amer. Chem. Soc.*, **102,** 2345 (1980).
225 D. Cram, T. Clarke and P. v. R. Schleyer, *Tetrahedron Letts*, **21,** 3681 (1980).
226 M. Rover-Kevers, L. Vertommen, F. Huys, R. Merenyi, Z. Janousek and H. G. Viehe, *Angew. Chem., Int. Edn*, **17,** 1023 (1981).
227 L. A. Pryor, A. Lee, C. E. Witt and G. L. Kaplan, *J. Amer. Chem. Soc.* **86,** 4229, 4234, 4237 (1964).
228 Z. Janousek, F. Huys, L. Rene, M. Masquelier, L. Stella, R. Merenyi and H. G. Viehe, *Angew. Chem., Int. Edn*, **18,** 615, 616 (1979).
229 L. Stella, Z. Janousek, R. Merenyi and H. G. Viehe, *Angew. Chem., Int. Edn*, **17,** 691 (1978).
230 R. W. Baldock, P. Hudson, A. R. Katritzky and F. Sati, *J. Chem. Soc., Perkin I*, 1422 (1974).
231 C. Rüchardt, *Top. Curr. Chem.*, **88,** 1 (1980).
232 J. M. Tedder, *Tetrahedron*, **38,** 313 (1982).
233 B. Giese, *Angew. Chem., Int. Edn*, **22,** 753 (1983).
234 J. M. Poblet, E. Canadell and T. Sordo, *Can. J. Chem.*, **61,** 2068 (1983).
235 F. Delbecq, N. T. Anh and J. M. Lefour, *J. Amer. Chem. Soc.*, **107,** 1623 (1985).
236 L. Kaplan, *Reactive Intermediates*, Vol. I, Wiley, New York, 1978, p. 163.
237 E. L. Eliel and R. V. Acharya, *J. Org. Chem.*, **24,** 151 (1959).
238 P. S. Engel, *Chem. Rev.*, **80,** 99 (1980).

16 Organic photochemistry

The absorption of a quantum of ultraviolet or visible light (200–800 nm wavelength range or approximately 600–150 kJ mol^{-1} (150–35 kcal mol^{-1})) by an organic molecule takes it from its lowest energy state (ground state) into a higher electronic state, accompanied by changes in electronic distribution, geometry and the uncoupling of electron spins. Although electronically excited molecules are normally very short-lived, their high reactivity is frequently sufficient for unique chemistry to occur. This may be ascribed to the large amount of energy which is available, and photochemical reactions frequently lead to high-energy products[1-5].

16.1 EXCITED ELECTRONIC STATES

16.1.1 Absorption of light by molecules

Light absorption is associated with a part of a molecule known as a chromophore (e.g. C=O, aryl), which may or may not encompass the whole molecule. At the simplest approximation, one electron from the HOMO of an organic molecule will be promoted to the LUMO (or higher unoccupied orbital) if it absorbs a quantum of radiation equal in energy to that of the transition, and Hückel approximations can be used to describe the electronic structure of the excited state in terms of two half-filled MOs (Fig. 16.1). For most purposes, however, this approximation is inadequate because of the involvement of σ- or n-electrons (unshared pairs) or because the changes in geometry upon excitation may be crucial to understanding the chemistry of the excited state. The molecular orbitals of the ground state are not identical to those in an excited state and it is more meaningful to describe molecules in terms of 'states' and their associated symmetries and multiplicities than to use ground state orbitals although, in making orbital symmetry correlations, the latter approach will give correct predictions (see Section 16.2.2).

16.1.2 Vertical and horizontal excitation

The time required for absorption of a quantum of ultraviolet radiation is in the region of 10^{-16} s, far shorter than times corresponding to vibrational

777

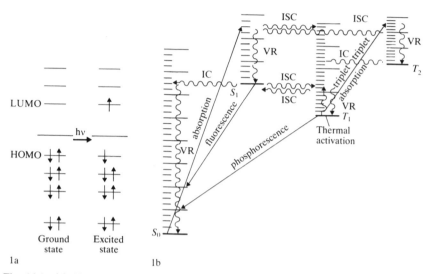

Fig. 16.1 (a) Simple Hückel MO representation of photochemical excitation. (b) Jablonski Diagram showing schematically the options available to ground and excited states. The short lines represent vibrational and rotational sublevels. ISC, intersystem crossing; IC, internal conversion; VR, vibrational relaxation.

frequencies, so the excited state is initially formed in the same conformation as the ground state although electronic reorganization has occurred — a vertical excitation. That electrons move much more rapidly than nuclei is known as the Franck–Condon Principle. However, within a few vibrational periods (*ca* 10^{-11} s), the molecule will relax to a new geometry of lower energy (a horizontal excited state) from which the subsequent chemistry will occur. Figure 16.2 shows equilibrium geometries of some excited molecules

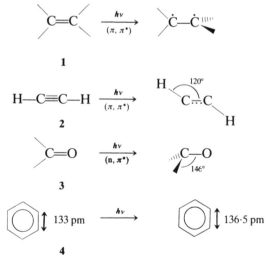

Fig. 16.2 Geometrical changes resulting from electronic excitation.

(1–4) inferred from their spectra. Electronic reorganization usually implies a change in the dipolar properties of the molecule and its polarizability and, in consequence, its solvation characteristics which are the basis of the E_T scale of solvation (Section 5.6.4).

16.1.3 Spin multiplicity: singlet and triplet states

The spin properties of electrons in organic molecules are not independent but interact even when they are associated with different MOs (they are said to 'correlate') so that the energy of the system is further reduced. Spin multiplicity describes the number of unpaired electrons in a molecule; the terms arise from the number of spin-inversion states with which each is associated, i.e. the number of spin energy levels into which the species splits when placed in a magnetic field:

No. of unpaired spins:	0	1	2	3 . . .
Designation:	singlet	doublet	triplet	quartet . . .

The ground state of most molecules is all spin-paired and hence is a singlet, denoted S_0. Free radicals are doublets. The two electrons in the formally singly-occupied orbitals of an excited state may have their spins in the opposite sense (a singlet state) or in the same sense (a triplet state). These will be of different energies and properties. There will normally be a series of both singlet and triplet excited states ('manifolds') of an organic molecule, denoted $S_1, S_2 \ldots$ and $T_1, T_2 \ldots$ respectively. There is a fundamental restriction upon changes in multiplicity which are said to be 'forbidden' (which, in practice, usually means very slow) so that singlet ground state molecules are initially excited to a singlet excited state and this then may be converted into a triplet, a process known as 'intersystem crossing'.

The relationships between states of different multiplicity are shown in Fig. 16.3b, a Jablonski Diagram. From this it will be seen that excited states will revert to the ground state if they are not intercepted by a chemical process and, in doing so, will emit their excitation energy in the form of a quantum of light. If this originates from a singlet state it is known as 'fluorescence', and if from a triplet 'phosphorescence'. The latter requires a change of multiplicity to return to the singlet ground state and hence is forbidden. Triplet states are therefore much long-lived than singlet states and much interesting chemistry is derived from triplet excited molecules (Table 16.1). Further pathways to the ground state include transference of the excitation energy to another molecule thereby raising it to its excited state ('radiationless transfer'), and non-radiative decay whereby the excited molecule passes successively through vibrational states of the excited and ground electronic states which may overlap in energy. Each of these processes, except energy

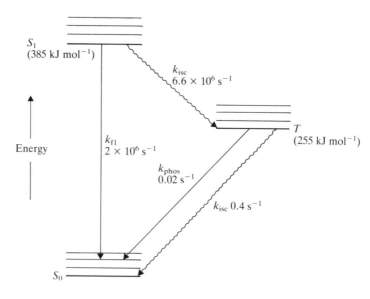

Fig. 16.3   Rate constants for some photochemical processes in napthalene.

transfer, are first order and may be characterized by rate constants as shown in Fig. 16.2 which reflect relative probabilities of the various processes.

16.1.4   Sensitization and quenching

The transference of electronic energy from one molecule to a second one is a means of obtaining the latter in an excited state, a technique known as sensitization, which is particularly useful when direct excitation or inter-system crossing in the desired species is inefficient. Sensitization may, since spin conservation is required, lead directly to either singlet or triplet states of the excited species, depending upon the availability of excited states of

Table 16.1   Lifetimes of some excited states

Molecule	Half-life/s	
	Singlet (25°C, solution)	Triplet (−196°C, glass matrix)
Pyrene	4.5×10^{-7}	0.5
Benzene	3×10^{-8}	6.3
Biphenyl	1.6×10^{-8}	4.6
Benzophenone	5.0×10^{-12}	6.0×10^{-3}
Acetone	2.0×10^{-9}	6.0×10^{-4}

Table 16.2   Excitation energies of some organic molecules

	$\Delta E, S_0 \rightarrow S_1$ /kJ mol^{-1} (kcal mol^{-1})	$\Delta E, S_0 \rightarrow T_1$ /kJ mol^{-1} (kcal mol^{-1})
Acetone	365 (87)	325 (78)
Formaldehyde	338 (81)	302 (72)
Benzophenone	318 (76)	291 (70)
Acrolein	310 (74)	290 (69)
Benzene	577 (137)	354 (84)
Naphthalene	415 (99)	255 (61)
Buta-1,3-diene	430 (103)	250 (60)
Stilbene	395 (94)	210 (50)

the sensitizer, S:

$$X(S_0) + S(S_1) \rightarrow X(S_1) + S(S_0)$$
$$X(S_0) + S(T_1) \rightarrow X(T_1) + S(S_0).$$

Obviously, the excited sensitizer must possess sufficient energy to bring about the desired transition; some sensitizer energies are given in Table 16.2. Ketones and aromatic species are frequently the choice for triplet sensitizers. Whereas the term 'sensitization' focuses attention on non-radiative transfer to the species of interest, the term 'quenching' refers to the same process but with regard to de-excitation of that species. This may be unfavorable for the achievement of photochemistry in that the excited reactant is returned to its ground state and the light energy lost. Quenching may also be desirable, however: for example, by the introduction of a quencher, either a singlet or a triplet state of the excited reactant can be selectively removed in order to bring about the desired chemistry of the unaffected state.

16.1.5   Techniques of photochemistry[6]

Relatively few organic compounds absorb in the visible region; those which do are colored. Most will absorb in the ultraviolet region to which air is transparent (200–400 nm). Mercury discharge lamps produce strong emission around 254 nm, a resonance line of the mercury atomic spectrum which, in lamps running at high pressure, is more intense and also broader in range than at low pressure. No exact matching of lamp emission to absorber is usually necessary since electronic absorption bands are usually quite broad due to coupling to vibrational transitions. Vessels for photochemical work need to transmit through the range of frequencies used and are normally of high-grade quartz.

16.2 PHOTOCHEMISTRY OF THE CARBON–CARBON DOUBLE BOND[7,8]

Simple alkenes absorb ultraviolet light with a λ_{max} around 180 nm which passes to longer wavelengths in conjugated systems with an increment of about 35 nm per successive double bond. The intense excitation is of the π,π^* type and in the far ultraviolet region. The excited singlet state, S_1, has an equilibrium geometry with the two halves of the double bond orthogonal, i.e. at 90°, rather than the planar disposition of the ground state. There is no net π-bonding in the S_1 state and little barrier to rotation. The same geometry is preferred for the triplet state, T_1; a result of this is that both fluorescence and phosphorescence from simple alkenes is weak since the excited states would need to regain their planar geometry before emitting a photon.

16.2.1 Geometrical isomerization[1,9]

Since the S_1 state has lost its memory of the original molecular geometry, it may revert to the ground state in either the Z or E forms with more or less equal probability. This would suggest that irradiation of either isomer of the alkene would lead to a 50:50 mixture of the two. One can in fact do something more interesting since the final photostationary mixture depends also upon the relative rates of excitation of the two isomers and these in turn depend upon their absorbances. If it were possible to choose a wavelength such that one isomer absorbed but the other did not, it should be possible to convert the whole sample into the non-absorbing species almost entirely. This may be done in the case of stilbene, **5**, the two isomers of which absorb with very different strengths, at 313 nm; $\varepsilon_E/\varepsilon_Z = 7\cdot1$. Consequently, on irradiation of either, a mixture of $Z:E = 93:3$ is obtained. The Z-isomer is thermodynamically the less stable so that photochemistry is the only way to obtain this species. Other examples of photochemical geometrical isomerization are shown in reactions **6–11**[10] and include rhodopsin, **7**, a visual pigment whose *cis–trans* isomerization at C-11 by visible light sets off a chain of events which results in stimulation of the optic nerve[11]. Catalyzed isomerization **10** may be brought about by adding traces of iodine, which avoids the necessity of far-ultraviolet radiation. Even

7%   93%

5

6

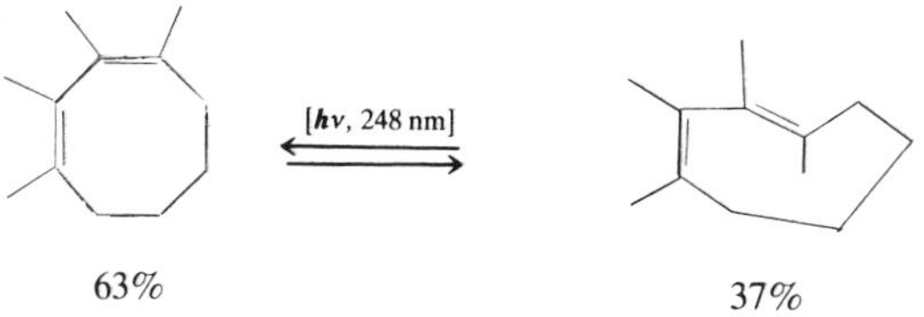

63%   37%

7

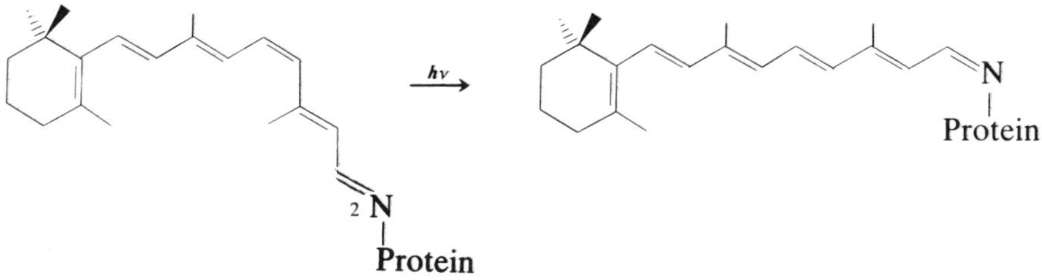

8

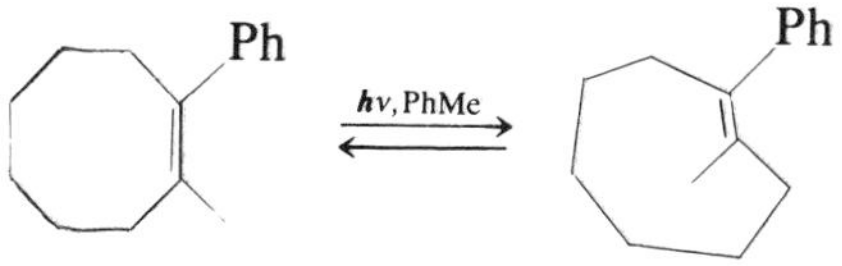

9

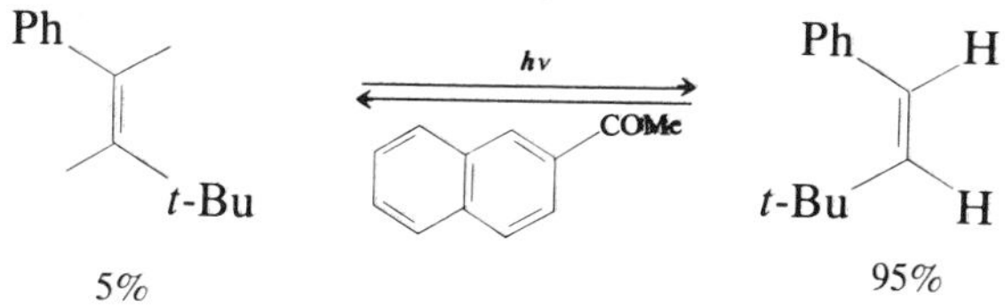

5%   95%

10

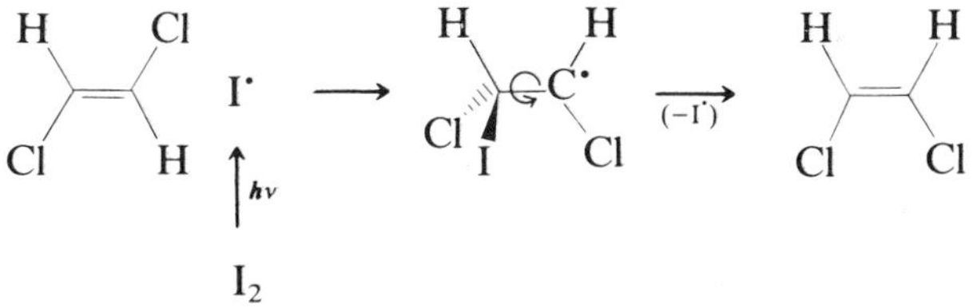

11

(mixed isomers)

Vitamin A

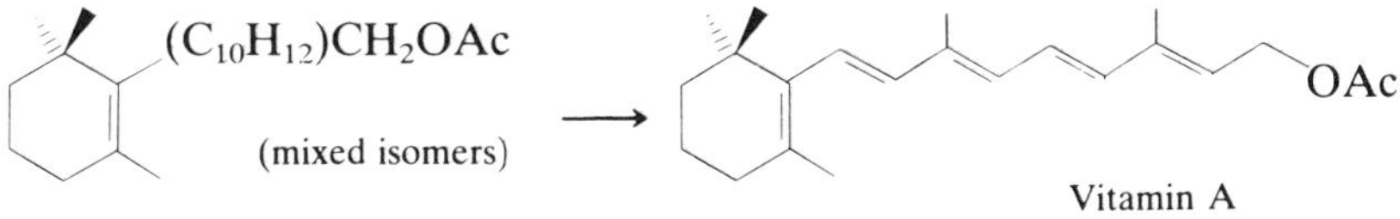

visible light will produce iodine atoms which add to form a radical which, after rotation, can expel the iodine atom and revert to the alkene in a ratio of isomers determined by their thermodynamic stabilities, a reaction similar in principle to acid-catalyzed isomerization (Section 12.1). These reactions can also take place through the triplet states of the alkenes [12], accessible through photosensitization, for example by use of an aromatic molecule as in **8** and **9**[13,14]. The same results can also be achieved by photo-induced electron transfer and rotation of the C—C bond in the radical anion, leading usually to the most stable isomer; an example is the commercial synthesis of Vitamin A acetate, **11**.

16.2.2 Photochemical pericyclic reactions

As mentioned in Chapter 14, many pericyclic reactions can take place from excited states. The number of electron pairs involved differs from that required for ground state reactions according to the Woodward–Hoffmann Rules (Section 14.2.4).

Cycloadditions between two π-systems may take place when one component is excited; orbital-symmetry-permitted processes now involve $4n$ electrons in contrast to the requirement of $4n + 2$ electrons in ground state cycloadditions, **12**. Example **13** indicates that reactions are stereospecific[15], a requirement for concerted processes which these presumably are. Photochemical $(2 + 2)$ cycloadditions are probably the best methods for synthesis of four-membered rings, **13–20** and can result in the formation of highly strained systems[16,17].

12

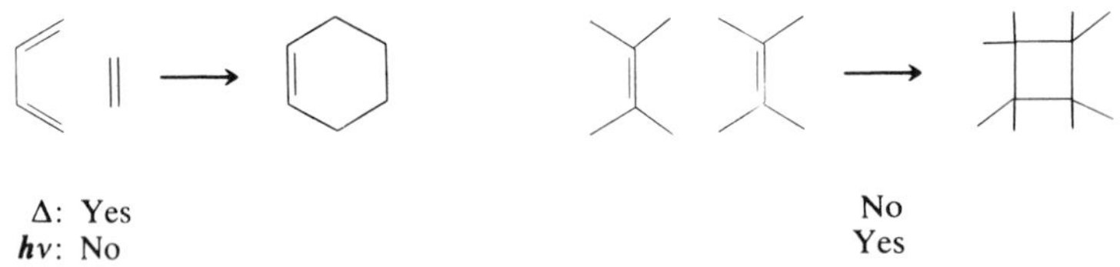

Δ: Yes
hv: No

No
Yes

13   2

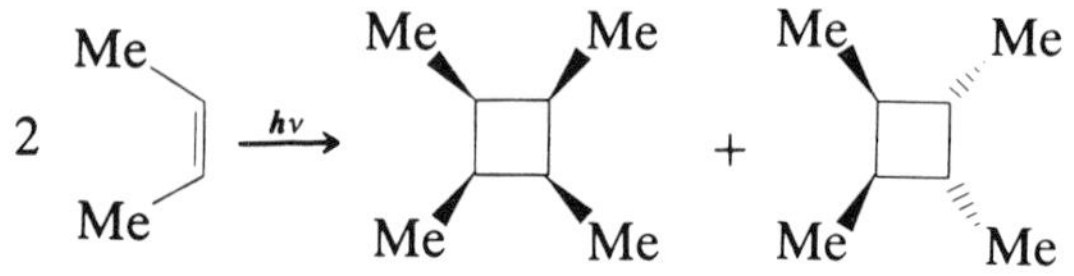

14

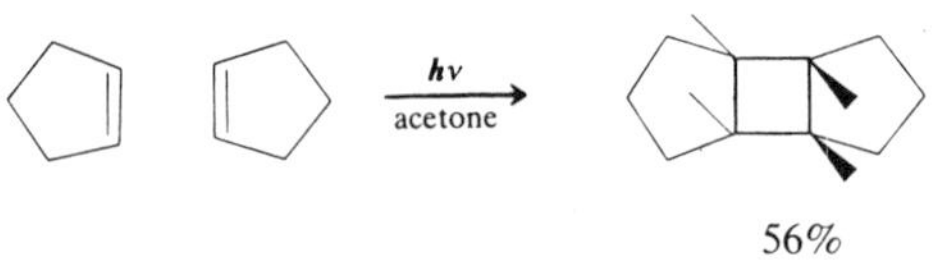

56%

15

55%

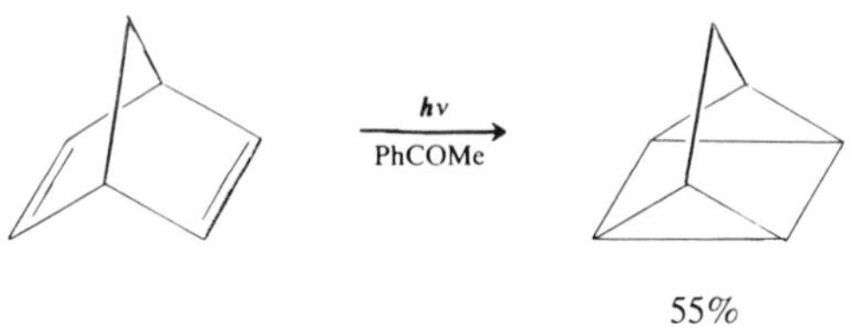

16 → hv / acetone → 92%

17 2 → hv → **17a** → + +

18 ← hv (sensitizer) → hv →

19 → hv CN (S) CN → [OEt S⁻]⁺ → OEt → OEt OEt

20 → hv methyl acetoacetate → 23% + 27% + 42%

Photodimerization of 1,3-dienes, **17**, by contrast is non-concerted and appears from the product distribution to take place via an intermediate diradical, **17a**[18]. This reaction competes with electrocyclic ring closure **18** (see below)[19]. Cycloadditions of nucleophilic alkenes such as vinyl ethers take place efficiently in the presence of electrophilic sensitizers, **19**, by way of an initial electron exchange and result in non-stereospecific addition of the cation-radical. Cycloalkenes may dimerize by prior photo-isomerization to the highly reactive *trans*-olefins, **28**, which, since the π-system is twisted, are able to dimerize thermally in the ground state (Section 14.4)[20].

Electrocyclic reaction, the orbital-symmetry-controlled interchange of cyclic and acyclic isomers, takes place under photochemical conditions with stereochemistry the reverse of that in the ground state. The butadiene–cyclobutadiene interconversion and other 4n processes (Section 4.6) now occur with disrotation, whilst $4n + 2$ electrocyclic reactions take place with conrotation (**21, 22**). This can be predicted by consideration of the symmetries of the higher half-filled MOs in the excited state in relation to the product orbital with which it corresponds. Electrocyclic reactions are, in principle, reversible since orbital symmetry considerations apply to reactions equally in both directions, but the interesting feature of these photoreactions is their ability to lead to the high-energy isomer. The cyclobutanes **23–25**, for instance, can be obtained in high yield, whereas thermal reactions tend to promote formation of the more stable butadienes[21,22].

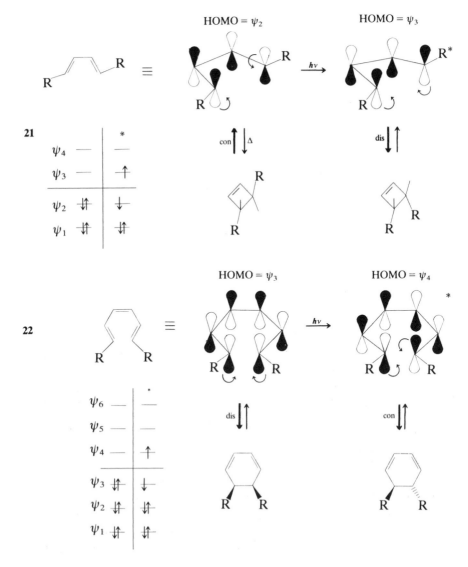

23          $\xrightarrow{hv}$

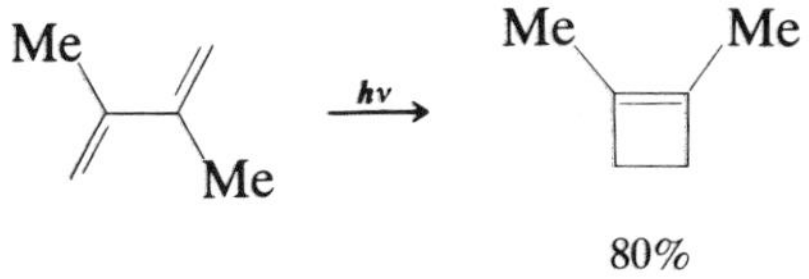

80%

24          $\xrightarrow{hv}$

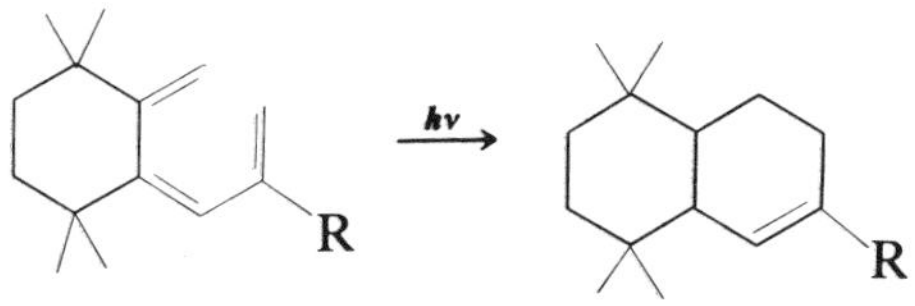

80%

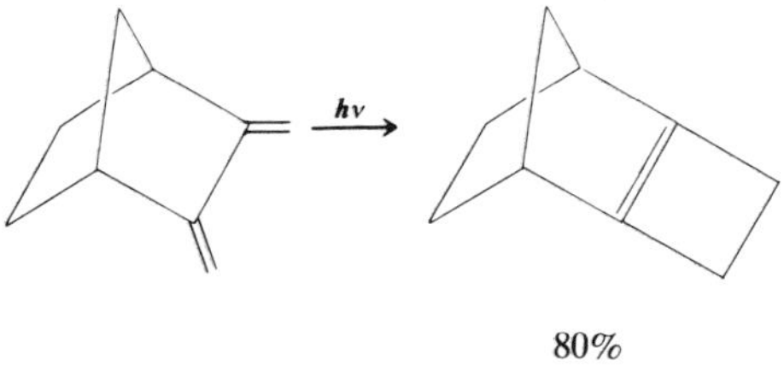

25          $\longrightarrow$

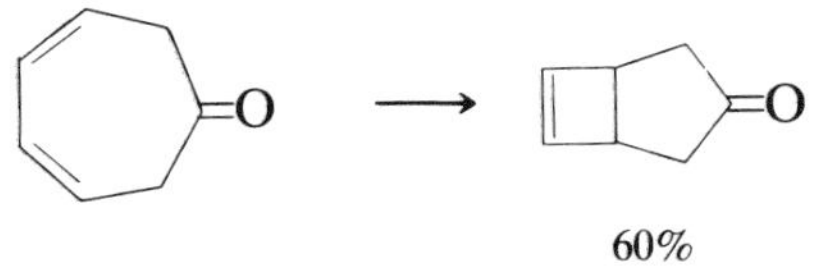

60%

26          $\xrightarrow{hv}$

27          $\xrightarrow{300\ nm}$

$\downarrow$ 254 nm

$\xrightarrow{\Delta}$

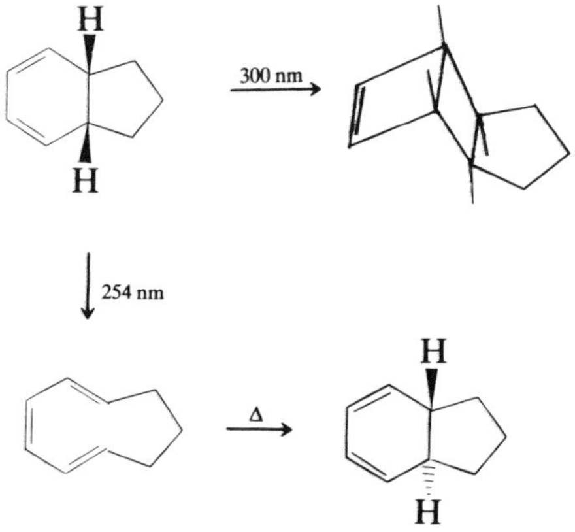

Photochemical (2 + 4) cycloadditions and the many examples of photo-chemical cyclization of arylalkenes **26–28** are electrocyclic reactions[23,24]. Sigmatropic reactions (Section 14.8) also may occur photochemically and this time it is the topology which is reversed with respect to that required for ground state reactions. 1,3-Shifts may now be envisaged taking place by movement of a hydrogen atom across the allylic system with respect to the

787

antibonding orbital ψ_3, **29a**. This is seen to be a suprafacial process and therefore one likely to occur. 1,5-Shifts from excited states **29b**, which are suprafacial and of frequent occurrence in thermal reactions, are now required to be antarafacial and so are usually rather difficult on steric grounds. Being suprafacial, 1,3- and 1,7-shifts are now facile and frequently observed in photochemical systems (**30–32**). Groups other than hydrogen

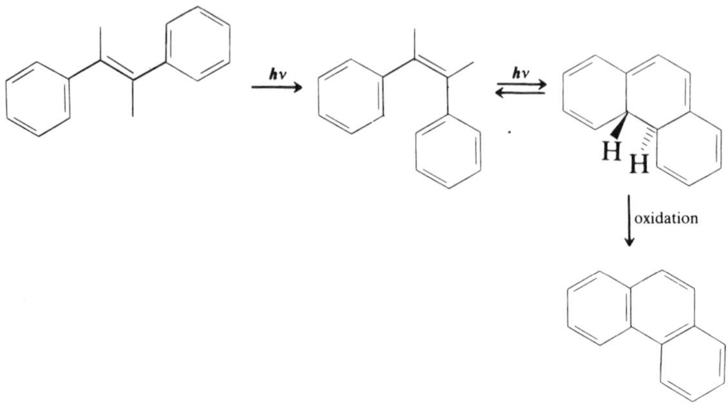

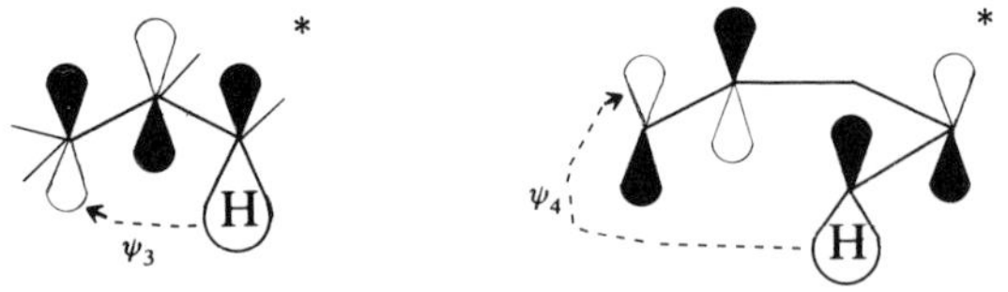

suprafacial
29a

antarafacial
29b

85%

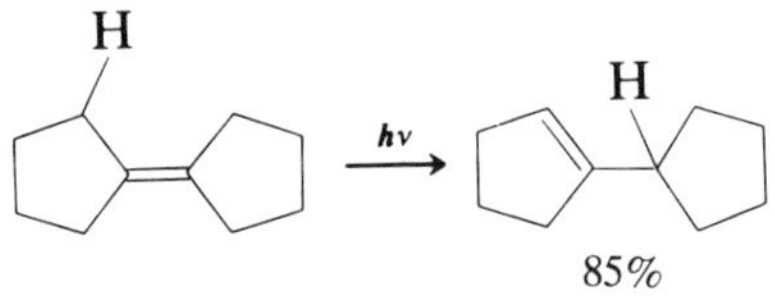

etc.

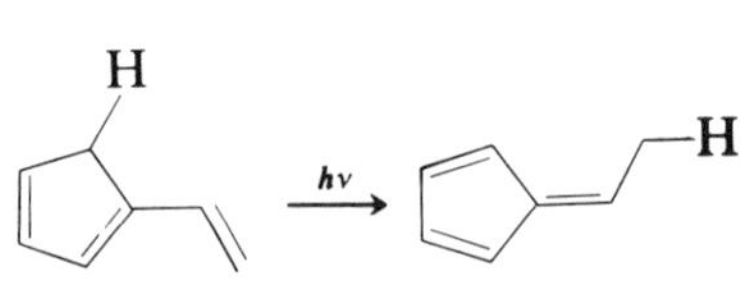

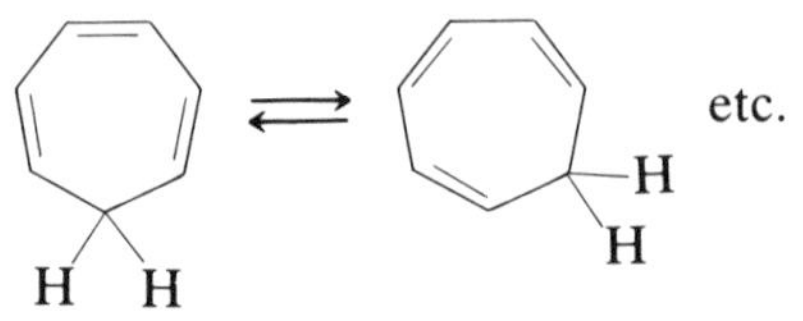

can migrate in sigmatropic processes, e.g. **33**, **34**[25]. An example of a 1,5-antarafacial shift is in the dimethylenecyclobutane, **35**, the non-planar geometry of which facilitates this rearrangement.

16.2.3   The Di-π-methane rearrangement[26]

This name is given to photochemical reactions of 1,4-dienes or 3-arylalkenes which lead to vinylcyclopropanes or phenylcyclopropanes, **36a,b**, **37**[27–31]. The reactions occur from the S_1 state of acyclic systems or from the T_1 state in six-membered ring precursors (the latter reached by sensitization) and can be seen as a 1,2-shift of the vinyl or aryl group coupled with cyclization. This reaction can lead to highly strained ring systems, **38–40**.

38 $\xrightarrow[\text{acetone}]{h\nu}$ 60%

39 $\xrightarrow[\text{(sensitizer)}]{h\nu}$

Ph Ph

40 $\text{ArCH}=\text{CH}_2 \xrightarrow[\text{H}_2\text{O, H}^+]{h\nu} \underset{\overset{|}{\text{OH}}}{\text{ArCH}}-\text{CH}_3$

16.2.4 Photoadditions to alkenes [14]

Excited states are frequently both more nucleophilic and more electrophilic than ground states and, owing to their high reactivities, will undergo addition reactions with a variety of protic reagents. The direct transfer of a proton followed by a nucleophile results in Markownikoff orientation via the more stable carbocation (**41, 42**). Addition in the presence of a triplet sensitizer often results in anti-Markownikoff addition (**43, 44**) and pre-

41 $\text{Ph}_2\text{C}=\text{CH}_2 \xrightarrow[\text{OMe}]{\text{MeOH, } h\nu} \text{Ph}_2\text{C} \overset{\text{OMe}}{\underset{\text{CH}_3}{<}}$

$\xrightarrow[\text{CN}]{h\nu}$

43 $\xrightarrow[\text{ACOH}]{h\nu}$

sumably occurs by another mechanism. This is possibly via initial electron transfer to give a radical ion from which the products are derived. Yet another route for cyclic alkenes is photoisomerization to the *trans* isomer, which is sufficiently reactive to add a protic solvent across the twisted double bond (**43**).

16.3 PHOTOREACTIONS OF CARBONYL COMPOUNDS[15]

Simple aliphatic aldehydes and ketones show strong absorption in the far-ultraviolet region (215–220 nm) due to a π,π^* transition, and also weak absorption around 280 nm due to a forbidden n,π^* transition to a singlet state which relatively rapidly undergoes intersystem crossing to the first triplet state. The latter is the transition usually employed for photochemical purposes although one may obtain either singlet- or triplet-derived products from the exceptionally rich excited-state chemistry of the carbonyl group. α,β-Unsaturated and aryl ketones absorb at longer wavelength (300–320 nm for the n,π^* transition) and the n,π^* and π,π^* states, which have different photochemical characteristics, are closer in energy than is the case for simple ketones. Indeed, the π,π^* state may be lowest in the triplet manifold and has considerable chemical resemblance to a biradical. Carboxylic acids tend to absorb only at very short wavelength and are much less photochemically reactive than ketones unless special features are present.

16.3.1 Carbon–carbon bond cleavage[32]

One of the earliest photochemical systems studied was the photolysis of acetone, which from the n,π^* state yields alkyl and acyl radicals which undergo their typical thermal reactions, **45**, **46**[33]. This process is known as a Norrish Type 1 reaction; n,π^* states tend to undergo reactions akin to those of radicals and this is an example of radical β-scission (Section 15.3.4).

Diaryl ketones do not fragment since their excitation energies are too low for there to be sufficient energy.

An excited carbonyl compound may bring about bond cleavage of a second molecule which it encounters by abstracting a hydrogen atom to give a radical pair (**47, 48**). This may achieve stability by several competing pathways: recombination resulting in addition to the C=O bond, **47a**; dimerization of the primary radical (e.g. pinacolic reduction, **47b**); or abstraction of a further hydrogen atom resulting in reduction, **47c**. Abstraction can occur intramolecularly when a six-membered transition state is available (**48**) ultimately forming a cyclobutane. The intermediate diradical may cleave forming a carbonyl compound so that the net result is an elimination known as a Norrish Type 2 reaction (**49, 50**)[34,35]. The lowest excited state of an α,β-unsaturated ketone is a triplet derived from a π,π^* transition which has less radical character than an n,π^* state. Geometrical isomerization, e.g. **51**[36], or a 1,3-sigmatropic shift, such as **52**[20], is frequently observed. Both processes tend to give the higher-energy isomer since in each case these have the lower absorbances.

Photoenolization is the primary process in **53**; the diene character of the product permits it to undergo Diels–Alder addition to a dienophile[37].

16.3.2 Cycloadditions[38–40]

The n,π* states of a ketone, both singlet and triplet, behave like electrophilic radicals and readily add to alkenes and alkynes with the formation of oxetanes, **54**, or oxetes, **55**, via biradical intermediates which can also be trapped by oxygen, **56**. These cycloadditions are therefore not usually stereospecific. An exception is the additions of maleonitriles, **57**[41], which are believed to occur via an intermediate 'exciplex', an adduct of the excited ketone and the addend, **57a**, which presumably has a definite

54

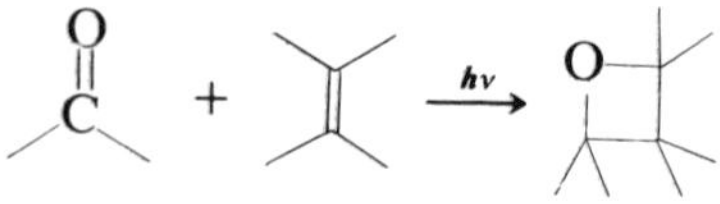

55

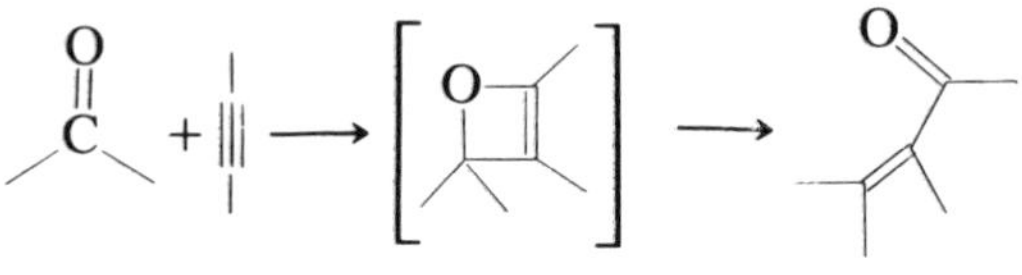

55a

56

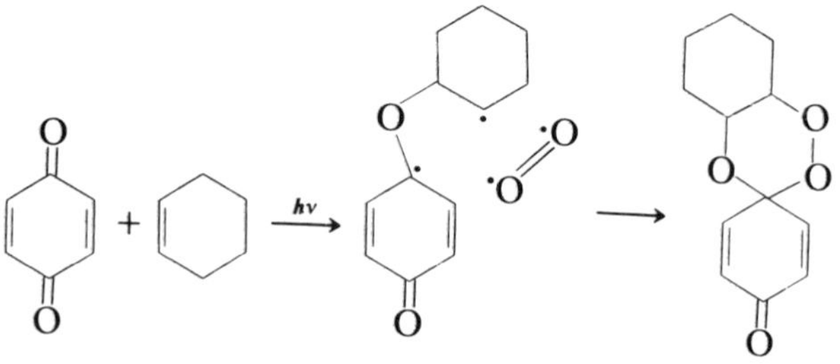

57

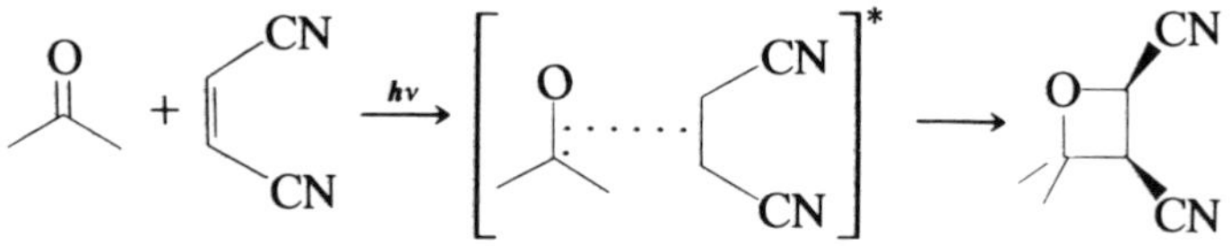

exciplex
57a

90%

58

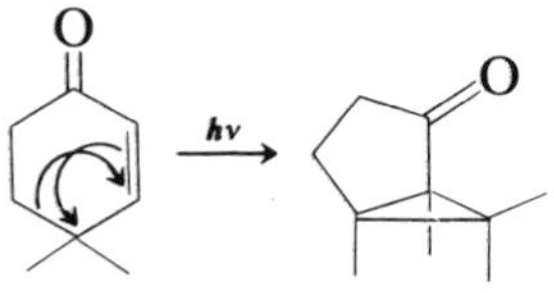

60%

59

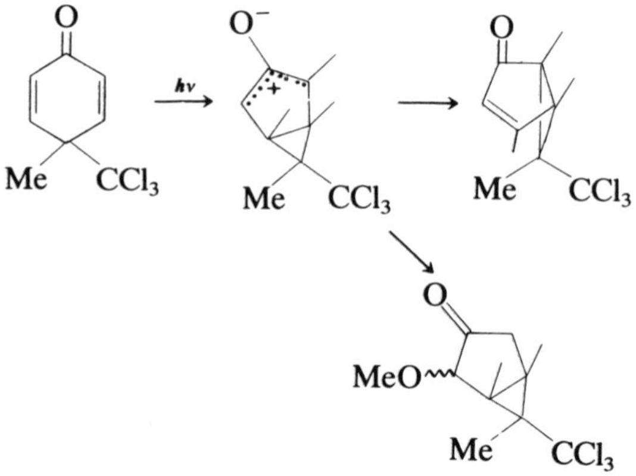

Me   CCl₃

Me   CCl₃

Me   CCl₃

MeO~

Me   CCl₃

geometry. The formation of oxetes, **55**, from photocycloadditions of ketones to alkynes, is usually inferred from the formation of the α,β-unsaturated compounds (**55a**) which are actually isolated and are known to form spontaneously by opening of the unstable four-membered ring. Cyclohexadienones substituted at C4 undergo unique rearrangements on photolysis leading to the formation of bicyclic products (**58, 59**)[42]. The reactions are stereospecific, suggestive of some type of pericyclic process.

16.4 PHOTOCHEMISTRY OF AROMATIC COMPOUNDS

Benzene and its simple homologues absorb strongly in the far-ultraviolet region at around 180 and 220 nm and weakly, due to a forbidden transition, around 260 nm. It is the latter band which is most accessible and excitation by the conveniently situated 254 nm emission line of the mercury lamp brings about some unique chemistry.

16.4.1　Photosubstitutions at the aromatic ring[43]

Excited aromatic molecules are subject to nucleophilic displacements of groups other than hydrogen such as methoxyl, nitro and chloro (**60–63**). The normal electronic effects for S_NAr reactions do not seem to apply. Whereas, in the ground state, a nucleophilic leaving group would be activated by a nitro substituent in the *para* position (Section 10.5), in photosubstitutions the methoxyl group which is displaced is *meta* to the nitro group in preference to *para* and an amino substituent which is normally deactivating towards nucleophilic displacement now activates. The mechanism is evidently complex and it seems likely that the substitution takes place from a radical cation. Calculations seem to agree that such species would show the orientational preferences observed. Radical cations from species such as naphthalene on irradiation may be observed and, in the presence of oxygen, displacements, even of hydride (**63**), may be carried out [44]. Photochemical displacements of halogen by carbanions, **64**, are useful synthetic procedures for attaching sidechains to the aryl nucleus [45].

16.4.2 The Photo-Fries rearrangement [46]

A sidechain-to-nucleus rearrangement of aryl esters, the Fries rearrangement, may be brought about by acid catalysis (Section 9.2.6) or by irradiation. This brings about homolysis of the acyl–oxygen bond and the two radical species then recombine at a ring position *ortho* or *para* to the oxygen, **65**. Cyclic and nitrogen analogs of this process are also known (**66**) [47].

85%

16.4.3   Valence isomerization[48]

Irradiation of benzene over a long period results in the formation of a small yield of the yellow isomer, fulvene (67), probably by way of the tricyclic intermediate 'prefulvene', 67a. There are many examples of the formation under ultraviolet irradiation of bi- and tri-cyclic isomers of aromatic systems, including heteroatom systems (72) and 'Dewar benzene' isomers (69, 74); even prismanes (70) can be formed from the irradiation of perfluorohexamethylbenzene[49,50]. The mechanisms of these reactions are obscure at present. The same kind of process is responsible for isomerization of substituent orientation which can occur on irradiation of alkylbenzenes, 75.

16.4.4 Photocycloadditions[51,52]

Alkenes may add photochemically (76) to benzene to give 1,2-, 1,3- or 1,4-adducts, 77. Of these, the 1,3-mode is perhaps the most commonly experienced, leading to a variety of tricyclo-octanes, 78[53]. 1,2-Addition is often preferred when the alkene bears either highly electron-withdrawing or electron-donating substituents (79). The photoadducts, being cyclohexadienes, may then add a further molecule of a dienophile in a normal Diels–Alder reaction or, in the case of the adducts from acetylene dicarboxylic esters, 80, undergo valence isomerization to the more stable cyclo-octatetraene derivative, 81[54]. Naphthalenes also prefer addition across the C1–C2 bond, 82[55]. 1,4-Cycloaddition is unusual and may be a minor pathway in some of the above cycloadditions. Polyacenes, including anthracene, will dimerize upon irradiation (83)[56].

16.4.5 Photo-oxidations with oxygen

Many species will undergo hydrogen transfer to molecular oxygen under the influence of ultraviolet light. This may be sensitized by mercury atoms which in their excited triplet state are reactive enough to abstract even from alkanes (**84, 85**)[57,58], or the reactive reaction may take place within a weak complex of the substrate and an oxygen molecule. The resulting radicals will then undergo their normal ground-state chemistry leading, for example, to hydroperoxides, a useful method of functionalizing alkanes. Photochemical generation of chlorine atoms from nitrosyl chloride in cyclohexane results in the formation of cyclohexanone oxime, **86**, which in turn is converted to caprolactam by a Beckmann rearrangement (Section 9.2.6) and thence to nylon-6, **87**. This is at present the largest-scale photochemical process used commercially[59].

PROBLEMS

1 What products would you expect to obtain from the following irradiations?

Organic photochemistry

a Me-C₆H₅ + OAc allyl acetate $\xrightarrow{hv}$

c $\xrightarrow{hv}$ (structure with N=N)

b (spiro structure with Ph) $\xrightarrow{hv}$

d Ar—N₃ $\xrightarrow{hv}$

2 Explain the origin of the products in the following reactions:

t-Bu—O—O—t-Bu (di-peroxide) $\xrightarrow{hv}$ acetone + C₂H₆

Ph—C(O)—CH(Ph)—Ph $\longrightarrow$ Ph—CH₂—CH₂—Ph + Ph—CH(Ph)—CH₂—Ph + Ph—CH(Ph)—CH(Ph)—Ph

3 *p*-Benzoquinone and cyclopentadiene add to give compound **A** which then undergoes a photochemical reaction to **B** having the same molecular formula; ¹H NMR spectra of **A** and **B** are given below. Elucidate the structures of **A** and **B**.

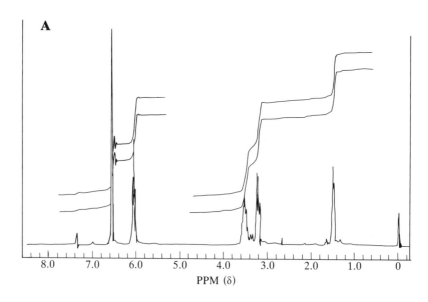

A

8.0   7.0   6.0   5.0   4.0   3.0   2.0   1.0   0
PPM (δ)

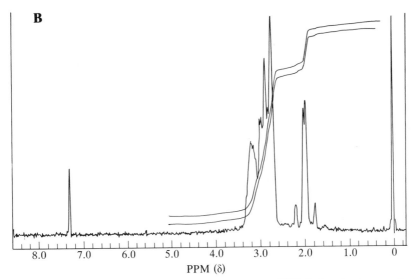

B

PPM (δ)

(Resonances at 0 and 7·3 ppm are due to SiMe₄ reference and to CHCl₃ in the solvent, respectively.)

REFERENCES

1   J. A. Barltrop and J. D. Coyle, *Excited States in Organic Chemistry*, Wiley, London, 1975.
2   J. D. Coyle, *Introduction to Organic Photochemistry*, Wiley, Chichester, 1986.
3   A. Padwa (Ed.), *Organic Photochemistry*, Dekker, New York, 1983.
4   J. G. Calvert and J. N. Pitts, *Photochemistry*, Wiley, London, 1966.
5   D. Bryce-Smith (Ed.), *Specialist Periodical Reports, Photochemistry*, Vols 1–16, Royal Society of Chemistry, London, 1970–1986.
6   W. M. Horspool, *Synthetic Organic Photochemistry*, W. M. Horspool (Ed.), Plenum Press, New York, 1984, Chapter 9.
7   A. Gilbert in *Photochemistry in Organic Synthesis*, Special Publication No. 57, Royal Society of Chemistry, London, 1986.
8   P. Wender in *Photochemistry in Organic Synthesis*, Special Publication No. 57, Royal Society of Chemistry, London, 1986.
9   J. Saltiel and J. L. Charlton, in *Rearrangements in Ground and Excited States*, Vol. 3, P. de Mayo (Ed.), Academic Press, New York, 1980.
10 W. G. Nebe and G. J. Fonken, *J. Amer. Chem. Soc.*, **91**, 1249 (1969).
11 K. Nakanishi, *Pure Appl. Chem.*, **49**, 223 (1977).
12 G. Beddard, *Light, Chemical Change and Life*, The Open University Press, Milton Keynes, 1982.
13 N. J. Turro, *Modern Molecular Photochemistry*, Benjamin, Menlo Park, California, 1978.
14 P. J. Kropp, *Org. Photochem.*, **4**, 7 (1979).
15 H. Yamazaki, R. J. Cvetanovic and R. S. Irwin, *J. Amer. Chem. Soc.*, **98**, 2198 (1976).
16 H. Prinzbach, *Pure Appl. Chem.*, **16**, 17 (1968).
17 R. R. Sauers and K. W. Kelly, *J. Org. Chem.*, **35**, 3286 (1970).
18 W. G. Herkstroeter, A. A. Lamola and G. S. Hammond, *J. Amer. Chem. Soc.*, **86**, 4537 (1964).
19 J. L. Charleton, P. de Mayo and L. Sketterbol, *Tetrahedron Lett.*, 4679 (1965).
20 R. G. Salomon, K. Folting, W. E. Streib and J. K. Kochi, *J. Amer. Chem. Soc.*, **96**, 1145 (1974).
21 J. M. Garrett and E. J. Fonken, *Tetrahedron Lett.*, 191 (1969).

22 D. H. Aue and R. N. Reynolds, *J. Amer. Chem. Soc.*, **95**, 2027 (1973).
23 W. G. Dauben and M. S. Kellogg, *J. Amer. Chem. Soc.*, **93**, 385 (1971).
24 W. H. Laarhoven, *Pure Appl. Chem.*, **565**, 1225 (1984).
25 C. A. Akhtar, J. J. McCullough, S. Vaitekunas, R. Faggliani and C. J. L. Lock, *Can. J. Chem.*, **60**, 1657 (1982).
26 H. E. Zimmerman, in *Rearrangements in Ground and Excited States*, P. de Mayo (Ed.), Academic Press, New York, 1980.
27 K. Dietliker and H. J. Hanson, *Chimia*, **35**, 52 (1981).
28 H. E. Zimmerman and A. C. Pratt, *J. Amer. Chem. Soc.*, **92**, 6267 (1970).
29 H. E. Zimmerman and G. L. Grunewald, *J. Amer. Chem. Soc.*, **88**, 183 (1966).
30 J. S. Swenton, J. A. Hyatt, T. J. Walker and A. L. Crumrine, *J. Amer. Chem. Soc.*, **93**, 4808 (1971).
31 P. S. Mariano and R. B. Steitle, *J. Amer. Chem. Soc.*, **95**, 6114 (1973).
32 R. F. Newton in *Photochemistry in Organic Synthesis*, Special Publication No. 57, Royal Society of Chemistry, London, 1986.
33 N. C. Yang, E. D. Feit, M. H. Hoi, N. J. Turro and J. C. Dalton, *J. Amer. Chem. Soc.*, **92**, 6976 (1970).
34 J. A. Barltrop and J. D. Coyle, *J. Amer. Chem. Soc.*, **90**, 6584 (1968).
35 P. E. Eaton and K. Liu, *J. Amer. Chem. Soc.*, **86**, 287 (1964).
36 R. R. Rando and W. v. E. Doering, *J. Org. Chem.*, **33**, 1671 (1968).
37 E. Block and R. Stevenson, *J. Chem. Soc., Perkin I*, 308 (1973).
38 H. A. J. Carless, in *Synthetic Organic Photochemistry*, W. Horspool, (Ed.), Plenum Press, New York, 1984.
39 H. A. J. Carless in *Photochemistry in Organic Synthesis*, Special Publication No. 57, Royal Society of Chemistry, London, 1986.
40 T. J. Katz, J. C. Carnahan, G. M. Clarke and N. Acton, *J. Amer. Chem. Soc.*, **92**, 734 (1970).
41 J. J. Beereboom and M. S. v. Wittenham, *J. Org. Chem.*, **30**, 1231 (1965).
42 D. I. Schuster and K. Liu, *J. Amer. Chem. Soc.*, **93**, 6711 (1971).
43 E. Havingea and J. Cornelisse, *Chem. Rev.*, **75**, 353 (1975).
44 M. Yasuda, C. Pac and H. Sakurai, *J. Chem. Soc., Perkin I*, 746 (1981).
45 R. A. Alonso, and R. A. Rossi, *J. Org. Chem.*, **43**, 4760 (1980).
46 C. E. Kalmus and D. M. Hercules, *J. Amer. Chem. Soc.* **96**, 449 (1974).
47 M. Fischer, *Tetrahedron Lett.*, 4295 (1968).
48 D. Bryce-Smith and A. Gilbert, in *Rearrangements in Ground and Excited States*, P. de Mayo, (Ed.), Academic Press, New York, 1980.
49 M. G. Barlow, R. V. Haszeldine and R. Hubbard, *Chem. Comm.*, 202 (1969).
50 Z. Yoshida, F. Kawamoto, H. Miyoshi and H. Ikikoshi, Jpn Kokai Tokkyo Koho 79, 138349.
51 D. Bryce-Smith and A. Gilbert, *Tetrahedron*, **33**, 2459 (1977).
52 P. A. Wender and A. Gilbert in *Photochemistry in Organic Synthesis*, Special Publication No. 57, Royal Society of Chemistry, London, 1986.
53 A. Gilbert and M. W. bin Samsudin, *J. Chem. Soc., Perkin I*, 1118 (1980).
54 D. Bryce-Smith and D. E. Lodge, *J. Chem. Soc.*, 695 (1963).
55 A. Akhtar and J. J. McCullough, *J. Org. Chem.*, **46**, 1447 (1981).
56 R. E. Borkman and D. R. Kearns, *J. Chem. Phys.*, **44**, 945 (1966).
57 J. M. Campbell, O. P. Strausz and H. E. Gunning, *J. Amer. Chem. Soc.*, **95**, 740 (1973).
58 N. Kulevsky, P. V. Sneeringer and V. I. Stenberg, *J. Org. Chem.*, **37**, 438 (1972).
59 M. Pape, *Fortschr. Chem. Forsch.*, **7**, 559 (1967).

Symbols and abbreviations

α	Coulomb integral (energy unit)
α	Brønsted coefficient
α	Taft solvation parameter
β	Resonance integral (energy unit)
β	Brønsted coefficient
β	Taft solvation parameter
β	Bohr magneton
Γ	parameter in Kaptein's equation
γ	activity coefficient (molal units)
Δ, δ	a difference between or change in (a quantity)
Δ	spectroscopic term index
Δ	pyrolysis (reaction condition)
δ	NMR chemical shift
ε	relative permittivity (dielectric constant)
ε	molar extinction coefficient (absorptivity)
ε	energy (occasional use)
ζ	mixed solvent composition (occasional use)
θ	angle
κ	transmission coefficient
λ	wavelength
λ	term in Marcus equation
λ	dual parameter reaction constant
μ	reduced mass
ν	frequency
$\bar{\nu}$	wave number
π	molecular orbital with plane of symmetry
π^*	an electronically excited π-orbital.
π^*	an empirical solvation parameter
ρ	reaction constants of the Hammett and related equations
ρ	spin density
Φ	empirical solvation parameter
ϕ	Bunnett–Olson parameter
ϕ	fractionation factor
ϕ	angle
ϕ	atomic orbital wavefunction

Symbols and abbreviations

σ	Hammett substituent constant
ψ	molecular orbital wavefunction
χ	empirical solvation parameter
A	Helmholtz free energy
A	antisymmetric (orbital)
A	anchimeric assistance, extent of
A	pre-exponential term in Arrhenius equation
Ad	adamantyl
Ad	addition (reaction type)
AN	acceptor number (solvation parameter)
AO	atomic orbital
a	hyperfine coupling constant, ESR
anti-	on opposite sides (cf syn-)
b	principal polarizability
B	hydrogen-bonding solvation parameter
B	magnetic flux density
Bu	butyl
bros	p-bromobenzenesulfonate
C	Coulomb (electrical charge)
C_p, C_V	Heat capacity (constant pressure, constant volume)
CIDNP	Chemically-induced dynamic nuclear polarisation
c	concentration (molar)
D	bond dissociation enthalpy
D	cohesive energy density
D_π^-	empirical solvation parameter
DNB	dinitrobenzoate
d	density
d	diffusion coefficient
E	energy
E	elimination (reaction type)
Et	ethyl
$E_q, E_{pol}, E_{exch}, E_{ct}$	electrostatic, polarisation, exchange and charge transfer energy
E_s, E_s^c	steric constants
E, E^+	a generalized electrophile
E, ES	(in structures) enzyme, enzyme-substrate complex
E_A	Arrhenius activation energy
E_T	empirical solvation parameter
E_N	Edwards nucleophilicity parameter
E_R	reaction field
EC	effective concentration
ESR	electron spin resonance (= electron paramagnetic resonance, EPR, PMR)
ERE	empirical rate equation
exo, endo	stereochemistry with reference to a component of

structure related to 'boat' cyclohexane:

F	force
F	field effect parameter
$\mathscr{F}$	Swain–Lupton field effect parameter
$f(f_0, f_m, f_p)$	partial rate factor (relating to ortho, meta, para positions of a substituted benzene)
f_ε, f_n	Kirkwood electrostatic solvation functions
f	activity coefficient (mole fraction units)
G	Gibbs free energy
ΔG_{tr}	free energy of transfer
G_A, G_B	constants in the Brønsted catalysis law
g	gyromagnetic ratio
H	enthalpy (heat)
$\boldsymbol{H}_{i,i}$	Coulomb integral
$\boldsymbol{H}_{i,j}$	Resonance integral
$\mathscr{H}$	Hamiltonian operator
ΔH_{at}	standard heat of atomization
HMO	Hückel molecular orbital
HOMO	highest occupied molecular orbital
ΔH_f	standard heat of formation
H_0, H_I, H_R, H_A, H_-	acidity functions
H_N	Edwards nucleophilicity parameter
$\boldsymbol{h}$	Planck's constant
I	indicator ratio
$+I, -I$	inductive effect (electron donating, electron withdrawing)
J_0	acidity function (see H_0)
K	Kelvin (temperature scale)
K	equilibrium constant
K_A	acid dissociation constant
K_M	Michaelis constant
k	specific rate constant: $k_1, k_2, k_3, \ldots$ are used both to denote rates of successive stages of a reaction and also to distinguish unimolecular, bimolecular, termolecular . . . processes.
k_{rel}	relative rate constant
k_H, k_D	rate constants for reaction of isotopic species containing H, D respectively
k_c, k_s, k_Δ	rate constants for components of solvolysis
$\boldsymbol{k}$	Boltzmann constant
L	Avogadro's number
L	a generalized hydrogen isotope (i.e. H, D or T)
L_+, L_-	localization energy

Symbols and abbreviations

LUMO	lowest unoccupied molecular orbital
l	nucleophilicity coefficient in Grunwald–Winstein equation
M	molar mass (molecular weight)
MO	molecular orbital
m	meter
m_s	polarity coefficient in Grunwald–Winstein equation
mes	methanesulfonate ($-SO_2CH_3$)
N	Newton (unit of force)
N, N_+	nucleophilicy parameters
Nu:, Nu$^-$:	a generalized nucleophile
NAD (NADH)	nicotinamide adenosine dinucleotide (reduced form)
NGP	neighboring group participation
n	refractive undex
n	nucleophilicity parameter
n-	unshared pair (electrons, MO)
n	an integer
$\mathbf{p}$	dipole moment
$\mathbf{P}_E$	total polarizability
P	empirical solvation parameter
PNP	paranitrophenyl
PKIE	primary kinetic isotope effect
P	total bond order
Ph	phenyl
Pr	propyl
p	pressure (vapor pressure)
p, p_π	partial (π) bond order
Q	partition function
Q	constant in McConnell's equation
q	electric charge
q	integer in Woodward–Hoffmann rule
q	heat
$\mathbf{R}$	gas constant
R–	a generalized unit of structure, usually an alkyl group
+R, –R	resonance effect (electron donating, electron-withdrawing, respectively)
R	molar refraction
$\mathcal{R}$	Swain–Lupton resonance effect constant
r	correlation coefficient
r	integer in Woodward–Hoffmann rule
r	Yukawa–Tsuno constant
$S_{i,j}$	overlap integral
S	entropy
ΔS_f	standard entropy of formation
ΔS_{tr}	standard entropy of transfer

S	selectivity
S	empirical solvation parameter
$\mathscr{S}$	empirical solvation parameter
S	symmetric
S_N, S_E	substitution (nucleophilic, electrophilic).
S	substrate (in enzymic reaction schemes)
SOH	a generalized protic solvent
syn-	stereochemical designation; on the same side (cf. anti-)
s	coefficient in nucleophilicy correlation (see *n*)
T	temperature
t	time
tos	*p*-toluenesulfonyl, CH_3—⟨ ⟩—SO_2O—
tert-Bu	tertiary butyl, $(CH_3)_3C$—
U	total internal energy
u	function in Bigeleisen equation
V	vibrational quantum number
V	volt
VB	valence bond
v	volume
v	velocity of reaction
w	work
w	parameter in Bunnett equation
X-	an electron-donating (resonance) substituent e.g. MeO-
X	empirical solvation scale
x	concentration, mole fraction
x	a fractional amount $(0 < x < 1)$
Y	empirical solvation scale
y	activity coefficient (molar units)
Z-	an electron-withdrawing (resonance) substituent e.g. $-NO_2$.
⌒→	notional movement of an electron-pair (note the arrow must be directed Nu:$\overset{\frown}{\ }$ E)
⌒	notional movement of a single electron
‡	pertaining to the transition state
*	pertaining to an electronically-excited state
⥮	spin-paired electrons
↑	unpaired electron
↔	resonance, symbol connecting VB contributing structures

MECHANISTIC DESIGNATIONS

A shorthand notation for the designation of reaction types is widely used in organic chemistry although it is not entirely systematic. In general, the symbol specifies the reaction as a substitution (S), addition (Ad) or elimination (E) followed by the type of process — nucleophilic (N), electrophilic (E) or homolytic (H) and the molecularity of the slow step (1, 2 and sometimes 3). Other symbols include A (acid catalysed), B (base catalysed), Ar (aromatic) and cb (conjugate base); for carbonyl-type substitutions there are in addition, Ac (acyl–oxygen fission), Al (alkyl–oxygen fission). The following schemes are intended to be a quick guide to the essential features of various mechanistic types including their supposed transition states, enclosed in brackets and denoted by the sign (‡). Nu:$^-$ and E$^+$ represent generalized nucleophiles and electrophiles while —Nu and —E in structures are nucleofugic and electrofugic groups (i.e. leaving groups); the latter is usually —H. —X and —Z are electron-donating (+ R) and electron-withdrawing (− R) substituents respectively, S stands for substrate (reagent) and SOH represents a protic (hydroxylic) solvent. The rate-determining step is also indicated by 'slow' for multistep reactions.

page

S_N1 **386**

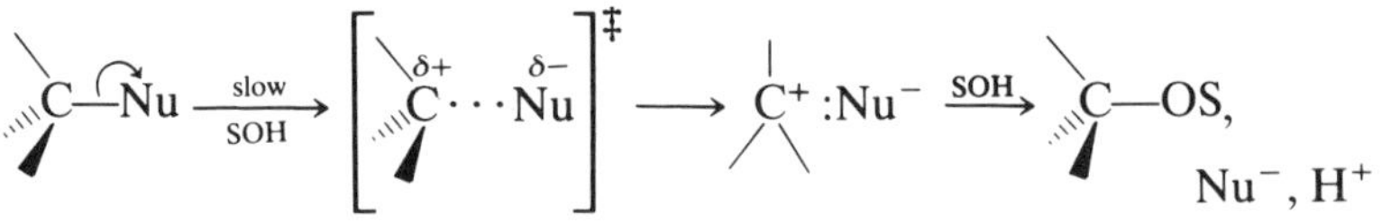

S_E1 **417**

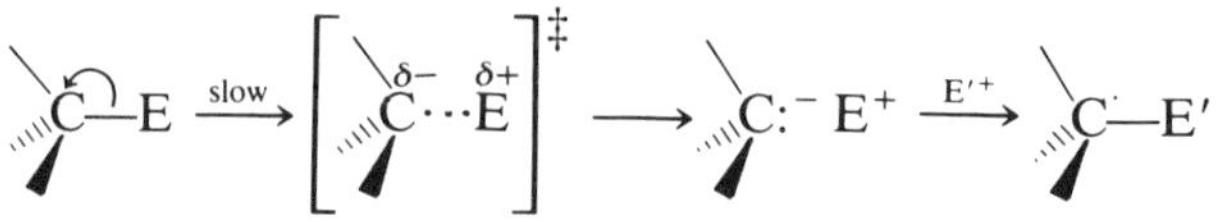

S_H1 **737**

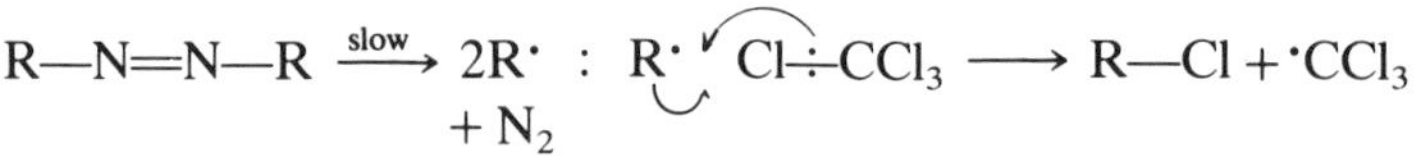

S_N2 **375**

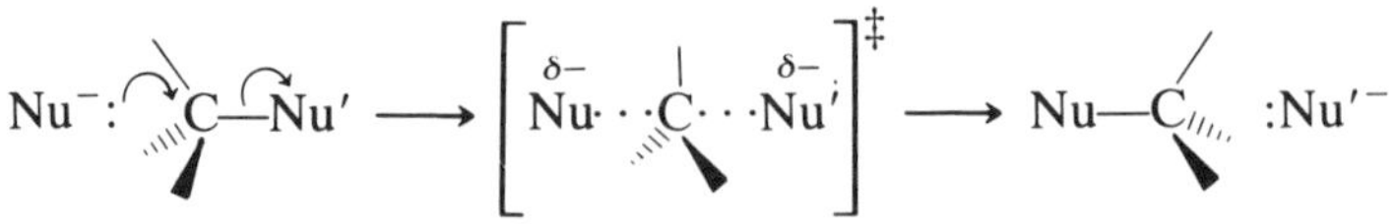

this type of synchronous transfer is not limited to reactions at carbon, for instance:

B:$\curvearrowright$H—$\overset{+}{O}$H$_2$ $\longrightarrow$ $\overset{+}{B}$—H :OH$_2$ S_N2 reaction at H

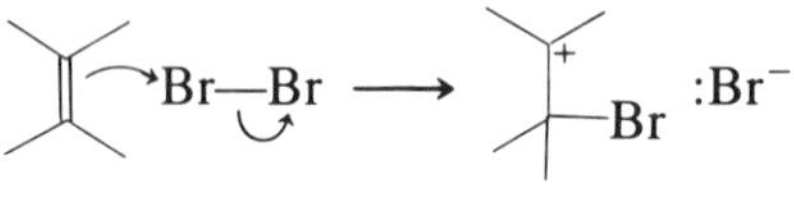

S_N2 reaction at Br

S$_E$2 417

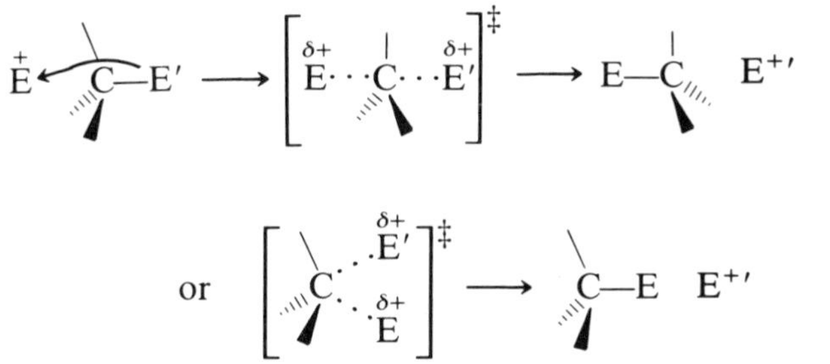

S$_H$2 737

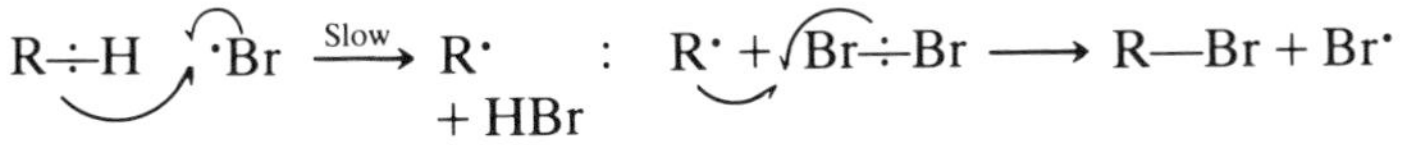

S$_E$2Ar 427

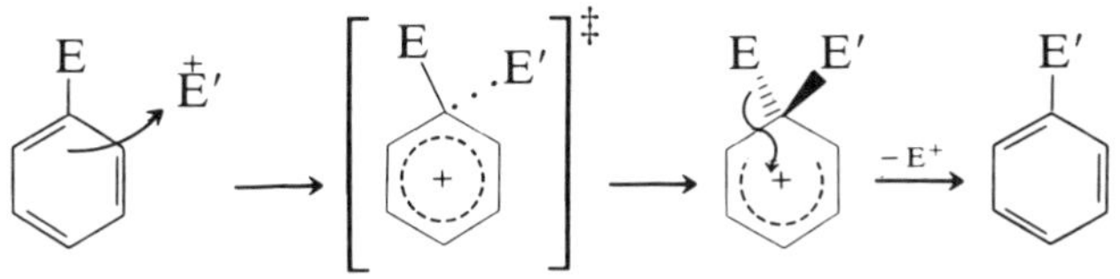

S$_N$2Ar 452

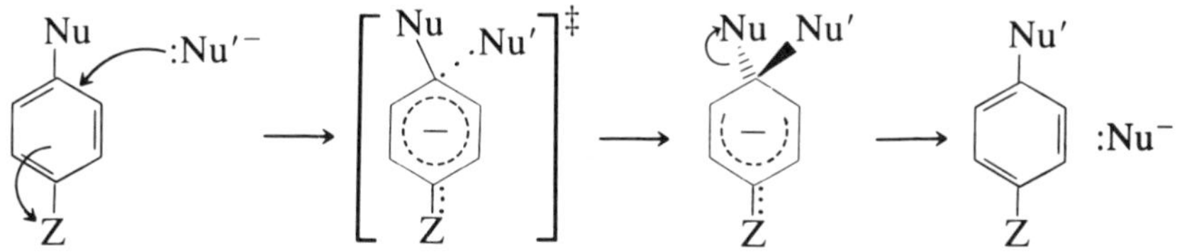

S$_N$i (intramolecular) 414

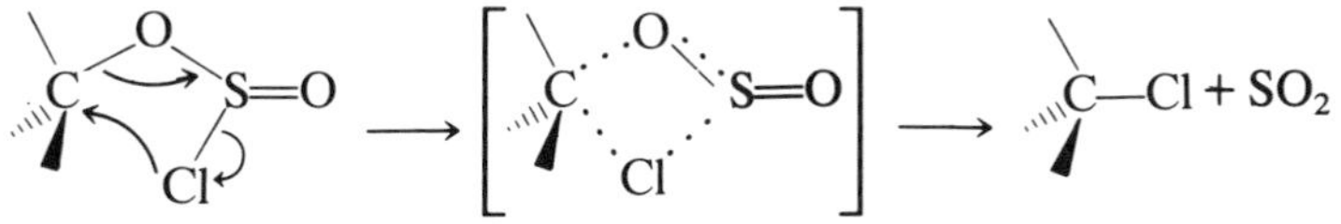

S$_N$1′ (S$_N$1 with allylic rearrangement) 400

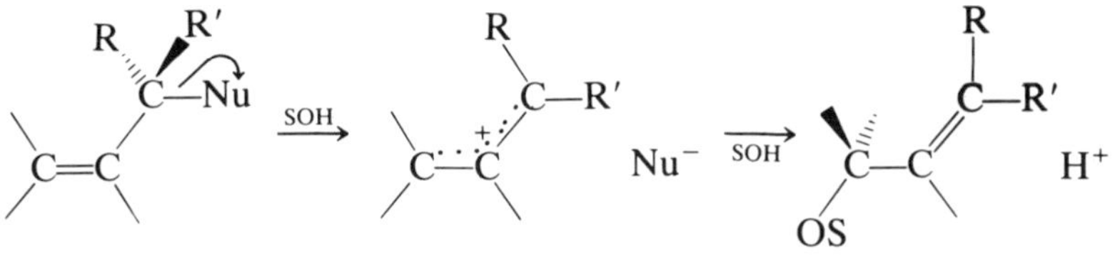

Symbols and abbreviations

S_N2' (S_N2 with allylic rearrangement)   **400**

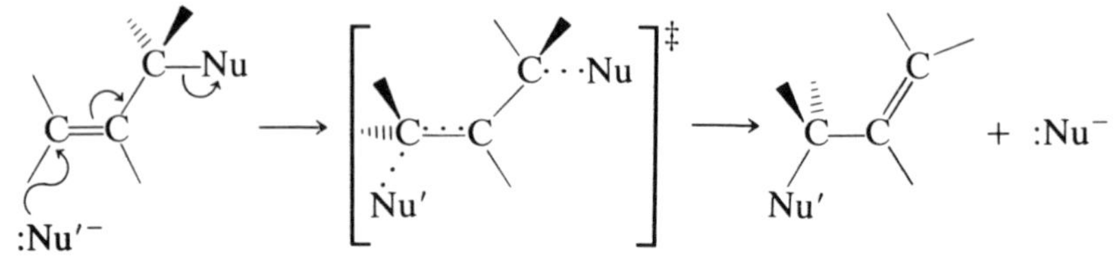

$S_{RN}1$, $S_{RN}2$ 1-electron reduction followed by S_N1 or S_N2 sequences (p. 749)
$S_{ON}1$, $S_{ON}2$ 1-electron oxidation followed by S_N1 or S_N2 sequences (p. 749)

A–S_N2,   **385**

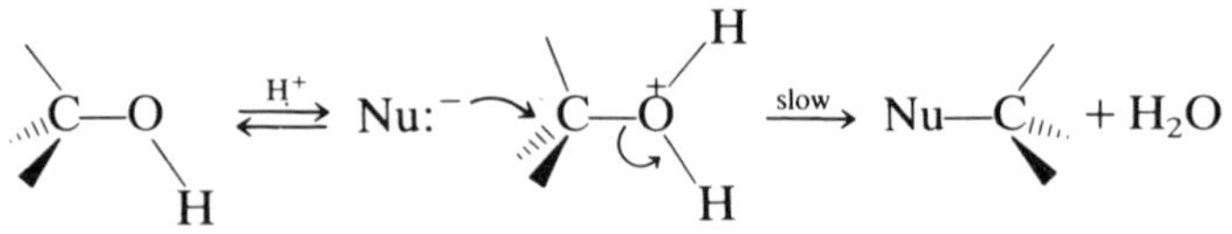

A1 (hydrolytic processes)   **334**

$$S + H\!-\!\overset{+}{B} \rightleftharpoons S\overset{+}{H} \quad :B$$

$$S\overset{+}{H} \xrightarrow{\text{slow}} \text{products}$$

A2 (hydrolytic processes)   **334**

$$SH^+ + H_2O \xrightarrow{\text{slow}} \text{products}$$

eliminations

E1   **504**

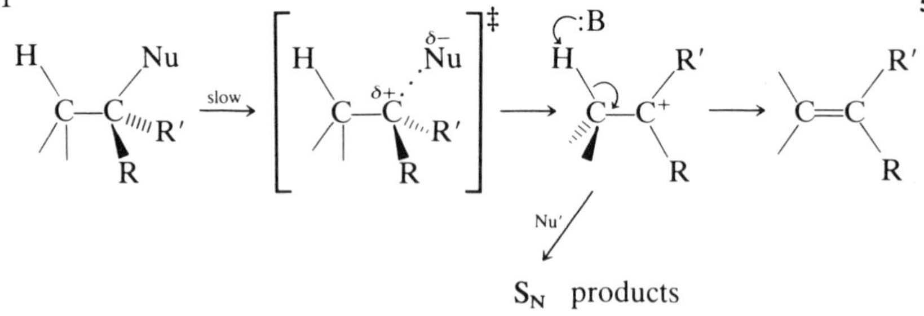

S_N products

E2 (E2H)   **504**

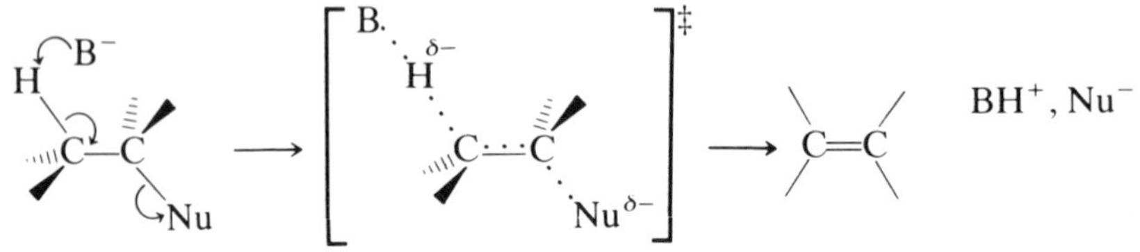

E2C like E2H but the 'soft' base interacts both with C and H   **504**

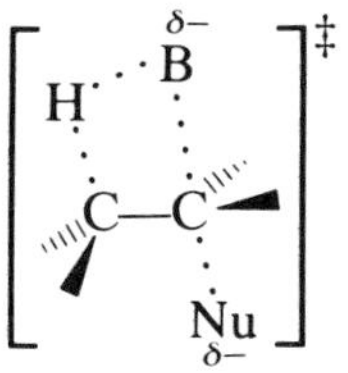

Elcb   **504**

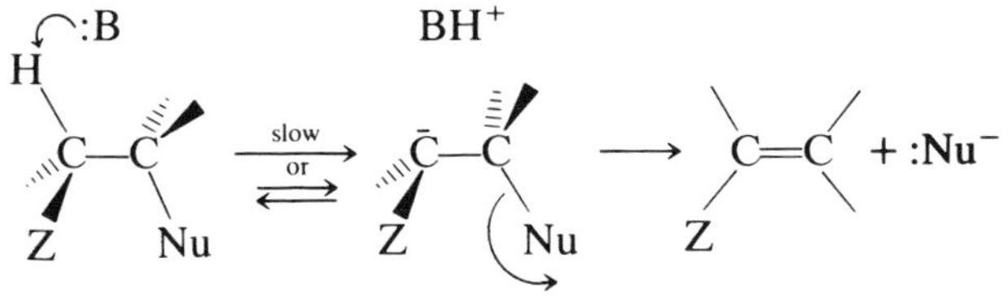

BH$^+$

slow
or

$+ :Nu^-$

additions

E$_i$ **533**

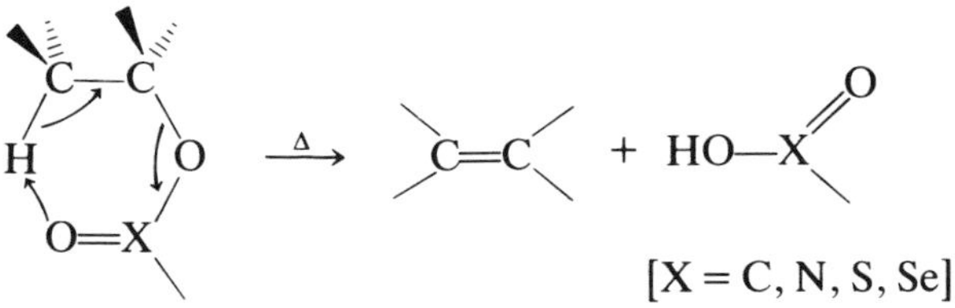

$\xrightarrow{\Delta}$ $+ \ HO-X$

[X = C, N, S, Se]

Ad$_E$2   **548**

slow

Ad$_N$2   **548**

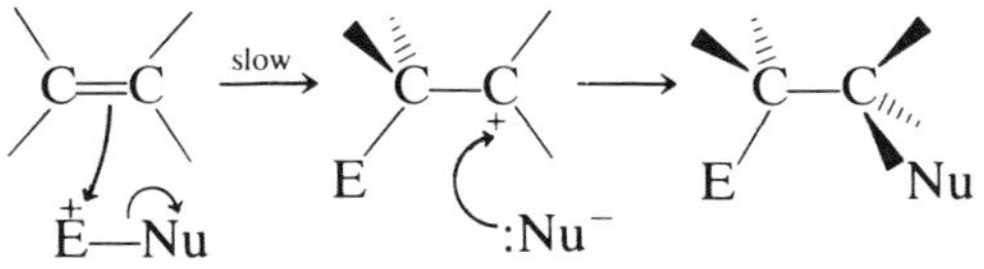

(H$^+$)

(H$^+$)

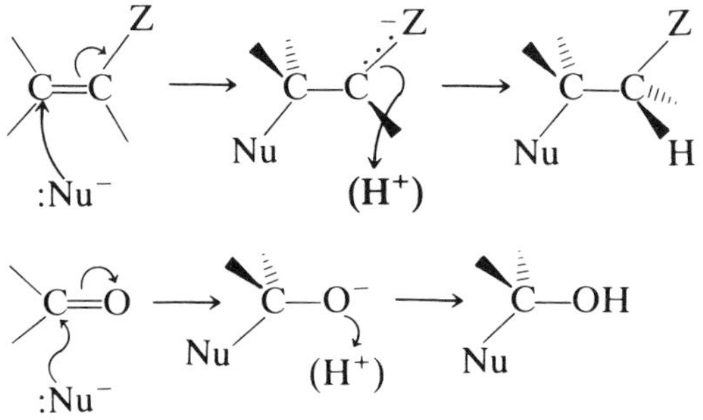

Ad$_H$2   **740**

·Br $\longrightarrow$

Br

H   Br

H$\div$Br

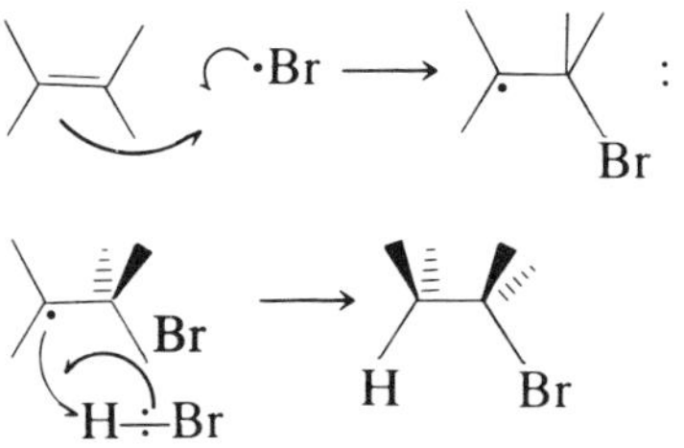

Ad—E: a sequence of addition followed by elimination, the net result being substitution   **423**

E—Ad: a sequence of elimination followed by addition resulting in substitution. See arynes.   **423**

carbonyl and related substitutions

$B_{Ac}2$ (an Ad_N—E sequence)   **466**

$A_{Ac}2$ **473**

$A_{Al}1$ (an S_N1 process)   **480**

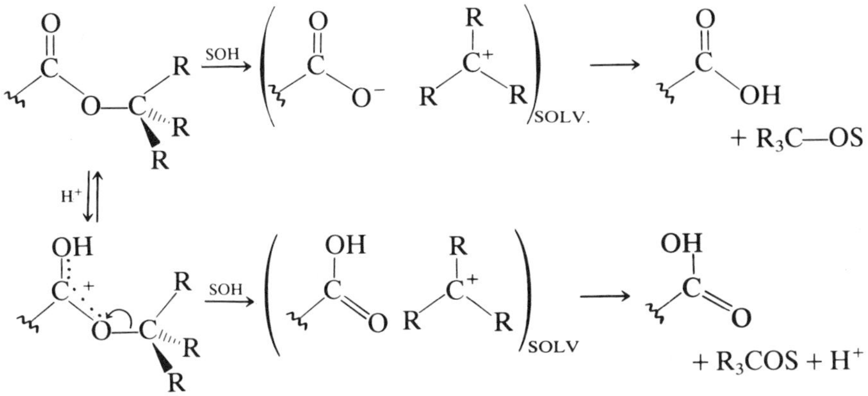

$A_{Al}2$ (an S_N2 process)   **482**

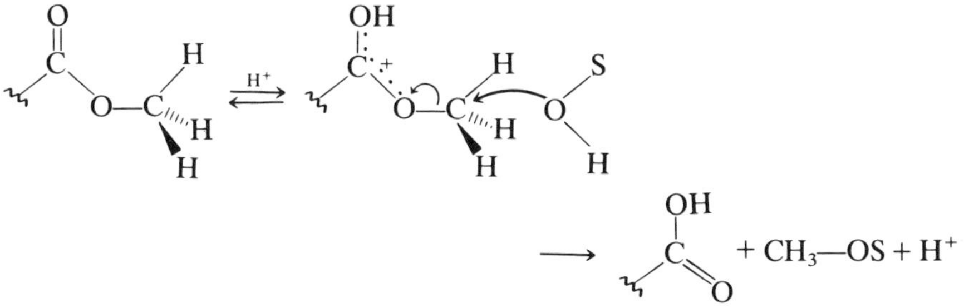

$A_{Ac}1$ **476**

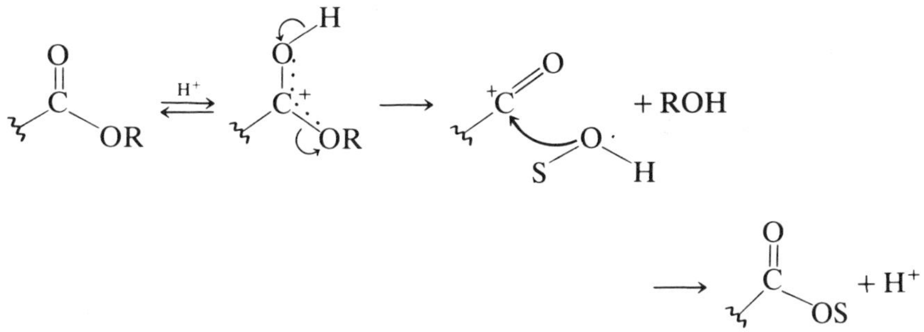

813

Index

α-effect, 218, 247
α-methyl effects, S_N reactions, 407
$A_{Ac}2$ mechanism, 336, 474
$A_{Al}1$ mechanism, 335, 481
$A_{Al}2$ mechanism, 483
A1, A2 mechanisms, 336
α-solvation scale, 198
A-solvation parameter, Table, 204, 205
ab initio M.O. methods, 20
abstraction reactions, 737
  rates, 738
acceleration, steric, 290
acceptor number (AN), solvation
      scale, 199
acetal hydrolysis, 277
  solvent isotope effect on, 278
acetaldehyde
  aldol reactions, 571
acetoacetate, alkylation of, 249
acetolysis, estimation of rates, 413
acetone, M.O.'s of, 23
  bromination of, 353
acetylene, acidity, 224
  electrophilic additions, 566
acids, strengths of, 211
  super-, 397, 448
acid-base dissociation, 210
acid dissociation constant, 210
  interpretation of, 214
  isotopic effect on, 274
  table of, 212
acidity, gas phase, 225
  ammonium ions, 218
  aqueous, table, 212
  carbon acids, 219
  table, 226
acidity function, H_0, 232
  and ester hydrolysis, 480
  table of, 233
  values of for H_2SO_4-H_2O, 234

acidity function, H_-, 236
  and alkene additions, 550
acrolein, M.O.'s of, 23
activation, 88
  effect of tunnelling, 272
  energy of, 89, 94
activity, 179
  in acid dissociations, 210
  of water in strong acids, 338
activity coefficients, 179
acyl halides, hydrolysis mechanisms,
      486
  aromatic substitution by, 432
adamantyl, solvolysis, 191, 388
addition (Ad reactions), 119
  to aromatic systems, 448
  to carbonyl compounds, 570
  electrophilic, 548 et seq.
  –elimination in vinyl compounds,
      579
  Michael, 568
  nucleophilic, 567
  photochemical, 790
  radical, 740
affinity, reaction, 77
alcohol dehydrogenase, 637
alcohol oxidation
  by chromic acid, 541
  conformational effects, 321
  by dimethyl sulfoxide, 543
  formation by hydride, 570
  reduction, Barton method, 747
aldehyde hydrates, dehydration, 358
aldol reaction, 571
alkane, energies of thermolysis, 753
alkenes
  additions to, 548 et seq.
  bromine addition, steric effects, 301
  hydration, 549
  hydration, catalysis of 359

alkenes (*cont.*)
  photochemical additions, 790
  photochemical dimerization, 784
alkyl chain, charge distribution in, 139
alkyl halides,
  eliminations, 504
  pyrolysis, 538
  solvolysis, 386
alkyl substituents, 143
allenes,
  cycloadditions, 679
  electrophilic additions, 566
allyl anion, M.O.'s of, 248
allyl lithium, spectrum of, 326
Alternant systems, 12
ambident nucleophiles, 248
amides
  conformations of, 312, 314
  dehydration, 370
  hydrolysis, 484
amines, dissociations of, 218
  inversions of, 315
amination, 460, 461
amine oxide pyrolysis, 533, 537
α-aminoacids, 625
aminoalkyl halides, solvolysis, 589
anchimeric assistance, 590, 593
anhydrides, hydrolysis mechanisms, 487
  aromatic substitution, 432
  in Perkin reaction, 572
ANRORC reactions, 461
antarafacial geometry, 655, 788
anti-aromatic systems, 14, 654
argon, radial distribution of liquid, 173
aromatic systems, 13
  additions, radical, 742
  electrophilic substitution, 428
  eliminations, 459
  nucleophilic substitution, 453
  pericyclic transition states, 654
  photochemical reactions, 795
  radical substitution, 743
aromaticity, 13
  acidity due to, 224
  in cycloadditions, 654
Arrhenius equation, 93
  non-linear, 271
arynes, 459
assistance, nucleophilic, 388, 589, *et seq.*
  electrophilic, 405
  intramolecular, 411

asymmetric induction (synthesis), 288, 323
autoxidation, 744
azides, thermolysis, 101
  dipolar cycloaddition, 687
aziridines, *N*-nitroso, thermolysis, 692
azo compounds
  CIDNP in, 729
  photolysis, 717
  rates and energies of thermolysis, 715, 754
azoisobutyronitrile (AIBN), thermolysis, 715
azomethineimines, 685, 687

β-solvation scale, 198
B-solvation parameter, 196
  table, 204, 205
$B_{Ac}2$ mechanism, 467
$B_{Al}2$ mechanism, 483
$B_{Al}1$ mechanism, 484
Baeyer–Villiger oxidation, 372
Bamberger rearrangement, 371
Barton reaction, 744
  alcohol oxidation, 747
basicity, gas phase, 225
Beckmann rearrangement, 361, 799
  abnormal, 371
Benesi–Hildebrand equation, 68
benzene, Hückel M.O.'s, 9
  photochemistry, 798
  valence bond description, 3
benzene-chlorine, interaction energy, 60
benzenium ions, 448
  stability, 450
  acidity, 224
benzidine rearrangement, 364
benzoate ester hydrolyses
  pressure effects on, 111
  substituent effects on, 130, 147
benzoic acids, acidity of, 131, 166
benzyl, M.O.'s of, 451
benzyl acetate, mechanistic change in hydrolysis, 480
benzyl chloride, hydrolysis kinetics, 110
BH_3–CO, interaction energy, 59
bicyclic compounds, strain energies, 284
Bigeleisen equation, 258, 263
bimolecular reactions, 89

biphenyl radical ion, spectra, 52
biphenyls, racemization,
   steric effects, 290, 324
   isotope effects on, 268
boiling point, 63
bond-angle strain, 289
bond angles, 27
bond dissociation enthalpies, 36
   tables, 39–43
bond lengths, table, 26
bond length, calculated, 19
bond order, 15, 18
   substituent effects on, 125
9-borabicyclo[3,3,1]nonane (9-BBN),
   562
bridgehead compounds, solvolysis, 407
bromine
   addition to alkenes, 554
   addition to diphenylethene, 584
   aromatic substitution, 433
p-bromophenol, rearrangement, 495
bromopropanoate, hydrolysis, 607
Brønsted Catalysis Law, 340
Brønsted plot, coefficients, 342
   bromination of acetone, 354
   carbonyl transfer reactions, 464
   in elimination, 512, 517
Brown–Okamoto equation, 149
Bunnett w-parameter, 337, 357, 480,
   482
   and alkene addition, 550
   benzenium ions, acidity of, 224
   and Olson ϕ parameter, 339, 355
   table, 338
Burke–Lineweaver plot, 627
butadiene, M.O.'s of, 4
   substituted, M.O.'s of, 578
but-2-ene, thermodynamic functions,
   85
tert-butyl chloride, solvolysis, isokinetic plot, 104
   definition of Y-parameter, 190

χ-solvation parameters, 194
Cannizzaro reaction, 570
   radical involvement, 751
cage effects, 726, 733
calciferol, isomerization, 699
captodative radicals, 757
carbenes, 539
   CIDNP in, 729
   cycloadditions, 692

carbenoid reagents, 540, 709
carbinolamine, dehydration of, 344
carbocations
   as intermediates, 405
   in addition to alkenes, 565
   aryl, 458
   in S_N1 reactions, 386
   stability, 389, 406
carbon acids
   acidities of, 219
   measurement, 221
   rates of dissociation, 225
   table, 220
carbon isotope effect, 269
carbonyl group
   nucleophilic reactions, 462
   photochemistry, 791
carboxypeptidase, 635
catalysis, 330 et seq
   acid-base, 331
   anhydride hydrolysis by metal ions,
      487
   of Diels–Alder reactions, 704
   electrophilic, 351
   general base, 490
   mechanisms, 334
   by micelles, 368
   neighboring groups, 490, 589 et seq
   by non-covalent binding, 367
   nucleophilic, 345, 489, 593
   table, 333
CH_3F–F, interaction energy, 62
chain reactions, 731
chain transfer, 737
charge densities, aromatic, 442
charge-transfer, complexes, 67
   energy, 57
   and solvation, 177
   table of properties, 69
charge type, reaction, 184
chelotropic reaction, 692
chemical potential, 83
chemically induced dynamic nuclear
      polarization (CIDNP), 725
   table, 729
chemiluminescence, 685
chlorambucil, 641
chloradamantane, 642
chlorine, additions to alkenes, 549, 585
chloromethane, V.B. structures, 2
chlorosulfonyl isocyanate, 679
cholesteryl derivatives
   solvolysis, 606, 618

chromic acid, 541
chromophore, 777
Chugaev reaction, 536
chymotrypsin, 632
'cine'-Substitution, 448
Claisen reaction, 571
Claisen rearrangement, 697
CNDO methods, 24
coalescence temperature, 305
cofactors, 635
cohesive energy density, 188
common ion effect, 399
compressibility of activation, 96
condensation reactions, 575
conformational analysis, 301
    effects on rates, 315
    table, 311
'Contra–Markownikoff' addition, 557,
    561
    peroxide effect, 741
    photochemical, 790
contributing structures, 2
Cope reaction, 533
Cope rearrangement, 696
Coulomb integral, 6
coupling, radicals, 735
covalency, 1
covalent radius, 26
'crossover' processes, 734
crown ethers, 511, 517
cumene, autoxidation, 746
cumyl chlorides, solvolysis, 146
Curtin–Hammett principle, 317, 524
Curtius rearrangement, 361
curved arrow notation, 121
cyanide, alkylation of, 250
cyanoalkenes, acidities of, 220
cyanobenzoate, hydrolysis, 643
cyanoethylation, 569
cybotactic region, 175
cyclic compounds
    conformations of, 310
    strain energies, 283
cyclization, energetics of, 609, 614
    rates of, 613
cycloaddition
    correlation table, 656
    1,3-dipolar, 685
    of ketones, photochemical, 793
    $(4 + 2)$, M.O. diagram, 650
    $(2 + 2)$ M.O. diagram, 651, 676
    photochemical, 784
    two-step, 680

cycloaddition $(2 + 2)$, solvent effects
    on, 200
cycloalkenes from elimination, 526,
    527
cyclobutenes, butadiene interconver-
    sion, 689
cyclobutenones, acidity of, 224
cyclodextrins, 641
cyclohexanes, conformations, 312, 313
    hexachloro, eliminations, 544
    rate effects in, 317
cyclooctene, participation in, 623
cyclopentadiene, dimerization
    acidity of, 220, 224
    effect of pressure, 95
    effect of solvent, 206
cyclopentadienide, M.O.'s of, 13
cyclopropane, M.O.'s of, 622
cyclopropanes, from carbenes, 540
cyclopropylmethyl derivatives sol-
    volysis, 621
$(2 + 2)$-cycloreversions, 682

D_π-solvation scale, 198
decarboxylation, isotope effect on, 262,
    269
    acid-base catalysis of, 358
Dewar benzene, 797
diazomethanes
    cycloadditions, 685
    photolysis, 695
    reaction with acids, 366
diazonium ion, decomposition, 456
    CIDNP in, 729
diborane, addition to alkenes, 561
β-dicarbonyl compounds, enolization,
    355
dichlorethane, conformations of, 303
dichromate, oxidations by, 321
dicyclohexylcarbodiimide, 542
Dieckmann reaction, 571
dielectric constant, 184
Diels–Alder reaction
    acid catalysis, 704
    of arynes, 460
    intramolecular, 673
    orbital symmetry conservation in,
        650
    reaction rates, 661
    retro, 670
    solvent effects on, 200
    steric effects on, 288

theory, 658
volume of activation, 112
1,3-dienes
   electrophilic additions, 566
   photochemical dimerization, 784
dienophiles, 659
diffusion control, 217
diimide, 676
β-diketones
   hydrogen bonding in, 67
   solvent effect on, 205
dioxetanes, thermolysis, 684
di-π-methane rearrangement, 789
diphenyldiazomethane
   mechanism, 365
   solvent effects on reactions, 204
   solvent isotope effect on reactions, 278
diphenylethenyl halides, substitution mechanism, 581
diphenylketene, 677
diphenylpicrylhydrazyl radical, 721, 736
dipoles, 49
dipole–dipole forces, 50
dipole moments, 31
   of charge-transfer complexes, 69
dipolar aprotic solvents, 171
disalicyl phosphate, solvolysis, 643
disiamylborane, 562
disproportionation radicals, 727, 737
donor number (DN), solvation parameter, 196
donor–acceptor interactions, 177
   solvation scales, table, 197
d-orbitals, effect on acidity, 223
Drude–Nernst equation, 102, 190
dual parameter correlations, 153
durosemiquinone radical anion, ion-pairing in, 48

E1 reactions, 119, 190, 505
E2 reactions, 119, 261, 269, 505 *et seq.*
E1cb processes, 119, 530 *et seq.*
   in ester hydrolysis, 532
E_s-Taft steric constant, 297
   table, 298
E_s^c-Hancock steric constant, 297
   table, 298
E_N-nucleophilicity parameter, 243
E_T-solvation parameter, 194

eclipsed conformation, 301
Edwards equation, 242
effective concentration (EC), 601
   table, 602
Eigen, model of acid dissociation, 229
electrocyclic reactions, 689
   photochemical, 786
electron densities of conjugated hydrocarbons, 12
electronic demands, 129
   reverse, 663
electronegativities
   atomic, 31
   of substituents, 163
electron spin resonance (ESR), 720
electrophilicity, measure of, 250
electrophiles, 116
electrostatic forces, 46
electrostriction, solvent, 102, 176
eliminations, 119, 504 *et seq.*
   α-, 539
   competition with substitution, 520
   intramolecular pyrolytic, 533, 697
   isotope effects in, 261, 266, 269
   kinetics, 507
   oxidative, 541
   structural effects, 508
   table, 521
empirical rate equation, 96
enamines alkylation, 355
endothermic reactions, 79
'ene' reaction, 674
energy profiles, 90, 591
enolate, M.O.'s of, 248
enolization, 423
enthalpy, H. 77
   temperature dependence, 87
entropy, S, 79
   contributions to, 83
   of liquids, 172
   molar, table of, 83
   of reaction, 102
entropy of activation
   acyl halide hydrolysis, 486
   E1cb reactions, 533
   ester hydrolysis, 469
   hydrolytic reactions, 340, 414
   Menshutkin reactions, 494
entropy of formation, 45
   table, 39–43
enzymes
   kinetics, 626
   structures, 624

enzymic reactions, 624
equilibrium, 84
  constant, 86
epoxidation, chiral, 324
  mechanism, 563
    esters, hydrolysis, basic, 467
    esters, hydrolysis, acidic, 474
esters, hydrolysis, $A_{Al}1$ mechanism, 477
esters, carboxylic
  hydrolysis mechanisms, 467
  pyrolysis, 533
ethane, conformations of, 302, 306
  stability–strain relationship, 760
ethanol, enzymic oxidation, 637
ethane, M.O.'s of, 22
ether cleavage
  by HI, 239
  by halides, 493
ethyl acetate hydrolysis kinetics, 467
ω-ethylenic acids, iodolactonization, 641
Evans–Polanyi equation, 756
excitation, electronic, 777
  energies, 781
excited molecules, geometry of, 778
exo–endo ratio, solvolysis, 620

Fenton's reagent, 719, 725
field effect, 138
  from gas phase equilibria, 165
  parameter, 301
  parameter of $\mathcal{F}$, 156
Fischer–Hepp rearrangement, 363
fluorenyl anion, ion pairing in, 49, 53
9-fluorenylmethanol, elimination, 545
fluorescence, 779
fluorobenzenes, substitution in, 495
force constants, 28, 256
  table, 30
fractionation factors, 273
  table, 274
Franck–Condon principle, 778
free energy, Gibbs, 81
  Helmholtz, 87
  of solution, table, 178
  of transfer, 179
free valence, 16, 19
Friedel–Crafts reactions, 432
Fries rearrangement, 363
  photo, 796
frontier orbitals, 21
  in additions, 578

aromatic substitution, 450, 457
  in cycloadditions, 652
  Diels–Alder reactions, 662
  in elimination, 529
  and the HSAB principle, 239
  in radical reactions, 763
fulvene, 797
  Hückel M.O.'s, 9
furoate esters, conformations of, 305

gas law, 80
gas phase reactions, S_N2, 415
gas-phase equilibria, substituent effects from, 164
gem-dimethyl effect, 611
Gibbs function G, 81
Gibbs–Helmholtz equation, 88
glucose
  acid–base catalysis, 331
  Brønsted plot, 341
  mutarotation, solvent isotope effect on, 278
glyceraldehyde, enzymic oxidation, 638
Grignard reagents, 323
Grunwald–Winstein equation, 242, 244
  in additions, 550

H_N-nucleophilicity parameter, 243
halogenation, aromatic, 432
halolactonization, 565
Hammett equation, 131
  deviations from, 146
Hammett indicators, 233
Hammett–Zucker postulate, 337
Hammond postulate, 104, 673, 759
hard-soft acids and bases (HSAB), 239
  table, 240
heat capacity, C_p, C_V, 87
heat capacity of activation, 94
  in solvolysis, 413
heats of formation, atoms and molecules, 38
heavy atom isotope effects, 268
hemiorthoesters, 487
  table, 488
heptane, chlorination, 759
heteroatom systems, M.O.'s of, 14
heterolysis, enthalpy of, 391
hexachlorocyclohexane, isomers of, 326
hexa-1,3,5-triene, Hückel M.O.'s, 9

hexatriene-cyclohexadiene
  interconversion, 690
HOMOgen, 115
HOMO–LUMO interactions, 122, 126,
  648
Hofmann's Rule, 522
homoallyl cation, 616
homotropyllium ion, 617
Hückel M.O., method, 3
  electrophilic aromatic substitution,
    450
  energies of conjugates systems, 10
  nucleophilic aromatic substitution,
    456
  pericyclic reactions, 649, *et seq.*
  radical reactions, 764
Hughes–Ingold rules, 183
Hunsdiecker reaction, 747
hydration of alkenes reactivities, 553
hydration carbonyl compounds, 336
hydration, energy of, 181
  effect of mixed solvents, 183
hydrazines, 576
  basicity of, 218
hydrazone formation, 358
hydride transfer
  boranes, 561
  Cannizzaro reaction, 570
  complex metal hydrides, 570
hydroboration, steric effects, 297
  mechanism, 561
hydrogen abstraction, selectivity in,
    107, 719
hydrogen-bond, 62
  and solvation, 176
  table, 64
hydrogen transfer, concerted, 676
hydrolysis, acid catalyses, 336
hydroperoxide, nucleophilicity, 247
hydrophobic bonding, 367
p-hydroxybenzoate hydrolysis, 533
2-hydroxyethylpyridine
  hydrogen bonding in, 66
hydroxylamine, 576
2-hydroxypyridine, solvent effect on
    tautomerism, 182
hyperconjugation, 143
  and isotope effect, 265

i-cholesteryl derivatives, 618
imidazole catalysis by, 491
imines, cycloaddition, 679

indene, hydrogen scrambling, 698, 710
index of solvation, empirical, 184
  relations between, 199
  specific, 196
  table, 180
indicator ratio, 210
inductive effect, 137
inductive isotope effects, 268
infrared frequencies, solvent effects on,
    196
inhibition, 733
initiator, 732
intersystem crossing, 779
intermediates, reactive, 91, 98
  in additions, 558
  in aromatic substitution, 447, 456
  in carbonyl substitutions, 487
  in eliminations, 504
  in NGP, 613
  in solvolysis, 396
intermolecular forces, weak, 56
intermolecular reactions, 589 *et seq.*
iodolactonization, 641
ion-cyclotron resonance, 164, 225
ion-dipole forces, 50
ion-pairs, 49
  association constants, 47
    solvent effects, 203
  in E2 reactions, 510
  forces between, 46
  in S_N1 reactions, 386
ipso addition, 449
isocyanate cycloaddition, solvent effect
    on, 207, 681
isokinetic relationship, 102
isonitriles, additions to, 577
isopropyl acetate, mechanistic change
    in hydrolysis, 479
isotopic effects
  effect of base strength, 264
  inductive, 268
  kinetic, 255 *et seq.*
  solvent, 272
  steric, 268
  table, 259, 261
  theory, 256
isotopic scrambling, ester hydrolysis,
    469
  table, 471

Jacobsen rearrangement, 361, 447
Jablonski diagram, 780

Kaptein's rules, 730
ketal family of compounds, 352
ketals, conformational effects in, 322
  hydrolysis, 356
ketene
  additions, 576
  cycloadditions, solvent effect on,
    201, 206, 706
  M.O.'s of, 577
  rates of cycloaddition, 673, 677
keto-enol equilibria, 355
ketones, acidities of, 220
  additions to, 570
  halogenation of, 353
ketyl radical, 721
kinetics, 77 *et seq.*
  aromatic substitution, 436
  elimination, 507
  enzymic, 626
  ester hydrolysis, 472
  neighboring group participation, 593
  radical reactions, 732
  solvolysis, 399
kinetic and thermodynamic control,
    107
Kirkwood function, 185
  in diazomethane reactions, 367
Kirkwood–Bauer–Magat equation, 196
Kirkwood–Onsager equation, 185
Knoevenagel reaction, 572

β-lactams
  formation from cycloaddition, 679
  hydrolysis mechanism, 486
  lactones, hydrolysis, 336
  thermolysis, 683
latent heat of vaporization, molar, 188
lead, tetramethyl, thermolysis, 715
leaving group (nucleofuge) effects
  in eliminations, 510
  in S_N2 reactions, 385
  in S_N1 reactions, 404
  in S_NAr reactions, 456
  table, 406
least motion, principle, 108, 425
Leffler–Hammond principle, 105
Lewis base, 115
limiting ionization, 388
linear free energy relationships
    (LFER), 129 *et seq.*
  acid dissociations, 216, 219
  additions, 551, 582
  aromatic substitution, 440

cycloadditions, 663
1, 3-dipolar cycloadditions, 686
elimination, 513
ester hydrolysis, 470
neighboring group participation, 596
radical reactions, 748
S_N reactions, 407
steric, 299, 761, 555
  table, 300, 302
liquids, structure, 172
localization energy, 16, 19
  in aromatic substitution, 443, 457
  and substituent effects, 164
lone pair substituents, 140
Lossen rearrangement, 361
Lucas' reagent, 386
luminol, chemiluminescence of, 684
LUMO, 21
LUMOgen, 115
lyonium ion, 214
lysozyme, 631

macrostates and microstates, 80
maleate ester hydrolysis, 604
Mannich reaction, 496, 575
Marcus, theory of proton transfer, 230
Markownikoff Rule, 556
matrix isolation, 725
Maxwell distribution law, 89
mechanism, 114 *et seq,* 120
Meisenheimer complexes, 457
Menshutkin reaction, solvent effect on,
    185, 189
  structural effects on, 494
mercury exchange, 418
metalloenzymes, 632
McConnell's equation, 722
methane
  combustion, 77, 79, 82
  formation, 78
  M.O.'s of, 22
ω-methoxyalkyl sulfonates, solvolysis,
    590
methyl cation, M.O.'s of, 398
2-methylbutane, abstractions at, 762
2-methyl-1,3-diphenylallyl anion, ion
    pairing in, 48
methylenecyclopropene, M.O.'s of, 17
methyl perchlorate, hydrolysis, 493
methyl radical, stereochemistry, 766
micelles, 368
Michaelis constant, 626

Michaelis–Menten kinetics, 626
  constants, table, 629
microscopic reversibility, 109
Möbius systems, 653
molecular mechanics, 292
  table, 293
molecular orbital, model, 3
  and substituent effects, 163
More O'Ferrall, Albery, Jencks diagram, 123
  in eliminations, 507
  in hydrolytic reactions, 348
  in sigmatropic reactions, 700
Morse curve, 24
  isotopic effect on, 257
morpholine, additions to alkenes, 585
mustard gas, 594, 641

N_+, nucleophilicity parameter, 244
n, nucleophilicity parameter, 242
naphthalene, radical ion spectra, 52
  e.s.r. spectrum, 54
neighboring group participation, 589
  activation parameters, 611
  in carbonyl transfer, 601
  factors influencing, 608
neophyl solvolysis, 191
  pyrolysis, 535
neopentyl compounds, 381, 393, 493
NH_3–BH_3 interaction energy, 58
nicotinamide adenine dinucleotide
  (NAD), 637
nitration, 431
  effect of pressure, 95
  and Hammond postulate, 105
  kinetics, 495
nitrenes, 695
nitrileimines, 685, 687
nitrile oxides, 685, 687
nitrile ylides, 685, 687
nitrite, M.O.'s of, 249, 250
  photolysis, 744
nitroalkanes, acidities of, 220
nitrocumyl chloride, solvolysis of, 147
nitronium ion, M.O.'s of, 250
N-nitroanilines, rearrangement of, 364
p-nitrobenzyl halides, radical reactions, 750
nitrones, 688
p-nitrophenyl acetate, hydrolysis, 345
  Brønsted plot, 347
nitrosamines, NGP by, 643
nitroso-isobutane, 725

nitroxyl radicals, 725
non-classical ions, 616
non-nucleophilic anions, 235
norbornenyl compounds, solvolysis,
  619
norbornyl solvolysis, 594, 619
2-norbornyl cation, 623
  energy profile, 623
Norrish reactions, 770, 791, 792
nuclear magnetic resonance, conformational studies by, 305
nucleofuge, 246
nucleophiles, 115
  ambident, 115
nucleophilicity, 242
  table of parameters, 243

orbital symmetry conservation, 649
orthocarbonates, hydrolysis, 370
ortho-substituent effects, 161, 285
  steric, 301
Orton rearrangement
  effect of pressure, 95
  mechanism, 363
overlap integral, 7
oxetans, 793
oxetes, 793
oxidation
  chromium-III, 279, 541
  photochemical, 799
  thallium-III, 584
oxime, formation, 358, 576
  rearrangement, 361, 799, 371
oxiranes
  hydrolysis, 333
  S_N2 reactions, 376
oxy-Cope rearrangement, 696
ozone, 685, 688

π^*-solvation parameter, 196
π-complexes, 449, 558
π-energy, 15, 18
π-polarization, 139
P-solvation parameter, 195
Paneth experiment, 715
'paraquat' radical, 721
partial rate factors, 443
  table, 442
participation, π-, 616
  σ-, 616
partition function, 257
Paterno–Buchi reaction, 707

Index

Pauling–Alfree electronegativities, 31
pentane, conformations of, 307
pericyclic reactions, 120, 648 *et seq.*
　photochemical, 784
Perkin reaction, 572
permittivity (relative)
　see dielectric constant
peroxides
　CIDNP in, 727
　table, 716
　thermolysis, 715
peroxide effect, 741
peroxycarboxylic acids, 563
perturbation M.O. theory, 241
pH-rate profile, ester hydrolysis, 470
phenacyl halides, 751
phenols, acidity of, 166
phenylethyl compounds, conformational energy, 293
　solvolysis, 597
phenyloxirane, reactions, 110
phosgene (carbonyl chloride), 497
phosphate ester hydrolysis, 496
phosphoranes, in Wittig reaction, 573
phosphorescence, 779
photochemical reactions, 777 *et seq.*
　cycloadditions, 656, 798
　geometrical isomerization, 782
photo-electron spectroscopy, 143
phthalate ester hydrolysis, 604
picrate, spectra of ion-pairs, 52
pinacol rearrangement, 361
piperidines, oxidation, 318
polarity, solvent, 176
polar reactions, 114
polarization, nuclear spin, 728
polarization energy, 56
polarizabilities, bond, 33, 194
　of leaving groups, 246, 456
polymerization
　cationic, 565
　radical, 732
Pondorff–Oppenauer reaction, 291
potential energy curve, bond, 24
　isotopic effect on, 257
pre-equilibrium, 99
pressure, effect on rates, 95
　effect on isotope effects, 271
Prévost reactions, 563
primary kinetic isotope effect (PKIE), 256
　in additions to alkenes, 555
　in eliminations, 514

in radical abstractions, 737, 739
in S_EAr reactions, 428, 434
in S_N1 reactions, 394
in S_N2 reactions, 382
transition states and, 263
prismane, 797
propagation, polymerization, 732
propene, conformations of, 308
protic solvents, 171
proton exchange, aromatic, 434
proton inventory, 275
proton relay mechanism, 217
proton transfer, H-bonding, 66
　rate of, 216
　theories of, 229
prototropic equilibrium, 355
pyridines, alkylation, steric effects, 285
pyridinium, radical displacement, 750
pyrrolidines, alkylation, conformational effects, 319

quenching, photochemical, 780
quantum yield, 734

radiationless transfer, 779
radicals, 116, 714 *et seq.*
　coupling, 735
　detection, 720
　fragmentation, 743
　generation, 714
　polar influences, 757
　reactions, 735
　reactivities of, 738, 752
　rearrangement of, 744
　stability of, 752
　stereochemistry of, 766
radiolysis, 717
Raoult's law, 63
rate constant, 97
rate-determining step, 99
rate equations, linear, table, 97
reaction constant, ρ, (ρ^0), 133
　acid dissociation, 145
　acyl halide reactions, 486
　addition-elimination, 582
　(2 + 2)-cycloadditions, 681
　dipolar cycloadditions, 686
　eliminations, 509
　　table, 513
　ester hydrolysis, 470
　interpretation of, 144
　ketal hydrolysis, 357
　pericyclic reactions, 686

S_N2 reactions, 380
  table of, 145
reaction constant, ρ^+, 150
  in alkene additions, 551
  Baeyer–Villiger oxidation, 372
  carbene additions, 540
  chlorination of alkynes, 167
  epoxidation, 563
  neighboring group participation, 596
  pyrolytic eliminations, 534
  radical reactions, 749
  S_N1 reactions, 389
reaction constant ρ^-, 151
  sulfonamide ionization, 151
  S_NAr reactions, 454
reaction constant ρ_I, 154, 156, 159,
    162, 552
  table, 160
reaction constant ρ_R, 154, 156, 159,
    162, 552
  table, 160
reaction constant ρ^*, 154
  acyl halides, 485
  alkene additions, 551
  neighboring group participation, 598
  in solvolysis, 389, 412
reaction isotherm, 86
reaction coordinate, 88
reaction field, 187
reactivity
  correlation with structure, 129 *et seq.*
  in Diels–Alder reactions, 662
reactivity index, 15
redox systems
  radical production, 717
  potentials, table, 718
refractive index, 186
refractivity, exaltation, 35
regioisomerism, 522
regioselectivity
  in additions, 569
regio-specificity in Diels–Alder reactions, 665
resonance effect, 136
  enhanced, 148
  variable, 152
resonance integral, 7
resonance, steric inhibition, 290
  steric enhancement, 291
retro-aldol process, 571
return, 398
ribonuclease, 634
ribose phosphate, 636

Ritchie nucleophilicity scale, 244
Ritter reaction, 372
RNA, enzymic hydrolysis, 636
Robinson annellation, 497
rotating cryostat, 725

S-solvation parameter, 195, 192
S-steric parameter, 295
S_N1 reactions, 202, 203, 386 *et seq.*,
    458, 481
S_N2 reactions, 118, 126, 202, 287, 299,
    320, 376 *et seq.*, 491, 589
S_E1 reactions, 418
S_E2 reactions, 118, 419
S_NAr reactions, 247
S_EAr reactions, 262, 429
S_{RN} reactions, 749
S_{ON} reactions, 751
S_R2 reactions, 118, 737
salicylate hydrolysis mechanism of, 276
salt effects, 385
  'special', 402
Saytzev's Rule, 523
s-character, bonds, 27
  and acidity, 224
Schleyer–Foote equation, 412, 599
Schmidt rearrangement, 361
Schrödinger equation, 3
secondary kinetic isotope effect
    (SKIE), 263
  conformational dependence, 267
  in cycloadditions, 678
    table, 702
  in eliminations, 515, 516
  in neighboring group participation,
    599
  in S_N reactions, 394
    table, 395
selectivity, 105, 106
  ortho-para ratio, 446
  in radical reactions, 756
selenoxide pyrolysis, 537
semicarbazide, 576
semicarbazone formation, 358
semiquinone, 721
sensitization, photochemical, 780
serine proteinases, 632
Sharpless'
  epoxidation reagent, 324
  selenium reagent, 544
sigmatropic reactions, 696
  photochemical, 787
silicon, displacements of, 493

singlet states, 779
singly occupied orbital (SOMO), 763
Simmons–Smith reagent, 540
solubilities, molar, table, 178
solutions, 173
solubility parameter, 188
solvation, 176
  effect on rates and equilibria, 182
  in ester hydrolysis, 473
  in gas phase, 181
  in NGP, 600
  in radical reactions, 759
  in S_N2 reactions, 384
  in S_NAr reactions, 458
solvation parameters, empirical, table,
  180
solvation parameters, empirical
  A, 204, 205
  α, 198
  AN, 199
  β, 198
  B, 196
  D_π, 198
  DN, 196
  E_T, 194
  f_ε, f_n, 185
  $\mathscr{S}$, 192
  χ_B, χ_R, 194
  S, 195
  Z, 193
  X, 192
  P, 195
  π^*, 196
solvent effects, 171, *et seq.*
solvent isotope effects, 272
  in acid catalysed reactions, 350
solvolysis, 386, 590
solvatochromism, 192
Sommelet–Hauser rearrangement, 371
spectra, ion pairs, 51
spin density, 722
  table, 723
  multiplicity, 779
spintrapping, 725
staggered conformation, 301
standard states, elements, 78
state, electronic, 777
  lifetimes, 780
steric hindrance, 285
  in borane-amine coordination, 288
  in cycloadditions, 288
  radical reactions, 757, 759
  in S_N2 reactions, 287, 381

stereospecificity in Diels–Alder reactions, 665
stereochemistry, NGP, 605
stereochemistry, radicals, 766
stereoelectronic factors
  in additions, 559
  in addition-elimination, 583
  in aliphatic substitution, 383
  in eliminations, 524
  ester hydrolysis, 475
  ketal hydrolysis, 357
  in radical reactions, 757
Stevens rearrangement, 729
stereospecificity, cycloadditions, 679
steroids, conformations, 320
steric selection, 288
steric effects, 282 *et seq.*
  on rates, 296
steric isotope effects, 268
steric parameter, E_s, 761, 155, 297
  E_s^c, 762, 297
  v, 299
  $\mathscr{S}$, 295, 700
Stieglitz rearrangement, 371
stilbene, *cis* isomer, 782
Stobbe condensation, 572
Stock–Brown relationship, 446
Stork reaction, 355
strain energy, 282
  B-, 291
  Baeyer, 285
  F-, 287
  I-, 289
  Pitzer, 285
  Prelog, 282
  table, 283
styrene, polymerization by carbanions,
  55
substituent constant, σ (σ°), 133
  table of, 134
  interpretation of, 140
  σ^+, 149, 360, 439
  σ^-, 151, 453
  σ_I, 154, 156
  σ_R, 154, 156
  σ^*, 154
substituent effects
  on acid dissociation, 223
  theory of, 135
substitution reactions, 375 *et seq.*
  at aromatic carbon (S_E), 428
  at aromatic carbon (S_N), 452
  α to carbonyl, 353

at carbonyl carbon (S$_N$), 462
photochemical, 795
radical, 737
at saturated carbon, 375
at vinyl carbon, 424
sulfolenes
formation equilibrium, 109
reactivities, 694
thermolysis, 693
sulfides, oxidation by I$_2$, 642
sulfonation, aromatic, 433
sulfonates
hydrolysis, 385
neighboring group participation, 590
sulfones, acidities of, 220
sulfoxide, dimethyl pyrolysis, 180, 191, 536
sulfoxide racemization, solvent effect on, 201
sulfur, isotope effect, 269
superacid
carbocations in, 448
enthalpies of ionization in, 391
'super-basic' media, 236
superdelocalizability, 16
suprafacial geometry, 655, 788
surface, potential energy, 91
Swain solvation parameters, 204
Swain–Lupton constants, 155
table, 157, 301
Swain–Scott equation, 242
sydnones, 686
syn, anti elimination modes, 511, 528

Taft–Ingold equation, 155, 297
table, 156
tautomerism, solvent effect on, 205
TCNE cycloadditions, solvent effect on (2 + 2) cycloadditions, 200
termination, polymerization, 732
tetrabutylammonium perchlorate, ion pairing in, 47
tetrafluoroethene dimerization, 681
tetrahedral intermediates, 488
see hemiorthoesters
thallium, additions of, 584
thermodynamics, 77 *et seq.*
first law, 78
second law, third law, 81
thiironium ion, 643
thiocarbonyl chloride, 497
Thornton's Rule, 123
Thorpe reaction, 572

tin alkyls, substitution, 420
radicals, 747
total energy, 77
trans-cycloalkenes
dimerization, 681
valence isomerism, 689
transfer functions, 177
transfer reactions, 118
at carbonyl carbon, 462
at saturated carbon, 375
transition state model, 93
for Diels–Alder reactions, 672
transmission coefficient, 93
triazenes, reaction with acids, 366
triplet states, 779
triosephosphate isomerase, mechanism, 640
trioxane, hydrolysis, effect of pressure, 95
triphenylmethyl radical, 721
dimerization, 736
tunnel effect, 269
in radical abstractions, 739
table, 270

valence bond model, 1
valence isomerization, 797
Van't Hoff equation, 51, 88
and proton transfer, 230
vibrational energy levels, 24
vinyl alcohol, 356
vinyl cations, 427
vinyl substitution, 579
vinyl substituent correlation, 160
Vitamin A, 783
volume of activation, 96
alkene hydration, 550
contributions to, 102
for cycloadditions, 672, 681
for hydrolytic reactions, 340
for S$_N$2 reactions, 379, 414

Wagner–Meerwein rearrangement, 361, 696
in solvolysis, 396
in terpenes, 371
water–gas reaction, 82
water, structure of liquid, 174
Winstein–Holness equation, 316
Wittig reaction, 573, 696
Woodward reaction, 563
Woodward–Hoffmann rule, 656
Wurtz–Grignard reaction, 729

X-groups (substituents), 140
X-solvation parameter, 192
xanthates, pyrolysis, 532
xenon fluoride, additions of, 585

Y-solvation parameter, 190
  and S_N reactions, 387
  table, 191
Yates–McClelland plot, 480
ylides, 223
  azomethine, 685, 687

carbonyl, 688
nitrile, 685, 687
phosphorus, 573
sulfur, 574
Ylide rearrangement, 696
Yukawa–Tsuno equation, 153
  table, 153

Z-groups (substituents), 144
Z-solvation parameter, 193
zeropoint energy, 256